Lecture Notes in Civil Engineering

Volume 597

Lecture Notes in Civil Engineering (LNCE) publishes the latest developments in Civil Engineering—quickly, informally and in top quality. Though original research reported in proceedings and post-proceedings represents the core of LNCE, edited volumes of exceptionally high quality and interest may also be considered for publication. Volumes published in LNCE embrace all aspects and subfields of, as well as new challenges in, Civil Engineering. Topics in the series include:

- Construction and Structural Mechanics
- Building Materials
- Concrete, Steel and Timber Structures
- Geotechnical Engineering
- Earthquake Engineering
- Coastal Engineering
- Ocean and Offshore Engineering; Ships and Floating Structures
- Hydraulics, Hydrology and Water Resources Engineering
- Environmental Engineering and Sustainability
- Structural Health and Monitoring
- Surveying and Geographical Information Systems
- Indoor Environments
- Transportation and Traffic
- Risk Analysis
- Safety and Security

To submit a proposal or request further information, please contact the appropriate Springer Editor:

- Pierpaolo Riva at pierpaolo.riva@springer.com (Europe and Americas);
- Swati Meherishi at swati.meherishi@springer.com (Asia—except China, Australia, and New Zealand);
- Wayne Hu at wayne.hu@springer.com (China).

All books in the series now indexed by Scopus and EI Compendex database!

Xiao Lin Zhao · Jian-Guo Dai · Siwei Liu ·
Soon Heng Lim
Editors

Proceedings of the Fourth World Conference on Floating Solutions: WCFS 2024; 2–4 December; Hong Kong

Springer

Editors
Xiao Lin Zhao
Research Institute for Land and Space
The Hong Kong Polytechnic University
Kowloon, Hong Kong

Siwei Liu
Department of Civil and Environmental Engineering
The Hong Kong Polytechnic University
Kowloon, Hong Kong

Jian-Guo Dai
City University of Hong Kong
Kowloon, Hong Kong

Soon Heng Lim
The Society of Floating Solutions
Singapore, Singapore

ISSN 2366-2557 ISSN 2366-2565 (electronic)
Lecture Notes in Civil Engineering
ISBN 978-981-96-4571-8 ISBN 978-981-96-4569-5 (eBook)
https://doi.org/10.1007/978-981-96-4569-5

This Springer imprint is published by the registered company Springer Nature Singapore Pte Ltd.
The registered company address is: 152 Beach Road, #21-01/04 Gateway East, Singapore 189721, Singapore

Committees

International Scientific Committee

Chair

Prof. Chien-Ming Wang, Australia

Members

Mr. Per Arild Åland, Norway
Dr. Andrew Chan, Hong Kong, China
Prof. Jianfei Chen, China
Dr. Rutger De Graaf-Vandinther, The Netherlands
Prof. Jian-Guo Dai, Hong Kong, China
Prof. Xiaoli Ding, Hong Kong, China
Prof. Peng Feng, China
Prof. Shixiao Fu, China
Mr. Paul Holthus, USA
Dr. Stefan Huebner, Singapore
Prof. Tomoki Ikoma, Japan
Prof. Bas Jonkman, The Netherlands
Dr. Kwanghoe Jung, South Korea
Mr. Rick Lam, Hong Kong, China
Mr. Soon Heng Lim, Singapore
Prof. Kenny Leung, Hong Kong, China
Mr. Terence Leung, Hong Kong, China
Prof. Ming Lin, China
Dr. Yifeng Lin, China
Prof. Hongbin Liu, Hong Kong, China

Emeritus Prof. Torgeir Moan, Norway
Mr. Robert Newman, UK
Prof. Dezhi Ning, China
Mr. Koen Olthuis, The Netherlands
Dr. William Otto, The Netherlands
Prof. Irene Penesis, Australia
Prof. Chi-Sun Poon, Hong Kong, China
Mr. Joe Quirk, USA
Ms. Tytti Sirola, Finland
Prof. Shigeru Tabeta, Japan
Prof. Jin-Guang Teng, Hong Kong, China
Mr. Anil Thapar, Singapore
Prof. Tomoaki Utsunomiya, Japan
Dr. Brydon Wang, Australia
Prof. Kahing Wong, Hong Kong, China
Prof. Yufei Wu, China
Prof. Hong Zhang, Australia
Prof. Xiao Lin Zhao, Hong Kong, China
Prof. Ronghua Zhu, China

Local Organising Committee

Chair

Prof. Xiao Lin Zhao

Co-chair

Prof. Jian-Guo Dai

Secretary

Dr. Siwei Liu

Members

Preface

We are pleased to present the conference proceedings of the 4th World Conference on Floating Structures (WCFS 2024), held in Hong Kong, 2–4 December 2024. In the context of Hong Kong, it is crucial to develop sustainable floating structure technology to create innovative and resilient means to foster the city's sustainable growth and enhance its liveability with strategic significance not just for Hong Kong but also for other coastal megacities, addressing the mounting threat of sea level rises and overpopulation. It is equally important to adopt floating solutions for renewable energy, food security, and food sustainability.

WCFS2024 featured four prominent keynote speakers: Prof. Torgeir Moan, the Norwegian University of Science and Technology, Norway, Prof. Chien-Ming Wang, the University of Queensland, Australia, Prof. Masashi Kashiwagi, Hyundai Engineering and Construction, South Korea, and Mr. Jacky Wu, Head of Sustainable Lantau Office, CEDD, Hong Kong.

These conference proceedings contain over 110 papers accepted for presentation at WCFS 2024 with a theme of "Floating solutions for sustainable ocean development and blue economy". The contributions from 18 countries cover a wide range of topics including

- Application and Prospect
- Construction and Design
- Environmental Enhancement
- Floating Tunnel and Mooring System
- Food, Energy, and Architecture
- Hydrodynamic Analysis
- Materials for Floating Structures
- Modular Floating Structures
- Policy, Social Acceptance, and Legal Matters
- Urban Planning
- Wave Tank Tests.

We are pleased to have received 23 Student Presentations which were presented by their lead authors.

We extend our heartfelt gratitude to the authors, reviewers, session chairs, as well as the volunteers who have generously invested their time and expertise to make WCFS 2024 a success.

WCFS® is a platform for promoting floating solutions worldwide, initiated by the Society of Floating Solutions (Singapore). It enables the confluence of ideas to mitigate the effects of climate change and combat the loss of biodiversity linked to land reclamation.

We are pleased to note that Bluet Floating Solutions has won the rights to host the 5th conference in this series in Finland. We wish them every success.

Kowloon, Hong Kong — Xiao Lin Zhao
Kowloon, Hong Kong — Jian-Guo Dai
Kowloon, Hong Kong — Siwei Liu
Singapore — Soon Heng Lim

Organised by

DEPARTMENT OF
CIVIL AND ENVIRONMENTAL ENGINEERING
土木及環境工程學系

RILS 土地及空間研究院
Research Institute for Land and Space

RESEARCH INSTITUTE
FOR SUSTAINABLE URBAN DEVELOPMENT
可持續城市發展研究院

RiFood
Research Institute
for Future Food
未來食品研究院

Society of Floating Solutions
(Singapore)

Non-financial Supporters

漁農自然護理署
Agriculture, Fisheries and Conservation Department

土木工程拓展署
Civil Engineering and Development Department

渠務署
Drainage Services Department

Marine Department
HKSAR

THE HONG KONG INSTITUTION OF ENGINEERS
香港工程師學會
Structural Division
結構分部

Sponsors

CITF 建造業創科基金

煌港特鋼
PEAK KONG SPECIAL STEEL

WO LEE GREEN SOLUTIONS Ltd.
和利綠色動力有限公司

Contents

Keynote Papers

Advancements in Offshore Aquaculture: Enhancing Environmental Sustainability and Elevating Seafood Production

Chien Ming Wang, Yunil Chu, and Huu Phu Nguyen

Abstract This keynote paper is concerned with recent research and developments on offshore aquaculture for enhancing environmental sustainability and increasing seafood production. We first discuss the motivation and challenges for moving offshore for the aquaculture industry. This is followed by highlighting technological advances and research needs in offshore fish farming infrastructure and offshore seaweed cultivation infrastructure.

Keywords Offshore aquaculture · Fish farms · Seaweed cultivation

1 Introduction

The global demand for sustainable seafood production continues to rise due to human population growth and improved living standards. Marine aquaculture has emerged as a crucial contributor in filling the gap between wild capture seafood supply and the rising demand. Farmed seafood production has increased 5.8 times reaching 29 million tonnes annually from 1990 to 2020, and it is expected to grow to 74 million tonnes by 2050 [1].

Most marine aquaculture farms are primarily sited in sheltered nearshore waters for safe operation, easy access, and quick transportation. However, in recent years, the annual growth rate of farmed marine production is slowing due to public and environmental opposition to water, seabed and shoreline pollution, and competing uses of sheltered sea space such as shipping, water leisure activities, conservation, and tourism. Additionally, more regulations and restrictions are increasingly requested for nearshore marine aquaculture to transform the traditional farming operations into more eco-friendly and sustainable practices. This includes prohibiting the use

C. M. Wang (✉) · Y. Chu · H. P. Nguyen
School of Civil Engineering, The University of Queensland, St Lucia, QLD, Australia
e-mail: cm.wang@uq.edu.au

Blue Economy Cooperative Research Centre, Launceston, Australia

X. L. Zhao et al. (eds.), *Proceedings of the Fourth World Conference on Floating Solutions: WCFS 2024; 2–4 December; Hong Kong*, Lecture Notes in Civil Engineering 597, https://doi.org/10.1007/978-981-96-4569-5_1

of antibiotics and taking social responsibility for using public waters. In response to these criticisms and constraints, offshore (exposed or open ocean) aquaculture is gaining traction as an alternative solution.

By moving aquaculture activities offshore, the potential for habitat destruction and water pollution due to fish wastes and uneaten feed on the sensitive nearshore water can be significantly reduced because of better water circulation and dilution from ocean waves and currents [2, 3]. Additionally, more spacious and pristine water column, cooler temperature, less pathogens and parasites promote healthier and faster fish growth without the need for antibiotics and other treatments as compared to nearshore farms [4, 5].

Offshore aquaculture, however, poses several challenges that include having to contend with a harsher sea environment, an uncertain environment for marine aquaculture, operational requirements regarding workers' safety and support vessels, and the lack of experience and standards in designing offshore aquaculture farms [6]. The most challenging aspect of offshore aquaculture is the survivability of farming infrastructure and the well-being of aquatic creatures during severe storms that are accompanied by huge surface waves, winds, and shear current actions.

2 Offshore Fish Farming

In recent years, there has been a remarkable upsurge in research and development in offshore fish cages, driven by the integration of cutting-edge technologies. This is evident from the design and construction of very large robust fish cage systems to withstand highly energetic offshore environments. Examples are Ocean Farm 1 [7] with a height of 69 m, a diameter of 110 m that is able to accommodate 1.2 million fish, and HavFarm 1 with a length of 380 m, width of 59 m and houses 10,000 tonnes of salmon [8, 9] as shown in Fig. 1a, b. However, these cages come with a substantial price tag, exceeding USD100 million each, which is not affordable by most fish farming companies.

(a) Ocean Farm 1

(b) HavFarm 1 (NSK Ship Design)

Fig. 1 **a** Ocean Farm 1, **b** HavFarm 1 (NSK ship design)

New submersible offshore fish cage concepts have attracted the attention of the fish farm operators due to their cheaper costs for sustainable and profitable offshore fish farming. Strong surface waves during storms can be avoided by submerging the fish cage to an appropriate water depth. Examples of such fish cages are SubFlex [10] (see Fig. 2a), SeaStation [11] (see Fig. 2b), Atlantis (see Fig. 2c) and Nautilus (see Fig. 2d). These submersible cages can unlock new offshore sites for aquaculture production where floating open net cages are inappropriate due to strong surface waves or seasonal variations in the marine environment. Tests have been conducted on some submersible fish cage designs, such as Atlantis, which can farm fish below the water surface during a stormy period. These fish cages have demonstrated their ability to produce healthy fish products [10, 12]. Also, there have been developments in autonomous submersible steel fish cages armed with motor valves controlling variable ballast tanks, housings for air compressors, control system, batteries, and seawater pumps [13]. More recently, a groundbreaking submersible offshore fish cage design known as the SeaFisher [14] (see Fig. 2e) was proposed. SeaFisher's modular frame structure is formed from bundled High-Density Polyethylene (HDPE) pipes that are held by custom brackets and connector pods. HDPE pipes are adopted due to their resistance to rotting and weathering, easy construction for diverse structural forms, good antibiofouling characteristics when compared to other materials, as well as an economical choice when procured in large quantities [15]. A single-point mooring (SPM) system is adopted for the SeaFisher to allow it to behave like a weathervane about its mooring point to reduce environmental loads [8, 16]. The SPM system also offers improved dispersal of fish waste and uneaten feed over a larger sea space.

Additionally, there is a new concept of offshore fish cages that are mobile. One of the famous mobile fish cages is the Chinese fish tanker named Guoxin 1 (see

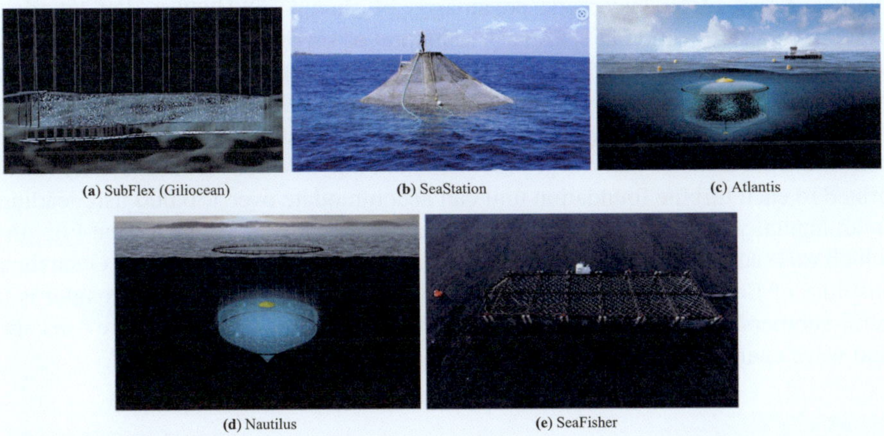

Fig. 2 **a** SubFlex (Giliocean), **b** SeaStation (https://photolib.noaa.gov/Collections/Fisheries/Other/emodule/1054/eitem/61652), **c** Atlantis (https://www.akvagroup.com/atlantis/), **d** Nautilus (https://www.akvagroup.com/nautilus/), **e** SeaFisher

(**a**) Guoxin 1 (**b**) Ocean Ark

Fig. 3 **a** Guoxin 1 (<https://www.youtube.com/watch?v=LqS6Y9HRGcs>), **b** Ocean Ark (https://oceansovereign.com/#solutions)

Fig. 3a). It has the size of an oil tanker and is expected to produce more than 3700 tons of fish annually [17]. Guoxin 2 and Guoxin 3 are currently being built and should be delivered in 2024. The Ocean Ark (see Fig. 3b) is another concept of a mobile aquaculture vessel developed by Ocean Sovereign. The vessel is a self-propelled, AI-assisted, low emissions trimaran, measuring 170 m long and 64 m wide and can farm up to 3900 tonnes of fish. The vessel design has received Approval in Principle from RINA, a ship classification society. Such mobile fish cages/tankers can be moved to a calmer sea environment to avoid large waves during storms. This type of mobile fish pens will appeal to land-locked nations that wish to farm fish in international waters. More recently, the New Zealand Food and Plants Research Institute proposed a conceptual design of mobile fish cage where a submerged cylindrical fish net tube is towed by an autonomous vessel to various locations for optimal conditions all year round [18, 19].

There is also a design trend to integrate offshore fish cages and renewable energy platforms that offers synergetic advantages of shared substructure and mooring system, better utilization of sea space, reduction of service and maintenance cost as well as to assist in the decarbonization of the fish farming industry. For example, the Ming Yu No.1 (see Fig. 4a) was successfully deployed 69 km off Guangdong's coast in water depths of 48 m in August 2023. The offshore installation comprises 40 large offshore wind turbines that can produce a total of 500 MW. The fish cage that is integrated to each turbine foundation unit can accommodate over 150,000 fish, leading to an annual yield of 75 tonnes [20]. Another example is the Penghu (see Fig. 4b) which was successfully deployed in Zhuhai, China in 2019. Developed by Guanzhou Institute of Energy Conversion of the Chinese Academy of Sciences, Penghu is a semi-submersible aquaculture platform that combines offshore aquaculture fish cage and wave energy conversion system [21].

(**a**) Ming Yu No.1 (**b**) Penghu

Fig. 4 **a** Ming Yu No.1, **b** Penghu

3 Offshore Seaweed Cultivation

Seaweeds have wide-ranging applications such as animal feed, biostimulants, food, cosmetics, pharmaceuticals, biofuels, bioplastics, fabrics, water treatment, construction materials (e.g. seabricks), and carbon dioxide sequestration [22]. While most seaweed farms are currently sited nearshore, interests in farming seaweeds in offshore waters are growing. However, there are challenges in offshore seaweed cultivation, and they include technological complexities, uncertainties in the performance of the seaweed, low nutrient concentrations and lack of clear legislations for offshore seaweed farming.

Early developments of offshore seaweed cultivation infrastructure started from the 1970s in the United States with the proposal and trial of inverted umbrella-like structure (see Fig. 5a) [23]. The structure comprises a buoyed upwelling pipe at its center to irrigate seaweed with nutrient-rich cold deep seawater. These structures were deployed at multiple sites off the coast of California to assess the feasibility and economic potential of offshore seaweed farming. Each 0.48-hectare farm incurred costs of hundreds of millions of USD. However, the experiment results were highly disappointing, with substantial infrastructure damage and slow kelp growth rates.

In 2010, Seaweed Energy Solutions AS in Norway trialed the Seaweed Carrier for kelp cultivation (see Fig. 5b) [23, 24]. This system employs a single mooring setup with a vertical cultivation net measuring 6.5 m in length and 5 m in depth, where the kelp grows. Data from the trial of the Seaweed Carrier is not available in the published literature, and the system has not been adopted for commercial seaweed farming.

In the 2010s, the BioArchitecture Lab introduced a new kelp cultivation infrastructure in Chile [23]. This system was designed as a rectangular grid, consisting of twenty-one 1-hectare modules, each with 99 horizontal lines of 100 m (see Fig. 5c). The highest production yield per module averaged 124 tonnes per hectare annually, with operational costs comparable to those of wild-harvested kelp [23]. Despite its commercial potential, the system has not yet been commercially implemented.

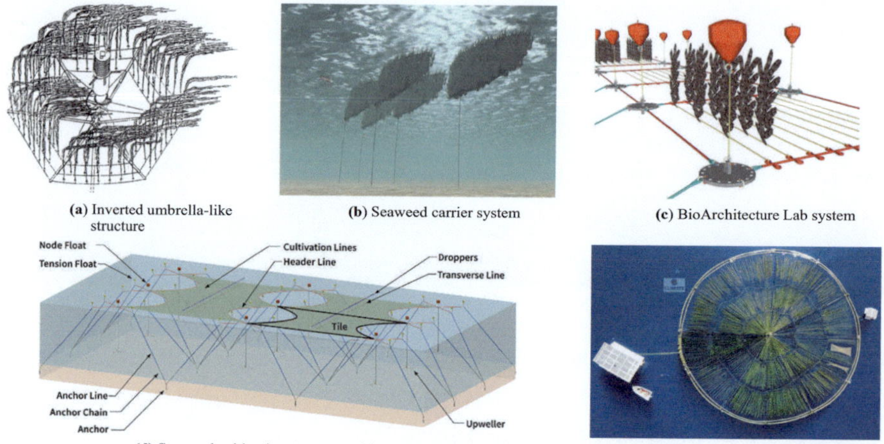

(**a**) Inverted umbrella-like structure
(**b**) Seaweed carrier system
(**c**) BioArchitecture Lab system
(**d**) Seaweed cultivation structure with modular "tiles"
(**e**) Seaweed cultivation infrastructure tested by Climate Foundation

Fig. 5 **a** Inverted umbrella-like structure, **b** seaweed carrier system, **c** BioArchitecture Lab system, **d** seaweed cultivation structure with modular "tiles", **e** seaweed cultivation infrastructure tested by Climate Foundation

Recently, a team from the United States introduced an innovative seaweed farming system featuring a lattice mooring design as shown in Fig. 5d [25]. This mooring system supports large, stable, horizontal arrays of modular components, each measuring 140 m by 70 m. Sea trials of this system improved understanding of its durability and functionality in offshore environments. It was found that accessibility, tension control, and precise anchor placement need improvements [25].

To further improve the safety and economic viability of offshore seaweed cultivation, substantial international efforts and collaborations are being invested in multi-million USD programs worldwide, such as the MARINER program in the United States [26]. Current R&D focuses on [27, 28]:

- developing automated seeding and harvesting facilities for offshore sites. An example of such facilities is the Standardized Production of Kelp (SPOKe) concept [24] that can be deployed from a normal service vessel for seeding and harvesting of circular seaweed platforms.
- farming seaweed next to fish farms or offshore renewable energy farms for synergistic benefits with respect to marine space usage, decarbonisation, and nutrient management. The seaweed cultivation infrastructure can be integrated to other-purpose offshore infrastructure or be placed next to each other. The latter appears to be more promising for applications in the near term due to their simplicity in infrastructure designs; whereas the former would require comprehensive review and modifications of infrastructure to accommodate different requirements for multiple purposes. The most notable co-located aquaculture farms cover 100 km^2 of the Sanggou Bay (China) [29]. In this huge integrated multi-tropic aquaculture (IMTA), fish, oysters, bivalves, scallops and seaweeds are farmed in an integrated

fashion to improve efficiency, reduce waste and provide ecosystem services such as bioremediation. A survey in Germany [30] found that co-locating seaweed and offshore wind farms is supported in principle by industry partners. Concerns about seaweed farms introducing additional structures and equipment into the marine environment that could potentially interfere with the operation and maintenance of wind turbines must be addressed.
- optimizing seaweed cultivation infrastructure comprising floating platforms, buoys, lines, tube nets, and anchors. Commercial offshore seaweed farms are expected to be large-scale, similar to those at nearshore sites. Therefore, innovative sea-keeping solutions are necessary for ensuring safety and minimizing mooring costs for large-scale offshore farming systems.
- developing submersible cultivation infrastructure to avoid strong surface waves during storm events and for accessing nutrient-rich deep waters to accelerate seaweed growth. Deep submergence of seaweed platforms to depths of at least 100 m, during night time and resurfacing them in day time for photosynthesis to take place, has shown to substantially accelerate seaweed growth rate even in conditions of low nutrient levels at the water surface [31].
- utilizing existing fish farming infrastructure with modifications for farming seaweeds. A promising submersible seaweed cultivation platform (comprising circular HDPE pipes and grow-out lines) held in place by a single-point mooring system and buoys was recently proposed and tested by The Climate Foundation in the Philippines [32] (see Fig. 5e). The platform is submerged during storms to avoid the strong surface waves and to soak the seaweed in cold nutrient-rich deep waters that help in accelerating the growth of seaweed. This demonstration showed how offshore seaweed cultivation may be carried out in a cost-effective manner. Such platform design has the potential for repurposing fish cages for farming seaweed, and for low-cost integration of seaweed farming into existing fish farming infrastructure.

4 Discussion and Closing Remarks

The potential benefits of offshore fish and seaweed farming are significant, and various offshore solutions have been suggested, with some already deployed for prototype testing. However, operational requirements under the offshore environment must be carefully considered and they should be aligned with the interests of stakeholders, encompassing the entire farming process such as breading, feeding, crowding, cleaning and harvesting. In addition, the divergent interests in integrated or co-located cultivation may be even more challenging to incorporate different operational design requirements as well as rapid growth in farming size posing a challenge to design mooring and sea-keeping systems.

Thereby, it is essential to address operational requirements and develop strategic plans that respect the specifications of offshore designs to create sustainable and economically viable offshore aquaculture models. By learning from ongoing prototype projects, it is expected to develop better strategies to mitigate risks and clarity uncertainties as offshore aquaculture progresses.

Acknowledgements The authors acknowledge the financial support of the Blue Economy Cooperative Research Centre (CRC), established and supported under the Australian Government's CRC Program, grant number CRC-20180101.

References

1. DNV (2021) Oceans' future to 2050. Marine aqiculture forecast
2. Sanz-Lazaro C, Casado-Coy N, Calderero EM, Villamar UA (2021) The environmental effect on the seabed of an offshore marine fish farm in the tropical Pacific. J Environ Manage 300:113712
3. Chu YI, Wang CM, Park JC, Lader PF (2020) Review of cage and containment tank designs for offshore fish farming. Aquaculture 519:734928
4. Tveteras R, Hovland M, Reve T, Misund B, Nystoyl R, Bjelland HV, Misund A, Fjelldal O (2020) Value creation potential and roadmap for offshore aquaculture. UiS: Stavanger, Norway
5. Morro B, Planellas SR, Davidson K, Adams TP, Falconer L, Holloway M, Dale A, Aleynik D, Thies PR, Khalid F et al (2021) Offshore aquaculture of finfish: big expectations at sea. Rev Aquac 14(2):791–815
6. Chu YI, Wang CM, Zhang H, Abdussamie N, Karampour H, Jeng DS, Baumeister J, Aland PA (2023) Offshore fish farms: a review of standards and guidelines for design and analysis. J Mar Sci Eng 11:762
7. Zhao Y, Guan C, Bi C, Liu H, Cui Y (2019) Experimental investigations on hydrodynamic responses of a semi-submersible offshore fish farm in waves. J Mar Sci Eng 7:238
8. Wang CM, Chu YI, Baumeister J, Zhang H, Jeng DS, Abdussamie N (2022) Offshore fish farming: challenges and developments in fish pen designs. In: Nazrul IM, Bartell SM (eds) Global Blue Economy. CRC Press, pp 87–128
9. Li L, Jiang Z, Ong MC (2017) A preliminary study of a vessel-shaped offshore fish farm concept. Int Conf Offshore Mech Arct Eng 57724
10. Milich M, Drimer N (2019) Design and analysis of an innovative concept for submerging open-sea aquaculture system. IEEE J Ocean Eng 44:707–718
11. Loverich GF (2010) A case study of an offshore SeaStation® sea farm. Mar Technol Soc J 44(3):36–46
12. Drimer N (2019) First principle approach to the design of an open sea aquaculture system. Ships Offshore Struct 14:384–395
13. Kim T, Hwang KS, Oh MH, Jang DJ (2014) Development of an autonomous submersible fish cage system. IEEE J Oceanic Eng 39(4):702–712
14. Wang CM, Chu YI, Baumeister J, Zhang H, Qiao YP, Karampour H, Jeng DS, Savage L (2023) SeaFisher—a submersible high-density polyethylene offshore fish pen. J Mar Sci Eng 11:1795
15. Beveridge MCM (2008) Cage aquaculture. Wiley vol 5
16. Goudey CA, Loverich G, Kite-Powell H, Costa-Pierce BA (2001) Mitigating the environmental effects of mariculture through single-point moorings (SPMs) and drifting cages. ICES J Mar Sci 58:497–503
17. The fish site: 100,000 tonne mobile fish farm docks in China, Fish Site (2022) Available online: https://thefishsite.com/articles/100-000-tonne-mobile-fish-farm-docks-in-china-guoxin-1. Last accessed 2024/07/26

18. Black S, Vennell R, Kregting L, Moran D (2024) Designing for the animal: reimagining offshore fish structures. In: Proceedings of the ASME 2024, 43rd international conference on ocean, offshore and arctic engineering, OMAE2024-125703, Singapore EXPO, Singapore
19. Pillai A, Ashton I, Johanning L, Cook D, Vennell R, Black S (2024) Navigating the future: strategic management of a mobile aquaculture system with receding-horizon control. In: Proceedings of the ASME 2024, 43rd international conference on ocean, offshore and arctic engineering, OMAE2024-121565, Singapore EXPO, Singapore
20. WeBridge (2023) The world's first integrated offshore wind and aquaculture system completed in Guangdong, China. Available online: https://windenergybridge.com/insights/the-world%27s-first-integrated-offshore-wind-and-aquaculture-system-completed-in-guangdong,-china. Last accessed 2024/08/02
21. Ding WW, Jiang JQ, Yue WZ, Li YZ, Wang WS, Sheng SW, Chen M (2023) Numerical study on hydrodynamic performance of a new semi-submersible aquaculture platform. Appl Sci 13(23):12652
22. Tullberg RM, Nguyen HP, Wang CM (2022) Review of the status and developments in seaweed farming infrastructure. J Mar Sci Eng 10:1447
23. Bak UG, Gregersen Ó, Infante J (2020) Technical challenges for offshore cultivation of kelp species: lessons learned and future directions. Bot Mar 63:341–353
24. Solvang T, Bale ES, Broch OJ, Handå A, Alver MO (2021) Automation concepts for industrial-scale production of seaweed. Front Mar Sci 8:613093
25. Moscicki Z, Swift MR, Dewhurst T, MacNicoll M, Chambers M, Tsukrov I, Fredriksson DW, Lynn P, Landon ME, Zotter B, MacAdam N (2024) Design, deployment, and operation of an experimental offshore seaweed cultivation structure. Aquacult Eng 105:102413
26. Agency-Energy (2022) Macroalgae research inspiring novel energy resources. Available online: https://arpa-e.energy.gov/technologies/programs/mariner. Last accessed 2022/01/31
27. Buck BH, Troell MF, Krause G, Angel DL, Grote B, Chopin T (2018) State-of-the-art and challenges for offshore integrated multi-trophic aquaculture (IMTA). Front Mar Sci 5:165
28. Nguyen HP, Wang CM, von Herzen B, Huang C (2023) Hydroelastic responses of a submersible ring structure for offshore seaweed cultivation under wave action. J Mar Sci Eng 11:2238
29. Li R, Liu S, Zhang J, Jiang Z, Fang J (2016) Sources and export of nutrients associated with integrated multi-trophic aquaculture in Sanggou Bay, China. Aquac Environ Interact 8:285–309
30. Wever L, Krause G, Buck BH (2015) Lessons from stakeholder dialogues on marine aquaculture in offshore wind farms: perceived potentials, constraints and research gaps. Mar Policy 51:251–259
31. Navarrete IA, Kim DY, Wilcox C, Reed DC, Ginsburg DW, Dutton JM, Heidelberg J, Raut Y, Wilcox BH (2021) Effects of depth-cycling on nutrient uptake and biomass production in the giant kelp Macrocystis Pyrifera. Renew Sustain Energy Rev 141:110747
32. Nguyen HP, Wang CM, von Herzen B, Huang C (2024) Hydroelastic responses of submersible seaweed cultivation platforms with single-point mooring systems. In: Proceedings of the ASME 2024, 43rd international conference on ocean, offshore and arctic engineering, OMAE2024-124766, Singapore EXPO, Singapore

Life Cycle Structural Integrity Management of Floating Structures with a Focus of Floating Bridges

Torgeir Moan

Abstract An overview of important developments regarding safety management of floating structures is given. The possible transfer of approaches from the oil and gas industry is briefly described. Based on relevant experiences with accidents and means to control the associated risks, are categorized from a technical-physical as well as human and organizational point of view. Structural risk relates to extreme environmental and accidental events, as well as structural degradation and can be controlled by use of adequate design criteria, especially to provide damage tolerance, as well as life cycle quality assurance and control. A risk analysis relating to ship impacts on a floating bridge is briefly presented and possible mitigation of the risk of road traffic disruption, is commented upon.

Keywords Life cycle · Floating structures · Integrity management · Ship impact risk

1 Introduction

Floating structures are used for different purposes in various sectors, like oil and gas industry, offshore renewable energy, fish farms as well as infrastructure like floating bridges (Fig. 1). In deep water they are used because the costs are less than for bottom fixed structures while they are used in more shallow water because of challenges with possible foundations on soft seabeds or simply because they provide some mobility, i.e. for use in different locations. While ships and rafts were the earliest floaters, support on discrete pontoons- (semi-submersibles, pontoon bridges), are commonly used for permanently located structures. They are moored by a connection to land, piles (in shallow water), or catenary, taut of tensioned mooring lines. Ship-shape

T. Moan (✉)
Norwegian University of Science and Technology, Trondheim, Norway
e-mail: tormo@ntnu.no

Faculty of Maritime and Transportation, Ningbo University, Ningbo, China

X. L. Zhao et al. (eds.), *Proceedings of the Fourth World Conference on Floating Solutions: WCFS 2024; 2–4 December; Hong Kong*, Lecture Notes in Civil Engineering 597, https://doi.org/10.1007/978-981-96-4569-5_2

(a) Oil & gas floating. (b) Wind turbines. (c) Aquaculture cages. (d) Coastal infrastructure and subseastructures (floating bridge).

Fig. 1 Examples of floating structures in various ocean sectors

facilities in harsh environments are positioned by a turret arrangement; i.e. the vessel is designed to rotate about a cylinder which is moored to the seabed by catenary mooring lines, while heading control is ensured by a DP system.

Floating structures in different sectors should in general be fit for their purpose and fulfill socio-political criteria that address the safety of personnel, aesthetics especially for near shore structures, environmental sustainability, budgetary and legal constraints. Moreover, they should not represent a hindrance to other activities, e.g. bridges should not hinder ship traffic unnecessarily.

The continuous innovation to deal with new serviceability and safety requirements in the oil and gas industry in the last decades, have led to an industry which has been in forefront of developing design and analysis methodology for safety assessment, including decision under uncertainty through reliability and risk analysis methods [1–3]. On the other hand, the cost of structures in this industry are relatively small compared to the income. Hence, transferring technology from this sector to other sectors should be done with cost reduction in mind.

In the paper the structural integrity management (SIM) in emerging ocean sectors, with emphasis on civil infrastructure, is discussed with reference to the comprehensive basis for principles and methods established in the oil and gas industry, a leading ocean sector.

2 General Requirements

The basic concepts of floating structures, initially created in the oil and gas industry, are semisubmersible with catenary or tension leg mooring system, spar and ship-shape floater/barge [1]. Those concepts have served as models or inspiration for floating structures in emerging ocean or coastal sectors. Floating structures should be designed to be serviceable, safe and, increasingly, sustainable, as well as fabrication friendly.

2.1 Serviceability

Serviceability requirements refer to behavior that hamper the function as an industrial facility or its function as a means in a transport system etc. The different types of floating structures considered herein; all have the function of carrying a payload. Some oil and gas platforms and fish cages also need to provide containment of oil and gas and live fish, respectively. Oil and gas platforms and wind turbines support industrial plants which call for limited motions (accelerations) to ensure an efficient operation of the machinery and mooring system. Floating bridges provide support for the public that should "always" be available, i.e. the disruption of road traffic should be limited.

2.2 Safety

General. The focus in this paper is on structural integrity management during the life cycle: design, fabrication/installation and operation. While all floating structures are subjected to hazards associated with payloads, metocean conditions and ship collisions, the latent energy in hydrocarbons in oil and gas platforms implies a particular risk potential. Safety requirements are specified to avoid fatalities, environmental damage and property damage, and are related to the following failure modes:

- overall, rigid body instability (capsizing) under permanent, variable and environmental loads
- failure of (parts of) the structure, foundation or mooring systems, considering permanent, variable, and environmental as well as accidental loads.

which ultimately can lead to fatalities, pollution or economic loss. Fatalities depend on the number of persons on the facility—on duty or public, evacuation options etc. Economic losses involve failure and damage expenses as well as the costs of being partly or completely out of service.

Service experiences. Safety is the absence of failures or accidents. Hence, experiences with accidents is an important basis for establishing a safety management approach. While accident statistics give an overview of the situation, more insight about the safety features can be obtained from more detailed investigation of individual accidents. Accident statistics for oil and gas platforms has been regularly compiled in WOAD [4]. Information about individual accidents or failures of oil and gas platforms, described e.g. in [2, 5–7]. Most accidents are due to human errors occurring at some stage in the lifecycle. Accidents statistics and causes for bridges are presented e.g. by Zhang et al. [8]. Design errors, construction mistakes, collisions and overloads make up more than 70% of the failures. It is noted that there is a global failure among the few existing bridges floating bridges (Hood Canal [9]). Statistics of wind turbine accidents, mainly related to turbines on land maybe found in [10, 11]. The focus is on relatively frequent electrical and mechanical failures, due to deterioration, lack of maintenance etc. "Blade failure" can arise from several possible sources, and often occur in areas with a drastic change of the geometry, e.g. at the blade root. Tower failure is attributed to lightning strikes, but also poor design, fabrication defects and lack of proper maintenance (which is difficult to perform) as well as underestimation of wind speeds, e.g. in typhoon prone areas. Moreover, tower failures have been caused by broken blades hitting the tower. Since there are very few floating wind turbines, the accident potential for the hull and mooring system needs to be inferred from that of similar oil and gas platforms with due account of the difference in life cycle integrity management. In the fish farming-industry the safety is focused on prevention of fish escape which implies environmental damage, and occupational safety [12]. Fish escape occurs due to operational errors and damage to the cage containing the fish associated with well boat operations or initial fabrication defects or deficient maintenance.

A common feature of accidents in the oil and gas industry and bridges is the contribution of human factors at some stage in the life cycle. The experiences with floating structures in the emerging sectors, is limited but the trend seems to similar as in the mentioned sectors. Operational errors are the primary cause of accidental events such as explosions, fires and ship impacts. Fabrication errors might cause a deficiency in strength. Design and analysis errors occur. The management of structural integrity to ensure acceptable safety is discussed in Sects. 2.3 and 2.4.

It is noted that it took several years of operation before deterioration especially in terms of fatigue cracks, was recognized as a hazard to ships and offshore structures.

Failure modes. Capsizing is assumed according to current practice, to occur for an intact freely floating structure about an axis where the overturning moment due to wind exceeds the stabilizing hydrostatic moment. It is noted that the motions induced by waves are not considered despite attempt to develop dynamic stability requirements. While for most floating structures the extreme environmental condition with a return period of 50 or 100 years, results in the governing overturning moment while for wind turbines it is an operational condition with so called rated wind speed which is most critical. For a floater with tension-leg mooring the tension provides the stability and slack in the mooring implies instability. For a floating bridge in an operational mode, the anchoring at both ends is intended to resist relevant torsional moments in the girder and prevent "capsizing".

While stability of the intact floater is an ULS criterion, stability of the floater in damaged condition is an ALS requirement which depends on the damage condition and the overturning wind moment. Damage is typically caused by ship impact which results in flooding and buoyancy loss, but technical or operational errors associated with a possible ballast system, could also lead to buoyancy loss. Some standards specify damage conditions in terms if one or two compartment flooding while others require risk analysis to determine relevant damage at a certain probability level.

It is interesting to note that capsizing of an intact floating platform or wind turbine is absent in the experience basis, global instability typically develop due to operational error of the structure causing abnormal ballast condition or structural damage condition, e.g. due to explosions, ship impacts or other external cause. For instance,

a) Alexander L. Kielland – fatigue failure, loss of leg, progressive failure and capsizing, North Sea, 1980

b) P - 36 explosion, flooding and capsizing, Brazil, 2001

c) The Typhoon Tension Leg Platform in the Gulf of Mexico, upside down following Hurricane Rita in 2005

d) Deepwater Horizon: explosion, flooding and sinking; blowout, Gulf of Mexico, 2010

Fig. 2 Selected catastrophic accidents in the oil and gas industry (oil platform disasters—Wikipedia)

the three catastrophic accidents in Fig. 2 resulted in capsizing/sinking after a sequence of events with structural damage and flooding [13].

Basically, structural failure occurs due to overload or deterioration due to fatigue or corrosion, when the resistance, R is less than the load effect, S or when the cumulative damage exceeds the failure limit. From a human factors perspective failure can be due to too small safety factors to account for the normal uncertainty and variability in the resistance or load effects relating to ultimate limit state (ULS) and fatigue limit state (FLS) criteria. But the main cause of actual structural failures is abnormal resistance or accidental loads due to human errors and omissions, not least in connection with fatigue which is the most important strength degradation process. Due to the small size of initial (fabrication) defects, even abnormal cracks (with a sizes > 0.1–0.5 mm depth) are difficult to detect, and such defects have been the cause of major failures [6].

The failure rate for mooring lines on mobile units in the North Sea was 0.04 per line year of. Chains made before 1985. When design procedures were improved, it was reduced to 0.01, e.g. [7, 14]. The high early failure rate was due to use of oversimplified methods for calculating the line tension as well as due to abnormal strength such as local bending at fairleads (which is not accounted for in design; fabrication defects, wear etc.) and damage during handling of the lines, The relatively high prevailing failure rate is partly due to the latter type of causes. The data for mooring systems for other floating structures is limited, but the failure rate for production platforms seems to be like that of mobile units [15]. Further clarification of the experienced failure rate is needed.

However, to fully explain accident event sequences, it is necessary to interpret them in view of the human and organizational factors (HOF) of influence, as reviewed in [2, 3, 5].

2.3 Structural Integrity Management for Oil and Gas Platforms

General. In Table 1 the causes of structural failures are categorized and the corresponding measures to control the accident potential are listed. Accidents with load carrying structures in most sectors are due to human errors that materialize in accidental loads or deficient strength [2, 3]. A rational integrity management approach should therefore be based on: goal-setting; not prescriptive; probabilistic; not deterministic; first principles; not purely experiential, integrated total; not separately; and balance of safety elements; not only hardware. In general, the measures include design criteria, quality assurance and control (QA/QC) relating to the engineering process as well as the hardware and operational procedures. With the increased digitalization of design, fabrication and operation of structures, software errors are of particular concern in QA/QC. Structural reliability analysis (SRA) can be used to establish ULS design criteria by calibration of partial safety factors to be consistent with a target level for the notional failure probability [16, 17]. It is noted that the random uncertainties in the ultimate strength commonly have limited effect on the reliability compared to that inherent in load effects. On the other hand, the systematic uncertainty, bias, in strength and load effects have the same effect on the reliability measure. Regarding fatigue the most important issue is to recognize this failure mode in design and inspection planning. Fatigue criteria are calibrated also to reflect the effect of inspection [7]. The SRA is based on normal variability and uncertainties in load effects and strength, and not the effect of human errors. Since the fatigue endurance depends on local geometry, which is hard to control, it is susceptible to fabrication "errors". The accidental loads and abnormal strength caused by human errors, are considered in the Accidental Collapse Limit State (ALS), and quantified (to the extent possible) by Quantitative Risk Analysis [18].

ALS approach to make structures robust (damage tolerant). Despite the efforts made to avoid error-induced accidental actions or abnormal resistance, they cannot be eliminated. For this reason, ALS criteria were introduced to ensure that "small

Table 1 Cause of structural failures and risk reduction measures [2]

Cause	Risk reduction measure
Less than adequate safety margin to cover "normal" inherent uncertainties	•Increase safety factors or margins in ULS, FLS •Improve inspection of the structure (FLS)[a]
Gross error or omission during: • design (d) • fabrication (f) • operation (o)	• Reduce probability and consequence of hazards, e.g. by • Improving skills, competence, self-checking (for d, f, o) •QA/QC of engineering process (for d) • Direct design for damage tolerance (ALS) (for events occurring during f, o) • Provide measures to detect, control and mitigate hazards at an early stage and initiate mitigation actions, e.g. by monitoring or Inspection/repair of the structure (for f, o)[b]
Unknown phenomena	• Research & development

[a] Measure by structural reliability analysis
[b] Measure by risk assessment

damages, which inevitably occur, do not escalate into disproportionate consequences" [7]. Such criteria were first introduced by the Norwegian Petroleum Directorate (Currently: The Norwegian Ocean Industry Authority) [19], and later adopted for instance in ISO standards for oil and gas platforms, bridges [20]. The design check is a two-step procedure, [21]:

(1) estimate accident scenarios for various hazards with an annual exceedance probability of the order 10^{-2} to 10^{-4}.
(2) demonstrate survival under permanent and environmental loads, by a ULS design check with partial safety factors equal to 1.0.

It is noted that the robustness that is provided by the ALS criteria is also needed to get the necessary damage-tolerance to rely on inspection of e.g. crack before they lead to rupture. Finally, it is noted that an ALS criterion was introduced to limit the risk associated with "abnormal" wave loading at an exceedance probability of 10^{-4}, say with a high crest or other unusual shape—which is not a simple "extrapolation" of the 10^{-2} event, with relevance for potential wave-load-in-deck scenarios.

Quality assurance and control (QA/QC). In view of the service experiences with accidents, QA/QC of the analysis and design and fabrication process as well as during operation, is crucial. The QA/QC in the design phase is particularly important for novel concepts and, with the increasing digitalization, also of computational tools and their use. The QA/QC of the engineering process should address errors and omissions of individuals, inadequate procedures, software and lack of robustness of the organizations. Lack of adequate training, time, and teamwork or back-up (disciplined process auditing procedures in the organization) has been the root cause of such errors [2, 5]. A main challenge in the design phase is to be aware of possible new phenomena, e.g. new load phenomena, dynamic response features and failure modes, especially associated with novel conceptual designs. Laboratory tests are crucial in this context. "Recently" discovered new phenomena such as ringing and springing of TLPs, and vortex induced motions of floating platforms, were observed in time before any catastrophic accident occurred, and the necessary precautions made. It is also interesting to observe that the failure mode of fatigue for ships and offshore platforms—was fully recognized in 1980-1990s despite earlier experiences [6]. Another issue is that fatigue depends on local structural quality and hence easily be susceptible to fabrication errors. And is hence dependent on adequate damage tolerance. The quality check of the structure during fabrication and operation is primarily done by visual or NDT inspection. Remote monitoring of structural components is carried out mainly performed to ensure the accuracy of methods used in predicting the behavior, while health monitoring to detect e.g. crack damage before it is critical, is challenging due to the extensive instrumentation needed to have an adequate representation of possible crack sites in large steel structures, and is not much in use for platforms. However, monitoring of possible leaks in floating structures is a viable approach to detect flooding before it escalates into more severe consequences.

Regulations, standards, guidelines. Standards provide a bridge between the state of knowledge (research front) and engineering design and operation. Moreover, standardization can help reduce costs since they can limit unnecessary alternatives and thereby make fabrication and assembly focused on optimal life cycle costs. While structures traditionally should fulfill serviceability and safety requirements, serviceability requirements are recommended in standards, but are decided by the owner. The focus in standards is hence of safety. Sustainability requirements in standards, are still at an early stage, and efforts to establish criteria and methods for assessment is needed.

The regulatory regimes for floating structures in various sectors, differ. There are international standards like ISO for oil and gas platforms [20]. IEC for floating wind turbines, [22]. Provisions for stability and other maritime issues are normally based on those of IMO and their corresponding national organizations. In addition, many national authorities and classification societies issue standards and recommended practice.

A main challenge with the ULS design criteria is the occurrence of new phenomena, especially associated with novel conceptual designs. "Recently" discovered new phenomena such as ringing and springing of TLPs, and vortex induced motions of floating platforms, were observed in time before any catastrophic accident occurred, and the necessary precautions made. Other phenomena, such as wave-in-deck load, caused severe damage on some platforms, before risk mitigation actions were taken [23].

2.4 Structural Integrity Management of Floating Structures in the Emerging Sectors

As mentioned above, knowledge transfer from structural engineering in the oil and gas sector should be made with due attention to the differences in function (Sect. 2.1), hazards and consequences and hence the required safety level as well as the cost of improving the safety level. In the following the focus is on safety aspects.

The offshore oil and gas industry has a large damage potential due to the large energy and pollution potential inherent in the hydrocarbons. All ocean structures are more or less subjected to ship impact risk and environmental loads.

The fatality risk varies with the personnel at risk—from none on floating wind turbines, small fish farms and solar power plants, to some manning on large fish farms, to public exposure on large on floating bridges. The potential environmental damage caused by failures of these emerging floating structures is limited compared to that for oil and gas platforms. But escape of fish from fish farms represents an environmental threat of concern.

A very important difference between the emerging sectors and the oil and gas sector is concerned with the level of costs and profit. While floating wind, solar power plants and fish farms, typically represent an economy in a smaller scale than oil and gas (production) platforms, floating bridges which are part of the national highway system might represent a large investment and a potential societal loss in case of failure.

These differences would have an influence on the use of various safety measures

- while in general SLS, ULS and FLS design criteria are applied for all types of structures while the applicability of, e.g. the ALS criteria, needs to be discussed. ALS requirements are used for floating bridge [24], to have at least the same safety level as oil and gas platforms.
- QA/QC is important for structures in the emerging sectors due to the novelty of function, layout etc. that can imply new load conditions and failure modes etc.
- Moreover, differences in loads and structural behavior modes might require different environmental data and analysis methods. For example, long floating bridges are subjected to inhomogeneous environmental conditions, and the behavior of floating wind turbines depends on hydro- and aerodynamic loads and active automatic control. Another example is the complex interaction between modular, elastic floating structures and a multi-line mooring systems.
- The limited operational experiences with the floating structures in the emerging ocean sectors, also need to be taken into account in the design of such facilities.

While the standards for oil and gas platforms have reached some maturity, the international standard, IEC 61400-3-2 for floating wind turbines is at an early stage and refer extensively to oil and gas platform practice. Provisions for stability and other maritime issues are based on those of IMO and their corresponding national organizations, even if consequences in terms of fatalities and environmental damage is very limited. The application of damage tolerance criteria (ALS) for overall stability, strength of the hull and mooring system is still open for discussion. For fish-farms and floating civil infrastructure like floating bridges, terminals etc. there are some national standards, but with quite general guidance and reference to other standards. For floating bridges, some regulations//standards, like N-400 [24] exist, while in practice, a specific design basis for each project is established. In due time, when more in service experiences have been collected, they can result in a standard.

In principle, application of all risk mitigation approaches in Table 1, should be considered for a balance in safety and lifecycle expenditure.

Public infrastructure is in a particular situation since a (global) failure might cause public fatalities and represent large costs, partly for repair/replacement, but mainly due to the costs of road traffic disruption. The Norwegian Public Road Administration [24], hence, implements all risk mitigation measures in Table 1, i.e. limit state criteria including ALS. Emphasis is placed on QA/QC (including risk analysis) to identify possible new hazards (load phenomena) and failure modes as well as measures to control the associated risk. The QA/QC performed for floating bridges in Norway, identified challenges that led to research projects, including extensive model testing, e.g. [25]. Moreover, several engineering design studies were conducted by two independent consortia of consulting companies. For example, in the Bjørnafjord floating bridge project in Norway the QA/QC process identified new phenomena, e.g. global dynamic instability of a curved floating bridge, hydrodynamic interaction between pontoons, the effect of inhomogeneity in environmental conditions, wave-current interaction and the important contribution of relatively "frequent", moderate ship impacts to the risk of road traffic disruption [25]. Moreover, there is a challenge to design the mooring system in view of the disturbingly high failure rate of mooring lines in the offshore industry. Rather than introducing extra safety margins to cover such "unknown" phenomena, it is crucial that sufficient numerical and experimental investigations are carried out during the OA/OC process to establish a basis for accounting for the possible effects.

2.5 *Structural Reliability and Risk Analysis*

The total efforts to ensure safety (design criteria, QA/QC) are supposed to ensure an acceptable safety level. In principle tools like structural reliability analysis (SRA), e.g. [26] and quantitative risk analysis (QRA), e.g. [18], provide means to quantify failure probabilities of different failure consequences, and hence the risk of failure. As mentioned in Sect. 2.3. SRA has been applied to calibrate the ULS requirements for floating structures in the oil and gas industry, but not yet in emerging ocean sectors.

Some regulations and standards for oil and gas platforms require a QRA to estimate the risk of fatalities, environmental damage and economic loss, e.g. [18], and its compliance with target levels. The contribution to the risk due to structural failure have commonly been estimated by assuming that the ALS criteria are fulfilled based on characteristic accidental loads at an annual probability level on e.g. 10^{-4} (Sect. 2.3) estimated by the risk analysis. This is, however, an approximation that does not give a good estimate of the risk, as discussed in Sect. 3.

Moreover, risk analysis is crucial at the initial stages of developing new concepts, e.g. floating structures in the emerging sectors—to identify hazards, failure modes and—consequences. The results can then be used to specify necessary research and development to cope with the identified challenges.

An illustration of the use of risk analysis is presented in Sect. 3 to use risk analysis to shed light on existing design criteria for floating bridges.

3 Risk-Informed Design of Floating Bridges Against Ship Impacts

3.1 *General*

Accidental loads are primarily handled by ALS criteria, which were introduced for offshore structures in 1984 to ensure more robust structures by a design check in two steps, as mentioned in Sect. 2.3 [19]. This approach is also adopted for bridges (e.g. [24]), considering relevant accidental loads. Ship impacts is the dominating accidental load and among the consequences: fatalities, environmental damage and economic loss, the latter and especially the cost of traffic disruption, is the governing design condition [9]. The accidental load should be determined by a risk analysis. As reviewed in [27], some standards for bridges (e.g. ASHTO; Eurocodes) refer to risk-based design of bridges. However, the existing risk analysis procedures have limitations with respect to application to floating bridges. To establish a basis for risk-informed of structural integrity relating to ship impacts, a comprehensive risk analysis has been conducted for an early design of the Bjørnafjord floating pontoon bridge.

3.2 *Ship Impact Risk Analysis Procedure*

The purpose of the present risk analysis is to estimate the ultimate consequence of ship impacts, namely: fatalities to users of the bridge as well as road traffic disruption caused by the impact. The ship collision risk analysis [9, 27] is carried out by estimating the

- Frequency versus impact energy and direction (traffic pattern-types and number of vessels, possible lanes etc., root causes of operational and technical faults, bridge layout, considering the mitigation measures.
- Consequences

  Immediate consequences in terms of structural damage, possible buoyancy loss and girder motions
  Ultimate consequences in terms of fatalities/injuries and roads traffic disruption

It turns out [9] that the impact on the pontoons makes the largest contribution to the ship impact risk for the current bridge. It is also found that the contribution to the fatality risk due to ship collisions, is small compared to the road traffic risk of fatalities on the bridge. Hence, the focus in this paper is on road traffic disruption.

A three—step analysis procedure is necessary to apply to determine the immediate consequences in terms of internal forces, possible permanent damage and the girder motions (accelerations) [9]:

(1) nonlinear shell analysis is carried out to determine the impact force–displacement (indentation) relationship in the impact region by using the finite element model in Fig. 3.
(2) global analysis of the impact is made. As illustrated in Fig. 4, the ship-bridge contact, is modeled by a mass-nonlinear spring system connected to the bridge, with a stiffness obtained by the force–deformation curve in step1. The default model of the other parts of the bridge is modelled by a linear elastic model but is replaced by a nonlinear shell sub-structure if the linear analysis indicates nonlinear material behavior.

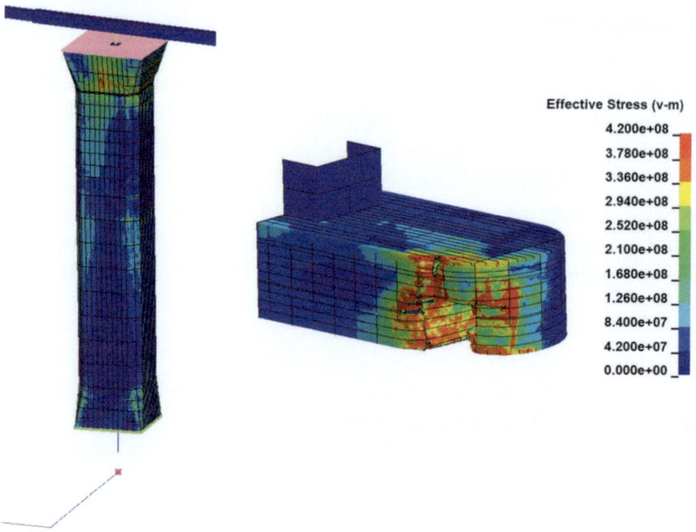

Fig. 3 Impact analysis model [9]

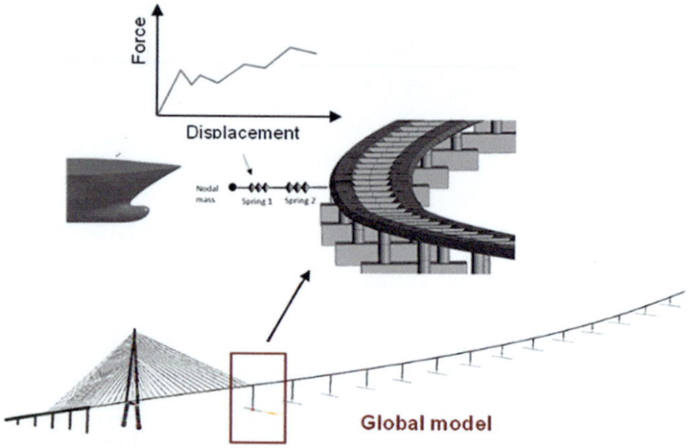

Fig. 4 Example of damage in a 125 MJ transverse impact on the pontoon of the tall column [9]

(3) Based on the results from these first two steps, the damage (indentation, fracture) and possible flooding is determined. Figure 3 shows an example of damage after the global analysis of a transverse impact on a pontoon.

The repair time—including mobilization of the equipment, i.e. by use of drydock for pontoon repair, was estimated to be between 2 and 20 months [9].

The traffic disruption time, when ULS and SLS criteria are nor fulfilled, was estimated by the annual frequency of each impact scenario, the immediate consequence (structural damage, flooding) and the time for damage and repair conditions based on the damage condition and choice of repair method.

3.3 Risk Estimation and Mitigation

It turned out that the expected annual downtime due to ship impacts was about 7 days in the initial study, with 90% of the downtime stemming from impacts on the pontoons with an energy up to 50 MJ. This result caused scrutinizing of the impact frequency analysis, which initially was focused on severe impacts (at an annual frequency of 10^{-4}) for the ALS check. Moreover, further ship traffic control measures to reduce the impact probability, are pursued. Another option that was proposed [9, 27], was to introduce a limit state in addition to the ALS to address the risk associated with frequent impacts, implying strengthening of the bridge girder and columns that did not fulfill the ULS or SLS criteria in damaged and repair conditions.

4 Concluding Remarks

Lifecycle structural integrity management is discussed in view of accident experiences for floating structures in various sectors. The main cause of accidents is human and organizational errors and omissions that cause accidental loads or abnormal strength. To achieve an acceptable safety level, therefore, requires fulfillment of ULS and especially FLS (and inspection plan) requirements, since the latter are often not recognized in early phases of the developments of new concepts. For high reliability/consequence class structures, fulfillment of damage tolerance (ALS) requirements, as well as QA and QC of the engineering process (including identification of hazard and failure modes); inspection, monitoring and repair of the structure, are important. QA and QC tasks (including assessment of digital tools and their use), are particularly important for novel concepts. Finally, a study of risk-informed design of a floating bridge under ship impacts, is presented, indicating that ("frequent") impacts with an energy in the range of 20–50 MJ, contribute more to the road traffic disruption than the impacts with an annual exceedance frequency of 10^{-4}, corresponding to an energy of about 150 MJ used for the ALS check. To ensure an acceptable risk of traffic disruption, one risk mitigation option that was proposed in [9, 27], was to introduce a limit state for design against frequent ship impacts in addition to the ALS requirement.

References

1. Chakrabarti S (ed) (2005) Handbook of offshore engineering, vol 1&2. Elsevier, Amsterdam

2. Moan T (2004) Safety of floating offshore platforms. Keynote lecture. In: 9th PRADS conference Travemünde-Lübeck, STG
3. Moan T (2014) Safety of Offshore Structures, Keynote lecture, OSRC Conf., API, 16–18 Sept, Houston, USA
4. WOAD: Worldwide offshore accident databank. Det Norske Veritas, Oslo, updated continuously
5. Bea RG (2005) Design for reliability: human and organizational factors, Chapter 12, Handbook of Offshore Eng, Chakrabarti S (ed). Elsevier, Amsterdam
6. Moan T (2007) Fatigue reliability of marine structures—from the Alexander Kielland accident to life cycle assessment of safety, ISOPE Keynote lecture, San Francisco. J ISOPE 17(1):1–21
7. Moan T (2009) Development of accidental collapse limit state criteria for offshore structures. Struct Saf 31(2):124–135
8. Zhang G, Liu Y, Lan S, Yang J (2022) Causes and statistical characteristics of bridge failures: a review. J Traffic Transp Eng 9(3):388–406
9. Moan T, Jin Y (2023) Risk Analysis of a floating bridge subjected to ship collisions. Report to NPRA, January 15
10. CWIF (2020) Summary of wind turbine accident data to 31. March 2020, www.caithnesswindfarms.co.uk. Accessed 24 April 2020
11. Santelo TN, de Oliveira CMR, Dias Maciel C, de A. Monteiro RB (2022) Wind turbine failures review and trends. J Control Autom and Electric Syst 33:505–521
12. Holen SM, Yang X, Utne IB, Haugen S (2019) Major accidents in Norwegian fish farming. Saf Sci 120: 32–43
13. OPD (2024) Oil platform disasters—Wikipedia. Accessed on 20.08.2024
14. Stiff J (2014) How reliable are reliability calculations? (Illustrated with station keeping examples). In: Proceedings of the offshore structural reliability conference, 16–18 Sept, Houston, USA
15. Smedley P, Petruska D (2014) Comparison of global design requirements and failure rates for industry long term mooring systems. In: Proceedings of the offshore structural reliability conference, API, 16–18 Sept, Houston, USA
16. Cornell CA (1995) Structural reliability—some contributions to offshore technology, Paper No. OTC 7753. In: Proceedings of the 27th off-shore technology conference, Houston
17. Moan T (1995) Safety level across different types of structural forms and material—implicit in codes for offshore structures, SINTEF Report STF70 A95210, Trondheim. Prepared for ISO/TC250/SC7
18. Vinnem JE (2014) Offshore risk assessment. Springer, London
19. NPD (1984) Regulations for load-carrying structures for extraction or exploitation of petroleum. The Norwegian Petroleum Directorate, Stavanger
20. ISO 2394 (2015) General principles on reliability for structures. International Standards Organization, London, UK
21. NORSOK N-001 (2021) Integrity of offshore structures. Norwegian Technology Standards, Oslo, Norway
22. IEC 61400-3-2 (2024) Design requirements for floating wind turbines. IEC, Geneva, Switzerland
23. Ma L, Swan C (2023) Wave-in-deck loads: an assessment of present design practice given recent improvements in the description of extreme waves and the nature of the applied loads. Ocean Eng 285:115302
24. NPRA (2024) Handbook N400: design of bridge (in Norwegian). Norwegian Public Road Administration, Oslo
25. Moan T, Økland OD, Øiseth OA (2024) Analysis and design of floating bridges—recent advances and challenges. In Wang, Wang, Weinert, de Graaf-van Dinther (eds) Large floating solutions: design, construction and legality of offshore structures and buoyant urbanism
26. Melchers RE (1999) Structural reliability analysis and prediction. Wiley, Chichester, UK
27. Moan T (2024) Estimation and mitigation of ship collision risk of a floating bridge. Submitted for publication

Expandable Floating City with Modular Floating Structures and Related Hydrodynamic-Elastic Analyses

Masashi Kashiwagi

Abstract Some examples of expandable floating city are introduced, consolidating box-type modular floating structures (MFS) with square or equilateral triangle plane shape. Since reduction of wave-induced motions of the floating city is important for safety and comfort to use it for residential purpose, an overview is written of existing methods available for accurate computation of hydrodynamic forces and resulting motions. Mathematical modeling is also described for mechanical interaction forces originating from the mooring, connector, and wave energy converter equipped with power take-off system. Effects of elasticity in the MFS system are briefly reviewed.

Keywords Floating city · Modular floating structures · Wave-induced response · Mooring and connector system

1 Introduction

It is said that about 2.5 billion people live in coastal cities at present, and this number will increase much more in the future [1]. Thus, the demand for extending the habitation space near coastal areas is on the rise but land availability is limited. Furthermore, due to global warming, the mean sea level is said to rise, and a big earthquake followed by huge tsunami is predicted to occur. In order to address these concerns, very large floating structures (VLFSs) have been considered, which is more promising and sustainable than the traditional land reclamation, owing to the advantages of low environmental impact, free site selection, and flexibility for expansion and decommission.

Although VLFSs have been intensively studied [2] and analysis methods for hydroelasticity have been developed [3], they are usually designed to serve a single

M. Kashiwagi (✉)
Hyundai Engineering & Construction, Seoul, Korea
e-mail: kashi@naoe.eng.osaka-u.ac.jp

Osaka University, Suita, Osaka, Japan

X. L. Zhao et al. (eds.), *Proceedings of the Fourth World Conference on Floating Solutions: WCFS 2024; 2–4 December; Hong Kong*, Lecture Notes in Civil Engineering 597, https://doi.org/10.1007/978-981-96-4569-5_3

purpose (such as airport), and not considered as a viable alternative for residential development. Due to its huge plane dimensions, a continuous VLFS tends to cause significant bending moments and shear forces. As an alternative idea for resolving this problem, a modular floating structure (MFS) system has been also considered. The concept of the MFS system is based on an array of relatively small floating modules consolidated with rigid or flexible connectors, which can facilitate future extension simply by adding new modules according to actual demand. Since the size of each module would be comparable to typical wavelengths in the ocean, it is effective to consider the rigid module and flexible connector (RMFC) method for hydrodynamic analysis in the preliminary design of the MFS system for urban use.

In considering feasibility of the MFS system for a residential area, its safety and comfort are of critical importance, and hence the wave-induced responses of MFS system must be as less as possible in magnitude. It may be achieved by use of breakwaters, but from the viewpoint of being eco-friendly and not restricted by water depth at the installation site, development of efficient floating breakwaters should be considered. In addition, recently the integration of response reduction and wave-energy power generation for MFS systems attracts attention. For instance, Nguyen and Wang [4] investigated heaving Wave Energy Converter (WEC)-type attachments to a pontoon-type VLFS for extracting wave energy and reducing the hydroelastic response of the VLFS. Each heaving WEC-type attachment is assumed to consist of a linear Power Take-Off (PTO) system connected to VLFS at one end and to the seabed at the other end.

In the MFS system, adjacent modules will be consolidated with connectors, and hence the motion characteristics of the MFS system change depending on the number of degrees of freedom (DOF) to be restricted by the connector system. Generally, the more DOF a connector restricts, the more the relative motion between two adjacent modules can be suppressed, but at the expense of increased load on the connector. Therefore, it is of critical importance to understand the characteristics of the connector system to be used and their effects on the response of the entire MFS system. Jiang et al. [5] presented a comprehensive review of connectors and mooring systems.

In this paper, an idea for realizing a floating city using the MFS system is introduced, in which the box-type floating structures with square or equilateral triangle plane shape are used as modular floats, and those will be consolidated by connectors. In order to guarantee the safety and comfort of this-type of floating city, wave-induced responses must be efficiently reduced by installing floating breakwaters, flexible connectors, WECs with PTO system, and so on. An overview of existing calculation methods for accurate prediction of wave-induced responses is described, with attention paid to how to construct the motion equations by taking account of hydrodynamic and mechanical interactions among multiple floats and connectors.

2 Examples of Expandable MFS

Due to global warming, it is said that the mean sea level will rise and 90% of the world's megacities are vulnerable to rising sea levels. A floating city under planning in Busan Metropolitan City, South Korea, designed by New York-based ocean tech company OCEANIX [6] can be a prototype that is resistant to rising sea levels. This plan was unveiled in 2022 at the Second UN Roundtable on Sustainable Floating Cities, and design details are in progress. Some unit floating structures will be serially connected to build up a floating city, which has the scalability and adaptiveness that can respond to the increase in residents and changes in needs.

As a modular unit structure, the box-type floats with square and equilateral triangle plane shapes are considered as shown in Fig. 1. Then, by combining these two box-type modules, a larger system of modular floating structures (MFSs) can be generated as shown in Fig. 2, which may be further connected and integrated to expand the area (see Fig. 3) or rearranged to adjust the size and layout, which can be used for the residence, comfort and commercial facilities, and so on.

Based on a similar idea, the Space@Sea project [7], funded by Horizon 2020, was conducted by the consortium of 17 European partners, which has studied and

Fig. 1 Modular unit floats

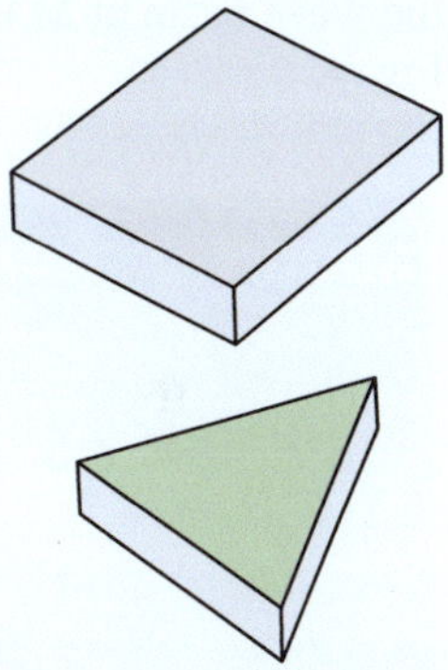

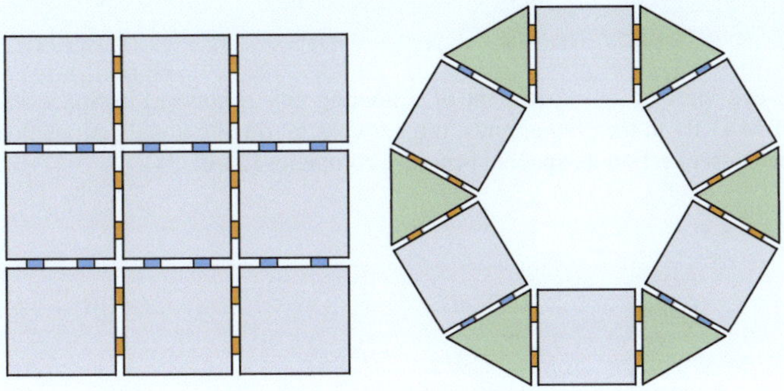

Fig. 2 Examples of consolidated modular floating structures

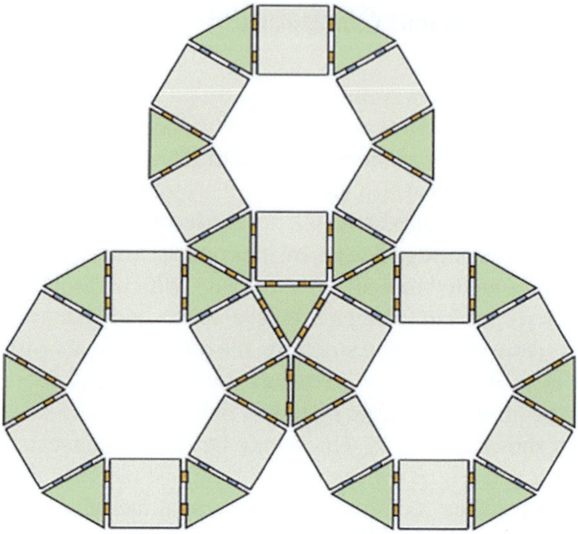

Fig. 3 Example of expanded floating city using modular floats

developed a system of interconnected floating pontoon-type island. Experiments in the wave basin at MARIN of the Netherlands were also carried out, as depicted in Fig. 4.

Fig. 4 Bird's-eye view of an experiment of a floating city conducted in Space@Sea project at wave basin of MARIN in the Netherlands. https://www.marin.nl/en/publications/floating-islands-one-step-closer-to-realisation-as-spacesea-project-completes (24/07/31)

3 Numerical Computations for Realization of Floating City

Of particular importance to realize a floating city with concept of the MFS system are the safety and comfort in a harsh environmental condition with large-amplitude waves. For residential purpose, the motion amplitudes in heave, pitch and roll and the resultant vertical acceleration must be suppressed as small as possible at some important local positions on the MFS to be considered. Therefore, we must understand correctly hydrodynamic characteristics of interacting multiple floating bodies in waves, and then be able to design optimized connectors between adjacent modular floats, develop a floating breakwater with excellent wave-reflection performance, and invent an efficient device for the integration of wave-energy absorption and motion reduction.

In order to perform those investigations, we must be able to compute the following items:

(1) Frequency–response characteristics of box-type floating bodies with square and equilateral triangle plan forms to be adopted as the modular unit float.
(2) Hydrodynamic interactions among a large number of modular floating structures oscillating with six degrees of freedom in regular incident waves.
(3) Modeling of connectors and mooring systems, and those mechanical interaction forces to be incorporated into the motion equations of the MFS system.
(4) Wave reflection characteristics of floating breakwaters for a practical range of wavelengths and their optimization in performance.
(5) Modeling of the WEC equipped with PTO systems and the absorbed wave-energy power, and the resulting reduction of the motion of integrated MFS system.

In the next section, with the premise that in-house computation codes will be used as much as possible, an overview is described on computation methods that can be used at present, some results obtained so far, the present status and improvement to be made, some points to be cautious, related researches done in the past, and so on.

4 Hydrodynamic-Elastic Analyses

4.1 Frequency–Response Characteristics of Multiple Floating Bodies

The basic unit float in the MFS system considered is a simple box-type module due to easy construction and cost reduction, and assumed to be rigid. As the first step, it is necessary to be able to calculate accurately the hydrodynamic forces on the modular floats, and to understand the frequency–response characteristics in waves and their physical reasons. When there are multiple floating structures, the frequency–response characteristics of hydrodynamic forces vary greatly due to hydrodynamic interactions among the modular floats, but the numerical calculation method can be basically the same as that for a single floating body. In the present floating-city project, a higher-order boundary element method (HOBEM) [8] is applied to obtain accurate results with a smaller number of unknowns. The underlying theory is the linear potential theory, and the velocity potential representing the flow field around floating bodies is expressed as

$$\phi(\boldsymbol{x}) = \mathrm{Re}\left[\frac{ig\zeta_a}{\omega}\left\{\phi_I(\boldsymbol{x}) + \phi_S(\boldsymbol{x}) + K\sum_{j=1}^{6}\sum_{n=1}^{M} X_j^n \phi_j^n(\boldsymbol{x})\right\}e^{i\omega t}\right] \tag{1}$$

where $\boldsymbol{x} = (x, y, z)$ is the position vector, the x–y plane is taken on the still water surface, and the z-axis in the Cartesian coordinates is positive vertically upward. $\phi_I(\boldsymbol{x})$ denotes the incident-wave potential, which is assumed known and made nondimensional with amplitude ζ_a, circular frequency ω, and gravitational acceleration g. The wavenumber $K = \omega^2/g$ is, in the case of finite water depth, different from the wavenumber $k_0 = 2\pi/\lambda$ of incident wave (λ is the wavelength), but they are correlated by the dispersion relation given as $K = k_0 \tanh k_0 h$, with h being the constant water depth. $\phi_S(\boldsymbol{x})$ represents the scattering potential and $\phi_I(\boldsymbol{x}) + \phi(\boldsymbol{x}) \equiv \phi_D(\boldsymbol{x})$ is referred to as the diffraction potential. $\phi_j^n(\boldsymbol{x})$ in Eq. (1) is the radiation potential due to the j-th ($= 1 \sim 6$) mode of motion of the n-th ($= 1 \sim M$) floating body, and X_j^n is its complex amplitude. This complex amplitude X_j^n will be determined by solving the motion equations of floating bodies. Thus, when considering independent free motions of M rigid floating bodies, there are $6 \times M$ unknowns, and hence we should consider $6 \times M$ coupled motion equations.

In order to calculate the hydrodynamic forces on floating bodies, it is necessary to determine the pressure, equivalently the velocity potentials, $\phi_D(\boldsymbol{x})$ and $\phi_j^n(\boldsymbol{x})$, on the wetted surface of floating body. This can be achieved by solving the integral equation for the velocity potential on the body surface, which can be derived by Green's second theorem in terms of the free-surface Green function [8]. To solve this integral equation with HOBEM, the wetted surface of floating body is divided into panels, and the geometric shape and the unknown velocity potential on each discretized panel are represented using the same shape function (called iso-parametric element). Then, the values of velocity potentials at the nodal points making up panels become new unknowns to be determined by solving the transformed integral equation.

Once the velocity potential on the body surface could be obtained, the hydrodynamic force, F_j^m, acting on the m-th floating body in the j-th direction can be computed and expressed as follows:

$$F_j^m = E_j^m - \sum_{k=1}^{6}\sum_{n=1}^{M} X_k^n \mathcal{F}_{jk}^{mn} - \sum_{k=1}^{6} X_k^m C_{jk}^m \tag{2}$$

where E_j^m denotes the wave-exciting force, $\mathcal{F}_{jk}^{mn}$ the radiation force exerted by the k-th mode of motion of the n-th floating body. Note that these are given in complex, but as shown in Eq. (1), the time dependency is assumed as $e^{i\omega t}$ and the real part of multiplication of complex quantities must be taken in the end. These hydrodynamic forces can be computed from

$$E_j^m = -\rho g \zeta_a \iint_{S_m} \phi_D n_j^m dS, \quad \mathcal{F}_{jk}^{mn}/(i\omega)^2 = -\rho \iint_{S_m} \phi_k^n n_j^m dS \equiv A_{jk}^{mn} + \frac{1}{i\omega} B_{jk}^{mn} \quad (3)$$

Here, A_{jk}^{mn} and B_{jk}^{mn} are the added mass and the damping coefficient, respectively; S_m the wetted surface of floating body m; n_j^m the j-th component of the normal vector on floating body m. Furthermore, C_{jk}^m in Eq. (2) denotes the restoring-force coefficients of floating body m, in which there is no coupling among the floating bodies.

Substituting these equations into the basic motion equations based on Newton's second law, we can obtain the $6 \times M$ coupled motion equations for M modular floating bodies, which may be expressed as follows:

$$\sum_{k=1}^{6} \left[\left\{ (i\omega)^2 M_{jk}^m + C_{jk}^m \right\} X_k^m + \sum_{n=1}^{M} \mathcal{F}_{jk}^{mn} X_k^n \right] = E_j^m \quad (4)$$

where $m = 1 \sim M$ and $j = 1 \sim 6$, and M_{jk}^m denotes the mass-matrix coefficients of floating body m.

This motion equation can also be expressed in a different form by separating the radiation forces into the component for floating body m alone (that is, when no interactions exist with other floating bodies), which is denoted as A_{jk}^m and B_{jk}^m, and the remaining component originating from hydrodynamic interactions due to the.

presence of other floating bodies. The resulting expression is written as

$$\sum_{k=1}^{6} \sum_{n=1}^{M} \left[D_{jk}^m \delta_{mn} + \mathcal{D}_{jk}^{mn} \right] X_k^n = E_j^m \quad \text{for} \begin{pmatrix} m = 1 \sim M \\ j = 1 \sim 6 \end{pmatrix} \quad (5)$$

where

$$D_{jk}^m = (i\omega)^2 \left\{ M_{jk}^m + A_{jk}^m \right\} + i\omega B_{jk}^m + C_{jk}^m, \quad \mathcal{D}_{jk}^{mn} = \mathcal{F}_{jk}^{mn} - \left\{ (i\omega)^2 A_{jk}^m + i\omega B_{jk}^m \right\} \quad (6)$$

and δ_{mn} means Kronecker's delta symbol, equal to 1 only for $m = n$.

It should be noted here that, when considering hydrodynamic forces on floating body m due to its own oscillatory motion, the waves generated by the motion of float m are reflected by the other floats and incident again on float m as an incident wave. Thus, the additional "scattering" force due to the reflected waves is expressed as $\mathcal{D}_{jk}^{mn}$, which must be never zero.

By the way, the hydrodynamic interaction forces due to multiple wave reflections among a large number of floating bodies may be calculated by the wave interaction theory of Kagemoto and Yue [9]. In particular, when the modular floating structures are identical in shape and size, the interaction forces can be calculated with relative ease, as long as the diffraction characteristics matrix of the unit floating body is obtained, which is effective in reducing the computation time, especially as the number of modular floats increases. However, when adjacent floating bodies are so close that there exists a region where the addition theorem for Bessel functions used in the wave interaction theory becomes invalid, the calculation accuracy decreases, as discussed by Kashiwagi [10]. Therefore, in those cases with close proximity between adjacent bodies, HOBEM for multiple floating bodies must be directly applied, even if it increases the number of unknowns and the computation time.

4.2 Modeling of Additional Forces Due to Mooring and Connector Systems

As shown in Fig. 5, we consider a gen- eral case where there is a mooring sys- tem or a device of connector consoli- dating with the adjacent floating body at an arbitrary position of 3-D multiple floating bodies, for which we obtain the expressions to be added to the motion equations as mechanical interaction forces due to those mooring and connector systems by referring to Matsuura and Ikegami [11].

First, we assign serial numbers such as m and n to multiple modular floating bodies. Let us express the motion displacement of floating body m with 6-DOF, considered in the coordinate system with the origin at the center of floating body m, in terms of vector notation as follows:

$$\{\xi^m\} = (\xi_1, \xi_2, \xi_3, \xi_4, \xi_5, \xi_6)^T = (\boldsymbol{\alpha}_T, \boldsymbol{\alpha}_R)^T \tag{7}$$

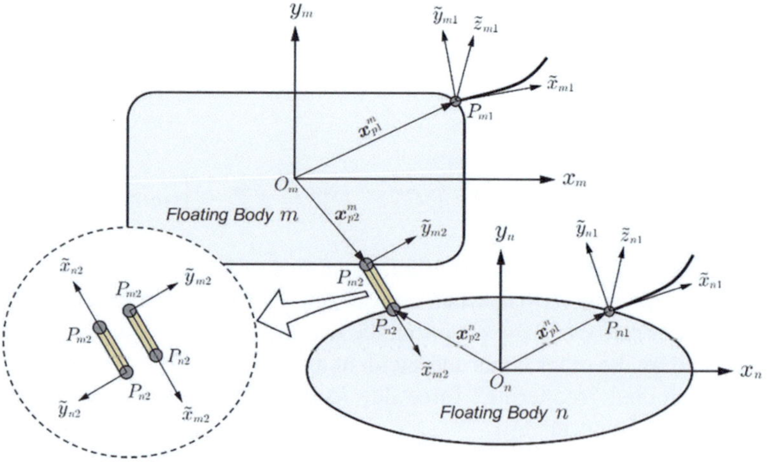

Fig. 5 Coordinate systems and notations for the analysis of modular connected multiple floating bodies

where $\boldsymbol{\alpha}_T = (\xi_1, \xi_2, \xi_3)$ and $\boldsymbol{\alpha}_R = (\xi_4, \xi_5, \xi_6)$ are the vectors of the linear and rotational motions, respectively.

Next, we consider the motion displacement vector $\left\{\xi_p^m\right\}$ at $\boldsymbol{x}_p = \left(x_p, y_p, z_p\right)$ where the mooring system, such as a mooring cable, is attached. The rotational motion is the same at any point considered, and the vector of linear motion is given by $\boldsymbol{\alpha}_T + \boldsymbol{\alpha}_R \times \boldsymbol{x}_p$. Therefore, by using the transformation matrix $[\Xi]$ representing these simple relations, the motion displacement vector can be expressed as $\left\{\xi_p^m\right\} = [\Xi]\{\xi^m\}$.

The mooring force, for example, given by the catenary theory is often more conveniently described in terms of local coordinates $\tilde{\boldsymbol{x}}_p$ with the origin placed at the mooring position. The relationship between $\tilde{\boldsymbol{x}}_p$ and $\boldsymbol{x}_p$ can generally be expressed by rotating the coordinate system using the Euler angles in a form of $\tilde{\boldsymbol{x}}_p = \left[\gamma\right]\boldsymbol{x}_p$, with the coordinate transformation matrix denoted as $\left[\gamma\right]$ (its details are omitted here). If we express the motion displacement vector in this local coordinate system as $\left\{\eta_p^m\right\}$, it can be correlated with $\{\xi^m\}$ as follows:

$$\left\{\eta_p^m\right\} = [\Gamma]\left\{\xi_p^m\right\} = [\Gamma][\Xi]\{\xi^m\} \equiv [T_m]\{\xi^m\} \tag{8}$$

where $[\Gamma]$ is defined as a 6×6 matrix, with 3×3 submatrix $\left[\gamma\right]$arranged as the diagonal element.

The mooring force and moment are modeled to be proportional to the 6-DOF motion displacement vector $\left\{\eta_p^m\right\}$ in the local coordinates $\tilde{\boldsymbol{x}}_p$ in the following form

$$\left\{F_p^m\right\} = -[K]\left\{\eta_p^m\right\} \tag{9}$$

where $[K]$ is defined as 6×6 matrix of proportional coefficients representing the restoring force due to the moor- ing system. The relationship between this vector of the mooring force and moment $\left\{F_p^m\right\}$ and the corresponding vector $\{F^m\}$ expressed in the original coordinate system with the origin at the center of floating body m can be obtained using the transformation matrix $[T_m]$ given in Eq. (8), and further expressed with Eqs. (8)–(9) as

$$\{F^m\} = [T_m]^T\left\{F_p^m\right\} = -[T_m]^T[K]\left\{\eta_p^m\right\} = -[T_m]^T[K][T_m]\{\xi^m\} \equiv -\left[H^m\right]\{\xi^m\} \tag{10}$$

We can see from this equation that $[H^m]$ is a 6×6 matrix of coefficients representing the additional force on floating body m by the mooring system connected at $\boldsymbol{x}_p$ of floating body m.

Next, let us consider the coefficient matrix of the additional force due to the connector consolidating float m and float n. The position of this connector is denoted as $\boldsymbol{x}_p^m$ in float m and $\boldsymbol{x}_p^n$ in float n, the vectors of 6-DOF motion displacement of each float are expressed as $\{\xi^m\}$ and $\{\xi^n\}$, and the vectors of motion displacement in the local coordinate system with origin at the connection point are expressed as $\left\{\eta_p^m\right\}$ and $\left\{\eta_p^n\right\}$. In this case, denoting the 6 × 6 matrix of the consolidation-force coefficients as $[K_C]$, proportional to the relative motion displacement between float m and float n, we can write the consolidation force acting on float m in the following form

$$\{F^m\} = -[T_m]^T[K_C]\left(\left\{\eta_p^m\right\} - \left\{\eta_p^n\right\}\right) = -[T_m]^T[K_C]([T_m]\{\xi^m\} - [T_n]\{\xi^n\}) \equiv -[E^{mm}]\{\xi^m\} - [E^{mn}]\{\xi^n\} \tag{11}$$

where

$$\left[E^{mm}\right] = [T_m]^T[K_C][T_m], \quad \left[E^{mn}\right] = -[T_m]^T[K_C][T_n] \; (n \neq m) \tag{12}$$

By taking account of the coefficient matrices representing the mooring force, $[H^m]$, the consolidation forces acting on float m due to the connector between float m and float n,$[E^{mm}]$ and $[E^{mn}]$, we can modify the coupled motion equations for M floating bodies as follows:

$$\sum_{k=1}^{6}\sum_{n=1}^{M}\left[\left(D_{jk}^m + H_{jk}^m\right)\delta_{mn} + \mathcal{D}_{jk}^{mn} + \mathcal{E}_{jk}^{mn}\right]X_k^n = E_j^m \quad \text{for} \begin{pmatrix} m = 1 \sim M \\ j = 1 \sim 6 \end{pmatrix} \tag{13}$$

Here, H_{jk}^m and $\mathcal{E}_{jk}^{mn}$ represent the matrix elements of the 6 × 6 matrices $[H^m]$ and $[E^{mn}]$, respectively.

As a method of consolidating multiple floating bodies, hinge connectors can be used between adjacent modular floating bodies. In hinge connections, the rotational motions at the hinge position are free, thus it is expected that the connecting force and moment due to the connector can be reduced, and then the motion displacement of the modular floats can also be reduced. Shi et al. [12] presented an analysis method for a "Flexible-Base Hinged Connector (FBHC)" in which the center of the connector is ball-hinged, and the consolidation force at the connection point has only three components proportional to the relative linear displacement in the x-, y- and z-axes between the modular floating body and the connection point because of free rotation. Details of the analysis are omitted here due to space limitation and should be referred to Shi et al. [12].

4.3 Absorption of Wave Energy and Response Reduction by a WEC-PTO System

As an attractive idea for reducing the vertical displacement of wave-induced motion of the MFS system, as Nguyen and Wang [4] investigated, it may be effective to install a heaving WEC system using an elastic cable attached to some of modular floating bodies on the upwave side at one end and to the seabed at the other end, with a PTO system equipped at an appropriate position to absorb wave energy.

As a mechanical model of the additional force due to this WEC-PTO system, we consider the linear restoring force by a spring, proportional to the vertical displacement at the mooring point $\boldsymbol{x}_p$, and the linear damping force by a dashpot, proportional to the vertical velocity. In this case, $i\omega\left(b_p\right)_{33} + \left(k_p\right)_{33}$ should be considered in the heave-related element, K_{33}, of the proportional coefficient matrix $[K]$ for the restoring force described in Eq. (9). Here, $(b_p)_{33}$ denotes the damping coefficient of dashpot and $(k_p)_{33}$ is the restoring-force coefficient of spring.

The time-averaged power produced by the PTO system can be calculated by

$$P_{av} = \sum_{i=1}^{N_W} \frac{1}{2}\left(b_p\right)_{33}\omega^2\left|\xi_{p3}\right|^2 \tag{14}$$

where ξ_{p3} denotes the vertical displacement by the heave motion and N_W denotes the number of PTO systems installed. Then, by evaluating the ratio of this power produced to the wave-energy power per unit width of incident wave, the power capture width can be obtained as an index of performance.

4.4 Effects of Elasticity of Module Float and Connector

In the analysis in the preceding sections, the consolidation force added to the modular floating bodies by the connector was modeled using the coefficient matrix $[K_C]$, which is proportional to the relative displacement of adjacent modular floats consolidated by the connector. This is equivalent to a spring or spring-dashpot approximation by considering the connector as an elastic member. If the connector is sufficiently stiff, the spring constant, which is an element of the coefficient matrix $[K_C]$, can be set to a very large value.

On the other hand, if both floating body and connector are considered elastic bodies, the analysis method for VLFS may be applied. However, since VLFSs studied so far have much larger plane dimensions than the draft, they are often approximated as an elastic zero-draft floating plate, and hence the flexural rigidity becomes relatively small so that the elastic response becomes dominant. In this case, as an additional restoring force due to the flexural rigidity of the plate, the stiffness matrix $\mathcal{S}_{jk}^m$ must be included in the motion equations for the vertical displacement. Thus, the motion equations may be written as follows:

$$\sum_{k=1}^{N}\sum_{n=1}^{M}\left[\left(D_{jk}^m + H_{jk}^m + \mathcal{S}_{jk}^m\right)\delta_{mn} + \mathcal{D}_{jk}^{mn} + \mathcal{S}_{jk}^{mn}\right]X_k^n = E_j^m \quad \text{for} \begin{pmatrix} m = 1 \sim M \\ j = 1 \sim N \end{pmatrix} \tag{15}$$

where the mode number k in the complex amplitude X_k^n is extended up to N ($\geq$ 7) to include elastic modes in the deflection of floating plate, with the elastic mode functions (including rigid modes) expressed as $w_j(x, y), j = 1 \sim N$. The calculation formula for $\mathcal{S}_{jk}^m$ for the m-th flexible floating plate with zero draft can be found in Kashiwagi [13], and $\mathcal{S}_{jk}^{mn}$ in Eq. (15) is the additional stiffness matrix due to a flexible connector with beam approximation [14], connecting adjacent two VLFSs, that is, float m and float n.

References

1. Sengupta D, Chen R, Meadows ME (2018) Building beyond land: an overview of coastal land reclamation in 16 global megacities. Appl Geogr 90:229–238
2. Watanabe E, Utsunomiya T, Wang CM (2004) Hydroelastic analysis of pontoon-type VLFS: a literature survey. Eng Struct 26:245–256
3. Ohmatsu S (2005) Overview: research on wave loading and responses of VLFS. Mar Struct 18:149–168
4. Nguyen HP, Wang CM (2020) Heaving wave energy converter-type attachments to a pontoon-type very large floating structure. Eng Struct 219:110964
5. Jiang C, Xu P, Bai X, Zhao Z, el Moctor O (2023) A review of advances in modeling hydrodynamics and hydroelasticity for very large floating structures. Ocean Eng 285:115319
6. OCEANIX Busan. https://oceanix.com/busan/ (Reference: 2024/07/31)
7. Flikkema M, Waals O (2019) SPACE@SEA The floating solution. Front Mar Sci 6. Article 553. https://doi.org/10.3389/fmars.2019.00553 (Reference: 2024/07/31)
8. Kashiwagi M, Endo K, Yamaguchi H (2005) Wave drift forces and moments on two ships arranged side by side in waves. Ocean Eng 32:529–555
9. Kagemoto H, Yue DKP (1986) Three-dimensional bodies in water waves: an exact algebraic method. J Fluid Mech 166:189–209
10. Kashiwagi M (2008) 3-D Calculation for multiple floating bodies in proximity using wave interaction theory. Int J Offshore Polar Eng 18:1–14
11. Matsuura M, Ikegami K (1994) Dynamic response analysis of composite floating bodies in waves (in Japanese). Technical Report of Mitsubishi Heavy Industries. Ltd. 31(4):266–269
12. Shi QJ, Xu DL, Zhang HC, Zhao H, Wu YS (2018) Optimized stiffness combination of a flexible-base hinged connector for very large floating structures. Mar Struct 60:151–164
13. Kashiwagi M (1998) A B-Spline Galerkin scheme for calculating the hydroelastic response of a very large floating structure in waves. J Mar Sci Technol 3(1):37–49
14. Lin X, Takaki M (1998) A B-Spline type direct method for predicting the wave-induced motion of a very large floating structure. J Jpn Soc Naval Architects Ocean Eng 184:253–261

Application and Prospect

Connector Stiffness Influence on Comfort in Large-Scale Floating Developments

Rui P. F. Gomes and Bart Roeffen

Abstract Coastal cities face imminent challenges of land scarcity due to overpopulation and rising sea levels. Floating structures offer a solution to both of these challenges by creating new living space that adapts to water level variations. While small-scale floating urban developments have been prevalent in sheltered areas, large-scale implementations in challenging sea conditions hold the key to addressing coastal city issues effectively. The design of such structures faces technical challenges related to internal loads and human comfort. Typically, large structures are subjected to high internal loads, while small structures are exposed to larger motions, resulting in a lower level of comfort. The application of compliant mechanical connectors between modules of large floating structures is an alternative for attaining an adequate balance between internal loads and comfort level. This study addresses the analysis of a multi-platform floating structure for urban applications subject to wave action. In particular, it focusses on how the platform comfort level is influenced by the connection bending stiffness. The interaction between waves and floating structures is computed using a radiation-diffraction boundary-element method. Each platform is represented as a six degree-of-freedom rigid body, which incorporates two buildings. The connectors are modelled as generic six degree-of-freedom springs located between neighboring platforms. Results demonstrate the dependency of comfort levels and bending moments on the connector bending stiffness and incident wave conditions. For an adequate combination of wave period and bending stiffness, the compliant connector provides a good balance between comfort and loads, when compared with flexible and rigid connectors.

Keywords Multi-module floating urban structures · Compliant connection system · Wave-body interaction

R. P. F. Gomes (✉) · B. Roeffen
Blue21 Group B.V., Paardenmarkt 1, 2611 PA, Delft, The Netherlands
e-mail: rui@blue21.nl

X. L. Zhao et al. (eds.), *Proceedings of the Fourth World Conference on Floating Solutions: WCFS 2024; 2–4 December; Hong Kong*, Lecture Notes in Civil Engineering 597, https://doi.org/10.1007/978-981-96-4569-5_4

1 Introduction

Coastal cities worldwide are dealing with severe challenges due to overpopulation and rising sea levels. As urban populations tend to increase, cities are facing significant problems in housing, infrastructure, and urban planning. Simultaneously, global warming-induced sea-level rise endangers low-lying areas, intensifying land scarcity and increasing risks to current developments [1]. These critical issues emphasize the need for innovative and sustainable solutions to ensure the resilience and growth of coastal urban environments. Floating structures are seen as promising solution for these problems, as they offer a flexible and adaptive approach to land scarcity by creating new living spaces on water, capable of adjusting to sea level variations [2]. Even though small-scale floating structures have found success in sheltered waters, the upscaling to large-scale developments in more challenging sea conditions remains relatively unknown and technically complex.

The engineering design of large-scale floating structures is fundamental in addressing these challenges. Engineering considerations must account for internal loads and human comfort, two critical factors that significantly impact the feasibility and sustainability of these structures [3, 4]. Large floating platforms are typically subjected to high internal loads, affecting their structural integrity and durability [5]. In contrast, smaller floating modules are more prone to motion induced by waves, which can reduce the occupant comfort of these spaces. Therefore, achieving a balance between structural integrity and comfort is essential for the success of large-scale floating urban developments.

Advanced modelling techniques play a crucial role in the engineering design of floating structures. Through detailed simulations and analyses, engineers can predict the behaviour of these structures under various conditions, allowing for optimized design solutions [6]. In this study, a multi-body radiation-diffraction boundary-element method is applied the analyse the interaction between the floating structure and waves. The study focuses on the importance of compliant mechanical connectors between the modules of large floating structures [7]. These connectors can mitigate internal loads while maintaining an acceptable level of comfort by accommodating relative motions between modules. The findings of this study emphasize the dependency of comfort levels and structural loads on the stiffness of the connectors. Results demonstrate that compliant connectors can effectively balance the demands of structural integrity and comfort. This balance is fundamental for the practical implementation of large-scale floating structures in coastal cities.

Characteristic	Value
Water depth [m]	12
Total length [m]	300
Total width [m]	150
Number of buildings	18
Number of single platforms	9
Number of floors per building	5
Height of building [m]	16
Total building floor area [m^2]	93,600
Total water covered area [m^2]	45,000
Total mass of platforms [ton]	81,000
Total mass of buildings [ton]	64,154
Length of single platform (l) [m]	100
Width of single platform (w) [m]	50
Height of platform (h) [m]	6
Height of center of gravity [m]	4.27
Draft of platform [m]	3.15

Fig. 1 Three-dimensional view of the 9-platform floating structure and 18 5-floor buildings and summary of the characteristics of the floating urban development

2 Methodology

2.1 Problem Definition

This study analyses a large scale urban floating development consisting of 9 platforms interconnected with a generic six degree-of-freedom spring. The platforms have no gaps between them. Each platform has two four-story buildings. The problem is solved using the urban design and simulation software HydroMEC+. A three-dimensional view of the floating structure and buildings and a summary of the characteristics of the floating structure are presented Fig. 1.

2.2 Numerical Modelling

The fluid–structure interaction problem is analyzed in this study by applying three-dimensional potential flow theory. Therefore, the formulation presented here considers irrotational flow and neglects viscous effects. Linear wave theory and linear hydrodynamic interactions are considered by assuming small wave amplitude and small body motions, respectively. The floating structure consists of an arrangement of several rectangular pontoon platforms, which are linked to neighboring platforms via compliant connectors, which allow a predominantly hinge-type motion, where the bending stiffness can be adjusted.

The rectangular platforms (length l, width w and height h) are modeled as rigid bodies, each with 6 degrees of freedom (3 translations—surge, sway and heave—and 3 rotations—roll, pitch and yaw) defined at the center of gravity of the corresponding body. Waves travel along the positive x-axis direction for a zero wave direction θ. The full system is modelled as having $6N$ degrees of freedom, where N represents the total number of platforms. The frequency-dependent wave radiation and diffraction hydrodynamic coefficients were computed using a three-dimensional boundary-element method (BEM) software [8].

All forces relevant to describe the system dynamics are assumed linear. Under this assumption, the system becomes time invariant and we may write the displacement x_i, velocity $\dot{x}_i$ and acceleration $\ddot{x}_i$ of each degree of freedom as

$$\{x_i, \dot{x}_i, \ddot{x}_i\}(t) = \mathrm{Re}\left(\left\{X_i, -\mathbf{i}\omega X_i, -\omega^2 X_i\right\} e^{-\mathbf{i}\omega t}\right) \tag{1}$$

where X_i is a complex displacement amplitude associated to the i-th degree of freedom and $\mathbf{i}$ is the imaginary unit. Note that dynamic forces can also be described the same way. All displacements of the system can be represented the vector of complex displacement amplitudes

$$\mathbf{x} = \left[X_1, X_2, X_3, X_4, X_5, X_6, X_7, \ldots, X_{6N-1}, X_{6N}\right] \tag{2}$$

where $[X_1, X_2, X_3, X_4, X_5, X_6]$ represents the 6 degree-of-freedom displacement complex amplitude vector of body 1 (surge, sway, heave, roll, pitch and yaw), $[X_7, X_8, X_9, X_{10}, X_{11}, X_{12}]$ is the corresponding vector for body 2, and so forth for the other bodies.

The equations of motion of the $6N$-degree-of-freedom system is given by

$$\left[-\omega^2(\mathbf{M} + \mathbf{A}) - \mathbf{i}\omega\mathbf{B} + (\mathbf{C} + \mathbf{C}_\mathrm{c} + \mathbf{C}_\mathrm{m})\right]\mathbf{x} = A_\mathrm{w}\mathbf{f}_\mathrm{d} \tag{3}$$

where $\mathbf{f}_\mathrm{d}$ is a column vector representing the force coefficient due to wave diffraction, A_w is the incident wave amplitude, $\mathbf{M} \in \mathbb{R}^{6N\times 6N}$ is the inertia matrix, $\mathbf{A} \in \mathbb{R}^{6N\times 6N}$ is the radiation added inertia matrix, $\mathbf{B} \in \mathbb{R}^{6N\times 6N}$ is radiation damping matrix, $\mathbf{C} \in \mathbb{R}^{6N\times 6N}$ is the hydrostatic stiffness matrix, $\mathbf{C}_\mathrm{c} \in \mathbb{R}^{6N\times 6N}$ is the connector stiffness matrix, and $\mathbf{C}_\mathrm{m} \in \mathbb{R}^{6N\times 6N}$ represents the mooring system stiffness matrix. In this case, mooring forces are applied directly in the center of gravity, and therefore $\mathbf{C}_\mathrm{m}$ becomes a diagonal matrix.

2.3 Connector Stiffness

The compliant connectors presented in Fig. 2 are modelled as generic six degree-of-freedom springs located between neighboring platforms. The relative surge, sway and heave displacements between neighboring platforms at the connector point are constrained by translational springs with stiffnesses K_x, K_y and K_z, respectively. Relative roll, pitch and yaw are restricted by rotational springs with stiffnesses K_{rx}, K_{ry} and K_{rz}, respectively. The relationship between the displacement vector of multi-platform system degrees of freedom and the restoring forces and moments induced at the center of gravity of each platform is given by matrix $\mathbf{C}_\mathrm{c}$. This matrix is obtained by applying Hamilton's principle [9] to describe the forces induced on the connectors

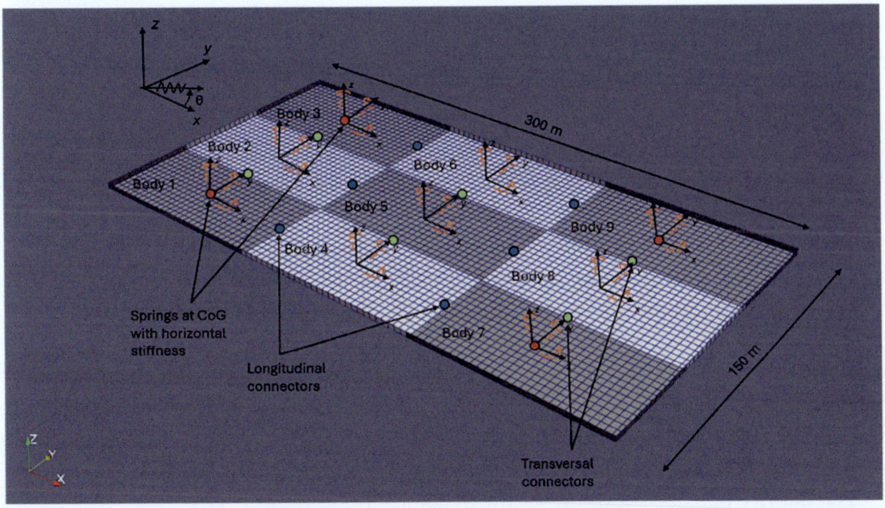

Fig. 2 Schematic representation of the model used to simulate multi-platform floating structure with 9 platforms in the 3-by-3 grid arrangement, and mesh used to discretize the floating structure wetted surface

by the motion of the bodies and by neglecting terms higher than first order [10]. The forces-moment vector acting on a connector is

$$\mathbf{f}_c = \mathbf{K}_c(\mathbf{x}_{CA} - \mathbf{x}_{CB}) \tag{4}$$

where $\mathbf{K}_c \in \mathbb{R}^{6\times6}$ is a diagonal matrix representing the connector stiffness ($\mathbf{K}_c = \text{diag}\left(K_x, K_y, K_z, K_{rx}, K_{ry}, K_{rz}\right)$), $\mathbf{x}_{CA}$ and $\mathbf{x}_{CB}$ are the position vectors of the connector point C due to the motion of bodies A and B, respectively.

2.4 Comfort Evaluation

The comfort evaluation is based on the computation of acceleration at critical points in the platform accessible to occupants. For a given point $\boldsymbol{P}$, the three coordinates of its displacement are given by the vector

$$\mathbf{x}_T^P = \mathbf{x}_T + \left(\mathbf{x}_R \times \mathbf{r}^P\right) \tag{5}$$

where $\mathbf{x}_T$ is a vector representing the three translational degrees of freedom of the corresponding platform, $\mathbf{x}_R$ is a vector representing the three rotational degrees of freedom (at the center of gravity) and $\mathbf{r}^P$ is a vector describing the position of point $\boldsymbol{P}$ relative to the platform center of gravity. Note that $\mathbf{x}_T^P$ is linear due to the small rotation assumption. Considering the formulation presented in Eq. (1), the acceleration vector at point $\boldsymbol{P}$ is

$$\mathbf{a}^{P} = -\omega^2 \mathbf{x}_T^P \tag{6}$$

The analysis of comfort requires the evaluation of the system response under irregular wave conditions, and therefore a stochastic approach is applied. This approach uses the linear response amplitude operators for the computation of statistic representative values of the floating structure dynamics. Let us consider a random sea state represented by the superposition of elementary regular wave components, each with its own amplitude and frequency. This assumption implies that the instantaneous free-surface elevation at a given position can be described as a stationary, ergodic and Gaussian process [11, 12]. To simulate irregular waves, a JONSWAP spectrum S_ω is considered, whose energy density variation over frequency ω is dependent on the significant wave height H_s, peak period T_p and shape factor γ [12].

The energy density spectrum of the acceleration at point $\boldsymbol{P}$ is

$$S_{\mathbf{a}^P}(\omega) = S_\omega(\omega)\left|\frac{\mathbf{a}^P(\omega)}{A_w}\right|^2 \tag{7}$$

Since the system is linear and the water elevation is described by a Gaussian probability density function, then any linear response S_x can be described by a Gaussian probability density function. The root-mean-square value of the acceleration at point $\boldsymbol{P}$ is

$$\mathbf{a}_{RMS}^{P} = \sqrt{\int_0^\infty S_{\mathbf{a}^P}(\omega)d\omega} \tag{8}$$

where the vector $\mathbf{a}_{RMS}^{P}$ represents acceleration values along the x-, y- and z-axis direction.

3 Results

3.1 Frequency-Domain Analysis

The frequency-domain approach described in Sect. 2 was applied to solve the 9-platform dynamic problem presented in Fig. 2. The computation of the hydrodynamic coefficients considers a regular mesh of the wetted surface with 3912 panels. Hydrodynamic results are obtained for a total of 81 equally-spaced wave frequencies, between 0.1 and 2.0 rad/s. The floating structure is moored via 4 spring elements at the center of gravity of bodies 1, 3, 7 and 9, to limit the horizontal motions of the platform. The stiffness coefficients of the spring elements are $K_x = K_y = 5 \times 10^8$ N/m and $K_{rz} = 10^{10}$ N m/rad. In order to simulate connector behaving as a hinge with adjustable bending stiffness, the spring stiffness of the other degrees of freedom is set with relatively high values. The stiffness coefficients of the longitudinal connectors are $K_x = K_y = K_z = 10^{13}$ N/m, $K_{rx} = K_{rz} = 10^{15}$ N m/rad and the K_{ry} is variable. The stiffness coefficients of the transversal connectors are $K_x = K_y = K_z = 10^{13}$ N/m, $K_{ry} = K_{rz} = 10^{15}$ N m/rad and K_{rx} is variable. All results presented here consider a wave direction equal to zero (aligned with the x -axis).

Heave and pitch response amplitude operators (RAOs) are shown in Fig. 3, for a range of wave periods between 4 and 16 s. Results are presented for the three middle longitudinal bodies for three different bending stiffness values. For $K_{ry} = 10^0$ N m/rad, the connectors have a negligible stiffness and the system behaves similarly to a hinge (flexible connector). For $K_{ry} = 10^{15}$ N m/rad, the connectors behave rigidly and restricts completely the displacement of the relative pitch displacement between platforms (rigid connector). The case with $K_{ry} = 10^{11}$ N m/rad represents an intermediate stiffness situation (semi-flexible connector—hinge with significant bending stiffness). Body 2, which is heading the waves, shows generally higher heave and pitch displacements. As expected, for most periods, the flexible-connector configuration shows the largest pitch and the rigid-connector configuration displays the smallest pitch.

Figure 4 shows the irregular-wave normalized bending moment root-mean-square-value at the connectors between bodies 2 and 5 and between bodies 5 and 8, for sea states with peak period (T_p) equal to 6, 8 and 10 s. The irregular-wave

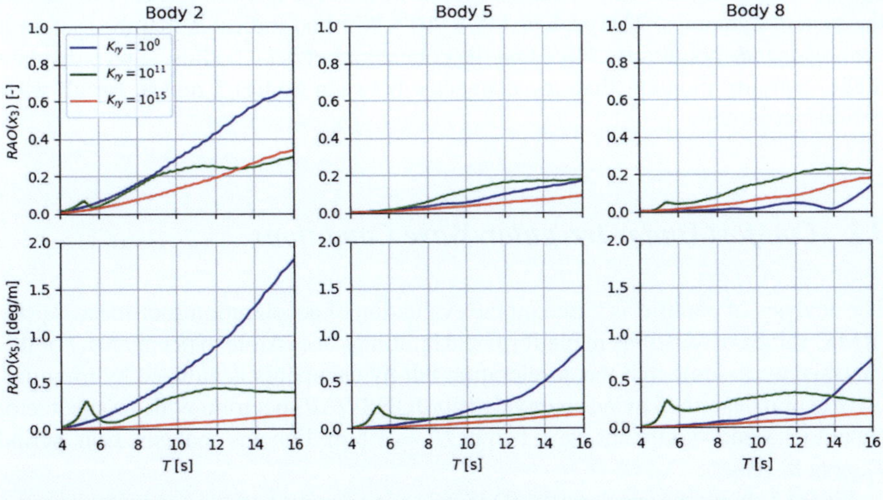

Fig. 3 Response amplitude operators (RAOs) of heave and pitch as a function of the incident wave period, for bodies 2, 5 and 8 and for different bending stiffness values

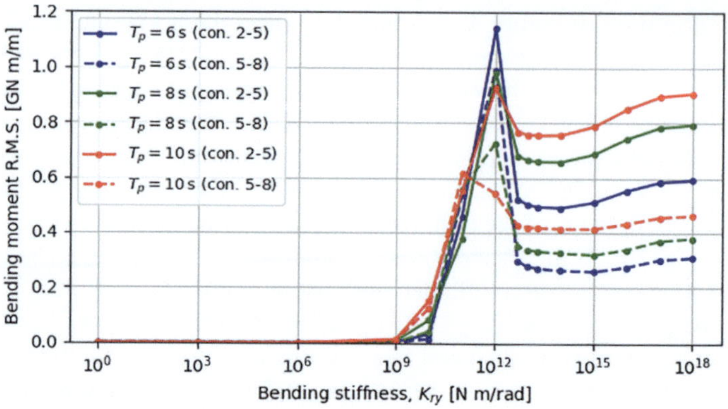

Fig. 4 Normalized bending moment root-mean-square value as a function of the bending stiffness, for the connectors between bodies 2 and 5 and between bodies 5 and 8, for 3 irregular-wave sea states. Bending moment is normalized with respect to the incident wave free-surface elevation root-mean-square value

sea states are described by a JONSWAP spectrum. The bending moment root-mean-square-value is computed through the stochastic approach presented in Eqs. (7–8), for the acceleration root-mean-square value calculation. For $K_{ry} < 10^9$ N m/rad, no significant bending moment is measured at the connections. For $K_{ry} > 10^{17}$ N m/rad, the bending moment is independent from K_{ry}, which indicates that the systems behaves rigidly. Note that for $K_{ry} > 10^{13}$ N m/rad, the response is weakly dependent on the stiffness K_{ry}. For $10^9 > K_{ry} > 10^{13}$, the bending moment varies significantly, suggesting that this intermediate stiffness range has a significant impact on the system dynamics. The peak at $K_{ry} = 10^{12}$ N m/rad indicates that the system is near resonance conditions. Note that the connector between bodies 2 and 5 displays higher bending moment than the connector between bodies 5 and 8, for all wave periods presented.

3.2 Comfort Under Irregular-Wave Conditions

The analysis of comfort is based on the evaluation of acceleration root-mean-square (RMS) values of locations in platform and buildings accessible to occupants. For this analysis, we assume that the acceleration felt by occupants is induced by irregular-wave action described by an energy density JONSWAP spectrum with a shape factor equal to 3.3 and significant wave height $H_s = 1.5$ m. For this analysis, peak period $T_p = 6, 8, 10$ s.

Figure 5 shows the acceleration RMS value as a function of the connector bending stiffness for the least comfortable location in the floating structure. The cases with flexible hinge-type connectors ($K_{ry} < 10^9$ N m/rad) shows maximum acceleration values independent from the bending stiffness value. The same is observed for rigid

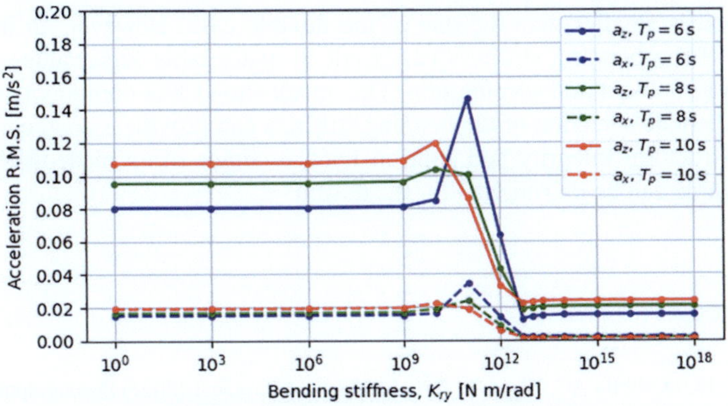

Fig. 5 Acceleration root-mean-square value as a function of the bending stiffness, for the point in the floating structure subject to the largest acceleration

connectors ($K_{ry} > 10^{13}$ N m/rad). For the intermediate stiffness range ($10^9 > K_{ry} > 10^{13}$), the acceleration values present significant variation, with the largest vertical acceleration RMS value occurring for $T_p = 6$ s and $K_{ry} = 10^{11}$ N m/rad, and the largest horizontal acceleration RMS value occurring for $T_p = 6$ s and $K_{ry} = 10^{11}$ N m/rad.

For this particular configuration, within a range of intermediate stiffness values, it is possible to achieve a better comfort level (see Fig. 5) than the one with the flexible connectors and a bending moment (see Fig. 4, for connector 2–5) smaller than the one with rigid connectors ($K_{ry} = 10^{18}$ N m/rad). This is the case for $5 \times 10^{12} < K_{ry} < 10^{15}$ N m/rad, for all peak periods. Resonance conditions for the bending moment (at $K_{ry} = 10^{12}$ N m/rad) and for the accelerations (at $K_{ry} = 10^{11}$ N m/rad) are observed for specific intermediate stiffness values.

4 Conclusions

A numerical model of a 300 m long and 150 m wide 9-platform floating structure with the characteristics of a large-scale urban development was simulated under wave action. Platforms were assembled in a 3-by-3 grid arrangement and neighboring platforms were connected via hinges. At the hinges, a bending stiffness was applied, and the effect of its magnitude on the dynamic response of the whole 9-body system was evaluated. Three bending stiffness regions were identified: a low stiffness region where connectors behave as hinges, a high stiffness region where connector are rigid, and an intermediate region where the system dynamics is highly sensitive to the bending stiffness value. The analysis of bending moments and comfort showed that for a given wave condition, the selection of an adequate bending stiffness provides a smaller bending moment than the rigid situation and simultaneously the maximum acceleration is smaller than the one of the flexible case. However, an inadequate selection of the bending stiffness can result in undesirable high values for either the bending moment and acceleration. This result shows that compliant connectors with an adequate selection of the bending stiffness can provide a good compromise between structural integrity and comfort but the choice of an inadequate bending stiffness value can be detrimental for the design.

References

1. Best K, He Q, Reilly AC, Niemeier DA, Anderson M, Logan T (2023) Demographics and risk of isolation due to sea level rise in the United States. Nat Commun 14(1)
2. IPCC (2022) Climate Change 2022: Impacts, adaptation, and vulnerability. In: Contribution of working group II to the sixth assessment report of the intergovernmental panel on climate change. Cambridge University Press, Cambridge
3. Wang G, Goldfeld Y, Drimer N (2019) Expanding coastal cities—proof of feasibility for modular floating structures (MFS). J Clean Prod 222
4. Wang G, Rosenfeld Y, Drimer N, Goldfeld Y (2021) Occupant comfort analysis for rigid floating structures—methodology and design assessment for offshore dwelling module. Ships Offshore Struct 16
5. Yoon J-S, Cho S-P, Jiwinangun RG, Lee P-S (2014) Hydroelastic analysis of floating plates with multiple hinge connections in regular waves, Mar Struct 36
6. Huang H, Chen X-J, Liu J-Y, Ji S (2022) A method to predict hydroelastic responses of VLFS under waves and moving loads. Ocean Eng 247
7. Jiang D, Tan KH, Wang CM, Dai J (2021) Research and development in connector systems for very large floating structures. Ocean Eng 232
8. Ancellin M, Dias F (2019) Capytaine: a Python-based linear potential flow solver. J Open Source Softw 4(36):1341
9. Shabana A (2020) Dynamics of multibody systems, 3rd edn. Cambridge University Press, Cambridge, UK
10. Sirigu SA, Boccanfuso V, Carapellese F (2024) Multi-body analysis of modular floating islands: optimisation of connector stiffness. In: Ikoma T, Tabeta S, Lim SH, Wang CM (eds) Proceedings of the third world conference on floating solutions. WCFS 2023. Lecture Notes in Civil Engineering, vol 465. Springer, Singapore
11. Sarpkaya T, Isaacson M (1981) Mechanics of wave forces on offshore structures. Van Nostrand Reinhold, New York, USA
12. Goda Y (2010) Random seas and design of maritime structures, 3rd edn. World Scientific, Singapore

Research and Development of Thermoelectric Modules and Their Application in Floating Cities

Jingming Cai, Yujin Yuan, and Zhiyang Pei

Abstract This paper investigates the development and application of thermoelectric (TE) modules, particularly focusing on their integration into floating city architectures to address unique energy demands and sustain-ability challenges. TE modules, which operate on the Seebeck effect, convert thermal gradients directly into electricity. Despite existing challenges, such as the cost and performance limitations of current TE materials, recent advancements have improved their properties, making them viable for practical applications. The focus of this study is on testing the performance of TE modules, including their mechanical properties, Seebeck coefficient, and electrical and thermal conductivity. These are all very important for improving the output power and energy conversion efficiency. The paper also investigates the innovative use of TE modules in floating cities to enhance energy independence through waste heat recovery and direct heat-to-electricity conversion, thereby supporting sustainable and autonomous urban ecosystems. The potential of TE modules for building materials and thermal management in floating cities is also dis-cussed, providing a path forward for resilient and energy-efficient urban development.

Keywords Thermoelectric (TE) modules · Floating cities · Energy sustainability

1 Introduction

In recent decades, the imperative to conserve energy and reduce emissions has driven the global agenda, leading to innovations in building materials and energy conversion technologies [1]. Floating cities, characterized by their need for self-sufficiency and

J. Cai (✉)
Faculty of Civil and Environmental Engineering, TU Darmstadt, Franziska-Braun-Straße, Darmstadt, Germany
e-mail: jingmingcai@seu.edu.cn

J. Cai · Y. Yuan · Z. Pei
Key Laboratory of Concrete and Prestressed Concrete Structures of Ministry of Education, Southeast University, Nanjing, China

X. L. Zhao et al. (eds.), *Proceedings of the Fourth World Conference on Floating Solutions: WCFS 2024; 2–4 December; Hong Kong*, Lecture Notes in Civil Engineering 597, https://doi.org/10.1007/978-981-96-4569-5_5

resilience, necessitate advanced energy solutions that can operate independently of traditional power infrastructure [2–4].

As shown in Fig. 1, adopting a variety of new energy sources can significantly enhance the self-sufficiency of floating cities, achieving zero external electricity purchases. Thermoelectric (TE) modules, which convert thermal energy into electrical energy through the Seebeck effect, present a promising solution for floating cities [5, 6]. These modules can harness thermal gradients to generate electricity, reducing dependency on external power sources. In floating cities, it is especially important to have energy systems that are compact, efficient, and environmentally benign because to limited space and environmental concerns.

Despite the potential advantages, current commercial TE modules face challenges related to cost, durability, and performance [7, 8]. Cement-based TE materials in building applications offer the significant advantages of cost-effectiveness and enhanced structural integrity, making them appealing for integration into energy-harvesting buildings and reducing reliance on expensive commercial TE modules [9]. However, they face challenges such as a lack of systematic research on their building structural strength and low TE performance with ZT values, which can severely impair their energy-harvesting capabilities. Recent advancements in TE materials, particularly the development of geopolymer-based TE composites, present a viable alternative [10]. These materials exhibit improved TE properties, making them suitable for application in floating cities.

In summary, the deployment of advanced TE modules in floating cities is a critical step towards achieving energy self-sufficiency and environmental sustainability. This paper explores the mechanical strength and ZT performance of TE modules, with a particular focus on their application in buildings. It highlights the potential of TE modules to meet unique energy demands and discusses their integration into building

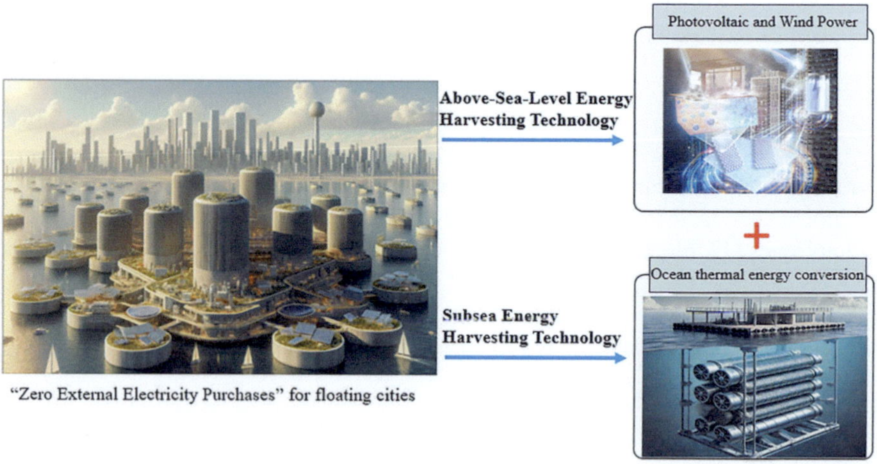

Fig. 1 Energy harvesting for floating cities

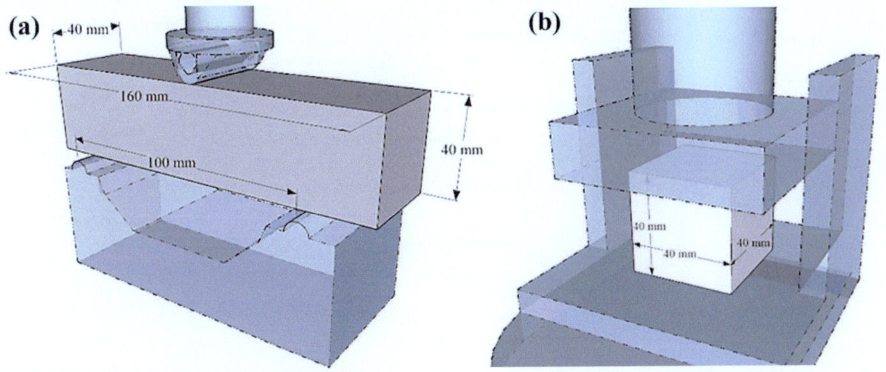

Fig. 2 Mechanical properties test **a** Flexural test. **b** Compressive test

materials and thermal management systems, emphasizing their suitability in floating urban environments.

2 TE Module Performance Test Method

2.1 *Mechanical Properties*

The compressive strength measurements were conducted in accordance with EN 1015-11 [11]. A universal testing machine performed standard prism specimen testing. For the bending tensile strength measurements, three prismatic specimens were subjected to a three-point loading assessment, and the compressive strength testing was implemented on six prismatic specimens obtained from the bending test (cf Fig. 2).

2.2 *Seebeck Coefficient*

Cylindrical molds measuring 50 mm in diameter and 100 mm in height were utilized to prepare the polymer samples for measuring the Seebeck coefficient. To measure the voltage differential (ΔV), a digital multimeter was used. Both ends of the module were surrounded by a copper sheet, and silver paint was applied to enhance the transmission of electricity between them (cf Fig. 3). The copper sheet interface was utilized to connect the digital multimeter. A digital thermometer equipped with a thermocouple was utilized to measure the temperature variation (ΔT) of the module. Using a resistance heater, a temperature difference was created between the two

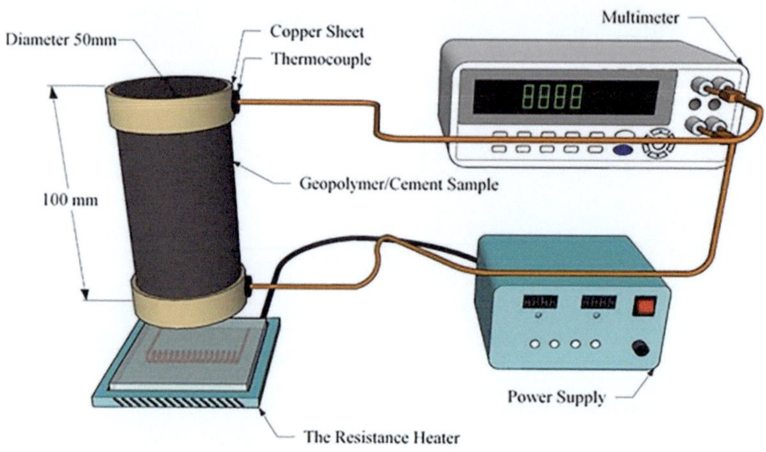

Fig. 3 Seebeck coefficient test set-up

surfaces of the sample, which produced a corresponding voltage difference that was recorded simultaneously.

2.3 *Electronic Conductivity*

To evaluate the electrical conductivity characteristics of the module, the impedance was measured using an impedance analyzer IM 3570, employing the double-probe method. The resistance obtained was subsequently substituted into Eq. (1) to calculate the electrical conductivity.

$$\rho = \frac{L}{SR} \tag{1}$$

where ρ represents the volume resistivity, *R* denotes the resistance, *S* is the transverse cross-sectional area of the sample relative to the direction of resistance measurement, and *L* indicates the distance between the electrodes (Fig. 4).

2.4 *Thermal Conductivity*

The thermal conductivity measurements in this paper were conducted using the LFA 467 laser flash apparatus from Netzsch, Germany, employing the transient method. The vertical configuration of the measurement system places the xenon lamp light source at the bottom, the sample at the center, and the signal detector at the top. The light source is controlled by software, allowing for adjustment of the pulse

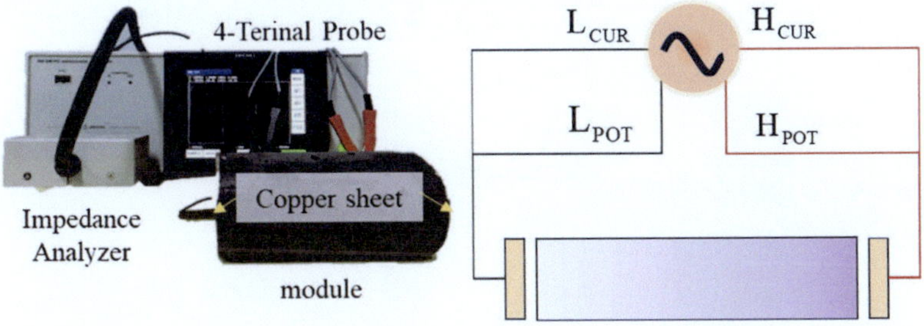

Fig. 4 Electrical conductivity measurement

width from 20 to 1200 μs. The xenon lamp uniformly heats the lower surface of the sample, while an infrared detector measures the temperature rise on the upper surface as a function of time. The test system, treated as an adiabatic system, enables the calculation of the thermal diffusivity coefficient, a (mm^2/s), as shown in Eq. (2).

$$a = 0.1388\frac{l^2}{t_{1/2}} \tag{2}$$

In the formula, l represents the thickness of the sample, and $t_{1/2}$ denotes the time required for the temperature to increase by 50%. In addition to the thermal diffusivity coefficient, the specific heat capacity c_p of the sample can also be determined using a reference specimen. The thermal conductivity (λ) at a given temperature can be calculated using Eq. (3).

$$\lambda(T) = a(T)\rho(T)c_p(T) \tag{3}$$

where ρ denotes the density of the material, while T represents the temperature achieved by the sample (Fig. 5).

To comprehensively assess the TE effect of this module, a dimensionless figure of merit (ZT) is utilized. This value, crucial for determining conversion efficiency, comprises the Seebeck coefficient (S), electrical conductivity (σ), thermal conductivity (κ), and the absolute measurement temperature (T). It is defined as follows.

$$ZT = \frac{S^2\sigma T}{\kappa} \tag{4}$$

The average value of the ZT for this module at ambient temperature has been calculated.

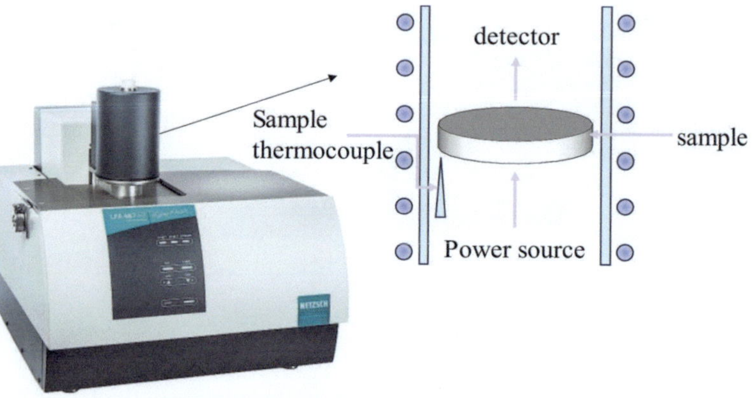

Fig. 5 A schematic of the thermal conductivity testing principle

2.5 *TE Module*

Materials structured into modules shape TE modules to convert thermal energy into electrical energy. By connecting p-type and n-type TE legs with metal electrodes, these modules improve thermal flow in a single direction perpendicular to the substrate. Typically, an arrangement combining electrical series and thermal parallel connections interconnects several -shaped modules to augment output power.

The ZT primarily determines the theoretical maximum efficiency of TE modules. However, during the design and fabrication processes, factors at the level, such as the topology, interface elements, and fabrication techniques, significantly influence the actual performance indicators. These indicators include energy conversion efficiency, maximum power output, and the stability and reliability of the modules.

Thermoelectric power generation (TEG) is based on the Seebeck effect. When a temperature difference is applied across the ends of a TE module—with the hot end and cold end set at temperatures T_h and T_c, resulting in a temperature difference $\Delta T = T_h - T_c$, TE voltage is generated. This voltage can drive load resistance. The two most critical performance parameters are the power generation efficiency and the maximum output power [12].

$$\eta = \frac{P}{Q_h} = \frac{T_h - T_c}{T_h} \frac{\sqrt{1 + \overline{ZT}} - 1}{\sqrt{1 + \overline{ZT}} + T_h/T_c} \tag{5}$$

The higher the ZT value of the material, the greater the conversion efficiency of the TE modules. Currently, the average ZT value of most TE materials systems discovered at room temperature remains below 1.0, which is a critical issue limiting the enhancement of power generation efficiency in TEG. The Seebeck voltage generated in the temperature difference circuit between the hot and cold ends determines the maximum output power. To achieve this maximum power, the external resistance

R_L must be matched with the internal resistance R of the module. At this point, the module can achieve its maximum output power, as shown in Eq. (6) [13].

$$P_{max} = \frac{{S_{np}}^2 \Delta T}{4R} \quad (6)$$

In this study, n-type and p-type architectural TE materials were initially prepared. Subsequently, these materials were used to construct 36 pairs of modules on a substrate. Mechanical strength and TE performance were considered as the basis for TE optimization, aimed at enhancing the power generation and efficiency of the modules.

3 The Potential Application of TE Modules in Floating Cities

3.1 Energy Demands and Challenges of Floating Cities

Floating cities face unique challenges, such as limited space and the need for sustainable energy. In this context, TE modules offer unique applications tailored to the environmental and spatial challenges inherent in such habitats. These modules, leveraging the direct conversion of heat into electricity, can be crucial in enhancing energy sustainability, optimizing thermal management, and supporting the complex infrastructure of floating cities.

Floating cities face unique challenges, such as limited space and the need for sustainable energy. TE modules are crucial in these settings, as they can convert the marine temperature differentials into a reliable energy source. This ability enhances the sustainability and energy independence of floating cities, making them more viable solutions to urban and environmental challenges.

In the context of floating city buildings, TE modules offer unique applications tailored to the environmental and spatial challenges inherent in such habitats. These modules, leveraging the direct conversion of heat into electricity, can be crucial in enhancing energy sustainability, optimizing thermal management, and supporting the complex infrastructure of floating cities.

4 Applications of TE Modules

Integration with Building Materials

TE modules can be embedded into structural elements of buildings, such as roofing panels or facades, as illustrated in Fig. 1. This integration allows the building itself to function as an energy harvester, utilizing the thermal energy from sunlight. Such applications not only provide power but also reduce heat gain from solar exposure, contributing to more stable internal temperatures.

TE Modules in Marine Structures

Floating city buildings can utilize TE modules to harness the thermal gradient between the warmer surface water and the cooler deep-sea water. This temperature differential can be a continuous source of energy, ideal for TE modules to generate electricity efficiently. This renewable energy source supports the floating city's power needs, reducing dependency on traditional energy supplies and minimizing environmental impact.

Efficient Thermal Management

In floating cities, managing indoor climate efficiently is paramount due to exposure to marine elements and limited space. TE modules can be integrated into the building envelopes or windows of floating city buildings to regulate temperature more dynamically. These modules can absorb excess heat from the interior and release it into the cooler marine environment or vice versa, maintaining comfort without extensive energy consumption.

Autonomous Energy for Safety and Monitoring Systems

TE modules can power critical safety and monitoring equipment continuously. In floating cities, where reliability is key due to isolation from mainland infrastructure, having autonomous power sources for emergency lighting, alarms, and structural health monitoring sensors ensures enhanced safety and operational stability.

Reducing Energy Footprint

By converting waste heat from residential, commercial, or industrial activities within the floating city into usable electricity, TE modules help reduce the overall energy footprint. This conversion contributes to the sustainability goals of floating cities by minimizing waste and optimizing resource use.

Overall, the application of thermoelectric modules in floating city buildings represents a forward-thinking approach to building design and energy management. These modules not only enhance energy efficiency and comfort but also align with the broader sustainability and resilience goals of floating urban developments.

References

1. Hassan Q, Viktor P, Al-Musawi TJ, Ali BM, Algburi S, Alzoubi HM, Jaszczur M (2024) The renewable energy role in the global energy transformations. Renew Energy Focus 48:100545
2. Nguyen TTT (2021) Promoting sustainability and resilience in Vietnam's floating community. Universität CottbusSenftenberg
3. Takahashi A, Lin F-Y, Yoshida I, In World conference on floating solutions. Springer, 101–120
4. Kirimtat A, Krejcar O, Kertesz A, Tasgetiren MF (2020) Future trends and current state of smart city concepts: a survey. IEEE access 8:86448–86467

5. Nandihalli N (2022) Thermoelectric films and periodic structures and spin Seebeck effect systems: facets of performance optimization. Mater Today Energy 25:100965
6. Jouhara H, Żabnieńska-Góra A, Khordehgah N, Doraghi Q, Ahmad L, Norman L, Dai S (2021) Thermoelectric generator (TEG) technologies and applications. Int J Thermofluids 9:100063
7. Zhou X, Carmeliet J, Derome D (2020) Assessment of risk of freeze-thaw damage in internally insulated masonry in a changing climate. Build Environ 175:106773
8. Bu Z, Zhang X, Hu Y, Chen Z, Lin S, Li W, Pei Y (2022) A record thermoelectric efficiency in tellurium-free modules for low-grade waste heat recovery. Nat Commun 13(1):237
9. Liu X, Jani R, Orisakwe E, Johnston C, Chudzinski P, Qu M, Yazawa K (2021) State of the art in composition, fabrication, characterization, and modeling methods of cement-based thermoelectric materials for low-temperature applications. Renew Sustain Energy Rev 137:110361
10. Cai J, Yuan Y, Xi X, Ukrainczyk N, Pan L, Wang Y, Pan J (2024) Unveiling the remarkable potential of geopolymer-based materials by harnessing manganese dioxide incorporation. Small 20(5):2305360
11. EN T (2000) Mortar testing method, part 11. Measurement of compressive and flexural tensile strength of mortar
12. Yang J, Caillat T (2006) Thermoelectric materials for space and automotive power generation. MRS Bull 31(3):224–229
13. Chen W-H, Lin Y-X, Wang X-D, Lin Y-L (2019) A comprehensive analysis of the performance of thermoelectric generators with constant and variable properties. Appl Energy 241:11–24

A Proposal of Floating Offshore Base for Wind Farm Deployed in EEZ

Mako Yamamoto, Saika Iwamatsu, Kazuhiro Iijima, Masaaki Shibata, and Hiroshi Tatsuhara

Abstract This study explores the feasibility of offshore floating bases for wind farms deployed in Japan's Exclusive Economic Zone (EEZ), aiming to achieve the national ambitious goal of installing 30–45 GW of offshore wind power by 2040. Construction of energy islands for wind farms in European waters is seriously discussed. Given the deep waters exceeding 200 m in Japan's EEZ, floating solutions are deemed more suitable than the artificial islands, contrasting with European practices. The research investigates three scenarios at varying distances from the coast (50 km, 100 km, and 300 km, respectively), considering environmental factors such as sea conditions, water depths, and wind characteristics. Referencing to the foregoing research of the energy islands, functional requirements for each floating offshore base are discussed and outlined, guiding the layout, area, and dimensions of the facilities. A semi-submersible structure supported by multiple vertical columns is finally selected from the viewpoint of its response characteristics in high waves and suitability for open ocean installation. The performance and structural safety under the design wave conditions are confirmed by numerical hydroelasticity simulations based on the finite element (FE) model.

Keywords Floating offshore wind turbine · Wind farm · Floating base · Hydroelasticity

1 Introduction

Aiming for installation of "30–45 GW of offshore wind power by 2040 yr" [1] as declared in Japan's Offshore Wind Industry Vision, discussions are ongoing regarding the utilization of Japan's exclusive economic zone (EEZ) for offshore wind power.

M. Yamamoto (✉) · S. Iwamatsu · K. Iijima · M. Shibata
Department of Naval Architecture and Ocean Engineering, Osaka University, Osaka, Japan
e-mail: yamamoto_mako@naoe.eng.osaka-u.ac.jp

H. Tatsuhara
Toda Corporation, Tokyo, Japan

X. L. Zhao et al. (eds.), *Proceedings of the Fourth World Conference on Floating Solutions: WCFS 2024; 2–4 December; Hong Kong*, Lecture Notes in Civil Engineering 597, https://doi.org/10.1007/978-981-96-4569-5_6

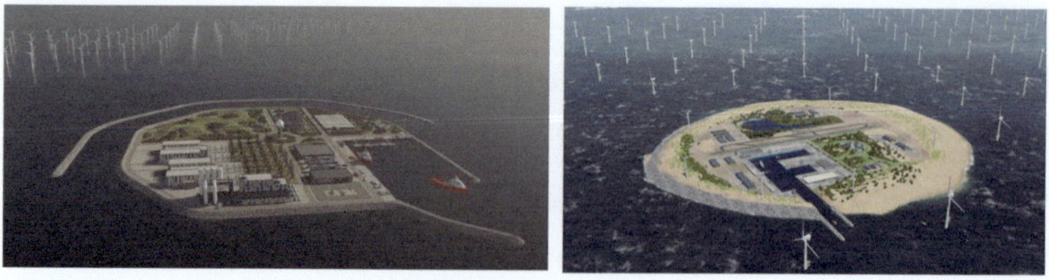

Fig. 1 Energy island of Denmark and Doggerbank [2]

Operation and maintenance (O&M) bases, construction bases, and other related industry bases in the sea areas is deemed necessary. In Europe, the concept of energy islands [2] has been planned and designed. Construction of one of such islands has even started. Considering the water depth exceeding 200 m in Japan's EEZ, floating solutions must be conceived. However, such research has not yet initiated in Japan. This study aims to explore the feasibility of floating bases to be installed in the EEZ.

2 Offshore Bases Conceptualized Overseas

As the reference for offshore bases, artificial islands are planned to be constructed in the North Sea of Denmark and Dogger Bank. In Fig. 1, the left side shows the artificial island in Denmark, and the right side shows the artificial island in Dogger Bank. Facilities to be developed on the islands include base ports, heliports, accommodation facilities, energy storage systems, Power-to-X systems, and high-voltage direct current (HVDC) transmission systems. The Danish artificial island is scheduled for completion in 2030 yr, and the Dogger Bank artificial island is set for 2035 yr.

3 Offshore Bases Needed in Japan's EEZ

When offshore bases suitable for Japan's EEZ is considered, floating solutions equipped with similar functionalities as European energy islands are discussed herein. Three installation scenarios at varying distances (50, 100 and 300 km) away from the coast of Japan's main island are brought up here. The offshore base located at 50 km from the coast are called Base1, that at 100 km from the coast are Base2, and that at 300 km from the coast are Base3. Based on the offshore wind condition map by New Energy and Industrial Technology Development Organization, Japan (NEDO) [3], eastern Japan on the Pacific side has strong annual average wind speeds, leading to the belief that many offshore wind farms will be installed in this area. The offshore bases will be set up in the sea areas on the Pacific side of eastern Japan. The design significant wave heights are determined by performing statistical analysis using the wave scatter diagram in the "Wave and Wind Database of Coastal Japan" [4] by the National Maritime Research Institute. The long-term distribution used for the long-term estimation is given by fitting the data to the Fisher-Tippet Type I distribution. Assuming a design life of 50 years, the design sea states for checking the structural safety is determined as given in Table 1. The operational limit wave height is separately determined by referencing to an exceedance probability of 10% and the significant wave height for checking the operational limit is set 3.5 m.

The functions needed for Base1, Base2, and Base3 are summarized in Table 2. Base1, located at 50 km from the coast, can be reached by boats within a day, so its primary function is to collect electricity from the wind farms, transmit it to land via subsea cables. Therefore, it has only the minimum facilities, such as a control room to manage the wind farms and a port for ships carrying personnel and goods. Base2, located at 100 km from the coast, mainly has transmission, assembly, and repair base functions. A significant feature is that assembly and commissioning can be carried out at the offshore base instead of making frequent trips by the vessels to build and construct the wind farm 100 km away. Base3, located at 300 km from the coast, in addition to the functions of Base2, has the feature of converting electricity into hydrogen or ammonia (P2X), with additional functions like solar panels and data centers for profitability. It not only has functions to operate wind farms but also to supplement new values as a floating structure for renewable energy like solar power generation, data centers, fishery bases and logistics bases. In this study, the solar power generation and data centers are considered.

For transmission facilities, wind power generation uses AC to freely raise and lower voltage, but AC has significant transmission losses. Therefore, when the distance from the coast exceeds 50 km, it is more advantageous to convert to DC before transmitting to land [5]. Hence, Base1 uses AC transmission while Base2 uses DC transmission. Base3 does not have transmission facilities because all electricity is converted to hydrogen or ammonia.

For accommodation facilities, considering the number of personnel working within a day [6], Base1, which is within a day trip distance, is assumed to have 10 people for O&M and transmission facilities. For Base2, 20 people for O&M, transmission facilities, and substations, and 20 people for assembly and repair are assumed. For Base3, 20 people for O&M, P2X, and substations, and 20 people for

Table 1 Conditions of base site

	Coastal distance (km)	Water depth (m)	Design significant wave height (m)
Base1	50	500	11.3
Base2	100	2000	12.7
Base3	300	5000	14.1
Ref (Doggerbank)	200	13–30	11.7
Ref (Denmark)	100	25–28	11–13

Table 2 Functions needed for Base1, 2 and 3

		Base1	Base2	Base3
Transmission and transformation	Transmission facility	HVAC	HVDC	–
	Substation	–	✔	✔
	Operation, Labo	✔	✔	✔
O&M	Accommodation	10	40	45
	Generator	–	✔	✔
	Warehouse	✔	✔	✔
Assembly	Assembly, commissioning (P&P)	–	✔	✔
Access	Port	✔ (small)	✔ (large)	✔ (large)
	Heliport	✔	✔	✔
	Runway	–	–	✔
P2X	Storage of electricity	✔	✔	–
	Factory of hydrogen, ammonia	–	–	✔
	Tank of hydrogen, ammonia	–	–	✔

assembly and repair are assumed. For the data center, it is assumed that one manager oversees 5000 servers, with 5 people managing 20,000 servers at the base.

For assembly and commissioning, typically, wind turbines are pre-assembled at an onshore base port and then towed to the installation site. However, for Base2 and Base3, which are located more than 100 km from the coast, not only assembly but also commissioning, including functional safety confirmation, connection of wind turbines to the grid, and the start of operation, can be carried out. The gantry crane used for assembly must move front, back, left, and right, and ensure that the object being lifted is firmly fixed when installing the nacelle and blades on the tower. Additionally, during assembly, when the rotor is hoisted in the "rabbit" state with the two blades up, the height is at its lowest, so a gantry crane with a height of at maximum 100 m is needed. Currently, detailed discussion on such cranes is not made, however, the development of new construction devices will be need in the future. The manufacturing of towers, nacelles, and blades are assumed to be done at the ports along the coast. These can also be towed from other floating factories. Such floating factory unit can be detached from the offshore base, towed to different locations once the wind farm at the site is completed.

Based on the above functions, the dimensions of the bases are set as follows: Base1 is 100 m × 100 m, Base2 is 600 m × 500 m, and Base3 is 1200 m × 600 m. Illustrations of Base1, Base2, and Base3 are given in Fig. 2.

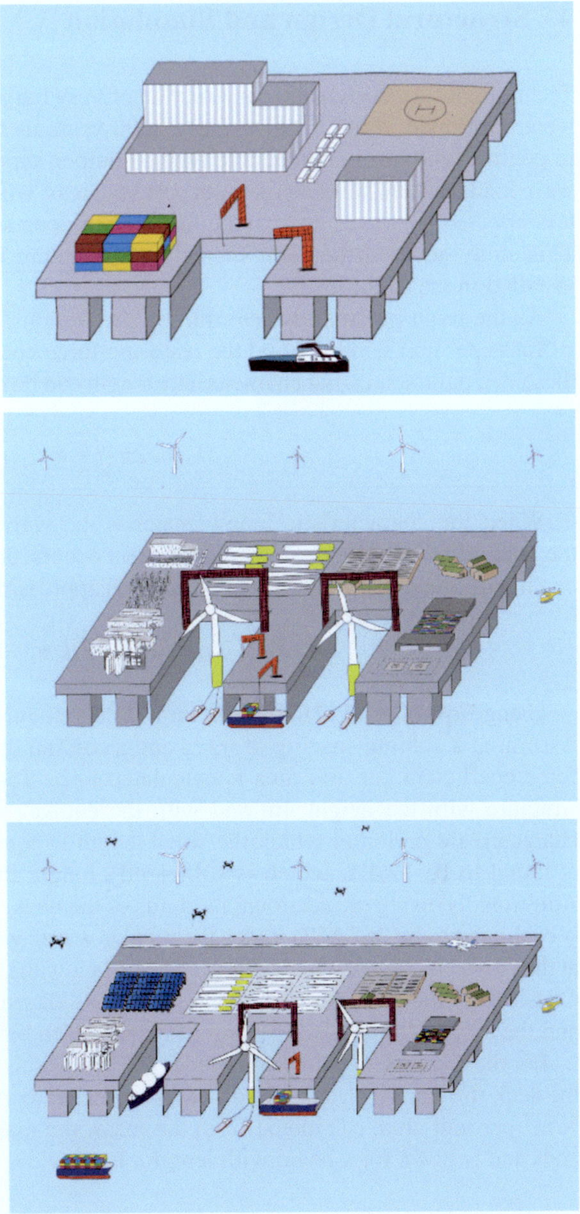

Fig. 2 Illustrative images of Base1 (top left), Base2 (top right) and Base3 (bottom)

4 Structural Design and Simulation

Technologies for design and construction of Very Large Floating Structures (VLFSs) were developed in late 1990 to early 2000 years in Japan. VLFS can be classified to pontoon and semi-submersible types. Pontoon types are relatively susceptible to wave loads and are suitable for calm port sea areas, while semi-submersible types are less affected by waves and can be installed in open seas with higher wave heights. This study adopted the semi-submersible structure for the floating offshore base installed in the EEZ.

As the main parameters determining the response of VLFS, there are the deck stiffness per unit width EI and the restoring force coefficient per unit area k_c. From these two parameters, the characteristic length can be defined as shown in Eq. (1).[7]

$$\lambda_s = 2\pi \sqrt[4]{\frac{EI}{k_c}} \# \tag{1}$$

When the characteristic length matches the wavelength of the incident wave, the maximum quasi-static response moment occurs, and this maximum longitudinal bending moment per unit wave amplitude is approximately given by Eq. (2).

$$M_{max} = \frac{1}{2}\sqrt{EIk_c} \# \tag{2}$$

Using Eqs. (1) and (2), the structural arrangement may be examined as follows. Assuming a column spacing of 60 m and a column diameter of 25 m, the restoring force coefficient per unit area is calculated to be 1340 N/m^2. Assuming the deck structure with the height 4m and with the equivalent plate thickness 25mm, the characteristic period at which the wave length is equal to the characteristic length is found to be 17.4 s, which is sufficiently longer than the average wave period. Additionally, using the sectional modulus of the deck, the bending stress at the deck is evaluated to be 262 MPa under the regular waves with a wave height of 14 m and at the period of 17.4 s, which is judged to be below the yield stress of steel, 355 MPa. Considering the mass per unit area is around 3000kg/m^2, the draft was set at 22.0 m, and the air gap for the severest wave is assumed to be 1.5 m.

For the local strength, a beam model with both ends clamped is considered. When the deck structure with a height of 4 m is subjected to a distributed load of $p = 19{,}612$ N/m^2 per unit area, calculated from the mass, the maximum moment occurs at the ends and is given for a beam with length l by Eq. (3).

$$M = \frac{pl^2}{12} \# \tag{3}$$

When $l = 60$ meters, the moment is calculated to be 5.9×10^6 Nm/m. Then, the local bending stress is found 59 MPa, which is well below the yield stress even though some stiffeners may have to be added locally.

This time, the structural feasibility of Base2 is further examined by using numerical simulations. For the evaluation, a numerical simulation tool based on frequency-domain hydroelasticity theory is used. It couples finite element for structural domain and potential theory for hydrodynamic domain, respectively and the theoretical details are given in [8]. As shown in Fig. 3, a fluid mesh model and an FE model consisting of beam elements are created, and response analysis in waves is conducted.

Figure 4 shows the stress response function near the center of the Base2. The characteristic frequency [7] is found at 0.36 rad/s, corresponding to the first peak. The frequency observed at 0.69 rad/s is the natural frequency of the two-noded vertical bending mode of the structural deformation, and the frequency observed at 0.91 rad/s is the natural frequency of the three-noded vertical bending mode. The resonance peak may further be reduced by inclusion of drag force terms, which however is not accounted for in the present simulations.

Using the stress response function and the spectrum of irregular waves (ISSC spectrum), a short-term prediction was made under sea conditions equivalent to the maximum wave height in 50 years. In the actual sea area, it is assumed that approximately 1000 waves persist under the same sea conditions which lasts two–three hours. Therefore, the 1/1000 maximum expected value can be considered the

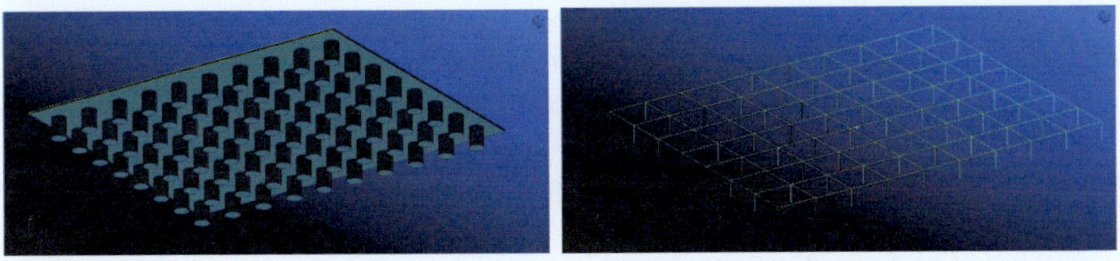

Fig. 3 Illustration of fluid model (left) and structural beam model (right)

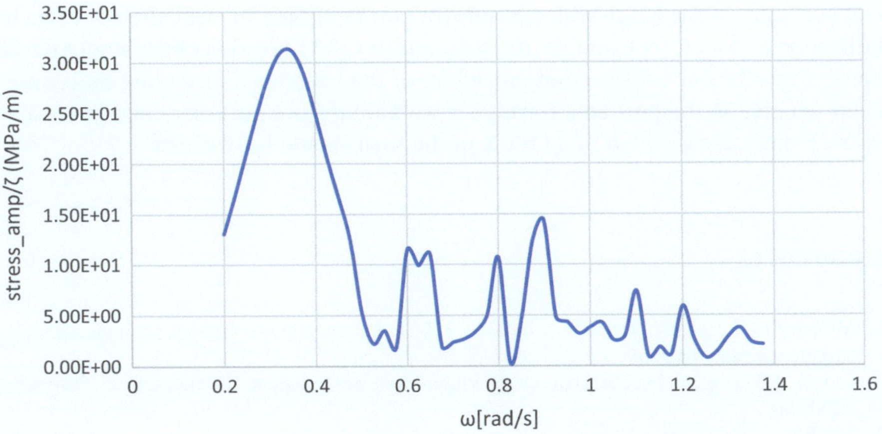

Fig. 4 RAO of stress at central part of the Base2

maximum value experienced under those sea conditions. The 1/1000 maximum response expected value is calculated to be 293.9 MPa, which is below the yield stress of steel (355 MPa), confirming the structural feasibility.

5 Conclusions

This study has aimed to establish a foundation for future discussions on offshore bases for wind farms deployed in EEZ through the following investigations:

- · The energy islands proposed in EU are examined and referenced for the facilities on the offshore bases.
- · The geographical and location conditions such as the distance from the coast (50 km, 100 km, 300 km, respectively), water depth, and wind conditions for three different offshore bases are assumed. The necessary facilities for each offshore base based on overseas proposals are examined to determine the layout, area, and dimensions of the bases, and to design the offshore bases. The primary functions include: collecting and transmitting electricity from the wind farm to land at Base1; transmission, O&M base, assembly and repair at Base2; and converting electricity to hydrogen or ammonia (P2X), solar panels, and a data center at Base3.
- · A large semi-submersible type, is selected for the floating structure of the base, by considering its response characteristics. The structural arrangements such as plate thickness, deck height, and column dimensions are determined.
- · Its feasibility from the structural engineers' viewpoint is confirmed based on the response functions of heave, pitch, and stress using a numerical simulation tool based on hydroelasticity theory.

Many future works need to be done. They may include: an improvement of the numerical model by giving appropriate viscous damping and reflecting more accurate geometry of the bases with cut-outs for port facilities; an evaluation of capacity utilization factor of, for example, the assembly or O&M function of the base; a development of efficient construction procedure of the base itself. Studying on creating calm sea area in the port area for higher workability is a necessary and interesting issue. Finally, the CAPEX and OPEX of the base should be clarified.

References

1. Ministry of Economy, Trade and Industry (2021) Green growth strategy through achieving carbon neutrality in 2050
2. COWI (2021) Cost benefit analyse og klimaaftryk af energiøer i Nordsøen og Østersøen, JANUAR
3. NeoWins (2021) https://appwdc1.infoc.nedo.go.jp/Nedo_Webgis/index.html. Last accessed 2024/07/16

4. Statistical Database od Winds and Waves around Japan (2024). https://cloud.nmri.go.jp/apps/wwjapan/service/serve/wsTDYiMbcx6XQhcnof6UXrnegvAICDEJV2TM0Y8lY4bBnzpGdqKdS4gvxJnYoMryE/. Last accessed 2024/07/16
5. https://www.meti.go.jp/shingikai/energy_environment/chokyori_kaitei/pdf/001_05_00.pdf. Last accessed 2024/07/16
6. https://www.jstage.jst.go.jp/article/jwea/35/3/35_85/_pdf/-char/ja, last accessed 2024/07/16
7. Suzuki H, Yoshida K (1995) A consideration on the dynamic behavior and the structural design of large scale floating structure. J Soc Naval Architects Jpn 178:473–483
8. Iijima K, Yoshida K, Suzuki H (1997) Hydrodynamic and hydroelastic analyses of very large floating structures in wave. Proc OMAE1997 4:139–145

Unlocking the Global Potential of Urban Development Offshore Floating Solutions in the Mediterranean Region

Gil Wang, Lieke Bikker, and Tomer Fishman

Abstract Recent research on floating solutions represents a significant advancement in promoting responsible human activities in the nearby marine environment and offshore regions. However, little is currently known about the actual scale of the potential for using floating solutions worldwide. The upscaling course of floating solutions is powered by the development needs of society and is restricted by natural environmental conditions and physical constraints. The main motivational drivers for floating solutions include land scarcity, population demographics, sustainability goals, recreational activities, and disaster mitigation. On the other hand, constraints primarily stem from issues such as wave energy, extreme weather events, zoning regulations, and bathymetric considerations. Thus, when considering both motivational drivers and constraints as layers of data, a model can be developed in a geographic information system (GIS) to help answer where floating solutions could be applied on a global scale. This study explores some of the methodological frameworks necessary for creating a global GIS model, focusing on the technical potential of floating solutions in the context of offshore dwelling.

Keywords Geographic information system (GIS) · Floating structures · Offshore urban development

G. Wang (✉)
CAMERI—Coastal and Marine Engineering Research Institute, Technion I.I.T, Haifa, Israel
e-mail: wang@technion.ac.il

L. Bikker · T. Fishman
Department of Industrial Ecology, Institute of Environmental Sciences (CML), Leiden University, Leiden, Netherlands

X. L. Zhao et al. (eds.), *Proceedings of the Fourth World Conference on Floating Solutions: WCFS 2024; 2–4 December; Hong Kong*, Lecture Notes in Civil Engineering 597, https://doi.org/10.1007/978-981-96-4569-5_7

1 Introduction

1.1 Background

Extending urban development into the adjacent marine environment has great potential for a more balanced relationship between man and planet, in the near and distant future [1]. The existing pressures in coastal urban regions, coupled with rising populations, threaten this equilibrium, leading to challenges that impact both people and nature [2]. Development needs often consume large amounts of natural resources and landscapes to accommodate new roads, infrastructures (e.g., energy, water, and communication), housing, industry, and commerce. This results in a massive loss of biodiversity and quality of life [3]. The tension between available space resources and development needs is further exacerbated in coastal areas suffering from land scarcity, flooding risks, and earthquakes [4]. Using a parametric GIS analysis model, our goal is to evaluate the worldwide feasibility of floating solutions as an alternative space for urban development [5, 6]. The current research focuses only on the Mediterranean Region showcasing the methodology and approach for global application to be used with a worldwide model.

1.2 Motivations and Technological Constraints

The GIS analysis is intended to address two primary questions. The first question addresses where floating solutions are a viable option for offshore expansion, based on a data set that examines motivational aspects. The second question addresses whether these locations can accommodate the associated technological constraints, or in other words if the threshold requirements with respect to the performance criteria of the floating structure can be met. The GIS model examines both questions to map the location with the greatest potential for the development of floating solutions. The model is a parametric investigation tool that may examine various design inputs reflecting the unique characteristics of any floating solution design inputted by the user.

The motivational aspects reflect the driving factors for offshore urban development which include sea level rise, land scarcity issues, limited potential growth, flooding threats, and the risk of earthquakes near major coastal communities. The constraints are dictated by the local environmental conditions such as wave climate and bathymetry along spatial zoning considerations such as shipping routes, and nature reserves to name only a few. Other constraints relate to the legal status of the potential location from municipal and regional perspectives, including borders and jurisdiction areas. The chosen floating solution layout (design and arrangement) is crucial for this analysis, as each design has distinct mechanical and structural properties for securing both ultimate and service limit states, thereby setting its performance criteria. In addition, the main particulars are also key features for safety (hydrostatics) and comfort (hydromechanics), along with the compliance to more shallow waters where draft is a limiting factor.

1.3 Demonstrator

The chosen demonstrator for this current study is the Modular Floating Structures (MFS) concept encompassing all equal modules, type Module9000 [7–10]. This layout has two principal configurations: 'MFS' and 'MFS + Floating Seawall'. The MFS is suitable for calm wave climate, up to a significant wave height of ca. 2 m [8], and the MFS + Floating Seawall is suitable for moderated wave energy with significant wave height up to 4 m, but it includes a survival condition of 1 to 100 years storm with a significant wave height of 8.5 m [9]. The latter is equipped with a protective Floating Seawall configuration made of floating breakwaters [11]. Figure 1. Shows the demonstrator layout, the chosen geographical location, and the Floating Seawall concept tested in the Ship Hydromechanics Laboratory in TU Delft.

2 Descriptive Methodology

2.1 Design of the GIS Model

This GIS model uses data layers of identified relevant factors as an input for establishing the technical potential. It combines them and generates maps showing on a predefined grid size where, if, and what configuration of MFS model can be used. The three options are color coded, and the grid size is equivalent to the schematic spatial arrangement given in Fig. 1. Using this data the technical potential calculations can be performed to estimate the maximal achievable area where MFS could potentially be implemented. The ArcGIS Project and Python code used for data preprocessing are available upon request.

2.2 Model Development, Verification and Validation

The GIS model is developed by combining the motivational factors for urban development offshore with the limiting factors related to the performance criteria of the MFS design. The environmental conditions limiting the technical potential of MFS include bathymetry, wave statistics, and wind climate. The zoning constraints include marine protected areas and shipping routes (military areas, and existing infrastructure are currently outside the scope of the model). The motivational factors for the potential of MFS include urban growth rates, topography (e.g., slopes in proximity

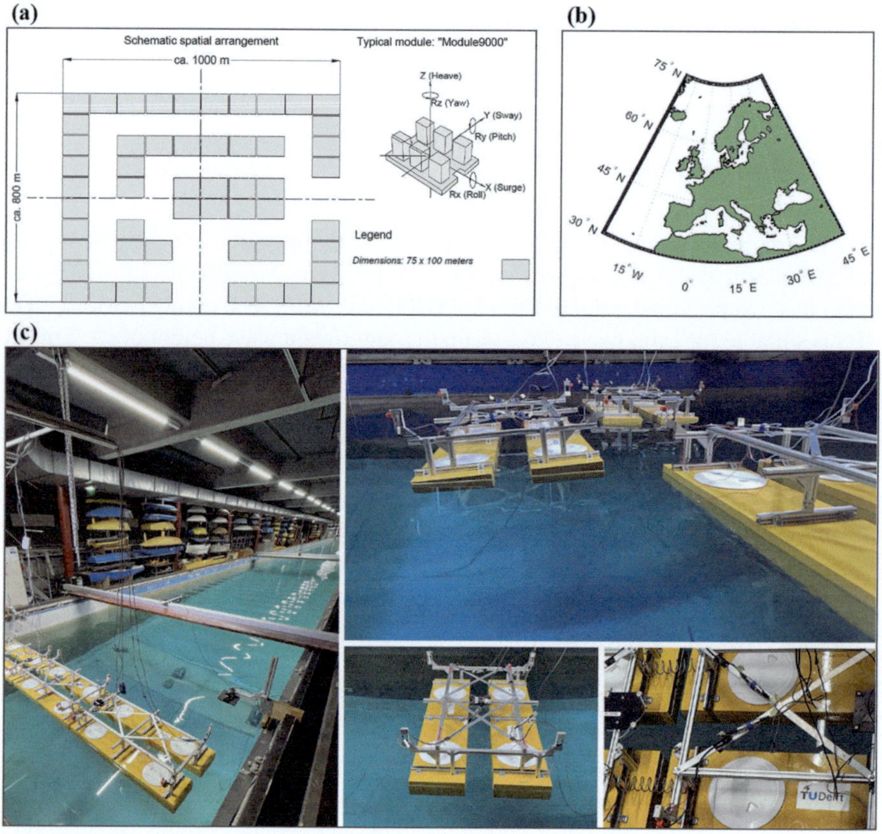

Fig. 1 The image is divided into three sections. The first section **(a)** shows a schematic spatial arrangement of a typical MFS city block, made from ca. 50 modules, all type "Module9000" [9]. The second section **(b)** is a map highlighting the Mediterranean Region which is the focus of the presented GIS analysis. The third section **(c)** shows a compilation of four photos from the physical testing of the Seawall concept [11] used for the alternative of MFS + Floating Seawall.

to coastal cities), land protected areas, and disaster mitigation such as exposure to sea-level-rise, flooding, and earthquakes. Data validation refers to the process of examining the quality, accuracy, and uncertainty of the used data sources before importing and processing the data. In the current study, data was collected from academic publications to ensure the quality. Ideally, raw data sets are used, and minimal data processing is applied within the GIS model. Accuracy can be affected by the level of resolution at which spatial data is made available. When data is provided with a coarser resolution, the divergence between the dataset value and the true value may be more pronounced, resulting in decreased accuracy—as is the case for the raw data extreme wave and wind values. Regarding data uncertainty, the use or inclusion of future projected data is important to consider. When dealing with future projections of urban growth and SLR, there are inherent uncertainties, as these projections are based on theoretical models. Data verification refers to the process of checking whether the data used accurately represents the actual data, mainly by cross-referencing it against published literature.

2.3 Data Preprocessing

After data validation and verification, the next step is to import the spatial data obtained in ArcGIS Pro (version 3.1.2) as individual layers. The datasets are obtained in different formats, resolution and geographical coordinate systems. The different formats include shapefiles, raster data, and tables. Following the definitions of Esri [12] shapefiles are simple formats storing the location, shape, and attributes of a geographic feature. These geographic features can be represented as points, lines, or polygons. A raster is a matrix of equally sized cells, arranged in rows and columns, where each cell contains a value representing specific information in the data set. Spatial resolution is associated with the raster data and refers to the size of cells in a raster. Lastly, the geographical coordinate system of the data defines where the data are located using coordinates. Both resolution and the coordinate system are important to consider, as these can determine how accurately and consistently spatial data can be measured, displayed, and analyzed.

When using multiple data layers in a GIS model or analysis, coordinate systems and resolutions should ideally be the same. As such, before analyzing these layers together, preprocessing of the individual layers was needed for coordinate alignment, cell resolution, and geographical coverage extent. The chosen geographical coordinate system used as the basis throughout the GIS analysis in this thesis was WGS84, with degrees as the unit of measurement. This choice is rooted in the fact that this is a widely recognized coordinate system, thus the coordinate system used for most of the obtained datasets, and to maintain consistency throughout the GIS analysis.

The resolution of raster layers used in the GIS model refers to the size of individual grid cells within a raster layer. Cell resolution is carefully selected with respect to the original raster data, to accommodate the global scale extent, while maintaining a desired level of detail for the coastal shoreline in a representative cell of 0.0125° by 0.0125°. This cell dimension choice is based on the schematic spatial arrangement of the MFS (Fig. 1). The resolution of 0.0125° by 0.0125° varies roughly between 1.3 by 1.3 km at 0 degrees latitude—at the equatorial—to 1.3 by 0.5 km at 67° latitude north/south—the arctic circle. The GIS model includes the geographical scope which includes territorial, internal, and archipelagic waters. A shapefile representing these waters was collected online and used as the spatial extent of the data layers.

3 Results

3.1 *Floating Urban Development in the Mediterranean Region*

The calculated suitable area in the GIS model is based on the motivational input and constraint layers outlined in Sect. 1.2 above and it accounts only for the geographical area close to the relevant coastal cities (cities with $\geqq$ 300,000 inhabitants), at a fixed radius of 35 km from the center of the city. The results of the global potential analysis in the Mediterranean Region show that the useable area for floating urban solutions is more than 18,000 km^2, an area the size of Slovenia or approximately half the size of Switzerland or Denmark. However, the key difference when comparing these areas is that the entire calculated area, in the model results, is within the metropolitan rim, adjacent to the main cities. Thus, this is the net area relevant for future urban development, excluding unusable surfaces such as mountain ridges, national parks, deserts, or even industrial terrain. Figure 2 shows the investigation region (including relevant coastal cities), and the results of the calculated potential.

3.2 *Taranto, Italy (40°28′0″ N, 17°14′0″ E)*

The vast potential for floating solution is further revealed when zooming in on a random or an explicit coastal city. Looking at Taranto, Italy, we can see that it has the potential of adding an additional 646 km^2 using MFS technology—three times its current size! The results of the GIS analysis are presented in Fig. 3a. Taranto is one of 12 Italian coastal cities analyzed in the GIS model, and it has the second highest potential suitable area for floating development after Pescara (with 766 km^2). Taranto was chosen randomly to evaluate the model outcome, and to explore the results using the methodology and approach outlined above. To validate whether Taranto is a suitable location from a technical perspective, the wave climate (which is a dominating factor for MFS feasibility) was cross-referenced with published research examining the wave energy at a 20 m water depth contour of the Mediterranean basin [13]. This research uses the Climate Data Store of the Copernicus Climate Change, while the GIS model uses the US Geological Survey (USGS) in partnership with Esri [12]. Good agreement was found between the two models showing a moderate wave climate with medium to low energy. This is a crucial point to ensure the performance criteria of the MFS with respect to occupant comfort in waves. Figure 3c provides the average MFS acceleration, in heave motion, used to determine occupant comfort criteria limits in waves [14, 15]. Closer examination of this specific location shows additional motivational factors beyond the scope of the GIS model. Taranto is located near the largest integrated cycle steel complex in Europe and has been declared an area at high risk of environmental crisis by the Italian Council of Ministers [16]. Thus, using floating urban development in the future could help improve some of the

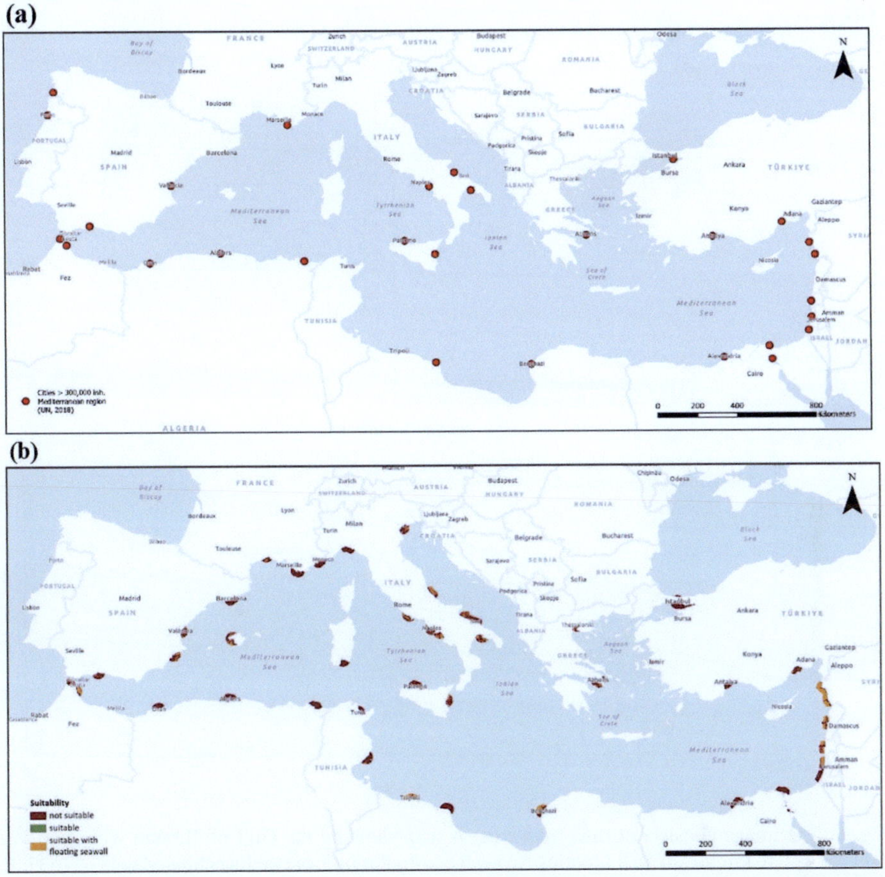

Fig. 2 Two maps of the Mediterranean region are shown. The first map **(a)** highlights cities with populations over 300,000 using red dots. The second map **(b)** illustrates the suitability of these areas for urban developments, using a color-coded system: green for suitable MFS, red for not suitable, and orange for suitable MFS + Floating Seawall. The maps provide a comparative analysis of urban density and floating development potential in the Mediterranean region.

local health and environmental conditions. For example, by creating additional urban green spaces and expanding the city towards the bay area, the exposure to long-term PM_{10} (particulate matter measuring $\leq 10\ \mu m$) can be reduced as shown in Fig. 3b.

4 Conclusions

Using our developed global GIS model in the Mediterranean region immediately reveals the vast potential of floating solutions for sustainable future growth. The total calculated potential area for urban MFS is 18,000 km^2, which can accommodate 100

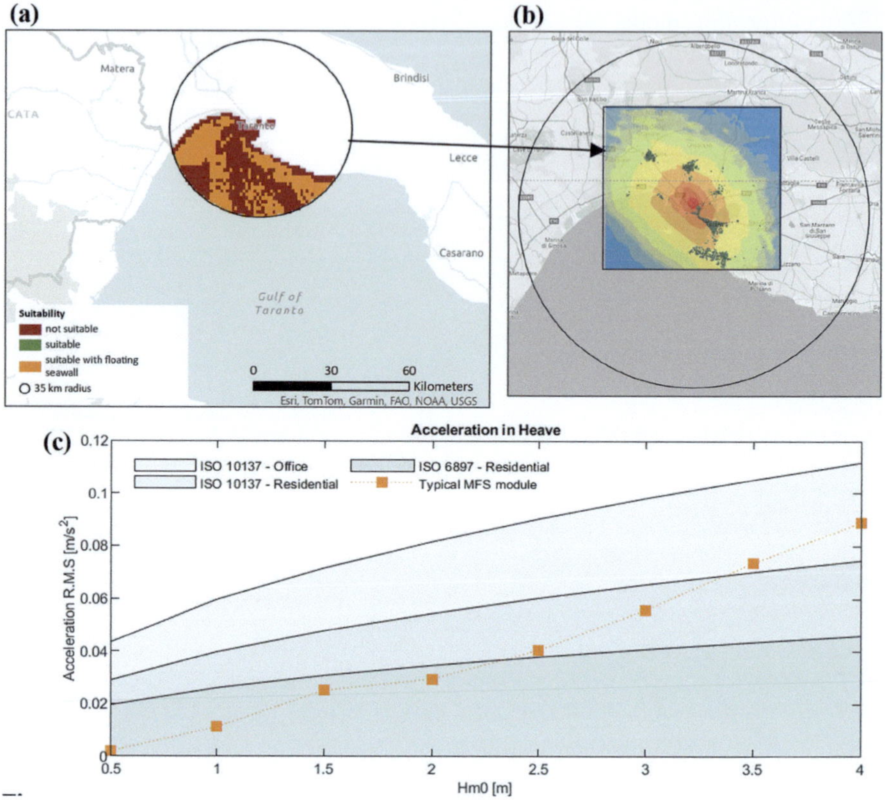

Fig. 3 The image consists of three parts: **(a)** A map showing the Gulf of Taranto with 646.407 suitable square km for MFS + Floating Seawall configuration from an investigated area of 1437.36 km^2. Part **(b)** shows a zoomed-in map section of long-term residential exposure to PM_{10}. Red-to-blue color coding indicates high-to-low PM_{10} concentrations (the dots represent residences) [16]. Part **(c)** shows an average MFS module acceleration in heave (with Floating Seawall) in significant wave heights of up to 4 m—corresponding to the local wave climate, presented on the main criteria limits for occupant comfort [8, 9]

million people, at a fix density of 6000 inhabitants per square km, and has the technical carrying capacity of over 300 million people [9]. A real game-changing figure in terms of reducing the tension on the available space resources and development needs in coastal areas. By performing a random spot check in Taranto, Italy, the model was validated against cross-referenced data to evaluate the main constraint for its performance, the local wave climate. In addition, site-specific insights provided an additional motivational driver for MFS, related to air quality and human health. Thus, using this GIS model enables researchers and planners to use it both for global strategies on a regional scale and tactical strategies at a specific site location. In both cases, the GIS model developed can be used to improve the design of floating solutions to conform to global or local requirements.

Hopefully, this model will provide a foundation for collaborations between researchers, industry, governments, and institutions looking to include the adjacent marine environment as a viable alternative for sustainable future urban growth.

References

1. Glavovic BC, Dawson R, Chow W, Garschagen M, Haasnoot M, Singh C, Thomas A (2022) Cross-chapter paper 2: cities and settlements by the Se. In: Climate change 2022: impacts, adaptation and vulnerability. Contribution of working group II to the sixth assessment report of the intergovernmental panel on climate change. Cambridge University Press, Cambridge, UK and New York, USA, pp 2163–2194
2. Cohen M (2017) A Systematic review of urban sustainability assessment literature. Sustainability 9:2048. https://doi.org/10.3390/su9112048
3. Simkin RD, Seto KC, McDonald RI, Jetz W (2022) Biodiversity impacts and conservation implications of urban land expansion projected to 2050. Proc Natl Acad Sci 119:e2117297119. https://doi.org/10.1073/pnas.2117297119
4. de Graaf-van Dinther R, Ovink H (2021) The five pillars of climate resilience. In: de Graaf-van Dinther R (ed) Climate resilient urban areas: governance, design and development in coastal delta cities. Springer International Publishing, Cham, pp 1–19. https://doi.org/10.1007/978-3-030-57537-3_1
5. Wang CM, Wang BT (2020) Floating solutions for challenges facing humanity. In: Reddy JN, Wang CM, Luong VH, Le AT (eds) ICSCEA 2019. Springer, Singapore, pp 3–29. https://doi.org/10.1007/978-981-15-5144-4_1
6. Wang CM, Watanabe E, Utsunomiya T (eds) (2007) Very large floating structures. CRC Press, London; New York
7. Wang G, Goldfeld Y, Drimer N (2019) Expanding coastal cities—proof of feasibility for modular floating structures (MFS). J Clean Prod. https://doi.org/10.1016/j.jclepro.2019.03.007
8. Wang G, Rosenfeld Y, Drimer N, Goldfeld Y (2020) Occupant comfort analysis for rigid floating structures—methodology and design assessment for offshore dwelling module. Ships Offshore Struct, 1–16. https://doi.org/10.1080/17445302.2020.1718267
9. Wang G, Drimer N, Goldfeld Y (2020) Modular floating structures (MFS) for offshore dwelling a hydrodynamic analysis in the frequency domain. Ocean Eng 216:107996. https://doi.org/10.1016/j.oceaneng.2020.107996
10. Wang S (2023) Simplified analytical solutions to the yaw dynamics of modular floating structures. Ocean Eng 276:114206. https://doi.org/10.1016/j.oceaneng.2023.114206
11. Wang G, Bar D, Schreier S (2024) The potential of end-of-life ships as a floating seawall and the methodical use of gap resonance for wave attenuation. Ocean Eng 298:117246. https://doi.org/10.1016/j.oceaneng.2024.117246
12. Esri (2023) What is GIS? https://www.esri.com/en-us/what-is-gis/overview
13. Simonetti I, Cappietti L (2023) Mediterranean coastal wave-climate long-term trend in climate change scenarios and effects on the optimal sizing of OWC wave energy converters. Coast Eng 179:104247. https://doi.org/10.1016/j.coastaleng.2022.104247
14. ISO-6897: 6897:1984 (1984) Guidelines for the evaluation of the response of occupants of fixed structures, especially buildings and off-shore structures, to low-frequency horizontal motion (0.063 to 1 Hz)
15. ISO-10137:10137:2007 (2007) Bases for design of structures—serviceability of buildings and walkways against vibration
16. Martín-Olmedo P, Ranzi A, Spadaro JV, Portier C, Pearce N (2023) Health impact assessment of steel plant activities in Taranto, Italy. WHO Regional Office for Europe, Copenhagen

Case Study Analysis: A Game-Changer in the Identification of Requirements for Floating Architecture

Livia Calcagni

Abstract The paper provides extensive case study research involving existing projects distributed across different scales, countries, water types, times, and functions. Identifying and cataloguing available and exemplary existing projects lays the foundations for creating an open-source archive of projects and solutions that allows the development of standard requirements and guidelines for floating building design. A preliminary screening based on specified criteria led to selecting 25 case studies, mainly designed for permanent inhabitation or public use, that are analysed following a requirement-structured rubric. The requirement-structured rubric and case study analysis protocol are evaluated by experts in different disciplinary domains and refined accordingly. The case studies illustrate the different technical and technological advancements in floating architecture, the challenges and opportunities associated with floating buildings and their aggregation in districts, and most importantly, their compliance with each requirement. A multi-criteria evaluation matrix returns best practices among the case studies according to their compliance with the requirement rubric and the priority distribution among requirements according to their fulfilment among the case studies. The multi-criteria matrix proved to be a helpful tool for comparing the different requirements and projects to make informed decisions in a structured and systematic manner.

Keywords Case study · Requirements · Floating building design · Multi criteria

1 Introduction

In the most virtuous European and international contexts, water-based development is gaining increasing attention and is becoming a component of city plans for sustainable development and climate adaptation [1, 2]. However, a significant gap in the field

L. Calcagni (✉)
Department of Planning, Design and Technology of Architecture, Sapienza University of Rome, Rome, RM, Italy
e-mail: livia.calcagni@uniroma1.it

X. L. Zhao et al. (eds.), *Proceedings of the Fourth World Conference on Floating Solutions: WCFS 2024; 2–4 December; Hong Kong*, Lecture Notes in Civil Engineering 597, https://doi.org/10.1007/978-981-96-4569-5_8

concerns the taxonomy and regulation of floating architecture. Since floating architecture is not considered a full-fledged building typology, it is not expected to comply with norms that regulate land-based buildings. There are no comprehensive guidelines for floating architecture, and the limited existing design frameworks primarily emphasize safety and comfort, driven by the demanding conditions of marine environments. Identifying and cataloguing available and exemplary existing projects lays the foundations for creating an open-source archive of projects and solutions that allows the development of standard requirements and guidelines for floating building design. Case study research provides information grounded in practice that has not necessarily already been taken into consideration within the theoretical and regulatory sectors [3]. The case studies analysis provides accurate information regarding highly significant requirements and the most used and widespread structural and technological components, materials, and construction processes, allowing the development of general considerations on state of the art within the practice field.

The review presented in this paper covers multiple cases and a comparative analysis to draw a set of cross-case conclusions. It is both descriptive and exploratory in purpose, as it aims at generating hypotheses for later investigation. In fact, on the one hand, it collects an extensive database of projects and presents them in a structured analytical format; on the other, it is intended for theory-building. More specifically, it serves to further develop and integrate an existing Performance-based Design Support Framework for floating building design, previously developed by the author [4], with inputs from the practice field. A multi-criteria matrix (MCM) allows the identification of best practices among case studies and of a priority order amongst requirements.

1.1 Methods

Workflow. The case study assessment protocol is articulated into six steps and uses a research-from-design approach to further integrate and update the framework for floating building design (PDSF).

Step 1 Definition of the objectives for the selection of case studies and sampling methods. Case studies were selected through selective sampling, as it represents the most appropriate method if only a limited number of primary data sources can contribute to the study, which is the case for floating buildings. The selection of the projects included a wide-ranging exploratory phase in which over 100 floating buildings built worldwide were considered. A first restriction was achieved by excluding cases with highly conditioned attributions and cases that did not adhere to the following conditions: built (still existing); realised over the last 20 years; located in inland and intermediate sheltered waters; compliant with minimum occupancy and indoor comfort conditions; data availability. A second boundary condition involved their compliance with at least four of the following objectives: ecosystem preservation, resource circularity, energy efficiency, wellbeing conditions, and economic feasibility. The 25 selected case studies represent a database snapshot considering the distribution between residential and non-residential projects. Ten housing projects - ranging from single housing units to multi-unit or entire districts – were selected to focus the attention on permanent inhabitation. The remaining 15 projects include 5 examples of temporary accommodations like touristic facilities and 10 non-residential projects hosting leisure, food production, infrastructure, tertiary, cultural-educational, and commercial functions. The goal of examining various functions, beyond residential buildings, is to create a framework for living and its essential services.

Step 2 Development of the requirement-structured rubric for the case study assessment. The projects are presented within the scope of the research with a particular focus on their (primary) function, scale, morphology, compliance to selection criteria, material and construction components of the critical technical units, and eventually, requirement compliance. An analysis sheet and a requirement-structured assessment rubric was developed to provide a simplified and structured reading method to facilitate the comparison between the different design solutions. The assessment form combines quantitative and qualitative analysis and a specific section is assigned to the project's compliance with each objective, schematically highlighted through the requirement-structured assessment rubric. The assessment rubric is based on the PDSF.

Step 3 Evaluation and validation of the case study assessment rubric. The case study assessment form was subjected to validation by a diverse group of scientific experts and practitioners involved in the fields of floating architecture, hydraulic and mechanic marine engineering, ocean engineering and ship technology, ecology and natural science, urban ecology, and environmental architecture. This collaborative validation process ensures that the assessment rubric is robust and comprehensive, reflecting the collective expertise of the research community.

Step 4 Refinement of the case study assessment rubric. Once the reviewing process ended, changes and variations were made to the case study assessment rubric according to the feedback received through the evaluation forms. The final requirement rubric can be found in Calcagni et al. [4].

Step 5 Case study assessment. The selected projects were analysed based on the refined case study assessment rubric. General information is acquired mainly through desk research integrated with field research (observational analysis and direct interviews with the main actors involved in the design and construction process). The data is used to provide general information about the project (identification data, site features, functional typology, dimensional aspects, and construction components) and to assess the project's compliance with the objectives defined in Step 1. Grey literature is also used for the evaluation of the project's compliance with each requirement. The requirement compliance is further assessed through observational research and direct interviews.

Step 6 Construction of the multi-criteria evaluation matrix. A MCM is built, placing the case studies on the x-axis and the classes of requirements on the y-axis. The requirements are included in each case study sheet within the section "requirement compliance.“ The construction of the matrix returns two results simultaneously. On one hand, it assesses and prioritises different projects on the basis of various criteria (performance requirements) helping decision-makers to objectively evaluate each project's benefits and drawbacks, and to identify best practices among the case studies according to their compliance with the requirements. On the other hand, by defining relevant criteria (presence/absence within selected projects), the matrix identifies a priority order amongst the classes of requirements according to their observance within the practice field, providing a percentage of fulfilment for each requirement and thus highlighting the importance of each requirement compared to the others.

2 Case Study Analysis

2.1 Case Studies

The residential case studies include: Ijburg, Amsterdam; Schoonship, Amsterdam; Tatami house, Amsterdam; Harnaschpolder, Delft; Harlem Shuffle, Haarlem; Urban Riggers, Copenhagen; Seattle floating home; Floating homes in Nassauhaven; Pvrok, Střelecký Island; Maldive floating house. The non-residential case studies include 10 buildings hosting public or private functions including infrastructure, commercial, cultural-education, food production, leisure, tertiary and greenery: GCA Floating Office, Rotterdam; Floating Pavillion, Rotterdam; Floating Farm, Rotterdam; Mansa Floating Hub - MFS IV, Mindelo Bay; Jellyfish barge, Florence; Meripaviljonki Sea Restaurant, Helsinki; Theater L'Île Ô, Lyon; Ferry Terminal, Terengganu; Joyous floating sauna, Oslo; Brockholes Visitor Centre, Lancashire. The other five include tourism related functions with a particular focus on hospitality in the sense of temporary living facilities: Marina Azzurra Resort, Lignano Sabbiadoro; DD16, Moscow; Ar-che, Brandenburg; Anthénea, Dora; Botel 2.0, Porto Ceresio. The analysis of such different examples highlights the diverse approach and common challenges encountered in floating building design, especially concerning the legal status and regulatory standards. The debate about the juridical and financial status of floating houses and their classification as either boats or houses is ongoing in several countries. In the Netherlands, floating houses are legally recognized as homes if they are moored to the ground (waterbed) in a specific location [5]. This has led to banks offering mortgages and insurance boosting market demand and trust among potential house buyers. Moreover, floating houses require an ordinary building permit and should fulfil ordinary building standards. Other countries, such as Finland, Norway, France, and Italy, lack specific codes for floating buildings, making permit acquisition complex and time-consuming, often resulting in structures being classified as houseboats, boats or barges. A common challenge regards the difficulties encountered in buying a plot of water or obtaining construction permits to build on water, that significantly extends project timelines.The construction of the Meripaviljonki Sea Restaurant, for instance, took up to ten years, mostly due to bureaucratic delays in obtaining permits.

Public floating buildings, like restaurants and museums, face stricter regulations than private ones, leading to their registration as buildings that comply with land-based laws, unlike residential floating units typically classified as houseboats. For instance, Botel 2.0 is registered as a navigation unit in northern Italy to secure legal permits, despite meeting land-based building requirements, while Anthenea is built with boat certification but does not comply with land-based rules.

The analysis revealed that only a few projects, like Botel 2.0 and the Floating Office in Rotterdam, include onboard instructions for safe building use, detailing occupancy limits, emergency procedures, and resource consumption. Climbing and holding devices are not provided if other measures like parapets along the perimeter, prevent falling into the water. Case studies show that tourist leisure accommodations or housing units prioritize water access for recreational purposes rather than safety reasons. Also climate and microclimate conditions strongly affect which requirements are observed. For instance, *thermal variation resistance of pipelines* is only met in areas prone to water freezing which require frost resistance, such as IJburg, Oslo sauna, or DD16.

Ultimately, the case studies have shown the great number of actors involved, directly or indirectly, in floating development, categorized into three groups: public actors, private actors, and society. Public actors, such as municipal committees and environmental agencies, operate at both the process and governance levels. Private actors include architects and developers directly involved in projects, alongside investors and port authorities indirectly. Society actors primarily encompass users and residents, as well as mortgage providers, housing corporations, and environmental activist groups at an indirect level.

2.2 Multi Criteria Analysis: Best Practices

Unlike conventional MCMs, this matrix can be read in both directions and all requirements are equally rated with a common weight of 1. According to the evaluation of the case studies (alternatives) based on the requirements (criteria) (Fig. 1), all case studies address more than 60% of the requirements.

2.3 Multi Criteria Analysis: Priority Order Among Requirements

The bee-swarm plot diagram in Fig. 2 displays the distribution of items (classes of requirements) over a continuous dimension (score). The only requirements met by all case studies (100%) are the following: *Mechanical resistance to static actions*, *Mechanical resistance to dynamic actions*, *Structural continuity with substructure*, *Structural fire integrity*, *Indoor temperature level and control*, *Indoor*

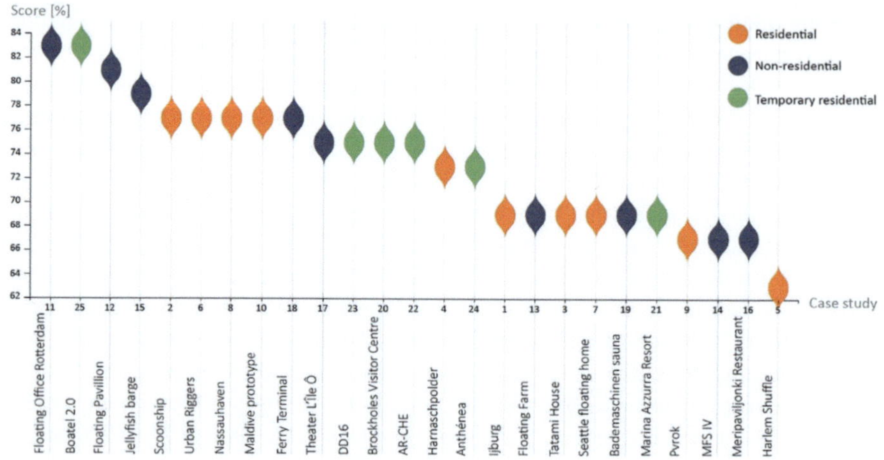

Fig. 1 A The graph shows the degree of compliance of the case studies (alternatives) with the design requirements (criteria)

humidity level and control, Ventilation control, Artificial illumination level and control, Occupancy rate, Ease of use and maneuver, Hygroscopicity, Freeboard, Watertight integrity, Mooring arrangements, Anchoring provisions and arrangements, Maintainability-repairability, Pipeline watertight integrity. These requirements mainly address safety, wellbeing, and buoyancy. Structural and fire safety are followed by all designers due to their mandatory nature across fields like architecture and offshore engineering. Wellbeing most conventional requirements like *thermal, acoustic, and visual comfort*, are met because they align with on-land building regulations and, therefore, followed by designers when building on water. *Motion comfort* is widely fulfilled (98%), while *biophilia, active design*, or *behavioural or community engagement* are generally not considered a priority in design and account respectively for 86%, 42%, and 38%. About 90% of projects address plant system maintainability, repairability, and watertight integrity.

Other essential requirements, accounting for 96%, are *Overtopping reduction (clearance above water)* and *Non-slip resistance* within the safety class, *Mobility (towing arrangements)* within usability class, *Atmospheric agents resistance* (management class), *Plant system integration* (integrability class), *Avoid interference with protected areas* (Environmental regeneration class), and *Adaptability of pipe-lines to water fluctuations* (plant system class). Requirements like *Overtopping reduction* (that prevents the threat of waves or tides washing over the deck) and *Non-slip resistance* (of the walking surfaces in contact with water) are observed by many projects because they are part of the safety class of demand despite being specifically related to the water environment. Requirement *Mobility (towing arrangements)* is undoubtedly crucial for the building's maintenance, repair, and relocation. Therefore, it is not surprising that most case studies are equipped with towing arrangements.

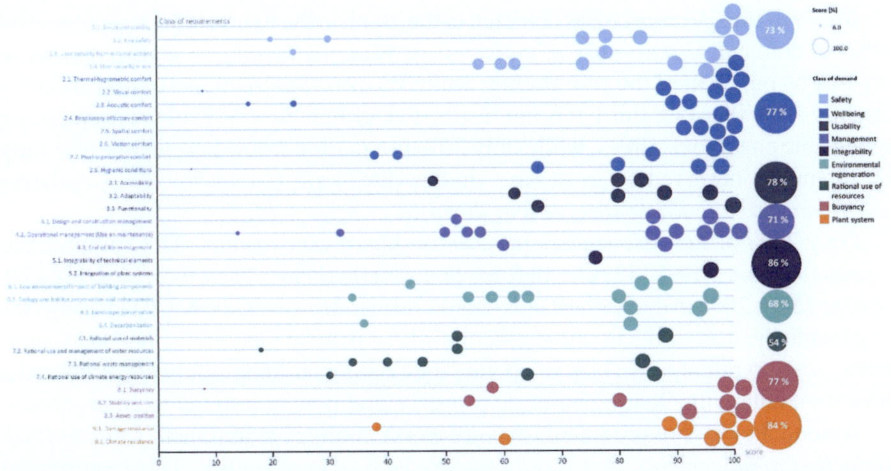

Fig. 2 Bee-swarm plot diagram displaying the fulfilment of the classes of requirements

Requirements like *Atmospheric agents resistance* (of exposed materials and components) and *Adaptability of pipelines to water fluctuations* are extremely important for floating buildings as the first ensures their long-term durability, and the latter their daily survival and adaptation to the changing site conditions. Adaptable pipelines are essential for ensuring the constant operation of the building's plant systems under all weather and climate conditions. Most case studies reasonably observe requirement *Plant system integration*, as it guarantees the overall efficiency of the building. Protected areas are regulated by specific regulations, which overrule building codes. Therefore, any construction must comply with the protected areas' regulations. Since the case studies are located in countries where regulations that protect these areas are in force, this requirement could be removed as it is implicit. However, since the framework could be applied anywhere, keeping the requirement as a reminder in the PDSF may be necessary.

Under keel clearance is revealed to be highly observed by designers (92% of case studies). This requirement is crucial as it is connected to several aspects, including safety, environmental impact, and mobility. It prevents the building from grounding, minimizes the impact of the building on the seabed and marine life, and allows it to manoeuvre the building easily and safely in shallow waters.

The requirements of the Environment regeneration class are accomplished by 68% on average. Only slightly more than one-third of the case studies address requirement Foster biodiversity. Between fifty and sixty percent of the case studies meet the requirements *Avoid impingement, entrainment, entanglement, and impairment of biostructure and aquatic vegetation, Avoid unnecessary reduction/obstruction and facilitate incoming sunlight in water*, and *Reduce underwater noise sources* (hydroacoustic energy). More attention is paid to *Reduce light pollution and avoid underwater illumination at night,* which is met by one-eighth of the case studies.

Rational use of resources, as depicted in the graph in the figure, is the least observed class of demand. *Circular use of materials* or *Water collection, treatment and reuse* are met by only half of the case studies, while *Waste-water treatment optimization* by slightly more than one-third. It is interesting to highlight how 64% of the projects use renewable energy resources, while only 30% use marine renewable energy. The only requirements observed by most case studies (86%) are *Use of bioclimatic passive solutions* and *Safe waste storage and disposal.*

Only 26% of the case studies have an on-board user manual. This requirement comes from the NTA 8111 rules in force in the Netherlands. Therefore, most projects realized after 2008 in the Netherlands should fulfil it. However, the NTA 8111 requires to provide the user of the floating structure with a manual containing the load limitations, with no reference to how to operate the floating building safely and efficiently, how to maintain the floating building properly, and how to respond to emergencies.

Among the other requirements that are worth noting is requirement *Collision risk reduction arrangements*, that is only met by 24% of case studies. This requirement comes from maritime regulations and refers to moving objects. The case studies that meet this requirement are equipped with fender devices mainly to allow easy boat docking rather than to protect the floating foundation bodies and decks from eventual collision with other structures or vessels. This is mainly due to two reasons. On one hand, most floating buildings are located in areas with low levels of boat traffic and thus lower risk of collision. On the other, the buildings are kept apart by mooring systems that limit the horizontal movements of the structures. Aside from this, several projects do not comply with any specific water-related regulation besides conventional architectural codes that reasonably do not include anti-collision arrangements.

Requirements *Emergency and signal lighting* and *Sink risk prevention indicators* are addressed in only 8% of the projects. According to building codes, the first is mandatory only for public buildings. This is the most likely explanation for it not being met by most case studies. The case studies that are public and meant for public use are only three: the Theater L'Île Ô, the Ferry Terminal, and the Brockholes Visitor Centre. As no information is available for the visitor centre regarding this requirement, it is only met with certainty by two case studies (8%). *Emergency and signal lighting* and *Sink risk prevention indicators* can be expensive to install and maintain. This can make them a less attractive option for some developers, especially those working on limited-budget projects and residential or private buildings that are not obliged to comply with these requirements according to on-land building regulations. *Sink prevention indicators* come from the shipping field. They could be extremely useful to ensure the safety of occupants and the structure by providing early warning and, thus, sufficient time to evacuate and take other precautionary measures. However, they are not considered by designers, as most projects comply with on-land-based building regulations that do not include this requirement as the buildings do not encounter this risk.

The least observed requirement is *Pest and dangerous animal prevention*, likely due to its conflict (mutual exclusion) with several other requirements like *Biophilia*,

Foster biodiversity, or *Toxic emission control of materials*. Designers and developers may prioritize sustainability and ecological design principles, leading to less emphasis on traditional pest and animal control methods in favor of alternatives that minimize environmental impact. Additionally, many case study sites are in sheltered waters near urban areas with low wildlife, resulting in a lower need for protection against dangerous animals. However, this doesn't diminish the importance of including this requirement in a design framework, as certain locations may attract insects and pests.

2.4 Theoretical Integrations from Practice

Despite being mentioned by very few regulations (UNI 8289; CLC SOR/2010–120), *Furniture integrability* (usability class) is often observed in projects (more than two-thirds of the case studies). *Functional/spatial flexibility*, which has no regulatory system supporting it but only scientific literature, has been proven extremely important as it is fulfilled by 80% of the case studies. Case study n°25 (Botel 2.0) has an automated domotic system that warns users when they exceed consumption limits. This system could be considered a digital version of the onboard user manual for the information related to the efficient energy operation of the building. This suggests how *Onboard user manual* could work with real-time monitoring systems requirement and integrated real-time information and alarms. The requirement *CO2 absorption design solutions* has been further integrated with examples from case studies that use photocatalytic materials and natural materials with embodied carbon properties. The case studies n° 2 (Schoonship), n°10 (Maldive floating city prototype), n° 11 (Floating Office Rotterdam), n° 12 (Floating Pavilion), n° 15 (Jellyfish barge), and n° 25 (Botel 2.0) have highlighted how, being in a water-based habitat, *Water collection, treatment, and reuse* should also embrace systems that extract and treat water from the surrounding environment. Desalination processes can remove high salt concentrations from seawater, while treatments like filtration, aeration, and disinfection can eliminate bacteria, algae, and chemicals from lake and river water to provide water for domestic use. Despite being mentioned only by certification systems and not by regulations, *Limit water consumption* is observed in 18% of the case studies. Several projects include water-saving taps equipped with technologies that limit water consumption, recirculating showers, and similar devices. This performance requirement, while not explicitly addressed in codes and guidelines, is of utmost importance within international and Europeans goals and objectives.

3 Conclusions

The database does not claim to provide a complete collection of all existing floating buildings. However, it sets up an initial repertoire of projects that can be further and constantly updated by other designers using the same methodology. Considering how floating buildings are relatively rare compared to traditional buildings, this collection provides a decent number of projects to draw significant conclusions. Overall, the case study analysis has confirmed the framework's requirements, leading to its validation within practice. The MCM proves to be an effective tool for comparing projects based on performance requirements. Overall, the case study research highlights 6 major issues.

1. **Legal and regulatory challenges**: floating buildings face significant regulatory and legal barriers, particularly regarding their classification as houses or boats. These challenges affect mortgage availability, insurance, and the time required to obtain construction permits.
2. **Sustainability and compliance**: non-residential and tourist accommodations generally have higher compliance with safety, accessibility, and sustainability requirements compared to residential buildings. Shared systems in larger developments also contribute to higher compliance rates.
3. **Material use and technical innovations**: projects employ a range of materials and innovations showing the potential for diverse and sustainable construction solutions in floating environments.
4. **Priority on safety and durability**: structural integrity, safety, and well-being, are the top priorities across the case studies, demonstrating how floating buildings are equivalent to land-based building in terms of compliance with these essential requirements.
5. **Environmental preservation**: while many projects meet energy efficiency and resource circularity goals, biodiversity protection and underwater noise reduction are less commonly fulfilled, indicating areas for improvement in integrating environmental considerations.
6. **Real-time monitoring**: some projects highlight the potential of real-time monitoring systems to enhance energy management and improve floating building operation efficiency.

These findings emphasize the need for further development of international standards and guidelines and the importance of drawing on practical, built examples to inform and develop these guidelines.

References

1. Ernst L, Dinther REDG, Peek GJ, Loorbach DA (2016) Sustainable urban transformation and sustainability transitions; conceptual framework and case study. J Clean Prod 112:2988–2999

2. Dal Bo Zanon B, Roeffen B, Czapiewska KM, de Graaf-Van Dinther RE, Mooij PR (2017) Potential of floating production for delta and coastal cities. J Clean Prod 151:10–20
3. Groat, L. N., & Wang, D. Architectural research methods. John Wiley & Sons (2013).
4. Calcagni, L., Hensel, D.S., Hensel, M.U., Battisti, A. A Performance-Based Design Framework for Floating Architecture. Trade-Offs and Correlations Between Requirements for Multiple Criteria Decision-Making Optimization. In: Ikoma, T., Tabeta, S., Lim, S.H., Wang, C.M. (eds) Proceedings of the Third World Conference on Floating Solutions. WCFS 2023. Lecture Notes in Civil Engineering, vol 465. Springer, Singapore (2024).
5. Vermande, H. (2009). Drijvende woningen en de bouwregelgeving, Handreiking voor ontwikkelaars, bouwers en gemeentelijke plantoetsers. Floating Houses and Construction Legislation, Guidelines for Developers, Contractors and Municipalities).

Floating Cities and Waste Management: A Case Study of a Waste Collection Vessel

Artur Karczewski, Maciej Reichel, Hanna Pruszko, and Ewa Marcinkiewicz

Abstract Floating cities, conceived as a remedy for overcrowding and the challenges posed by rising sea levels, present distinct hurdles, particularly in waste management. The case study of a Waste Collection Vessel (WCV) seeks to examine these associated challenges. It delves into the assumptions, limitations, and risks involved in addressing garbage management in large floating habitats. The analysis encompasses both formal situations and technical solutions. The culmination of this study is the proposal for a "floating garbage truck"—a zero-emission autonomous catamaran designed to be an intelligent vessel equipped with advanced technologies. It will be highly flexible and ideally suited for applications in floating cities, capable of tasks such as removing surface garbage, collecting garbage containers from habitats, addressing oil spills, and cutting weeds. The design of WCV ship was developed because of the ZeroWastePorts project.

Keywords Floating structure · Waste management · WCV · Autonomous technologies

1 Introduction

In the face of global challenges such as climate change, rising sea levels, and environmental pollution, the concept of floating cities is gaining significance. Floating cities are visionary solutions to global problems like overpopulation, rising sea levels, and land scarcity [1]. These innovative, floating habitats will offer adaptable, sustainable living spaces that open new possibilities for habitation, development, and environmental protection [2, 3].

A. Karczewski (✉) · M. Reichel · H. Pruszko
Faculty of Mechanical Engineering and Ship Technology, Gdansk University of Technology, Gdansk, Poland
e-mail: artkarcz@pg.edu.pl

E. Marcinkiewicz
Seatech Engineering LTD., Gdansk, Poland

X. L. Zhao et al. (eds.), *Proceedings of the Fourth World Conference on Floating Solutions: WCFS 2024; 2–4 December; Hong Kong*, Lecture Notes in Civil Engineering 597, https://doi.org/10.1007/978-981-96-4569-5_9

Despite many advantages, like any major human settlement, floating cities will have to address the issue of waste management also. That includes waste segregation and their collection, transportation, processing, and recycling [4]. Waste will accumulate in the inner parts of floating cities, as well as in ports, marinas, and associated water infrastructure. Additionally, there is a heightened risk of oil spills due to operational accidents during loading, unloading, and bunkering, and heavy ship traffic complicates cleaning efforts in these areas.

These factors indicate the need for an ecological vessel dedicated to cleaning the aquatic environment, both in the waters surrounding the city's edges and in ports. This vessel's tasks would include regular cleaning of floating debris and pollutants, cutting surface-level weeds, and emergency actions such as neutralizing or eliminating oil spills caused by leaks.

This article presents the conceptualization process of such a vessel, the Waste Collecting Vessel (WCV). It is an autonomous catamaran, 14 m in length, with zero emissions. This intelligent vessel will be equipped with advanced technologies, highly flexible, and perfectly suited for sheltered terminal and port applications. It will be capable of rapid response and water cleaning from pollutants, as well as recovering any oil contaminants.

2 Methodology

A standard method for designing a small craft was adopted, supplemented at the preliminary stage by analyses of the following unusual elements of the design process [5–7]:

- Modularity: The ship should have the ability to be reconfigured depending on the type of task.
- Modes of exploitation: The ship performs contrasting functions, several ships should be designed in one, like a multitasker.
- Degree of autonomy: The ship should exhibit the highest possible degree of autonomy.
- Task grouping: The ship's tasks should be grouped according to its modularity.

2.1 Modularity

Depending on the type of task to be performed, the operating ship will consist of a different number of modules:

- A single middle module (SET 1).
- Bow and middle modules (SET 2).
- Middle and aft modules (SET 3).
- A complete set of three modules (SET 4).

The bow module has been designed in three variants—1a (floating weir oil skimmer), 1b (smooth drum type oil skimmer), 1c (weed harvest option), therefore SET 2 will be available in the variants SET 2a, SET 2b and SET 2c. Only one of them can be used at a time. The middle and stern modules are designed in only one variant—for seaweed collection.

2.2 *Modes of Exploitation*

Ship should be designed for the following modes of exploitation.

- Waste collection from water (TYPE 1): Vessel designed to collect waste and garbage from water surface (to a depth of approx. 20 cm) while driving autonomously. For this operation ship will need SET 1 and occasionally SET 3 (if there is a lot of waste on a water surface).
- Oil spill recovery from water surface (TYPE 2): Vessel designed to collect oil spills from water surface. Depending on area depth and size of oil spill ship is to use one of two skimmers, variants a or b. For oil spill recovery there is need to use SET 2 (skimmer module).
- Seaweed harvesting (TYPE 3): Vessel designed to cut and/or collect seaweed in port area. For this operation there is need for SET 2c (harvest module) and occasionally SET 4.
- Collecting garbage from other ships (TYPE 4): Vessel designed to collect garbage and waste from other ships. For this operation there is need for SET 3. Ship will park side-to-side to the next vessel so it can throw waste into cargo boxes on the deck of the ship.

2.3 *Degrees of Autonomy*

The vessel, is designing, should operate in one of four degrees of autonomy [8]:

- Degree One (M): Ship with automated processes and decision support: Seafarers are on board to operate and control shipboard systems and functions. Some operations may be automated and at times be unsupervised but with seafarers on board ready to take control.
- Degree Two (R): Remotely controlled ship with seafarers on board: The ship is controlled and operated from another location. Seafarers are available on board to take control and to operate the shipboard systems and functions.
- Degree Three (RU): Remotely controlled ship without seafarers on board: The ship is controlled and operated from another location. There are no seafarers on board. Some vessels functions will be remotely controlled, and other become automated.

- Degree Four (A): Fully autonomous ship: The operating system of the ship can make decisions and determine actions by itself.

The extent of support from on-board personnel and the level of automation will depend on the task to be performed. The vessel will have a lower level of autonomy in certain operations, remaining under human control.

2.4 *Task Grouping*

To optimize the number and types of modules for the designed ship, tasks were grouped accordingly. The number of modules used will depend on the planned amount of garbage to be collected and the intended degree of autonomy for the ship. Considering the possible types of operations, module configurations, and degrees of autonomy, the ship will be able to use for 26 tasks listed below. Each task will be appropriately programmed, and depending on the selected task, the corresponding ship reconfiguration procedure will be activated.

- Waste collection from water:
  - TASKS 1–4

If the ship will be used to collect incidental waste from a specific area daily, it can be expected that only the middle module (SET 1) will be sufficient for such a task. This middle module will be a self-contained unit equipped with all the devices needed for independent navigation and waste collection facilities with a small tank for the collected waste. Depending on the needs the middle module (or the middle and aft modules) could be operate in any of the 4 levels of autonomy. In the event of a failure a single module as a fully autonomous ship is, due to its smallest size, less dangerous for other users than a unit in other configurations.

- TASKS 5–8

If it is necessary to collect more garbage than the permissible volume of the tank of the middle module, the ship will work as the middle module + stern module configuration (SET 3). The set consisting of the middle and stern modules will be able to operate in all 4 degrees of autonomy, like TASK 1–4.

- Oil spill recovery:
  - TASKS 9–14

Oil spill removing will be possible using the middle module connected to the bow module (SET 2), on which oil collection devices will be mounted 1a or 1b.

In the case of collecting oil spills, the ship will be able to operate in 3 degrees of autonomy:

human-operated ship with automated processes and decision support,
remotely controlled ship with seafarers on board,

remotely controlled ship without seafarers on board.

For safety reasons, the ship will not be able to operate as an unmanned autonomous vessel for these tasks.

- Weeds cutting and collecting:
  - TASKS 15–18

For weeds cutting task will be used the middle module and the bow module (SET 2) equipped with dedicated devices or all three modules (SET 4) in case when there are a lot of weeds to be harvested. In these tasks the vessel can operate as:

conventionally human-operated ship,
remotely controlled ship from the deck or from the quay (or operation station).

Due to the elevated risk of damage caused by the bow module, fully automatic work, and remotely controlled ship without seafarer on board is not recommended in this task.

- Collecting garbage from other ships:
  - TASKS 19–26

    Garbage collecting from other ships will be possible using only middle module (SET 1) or middle module connected with aft module (SET 3). In this task, ship can operate in all 4 autonomic levels of control.

A summary of all tasks with respect to the degree of autonomy, the number of intended modules and the type of operation to be done is shown in Table 1.

3 Results—Waste Collection Vessel

The result of the design work is zero emission autonomous catamaran, 14 m in length. The designed ship consists of three separate modules: the bow module (1), the middle module (2), and the stern module (3) (see Fig. 1). Only the middle module has its own propulsion. The other two modules, when connected to the middle module, are either towed or pushed. It is possible to control the ship in various module configurations, but always using the middle propulsion module. All working equipment on the different modules will be directly accessible from the deck. Each module is designed with a single flat deck, allowing personnel to move around easily and have access to all onboard and operational equipment.

The middle module (2) will serve as the propulsion module for the entire ship assembly. Below deck, it will space of the propeller, battery rooms, and electrical room. This module will also be equipped with waste collection equipment, a conveyor belt for transporting collected pollutants, and a small onboard tank, making it capable of collecting surface waste as an independent unit.

Table 1 A summary of all assumptions in range waste collection in design process

Modes of exploitation	Task number	Set				Autonomy					
		1	2			3	4	M	R	RU	A
			a	B	c						
TYPE 1	1	+						+			
	2	+							+		
	3	+								+	
	4	+									+
	5					+		+			
	6					+			+		
	7					+				+	
	8					+					+
TYPE 2	9		+					+			
	10		+						+		
	11		+							+	
	12			+				+			
	13			+					+		
	14			+						+	
TYPE 3	15				+			+			
	16				+				+		
	17						+	+			
	18						+		+		
TYPE 4	19	+						+			
	20	+							+		
	21	+								+	
	22	+									+
	23					+		+			
	24					+			+		
	25					+				+	
	26					+					+

The unpowered stern module (3) will be used exclusively for transporting the tank containing surface-collected waste, gathered algae, or waste collected from other ships. It is equipped with an additional tank, which, combined with the tank in the middle module, doubles its capacity.

The ship's superstructure is located on the middle module. It includes a control station equipped with a seat and a console panel for direct manual control. The station is covered to ensure operator comfort and will be mechanically and electronically secured against unauthorized access.

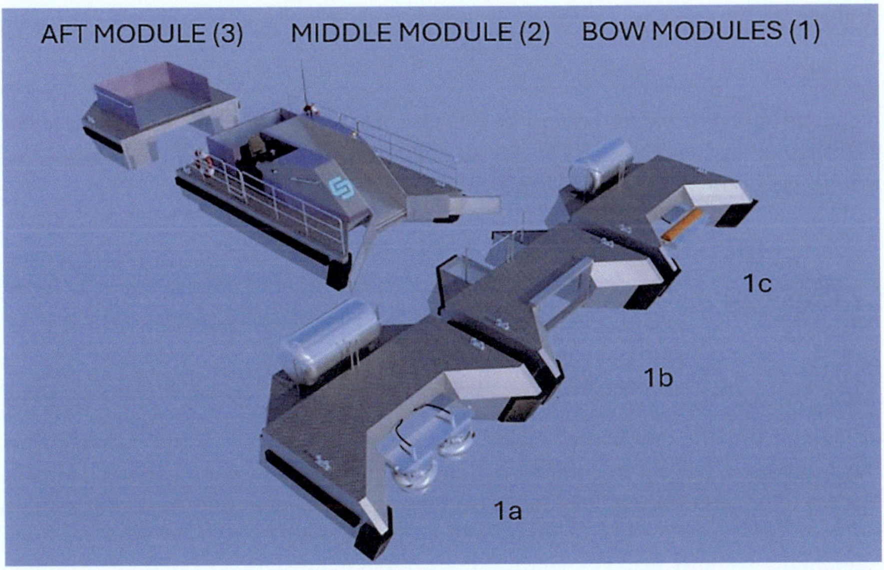

Fig. 1 General view of the ship configuration and modules

4 Discussion

Integration of modular design principles in shipbuilding processes represents a significant advancement in marine engineering. This study utilizes the flexibility and adaptability of modular construction to meet diverse operational requirements and environmental challenges. The ability to reconfigure the ship according to task type, as described in the article, underscores the importance of modularity in achieving versatile and multifunctional marine units.

The concept of modularity, particularly the ability to switch between different configurations such as SET 1, SET 2, SET 3, and SET 4, demonstrates an important level of innovation. This approach not only enhances operational efficiency but also promotes sustainable development by reducing the need for multiple specialized units. The specific design of bow, middle, and stern modules tailored for tasks such as waste collection, oil spill removal, and algae harvesting provides a practical example of modularity applied in real-world scenarios.

Furthermore, discussing levels of vessel autonomy in accordance with [8], highlights the evolution of maritime operations towards increased automation. The four autonomy levels provide a structural framework to understand how ships can transition from manual operations to fully autonomous systems. Gradual integration of autonomous capabilities ensures operational safety and reliability while leveraging technological advancements.

Task grouping based on operational requirements and autonomy levels is another key aspect discussed in the article. By categorizing tasks such as waste collection and oil spill removal, the study effectively matches modular configurations to intended functions and autonomy levels. The task grouping strategy optimizes ship performance and enhances its ability to adapt to diverse marine environments and operational demands.

5 Conclusions

Floating cities are emerging as a visionary solution to global challenges such as climate change, rising sea levels, and land scarcity. However, like any large human settlement, floating cities must tackle common issues such as waste management. This includes segregation, collection, transport, processing, and recycling of waste. Waste will accumulate on solid areas of floating cities, as well as in ports, marinas, and associated water infrastructures. These premises underscore the need for environmentally friendly water cleaning ships, both in the waters surrounding city edges and in ports. Their task would involve regular, periodic cleaning of floating debris and pollutants, surface weed cutting, as well as intervention actions when necessary, such as neutralizing or eliminating oil spills. These ships must be advanced and innovative to meet future challenges. Therefore, this article proposes an approach to ship design based on modularity, combined with various levels of autonomy, presenting a comprehensive solution to contemporary challenges in ship design. To promote efficiency and sustainable development, new standards in ship design and operations must be established. The detailed analysis and practical examples presented in this research serve as valuable reference points for future advancements in this field.

Acknowledgements The research was supported by the National Center for Research and Development (NCRD) within the project ZeroWastePorts: Zero Emission Waste Collecting Vessel to Use in Ports and Close-to-Shore Areas. Agreement no PL-TW/IX/43/ZEROWASTEPORTS/2022. The project visualization was supported by the Academic Computer Centre in Gdansk (CI TASK). All support is highly appreciated by the authors. Authors thank the reviewers for their valuable comments.

References

1. Neumann B, Vafeidis A, Zimmermann J, Nicholls R (2015) Future coastal population growth and exposure to sea-level rise and coastal flooding—a global assessment. PloS ONE 2015
2. Blue21 (2019) Sustainable floating cities as a future solution to sea level rise. In: Sustainable development goals report 2019
3. Fornari D, Landolfi G, Paci G, Ceccacci S (2018) Floating architecture: the challenge of a new urban development on water. Sustainability 2018
4. Smith J, Lee A, Kumar S (2022) Waste management in floating cities: challenges and opportunities. J Environ Manag 2022

5. Gełesz P, Karczewski A, Kozak J, Liwin W, Piatek Ł (2017) Design methodology for small passenger ships on the example of the Ferryboat Motława 2 driven by hybrid propulsion system. Pol Marit Res, pp 67–73
6. Choi M, Lee J (2024) Modular ship design: rapid prototyping and enhancing efficiency through design modules. In: 15th international marine design conference (IMDC-2024), Amsterdam, 2024
7. Erikstad SO (2019) Design for modularity: volume 1: optimisation of ship design and operation life cycle. In: A holistic approach to ship design, Springer, pp 2–28
8. IMO (2021) MSC.1/Circ.1638 Outcome of the regulatory scoping exercise for the use maritime autonomous surface ships (MASS). IMO, London

Sustainable Floating City as a Solution to Rising Sea Levels: An Integrated Project for Northwest Coast of Brazil

Allan Gomes, Fernando Borguetti, Guilherme Moraes, Sena Takahashi, Tomoki Ikoma, and Kazuo Nishimoto

Abstract This work presents an innovative integrated project for a floating city designed to address the challenges posed by rising sea levels, particularly in Recife, Brazil. The project encompasses comprehensive naval engineering concepts, with a focus on hydrostatics, hydrodynamics, anchoring, and renewable energies. It draws from a diverse range of sources for the various topics covered and involved sourcing suppliers for the city's components. Additionally, a partnership with Nihon University was established for in-depth structural and hydrodynamic analyses. Using the Oceanix model as a starting point, this research delves deeper into the essential engineering aspects necessary for developing this complex system. The primary focus is on integrating solutions, analyzing technical aspects, and ensuring the sustainability of the floating city. It is noteworthy that this work is open to future enhancements, aiming to conduct economic feasibility analyses and further explore the individual considerations of each component addressed in this document.

Keywords Floating city · Naval architecture · Sea level rise · Sustainability · Hydrostatics · Hydrodynamics · Renewable energies · Integrated solutions · Recife · Brazil

---

A. Gomes · F. Borguetti · G. Moraes · K. Nishimoto
Polytechnic School of the University of Sao Paulo, Sao Paulo, Brazil

S. Takahashi
Department of Architecture, Waseda University, Tokyo, Japan

T. Ikoma (✉)
Department of Oceanic Architecture & Engineering, CST, Nihon University, Chiba, Japan
e-mail: ikoma.tomoki@nihon-u.ac.jp

X. L. Zhao et al. (eds.), *Proceedings of the Fourth World Conference on Floating Solutions: WCFS 2024; 2–4 December; Hong Kong*, Lecture Notes in Civil Engineering 597, https://doi.org/10.1007/978-981-96-4569-5_10

1 Introduction

This paper introduces a strategic response to mitigate the adverse effects of rising sea levels, with a particular focus on the feasibility of implementing a floating city as a potential housing solution for future displaced populations. The objective is to apply floating systems engineering principles to examine the technical feasibility of a Very Large Floating Structure (VLFS) in Recife, one of the Brazilian cities most affected by this phenomenon. By investigating the feasibility of floating infrastructure, this study aims to provide a sustainable and innovative solution that can adapt to the dynamic and increased impacts of climate change on coastal urban areas.

The paper will cover real case studies, Oceanix Busan, deep case analysis, problem development, project requirements, weight and center analysis, stability assessments, energy module considerations, anchoring strategies, hydrodynamic analyses, and the final presentation of results. The multifaceted methodology, integrating naval architecture principles with global concepts, underscores the comprehensive nature of the study. This approach is pivotal in effectively addressing to several challenges associated with rising sea levels in Recife.

2 The Problem Formulation

Sea-level rise has become a major global concern due to global warming, intensified by anthropogenic effects. On a global scale, tangible consequences have already been observed:

- A sea level rise of 20 cm since 1900, with the potential to reach 1 m in the next 50 years.
- Increased frequency and intensity of coastal flooding, coastal erosion, and soil salinization.
- The World Bank estimates that coastal cities could incur losses exceeding 1 trillion USD by 2050.

In the Brazilian context, the Ministry of the Environment estimates that approximately 60% of the Brazilian population could be affected by the effects of sea-level rise due to living in coastal regions. Additionally, a study by the Federal University of Pernambuco (UFPE) highlighted that Recife is one of the most vulnerable cities in the country due to its low average altitude of only 4 m above sea level. In a scenario of a 0.5-m sea-level rise, it is estimated that the flooded area could be around 25.38 km^2, potentially affecting 175,000 people based on census density, highlighting the importance of finding a pivotal solution for this worst-case scenario.

Table 1 Estimated weight of functional modules

Module	Weight (ton)	KG (m)
A1—Freshwater production, treatment and storage	5.257	2.95
A2—Food production and storage	87	1.64
A3—Waste treatment	234	0.69
A4—Residential towers	14.726	14.99
A5—Floating structure	5.308	3.56
Total	25.611	9.97

3 Weights and Centers of General Arrangement

The weights and centers estimations were made by breaking down the different functional modules designed for the floating city (general arrangement). For each module, a should-weight analysis was made to build up the estimated weight by its subcomponents. Module "A2—Food Production and Storage", for example, followed a rationale of defining the average daily food intake per person, the crop yield of different crops and the weight of aeroponic equipment. The results are listed below (Table 1):

4 Structural Arrangement Analysis

The structural analysis for the studied case should start with a thorough analysis of the classification societies, the suitability of the existing rules for the context of the floating city, and their applicability to the design scenario of the primary, secondary, and tertiary structures. With this analysis, it was possible to determine the estimated number of structural components, thicknesses, dimensions, and the approximate weight of the structures.

The support structure for the designed neighborhood includes four main components: a lower hexagonal structure, internal support beams, bulkheads, and an upper deck shaped like an equilateral triangle (Fig. 1).

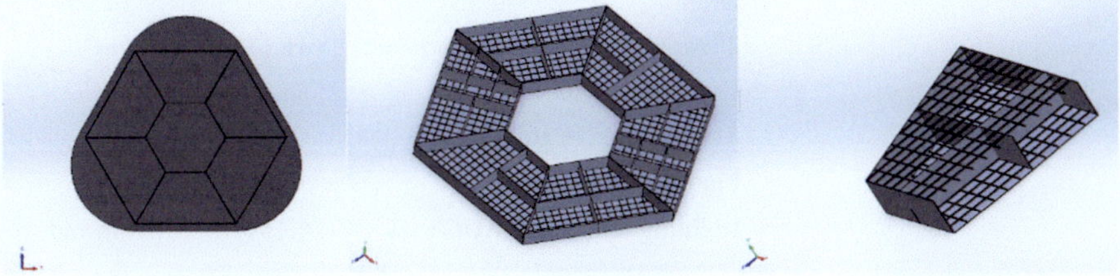

Fig. 1 Main structural components

5 Energy Generation

Aiming to establish a clean and renewable energy matrix, several energies generation sources for the city location were analyzed: solar, wind, tides, waves, and currents. Estimations based on local data are crucial to determine the optimal energy matrix for the floating city. Selecting an energy source for the floating city requires understanding specific environmental conditions. This section analyzes geographical, climatic, and oceanographic factors that affect energy source viability and efficiency.

5.1 Local Environmental Conditions and Energy Matrix Selection

The coastal Northeast region, where the floating city is proposed, has favorable wind conditions with speeds between 2.5 and 3.0 m/s and peaks of 12.0 to 15.0 m/s. Solar energy is also viable, with annual solar radiation ranging from 5000 to 5500 Wh/m^2 per day. However, tidal and tidal stream energies are limited due to low tide variations and insufficient wave amplitudes.

Considering Recife's total energy consumption and population, necessary daily energy are calculated. A hybrid energy approach, combining wind and solar generation with energy storage, is proposed to supply the city's energy requirements.

6 Intact and Damaged Stability

Since the structure is floating system, one crucial safety and comfort requirement for the floating city is its stability. Therefore, the **initial, intact and damaged stability (by lost buoyance method)** were evaluated for the project. The key metrics evaluated were in Table 2.

Table 2 Estimated weight of key functional modules

Module	Weight (ton)	KG (m)
Draft	5.257	2.95
2° Moment of Area (Waterplane)	16.688.488	m^4
Displaced volume (∇)	24.551	m^3
Metacentric Height (GM)	670.72	m
Flooding angle (θ_f)	3.36	°
Righting lever (GZ) AUC to θ_f	1.15	m-rad

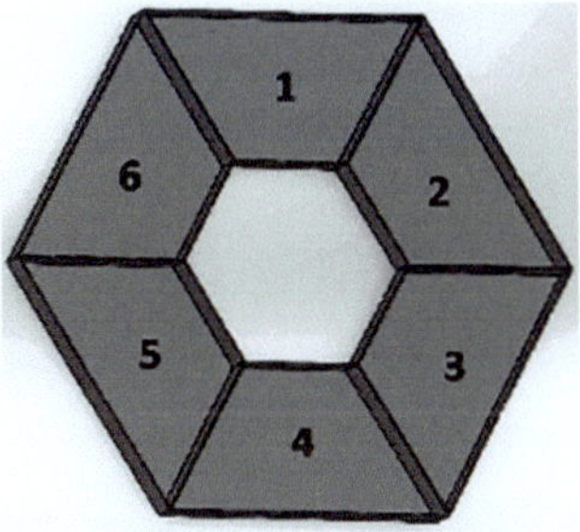

Fig. 2 Compartmentalization

For the damaged stability, the complete flooding of each compartment (Fig. 2) was evaluated. Due to the high value of the metacentric height, the maximum heel/trim angle was 0.95° (when compartment 1 was flooded), which is less than the flooding angle $\theta_f = 3{,}36°$.

7 Hydrodynamics

7.1 Wave Spectrum Consideration

For modeling the wave spectrum observed in the region of Recife (PE), it was necessary to consider a standardized sea spectrum that determines the relationship between wave height and period. For this project, the frequency domain spectrum was considered due to the ease of wave modeling with fewer parameters, using JONSWAP ("Joint North Sea Wave Observation Project") as it is the most universally used type, commonly employed for modeling on the Brazilian coast.

This type of spectrum model for waves uses only two parameters: significant wave height and peak period. Based on historical observations in the Recife region, approximately 10 km from the coast, these parameters have been estimated at 1–3.5 m for wave height and 7.5 s for the peak period. The Python implementation of the JONSWAP spectrum is illustrated in Fig. 3.

7.2 Transfer Function

With the wave spectrum in the frequency domain obtained, it was established the transfer function (spectral crossing) that will be responsible for determining the response curve of the floating unit. This response function can be obtained using the module of the Response Amplitude Operator (RAO).

For the initial analysis of the RAO of the structure, modeling was conducted using WAMIT software considering rigid body. This resulted in frequency domain

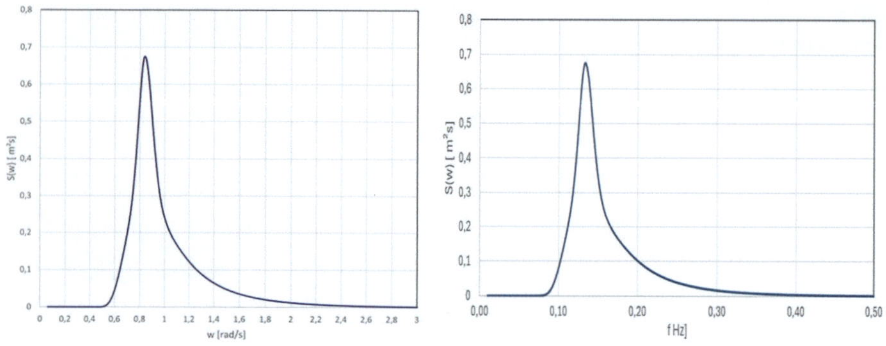

Fig. 3 Angular and frequency domain spectrum

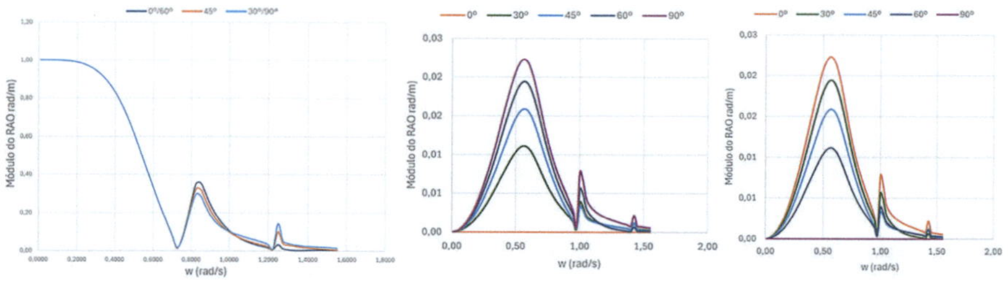

Fig. 4 RAOs for heave, roll and pitch

curves considering the heave, pitch, and roll movements of each floating parts. It is important to note that this preliminary analysis considered only the floating neighborhood, representing a single floating unit and the damping effect generated by the anchoring system was considered negligible, as its primary function was initially directed solely towards preventing the drift movement of the structure. Considering this, the following RAO curves in the frequency domain were obtained, considering 5 different incidence angles, as illustrated in Fig. 4.

7.3 Results Analysis

The analysis of response spectra allows us to conclude that:

- **Heave**: The response spectrum showed little variation in heave motion due to the high restoring coefficient of the structure, with a larger variation in vertical motion equal to 0.71 m. However, it was emphasized the importance of creating an anchoring system that ensures the natural period is outside any risk of resonance between the structure and the wave period of the region.

- **Roll and Pitch**: As explained in the stability chapter, one of the systemic concerns of the designed neighborhood was to ensure that sea conditions would not cause significant angulation. In this context, the response spectra showed that, despite the wave period of the region falling within a range of maxima, the values are negligible, with maximum variation values reaching 0.013 rad (0.771°) due to the large proportion of the floating structure.

In summary, hydrodynamic analyses have shown that the structure remains stable dynamically under the hydrodynamic conditions of the Recife region and 10 km from the coast. However, since the structure is very large, the hydroelasticity analysis will be required when several module will be connected each other.

8 Mooring System

8.1 Anchoring System Selection and Mooring Line Sizing

Designing large floating structures involves critical decisions, particularly in selecting the anchoring system. Challenges include accommodating their significant size, environmental loads, shallow depths, limited mooring cable space, and anchor installation complexities.

For the floating city, the Taut Legs system was chosen due to the shallow depth and proximity to the coast. This system employs angled ties and tensioned chains to minimize the anchor footprint, which is crucial in areas with high vessel traffic. An alternative catenary arrangement, where part of the mooring line rests on the seabed, was considered but deemed less efficient due to its larger footprint. The chosen mooring system must withstand environmental forces, primarily constant northeastern winds in the selected coastal region, with minimal impact from waves and currents.

A simplified approach involves sizing the mooring lines to account for the forces acting on the city. The mooring system is pre-tensioned to absorb energy and reduce movement. The primary consideration in the analysis was the horizontal wind force combined with the pre-tension established in the moorings. The estimated force was $F = 7.377$ kN, leading to the use of 12 moorings attached to the vertices of the hexagon to withstand this force. Choosing anchors for floating structures involves assessing various loading conditions and seabed soil properties, requiring thorough geotechnical analyses. For the floating city prject, driven piles with diameters of 1–3 m are selected for their strong resistance, capable of supporting over 5000 tons. These piles, ideal for systems like Taut Legs, are installed using pile-driving hammers in shallow depths. Anchor fixation is achieved through shear with the soil, utilizing the piles' hollow design effectively (Fig. 5).

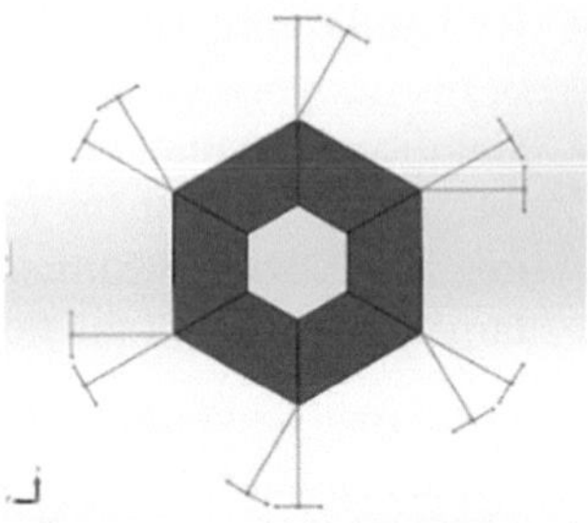

Fig. 5 Final mooring line diagram

Table 3 Calculation conditions

Floating body material	High strength concrete (cheaper than steel)
Floating body thickness B (mm)	1000
Young's modulus (N/mm^2)	4.0×10^4
Poisson's ratio	0.2
Mass density (N/mm^2)	2.5×10^{-10}
Depth of water (m)	15.0
Angle of incident waves (°)	0.0
Incident wave period (s)	4.0–20

9 Hydroelasticity Response

9.1 Analysis Method

In this study, eigenvalue analysis was carried out using the finite element structural analysis software Femap with NX Nastran®, and flow force elastic response analysis was carried out using eigenfunctions and eigenvalues in the manner of Sato et al. 6). The computational conditions for the unit-connected large floating structure used in the analysis are shown in Table 3 and an overview of the unit-connected floating structure is shown in Fig. 6.

The computational floating body mesh partition is shown in Fig. 7. The eigenfunctions and eigenvalues obtained from the eigenvalue analysis were used to analyse the hydroelastic response at incident wave periods from 4 to 20 s.

9.2 Vertical Displacement Calculation Points Overview

To evaluate the elastic deformation in waves, 16 points were selected from the nodes of the mesh data and flow force elastic response analysis was carried out using a calculation programmed for 8-node isoperimetric elements. 6 of the 16 points, A-F,

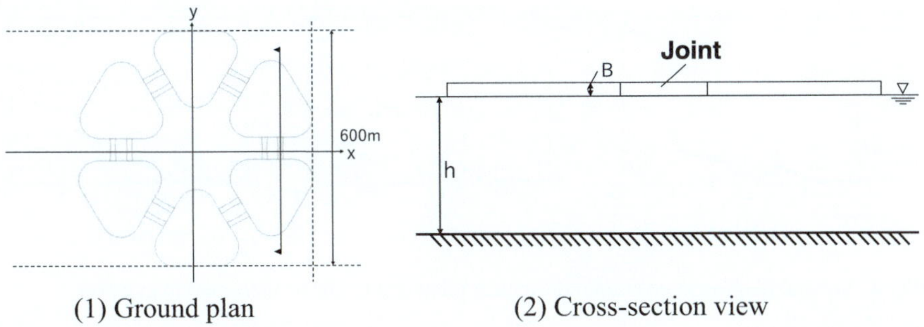

Fig. 6 Outline of unit-connected floating structure

Fig. 7 Calculation mesh model

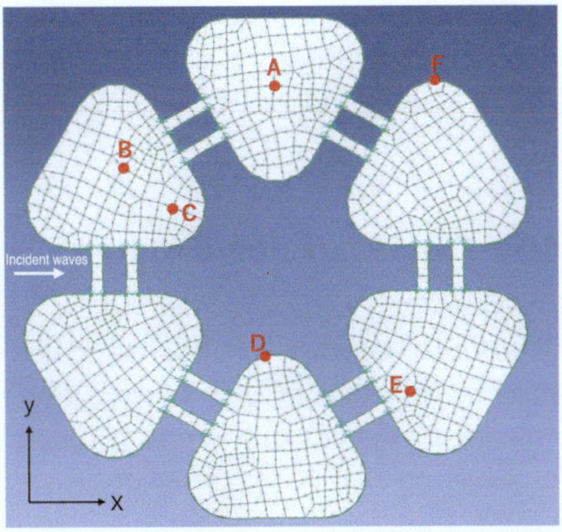

were selected, including points with high loads and the inside and outside of the connecting unit. The six evaluation points are shown in Fig. 7.

The maximum values of vertical displacement at each of the A to C assessment points are shown in Fig. 8. The point D behaved similar response with A and the vertical motion at E and F were similar because of almost symmetric positions. The vertical axis of the graph is shown against the dimensionless value of the deformation, while the horizontal axis is shown as the ratio of wavelength to the length of the floating body. Here, λ: incident wave wavelength, L: total length of the floating body, a: incident wave amplitude.

Similar changes were observed in the wavelength-to-float length ratio at points A and D, E and F respectively. Particularly large deformation was observed at point B, with a deformation of 0.025 at $\lambda/L = 0.39$. This result can be attributed to the fact that point B is the evaluation point located furthest in front of the incident wave. The next largest deformation was observed at point FH, with a deformation of 0.011 at $\lambda/L = 0.39$. The results show that similar changes were observed at points A and D, E and F on the outer side of the floating structure for each wavelength range, suggesting that elastic deformation in waves depends on the direction of the incident wave, and

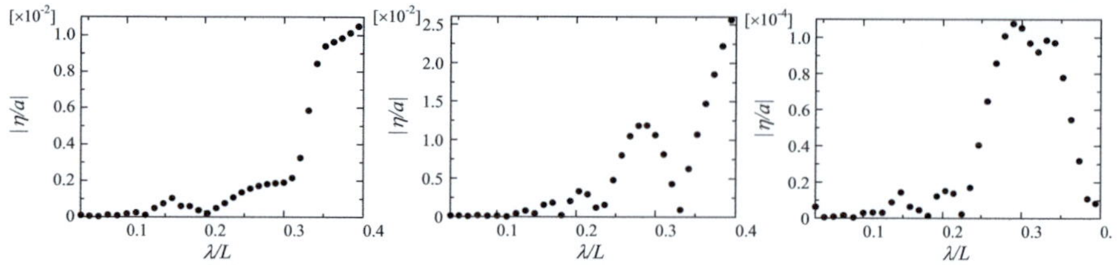

(a) Point A (b) Point B (c) Point C

Fig. 8 Vertical displacement amplitude at each point due to elastic deformation in waves

that the shallow water depth of 15 m in the planned area of this study makes the wavelength relatively small compared to the total length of the floating structure, so the motion The water depth in the planned study area is shallow (15 m), so the wavelength is relatively small compared to the total length of the floating structure, which is thought to have resulted in the micro-scale motion.

10 City Planning Proposal

The unit-connected floating structure proposed in this study was found to have minimal displacement at key locations due to elastic deformation. Therefore, it is considered that the discomfort people feel when living at sea is not significant. Based on these results, an overview of the actual urban design is shown in Fig. 9.

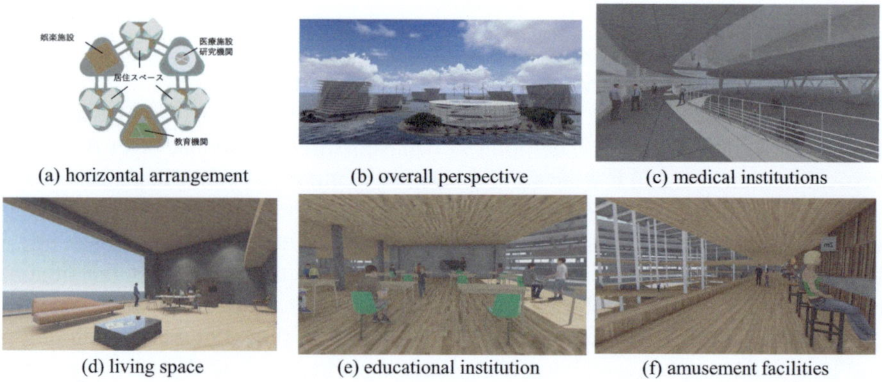

(a) horizontal arrangement (b) overall perspective (c) medical institutions

(d) living space (e) educational institution (f) amusement facilities

Fig. 9 Urban planning proposal

11 Conclusion

In summary, the system was proved to be feasible in the naval architecture aspects and safety aspects, considering the weights and centers, structural analysis, energy generation system, stability, hydrodynamics- hydroelasticity system, and mooring system.

References

1. Author F (2016) Article title. Journal 2(5):99–110
2. Author F, Author S (2016) Title of a proceedings paper. In: Editor F, Editor S (eds) Conference 2016, LNCS, vol 9999. Springer, Heidelberg, pp 1–13
3. Author F, Author S, Author T (1999) Book title. 2nd edn. Publisher, Location
4. Author F (2010) Contribution title. In: 9th International proceedings on proceedings. Publisher, Location, pp 1–2
5. LNCS Homepage. http://www.springer.com/lncs. Last accessed 2016/11/21
6. Sato F, Ikoma T, Masuda K, Kaneko T (2011) Study on the behaviour characteristics in waves of a large circular floating structure with a moon pool, conference proceedings, pp 385–386 (Marine Architecture 10006)
7. Masuda K, Ikoma T, Sato F, Kuroyanagi A, Kobayashi A, Hiroumi J, Nobukawa Y, Nakanishi M, Umemura Y, Saito T (2012) Proposal of floating plan planning considering response in waves of a multifunctional large floating body NU-FLOAT. In: 23rd Symposium on ocean engineering, Japan Society of Marine Engineering and Ocean Technology

Construction and Design

Removal Demonstration of 2.1-MW Hybrid-Spar Floating Offshore Wind Turbine in Actual Sea Area

Koji Tanaka and Tomoaki Utsunomiya

Abstract Floating Offshore Wind Turbines (FOWTs) are being researched, designed, and developed in many regions to achieve a global energy transition that meets the world's climate goals. These results have been published in the available literature and have contributed significantly to the spread and expansion of FOWTs. Especially in Europe, small-scale floating offshore wind farms with multiple types of floating substructures are in operation. On the other hand, it is not clear how these FOWTs will be removed. TODA CORPORATION are constructing Japan's first commercial-scale floating offshore wind farm (2.1-MW × 8) from 2021 which will start commercial operation in January 2026. For this project, a hybrid-spar floating substructure in which upper part is made of steel and lower part is made of concrete was selected. We have removed one of the FOWTs installed in this project at the end of December 2023 using a removal method we developed. Inside of this floating substructure, there exist gravel ballast and seawater ballast. By draining the seawater ballast to the outside of the floating substructure, the tower and floating substructure were laid down while remaining connected. The floating substructure was then pulled onto a submerged semi-submersible barge and successfully transported to onshore quay. In this paper, we describe the removal procedure of 2.1-MW hybrid-spar FOWT in actual sea area and a comparison of the laying operation of the floating substructure during removal was described between the values calculated from the static equilibrium analysis using potential energy and the measured values.

Keywords Floating offshore wind turbine · Hybrid-spar · Lay down

---

K. Tanaka (✉)
TODA CORPORATION, Chuo-Ku Tokyo, Japan
e-mail: kouji.tanaka@toda.co.jp

T. Utsunomiya
Kyushu University, Nishi-Ku, Fukuoka, Japan

X. L. Zhao et al. (eds.), *Proceedings of the Fourth World Conference on Floating Solutions: WCFS 2024; 2–4 December; Hong Kong*, Lecture Notes in Civil Engineering 597, https://doi.org/10.1007/978-981-96-4569-5_11

1 Introduction

The global energy transition towards net-zero greenhouse gas emissions by 2050 is a common challenge that the world must solve. Global Wind Energy Council predicts that floating offshore wind turbines (FOWTs) accelerate rapidly after 2030, as 80% of the world's offshore wind resource potential is located in region deeper than 60 m of water depth [1].

Various types of floating substructures technologies have been demonstrated in countries around the world [2], among which spar and semi-sub types have moved into mass-produced commercial operation [3, 4]. Regarding these technology demonstrations, papers have been published that compare actual measured floating substructure motions with simulations [5–7], whereas there is almost no published literature that describe detailed methods for removal of floating substructures.

In Japan, spar-type floating substructures are in operation at Sakiyama 2-MW floating offshore wind turbine (2-MW × 1 unit) [8] and under construction at Goto City Offshore Wind Power Generation Project (2.1-MW × 8 unit) [9]. Therefore, in the future, there is a possibility that these floating substructures will be transported to onshore quay for major repairs or removal. At the end of December 2023, TODA CORPORATION conducted a removal demonstration in actual sea area using the floating substructure of the Goto City Offshore Wind Power Generation Project. In this paper, we describe the removal procedure of 2.1-MW hybrid-spar FOWT in actual sea area and a comparison of the laying operation of the floating substructure during removal was described between the values calculated from the static equilibrium analysis using potential energy and the measured values. A good agreement between the measured value and the calculated values has been obtained.

2 Demonstration Site

Figure 1 shows the demonstration site. This demonstration was conducted on the assumption of the removal of floating substructures in the Goto City Offshore Wind Power Generation Project currently under construction off the coast of Goto City, Nagasaki Prefecture. In this project, construction is conducted at three areas: onshore quay which is construction of the floating substructures and loading to a semi-submersible barge (Area 1), mounting wind turbines site (Area 2), installation of the floating substructures site (Area 3). This floating substructure has been designed, including riggings, with the assumption that it will be removed, and the currently planned removal procedure is as follows: First, remove the mooring lines from the floating substructure in Area 3 and wet tow it to Area 2; Next, in Area 2, the blade and nacelle are removed and the seawater ballast inside the floating substructure is drained, thereby the floating substructure is laid down with the tower connected; Finally, the floating substructure which is laid down on the sea surface is loaded onto the semi-submersible barge, and transported to Area 1 and unloaded.

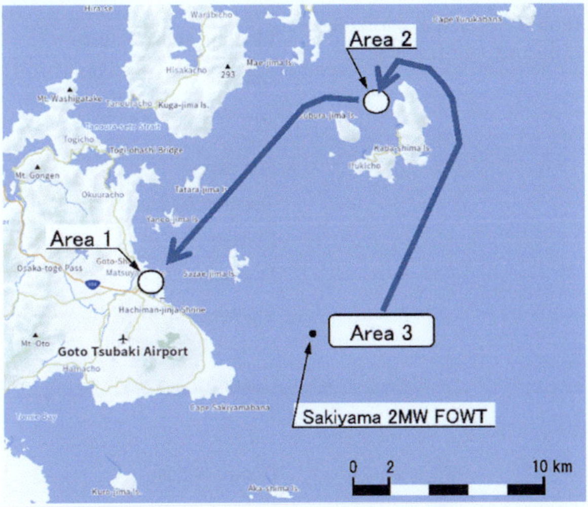

Fig. 1 Demonstration site

This demonstration covered the removal of the blade and nacelle in Area 2, and the draining of seawater ballast in the floating substructure to transport and unloading in Area 1.

3 Floating Substructure

Table 1 shows principal particulars of floating substructure. The floating substructure is 88 m long and is a hybrid structure with the upper 50 m made of steel and the lower 38 m made of concrete. The steel part consists of cylindrical members with outer diameters of 4.8 m and 7.8 m connected by conical cylinders. Various riggings such as platform, fenders, boat landing ladder and sacrificial anodes are attached on the outside of the floating substructure, moreover, gravel ballast and seawater ballast are included on the inside of the floating substructure. The volume of gravel ballast is 360 m^3 with a dry density of 2050 kg/m^3 and a saturated density of 2500 kg/m^3. In this paper, the density in natural state is referred to as the dry density. The seawater ballast can be drained by a drain pump as needed.

In this demonstration, a 50-m long temporary tower imitating a real tower was connected to the floating substructure for removal. A 70-ton weight was attached to the base of the temporary tower at 180° to the hatch of the entrance to the floating substructure to prevent the hatch from going underwater when the floating substructure was laid down. The center of gravity of the floating substructure is 36.7 m above the bottom and 0.11 m from the axial centerline toward the weight.

Figure 2 shows outline of the floating substructure. In this paper, the tilt angle θ refers to the angle around the x-axis shown in Fig. 2, where the upright state is 90° and the horizontal state is 0°.

Table 1 Principal particulars of floating substructure

Floating substructure	
Length	Total 88 m Steel 50 m Concrete 38 m
Diameter	4.8–7.8 m
Temporary tower	
Length	50.0 m
Diameter	2.5–4.8 m
Floating substructure with temporary tower	
Weight	1840 ton
Center of gravity	x:0 m y:− 0.11 m z:36.7 m
Gravel ballast	
Volume	360 m^3
Dry density	2050 kg/m^3
Saturation density	2500 kg/m^3

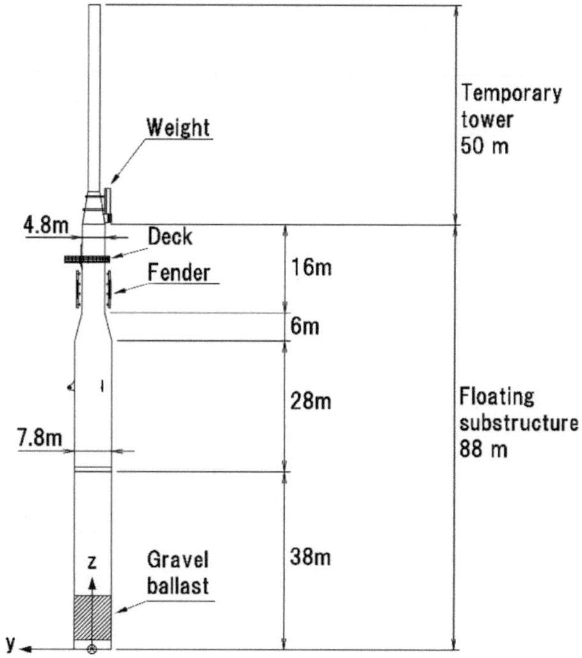

Fig. 2 Outline of floating substructure

4 Numerical Analysis of Floating Substructure Lay Down

Prior to the removal demonstration, the posture change of the floating substructure during the laying operation was calculated by static equilibrium analysis using the potential energy formulated by Tatsumi et al. [10]. The total potential energy U in the laying operation is shown in Eq. (1).

$$U = U_G + U_B + U_S + U_W \tag{1}$$

$$U_G = MgZ_G \tag{2}$$

$$U_B = -\rho V_B g Z_B \tag{3}$$

$$U_S = \rho V_S g Z_S \tag{4}$$

$$U_W = \rho V_W g Z_W \tag{5}$$

where U_G is the potential energy due to the weight of the floating substructure acting on the center of gravity of the floating substructure, U_B is the potential energy due to buoyancy acting on the center of buoyancy of the floating substructure, U_S is the potential energy due to the weight of gravel ballast acting on the center of gravity of the gravel ballast, and U_W is the potential energy due to the weight of the seawater ballast acting on the center of gravity of the seawater ballast. The potential energies U_B, U_S, and U_W are expressed as functions of the posture and draft of the floating substructure, which change with the drain of seawater ballast and the respective centers of gravity, which change with the distribution of gravel ballast and seawater ballast.

Each potential energy is shown in Eqs. (2)–(5), where M is the mass of the floating substructure, g is the acceleration of gravity, ρ is the density of seawater (1025 kg/m^3), Z is the center of gravity for each subscript in the spatial coordinate system, and V is the volume for each subscript. The potential energy was calculated using a spatial coordinate system with the origin at the intersection of the center line along the axis of the floating substructure and the water surface (Fig. 3). In calculating the potential energy U_W, the water surface of the seawater ballast was assumed to remain horizontal regardless of the posture of the floating substructure. On the other hand, the surface inclination of gravel ballast differs depending on the posture of the floating substructure: the surface remains in the initial state if $\varphi \geq 90—\theta$, where φ is the angle of repose; the surface collapses and stabilizes with the angle of repose if $\varphi < 90—\theta$. Figure 4 shows an overview of the posture of the floating substructure and the surface of the gravel ballast. In this paper, the angle of repose of the gravel ballast is set to 35°, which has been obtained in tests.

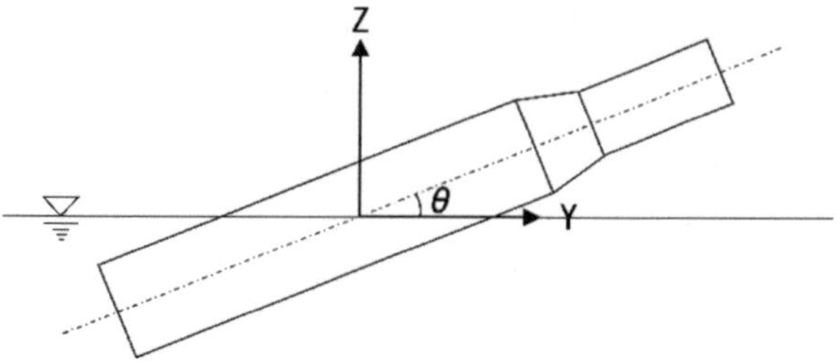

Fig. 3 Coordinate system

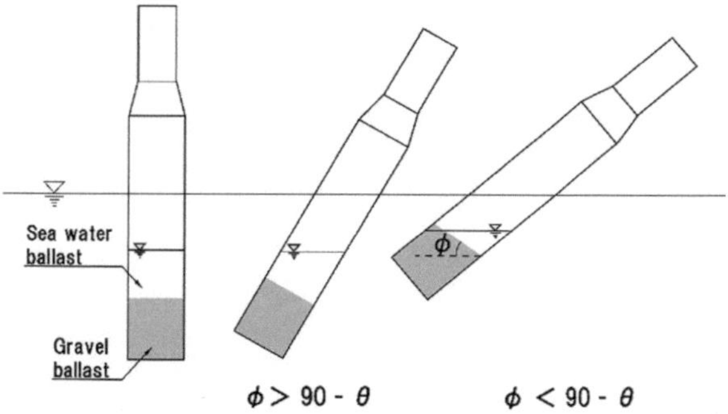

Fig. 4 Overview of the posture of the floating substructure and the surface of the gravel ballast

The equilibrium point in the analysis was determined by calculating the potential energy U for the tilt angle of the floating substructure for each volume of seawater ballast, and using the tilt angle that was the minimum point of this value. If more than one minimum point existed, the minimum point with the tilt angle closest to 90° was adopted as the equilibrium angle, the same as Tatsumi et al. [10].

5 Removal Demonstration

5.1 Blade and Nacelle Removal

After removing the mooring lines in the area where the floating substructure was installed, it was towed by tugboats to the Area 2 where the wind turbine was removed (see Fig. 1).

After arriving at the wind turbine removal area, a temporary mooring line was connected to the floating substructure. The procedure for removing the nacelle is as follows: First, the rotor was rotated so that one blade was horizontal, and the horizontal blade was removed using a floating crane; Next, the rotor was rotated so that the remaining two blades formed a V-shape, and the nacelle with the two blades attached was removed from the tower all at once (Fig. 5). It took one day each to remove the blade and nacelle, for a total of four days including preparation. Although

Fig. 5 Removal of nacelle

the plan for actual removal is to remove the nacelle and then mount weights on the top of the tower to lay the floating substructure, in this demonstration the tower was removed and a temporary tower was attached instead. After that, the temporary mooring line was removed. It took two days to remove the mooring line.

5.2 Laying Down and Loading

After attaching ropes to the floating substructure to prevent it from drifting, seawater ballast was drained by a pump until the floating substructure was laid down to the sea surface. Figure 6 shows draining of seawater ballast and Fig. 7 shows the floating substructure after draining. The floating substructure slowly tilted toward the weight attached to the base of the temporary tower as planned.

The semi-submersible barge "FLOAT RAISER [11]" was used to load the floating substructures. The semi-submersible barge has a total length of 110 m and a total width of 43 m. Islands are installed at the four corners of the deck, and each is equipped with a 25-ton winch and a 50-ton winch. Moreover, the deck of the FLOAT RAISER can be submerged up to 7.4 m below the sea surface. The loading procedure is as follows: First, the semi-submersible barge was submerged when the floating substructure was laid down. At that time, the floating substructure was tilted so that its lower end was sinking due to the gravel ballast, therefore the semi-submersible barge was also submerged in a downward posture from the bow to the stern. After the laying operation was completed, the winch installed on the island of the semi-submersible barge and the floating substructure were connected with a rope; Next, the rope was rolled up with a winch and the floating substructure was moved onto the deck of the semi-submersible barge; Finally, the semi-submersible barge was floated and the floating substructure was loaded onto the support frame on the deck. In this demonstration, we tested a removal method of lifting the lower end of the floating

Fig. 6 Draining of seawater ballast

Fig. 7 Lay down

substructure using a floating crane, assuming that the inclination of the floating substructure which was laid down was not tilted relative to the plan (Fig. 8). It took one day to prepare for laying operation, and another day from laying operation to loading. Including the next day's transportation, the work was completed in a total of three days.

Fig. 8 Loading of floating substructure

5.3 *Landing*

After loading the floating substructure onto the semi-submersible barge, it was transported to the onshore quay of Area 1 (see Fig. 1). Figure 9 shows appearance of floating substructure during transportation. At the onshore quay, a self-propelled modular transporter was entered under the support frame, then the floating substructure was unloaded in the same method as it was loaded (Fig. 10). Although an investigation of the unloaded floating substructure revealed damage to some of the sacrificial anodes attached to the outside of its steel part, relatively large riggings such as the platform and fenders were not damaged.

Fig. 9 Shipping of floating substructure

Fig. 10 Unloading of floating substructure

6 Comparison of Measured and Calculated Tilt Angle

During the laying operation, a steel cover plate was attached to the top of the temporary tower, and an accelerometer was attached in the center of the cover plate to measure the posture of the floating substructure. Figure 11 shows a comparison of the measured and analyzed values of the tilt angles. In actual work, when the tilt angle was approximately 25°, the winch installed on the island of the semi-submersible barge and the floating substructure were connected with a rope, and the draining work was interrupted for approximately 30 min. Therefore, the graph was corrected by removing that time period. The measured values show that the tilt changed slowly from 90° to around 80° over a period of about two hours, and that the tilt changed suddenly from 55° to around 35°, where the gravel ballast is thought to have collapsed. The analyzed values were able to represent the changes described above, and showed a good agreement with the measured values from the start to the end of the laying operation.

Figure 12 shows the tilt angle from 8000 to 9000 s. Although the analyzed value shows that the tilt angle changes continuously up to 55°, there is no equilibrium point between 55° and 35° of tilt angle, and the equilibrium point jumps to 35°. On the other hand, the measured value shows that the tilt angle changes more gradually from the 55° to around 25° than the analyzed value. The reason for this phenomenon is thought to be that it actually takes time for the gravel ballast surface to collapse, and that the motion is attenuated by the fluid resistance acting on the floating substructure.

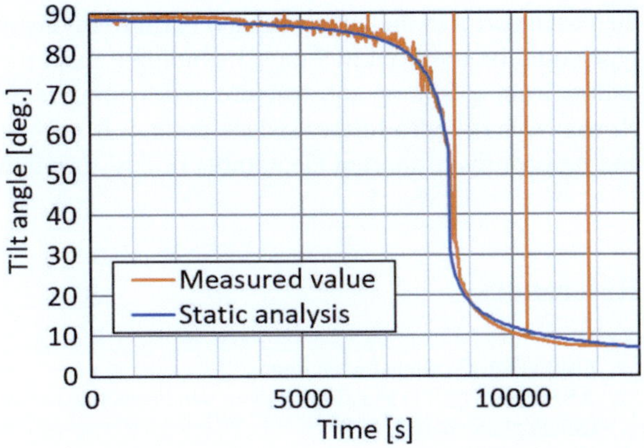

Fig. 11 Tilt angle of floating substructure

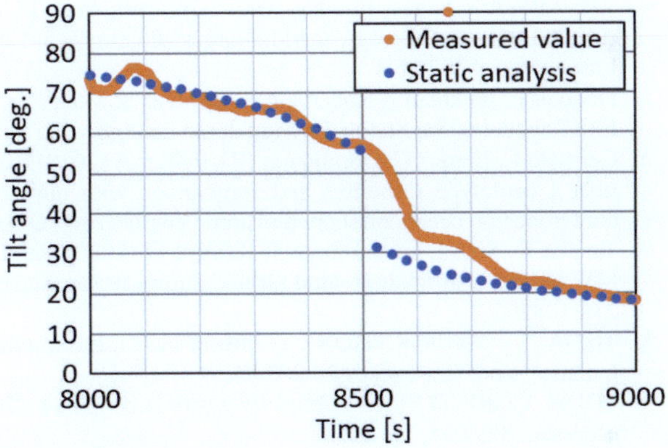

Fig. 12 Tilt angle from 8000 to 9000 s

7 Conclusions

In this paper, the removal demonstration of the 2.1-MW hybrid-spar type FOWT was presented, including the removal procedure. In addition, a comparison of the laying operation of the floating substructure during removal was described between the values calculated from the static equilibrium analysis using potential energy and the measured values.

In the static equilibrium analysis, a good agreement between the analyzed and measured values was obtained by appropriately setting the distribution of seawater ballast and gravel ballast and the density of the gravel ballast, dry density and saturated density, which changed with the posture of the floating substructure. Immediately after the gravel ballast surface exceeded the angle of repose, a discontinuous change of the equilibrium tilt angle was predicted, which was shown to be a point to note in the removal process.

When the water depth of harbors is not deep such as Japan, it is necessary to remove the blades and nacelles at sea, load the floating substructure onto a barge, and then transport it to the onshore quay. In this demonstration, we showed that the spar type could be removed in 9 days including transportation, by draining the seawater ballast from a floating substructure and loading it onto a semi-submersible barge. However, in order to further reduce costs, it is necessary to study efficient methods and procedures to shorten the number of days of offshore work in the future.

References

1. GWEC (2023) Global wind report
2. ABS Group (2021) Floating offshore wind turbine development assessment—final report and technical summary
3. Equinor (2024) The world's largest floating offshore wind farm officially opened. https://www.equinor.jp/en/news/hywind-tampen-opened. Last accessed 2024/5/5
4. ABS, World's largest floating wind farm built to ABS class. https://news.cision.com/american-bureau-of-shipping/r/world-s-largest-floating-wind-farm-built-to-abs-class,c3695668. Last accessed 2024/5/5
5. Driscoll F, Jonkman J, Robertson A, Sirnivas S, Skaare B, Nielsen FG (2016) Validation of a FAST model of the Statoil-Hywind demo floating wind turbine. Energy Procedia 94:3–19
6. Cermelli C, Leroux C, Dominguez SD, Peiffer A (2018) Experimental measurements of Windfloat 1 prototype responses and comparison with numerical model. In: 37th International conference on ocean, offshore and arctic engineering, OMAE2018-77057
7. Tanaka K, Sato I, Utsunomiya T, Kakuya H (2020) Validation of dynamic response of a 2-MW hybrid-spar floating wind turbine during typhoon using full-scale field data. Ocean Eng 218:108262
8. TODA CORPORATION. https://www.toda.co.jp/business/ecology/special/pdf/sakiyama2mw_e.pdf. Last accessed 2024/5/5
9. TODA CORPORATION. https://www.toda.co.jp/news/2022/20221007_003122.html. Last accessed 2024/5/5
10. Tatsumi K, Utsunomiya T, Sato I (2022) Experiment and analysis on upending and toppling behavior of spar with stone ballast. In: Conference proceedings The Japan Society of Naval architects and ocean engineers, vol 34, pp 535–541
11. TODA CORPORATION. https://www.toda.co.jp/business/ecology/special/pdf/floatraiser_e.pdf. Last accessed 2024/5/5

Design and Long-Term Dynamic Analysis of a Semi-Submersible Concrete Platform for 15 MW Wind Turbines

Shangpei Liu, Shunyun Zheng, Chao Li, and JinPing Ou

Abstract In the design and construction of floating wind turbine platforms, the use of steel is limited by cost and corrosion concerns. In contrast, concrete materials have gained recognition and popularity due to their affordability and durability. In this paper, a concrete floating platform for a 15 MW wind turbine is designed and compared with the VolturnUS-S. A surrogate model is established to predict the long-term dynamic response of the platform. The results show that the concrete platform in this paper exhibits advantages in hydrodynamic performance and satisfies the relevant requirements of floating wind turbines. Moreover, the cost estimation indicates that the use of concrete can substantially reduce the construction costs.

Keywords Semi-submersible wind turbine platform · Reinforced concrete · Long-term dynamic response · Surrogate model

1 Introduction

As offshore wind power development advances into remote ocean areas, floating wind power technology has made significant advancements. Prototype experiments have been built and tested in Europe, the United States, Japan, and China [1, 2] successfully. However, Floating Offshore Wind Turbines (FOWTs) still face challenges in large-scale commercial application due to the levelized cost of energy (LCOE) not meeting market demands.

Concrete materials are increasingly emphasized in the design of floating wind turbine platforms due to their durability, low material cost, and insensitivity to fatigue. Notable examples include projects such as FOATGEN [3], VolturnUS [4], and OO-Star [5]. Walia [6] designed a tension leg platform using prestressed ultra-high-performance concrete (UHPC) for floating wind turbines, demonstrating a potential cost reduction of 90%. Liu [7] developed a wind turbine system model based on the

S. Liu · S. Zheng · C. Li (✉) · J. Ou
School of Civil and Environmental Engineering, Harbin Institute of Technology, Shenzhen, China
e-mail: lichaosz@hit.edu.cn

X. L. Zhao et al. (eds.), *Proceedings of the Fourth World Conference on Floating Solutions: WCFS 2024; 2–4 December; Hong Kong*, Lecture Notes in Civil Engineering 597, https://doi.org/10.1007/978-981-96-4569-5_12

OC4-DeepCwind platform and studied the selected design load cases and dynamic response of the system using the FAST software. Yang [8] developed an aero-hydro-servo-elastic coupling software called F2A for the dynamic response analysis of floating offshore wind turbine systems. Zhou [9] utilized surrogate models to predict the long-term response of floating wind turbines and conducted fatigue calculations for the tower base and fairleads. Oh [* MERGEFORMAT 10] designed a prestressed concrete semi-submersible floater for 10 MW wind turbines and compared it with the traditional steel-based floater. Li [11] investigated the dynamic performance of two floating wind turbine systems mounted on concrete and steel Y-shaped semi-submersible platforms, respectively, using coupled multi-body simulations and conducted wind and wave model tests.

In this paper, a four-column semi-submersible floating platform for the 15 MW wind turbine is studied. Preliminary design is conducted for the concrete floating wind turbine foundation. Surrogate models are used to predict long-term dynamic responses, and a comparison of long-term dynamic responses is made with VolturnUS-S. The results indicate that the concrete foundation exhibits superior performance in terms of hydrodynamics and cost compared to the steel structure.

2 Description of the Platform

2.1 The IEA Wind 15-Megawatt Offshore Reference Wind Turbine

In this paper, the wind turbine used for the concrete floating platform is the IEA Wind 15-Megawatt Offshore Reference Wind Turbine. Detailed information about this turbine can be found in [* MERGEFORMAT 12]. The key parameters are given in Table 1 below.

Table 1 Key parameters for the IEA Wind 15-Megawatt Offshore Reference Wind Turbine

Parameters	Units	Value
Power Rating	MW	15
Rotor Diameter	m	240
Hub Height	m	150
Cut-in / Rated / Cut-out Wind Speed	m/s	3 / 10.59 / 25
Tower Mass	t	1,263
Rotor Nacelle Assembly Mass	t	991

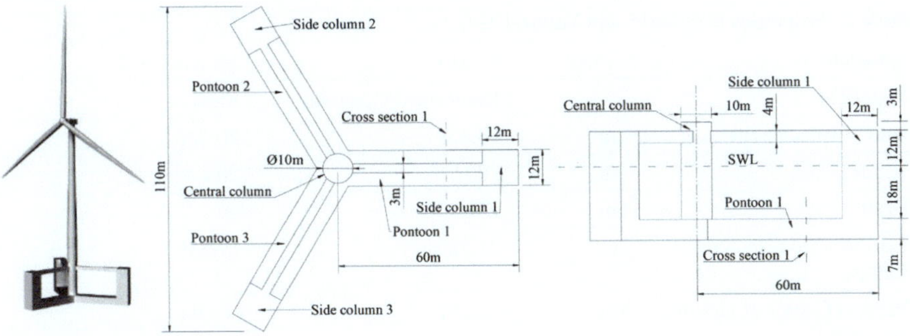

Fig. 1 Layout of the Con15

2.2 The Concrete Semi-Submersible Platform

The concrete platform for the 15 MW wind turbine described in this paper adopts a design comprising three square outer columns and a central circular column, using C50 prestressed concrete, which will hereinafter be referred to as Con15. The specifications of Con15 are illustrated in Fig. 1. The primary parameters of the concrete platform and the three-line catenary mooring system are provided in Table 2, with a comparison to VolturnUS-S [* MERGEFORMAT 13] included. The primary loads on the platform include gravity, hydrostatic pressure, and the forces of wind, waves, and currents. Considering factors such as bending resistance and construction requirements, a rectangular cross-section has been selected for the side columns. However, Con15 is designed without optimization or strength verification, thus the wall thickness is initially set at 0.45 m.

2.3 Service Location of Wind Turbine

The floating wind turbine studied in this paper is assumed to be deployed in the South China Sea at the specific coordinates of N22.0°, E114.5°. The data of wind and wave for the target location are sourced from ARGOSS, a subsidiary of BMT specializing in Earth observation and consulting services. The data used in this study are based on the average wind and wave data within a 50 km radius centered at N22.0°, E114.5°, covering a complete 25-year dataset from 1992 to 2016. The dataset has a sampling rate of 3 h per data point, resulting in a total of 73,052 samples. The data include detailed information such as significant wave height, peak wave period, wave direction, and average wind speed and direction at a height of 10 m. The probability density functions of wind speed, wave height, and spectral peak period are presented in Fig. 2. 80% of the wind speeds at a height of 10 m are distributed in the range of 3.2 m/s to 10.7 m/s, 80% of the significant wave heights are distributed in the range of 0.8 m to 3.1 m, and 80% of the peak wave periods are distributed in the range of

Table 2 Properties for Con15 and VolturnUS-S

Parameters	Units	Con15	VolturnUS-S
Materials	-	Reinforced Concrete	Steel
Hull Displacement	m^3	23,839	20,206
Water Depth	m	200	200
Draft	m	25	20
Platform Mass include Ballast	t	21,380	17,839
Vertical Center of Gravity from SWL	m	-13.38	-14.94
Roll Inertia about Center of Gravity	$kg{\cdot}m^2$	2.015E + 10	1.251E + 10
Pitch Inertia about Center of Gravity	$kg{\cdot}m^2$	2.015E + 10	1.251E + 10
Yaw Inertia about Center of Gravity	$kg{\cdot}m^2$	3.261E + 10	2.367E + 10
Mass/Unit Length of Cable	kg/m	685.00	685.00
Cable Length	m	850	850
Cable Stiffness (EA)	N	3.27E + 09	3.27E + 09
Fairlead Points Coordinates	(x, y, z)	(-58.0, 0, -14.0) (29.0, 50.2, -14.0) (29.0, -50.2, -14.0)	(-60.0, 0, -25.0) (30.0, 52.0, -25.0) (30.0, -52.0, -25.0)
Anchor Points Coordinates	(x, y, z)	(-837.6, 0, -200.0) (418.8, 725.4, -200.0) (418.8, -725.4, -200.0)	(-837.6, 0, -200.0) (418.8, 725.4, -200.0) (418.8, -725.4, -200.0)

5.2–8.2 s. Figure 3 depicts the wind and wave rose diagram, indicating that the wind and wave directions are primarily distributed eastward with a strong correlation.

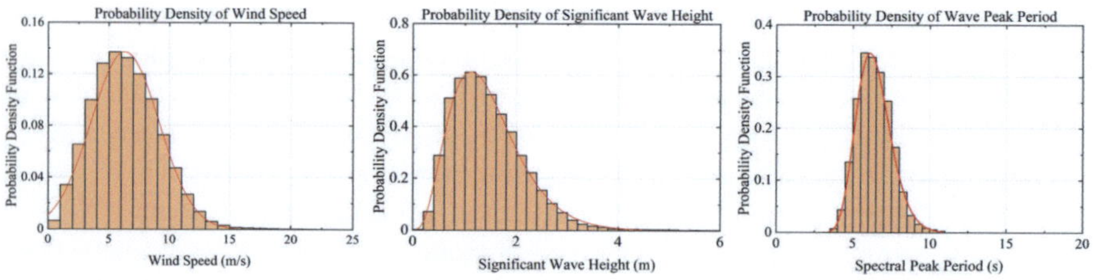

Fig. 2 Probability density of wind and wave environment at target locations

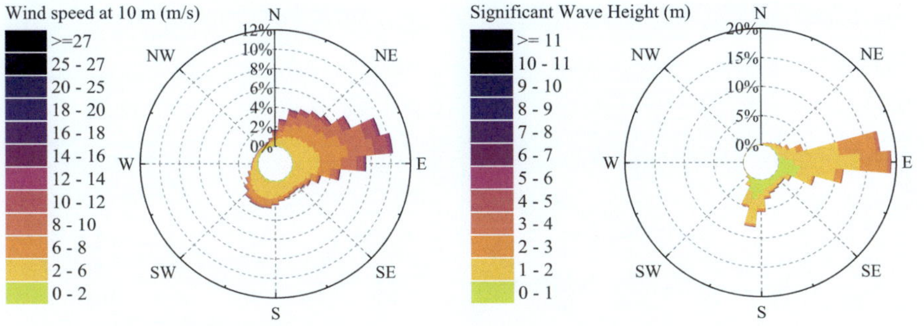

Fig. 3 Rose diagram of wind and wave environment at target locations

3 Long-Term Dynamic Analysis Method

3.1 Surrogate Model

Evaluating the full lifecycle response characteristics of a wind turbine system is crucial during the design phase. Given the current service lifecycle requirements for offshore wind turbine systems, typically span 25 years, focusing solely on short-term sea states is insufficient to comprehensively reflect the dynamic performance of floating wind turbine systems over their entire lifecycle. However, calculating all the sea conditions is clearly inefficient. Therefore, a surrogate model has been introduced to incorporate the statistical characteristics of sea states over the entire lifecycle of the target sea area into the response analysis. This approach enables the prediction of the long-term dynamic response of the floating wind turbine system, providing a basis for subsequent design.

To conduct long-term response analysis using the surrogate model, the selection of characteristic sea states must ensure an even distribution across the entire sample space, including the sample points on the outer contour of the total sample space. The Maximum Dissimilarity Algorithm (MDA) [* MERGEFORMAT 14] is used to select a certain number of short-term characteristic sea states from the sea state data covering the entire lifecycle of the target sea area. Subsequently, OpenFAST is used to perform short-term dynamic simulations on these selected sea states. Finally, the selected sea state data and the results of the short-term response are input into a Kriging surrogate model, which is more suitable for such nonlinear problems [* MERGEFORMAT 9]. The Kriging model then outputs the long-term dynamic response predictions for the entire sea states. In this paper, wind speed, wind direction, wave period, significant wave height, and wave direction are used as inputs for Kriging surrogate model, while platform heave, platform tilt, nacelle acceleration, tower base bending moment, and the fairleads tension are used as outputs.

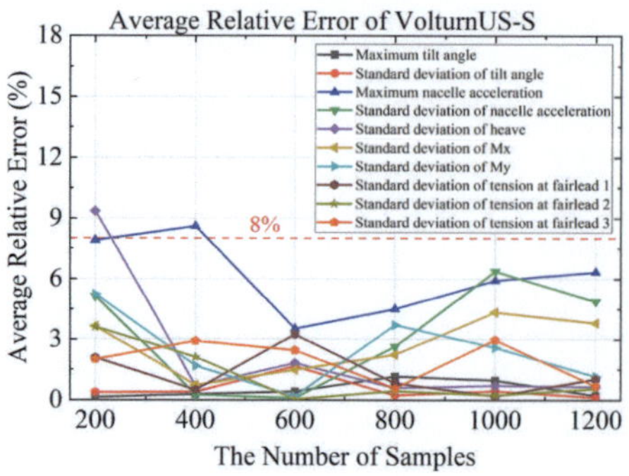

Fig. 4 Average relative errors of the surrogate model for Con15 and VolturnUS-S

3.2 Model Verification

To establish an effective and cost-efficient surrogate model, a set of 200, 400, 600, 800, 1000, and 1200 sea states, along with their corresponding dynamic responses, are selected as training samples. Moreover, a total of 2525 samples from the 2016 results are chosen as references. The accuracy of the surrogate model is evaluated through the comparison of average relative errors in Fig. 4. In this paper, the average relative error of the surrogate model is constrained to not exceed 8%. As the number of samples increases, the error of the surrogate model tends to decrease. At a training sample size of 1000, the average relative errors of all parameters are below 10%, which meets the requirements of optimization design.

4 Prediction of Long-Term Dynamic Responses

With the established surrogate model, the response of the wind turbine system under all sea states over the 25-year period can be quickly estimated. The platform's inclination angle under wind and wave load is a crucial indicator of the wind turbine system's dynamic performance. Significant platform inclination is detrimental to the long-term stable operation of the turbine. Nacelle acceleration is closely related to the load levels on drivetrain components and is an important consideration in the safe operation of floating wind turbines. Therefore, the maximum tilt angles and nacelle acceleration of the turbine over its entire lifecycle are predicted and presented in Table 3. The extreme response values of platform tilt angle and nacelle acceleration are obtained by selecting the values with an exceedance probability of 10%. A comparison of the extreme responses between Con15 and VolturnUS-S reveals a decrease in the maximum tilt angle. While the tilt angles for both platforms remain below the 10° constraint, VolturnUS-S exhibits a nacelle acceleration exceeding 3 m/s^2.

Table 3 Comparison of the long-term extreme response results

Parameters	Maximum response of platform tilt angle (°)	Maximum response of nacelle acceleration (m/s^2)
Con15	8.0291	2.9643
VolturnUS-S	9.5992	4.8504
Constraint Values	10	3

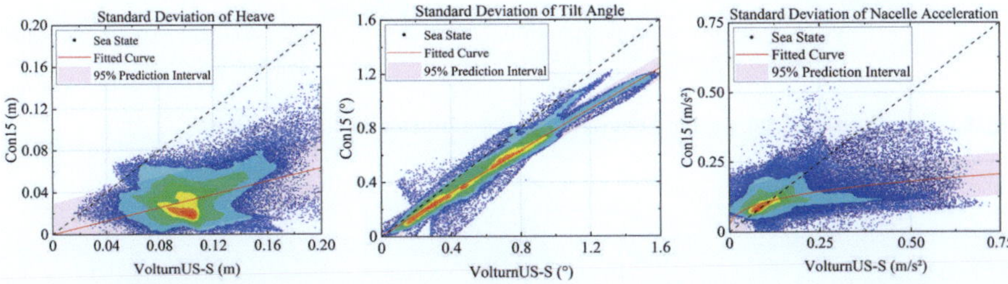

Fig. 5 Comparison of the standard deviation of motion

The standard deviations of Con15 and VolturnUS-S across all 73,052 sea states are compared in Fig. 5. The concrete platform, Con15, exhibits significant advantages in terms of heave motion. The standard deviations of the tilt angle responses for both types of foundations mainly fall within the range of 0.1° to 1.6°, indicating superior inclination motion performance, ensuring efficient power generation for floating wind turbine systems. Additionally, the standard deviations of their accelerations are concentrated within 0.5 m/s^2.

The base of the tower is one of the most vulnerable parts of the entire floating wind turbine platform. Figure 6 presents a comparison of the standard deviation of the tower base bending moment between Con15 and VolturnUS-S across all actual sea conditions. Overall, the standard deviation fitting curve of the tower base bending moment is situated below the diagonal line, and in most sea states, Con15 exhibits a smaller tower base bending moment response compared to VolturnUS-S. This indicates that the dynamic load performance of the tower base of Con15 is superior to that of VolturnUS-S in the long term. The moment at the tower base is mainly related to the tilt angle of the platform. Due to the smaller platform tilt angle of Con15 compared to VolturnUS-S, the moment is accordingly reduced. This reduction is advantageous in mitigating fatigue damage to the tower base.

The mooring cable at the fairleads experience maximum tension. Figure 7 presents a comparison of the standard deviation of the cable tension response at the three different fairleads between Con15 and VolturnUS-S under actual sea conditions. The tension of Con15 is higher than that of VolturnUS-S due to the greater mass of the concrete platform compared to the steel platform. The higher tension at fairlead will increase the fatigue damage.

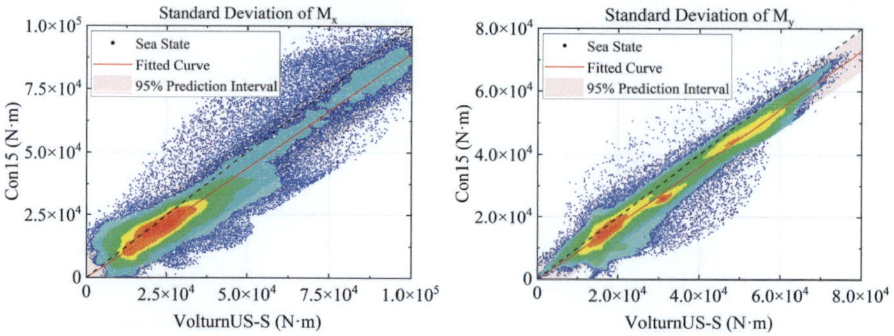

Fig. 6 Comparison of the standard deviation of the moments at the tower base

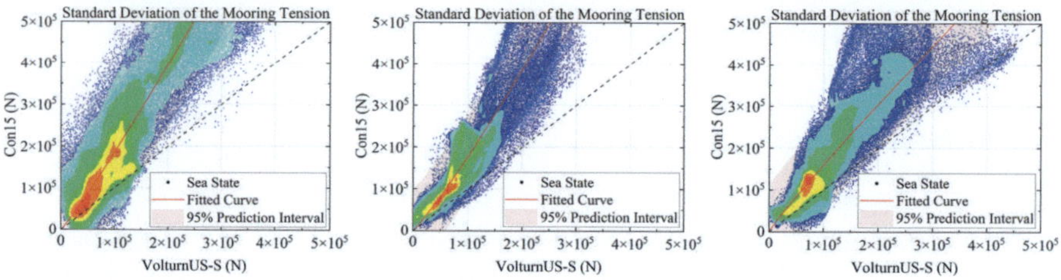

Fig. 7 Comparison of the standard deviation of mooring tension at fairleads

5 Comparison of Material Costs

In the design of a floating wind turbine, construction cost serves as a vital indicator for evaluating overall performance. In this section, the material costs of Con15 and VolturnUS-S are compared, specifically the costs of concrete and steel materials. Reinforced concrete is priced at €0.063/kg [* MERGEFORMAT 15] and steel is priced at €0.524/kg [* MERGEFORMAT 16]. The ballast is primarily composed of seawater and a small amount of iron-ore-concrete, and the cost of ballast is not considered in this paper. Table 4 provides an overview of the material costs for the two types of platforms. Despite the substantial reduction in platform weight achieved by VolturnUS-S using steel materials, Con15 surpasses half of the material cost savings due to the cost-effectiveness of concrete.

Table 4 Comparison of the material costs

Parameters	Unit	Con15	VolturnUS-S
Material	-	Reinforced Concrete	Steel
Hull Mass	t	14,840	3,914
Price	€	934,920	2,050,936

6 Conclusions

In this study, a concrete semi-submersible platform for a 15 MW wind turbine is designed. The surrogate model is used to predict the long-term dynamic response for a 25-year period. The results are compared with VolturnUS-S. The following conclusions are obtained:

The concrete platform Con15 has significantly increased mass and displacement compared to the steel platform VolturnUS-S due to the substantial differences in density and strength between concrete and steel materials. The long-term dynamic response prediction of Con15 shows superior performance in heave, platform inclination, and nacelle acceleration compared to VolturnUS-S. The bending moment at the tower base also decreases as a result of the reduction in platform tilt angle, which is beneficial for reducing fatigue damage. However, the mooring tension at the fairleads of Con15 is higher than that of VolturnUS-S, due to the increased mass of Con15, leading to increased fatigue damage.

References

1. Roddier D., Cermelli C., Aubault A., Weinstein A.: WindFloat: A floating foundation for offshore wind turbines. Journal of Renewable and Sustainable Energy, 033104 (2010).
2. Viselli A M., Goupee A J., Dagher H J., Allen C K.: Design and model confirmation of the intermediate scale VolturnUS floating wind turbine subjected to its extreme design conditions offshore Maine. Wind Energy, 1161–1177 (2016).
3. Jestin N., Modiano P., Bigard R., Crémona C., Dumas B.: FLOATGEN – Design and construction of the first floating wind turbine. IABSE Symposium, Nantes 2018: Tomorrow's Megastructures, pp S15–59.International Association for Bridge and Structural Engineering (2018).
4. Borisade F, Choisnet T, Cheng PW (2016) Design study and full scale MBS-CFD simulation of the IDEOL floating offshore wind turbine foundation. J Phys: Conf Ser 753:092002
5. Berthelsen P A.: Qualification of innovative floating substructures for 10MW wind turbines and water depths greater than 50m. University of Stuttgart (2015).
6. Walia D., Schiinemann P., Kuhl M., Adam F., Hartmann H., Großmann j., Ritschel U.: Prestressed ultra high performance concrete members for a TLP substructure for floating wind turbines. The 27th International Ocean and Polar Engineering Conference. ISOPE-I-17–304 (2017).
7. Liu J, Thomas E, Manuel L, Griffith D, Ruehl K, Barone M (2018) Integrated system design for a large wind turbine supported on a moored semi-submersible platform. Journal of Marine Science and Engineering 6(1):9
8. Yang Y, Bashir M, Michailides C, Li C, Wang J (2020) Development and application of an aero-hydro-servo-elastic cou-pling framework for analysis of floating offshore wind turbines. Renewable Energy 161:606–625
9. Zhou S, Li C, Xiao Y, Cheng P (2020) Importance of platform mounting orientation of Y-shaped semi-submersible floating wind turbines: A case study by using surrogate models. Renewable Energy 156:260–278
10. Oh S (2021) Comparison of concrete and steel semi-submersible floaters for 10MW wind turbines. J Phys: Conf Ser 2018(1):012029

11. Li C, Zhou S, Shan B, Hu G, Song X, Liu Y, Hu Y, Xiao Y (2022) Dynamics of a Y-shaped semi-submersible floating wind turbine: a comparison of concrete and steel support structures. Ships and Offshore Structures 17(8):1663–1683
12. Gaertner E., Rinker J., Sethuraman L., Zahle F., Anderson B., Barter G., Abbas N., Meng F., Bortolotti P., Skrzypinski W., Scott G., Feil R., Bredmose H., Dykes K., Sheilds M., Allen C., Viselli A.: Definition of the IEA 15-Megawatt offshore reference wind turbine. International Energy Agency (2020).
13. Allen C., Viselli A., Dagher H., Goupee A., Gaertner E., Abbas N., Hall M., Barter G.: Definition of the UMaine VolturnUS-S reference platform developed for the IEA wind 15-Megawatt offshore reference wind turbine. International Energy Agency (2020).
14. Kennard RW, Stone LA (1969) Computer aided design of experiments. Technometrics 11(1):137–148
15. Filgueira-Vizoso A., Cordal-Iglesias D., Puime-Guillén F., Lamas-Galdo I., Martínez-Rubio A., Larrinaga-Calderón I., Castro-Santos L.: Sensitivity study of the economics of a floating offshore wind farm. The case study of the SATH® concrete platform in the Atlantic waters of Europe. Energy Reports, 9: 2604–2617 (2023).
16. Castro-Santos L, Filgueira-Vizoso A, Carral-Couce L, Formoso JÁF (2016) Economic feasibility of floating off-shore wind farms. Energy 112:868–882

Sustainable Solutions by Design on Offshore Reclamation Construction

Ying Li, Marcel 't Hart, Hans de Wit, Wouter de Hamer, and Eric Smit

Abstract Reclamations as well as floating structures are used to create new coastal areas for development of housing, industries, or recreation. Subsea tunnels or bridges are normally applied as offshore traffic connections to provide smooth access to these new areas. The mitigation of environmental impacts and reduction in the use of natural resources are important issues for consideration of these waterfront developments. This paper provided two types of sustainable solutions on the reduction of environmental impact of large waterfront developments by design: first is the reuse dredged material from the Immersed tunnel trenches for land reclamation and nature development; and second is a polder solution as alternative to the traditional land reclamation design. Three hallmark projects are presented to demonstrate the solutions that are successfully applied.

Keywords Reclamation · Immersed Subsea tunnel · Material re-use

1 Introduction

Waterfront developments are booming in recent years. Reclamations as well as floating structures are used to create new coastal area for development of housing, industries or recreation. Subsea tunnel or bridges are normally applied as offshore traffic connection to provided smooth access to these new areas. An example can be the Kau Yi Chau Artificial Islands project planned in Hong Kong [1]. The reduction of environmental impacts and reduction in the use of natural resources are important issues for consideration of those waterfront developments. Combination of immersed tube tunnels (IMT) and reclamations are a popular solution in recent years. The Dutch "polder" solution is also an innovation that gains interest as it limits

Y. Li (✉) · M. 't Hart · H. de Wit
Tunnel Engineering Consultants, Utrecht, The Netherlands
e-mail: ying.li@rhdhv.com

Y. Li · M. 't Hart · H. de Wit · W. de Hamer · E. Smit
Royal HaskoningDHV, Utrecht, The Netherlands

X. L. Zhao et al. (eds.), *Proceedings of the Fourth World Conference on Floating Solutions: WCFS 2024; 2–4 December; Hong Kong*, Lecture Notes in Civil Engineering 597, https://doi.org/10.1007/978-981-96-4569-5_13

the use of fill material for reclamation. This paper will introduce the main idea of these two solutions following by 3 case studies and some conclusions.

2 Two Ways of Reducing Environment Impact by Design

To create new waterfront developments by reclamation, one of the concerns is the environmental impact pollution during construction activities. A lot of measures have been developed already to mitigate the environmental impact of such projects. Examples during construction are the use of more advanced dredging equipment, measures to contain the silt distribution locally (e.g. adapting the filling techniques and using silt screens), the use of real-time monitoring measures. For the long-term impacts designers propose more and more circular solutions, apply nature-based approaches and introduce ecological enhancements to minimize negative impacts. The focus on circularity and nature-based approach during the design phase, instead of mitigating the impacts during construction phase, is a more effective solution as the impact are mitigated upfront. This paper will focus on two ways to reduce environmental impact of large waterfront developments by design: the reuse dredged material from the IMT trenches for land reclamation and nature development; and a polder solution as alternative to the traditional land reclamation design.

2.1 Reuse Material for Reclamations by Combination with Immersed Tube Tunnel

Subsea tunnels or bridges are normally applied as offshore traffic connections to provide smooth access to these new areas. IMT is becoming more and more attractive for offshore connections compared to bridge and bored tunnels, as it cleverly utilizes the function of water for floating transportation and installation of the tunnel structures and other attracts mentioned below.

Attracts of Immersed Tube Tunnel (IMT) IMT as traffic connection crossing water has a lot of advantages compared with bridge and other tunnel solutions (Fig. 2):

- Shorter longitudinal alignment,
- More flexible cross section with highway and railway together,
- Not like bored tunnel with deep buried depth under the ground, IMT is laying on the seabed and easer to connect with new floating development areas or land reclamations due to more equivalent vertical elevations, and with less blockage of nature water flow by its transition structure between tunnel and its landing point.
- More flexible construction due to parallel construction of tunnel element production and floating transport & immersion,
- IMT is more attractive from sustainability point of view than bored tunnels (Jonathan, 2024).

Fig. 1 Kau Yi Chau Artificial Islands project planned in Hong Kong (an example of reclamation project) [1]

Reuse Material for land reclamations or nature development. However, IMT has its disadvantages as well, since the tunnel foundation construction requires dredging activity which may have environmental impacts. However, reclamation areas can be created by reusing the dredged materials from the tunnel trenches, which will give the waterfront development opportunities to enhance the area. This enhancement can be through new artificial islands as part of the large development (as done for Øresund Link) or for nature development (as doing now for the Fehmernbelt Link).

Note that the opportunity for re-use of dredged materials should be considered as part of the project's planning by client's conscious decision in very early project phase [2].

2.2 *Polder for Reclamation*

A polder is an area of low-lying land, that has been reclaimed from a body of water and is protected by dikes. It is especially popular in the Netherlands, where polders have been created since the fourteenth century (see Fig. 3). Polders have a surface level below the surrounding water level. Water therefore infiltrates the low-lying polder, which usually means that the polder has an excess of water, which is pumped out or drained by opening sluices at low tide.

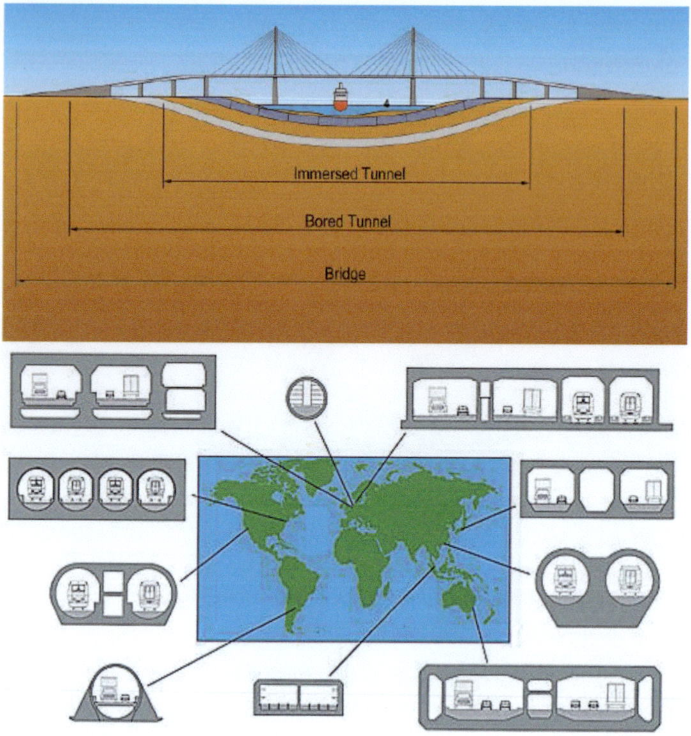

Fig. 2 Left: different connection methods crossing water; Right: examples of flexible cross-sections of IMTs

The polder concept was initially applied to create fertile agricultural lands from mainly low-laying inland marches. However, in the twentieth century the polder concept was also used to reclaim large areas from the sea. For the construction of these polders large dikes have been built on top of the existing seabed, and as such only limited fill material was used to reclaim the lands.

3 Case Study

Two cases of the re-use of dredged material from IMT trenches and one case on a recent polder concept solution are presented.

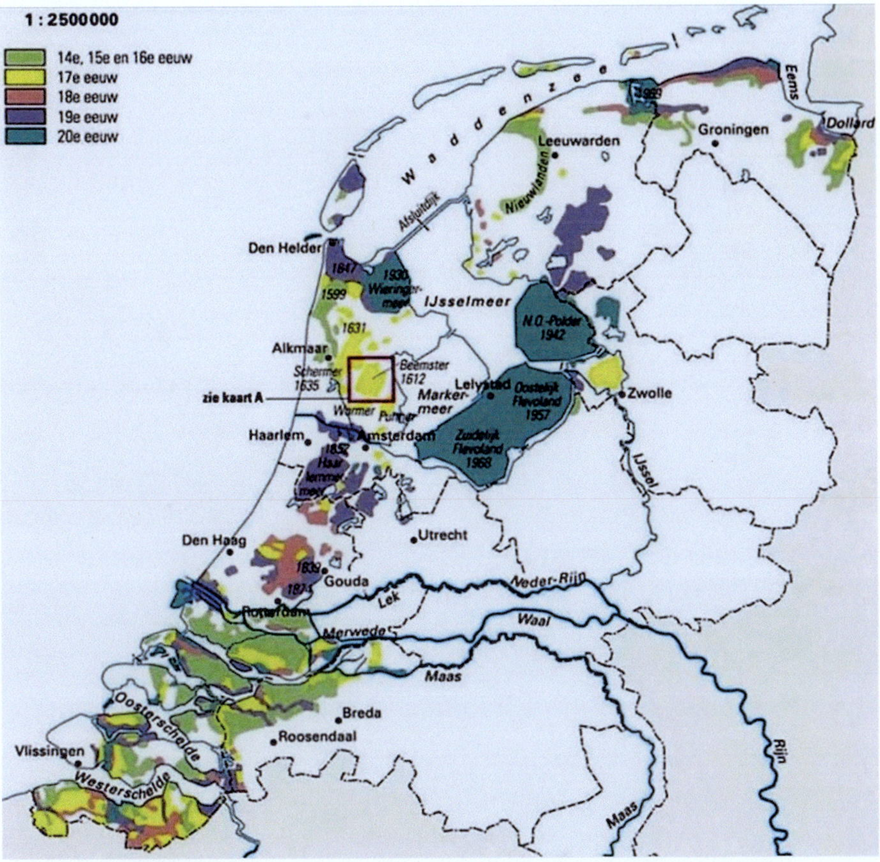

Fig. 3 The polders have been created since the fourteenth century in the Netherlands

3.1 *Øresund Link*

The 16 km fixed motorway and railway link across the Øresund connects Kastrup on the Danish coast with Lernacken on the Swedish coast. The main components of the Øresund Link is shown in Fig. 4.

The project has been divided into three main contracts: the tunnel sections; the dredging and reclamation; and the bridge. The dredging and reclamation work encompasses the excavation of a 3,500 m long immersed tunnel trench. Then, using the dredged material, a 4,000 m long artificial island complex is being reclaimed to form the connection between the tunnel and the bridge. The works also includes the deepening of two canals and the islands' shore protection [3]. This approach, dictated by the project client, enabled full re-use of the dredged material. Major sea crossings of this type that feature artificial islands particularly lend themselves to this re-use philosophy [2].

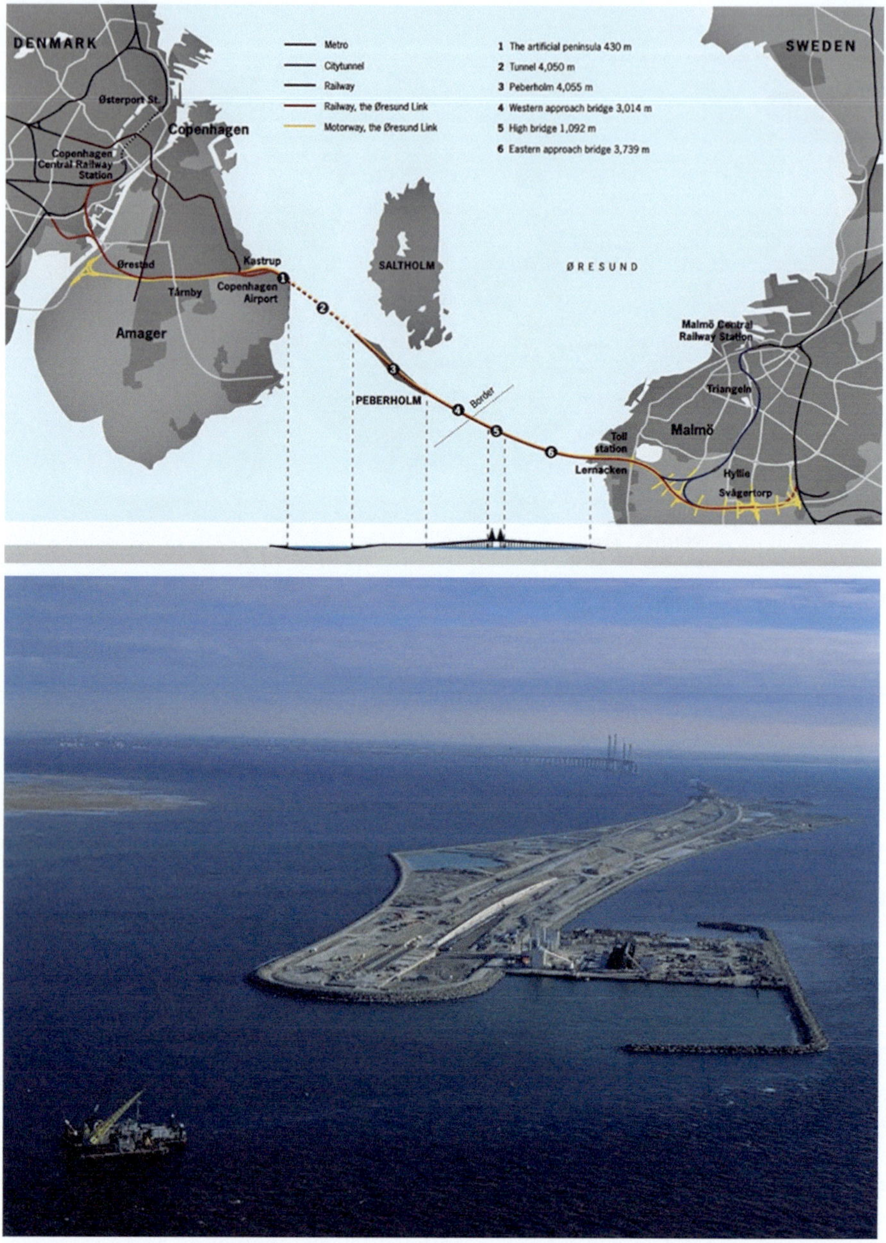

Fig. 4 Top: plan view of Øresund Link [4]; below: artificial islands under construction [5]

All dredged seabed material is reused in the reclaimed areas, which include sedimentation basins. This entails the following advantages with respect to environmental impact [3]:

- Minimal sediment spillage;
- No need for dumping sites; and
- Reduced need for import of sand and gravel from borrow areas.

3.2 Fehmarnbelt Link

Fehmarnbelt Link, an 18-km-long immersed tunnel between Denmark and Germany, which will be the longest immersed tunnel in the world combining a dual railway and motorway connection (see Fig. 5). The negative environmental impacts of such a mega project are to be mitigated through design, careful planning and project execution by the client.

The Fehmarnbelt tunnel construction will generate some 19 million m^3 of dredged sand and clay materials that will be stored and reused [2]. The impacted areas on Lolland will be replaced by nature areas of at least twice the size of the Femern industrial areas for the benefit of animal and plant life. Also, completely new nature areas will be established on the south coast of Lolland using materials dredged from the tunnel trench. These land reclamation areas in the immediate vicinity of the project area at Rødbyhavn on the Danish side will be established as nature areas with green corridors that support coherent, diversified nature, where insects and protected species and plants can thrive [6]. The large nature development by the reused material can be seen in Fig. 6.

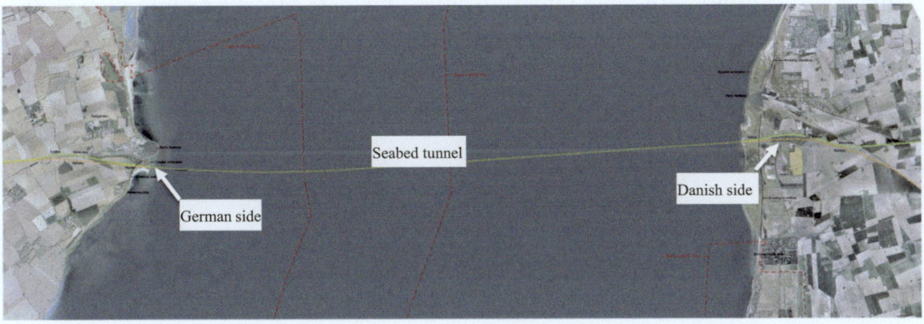

Fig. 5 Plan view of 18 km long Fehmarnbelt Link alignment;

Fig. 6 Left: Danish side of Fehmarnbelt link; right: German side of Fehmarnbelt link (red dashed line: existing coastline)

3.3 Pulau Tekong Pilot Polder in Singapore

The design of the polder development of Pulau Tekong in Singapore by Royal HaskoningDHV, has won the inaugural Innovative Engineering Award (Design) by Singapore's Housing & Development Board (HDB).

The polder concept was concluded to be technically and financially feasible due to the reduced volume of fill material required as compared to the traditional land reclamation method. By building dikes and installing pumping systems it is possible for the area currently below low water to be converted into land. Figure 7 provides an aerial view of the shoreline configuration and reclamation during the construction phase. Traditionally, sand is used to fill the area to be reclaimed above sea level. This new approach involves building a dike around the area to be reclaimed and draining the water from it with a pumping. This will allow the reclaimed area to be built at a lower level and hence significantly reduce the amount of sand needed for land reclamation.

The design is an innovative and cost-effective reclamation solution. Traditional drainage philosophy with large retention areas and state of the art pumping stations, make this design future-proof and good to operate and manage. The polder dike has a length of 10 km and a height of 7 m above polder land. To maintain the biodiversity of the area, the dike is aligned in such a way that the habitat around Pulau Unum is perserved.

4 Conclusions

To mitigate the long-term impacts of large waterfront development projects, designers propose more and more circular solutions, apply nature-based approaches, and introduce ecological enhancements to minimize negative impacts. The focus on circularity and nature-based approach during the design phase is the most effective solution, then mitigating the impacts during construction phase, as the impacts are mitigated upfront.

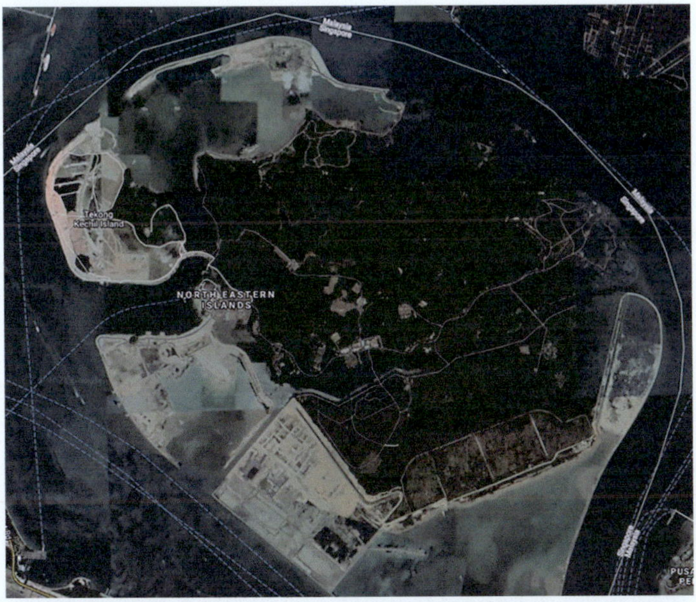

Fig. 7 Left:State of Shoreline Configuration and Reclamation at Pulau Tekong (Google maps 2021)

Fig. 8 Right: Artist impression of finished polder at Pulau Tekong (see: https://youtu.be/17BZB6ko_Nc)

This paper provided two types of solution on the reduction of environmental impact of large waterfront developments by design: first the reuse dredged material from the IMT trenches for land reclamation and nature development; and second a polder solution as alternative to the traditional land reclamation design.

If the reclamation is combined with an IMT which is a popular way now, reusing of materials dredged from the IMT into the reclamation can gain good benefits. However some attentions shall be paid to:

- It shall be considered and well planned for in early phase and supported by the project client.
- The ability to reuse dredged material shall be checked carefully such as the material's suitability, contamination, temporary storage, extra efforts other than just dumping it into designated area.

References

1. Kau Yi Chau Artificial Islands homepage, https://www.centralwaters.hk/en/, last accessed 2024/05/31
2. Jonathan B.: Achieving sustainable immersed tunnel projects. In: Yan, Celestino, Thewes & Eberhardt (eds.), ITA CONFERENCE on Tunnelling for a Better Life 2024, pp. 2333–2341. CRC Press (2024).
3. Maurice de K., Wouter D. and Rienk H.: The Øresund Fixed Link: Dredging Reclamation. Terra et Aqua, No.66, (1997).
4. ØresundsBro konsortiet, 10 years The Øresund Bridge and Its Region, June 2010.
5. Helena R. (author), Ying L.(translator): Partnership Pays: Project Management the Øresund Way (in Chinese), China Communications Press, Beijing (2017).
6. Jens Brenner L., Elisa Christina R., Søren Mikkel B., Kim Smedegaard A.: Sustainability of the Fehmarnbelt Tunnel, Proceedings of ITA-AITES World Tunnel Congress 2022 and 48th General Assembly Bella Center, Copenhagen (2022).

Optimization of a Hexagon-Shaped Floating Unit Based on the Response Surface Method

Ziyi Zhang, Yufei Wu, and Xiaoxu Huang

Abstract A hexagon-shaped floating unit (HS-FU) was proposed, and the optimization was performed with respect to the geometric dimension based on the response surface method (RSM). A multi-objective optimization model considering the total weight and heave response was established to achieve a satisfactory balance between economy and safety. Subject to the constraints of the roll response, stability height, freeboard, and collision issue, five independent design variables were optimized to achieve the minimum total weight and heave response. To reduce the computational burden and improve the optimization efficiency, surrogate models of output variables were constructed based on the RSM. The results show that, opposite to the heave response, the total weight of the floating unit is positively proportional to the side lengths of the water tank and living area; the stability performance can be improved by increasing the water tank height and the living area side length. The combination of weighting factors with a value of 0.4 for total weight and 0.6 for heave response was selected as one of the optimal designs for the floating unit. Compared with the initial design, the optimal design shows better performance, with the heave response decreasing by 22.51% and the stability increasing by 28.30%.

Keywords Hexagon-shaped floating unit · Multi-objective optimization · Response surface method · Stability performance · Heave response

1 Introduction

To obtain better motion performance and lower manufacturing costs, optimizing offshore floating platforms has become critical and widely studied over the past few decades. To craft a semi-submersible unit for oil recovery, a fully automated optimization procedure was proposed by Park et al. [1]: the extreme value of a three-hour heave and the structural weight were investigated by an annealing algorithm, and

Z. Zhang · Y. Wu · X. Huang (✉)
School of Civil and Transportation Engineering, Shenzhen University, Shenzhen, China
e-mail: xxhruby@szu.edu.cn

X. L. Zhao et al. (eds.), *Proceedings of the Fourth World Conference on Floating Solutions: WCFS 2024; 2–4 December; Hong Kong*, Lecture Notes in Civil Engineering 597, https://doi.org/10.1007/978-981-96-4569-5_14

four optimal solutions with the weight coefficients according to different objectives were obtained. Jang et al. [2] developed evaluation modules for the internal strength and limit states, which were used as constraints to optimize the structures. Emami and Mostafa [3] constructed a triple objective function with weight, heave response and pitch response as objectives and proposed a grid search algorithm to reduce the motion response while minimizing the floating structural mass.

With the complication of marine engineering, most engineering problems show a high nonlinear relationship between optimization objectives and design variables. Currently, RSM has been widely used in the optimization design of marine engineering. Tabatabaei [4] used the 3D modified squid model as a basis, and achieved the optimization of geometric parameters by RSM and optimization algorithm, which improved the pressure, viscous and total drag force of the model. Tang et al. [5] investigated the bearing capacity reliability of monopile foundations of offshore wind turbines based on the RSM, and verified the accuracy of the model through practical cases.

From the above studies, it can be seen that structural weight and motion response have been the predominant focuses when optimizing offshore structures: weight represents structural cost, while motion response represents structural safety and stability. In addition, semi-submersible platforms and other structures have been well developed and examined in current research, but new types of floating units have not received corresponding attention.

By imitating a honeycomb architecture, a hexagon-shaped floating unit (HS-FU) was proposed [6] and investigated in this paper. As shown in Fig. 1, the HS-FU is composed of a living tank, mechanical tank, and water tank. It is noted that the proposed HS-FU is mainly used in multi-floating unit combinations, the balancing of the safety and economy of individual floating unit plays a decisive role in the overall platform after combination. Therefore, this paper mainly focuses on the optimization of individual floating unit, providing useful suggestions for the subsequent design in practical engineering.

2 Design Variables

To determine an economically feasible and safe design, the size of the hexagon-shaped floating unit (HS-FU) needs to be optimized by considering the corresponding functional requirements. The economy of the HS-FU is represented by the total weight, and the safety of the HS-FU is represented by its stability and motion response. The freeboard of the HS-FU is constrained to ensure the connection between floating units and the barging vessel. In addition, considering the manufacturing technology and collision issues, the shape and size of the HS-FU are also constrained.

As shown in Fig. 1, seven design variables related to geometric dimensioning for HS-FUs are considered here, including the heave plate side length L_1, heave plate thickness H_1, water tank height H_2, water tank side length L_2, mechanical area height

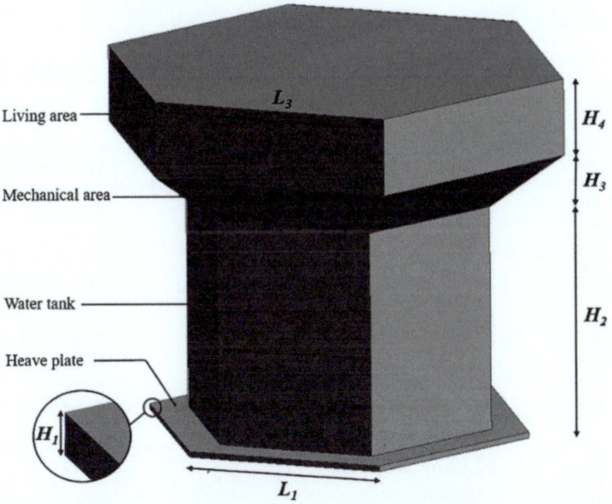

Fig. 1 Schematic of the hexagon-shaped floating unit

H_3, living area height H_4, and living area side length L_3. Due to the technique of compression casting and prefabrication, it is not convenient to change the thickness of each concrete plate; thus, the thickness of all the concrete plates is fixed at 0.3 m (i.e., $H_1 = 0.3$ m). The side length of the heave plate L1 is related to the living area side length L_3 and the water tank side length L_2 in the design process with the following assumptions. To avoid the collision issue of floating units, the heave plate side length L_1 is assumed to be at least 1.5 m shorter than the living area side length L_3. In addition, to avoid an unreasonable loading pattern and the deposition of marine organisms caused by the excessive overhang length, the heave plate side length L_1 should not exceed the water tank side length L_2 by more than 2 m. Thus, five independent variables are considered in the design. The initial values and limit values for each design variable are listed in Table 1.

3 Optimization Objectives

3.1 Weight Estimation

Weight is an important index for reflecting the economy of the floating unit. Based on the configuration of the HS-FU (see Fig. 1), it is clear that the total weight of the floating unit consists of four main parts, including the internal bulkhead, the top deck, the heave plate, and the frame slab of each layer. The design load of each layer is determined based on the Chinese building design code [7, 8]: the variable loads applied to the living area, the machinery area, and the upper deck are 2.0 kN/m^2, 7.0 kN/m^2, and 3.0 kN/m^2, respectively. Thus, the total weight of the floating unit can be evaluated analytically.

Table 1 Design Variables of the floating unit

Variables	Definition	Initial value (m)	Lower bound (m)	Upper bound (m)
H_1	Heave plate height	0.3	0.3	0.3
H_2	Water tank height	12	10	14
H_3	Mechanical area height	3	2	4
H_4	Living area height	4	3	5
L_1	Heave plate side length	10	min[L_3-1.5, $L_2 + 2$]	min[L_3-1.5, $L_2 + 2$]
L_2	Water tank side length	8	6	10
L_3	Living area side length	12	10	14

3.2 Heave Response

In an actual marine environment, the equilibrium state of the platform will be disrupted when it is subjected to wind, waves, currents, and other factors, leading to translational or rotational movements. Among the six degrees of freedom of the platform, the vertical heave motion and the roll motion can better reflect the hydrodynamic performance of the platform [9]. The heave motion of the platform is less affected by the model parameters and is difficult to change drastically, while the roll motion can be adjusted by changing the gravity centre. Thus, the heave response of the floating unit is selected as one of the optimization objectives.

4 Design Constraints

4.1 Freeboard Control

Combined with relevant specifications, the freeboard H can be calculated based on the rule of general layout (Ministry of Transport of the People's Republic of China, 2013):

$$H = HWL + \Delta \tag{1}$$

where HWL is the extreme high water level, generally taken as the wave height with an accumulated frequency of 1% of the wave column; Δ is the affluent height, generally taken as 0 ~ 0.5 m. It is assumed that the floating unit is applied in a

Table 2 Requirement of the stability height for different standards

Regulation	DNV	CCS	IMO	ABS	NMD
Stand value	≥ 0.3 m	≥ 0.15 m	≥ 0.15 m	≥ 0.0 m	≥ 1.0 m

Note: DNV—Det Norske Veritas; CCS—China Classification Society; IMO—International Maritime Organization; ABS—American Bureau of Shipping; NMD—Norwegian Maritime Directorate

wave condition with a maximum wave height of 4 m. Therefore, the freeboard of the HS-FU should be controlled above 4 m in this study.

4.2 Stability Performance

Under the effect of external forces (e.g., wind, waves, and current), the equilibrium state of the floating unit will be lost, leading to the movement of the unit. When the unbalanced external force disappears, the ability of the floating unit to recover to the equilibrium state automatically is called stability performance. Stability is an important indicator of the performance of floating structures.

GM is an important factor in the design of floating structures:M is the centre of transverse stability. G is the centre of gravity. when point M is higher than point G (i.e., GM > 0), the direction of the restoring moment is opposite to the heeling direction, which resists the heeling of the floating unit; when point M is lower than point G (i.e., GM < 0), the direction of the restoring moment is the same as the heeling direction, accelerating the tilt of the floating unit until it overturns.

Table 2 lists the requirements of the stability height for different specifications. In general, considering the engineering operation needs, the actual value is usually larger than the required value suggested in the standards. In this study, the stability height of the floating unit is assumed to be larger than 1 m to ensure stability.

4.3 Roll Response

The degrees of freedom and the method to evaluate the motion response of a floating structure have been described in Sect. 3.2 in detail. The roll response of the floating unit is related to the inherent period of the floating unit in the degrees of freedom of the roll. When the roll inherent period of the unit is much larger than that of the external forces, the rocking amplitude is small. However, when the roll inherent period of the unit is close to or equal to that of the external forces, the floating unit and the external forces resonate, and the rocking angle of the floating unit increases significantly. To avoid excessive shaking that affects the normal operation of the HS-FU, the roll response is limited to 5° in the following optimization issue.

4.4 Collision Issue

Considering the redundancy in the design process, the length gap between L_1 and L_3 should be larger than 1.5 m. As mentioned in Sect. 2, the length of L_1 depends on L_2 and L_3, thus, the constraint of the collision issue is expressed as the relationship between L_2 and L_3: the length of L_3 should be greater than the length of L_2 by 3.5 m.

5 Multi-Objective Optimization Model

5.1 Multi-Objective Function

In this paper, from the perspective of safety and economy, minimizing the total weight and the heave response of the floating unit are selected as the two optimization objectives. The mathematical model of the optimization issue is expressed as follows:

$$\mathrm{Min}.F(X_1, X_2 ... X_5) = \omega_M \left[\frac{W_e}{Ref.W_e} \right] + \omega_p \left[\frac{H_e}{Ref.H_e} \right]$$

$$\begin{aligned} & s.t.\, GM \geq 1\mathrm{m} \\ & H \geq 4\mathrm{m} \\ & R_O \leq 5^\circ \\ & L_3 - L_2 \geq 3.5\mathrm{m} \end{aligned} \tag{2}$$

where W_e and H_e are the mass and the heave response of the floating unit with the corresponding weighting factors denoted as ω_M and ω_p ($\omega_M + \omega_p = 1$), respectively; and *Ref* represents the reference value of each target. The reference value is selected as the optimal result of each objective in single-objective optimization (see Table 3).

Table 3 Single-objective optimization results

Item		Weight (kg)	Heave (m)
Objective	Min W_e	1,401,403	1.8012
	Min H_e	2,432,849	0.8953
Reference value		1,401,403	0.8953

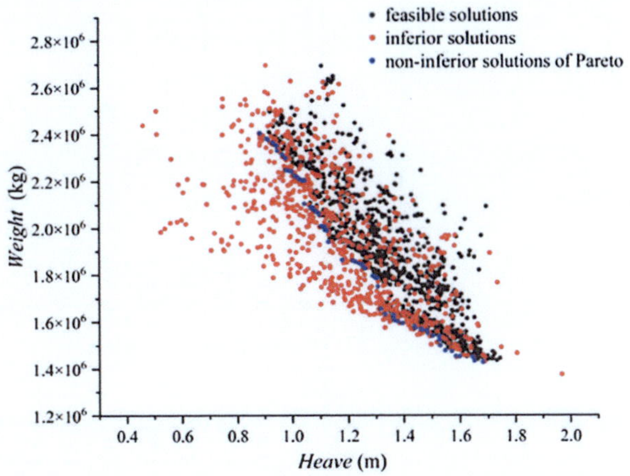

Fig. 2 Pareto set of the optimization

6 Results and Discussion

6.1 Optimization Result

Based on the initial design model in Table 1, after 2000 iterations, the Pareto set can be obtained as shown in Fig. 2. The red dots represent inferior solutions that do not satisfy the constraints. The black points are feasible solutions that satisfy the constraints. The blue dots not only satisfy the constraints but also outperform the black dots, which constitute the Pareto solution set. The total weight is inversely proportional to the heave response. Therefore, a balance between these two items is necessary. Considering different combinations of the weighting coefficients for the heave response and total weight, the corresponding optimization results are shown in Table 4. The optimization results shown in Table 9 all meet the constraints mentioned in Sect. 6.1. It can also be found that with the increase in ω_M, the weight of the floating unit decreases gradually, and the heave response increases gradually.

6.2 Optimal Design

To verify the accuracy of the optimization results, the discrepancy of the output parameters (i.e., weight, *GM*, roll, heave) obtained by the optimization model and the numerical model was evaluated. The numerical model was built with the same geometric dimensions in Table 4, and the four output parameters can be obtained by the software AQWA.

According to Fig. 3, a typical optimal solution with minimum errors is selected by considering weight and motion performance. When ω_M and ω_p take the values of 0.4 and 0.6, respectively, the error of each output parameter is relatively small and is

Table 4 Optimization results for different combinations of weighting factors

Item	Number of combinations								
	1	2	3	4	5	6	7	8	9
ω_M	0.1	0.2	0.3	0.4	0.5	0.6	0.7	0.8	0.9
ω_p	0.9	0.8	0.7	0.6	0.5	0.4	0.3	0.2	0.1
H_2 (m)	12.703	13.012	13.034	12.682	12.557	11.792	11.493	12.563	11.92
H_3 (m)	3.370	3.532	3.484	3.554	3.442	3.767	3.886	3.442	3.133
H_4 (m)	4.275	4.496	4.453	4.533	4.541	4.595	4.782	4.262	4.470
L_2 (m)	9.135	9.337	9.075	8.744	8.179	7.342	7.022	6.700	6.028
L_3 (m)	12.931	13.168	12.943	12.85	12.02	10.845	10.541	10.639	10.016
W_e (kg)	2,435,590	2,550,770	2,454,684	2,367,736	2,112,422	1,769,533	1,670,529	1,618,525	1,407,281
GM (m)	1.041	1.032	1.092	1.340	1.093	1.262	1.311	1.114	1.422
H (m)	4.034	4.028	4.012	4.010	4.008	4.006	4.074	4.002	4.090
H_e (m)	0.968	0.891	0.936	1.017	1.107	1.302	1.381	1.569	1.807
R_O (°)	3.055	3.178	3.194	3.155	3.200	3.741	3.878	3.772	3.895

within a reasonable range: the errors of weight, stability height, heave response, and roll response are 0.34%, 3.72%, -0.88%, and 6.66%, respectively. Herein, the result of this combination of weighting factors is considered one of the optimal designs for the floating unit, as listed in Table 5. Although the total weight and roll response increased by 17.57% and 6.94% after the optimization, respectively, the motion performance of the floating unit was improved: the heave response was reduced by 22.51%, and the stability performance was increased by 28.30%.

In addition, the iteration process of each variable in the optimization is shown in Fig. 12. After 2000 iterations, H_2, H_3, and H_4 show an obvious tendency to converge. The distributions of L_2 and L_3 are more scattered and lack regularity, while the ratio of L_2 to L_3 gradually converges as the number of iterations increases.

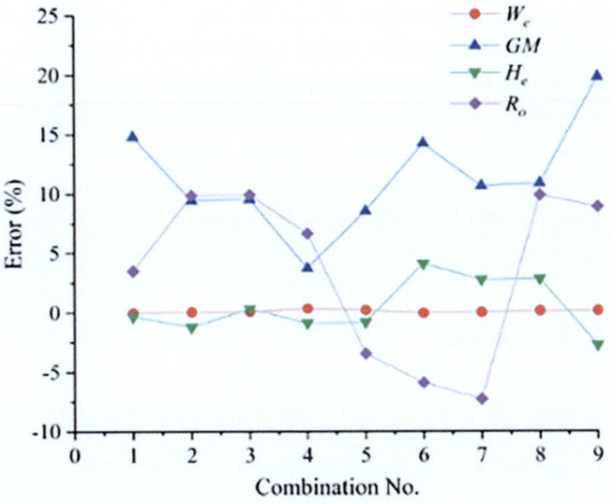

Fig. 3 Error evaluation for different combinations of weighting factors

Table 5 Comparison of optimal design and initial design

Item		Before optimization	After optimization	Change ratio
Design Variables	H_2 (m)	12	12.682	5.68%
	H_3 (m)	3	3.554	18.47%
	H_4 (m)	4	4.533	13.33%
	L_2 (m)	8	8.744	9.30%
	L_3 (m)	12	12.85	7.08%
Objectives	W_e (kg)	2,007,000	2,359,700	17.57%
	H_e (m)	1.324	1.026	-22.51%
Constraints	H (m)	4.000	4.010	2.50%
	GM (m)	1.007	1.292	28.30%
	R_O (^{O})	2.766	2.958	6.94%

7 Conclusions

In this paper, a hexagon-shaped floating unit (HS-FU) was proposed, and optimization was performed with respect to the geometric dimensions. To achieve a satisfactory balance between safety and economy for the floating unit, a multi-objective optimization model was constructed. The accuracy of the optimization results was verified by error estimation, and the combination of weighting factors with a value of 0.4 for total weight and 0.6 for heave response was selected as one of the optimal designs for the floating unit. Compared with the initial design, the floating unit with the optimal design shows better performance, with the heave response decreasing by 22.51% and the stability increasing by 28.30%.

In this study, only the optimization of an individual floating unit was focused on without consideration of the interactions between the multi-floating units, leading to the discrepancy from the actual service status. For a deeper understanding of the proposed floating unit in marine engineering, the optimization design considering the interaction caused by multi-floating units should be conducted in future studies. In addition, for a comprehensive design of HS-FU, the load-bearing capacity of the floating structure can be added into consideration by combining the finite element analysis and hydrodynamic analysis.

References

1. Park Y, Jang BS, Du Kim J (2015) Hull-form optimization of semi-submersible FPU considering seakeeping capability and structural weight. Ocean Eng 104:714–724
2. Jang BS, Kim JD, Park TY, Jeon SB (2019) FEA based optimization of semi-submersible floater considering buckling and yield strength. International Journal of Naval Architecture and Ocean Engineering 11(1):82–96
3. Emami A, Mostafa Gharabaghi AR (2020) Introducing a simple and reliable multi-objective optimization method to estimate hull dimensions of a semi-submersible rig. Journal of Marine Engineering 16(31):29–40
4. Malazi MT (2023) Design optimization of a longfin inshore squid using a genetic algorithm. Ocean Eng 279:114583
5. Tang S, Zhai J, Yin Q, Wang W, Xu S (2023) Bearing capacity reliability analysis using the coupled FEM-BPNN-RSM for monopile-supported OWTs. Ocean Eng 276:114228
6. Zhang, Q., Ding, Z., Pre-cast floating tubes for the combined construction of large floating platforms on water. China. CN107628208B. 2019–03–26.
7. Ministry of Housing and Urban-Rural Development of the People's Republic of China: Load Code for the Design of Building Structures. MOHURD, Beijing (2012).
8. Ministry of Transport of the People's Republic of China: Design Code of General Layout for Sea Ports. MOT, Beijing (2013).
9. Zhang X, Song X, Yuan Z, You Y (2017) Global motion and airgap computations for semi-submersible floating production unit in waves. Ocean Eng 141:176–204

The Challenges of Designing and Constructing Floating Buildings

Virginie Segura

Abstract As pressure on urban space around the world increases, councils and local governments look to innovative solutions to provide facilities to tackle social, economic and environmental issues in their jurisdictions. Land is being reclaimed from the sea in some countries and islands created in others, while many are opting to build floating structures to maximise space and the opportunities these brings. SPISEA is an engineering firm specialized in floating structure. We engineer small to medium size floating structures.

Keywords Regulation · Standards · Construction

1 Design Challenges

1.1 The Lack of Regulation

While the concept of constructing floating structures is not novel, there remains a notable absence of cohesive regulations and building standards governing such endeavors. The classification of floating buildings varies significantly across the globe. For instance, in the Netherlands, these structures are regarded as houses, whereas in Singapore, they are classified as boats and are required to possess an engine.

In France, the classification differs between inland and coastal areas—on inland waterways, a floating building is deemed a boat, whereas on seawater, it is considered a house. In some countries, such constructions are entirely prohibited.

V. Segura (✉)
Management & Construction of Floating Structures, SPISEA, Port, France
e-mail: v.segura@spi-sea.com

X. L. Zhao et al. (eds.), *Proceedings of the Fourth World Conference on Floating Solutions: WCFS 2024; 2–4 December; Hong Kong*, Lecture Notes in Civil Engineering 597, https://doi.org/10.1007/978-981-96-4569-5_15

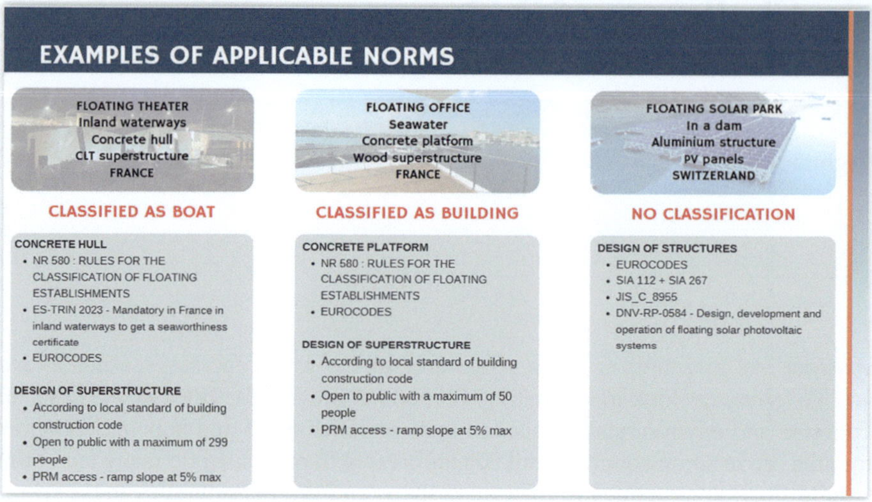

Fig. 1 Examples of applicable norms

The absence of a standardized framework hinders development companies from effectively addressing the challenges outlined. There is a pressing need for a comprehensive roadmap—a uniform set of regulations governing projects that incorporate floating buildings or structures.

In France's inland waterways, for example, we refer to ES-TRIN 23, which sets forth the technical requirements for vessels on interior waters; however, this pertains to boats and propulsion systems rather than to floating buildings that remain anchored in place.

SPISEA anticipates an increase in projects featuring floating buildings, as population pressures compel governments to adopt innovative solutions for providing amenities to their citizens. To ensure the safety and longevity of these structures, it is imperative that we establish a common standard for universal adoption, necessitating global collaboration to achieve this goal (Fig. 1).

2 Which Codes Can Be Used?

The construction of floating buildings requires compliance with a unique set of regulatory codes that cover both maritime and architectural standards. Given that these structures exist at the intersection of land and water, regulatory requirements often include aspects from both building codes and maritime regulations.

One key regulatory consideration is structural integrity and safety. Floating buildings must adhere to rigorous engineering codes to ensure they remain stable and secure under various water conditions. This includes compliance with buoyancy and stability standards similar to those used in shipbuilding, as well as wind, seismic, and load-bearing codes that apply to traditional buildings. These codes help to address challenges such as wave impact, water currents, and the potential for flooding or extreme weather.

3 NR 580: Rules for the Classification of Floating Establishments

- Bureau Veritas
- Refer to Eurocodes for design and cover the entire floating structure: hull and superstructure

3.1 Es-Trin 2023

- INLAND - EU DIRECTIVE 2016/1629/EC Rhines Regulations - ES-TRIN
- Provisions on inland navigation vessel construction, arrangement and equipment - mainly used in inner waterways to design floating building in steel
- Available in 4 languages (German, English, French, Dutch)

3.2 Eurocodes

- Eurocode 1: Actions on structures
- Eurocode 2: Design of concrete structures
- Eurocode 3: Design of steel structures
- Eurocode 7: Design of aluminium structures

3.3 Design of Superstructure

- According to local standard of construction
- Depending on the building destination: open to public or not?
- PRM access? Person with reduced mobility

4 Construction Challenges

4.1 Project Management as a Key Factor of Success

Effective project management is essential for the successful construction of floating structures, which require an extraordinary level of coordination, precision, and innovation. The unique challenges of these projects—ranging from engineering complexities to environmental considerations—demand a well-structured project management approach to ensure that all aspects are meticulously planned and executed.

One of the primary responsibilities of project management in floating construction is to oversee the integration of multidisciplinary teams. Architects, engineers, environmental consultants, and regulatory specialists must work together seamlessly to address the various technical, regulatory, and environmental requirements. Clear communication and collaboration facilitated by project management are crucial in preventing delays and ensuring that all team members are aligned on project goals, timelines, and responsibilities.

Risk management is another critical component. Floating structures face distinct risks, such as water-level changes, weather impacts, and maritime logistics challenges. Project managers must identify potential risks early, implement risk mitigation strategies, and adjust plans as necessary. This includes contingency planning for unexpected conditions like severe storms or supply chain delays that can impact the construction timeline.

Budget and resource management are equally vital. Given the specialized materials and techniques involved in floating structures, cost overruns are a common risk. Effective project managers monitor budgets closely, ensuring that resources are allocated efficiently and that spending remains within scope. They also coordinate the procurement of materials, some of which may require customization for water resistance and durability, which adds complexity to sourcing and budgeting.

Moreover, project management plays a key role in navigating regulatory compliance. Floating structures must adhere to stringent building, maritime, and environmental regulations. Project managers ensure that all necessary permits and approvals are obtained in a timely manner and that construction aligns with safety and environmental standards. This not only reduces the risk of legal issues but also enhances the project's sustainability and public acceptance.

In conclusion, project management is a cornerstone of successful floating structure construction. By managing risks, coordinating multidisciplinary teams, maintaining budget control, and ensuring regulatory compliance, effective project management ensures that floating structures are built safely, sustainably, and within the defined scope. It's clear that without strong project management, the unique challenges of floating construction would be much harder to overcome, making it an indispensable factor in the success of these innovative projects.

5 The Integration of Construction and Naval Architecture in Floating Structures

The design and construction of innovative floating structures require a unique synergy between two distinct professional disciplines: construction and naval architecture. This interdisciplinary collaboration presents a series of unique challenges that must be navigated carefully.

One significant challenge faced by construction companies involves the lack of a conventional reference point that is typically available in terrestrial projects. In the context of floating structures, project managers must continuously seek critical information such as the center of gravity. This requires meticulous consideration of the weight of equipment and individual components installed within the floating building.

Additionally, environmental factors—such as wave action, water level fluctuations, and currents—must be taken into account, as these are not considerations in traditional land-based construction.

From a project management standpoint, the initial step in proposing a floating structure is to conduct a comprehensive assessment of the overall weight of the building. Understanding the loads imposed on the floating platform is essential, as these factors directly influence the stability and buoyancy of the structure. This assessment must occur during the conceptual phase to provide a reliable estimate of the project's cost and complexity.

Following the conceptual design phase, greater precision is required regarding the technical specifications—such as electrical systems, plumbing, heating, ventilation, and cladding materials—that impact the building's weight. At this stage, it becomes possible to perform a stability assessment in accordance with established standards, such as ES-TRIN 2023 or NR580. Only upon confirming that the stability criteria are met can the project move forward to the construction phase.

Thus, it is evident that a significant amount of accurate information from construction teams is necessary before a floating project can be deemed viable at a specific location. This collaborative process represents a markedly different approach to project management and execution, highlighting the complexities inherent in the development of floating structures (Fig. 2, 3 and 4).

Fig. 2 2024 – Hong Kong – Tsui Ping River public area

Fig. 3 2022 – France – Maison de la Mer

Fig. 4 2024 – France – Private lodges

Worker Tracking Using Real-Time LiDAR Scanning for Offshore Construction Safety

Mingyu Zhang, Yushu Yang, and Heng Li

Abstract Offshore floating residences offer a future sustainable solution for coastal cities with limited land availability. However, the premise of floating residences relies on safe and efficient construction, which presents significant challenges due to the significant difference from on-land construction. The complex environmental factors such as waves, wind, and humidity make offshore construction highly difficult. Construction activities are disrupted by these intricate environmental factors, resulting in more safety risks and uncertainties compared to on-land construction. Therefore, worker activities in offshore construction require extensive monitoring and tracking. Traditional worker positioning and tracking methods, such as various tag sensors, may be difficult to deploy effectively on offshore platforms. This study proposes a real-time worker tracking method in offshore construction scenarios using LiDAR scanning. LiDAR scans the entire construction site in a contactless manner and identifies and locates multiple workers without requiring them to carry any positioning devices. An algorithm for worker identification, positioning, and tracking is developed based on LiDAR scanning data. The proposed method has been validated through on-site experiments, with results showing a positioning accuracy of 0.05 m based on real-time LiDAR scanning. This research provides essential safety monitoring capabilities for offshore construction and can further enhance worker safety in complex construction scenarios.

Keywords Offshore construction · Construction safety · Worker tracking

M. Zhang (✉) · Y. Yang · H. Li
Department of Building and Real Estate, The Hong Kong Polytechnic University, Hong Kong, China
e-mail: mingyu.zhang@connect.polyu.hk

X. L. Zhao et al. (eds.), *Proceedings of the Fourth World Conference on Floating Solutions: WCFS 2024; 2–4 December; Hong Kong*, Lecture Notes in Civil Engineering 597, https://doi.org/10.1007/978-981-96-4569-5_16

1 Introduction

The rapid urbanization and population growth in coastal cities have increased the demand for innovative solutions to address the issue of limited land availability, particularly in areas like Hong Kong where geographical constraints, such as mountains, preclude further land development. To mitigate these challenges, the concept of offshore floating residences [1–3] has garnered considerable attention as a forward-looking and sustainable alternative. Offshore floating residences are engineering structures designed to float on water bodies, providing habitable spaces without the need for traditional land-based construction [4]. These floating platforms offer a novel solution to the land scarcity problem, thus contributing to the sustainability of urban development. However, the construction of such offshore structures introduces challenges in terms of safety, viability, and efficiency of these projects.

The construction of offshore floating residences differs significantly from conventional on-land construction due to the complex environmental factors. Unlike land-based construction, which operates in relatively stable and predictable conditions, offshore construction involves a dynamic environment characterized by frequent severe fluctuations. For instance, wave action can apply substantial forces on floating platforms [5], which complicates construction activities that require precision and stability. These environmental factors not only increase the complexity of construction processes but also introduce safety risks for workers. In offshore environments, the margin for error is significantly reduced, and any lapse in safety protocols can lead to worker injuries and fatalities.

Given these risks, effective monitoring and tracking of worker activities in offshore construction become necessary. Traditional methods of worker positioning and tracking, commonly used in on-land construction, typically include tag-based sensors [6, 7], GPS devices, or RFID (Radio-Frequency Identification) systems. These devices are often carried by workers for continuous tracking. However, the deployment of these technologies in offshore settings has challenges. For example, the harsh marine environment can interfere with wireless communication, leading to inaccuracies in worker positioning data. Additionally, the constant movement of the floating platform can cause these devices to lose connectivity. Furthermore, ensuring that all workers consistently carry these devices can be difficult, especially in an environment where personal safety equipment is already cumbersome. These limitations suggest that traditional tracking methods may not be sufficiently reliable or effective for offshore construction scenarios, necessitating the development of alternative approaches.

In response to these challenges, this study proposes an innovative solution for real-time worker tracking in offshore construction environments by leveraging LiDAR (Light Detection and Ranging) technology [8]. LiDAR is a remote sensing method that utilizes laser beam to measure distances and generate precise 3D maps of an environment. In the context of offshore construction, LiDAR can be deployed to scan the entire construction site, creating a detailed spatial representation of the area. Unlike traditional tracking methods, LiDAR operates in a contactless manner, meaning that it does not require workers to carry any tracking devices. Existing studies have explored the application of LiDAR technology in construction. For example, LiDAR has been used in on-land construction for tasks such as site surveying [9], equipment monitoring [10], and object detection [11]. In these applications, LiDAR has demonstrated its capability to provide accurate and comprehensive data to optimize construction processes and improve safety. However, the use of LiDAR for worker tracking, particularly in offshore environments, remains a relatively unexplored area.

This research aims to develop an innovative real-time tracking system using LiDAR technology. To overcome the limitations of traditional tracking methods, which often struggle in dynamic marine environments, our study focuses on providing a robust, contactless solution to monitor workers. The methodology begins with preprocessing LiDAR data to filter out static background. This is followed by detecting and classifying workers among other moving objects. By leveraging advanced clustering techniques, we can accurately identify individual workers without requiring them to carry any additional devices. The final step involves ensuring precise tracking of these identified workers across time, maintaining their identity and trajectory even in challenging conditions. By integrating LiDAR technology, we offer a comprehensive and reliable method to enhance safety and operational efficiency, ultimately paving the way for further advancements in offshore construction management.

2 Methodology

2.1 Data Preprocessing

The large volume of LiDAR point cloud of an offshore platform poses a challenge for low operation latency. To address this issue, we first filter out the background that accounts for most of the point cloud in advance. Although the offshore platform is floating on the sea, the LiDAR is installed without relative displacement to the platform. Therefore, the background of the point cloud can be regarded as stationary. The distance of background points should have a small variance over time, close to the hardware measurement error of the LiDAR, while the point reflected from a worker has a large variance in range. Therefore, a statistical model can be utilized to model the measured distance associated with each laser beam. Specifically, a Gaussian mixture model [12] is applied to determine the position of the maximum probability distribution of the measured distance over an extended duration (beyond 60 s). Once the estimated background is obtained, the foreground points can be extracted. For each point, if its distance is within a predefined tolerance of the static background distance, the point belongs to the background, otherwise, it belongs to foreground objects. The resultant point clouds that have undergone background filtering are ready for detection.

2.2 Worker Detection

After the preprocessing stage, the foreground data contains workers, some unfiltered debris like moving equipment, and noise points. The primary goal of the worker detection module is to determine the positions of all workers within this foreground point cloud. Initially, a density-based clustering [13] technique is employed to segment the foreground point cloud into several clusters. However, closely adjacent workers may still be grouped into a single cluster due to the connected nature of point clouds. To address the issue, 2D kernel density estimation [14] is used to compute the spatial density distribution of the 2D projected point cloud. Given that each worker should align with a high-density area, clustering is guided by these density peaks. If a cluster contains multiple density peaks (more than one), a Gaussian mixture model [12] is employed with the same number of Gaussian kernels to segment the cluster. These steps ensure that each worker corresponds to a distinct point cloud cluster.

However, not all clusters generated are workers; thus, it is crucial to identify which clusters represent workers. We introduce an efficient and lightweight neural network to simultaneously classify clusters and regress their central coordinates. For computational efficiency, we utilize a voxel data format, converting unstructured point clouds into structured 3D grids. The network architecture comprises multiple layers of 3D convolution, incorporating ResNet-style [15] residual connections to bypass certain intermediate layers. The network is configured with six layers and an eightfold downsampling. Features extracted through the 3D convolutional neural network are finally processed by two fully connected layers to yield classification scores and central coordinates. This constitutes a binary classification problem, determining whether a cluster represents a worker. The worker detection module outputs only the results corresponding to clusters classified as workers.

2.3 Worker Tracking

The tracking module uses the detections obtained from the detection module. In general, a frame may output many detections with varying scores. To reduce the significant number of low-quality tracks, a score threshold is applied to discard the detections with lower confidence scores. Additionally, an enhanced non-maximum suppression (NMS) technique is applied following [16], which can eliminate the substantial quantity of overlapping low-quality detections. After this pre-processing step, the detection results are prepared for tracking. The motion model predicts the trajectory's state using the previous state and updates the state upon a successful match. The instance association step is the core component of tracking, as it matches the detections with the tracks. Due to the presence of new workers and the disappearance of previous workers in the scene, the tracks are constantly changing. Track management [17] is responsible for controlling the birth and death of the tracks, ensuring efficient tracking of the system.

The instance association part associates the set of N_T predicted trajectories with the N_D detection results. First, an affinity matrix with a dimension of $N_T \times N_D$ was established by computing the affinity score for each trajectory and detection pair. We simply employ Euclidean distance between the centers of two objects to compute affinity. Compared to other IoU-based metrics [18], it has a smaller computational overhead and consistent performance regardless of the size and rotation of the object. Then the affinity matrix is solved by the Hungarian algorithm [19], which finds the best match between trajectories and detection with the lowest total cost. In the experiment, it was found that the Hungarian algorithm performs better than the greedy algorithm. A match threshold is employed to constrain the matches to have an affinity score lower than the threshold.

3 Experiment

The LiDAR system utilized in this study is the Robosense RS-M1 [20]. It offers a horizontal field of view of 120 degrees and a vertical field of view of 25 degrees (± 12.5°). Both horizontal and vertical resolutions are 0.2°, with an operational frame rate ranging from 10 to 20 Hz. The device can measure distances up to 200 m, maintaining a measurement error of 0.05 m within a 100 -m range. For data collection, the LiDAR's frame rate was configured to 10 Hz.

We conducted field experiments to evaluate the tracking performance of the proposed method by collecting data in a real-world setting. A transport ship served as the test site for these experiments. Anchored in the To Kwa Wan typhoon shelter in Hong Kong, approximately 500 m from the shore, the ship measures 50 m in length, 20 m in width, 3 m in depth, with a draft of 1.5 m, shown in Fig. 1. Several steel pipes were arranged on its deck to simulate load and construction scenarios. On the day of testing, weather conditions were typical, with wind speeds and sea waves at average levels for Hong Kong. As a result, individuals standing on the deck could perceive the ship's periodic motion without experiencing dizziness or imbalance. These conditions effectively simulate offshore construction environments. The LiDAR was positioned 4 m above the deck, tilted downward at an approximately 10° angle. Apart from a 5 m radius around the Lidar, the entire deck was within the scanning range. During the experiment, 3–5 workers were instructed to move freely on the deck and simulate various construction activities, such as material transportation, assisting with lifting operations, and ground surface installation. Ultimately, two representative Lidar sequences were extracted from the collected scans for evaluation, details of which are summarized in Table 1 (#1 and #2). Additionally, to assess environmental impacts on performance, similar experiments were conducted in an on-land setting, with results also summarized in Table 1 (#3).

Fig. 1 The experiment platform (left) and layout (right) with a demonstration of tracked trajectory

Table 1 Information of collected data

LiDAR sequence	Environment	Sequence duration (Second)	No. trajectories	Ave. distance to LiDAR (m)	Total trajectory length (m)
#1	Offshore	90	6	26.6	135.2
#2	Offshore	90	5	23.1	109.5
#3	On-land	90	6	36.8	115.9

4 Results

We assessed tracking performance using the following metrics: (1) Average Multi-Object Tracking Accuracy (AMOTA), a well-established metric for evaluating multi-object tracking, as referenced in relevant literature [21]; (2) Average Positioning Precision (APP), which quantifies the average positioning error across all workers; (3) Average Tracking Recall (ATR), reflecting the average successful tracking rate for all trajectories; and (4) Identity Switches (IDS). The test results are detailed in Table 2.

The proposed method achieved AMOTA scores of 0.951 and 0.954 in the offshore platform tracking tests, demonstrating remarkably high tracking performance. Concerning positioning accuracy, our approach achieved a precision of 0.039 m. Considering that a worker's typical body width ranges from 0.6 to 0.8 m, this

Table 2 Results of tracking test

LiDAR sequence	AMOTA	APP (m)	ATR	IDS
#1	0.951	0.035	0.92	3
#2	0.954	0.042	0.90	2
Ave. #1 and #2	0.953	0.039	0.91	3
#3	0.958	0.036	0.94	2

error is significantly smaller than the scale of a worker's body. Furthermore, when compared to existing radio frequency-based tracking methods such as UWB (0.3 m) and BLE (0.5 m), our method offers a substantial advantage in positioning accuracy. The ATR for the offshore construction scenarios exceeded 0.9, indicating comprehensive coverage of worker trajectories. A minimal number of IDS were observed, highlighting the system's robustness in maintaining identity consistency. Figure 1 illustrates one example of a tracked worker trajectory. Moreover, the results from tests conducted in on-land environments are also included in Table 2. These findings illustrate that our method performs consistently across both on-land and offshore contexts, achieving similar scores across all evaluated metrics. This consistency highlights the method's resilience, as its tracking performance remains unaffected by the dynamic and challenging conditions typical of offshore environments.

5 Conclusion

This study presents a pioneering approach to real-time worker tracking in offshore construction environments using LiDAR technology. The proposed method effectively addresses the limitations of traditional tracking systems by providing a contactless solution that does not require workers to carry tracking devices. The proposed system demonstrates high tracking accuracy, positioning precision, and identity consistency, substantially outperforming conventional radio frequency-based methods. The experimental results highlight the system's robustness and reliability in both offshore and on-land scenarios, highlighting its potential for broader application in various construction environments.

The integration of LiDAR technology in worker tracking not only enhances safety protocols but also contributes to the sustainable development of innovative urban solutions such as offshore floating residences. Future work could explore the scalability of this system for larger offshore projects and its integration with other emerging technologies to further enhance construction site safety and efficiency.

Acknowledgements This work was supported by the RISUD Interdisciplinary Research Scheme [grant number 1-BBWV].

References

1. Lamas-Pardo M, Iglesias G, Carral L (2015) A review of very large floating structures (VLFS) for coastal and offshore uses. Ocean Eng 109:677–690. https://doi.org/10.1016/j.oceaneng.2015.09.012
2. Wang CM, Tay ZY (2011) Very large floating structures: applications, research and development. Procedia Eng 14:62–72. https://doi.org/10.1016/j.proeng.2011.07.007

3. Wang G, Goldfeld Y, Drimer N (2019) Expanding coastal cities—proof of feasibility for modular floating structures (MFS). J Clean Prod 222:520–538. https://doi.org/10.1016/j.jclepro.2019.03.007
4. Wang G, Rosenfeld Y, Drimer N, Goldfeld Y (2021) Occupant comfort analysis for rigid floating structures—methodology and design assessment for offshore dwelling module. Ships Offshore Struct 16:184–199. https://doi.org/10.1080/17445302.2020.1718267
5. Ren N, Zhang C, Magee AR, Hellan Ø, Dai J, Ang KK (2019) Hydrodynamic analysis of a modular multi-purpose floating structure system with different outermost connector types. Ocean Eng 176:158–168. https://doi.org/10.1016/j.oceaneng.2019.02.052
6. Huang Y, Hammad A, Zhu Z (2021) Providing proximity alerts to workers on construction sites using Bluetooth Low Energy RTLS. Autom Constr 132:103928. https://doi.org/10.1016/j.autcon.2021.103928
7. Li H, Lu M, Hsu S-C, Gray M, Huang T (2015) Proactive behavior-based safety management for construction safety improvement. Saf Sci 75:107–117. https://doi.org/10.1016/j.ssci.2015.01.013
8. Alaba SY, Ball JE (2022) A survey on deep-learning-based LiDAR 3D object detection for autonomous driving. Sensors 22:9577. https://doi.org/10.3390/s22249577
9. Kaartinen E, Dunphy K, Sadhu A (2022) LiDAR-based structural health monitoring: applications in civil infrastructure systems. Sensors 22:4610. https://doi.org/10.3390/s22124610
10. Yin C, Wang B, Gan VJL, Wang M, Cheng JCP (2021) Automated semantic segmentation of industrial point clouds using ResPointNet++. Autom Constr 130:103874. https://doi.org/10.1016/j.autcon.2021.103874
11. Zhang M, Wang L, Han S, Wang S, Li H (2024) Deep learning framework with local sparse transformer for construction worker detection in 3D with LiDAR. Comput Aided Civ Infrastruct Eng. https://doi.org/10.1111/mice.13238
12. Rasmussen C (1999) The infinite Gaussian mixture model. In: Advances in neural information processing systems. MIT Press. https://proceedings.neurips.cc/paper/1999/hash/97d98119037c5b8a9663cb21fb8ebf47-Abstract.html
13. Ester M, Kriegel HP, Xu X (1996) A density-based algorithm for discovering clusters in large spatial databases with noise
14. Węglarczyk S (2018) Kernel density estimation and its application. ITM Web Conf 23:00037. https://doi.org/10.1051/itmconf/20182300037
15. He K, Zhang X, Ren S, Sun J (2016) Deep residual learning for image recognition. In: 2016 IEEE conference on computer vision and pattern recognition (CVPR), IEEE, Las Vegas, NV, USA, pp 770–778. https://doi.org/10.1109/CVPR.2016.90
16. Li X, Xie T, Liu D, Gao J, Dai K, Jiang Z, Zhao L, Wang K (2023) Poly-MOT: a polyhedral framework for 3D multi-object tracking. http://arxiv.org/abs/2307.16675
17. Benbarka N, Schröder J, Zell A (2021) Score refinement for confidence-based 3D multi-object tracking. In: 2021 IEEE/RSJ International conference on intelligent robots and systems (iros), pp 8083–8090. https://doi.org/10.1109/IROS51168.2021.9636032
18. Pang Z, Li Z, Wang N (2023) SimpleTrack: understanding and rethinking 3D multi-object tracking, in: computer vision—ECCV 2022 workshops. Springer Nature Switzerland, Cham, pp 680–696. https://doi.org/10.1007/978-3-031-25056-9_43
19. Wang J, He P, Coo W (2007) Study on the Hungarian algorithm for the maximum likelihood data association problem*. J Syst Eng Electron 18:27–32. https://doi.org/10.1016/S1004-4132(07)60045-0
20. Robosense, Automotive Grade LiDAR—RS-LiDAR-M1—RoboSense—RoboSense—LiDAR for Autonomous Driving, Robots, V2X, Accessed 5 May 2023 (2023). https://www.robosense.cn/en/rslidar/RS-LiDAR-M1
21. Weng X, Wang J, Held D, Kitani K (2020) 3D Multi-object tracking: a baseline and new evaluation metrics. http://arxiv.org/abs/1907.03961. Accessed 13 Nov 2023

Topological Optimized Offshore Floating Wind Turbine Platform Design

Nouman Saeed, Jingliang Gong, Yuejia Wan, Ayesha Saeed, Haijun Zhou, and Lixiao Li

Abstract Diverse offshore floating platform structures are constructed for multi-purpose to serve in specific environments and depths of the ocean all over the globe. The practical applications of these offshore structures demand distinct appropriate approaches platform structures, equipment filtering, and drilling/production techniques for long term stability. This study proposes a novel conceptual structure design for floating wind turbine platform to harvest offshore energy by employing the structural optimization. Topology optimization (TO) have already been practiced in numerous engineering fields to achieve the desired sustainable performance objectives such as maximizing stiffness, minimizing compliance or improving structural strength. Initially, TO suggest obtaining the novel lightweight conceptual platform structure design with minimum compliance subjected to the volume minimization. Variable density based TO method is performed for novel platform design by removing the unnecessary material from the design domain under the applied wind turbine and hydrodynamic and mooring lines forces. The obtained topological optimized floating platform structure is a lightweight design structure that also makes sure long term structure strength integrity.

Keywords Offshore energy · Offshore floating platform · Topology optimization

N. Saeed · J. Gong · Y. Wan · H. Zhou · L. Li (✉)
College of Civil and Transportation Engineering, Shenzhen University, Shenzhen, China
e-mail: lilixiao@szu.edu.cn

H. Zhou · L. Li
Shenzhen Key Laboratory of Structure Safety and Health Monitoring of Marine Infrastructures, College of Civil and Transportation Engineering, Shenzhen University, Shenzhen, China

A. Saeed
State Key Laboratory of Alternate Electrical Power System With Renewable Energy Sources, North China Electric Power University, Beijing, China

X. L. Zhao et al. (eds.), *Proceedings of the Fourth World Conference on Floating Solutions: WCFS 2024; 2–4 December; Hong Kong*, Lecture Notes in Civil Engineering 597, https://doi.org/10.1007/978-981-96-4569-5_17

1 Introduction

The current world population of 6 billion people is expected to increase 9 billion by 2050 [1]. With increases primarily conspicuous in emerging economies, which are expected to dramatically increase their energy consumption as they upgrade their infrastructures, the global energy challenge to climb by more than 50% by 2030 [2]. The Global Wind Energy Council (GWEC) projects that worldwide wind capacity would rise from 35GW in 2022 to 380GW by 2030. Many nations' net-zero aspirations rely on enormous and rapid deployment of offshore wind power [3]. Offshore wind energy is more powerful and consistent in deep sea environments; with 80% of wind rich resources located over 60 m [4]. Most offshore wind turbines are fixed; they are supported by a framework extending from the turbine tower's base to the seabed [5]. Fixed structures are impractical to construct and uneconomic or technologically unfeasible in deep seas, there is open space available for the floating offshore wind turbines (FOWTs). An offshore floating platform is a structure designed to floats at the surface of the seas or oceans. The depth of the water is subdivided into the following three categories, shallow waters (0–30 m), transitional waters (30–50 m), and deep water (50–200 m) [6]. Floating platforms can operate in a wide range of water depths. They are particularly beneficial in deep water, as they eradicate the need for extensive seabed foundation work, reducing massive construction costs, and are generally easy to install and decommission compared to fixed structures, and can be moved from one place to other place, offering greater flexibility in project execution. These platforms are versatile and serve various purposes across different industries.

Offshore floating multifunctional platforms are a reasonable choice to harvesting energy in shallow and deep waters. However, floating wind turbine platforms are still in the beginning of development, with various platform designs possible. Lightweight and structural stability are fundamental factors considered for the floating platform structure design to assure load-bearing capability. Cost optimization is essential for the design of floating platforms. The platform's cost is mostly determined by the floating infrastructure design parameters. TO is employed to attain the innovative structure design having the best material distribution by applying the loads and boundary condition on geometry design domain [7]. TO has been demonstrated to be an effective use in many engineering aerospace, automobile, wind energy, civil and mechanical fields to providing the desired structure strength, durability robustness and stiffness of the subjected infrastructure design within boundary conditions and a variety of loads [8]. Bendsøe and Kikuchi was pioneer to propose the topology optimization in 1988 [9]. Subsequently, a number of TO methods have evolved, evolutionary structural optimization, moving morphable components [10], level set [11] and the variable density-based method [12]. Yang et al. [13] developed a reliability-based optimization approach for the optimal OWT tripod substructure design to meet dynamic response. Gentils et al. [14] employed a genetic algorithm and finite element (FE) analysis to obtain an integrated optimization solution for OWT substructures. Sandal et al. [15] implemented a size optimization technique and demonstrated its application to the OWT jackets structure design. Motlagh et al. [16] introduced a genetic method to optimizing the jacket platforms based on design constraints such as associated thickness and outer diameter. Zhang et al. [17] first acknowledged the importance of topological features on the OWT jacket structure design by emphasizing the X-type's superior bearing capability. Tian et al. [18, 19] suggested topology optimization approach to construct a lightweight jacket structure that incorporates size, shape, and topology. Zhang et al. [20] presented TO approach for 5 MW lightweight jacket structure by utilizing the weight coefficient at the early design stage. Lu et al. [21] applied TO to achieve the innovative lightweight 5 MW tripod substructure with minimum compliance. Karimi et al. [22] employed multi-objective optimization method to optimize the 5 MW FOWT cost. Song et al. [23] adopted a TO method to construct a conceptual optimal design of the hybrid 10 MW FOWTs platform structure.

In this study, we proposed a innovative lightweight offshore floating platform structure design is suggested, leveraging the density-based TO methodology for the coupled production and exploitation of solar and wind energy. TO methodology has employed to carry out the innovative multifunctional offshore floating platform lightweight robustness structure design with minimum compliance and maximum stiffness under multiple loads of solar penal, mooring lines and three 10 MW DTU wind turbines. According to the harsh environment of the South China Sea, the minimal compliance subject to the specified volume fraction is established and a novel platform design is implemented by interfacing the variable density approach in the OPTISTRUCT with HYPERMESH software.

Section 2 explains the methodology that applied to get desired layout. Section 3 describes the final optimized layout using the Optistruct software respectively.

2 Problem Statement

2.1 Topology Optimization Formulation

Topology optimization (TO) is a sophisticated powerful computational design approach used in structural optimization that allows engineers to explore the innovative design concepts, minimize material use, and enhance the performance of the components and structures. The goal of TO is determine the most optimum configuration of material distribution, the innovative customized lightweight, stiffer, and more structurally efficient design along particular recommended design domain with respect to the specific boundary conditions and multiple loads. Optimization algorithm iteratively improves the design to meet the optimization objective and constraints. Once the optimization process is complete, engineers can review and analyze the optimized design to understand how material has been redistributed and to validate its structural performance. A density-based TO approach is employed to minimize the volume fraction subjected the constraints, and mathematically may express as follows:

$$\begin{aligned}
&\text{find}: \ \boldsymbol{\rho} \\
&\min: \ V_f = \frac{\sum_{m=1}^{N_m} v_m \overline{\rho}_m}{\sum_{m=1}^{N_m} v_m} \\
&\text{s.t.} \ \mathbf{K}(\overline{\rho})\mathbf{u} = \mathbf{F} \\
&\quad 0 \leq \rho_m \leq 1 \qquad (m = 1, 2, \cdots, N_m)
\end{aligned} \tag{1}$$

where $\boldsymbol{\rho}$ is a design variable vector; v_m prescribes volume solid structure of element *m*, N_m is total mesh grid finite elements, and V_f represents volume fraction of the recommended domain, respectively. $\boldsymbol{K}$ denote a global stiffness matrix, $\boldsymbol{u}$ denote global displacement vector, and $\boldsymbol{F}$ refers to the force respectively. The symbol $\overline{\rho}_m$ conveying the physical density interrelated to every element ρ_m implies the design variable of the *m*th element vary among 0–1.

2.2 SIMP Scheme

Solid isotropic material with penalization (SIMP) is highly versatile, capable with large number of design constraints, multiple loads and for large system. It operates on fixed domain of finite element to minimize the objective function to identifying the each element must be towards their binary bound 0 or 1. The well-established SIMP interpolation exercised while the Young's modulus may defined as follows [24]:

$$E(\rho_m) = \overline{\rho}_m^p E_0 \tag{2}$$

where E_0 is the solid material Young's modulus and P is the penalization factor whose suggested value is 3.

3 Numerical Analysis

The primary geometry of the multipurpose floating hexagonal platform is depicted in Fig. 1, which has a height of 33 m and a regular hexagonal length of 215 m on each side. The offshore proposed multipurpose large floating platform structure bears not just environmental loads same time it is also associated with to the transferred loads from the platform deck, such as, solar panels, turbines' weight, bending moments and shear. The free FAST program was used to simulate the three DTU 10 MW WTs loads [25, 26]. Load data is simulated 700 s with ranging from 100 to 700 s by open FAST, and severe load values at tower bottom are given in the following Table 1.

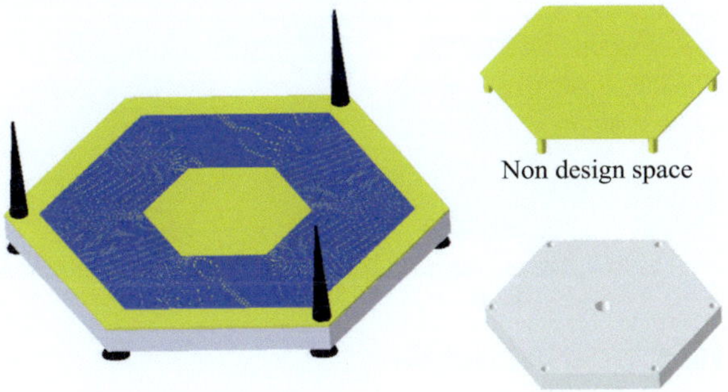

Fig. 1 Design domain, boundary and load conditions

Table 1 Tower bottom loa

x-direction load	y-direction load	z-direction load	Moment at x	Moment at y	Moment at z
4152kN	690.5 kN	– 20780 kN	– 65010 kN m	300800 kN m	– 7552 kN m

The symmetrical hexagonal platform geometry has been divided into design and non-design spaces, with non-design space not being considered for topology optimization. The upper deck plane features photovoltaic panels, while columns provide buoyancy for floating on water surfaces.

3.1 Linear Static Analysis

In Optistruct, linear static analysis is a sort of finite element analysis (FEA) that simulates the structure design behavior under applied loads, with the response assumed to be linearly connected to the loads. It identifies high stress areas, and defines constraints for optimization. It guides the process towards effective material distribution and redistributing. In this analysis, material characteristics are considered linear, and the structural behavior (high stress and potential failure area) is determined by linear equations. The equilibrium equations (Newton's second law) are solved to calculate the strains, displacements, and stresses, inside the structure under the applied loads. Optistruct computes the structural response using efficient algorithm.

The platform is meshed using a total of 4,330,459 3D finite elements, each having a size of 1.1 m. The total number of degrees of freedom is 2,442,852. After meshing the platform model, 2.51×10^5 Pa wave surface pressures is acting at outer surface of the platform. We applied Photovoltaic solar loads acting in $z = -2.4902e^{08}$N, mooring lines loads in $z = -9.61e^{07}$N and the WT loads in $x = 4.152e^{06}$ $y = 6.905e^{05}$N $z = -2.078e^{07}$N directions, at platform. The platform material density is $7.9e^{09}$kg/m^3, shear modulus is $8.07692e^{05}$, Poisson's ratio is 0.3, and the elastic

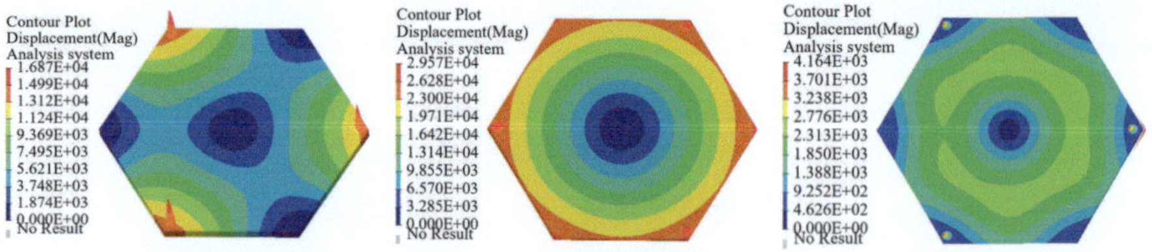

(a) Wind turbine, mooring line loads (b) PV solar, mooring line loads (c) Combined loads

Fig. 2 Linear static analysis

modulus is 2.1e^{05}Pa, respectively. In following Fig. 2 represents the linear static analysis of the proposed platform under different loads.

3.2 Optimized Layout

In the TO setting, the variable density-based TO approach is utilized to minimize the compliance and the objective function is volume fraction according to the Eq. 1. Optistruct solves topology optimization problem using the material density of each element. Figures 3 and 4 depict the TO of the platform model layout along couple load combination. The optimized layouts represent the distributed material close to the central column of platform is a force transmission path.

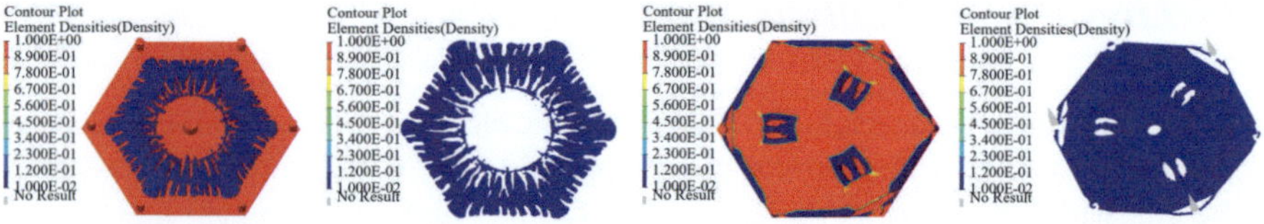

(a) Solar optimized platform (b) Solar optimized design (c) Wind turbine optimized platform (d) Wind turbine optimized design

Fig. 3 Topology optimized layout

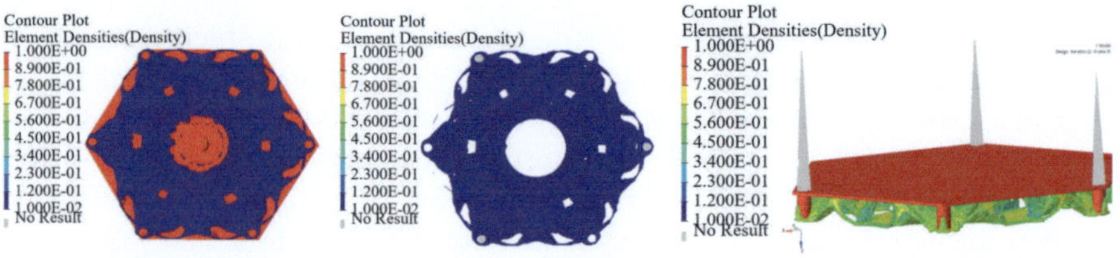

(a) Optimized platform (b) Optimized design space (c) Topology Optimized platform

Fig. 4 Final topology optimized multifunctional platform result

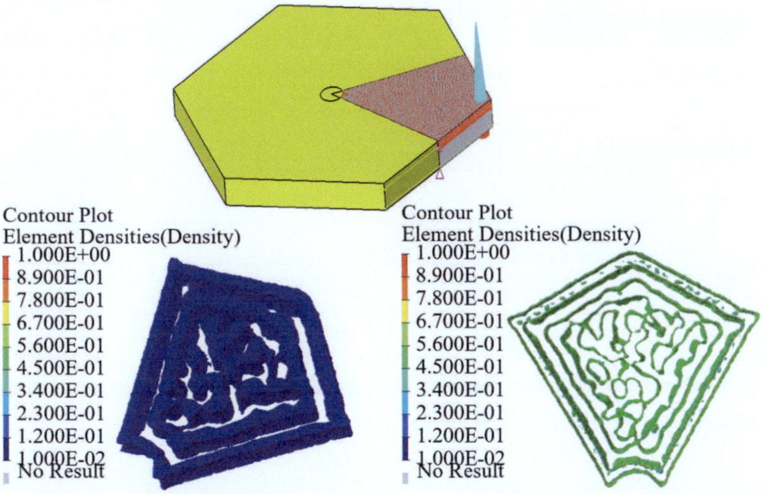

Fig. 5 Optimized result

Sometimes the investigations of TO lead to designs that are too complicated or unsmooth to be produced effectively. Optistruct suggests specific manufacturing constraints, such as maximum element size, draw direction (DVs with PSOLID), extrusion (DVs with PSOLID), pattern grouping, and pattern repetition, to obtain smooth and efficient optimized layout for fine manufacturing. At this stage, we utilized maximum element size 3 m, and Draw direction constraints to obtained fine and better optimized results. Figure 4 depicts the final topology optimized platform result.

The hexagonal offshore multipurpose floating platform size is quite big, it is not easy to manufacture even optimized. We divided the regular hexagonal platform geometry into six symmetrical parts, utilized density-based TO approach to optimized and optimized result is shown in Fig. 5.

Figure 6 depicts a schematic flowchart of the projected floating platform integrated with the TO.

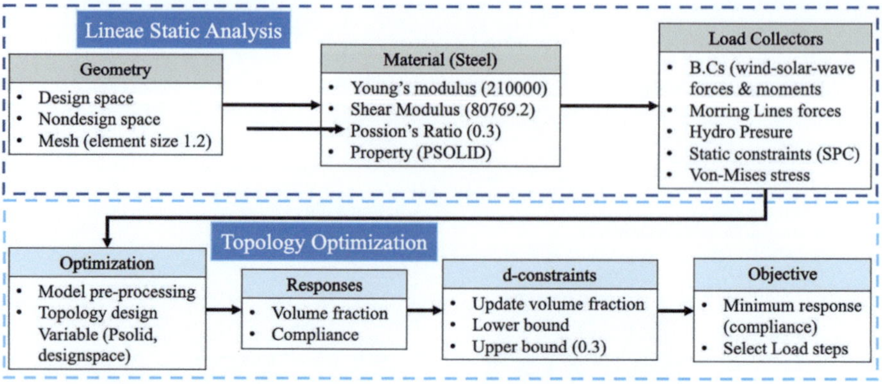

Fig. 6 Flowchart of the TO floating platform

4 Conclusion

Intending to achieve coupled energy development and utilization in the deep ocean, the primary aim of this study has been reveals a conceptual design for a multifunctional floating optimized platform structure. This proposed platform structure incorporates three DTU 10 MW WTs, and solar energy generation devices to produce power through wind and solar energy. The variable density-based topology optimization approach is applied to determine the best material distribution throughout the suggested multifunctional platform to achieve a lightweight, durable design and maximize stiffness, considering the environmental conditions of the South China Sea. A more definite force transfer path can be seen in the topological optimization layout of cells with densities between 0.6 and 1. Compared with the original given design domain, the newly obtained optimized platform structure reduced the total weight by 70% without promising the platform structure stability and improving the bearing capacity of the structure. These promising layouts encourage to the researchers to capitalize and extend TO approaches in their work to get innovative, durable structure.

Acknowledgements The support for this research has been provided by the National Natural Science Foundation of China (Grant No. 52178491), the Natural Science Foundation of Shenzhen (Grant No. JCYJ 20210324093809023) and Shenzhen Key Laboratory of Structure Safety and Health Monitoring of Marine Infrastructures (Grant No. ZDSYS20201020162400001) Shenzhen Science and Technology Program (Grant No. KQTD 2020 0820 1130 04005) are gratefully acknowledged.

References

1. Béné C, Barange M, Subasinghe R, Pinstrup-Andersen P, Merino G, Hemre GI, Williams M (2015) Feeding 9 billion by 2050–Putting fish back on the menu. Food Security 7:261–274
2. Cheng Z, Kai Z, Zhu S (2023) Does green finance regulation improve renewable energy utilization? Evidence from energy consumption efficiency. Renew Energy 208:63–75
3. Jenkins JD, Mayfield EN, Larson ED, Pacala SW, Greig C (2021) Mission net-zero America: the nation-building path to a prosperous, net-zero emissions economy. Joule 5(11):2755–2761
4. deCastro M, Costoya X, Salvador S, Carvalho D, Gómez-Gesteira M, Sanz-Larruga FJ, Gimeno L (2019) An overview of offshore wind energy resources in Europe under present and future climate. Ann N Y Acad Sci 1436(1):70–97
5. Yeter B, Garbatov Y (2022) Structural integrity assessment of fixed support structures for offshore wind turbines: a review. Ocean Eng 244:110271
6. Tayfun MA, Alkhalidi MA (2020) Distribution of sea-surface elevations in intermediate and shallow water depths. Coast Eng 157:103651
7. Saeed N, Long K, Ansari JA, Jaffri NR, Abrar U (2021) Improved efficient projection density function based on topology optimization. J Math 2021(1):1077990
8. Saeed N, Long K, Rehman A (2020) A review of structural optimization techniques for wind turbines. In: 2020 3rd international conference on computing, mathematics and engineering technologies (iCoMET). IEEE, pp 1–8
9. Bendsøe MP, Kikuchi N (1988) Generating optimal topologies in structural design using a homogenization method. Comput Methods Appl Mech Eng 71(2):197–224
10. Zhang W, Li D, Yuan J, Song J, Guo X (2017) A new three-dimensional topology optimization method based on moving morphable components (MMCs). Comput Mech 59:647–665
11. Van Dijk NP, Maute K, Langelaar M, Van Keulen F (2013) Level-set methods for structural topology optimization: a review. Struct Multidiscip Optim 48:437–472
12. Saeed N, Long K, Ansari JA, Jaffri NR, Abrar U (2021) Implementation of structural assemblies by simultaneous projection and density-based topology optimization. J Math 2021(1):8603046
13. Yang H, Zhu Y, Lu Q, Zhang J (2015) Dynamic reliability based design optimization of the tripod sub-structure of offshore wind turbines. Renew Energy 78:16–25
14. Gentile L, Greco C, Minisci E, Bartz-Beielstein T, Vasile M (2019) Structured-chromosome GA optimisation for satellite tracking. In: Proceedings of the genetic and evolutionary computation conference companion, pp 1955–1963
15. Sandal K, Verbart A, Stolpe M (2018) Conceptual jacket design by structural optimization. Wind Energy 21(12):1423–1434
16. Motlagh AA, Shabakhty N, Kaveh A (2021) Design optimization of jacket offshore platform considering fatigue damage using Genetic Algorithm. Ocean Eng 227:108869
17. Zhang P, Li J, Gan Y, Zhang J, Qi X, Le C, Ding H (2020) Bearing capacity and load transfer of brace topological in offshore wind turbine jacket structure. Ocean Eng 199:107037
18. Tian X, Wang Q, Liu G, Liu Y, Xie Y, Deng W (2019) Topology optimization design for offshore platform jacket structure. Appl Ocean Res 84:38–50
19. Tian X, Sun X, Liu G, Deng W, Wang H, Li Z, Li D (2022) Optimization design of the jacket support structure for offshore wind turbine using topology optimization method. Ocean Eng 243:110084
20. Zhang C, Long K, Zhang J, Lu F, Bai X, Jia J (2022) A topology optimization methodology for the offshore wind turbine jacket structure in the concept phase. Ocean Eng 266:112974
21. Lu F, Long K, Zhang C, Zhang J, Tao T (2023) A novel design of the offshore wind turbine tripod structure using topology optimization methodology. Ocean Eng 280:114607
22. Karimi M, Buckham B, Crawford C (2019) A fully coupled frequency domain model for floating offshore wind turbines. J Ocean Eng Mar Energy 5:135–158
23. Song J, Chen J, Wu Y, Li L (2022) Topology optimization-driven design for offshore composite wind turbine blades. J Mar Sci Eng 10(10):1487
24. Bendsøe MP, Sigmund O (1999) Material interpolation schemes in topology optimization. Arch Appl Mech 69:635–654

25. Bak C, Zahle F, Bitsche R, Kim T, Yde A, Henriksen LC, Hansen MH, Blasques JPAA, Gaunaa M, Natarajan A (2013) The DTU 10 MW reference wind turbine. In: Danish wind power research
26. Saeed N, Gong J, Wan Y, Long K, Saeed A, Mei L, Xiong C, Long W, Zhou H, Li L (2024) A novel design of multifunctional offshore floating platform structure based on topology optimization. Eng Struct 306:117782

Effect of Equivalent Design Wave Selection on the Stress Prediction of FOWT Semi-Submersible Hull

Lei Yang and Binbin Li

Abstract Equivalent design wave (EDW) and spectral analysis are two methods to predict the stress in the Semi-submersible hull of floating offshore wind turbine (FOWT). Compared with spectral analysis, EDW save the cost of finite element calculation, while its accuracy depends on the selection of global response and the method used to select EDWs. To investigate the effect of equivalent design wave selection on the stress prediction, EDW analysis with different selection of governing loads and spectral analysis are conducted for FOWT Semi-submersible (Semi-sub) hull. In this study, a typical 5 MW Semi-sub FOWT is selected and the finite element (FE) model of hull is constructed. Hydro-structure interaction analysis of the hull is conducted, with the hydrodynamic pressure recalculated at structural points, and different pressure components separately transferred from the hydrodynamic to structural model. The frequency, amplitude, wave heading of EDWs are selected according to the long-term spectral analysis results of the global response. The results indicate that in practical EDW analysis, in addition to selecting internal load and inertial load as governing loads according to the specifications, the pressure, velocity, and hydrodynamic coefficients should also be considered to enhance the accuracy of EDW predictions. Most of the EDWs are concentrated around a specific frequency and heading, which is related to structural type and geometric dimension. These findings will provide a reference for the selection of design waves and the design of geometric dimension.

Keywords Equivalent design wave · Hydro-structural interaction analysis · Long-term spectral analysis · Global structural response

L. Yang · B. Li (✉)
Shenzhen International Graduate School, Tsinghua University, Shenzhen, China
e-mail: libinbin@sz.tsinghua.edu.cn

X. L. Zhao et al. (eds.), *Proceedings of the Fourth World Conference on Floating Solutions: WCFS 2024; 2–4 December; Hong Kong*, Lecture Notes in Civil Engineering 597, https://doi.org/10.1007/978-981-96-4569-5_18

1 Introduction

Wind energy has been recognized as a key climate change mitigation technology, which has become increasingly cost-effective and available [1]. In the offshore market, 10.8 GW of new offshore wind was commissioned in 2023, bringing total global offshore wind capacity to 75.2 GW [1]. As the most prominent offshore wind technology, FOWTs (floating offshore wind turbine) are considered to have significant incremental potential, with an installed capacity of only 66.4 MW in 2022, projected to reach 16.5 GW by 2030 [2]. However, high costs are the main challenge limiting the development of FOWTs, with their costs reaching up to five times those of fixed offshore wind turbines [3]. In the cost breakdown of FOWT, the substructure and hull are predicted to account for 37.5% of capital expenditures, making it the highest proportion of all costs [4]. Therefore, cost-effective design of the hull is meaningful for cost reduction in FOWT.

A cost-effective design involves performing finite element (FE) analysis on the hull. However, most research on FOWTs has primarily concentrated on their motion response, leaving the structural response relatively unexplored. Although recent studies have begun to address the structural response characteristics of FOWT hulls [5, 6], these efforts only focused on the impacts of corrosion, structural elasticity, and material properties on the structural response. Lack of understanding about the sensitive wave parameter results in conservative structural design of the floating foundation.

There are two design approaches to conduct structural design for floating foundation, which are spectral analysis and equivalent design wave (EDW). The spectral analysis approach requires performing spectral analysis on each element individually, which is usually quite time-consuming. The EDW approach is widely used in the industry, it significantly reduces the computational cost, and allows to derive simplified formula for hydrodynamic loading [7].

However, the accuracy of the EDW approach is determined by selection of governing loads. To the best of the authors' knowledge, the structural response and governing loads of FOWT Semi-sub hull is unclear, which is difficult to obtain accurate results using EDW approach. So far, the governing loads recommended by standards for EDW analysis mostly pertain to quadrilateral-shaped Semi-submersible platforms. For triangular-shaped platforms, the governing loads are not well understood. It does inspire the investigation of governing loads for FOWT Semi-sub hull, which will provide a reference for EDW selection. This study explores the effect of governing loads selection on the EDW results. To validate the accuracy of EDW results, full spectral analysis is also conducted as for comparison purpose. The theory and methodology of linear quasi-static hydro-structural interaction analysis, short-term and long-term spectral analysis, and EDW analysis are introduced in Sect. 2. In Sect. 3, OC4-DeepCwind FOWT is selected as a case study to conduct linear quasi-static hydro-structure interaction analysis. Extreme von Mises stress of each element is predicted using von Mises stress response amplitude operator (RAO) and the wave scatter diagrams of typical coastal area of Guangdong Province using spectral analysis and EDW analysis. The comparison of extreme von Mises stress between full spectral analysis and EDW analysis are presented in Sect. 4. The governing loads for Semi-sub hull and the effect of different governing loads selection on the EDW results are also discussed. Finally, conclusions are drawn in Sect. 5.

2 Theory and Methodology

2.1 *Linear Quasi-Static Hydro-Structure Interaction Analysis*

Concerning hydro-structure issues, fully consistent non-linear hydro-structure calculations are not practically possible within the reasonable combination of CPU time and accuracy [8]. Therefore, different levels of approximation methodology are proposed to simplify the hydro-structure issue. This study adopts the linear quasi-static hydro-structure interaction analysis, which is the most general and efficient method to predict the extreme von Mises stress. Linear seakeeping analysis and boundary value problems are the basis of hydro-structure interaction analysis, which are explained in many studies [9, 10]. In order to transfer the hydrodynamic pressure in implicitly balanced model, the potential at each point in the structural model can be recalculated according to the hydrodynamic point.

$$\varphi(x_s) = \iint\limits_{S_B^H} \sigma(x_h) G(x_h; x_s) dS \tag{1}$$

where x_s denotes the structural point and x_h denotes the hydrodynamic point. G and σ are Green's function and unknown source strength, respectively. In this way, the hydrodynamic pressure at the structural model are recalculated and different pressure components are separately transferred from hydrodynamic to structural model [8]. The proposed transferring method can obtain implicit balanced model compared with the interpolation of hydrodynamic pressure. To improve the efficiency of computation, Gaussian Quadrature integration is used to integrate the pressure over the structural mesh to obtain the hydrodynamic coefficients. The motion response in the frequency domain can be obtained by

$$\left(-\omega^2\left([M] + [A]^S\right) - i\omega[B]^S + [C]^S\right)\{\xi\} = \left\{F^{DI}\right\}^S \tag{2}$$

where $\left\{F^{DI}\right\}^S$, $[B]^S$, $[C]^S$ are the incident and diffracted forces, added mass coefficients, and radiation damping coefficients, respectively. These hydrodynamic coefficients can be found by integrating the pressure over the structural mesh. The inertial force of each element and the hydrodynamic pressure of wet surface can be obtained according to motion response and acceleration. The real and imaginary components of the nodal force corresponding to each frequency and wave heading are implemented to the FE model for quasi-static analysis. Therefore, the von Mises stress RAO are obtained to conduct spectral analysis and EDW analysis.

2.2 Short- and Long-Term Spectral Analysis

A long-term spectral analysis is performed using wave scatter diagrams and von Mises stress RAOs to predict extreme stress. This prediction is then utilized for strength verification and to investigate the distribution of extreme stress. The long-term extreme stress distribution can be obtained by cumulating the results from the short-term analysis. In the short-term analysis, the variable of stress is considered to follow a Rayleigh distribution, which is stationary. The parameter of Rayleigh distribution has a relationship with the spectral moment of the structural response, which can be found by

$$m_n = \int \omega^n RAO^2(\omega, \beta_0) S_{\eta\eta}(\omega) d\omega \tag{3}$$

where $\beta_0, RAO(\omega, \beta_0), S_{\eta\eta}(\omega)$ are the selected wave direction, RAO of selected wave frequency and wave heading, given wave spectrum, respectively. The short-term extreme stress can be predicted by Rayleigh distribution with a specific probability.

In the long-term spectral analysis, the scatter diagram provides the probability of occurrence for each state $prob(SS)$. The cycles of each sea state can be calculated by the total cycles and the probability. The number of cycles exceeding X in each sea state can be written as follows:

$$n_{ex}(X) = \sum_{SS=1}^{SS=N_{SS}} n_{SS}(1 - P_{SS}(X)) \tag{4}$$

where N_{SS} and $P_{SS}(X)$ is the total number of sea states and the probability density function in specific sea state, respectively. Battjes wave counting method is used to obtain the long-term cumulative distribution of von Mises stress. The long-term extreme value of structural response can be determined by a specified probability level. According to the reference [11], the probability in calculating the long-term extreme value is taken as a probability of 50-year return period.

For each sea state in wave scatter diagrams, all wave directions are summed to have the same probability. Therefore, wave directions can be divided into several segments at regular intervals. The distribution of long-term extreme structural response in each wave direction can also be calculated. Combined with the specific probability level, the contribution of each wave direction to the long-term extreme value can be obtained. The maximum contribution of wave direction provides a reference to select the wave direction of EDW.

2.3 Equivalent Design Wave Analysis

EDW analysis is conducted according to results of long-term spectral analysis, as mentioned in Sect. 2.2. The EDW is chosen as the wave direction that makes the maximum contribution to the long-term extreme value [12]. Under the wave direction that makes the maximum contribution, the frequency contributing the most to the structural response RAO is selected to be the wave frequency of EDW. The amplitude and phase of EDW can be calculated by

$$A = \frac{X_{lt}}{|RAO(\omega, \beta)|} \tag{5}$$

$$\phi = -RAO_{\phi}(\omega, \beta) \tag{6}$$

where ω and β are the selected wave frequency and direction of EDW, respectively. EDW condition can generate the response stress equivalent to the long-term predictions values of stresses. EDW is chosen based on the long-term spectral analysis of the governing loads, which is suggested by several specifications, such as ABS and DNV specification.

3 Numerical Modeling

Among various types of Semi-sub hulls, the OC4-DeepCwind is the most extensively studied and representative [13]. Therefore, the classic OC4-DeepCwind Semi-sub hull of FOWT is selected as a case study. The Semi-sub hull is modeled using a combination of beam and shell elements: shell elements for the column shell and bulkhead, and beam elements for the stiffeners and pontoons. Three nodes are constrained, with two nodes restricted in Y and Z translations, and one node restricted in X and Z translations. The rotor nacelle assembly and tower are simplified into a single concentrated mass point, defined using a 6 × 6 mass matrix. The water ballast is modeled as a fully filled liquid tank, without accounting for the sloshing effect. The structural model of the FOWT Semi-sub hull is presented in Fig. 1a.

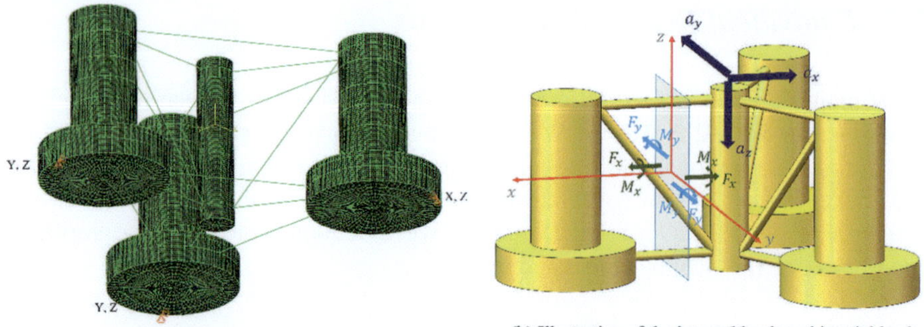

(a) Structural model of FOWT Semi-sub hull.

(b) Illustration of the internal loads and inertial loads

Fig. 1 Structural model of FOWT semi-sub hull

4 Governing Loads of FOWT Semi-Sub Hull

4.1 Comparison of the Extreme Von Mises Stress Using Spectral Analysis and EDW Analysis

In order to investigate the effect of EDW selection on the predicted wave-induced von Mises stress of FOWT Semi-sub hull, the comparison of extreme von Mises stress using different governing load selection are conducted. To validate the EDW results, the long-term predicted value based on spectral analysis is also calculated as a reference. In light of the governing loads of quadrilateral-shaped Semi-sub hull, this study uses similar internal loads such as split force, shear force, torsion moment, and inertia loads induced by topside acceleration. Figure 1b illustrates the internal loads and inertial loads of the FOWT Semi-sub hull. It is worth noting that FOWT has three internal force sections, which are equivalent. For simplicity, only one of these internal forces of section is shown in Fig. 1b. In the EDW analysis, the internal forces of all three sections will be considered. In addition to these internal loads and inertial loads, the global velocity response, hydrodynamic pressure near still water level, and hydrodynamic coefficients are also selected.

Figure 2 presents the comparison of maximum von Mises stress using different EDW selections, which is validated by full spectral analysis. The scatter plot of acceleration of topside EDW condition most closely aligns with $y = x$ line, which indicates that the topside acceleration RAO best reflects the variation trend of the RAO of stresses in most elements. Therefore, the inertial force induced by topside acceleration has the greatest impact on the von Mises stress, and should be the governing load for FOWT Semi-sub hull.

To quantitatively analyze the impact of different governing loads on the EDW analysis results, Table 1 lists the mean error, root mean square (RMS) error, and the number of elements with an error less than 5% for various EDW conditions.

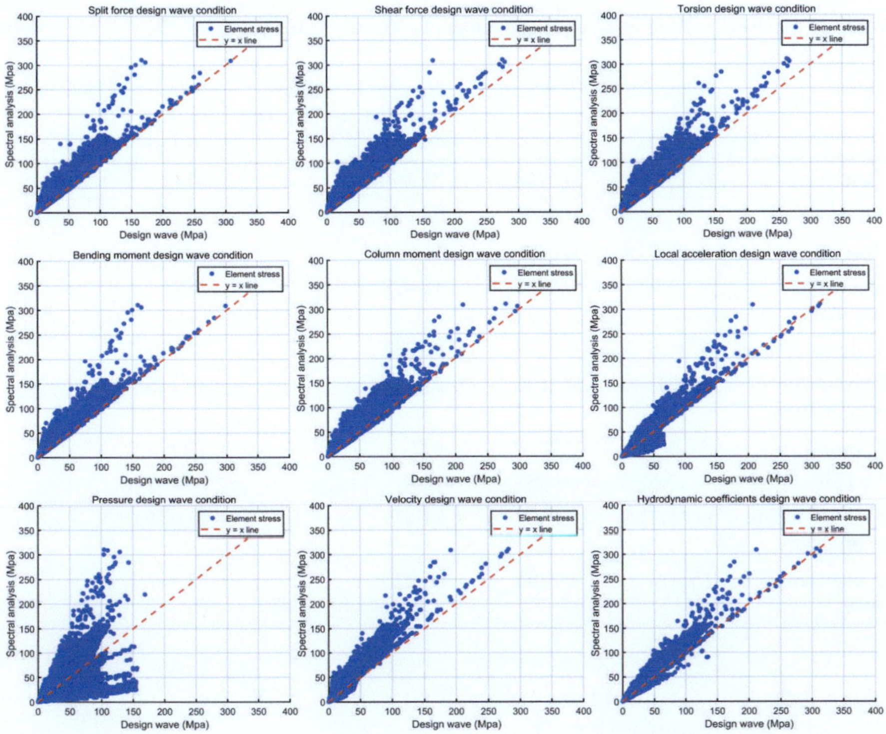

Fig. 2 The comparison of maximum stress between EDW condition and full spectral analysis

According to Table 1, the mean error of the topside acceleration is smallest, indicating that selecting these EDW conditions yields a better alignment with the spectral analysis results. In addition, the mean error of characteristic loads such as the internal loads and global motion are consistently less than zero, which means that these EDW condition underestimate the extreme von Mises stress. The RMS errors of the internal loads such as split force, shear force, torsion moment, bending moment, and column moment are the smallest, which also indicates that selecting these EDWs enables better prediction of extreme von Mises stress. Pressure EDW condition is worth noting that the predicted errors are relatively large, but most of the predicted results are higher than those from spectral analysis. Therefore, choosing these EDW conditions leans towards conservative considerations. The global velocity and hydrodynamic coefficients EDW condition show relatively good performance, with errors similar to the global structural response. Therefore, it is recommended to use these EDW conditions in the designing process.

In general, none of the governing load can accurately predict extreme von Mises stress for all elements using EDW analysis. In practical EDW analysis, achieving better predictive results often involves combining different governing loads. This

Table 1 Mean error, RMS error, and the number of elements with an error less than 5% for various EDW conditions

Items	Mean error	RMS error	Number of elements with an error less than 5%	Number of elements ($y < x$)
Split force	– 22.72%	0.0788	21,646	6242
Shear force	– 21.64%	0.0801	28,958	10,863
Torsion	– 23.97%	0.0886	16,975	4923
Bending moment	– 23.90%	0.0841	15,828	2185
Column moment	– 20.51%	0.0738	30,238	20,371
Topside acceleration	3.08%	0.2198	21,719	48,175
Pressure	88.22%	3.2810	4027	95,793
Velocity	– 19.61%	0.0958	7241	20,209
Hydrodynamic coefficients	– 15.71%	0.0402	24,476	13,834
All governing loads	15.60%	0.2395	66,220	75,497

entails selecting the maximum values from various EDW conditions to predict extreme von Mises stress (Fig. 3).

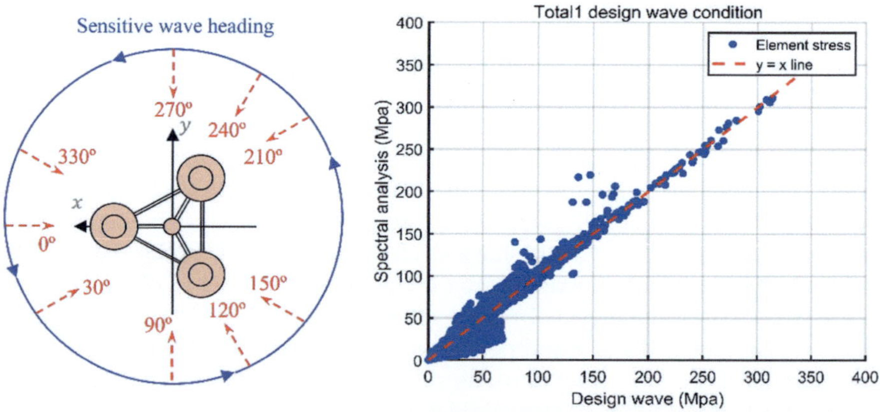

(a) Illustration of sensitive wave headings (b) Comparison of maximum stress using spectral analysis and EDW analysis considering all governing loads.

Fig. 3 Illustration of sensitive wave headings and the comparison of maximum stress using spectral analysis and EDW analysis considering all governing loads

4.2 Equivalent Design Wave Analysis Results

Table 2 presents the EDW analysis results of different governing load. The results show that the EDW amplitudes mostly range between 4 and 4.5 m, with the largest EDW amplitude being 4.827 m, obtained from the hydrodynamic coefficients EDW condition. However, the maximum EDW amplitude does not imply the maximum von Mises stress, and the maximum von Mises stress occurs under the topside acceleration EDW condition. Most of the EDW frequencies are around 0.80 rad/s. According to deep-water wave theory, this frequency corresponds to a wavelength approximately twice the height of the triangular configuration. Therefore, the sensitive frequency of von Mises stress is often related to the main dimensions of the structure. Reasonable main dimension design can be used to avoid the frequencies where wave energy is concentrated. Each governing load determines a different EDW direction, but most of them are near the three principal directions (0°, 120°, 240°), corresponding to the directions along the three pontoons (Fig. 3).

Table 2 EDW analysis results of different governing load

Governing loads	EDW direction(°)	EDW frequency(rad/s)	EDW amplitude (m)	Maximum von Mises stress (Mpa)
Split force	0	0.800	4.277	213.20
	330	0.800	4.274	259.35
	30	0.800	4.275	308.33
Shear force	230	0.825	4.295	248.95
	110	0.80	4.306	278.76
	250	0.80	4.305	275.15
Torsion	130	0.825	4.283	264.69
	110	0.800	4.115	266.43
	250	0.800	4.105	262.35
Bending moment	0	0.800	4.084	203.58
	150	0.775	4.302	298.46
	210	0.775	4.296	280.37
Column moment	110	0.800	4.591	297.24
Topside acceleration	270	0.725	4.734	313.96
Pressure	80	0.375	3.492	156.18
Velocity	270	0.675	4.816	281.69
Hydrodynamic coefficients	80	0.700	4.827	298.88

Considering EDW analysis for all governing loads, the results are presented in Table 1 and Fig. 2b. It can be observed that a better predictive performance is achieved by considering all governing loads.

5 Conclusion

This study conducts spectral analysis and equivalent EDW analysis to predict the extreme von Mises stress of FOWT Semi-sub hull. The extreme von Mises stress obtained from the spectral analysis are compared with that from EDW analysis, which selected by different governing loads. The local wave-induced stress are obtained by adopting the linear quasi-static hydro-structure interaction analysis. A novel load transferring method is adopted, which involves in the recalculation of hydrodynamic pressure at the structural model and separately transferring of different pressure components from hydrodynamic to structural model. The frequency, amplitude, wave heading of EDWs are selected according to the long-term spectral analysis results of the global response. The main conclusion are drawn as follow:

- Internal load and inertial load are the governing load for the FOWT Semi-sub hull, including the split force, shear force, torsion moment, and topside acceleration and so forth.
- In addition to internal load and inertial load, the pressure, velocity, and hydrodynamic coefficients should also be selected as governing loads in the practical EDW analysis. Velocity and hydrodynamic coefficients EDW condition achieve good performance in predicting the maximum extreme stress. Pressure EDW condition often overestimates extreme stress values, which leans towards conservative considerations.
- Most of the EDW frequencies are concentrated around a specific frequency, corresponding to a wavelength approximately twice the height of the structural triangular configuration. EDW headings are almost near the three principal directions, corresponding to the directions along the three pontoons.

However, this study only considers regular EDW selection methods during the EDW selection process. More advanced EDW selection methods, such as irregular EDW selection, remain to be considered.

References

1. Jonathan C, Ben B, Girish T, Mohamed JAR (2023) Global Wind Report 2024. Global Wind Energy Council: Brussels, Belgium
2. Rebecca W, Alvaro M (2023) Global offshore wind report 2023. Global Wind Energy Council: Brussels, Belgium 19:10–12
3. Remi E(2023) Energy transition outlook 2023. Høvik, Norway, Det Norske Veritas

4. Stehly T, Duffy P (2022) Cost of wind energy review. National Renewable Energy Lab. (NREL), Golden, CO United States
5. Gao S, Zhang L, Shi W, Wang W, Wang B, Li X (2024) Dynamic analysis of a 10 MW floating offshore wind turbine considering the tower and platform flexibility. J Ocean Univ China 23(2):358–370
6. Wang S, Moan T, Gao Z (2023) Methodology for global structural load effect analysis of the semi-submersible hull of floating wind turbines under still water, wind, and wave loads. Marine Struct 91
7. De Hauteclocque G, Derbanne Q, El-Gharbaoui A (2012) Comparison of different equivalent design waves with spectral analysis. Int Conf Offshore Mech Arct Eng
8. Malenica Š, Derbanne Q, Sireta FX, Bigot F, Tiphine E, De Hauteclocque G, Chen XB (2013) HOMER–integrated hydro-structure interactions tool for naval and offshore applications. Int Conf Comput Appl Shipbuild
9. Chen XB (2004) Hydrodynamics in offshore and naval applications-Part I. In: Keynote lecture of 6th international conference on hydrodynamics, Perth Australia
10. Lee CH, Newman JN (2005) Computation of wave effects using the panel method. WIT Trans State-of-Art Sci Eng 18
11. Igba J, Alemzadeh K, Durugbo C, Henningsen K (2014) Through-life engineering services: a wind turbine perspective. Procedia CIRP 22:213–218
12. Liu Y, Wu J, Zhu W, Li P (2018) Research on equivalent design waves of container vessel based on stress spectrum analysis. In: The 28th international ocean and polar engineering conference
13. Gueydon S (2019) Effects of variations on the experimental set-up on the motion response of a floating wind semisubmersible (OC4 Type). In: International conference on offshore mechanics and arctic engineering, vol 59353. American Society of Mechanical Engineers

Hydroelastic Analysis of a Novel Offshore Fish Cage Design

Xinyu Zhang, Hong Zhang, Chien Ming Wang, Yunil Chu, and Mingyuan Ma

Abstract A flexible, modular, submersible fish cage is an attractive design concept for offshore fish farming due to its cost-effectiveness in construction and survivability during extreme weather events when submerged at an appropriate water depth. However, under some intensive circumstances, the structure would experience extreme weather conditions at the water surface, which leads to potential failure. This paper investigates the hydroelastic behaviour of a simplified design of a newly proposed modular fish cage, named the SeaFisher, under various wave conditions, including waves in severe weather conditions at sea level. The software AquaSim is employed to model the cage frame, nets, and mooring line system and perform the hydroelastic analysis to assess the strength, stiffness, and stability of the structure in floating conditions. The study highlights the need for an optimal design to prevent large deformations and potential failure of the fish cage. As wave height and period increase, the axial compressive forces within the structural members also rise.

Keywords Offshore fish farming · Flexible frame · Hydroelastic behaviour analysis

1 Introduction

Moving fish farms offshore offers several advantages: access to larger sea space for large-scale farming, better water quality with lower temperatures, fewer parasites and pathogens, better dispersal of fish wastes and less interaction with other human activities than nearshore fish farming. However, the most significant challenge faced

X. Zhang (✉) · H. Zhang · M. Ma
School of Engineering and Built Environment, Griffith University, Gold Coast, QLD, Australia
e-mail: xinyu.zhang6@griffithuni.edu.au

C. M. Wang · Y. Chu
School of Civil Engineering, The University of Queensland, St Lucia, QLD, Australia

Blue Economy Cooperative Research Centre, Launceston, TAS, Australia

X. L. Zhao et al. (eds.), *Proceedings of the Fourth World Conference on Floating Solutions: WCFS 2024; 2–4 December; Hong Kong*, Lecture Notes in Civil Engineering 597, https://doi.org/10.1007/978-981-96-4569-5_19

by offshore fish farming is maintaining structural integrity and fish well-being under rough environmental conditions, including strong waves and faster currents [1].

Existing offshore fish cages such as Shenlan 1 and Havfarm 1 have huge steel frames to resist the high environmental loads on the structure. However, such designs come with a considerable construction cost of over 100 million US dollars [2]. A flexible, smaller, submersible fish cage design reduces construction costs as intense surface wave loads can be avoided.

Arising from a recent Blue Economy CRC project in Australia, the team developed a novel offshore fish cage called the SeaFisher [3]. This submersible modular cage is constructed using bundled high-density polyethylene (HDPE) pipes. Compared with the aforementioned offshore steel cages, the high-density polyethylene (HDPE) frame has a lower construction cost and is easier to transport and construct. Although HDPE is durable in seawater environments and an excellent anti-biofouling material, it has some drawbacks, such as its lightness and flexibility. Due to its lower stiffness, the flexible HDPE structure is more likely to fail under severe environmental conditions. Under some circumstances, the fish pen structure is possibly exposed to severe environmental load conditions at the water surface. When the wave frequency is closer to the structure's natural vibration frequency, resonance may cause the structure to deform excessively, leading to structural failure.

The present study evaluates the new HDPE structure fish cage design by examining the hydroelastic behaviour of a simplified SeaFisher design under different wave heights and wave periods while the structure is under floating conditions.

2 Hydroelastic Analysis

In this study, a simplified design of the SeaFisher is used. A single equivalent HDPE pipe was used for the structural member. The equivalent pipe has the same material properties and stiffness. The front bow structure was also modified by removing the shield mat to decrease the computational time and resources required. Since this study aims to evaluate the relationship between hydroelastic deformation and various environmental load conditions, the simplification has minor effects on the hydroelastic behaviour of the structure. Figure 1 presents the plan and side views of the simplified fish cage model. HDPE material was used for the main structure frames of the model, including top, bottom, vertical and bow frames. HDPE material has a density of 958 kg/m^3, Young's modulus of 1 GPa, and Poisson's ratio of 0.4. A multiple-point mooring system is adopted to capture better the interaction between the mooring line and the flow. The detailed design parameters of the model are given in Table 1 and Table 2. In this study, the model floats at a water depth of 70 m, with the free water surface as the reference for the Z coordinate. The Z value indicates the depth of each component from its centre to the free water surface. Submerged states of the fish cage at various depths were not considered.

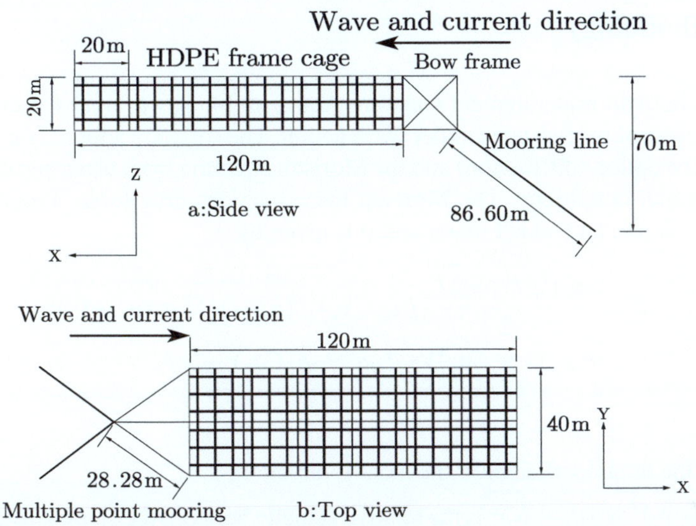

Fig. 1 Side view (**a**) and Top view (**b**) of the fish cage model

Table 1 Model fame structure

	Outer diameter (m)	Thickness (m)	Length (m)	Ballast	Element type
Top longitudinal frame	0.956	0.095	120	Yes	Beam
Top Transverse frame	0.956	0.120	20 per cubic cage	No	Beam
Vertical Frame	0.956	0.095	20 per cubic cage	No	Beam
Bottom Frame (including longitudinal and transverse members)	0.956	0.095	20 per cubic cage	No	Beam
Bow Frame	0.600	0.085	10–22 m	No	Beam

Table 2 Model mooring line and net structures

	Thickness (m)	Thread diameter (m)	Young's modulus (GPa)	Density (kg/m^3)	Poisson ratio	Element type	Area (m^2)
Net	0.002	0.002	1.0	1025	–	Membrane	–
Hawser	–	–	2.7	788	–	Truss	0.022

3 Methodology

The time-domain finite element software AquaSim [4, 5] was used for the hydroelastic analysis of the fish cage under wave action. The software adopted the hydrodynamic force option (diffraction) and the Morison equation from wave and current for environmental conditions. The Morison force includes drag force, Froude-Krylov diffraction force, and added mass, and it is given by

$$F = \frac{\rho_w C_{dy} Diam_N L_0}{2}\left(u_y - \dot{v}_y\right)\sqrt{\left(u_y - \dot{v}_y\right)^2 + \left(u_z - \dot{v}_z\right)^2} + \rho_w\left(1 + Ca_y\right)V_{2D}L_0 a_y - \rho_w Ca_y V_{2D} L_0 \ddot{v}_y \quad (1)$$

where

- C_{dy} is the drag coefficient in the local y-direction;
- $\sqrt{\left(u_y - \dot{v}_y\right)^2 + \left(u_z - \dot{v}_z\right)^2}$ is the relative velocity between the element and the fluid in the cross-section plane;
- $u_y = u_{ywave} + u_{ycurrent}$, where u_{ywave} is the fluid velocity due to wave, $u_{ycurrent}$ is the current velocity in the local y-direction;
- $\dot{v}_y$ is the velocity of the element in the local y-direction;
- a_y is the fluid acceleration in the local y-direction;
- C_{ay} is the added mass coefficient in the local y-direction;
- $Diam_N$ is the cross-section diameter in the direction of the relative velocity;
- ρ_w is the seawater density;
- V_{2D} is the 2D volume.

We have selected the time-domain finite element software AquaSim because the small diameters of the structural and mooring elements and the smooth HDPE frame have less effect on the surrounding flow field. In contrast, the analytical method makes it hard to capture the elastic behaviour of the fish pen. AquaSim provides suitable elements for modelling various structural components: membrane elements, beam elements, truss elements, and springs. Beam elements were used for the HDPE pipes forming the main structural frame and bow structure to account for bending moments and shear forces. Truss elements were used for the hawser and anchor lines, excluding compressive forces. The Kikkonet was modelled with a membrane-type element, automatically generating the net and allowing adjustments to the net's solidity by changing the thread diameter and mesh width. Biofouling effects were excluded. The drag coefficients for the hawser, net, and frame were assumed to be 1.2 per the default setup [4, 5].

Table 3 Load cases used to identify the critical element

Load case	Wave height (m)	Wave frequency (Hz)
1	3	0.125
2	4.5	0.100
3	6.5	0.083
4	8	0.071
5	8.5	0.063

Table 4 Load cases used for divided elements

Wave height (m)	Wave frequency (Hz)				
	$J_1 = 0.125$	$J_2 = 0.100$	$J_3 = 0.083$	$J_4 = 0.071$	$J_5 = 0.063$
$C_1 = 2$	(C_1, J_1)	(C_1, J_2)	(C_1, J_3)	(C_1, J_4)	(C_1, J_5)
$C_2 = 3$	(C_2, J_1)	(C_2, J_2)	(C_2, J_3)	(C_2, J_4)	(C_2, J_5)
$C_3 = 4$	(C_3, J_1)	(C_3, J_2)	(C_3, J_3)	(C_3, J_4)	(C_3, J_5)
$C_4 = 5$	(C_4, J_1)	(C_4, J_2)	(C_4, J_3)	(C_4, J_4)	(C_4, J_5)
$C_5 = 6$	(C_5, J_1)	(C_5, J_2)	(C_5, J_3)	(C_5, J_4)	(C_5, J_5)

4 Environmental Conditions

Several offshore environmental load cases were considered to capture the maximum compressive axial force distribution across the cage structure with the view of identifying the most critical member that is most likely to fail. Table 3 shows the load cases used to determine the critical element. The wave profile was generated according to de Hauteclocque et al. [6]. Based on the identified critical element, the relationship between the compressive axial force in each segment under various wave heights and wave periods was observed, and the combinations of wave height and period used are listed in Table 4. For all the environmental load cases considered, a 0.5 m/s current speed was assumed. The current and wave are assumed to be heading in the same direction toward the model's front end. The water depth is taken as 70 m.

5 Results and Discussion

5.1 Identifying the Critical Elements

During the simulated time period, the computed maximum axial compressive force distribution of the cage structure is presented in Fig. 2. The axial compressive force distribution scale is consistent across all load cases, with red indicating the highest compressive axial force and blue indicating the lowest. In AquaSim, the maximum axial compression force is displayed as a positive value. The distribution

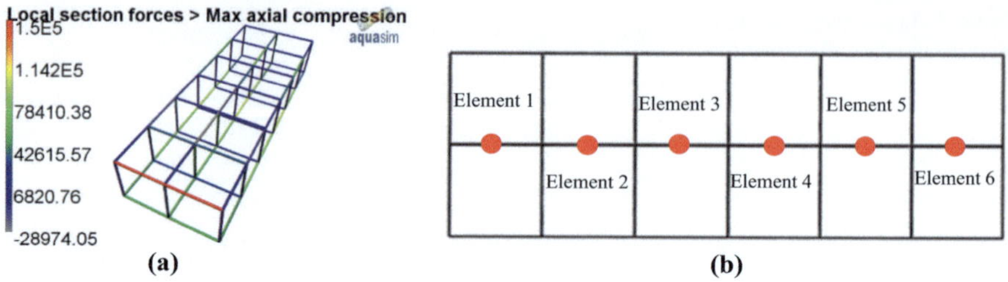

Fig. 2 Maximum axial compressive force distribution for Load Case 1 (**a**), divided element number and observation points marked as red dots (top view of the structure) (b)

shows that the bottom structural members experience higher compressive forces than other components, with the bottom-middle longitudinal element having the highest compressive force. The maximum axial compressive force of the entire structure increases under more severe environmental conditions, making the bottom-middle longitudinal element the critical component with the highest risk of failure.

The bottom-middle longitudinal element was divided into six segments for further analysis. The maximum axial compressive force observation point along each divided element was numbered to obtain a detailed axial compressive force distribution along the whole element, as shown in Fig. 2.

Figure 3 shows the maximum axial compressive force F divided by the cross-sectional area of the HDPE pipe A and elastic modulus E at the observation points as the axial compressive strain for the five extreme environmental conditions listed in Table 3. The Elastic modulus used for the HDPE Pipe is 1GPa, and the diameter of the HDPE pipe can be found in Table 1.

In each case, the middle part of the cage shows higher axial compressive strain distribution, while the front and back have lower axial compressive strain, which presents the elastic behaviour of the HDPE structure. Case 1 exhibits the lowest overall axial compressive strain distribution, whereas Case 4 has the highest. The axial compressive strain distributions for Load Case 4 are the highest. The middle elements, specifically elements 3 or 4, have the highest axial compressive strain.

Fig. 3 Axial compressive strain (F/EA) along the bottom middle longitudinal pipe

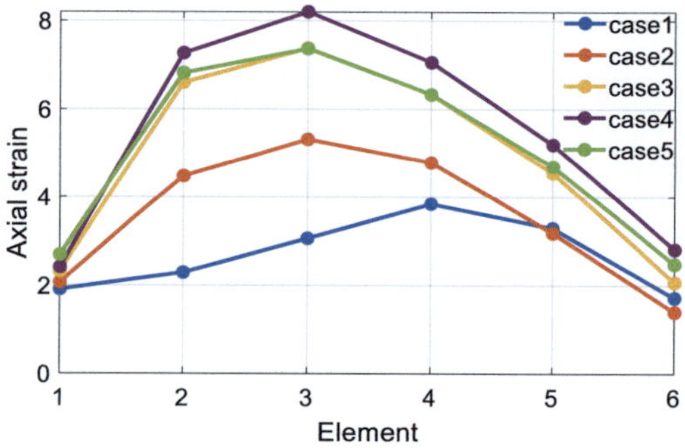

However, the axial compressive strain distribution pattern across elements varies with different environmental conditions, indicating a need for further study.

The maximum axial compressive strain distribution of each divided segment along the bottom middle element has been plotted using selected environmental condition combinations (Table 4) to investigate the relationship between the axial compressive strain distribution of each segment with wave height and wave frequency, with either the same wave height at different wave frequencies or the same wave frequency at different wave heights—Figs. 4 and 5 display the results for elements 2, 3, 4, and 5. Element 2 is at the front of the cage, element 3, which has the highest axial compressive force, is in the middle, and element 5 is toward the end of the cage. Figure 4 illustrates the relationship between F/EA and wavelength normalised by structure length. Figure 5 shows the relationship between F/EA and wave height, demonstrating the relationship between HDPE pipe axial compressive strain and various wave heights and frequencies.

Figure 4 illustrates the divided segments under the same frequency and current speed, and Fig. 5 demonstrates that for each divided segment under the same wave height and current speed. The maximum axial compressive strain in each element increases with higher wave heights, which indicates that the structure is more likely to fail. Under consistent wave height and current speed, the axial compressive strain distribution increases with higher frequencies. However, elements 2 and 3 at the front show a different pattern under high-frequency conditions. Moreover, when

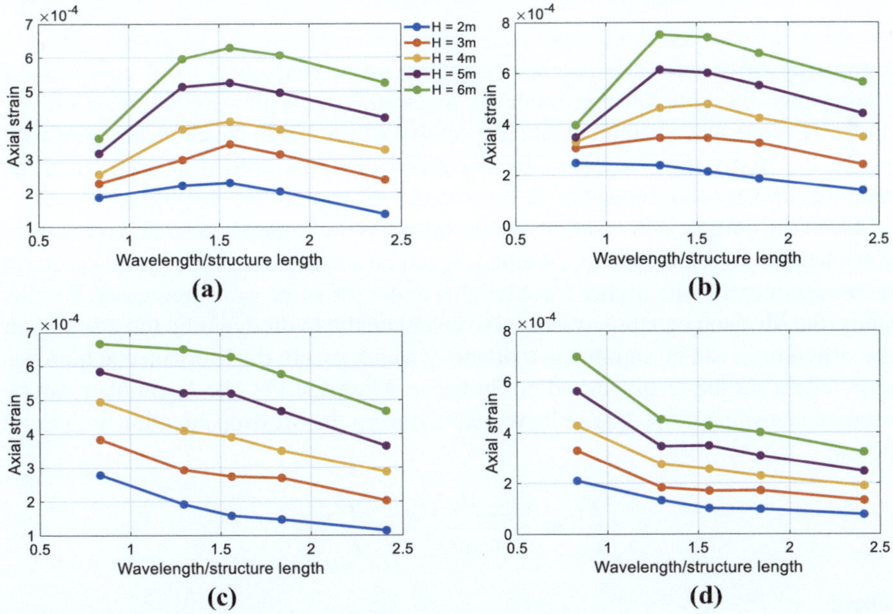

Fig. 4 Relationship between axial compressive strain and wavelength normalised by structure length for **a** element 2, **b** element 3, **c** element 4 and **d** element 5

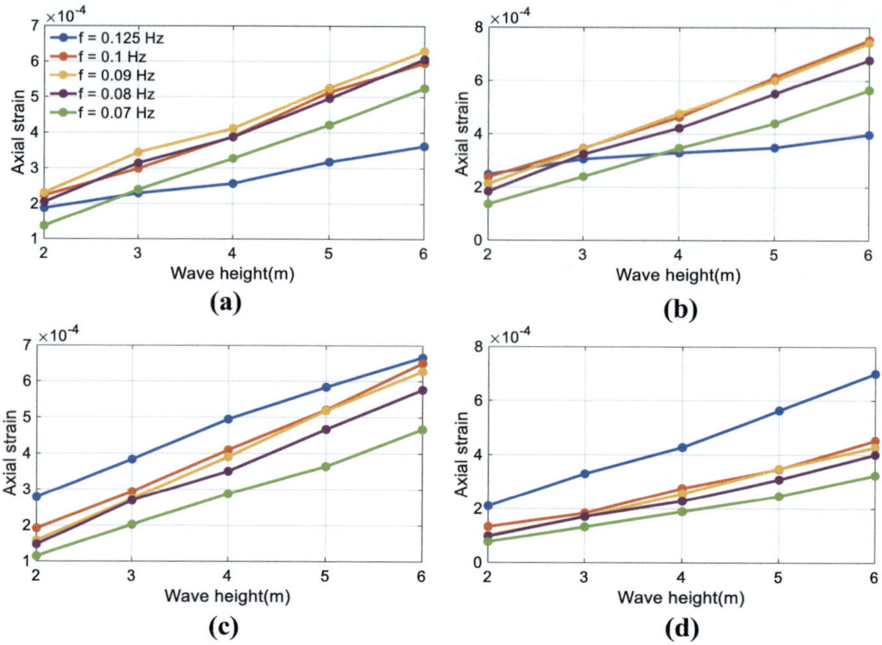

Fig. 5 Relationship between axial compressive strain and wave height for **a** element 2, **b** element 3, **c** element 4 and **d** element 5

comparing different wave heights under the same frequency, axial compressive strain generally shows an approximately linear increase with wave height. However, under the same wave height, axial compressive strain within the element varies with increasing frequency. Therefore, further investigation is needed to understand the relationship between frequency and elastic deformation of the flexible structure.

From the comparison result above between axial compressive strain and various wave heights for elements 2, 3, 4 and 5, it can be seen that axial compressive strain increases linearly with higher wave heights under the same wave frequency. For this study, the Morison equation was used to calculate the hydrodynamic force acting on the structure; it can be significant to identify which part of the Morison equation has more effect on the result. Based on Sumer and Fredsøe [7], the K_c number can be used to determine if the drag or inertia parts govern the Morison equation with Eq. 2 below.

$$\frac{F_I}{F_d} = \frac{C_I \pi d^2 \omega U_m}{2C_d d U_m^2} = \frac{\pi^2}{K_c} \frac{C_I}{C_d} \tag{2}$$

where

- F_I is the inertia force part of the Morison force;
- F_d is the drag force part of the Morison force;

- C_I is the inertia coefficient, which is equal to $1 + C_a$;
- C_a is the added mass coefficient;
- C_d is the drag coefficient from the Morison equation;
- d is the characteristic length;
- U_m is the maximum flow velocity calculated using linear wave theory;
- K_c is the Keulegan-Carpenter number.

Where K_c number is calculated by using the following Eq. 3:

$$K_c = \frac{U_m T}{D} \tag{3}$$

where U_m is the maximum flow velocity, T is the wave period; D is the characteristic length.

Figure 6 shows the resulting inertia and drag force ratio for the HDPE pipe, net, and mooring line under different wave conditions.

The above Fig. 6 shows that the inertia force of the Morison equation governs the main structure and front barrier. Where for the net and mooring of the fish pen, the drag force governs the force on the component. Under the same wave height,

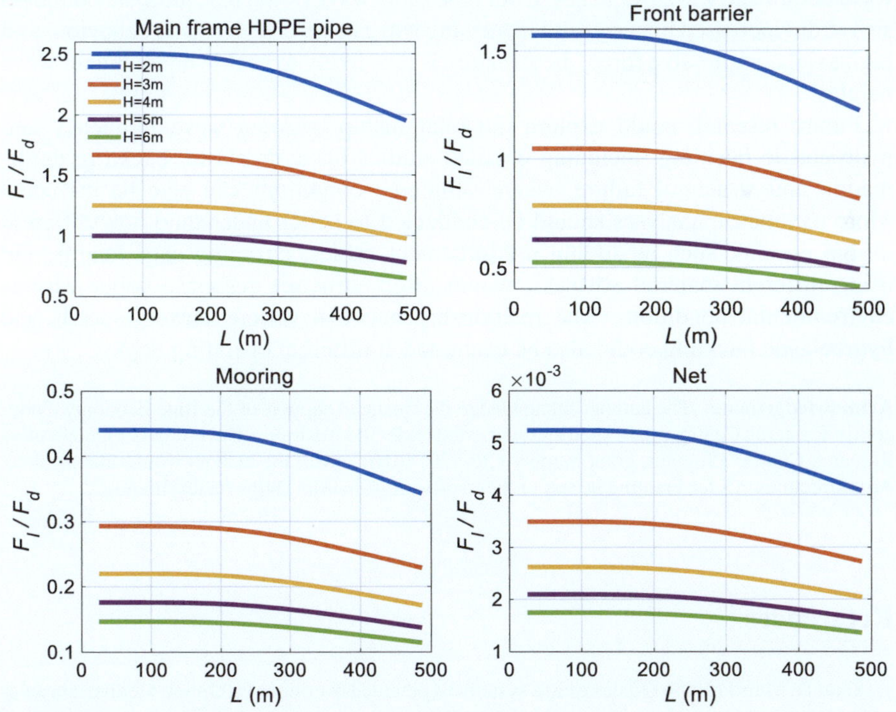

Fig. 6 Inertia force and drag force ratio with different wavelength (L) and wave height for each component

the inertia and drag force ratio have a non-linear descending trend, which clearly shows that with wavelength longer than the structure length, the Morison equation is more governed by drag force, while for short waves less than the structure length the inertia and drag ratio does not vary clearly with increasing wavelength. Also, for the main frame HDPE Pipe, under higher wave heights such as 5 m and 6 m, the ratio between inertia and drag force is lower than one, which could indicate that the drag force part is governing the Morison equation, whereas under lower wave heights such as 4, 3 and 2 m, the inertia force part of the Morison equation will be dominant the equation.

6 Conclusion

This study examined the hydroelastic behaviour of a simplified HDPE offshore fish cage design at sea level under various extreme weather conditions using the software AquaSim, aiming to investigate the structural internal forces caused by waves and changes in wave conditions when the structure is floating on the water surface. It is found that the axial compressive strain in structural members increases with higher wave height and wave frequency. Under the same wave frequency, the axial compressive strain increased linearly with increasing wave height. With elastic behaviour and high axial compressive force, the structure is also likely to buckle under higher wave heights.

Future research could explore the relationship between wave frequency and hydroelastic buckling, including detailed studies on hydroelastic buckling deformation and structural failure, where submerge conditions can also be included. More parametric analyses should be conducted to better understand how different design options, such as adding reinforcement cables, changing cage length, and using different material stiffness, would affect structure response under various environmental conditions. The relationship between various current speeds and hydroelastic buckling could also be examined at different operating depths.

Acknowledgements The authors acknowledge the financial support of the Blue Economy Cooperative Research Centre, established and supported under the Australian Government's Cooperative Research Centres Program, grant number CRC-20180101. Also, the authors would like to thank Aquastructures AS for granting access to their software AquaSim (https://aquasim.no).

References

1. Scott D, Muir J (2000) Offshore cage systems: a practical overview. In: Option Mediterranean's-International Centre for advanced Mediterranean agronomic studies, pp 79–89
2. Wang CM, Chu YI, Baumeister J, Zhang H, Jeng DS, Abdussamie N (2022) Offshore fish farming: challenges and developments in fish pen designs. Glob Blue Econ Anal Dev Challenges 87–128

3. Wang C-M et al (2023) Seafisher—a submersible high-density polyethylene offshore fish pen. J Mar Sci Eng 11(9):1795
4. Berstad A (2013) The AquaSim package theory user manual
5. Aquastructures AS (2022) AquaEdit user manual
6. de Hauteclocque G, Maretic NV, Derbanne Q (2023) Hindcast based global wave statistics. Appl Ocean Res 130:103438
7. Sumer BM, Fredsøe J (1997) Hydrodynamics around cylindrical structures. World Scientific, Singapore

Environmental Enhancement

Studying Water Quality Impacts of Large-Scale Floating Structures: Insights from Floating Projects

Rui L. P. de Lima, Rutger E. De Graaf, and Floris C. Boogaard

Abstract Floating projects have emerged in the last two decades as a climate resilient solution to address land scarcity in urban areas, offering opportunities for energy, food production, and living environments. However, these projects may significantly influence various aspects of water bodies, with potential positive or negative effects that are not yet completely understood. In the Netherlands, water quality research on smaller-scale floating projects has been conducted and well documented in recent years. This study provides an overview of various studies that were done in the past decade in the Netherlands, where critical water quality parameters beneath the floating projects were compared with reference conditions in open water for multiple floating housing and floating solar projects. Preliminary findings suggest minimal differences in water quality parameters, with slight decreases in dissolved oxygen concentrations and temperatures. Biofouling layers and ecosystems develop rapidly after construction, indicating potential for ecosystem restoration and biodiversity enhancement. These results may not be universally applicable to other water system types or floating platform sizes and shapes. Innovative monitoring strategies and tools were also implemented (e.g. underwater drones). Furthermore, the article describes the limitations of current studies for the scaling-up of floating solutions and describes how a new research initiative from the Netherlands "Floating Future'

R. L. P. de Lima (✉)
Utrecht University, Utrecht, The Netherlands
e-mail: rui.plima@indymo.nl

Indymo, Delft, The Netherlands

MARE—Marine and Environmental Sciences Centre, Coimbra, Portugal

R. E. De Graaf
Blue21, Delft, The Netherlands

Blue Revolution Foundation, Delft, The Netherlands

R. L. P. de Lima · F. C. Boogaard
Hanze University of Applied Sciences, Groningen, The Netherlands

F. C. Boogaard
Deltares, Delft, The Netherlands

X. L. Zhao et al. (eds.), *Proceedings of the Fourth World Conference on Floating Solutions: WCFS 2024; 2–4 December; Hong Kong*, Lecture Notes in Civil Engineering 597, https://doi.org/10.1007/978-981-96-4569-5_20

will address this knowledge gap. This research project aims to further investigate this topic, optimizing the scale of large-scale floating infrastructure in alignment with receiving system characteristics. Additionally, methods for optimizing below- and above-water designs will be developed to maximize ecological benefits and meet engineering demands.

Keywords Water quality impacts · Large-scale floating photovoltaic plants · Underwater drones

1 Introduction

1.1 Floating Projects and Solar Parks

Floating projects have emerged in the last two decades as a climate resilient solution to address land scarcity in urban areas, offering opportunities for energy and food production, as well as for sustainable living environments such as housing. Floating solutions are considered as robust adaptive solutions that can withstand climate related challenges such as water level rise, or flooding conditions. The size of new projects keeps increasing, with solar parks already built that cover over hundreds of hectares of the water surface of lakes, or full neighborhoods with floating buildings.

With the increase of scale of these projects larger areas of water are covered and blocked from sunlight penetration. This leads to various aquatic processes and biological, physical and chemical or ecological characteristics of water bodies being increasingly affected by the presence of the floating structures, generating potential positive or negative effects that are not yet completely understood.

Studies that evaluate potential impacts are scarce, and there are currently no established guidelines for mitigation actions or for the design and construction of floating developments. The current knowledge on the effects of floating structures is often based on literature, assumptions, and unvalidated models rather than in-situ data, leaving many questions from water authorities unanswered. This article provides an overview of various studies that were done in the past decade in the Netherlands. Furthermore, the article describes the limitations of current studies for the scaling-up of floating solutions and describes how a new research initiative from the Netherlands, "Floating Future' research project will address this knowledge gap.

2 Methodology

2.1 *Review of Monitoring Studies in the Netherlands*

In the Netherlands, water quality research on smaller-scale floating projects has been conducted and well-documented in recent years. To gain further insights into large-scale implementation of floating structures, the numerous large floating photovoltaic plants (FPVs) also represent compelling and relevant case studies. This study aims to provide an overview of recent advancements in water quality monitoring across multiple existing floating projects in the Netherlands.

Most of the studies highlighted in Sect. 3 are based on the comparison of critical water quality parameters beneath floating structures with reference conditions in open water, where various monitoring strategies and tools were employed, including dynamic monitoring using underwater drones equipped with sensors and static monitoring (designed to measure water quality and collect images of underwater ecosystems), or by installing permanent in-situ sensors and data loggers installed under the floating structures (measuring continuously at short intervals).

3 Results and Discussion

3.1 *Review of Past Studies*

Past and ongoing research are investigating the effects of floating urbanization on water quality and ecology. These studies are looking into the effects not only underneath floating solar panels, but also for floating houses, or pavilions. Water quality data collected under/near floating structures is frequently compared with the corresponding parameters at open water locations. Table 1 summarizes the main findings of the studies conducted by the authors in this topic since 2014.

3.2 *Discussion of Results and Findings*

Most listed studies (Table 1) report small differences in the measured water quality parameters below the floating projects. Lower dissolved oxygen concentrations and slightly lower temperature were measured underneath the floating structures, with less fluctuations. Underwater images collected by underwater drones show that biofouling layer and filter feeders rapidly develop into lively ecosystems that can emerge shortly after the construction at most locations. Findings also suggest minimal differences in water quality parameters in studies aimed at floating solar parks, with slight decreases in dissolved oxygen concentrations and temperatures. Biofouling

Table 1 Review of multiple studies aimed at monitoring of water quality and ecology underneath existing floating structures in the Netherlands

Reference Year	Type of project water system	Approach and main findings/highlights
[1]	Multiple floating houses, pavilions of different sizes and configurations. Various types of water system	Underwater drone images reveal diverse underwater life, including the presence of fish. Mussel accretion is seen as positive, as habitat provision for mussels and refuge for fish can aid in water purification. A decrease in oxygen levels is concerning, especially in shallow, still waters with silt layers, which may be less tolerant to additional shade. In any case, the collected data on dissolved oxygen did not go below the value defined in standards at any of the locations studied. Potential effects depend on the characteristics of the water bodies: in highly stressed water, further research or refraining from floating construction is advised. Conversely, in good quality water, sufficient depth (> 2 m) and flow a quantity of floating objects up to 18%, floating objects seem to have minimal negative impact or may even be beneficial
[2]	Floating houses, floating Pavilion Rotterdam	Underwater images collected by the underwater drone show that floating houses create habitats for fish, mussels, and zooplankton. Roach and perch were spotted sheltering at different locations, and using the floating structures for cover. Mussels, filamentous algae, and zooplankton thrive under Rotterdam's floating pavilion
[3]	Multiple floating houses, pavilions of different sizes and configurations. Various types of water system	Small-scale floating structures have minimal impact on water quality, with measured parameters like dissolved oxygen remaining at acceptable levels. Underwater footage reveals abundant aquatic life, including mussels and fish, attached to and swimming beneath the structures. If considered during the design, floating constructions can be optimized to enhance aquatic life and biodiversity by creating habitats and shelter for smaller and juvenile fish
[4]	Group of 5 floating houses in a row. Small and shallow water system with little flow	At the location monitored, floating houses seem to slightly reduce dissolved oxygen (~10%) in the upper water column, affecting levels and temperature through shading. A wind tunnel effect between the houses seems to contribute for local aeration. Despite reduced air–water exchange with the atmosphere, dissolved oxygen levels remained above critical conditions (above 5 mg/L). In this shallow and turbid pond with low water circulation, water quality parameters were not noticeably impacted. Continued monitoring is recommended due to the complex interactions among water quality parameters

(continued)

Table 1 (continued)

Reference Year	Type of project water system	Approach and main findings/highlights
[5]	Floating neighborhood, Amsterdam with multiple floating houses in the same water system	Research on the impact of floating structures on freshwater benthic ecosystems was conducted near four floating houses in Ijburg, Amsterdam. Underwater drones recorded videos and measured dissolved oxygen levels, while sediment samples assessed organic carbon, nitrogen content, and macroinvertebrate diversity. Results showed that the platforms were colonized by filter feeders, lacked macrophytes underneath, and had lower dissolved oxygen levels. Sediment samples under the platforms had higher organic matter and respiration rates, but no differences in redox potential or macroinvertebrate structure were found. A light intensity model suggested design optimizations to reduce low-light areas. A conceptual model highlighted that platform effects depend on design, location, and ecosystem sensitivity, leading to recommendations for minimizing negative impacts and the need for further research
[6]	French Polynesia, considers potential large-scale floating infrastructure and housing of different sizes and configurations	This report assesses the environmental impact of floating infrastructures of different sizes for the specific case study in French Polynesia. Different scenarios were evaluated, and a risk analysis/ description was performed. In addition to the negative impacts associated with the construction and installation, other threats include the impact of artificial light originating from the platforms, wastewater system malfunctions, littering, noise pollution, or increased boat traffic. Mitigation involves ensuring adequate light penetration by adjusting platform size and configurations, selecting low-impact materials, and using careful and sustainable construction practices. Opportunities are also identified, such as promoting marine wildlife colonization to improving water quality. Further research and adaptive management strategies will contribute to support new guidelines and design principles to ensure successful and sustainable future designs that consider ecological sensitivity and improve ecosystem resilience

(continued)

Table 1 (continued)

Reference Year	Type of project water system	Approach and main findings/highlights
[7]	Group of 7 floating houses, installed at an excavated water system, connected to a stream	A measurement campaign assessed the impact of floating houses on water temperature and dissolved oxygen content. Using multi-parameter probes, CTD-Divers, and an underwater drone, data were collected over 24–72 h, over 10-day periods. Results showed that water temperature and dissolved oxygen levels are highly weather-dependent. Floating houses seen to act as temperature buffers in weak winds but mix with the water column in strong winds. In shaded areas, water temperature was higher than in open water, particularly in autumn. Dissolved oxygen content was lower in shaded areas, especially at the bottom, which was attributed to reduced photosynthesis. This thesis also briefly discusses advantages of water-based constructions when compared to land-based infrastructure in terms of environmental impacts
[8]	Floating solar park, installed in sand pit lake (up to 35 m deep)	This study investigates the potential impacts of floating solar panels on water quality parameters. Data was collected at one of the largest floating solar parks in Europe, Lake Bomhofsplas. Statistically significant differences between parameters under the panels and parameters at open water conditions, across the water column were reported). Values showed strong correlation and followed seasonal and day/night variations. Dissolved oxygen concentrations were consistently lower under the panels, with temperature differences observed mainly in surface layers. The shading effect of the panels likely contributes to these differences, but further research is needed to assess other potential drivers. The design of the solar panels allows some wind and sunlight to reach the water surface, suggesting different designs could have varied impacts. The presence of biofouling and bivalve colonization on the panels indicates potential for ecological enhancement alongside solar energy production

(continued)

Table 1 (continued)

Reference Year	Type of project water system	Approach and main findings/highlights
[9]	Multiple floating houses, pavilions of different sizes and configurations. Various types of water system	The monitoring campaign studied 18 locations with various floating buildings in different water bodies and seasons. Results showed water quality differences based on system characteristics like hydrodynamics and depth. Lower dissolved oxygen levels were found under floating buildings due to reduced light and photosynthesis, though levels remained healthy for aquatic life. Temperatures were sometimes higher under structures, likely due to heat retention. Continuous measurements showed variations depending on weather and daily cycles. Underwater drone footage revealed colonization by filter feeders, fish near structures, reduced macrophyte growth in shaded areas, and accumulation of dead shells. The study recommends further research to understand interactions under floating structures and suggests the findings can guide policies for their construction and management
[10]	Multiple floating houses, pavilions of different sizes and configurations. Various types of water system	This study reports that variations in concentrations of water quality parameters between open water and under/near floating structures were detected, but consistently low. The characteristics of water bodies are decisive factors for the extent of impact of water quality structures on water quality, as currents and mixing capacity highly influence the renewal rate of water under the structures. No clear relationships between water quality and structure characteristics were established. Highlights the need for continuous water quality data collection, and the need for integrated studies with the characteristics water body where the floating structure is built (e.g. currents, flow velocities), as well as the inclusion of new parameters. Introduces opportunities for adjusting designs, and for building with nature, considering all the new habitat that was unveiled by the underwater images

layers and ecosystems develop rapidly after construction, indicating potential for ecosystem restoration and biodiversity enhancement.

These research studies were limited to the available small scale floating structures that currently exist in the Netherlands. Results may not be universally applicable to other water system types or floating platform sizes and shapes. Incorporation with modelling studies would bring value and additional knowledge. Moreover, in The Netherlands, different projects focus on improving existing unmanned monitoring methods (underwater, surface and aerial vehicles) in terms of cost-efficiency and features (e.g. autonomous navigation, real-time availability of collected data, intelligent data interpretation), and to develop new systems/methods to monitor aspects that were not possible to monitor so far.

4 Conclusions

4.1 Opportunities and Monitoring

Rapid innovation often leaves little time, budget, or knowledge for thorough research into environmental implications. It is equally important to learn how to design and construct floating projects in a way that positively impacts ecology, the environment, and water quality, while also preventing issues that could lead to disinvestment in this promising green technology.

To better understand the environmental impact of floating solutions it is crucial to further develop tools to efficiently monitor water quality and ecology, as well as to uncover new insights into the underwater impact of installing floating projects. Most past studies emphasize the urgent need for long-term, continuous, holistic monitoring from various disciplines, such as ecology, focusing on specific areas like birds and fish. There is often insufficient time for comprehensive research into these effects on ecology and water quality. Additionally, there is a lack of adequate tools, which are primarily based on point-sample collection and laboratory analysis, for conducting comprehensive, efficient, and cost-effective monitoring under and around the floating solar panels. Therefore, acquiring field data using innovative monitoring technologies and intelligent data interpretation techniques is urgently needed.

Leaching heavy metals or other toxic substances from the floating structures into the water during the life span of the solar park may also impact the environment. Further studies should contemplate chemical analysis of water quality to detect the presence of leached metals or plastics/chemicals. Potential sources include lead used for soldering, aluminum frames, or release of microplastics from floats. Another common problem is caused by animals (e.g. birds, snakes) that use the floating structures as habitat, producing waste that pollutes the water.

To ensure that large-scale projects do not negatively impact the environment, further research is essential to develop recommendations and best practices for extensive floating urbanization. Floating projects present an opportunity to integrate environmentally friendly features into their designs. This includes submerged objects or floating gardens that promote ecosystem growth by enhancing nutrient exchange and attracting wildlife beneath the floating structures. Furthermore, these eco-friendly additions can help dampen waves, protecting floating structures such as houses and solar panels, and improve the overall landscape. Future designs should incorporate features like floating wetlands and other solutions that foster underwater habitats. Moreover, creative solutions such as utilizing the generated solar energy to continuously power water aeration systems, or to provide artificial light underneath the floating solar panels to compensate for the shading caused by the floating structure, could be implemented to mitigate undesired effects if necessary. These measures can improve water quality and biodiversity, offer opportunities for green solutions, and strengthen the connection with their aquatic surroundings.

4.2 Floating Future

The objective of the interdisciplinary Floating Future Research Project (https://floating-future.nl/) is: Scaling up floating structures as climate-proof space-creating solution enabling societal, industrial, and ecological win-wins'. This research project from the Netherlands consists of a broad research consortium including universities, universities of applied sciences, research institutes, public authorities, businesses, and NGOs. The overall objective is to develop knowledge about the societal acceptance, the required governance and legal arrangements, the ecological impacts and the technological challenges of building large-scale floating islands. Floating Future will also further explore the ecological impact and water quality impact of large-scale floating structures, optimizing the scale of large-scale floating infrastructure in alignment with receiving system characteristics. Additionally, methods for optimizing both below- and above-water designs will be developed to maximize ecological benefits and meet engineering demands. Floating Future will further analyze large-scale implementation of floating cities and infrastructure, by addressing different disciplines, including societal, environmental, governance, or technology aspects.

References

1. Boogaard FC, de Graaf R (2014) Kennis En Marktontwikkeling Drijvend Bouwen-Effecten Drijvende Constructies Op Waterkwaliteit
2. Kroes, Martin, Boogaard F, de Graaf R (2015) In eerste instantie wordt bij de term ‘drone’ vaak gedacht aan onbemande, op afstand bestuurbare, vliegtuigjes. Minder bekend is dat er ook drones bestaan die onder water inzetbaar zijn. Deze ‘submerged drones’ kunnen worden gebruikt voor aquatisch ecologisch onderzoek
3. Lima RLP de, Boogaard FC, De Graaf R, Dionisio Pires LM, Sazonov V (2015) Monitoring the impacts of floating structures on the water quality and ecology using an underwater drone. In: 4. IAHR World Congress, Amsterdam, The Netherlands
4. Foka E, Rutten M, Boogaard FC, de Graaf RE, de Lima RLP, van de Giesen N (2015) The effect of floating houses on water quality. In: International Water Week. https://www.researchgate.net/publication/283502222_The_effect_of_floating_houses_on_water_quality.
5. Härtwich H (2016) The impact of floating platforms on the benthic community structure in Dutch freshwater ecosystems.pdf. Universiteit van Amsterdam, Amsterdam, The Netherlands
6. Czapiewska K, Bart R, Barbara Dal Bo Zanon, Fen-Yu (Vicky) Lin, Härtwich H (2017) Environmental assessment framework for floating development (French Polynesia). Blue21; BlueFrontiers. https://www.blue21.nl/wp-content/uploads/2020/04/EIA.pdf
7. Lenz J (2018) Impact assessment of floating houses on water temperature and dissolved oxygen in Himpenser Wielen, Leeuwarden (Netherlands)

8. Lima RLP de, Katerina P, Boogaard FC, Olof A, Fen Yu Lin (2021) In-situ water quality observations under a large-scale floating solar farm using sensors and underwater drones. *Sustainability* 13(11): 6421. Switzerland. https://doi.org/10.3390/SU13116421/S1
9. Lima RLP de, de Graaf-van Dinther RE, Boogaard FC (2022) Impacts of floating urbanization on water quality and aquatic ecosystems: A study based on in situ data and observations. J Water Clim Chang 13(3): 1185–1203. https://doi.org/10.2166/wcc.2022.325
10. De Lima, Rui LP, Boogaard FC, Sazonov V (2022) Assessing the influence of floating constructions on water quality and ecology. In: WCFS2020, Łukasz P, Lim SH, Wang CM, Graaf-van Dinther, RD (ed) Lecture notes in civil engineering. Singapore: Springer Singapore. 158:397–406 https://doi.org/10.1007/978-981-16-2256-4_24

Designing Blue Infrastructure to Facilitate Native Species and Prevent Marine Invasion

Katherine Dafforn, Melanie Bishop, Tegan Fuchert, Aria Lee, Mariana Mayer-Pinto, Nina Schaefer, and Maria Vozzo

Abstract Global increases in development have resulted in severe habitat modification along coastal zones and offshore. Marine built structures can impact on native marine species. The provision of extra hard surface presents additional free space, and research suggests invasive marine species may be able to exploit floating structures such as pontoons more effectively than native species. Floating structures then act as stepping stones to facilitate establishment and regional spread. We investigated whether (1) increasing the complexity of surfaces, (2) including natural materials in substrates, (3) transplanting native species and (4) increasing daytime lighting could enhance native species recruitment and growth over invasive species. We found that increasing complexity had different effects depending on tidal elevation. Specifically, increased complexity facilitated more invasive species at lower tidal elevations and particularly in moist, shaded microhabitats. We found that including natural materials had little effect on invasive species and that transplanting some native species could inhibit recruitment by invasive species. Overall, daytime lighting was most effective at encouraging growth by native seaweeds that then excluded invasive invertebrate species. These findings are significant because they identify factors that promote native over invasive species and can assist in designing nature-based approaches for infrastructure to prevent marine invasion.

Keywords Invasive species · Pontoon · Nature-based approach · Biosecurity · Biodiversity

K. Dafforn (✉) · M. Bishop · T. Fuchert · N. Schaefer
Macquarie University, North Ryde, NSW, Australia
e-mail: katherine.dafforn@mq.edu.au

A. Lee
Sydney Institute of Marine Science, Mosman, NSW, Australia

M. Mayer-Pinto
University of NSW, Sydney, NSW, Australia

M. Vozzo
CSIRO, St Lucia, QLD, Australia

X. L. Zhao et al. (eds.), *Proceedings of the Fourth World Conference on Floating Solutions: WCFS 2024; 2–4 December; Hong Kong*, Lecture Notes in Civil Engineering 597, https://doi.org/10.1007/978-981-96-4569-5_21

1 Introduction

1.1 The Spatial Extent of Marine Built Structures Is Increasing

Global increases in coastal and offshore development have resulted in severe habitat modification in the marine environment. Built structures such as pilings, pontoons and seawalls support burgeoning blue economy activities including ports and offshore renewables developments. The physical footprint of built structures is now estimated at > 32,000 km^2 while more than 1 M km^2 of seafloor is estimated to be impacted by stressors associated with these structures such as noise, light and contamination.

1.2 Floating Structures Create a Unique Environment that Supports Invasive Species

Comparisons of built structures and natural habitats have revealed distinct differences in the species able to colonise and persist. Intertidal communities on built structures tend to be distinct from those on rocky reefs, supporting fewer species and being dominated by algae [1]. Subtidally, assemblages on built structures may have fewer taxa and more coralline algae than natural habitats [2]. More specifically, built structures such as pontoons tend to support more bryozoans, ascidians and mussels than do rocky reefs [3, 4].

Differences between built structures and natural habitats arise due to their physical characteristics, including substrate composition [5–7], age [8], orientation [9–11], predation levels [12], illumination levels [13] and due to the fixed nature of some structures such as pilings, seawalls and natural habitats versus the movement of pontoons [14, 15]. Structures such as pontoons, typically support less biodiversity than the natural habitats they replace and can harbour invasive species [16]. This threatens ecosystems, as invasive species have the potential to reduce native biodiversity through competition for resources, habitat alteration or predation [17].

Marine life that can survive on floating structures are responding to several design factors including the material used in construction and availability of micro/macro habitats as well as local environmental conditions such as light and wave energy. Our understanding of how these factors might influence the types of marine life found on artificial structures is increasingly being used to restore native biodiversity in a practice commonly referred to as "ecological engineering" [18].

Fig. 1 Living seawalls add mimics of microhabitats like rockpools that are missing from built infrastructure

1.3 Eco-Engineering Can Be Used to Design Built Structures that Facilitate Native Over Invasive Species

Eco-engineering projects like the Living Seawalls project (www.livingseawalls.com.au) enhance biodiversity by adding complexity and surface area [19]. It is, however, unclear whether eco-engineering could be used to design biosecure marine built structures (Fig. 1).

2 Materials and Methods

We investigated whether the design of marine built structures could facilitate native species and reduce biosecurity risk with three studies in Sydney Harbour.

2.1 Surface Complexity

We manipulated surface complexity by installing complex and flat habitat enhancing panels. We assessed (1) whether colonisation of invasive species was enhanced on the complex panels when compared to the flat panel, and (2) whether colonisation patterns differed among tidal elevations.

2.2 Materials

We manipulated materials by including recycled oyster shells and sandstone in habitat enhancing panels, and we transplanted native seaweed onto the panels. We assessed (1) whether colonisation of invasive species was reduced on the natural materials when compared to concrete, and (2) whether the presence of seaweed transplants inhibited invasive species.

2.3 Light

We experimentally manipulated daytime lighting in a gradient from total shade to light levels reflecting midday sunlight. We deployed panels to investigate what would grow under this gradient of lighting. We assessed (1) whether colonisation of native seaweeds increased with increasing light intensity, and (2) whether growth of native seaweeds was associated with reduced growth of other species including invaders.

3 Results

3.1 Surface Complexity

At high and mid intertidal elevations, the contribution of invasive species to total abundance and richness was generally very small on both complex and flat panels (Fig. 2). At the low intertidal elevation by contrast, invasive species contributed approximately 75% sessile cover, 50% richness and were in some instances 50% more abundant and diverse in growth on the complex rather than flat panels. Within the panels, invasive species were particularly abundant in moist, shaded microhabitats.

3.2 Materials

At intertidal elevations, the cover of non-indigenous species was greater on sandstone than oyster or concrete materials (Fig. 3). At the low and deep subtidal elevations, concrete panels generally had greater cover of non-indigenous species than natural materials, but this was not significant.

At low subtidal elevations, covers of native and non-indigenous species were similar on panels with and without seaweed transplants (Fig. 4). Native species richness increased on panels with seaweed transplants while non-indigenous species richness remained similar to panels without transplants.

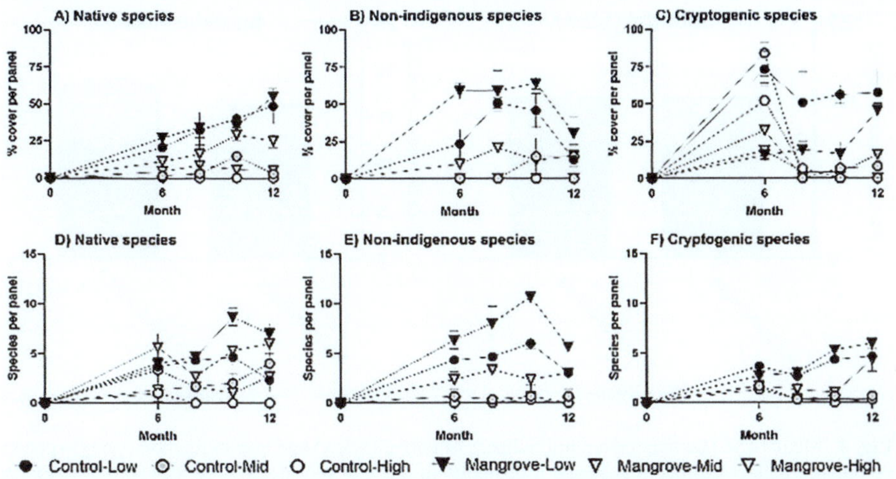

Fig. 2 Mean (+SE) percent covers of species colonising complex and control panels over time across three intertidal elevations. Cover and richness increased over time for all classifications, but varied between complex and flat panels, and between tidal levels. Complex panels may facilitate more non-indigenous species in the low intertidal than other tidal levels

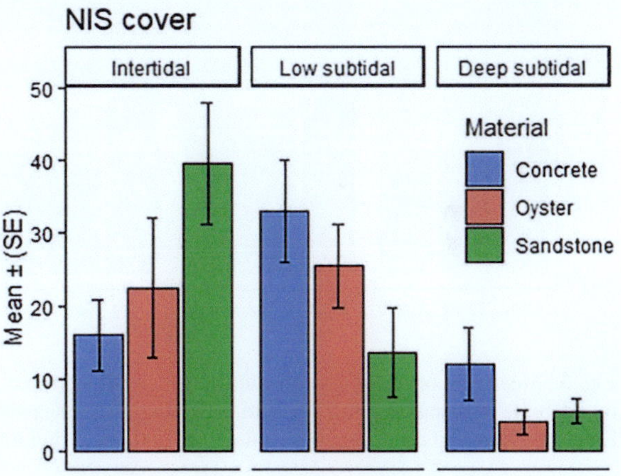

Fig. 3 Mean (+SE) percent covers of non-indigenous species colonising natural (sandstone, oyster shell) and concrete panels across three intertidal elevations (intertidal, low and deep subtidal). Cover varied between natural and concrete panels at all tidal levels. Concrete panels may facilitate more non-indigenous species in the low and deep subtidal than other tidal levels [4]

3.3 Light

Seaweeds were more abundant under all light treatments and virtually absent under the shaded treatment (Fig. 5). Conversely, invertebrates were more abundant in the low light treatments. Predation reduced the cover of both groups.

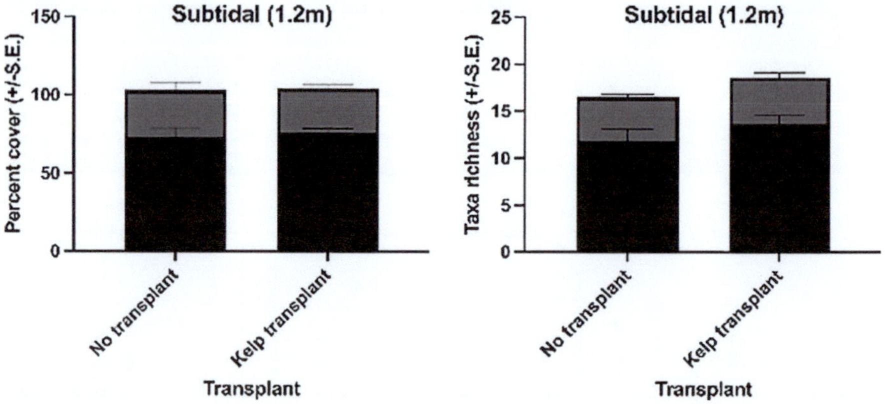

Fig. 4 Mean (+SE) percent cover and richness of native (black) and non-indigenous (grey) species colonising panels following seaweed transplants in the low subtidal. Cover was similar between treatments, but native species richness increased with seaweed transplants. Seaweed transplants may increase native biodiversity without facilitating non-indigenous species

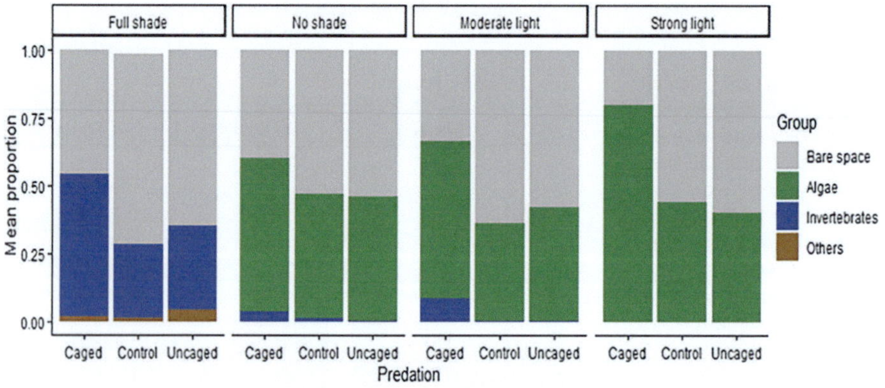

Fig. 5 Mean proportion of taxa on panels exposed to a gradient of light. Communities shifted from invertebrate dominance in full shade to increasing native seaweed dominance with increasing light. Enhanced seaweed covers may exclude other taxa including non-indigenous species present [20]

4 Discussion and Conclusions

Marine built structures such as pontoons, typically support less biodiversity than the natural habitats they replace and can harbour invasive species. Knowledge of the factors that support invasive species colonisation will assist in designing future eco-engineering interventions that reduce biosecurity risk.

References

1. Bulleri F, Chapman MG (2004) Intertidal assemblages on artificial and natural habitats in marinas on the north-west coast of Italy. Mar Biol 145:381–391
2. Connell SD, Glasby TM (1999) Do urban structures influence local abundance and diversity of subtidal epibiota? A case study from Sydney Harbour Australia. Mar Environ Res 47:373–387
3. Glasby TM (1999) Differences between subtidal epibiota on pier pilings and rocky reefs at marinas in Sydney, Australia. Estuar Coast Shelf Sci 48:281–290
4. Connell SD (2001) Urban structures as marine habitats: an experimental comparison of the composition and abundance of subtidal epibiota among pilings, pontoons and rocky reefs. Mar Environ Res 52:115–125
5. Anderson MJ, Underwood AJ (1994) Effects of substratum on the recruitment and development of an intertidal estuarine fouling assemblage. J Exp Mar Biol Ecol 184:217–236
6. Glasby TM (2000) Surface composition and orientation interact to affect subtidal epibiota. J Exp Mar Biol Ecol 248:177–190
7. Chapman MG, Bulleri F (2003) Intertidal seawalls—new features of landscape in intertidal environments. Landsc Urban Plan 62:159–172
8. Perkol-Finkel S, Shashar N, Barneah O, Ben-David-Zaslow R, Oren U, Reichart T, Yacobovich T, Yahel G, Yahel R, Benayahu Y (2005) Fouling reefal communities on artificial reefs: does age matter? Biofouling 21:127–140
9. Connell SD (1999) Effects of surface orientation on the cover of epibiota. Biofouling 14:219–226
10. Glasby TM, Connell SD (2001) Orientation and position of substrata have large effects on epibiotic assemblages. Mar Ecol Prog Ser 214:127–135
11. Saunders RJ, Connell SD (2001) Interactive effects of shade and surface orientation on the recruitment of spirorbid polychaetes. Austral Ecol 26:109–115
12. Knott NA, Underwood AJ, Chapman MG, Glasby TM (2004) Epibiota on vertical and on horizontal surfaces on natural reefs and on artificial structures. J Mar Biol Assoc UK 84:1117–1130
13. Nydam M, Stachowicz JJ (2007) Predator effects on fouling community development. Mar Ecol Prog Ser 337:93–101
14. Glasby TM (1999) Effects of shading on subtidal epibiotic assemblages. J Exp Mar Biol Ecol 234:275–290
15. Connell SD (2000) Floating pontoons create novel habitats for subtidal epibiota. J Exp Mar Biol Ecol 247:183–194
16. Holloway MG, Connell SD (2002) An introduced polychaete affects recruitment and larval abundance of sessile invertebrates. Ecol Appl 12:1803–1823
17. Chapman JW, Carlton JT (1991) A test of criteria for introduced species: The global invasion by the isopod *Synidotea laevidorsalis* (Miers, 1881). J Crustac Biol 11:386–400
18. Chapman MG, Underwood TJ (2011) Evaluation of ecological engineering of "armoured" shorelines to improve their value as habitat. J Exp Mar Biol Ecol 400:302–313
19. Bishop MJ, Vozzo ML, Mayer-Pinto M, Dafforn KA (2022) Complexity-biodiversity relationships on marine urban structures: reintroducing habitat heterogeneity through eco-engineering. Philos Trans R Soc B 377:20210393
20. Schaefer N, Bishop MJ, Bugnot AB, Foster-Thorpe C, Herbert B, Hoey AS, Mayer-Pinto M, Sherman CDH, Vozzo ML, Dafforn KA (2021) Ecological engineering of marine infrastructure for —phase 2. In: Report prepared for the department of agriculture, water and environment, p 105

Reconstructing Historical Water Quality and Biofouling Dynamics in Tokyo Bay: A Numerical Simulation Approach for Sustainable Floating Infrastructure

Teng Tu, Jinxin Zhou, Daisuke Kitazawa, Akane Takahashi, Yoshinobu Yoneyama, and Ikuo Yoshida

Abstract The rapid growth of cities along coastlines has increased the need for maritime structures, highlighting the importance of studying biofouling on floating infrastructure. Biofouling, the buildup of organisms on underwater surfaces, can significantly impact the stability, lifespan, and expenses of these structures. We conducted a thorough numerical analysis to replicate the water quality of Tokyo Bay in 2000 and study the biofouling dynamics on a large floating platform at Yokosuka Shipyard in Yokosuka City. Our study utilized a specialized biofouling submodel to simulate the conditions around the mega-float during that time. We used numerical simulations to reconstruct water quality parameters, which closely aligned with observed data, ensuring the reliability of our findings. We studied biofouling dynamics under four scenarios with varying mortality rates, a key variable with significant uncertainty. Our results indicated minimal differences in environmental effect of biofouling dynamics across these scenarios, suggesting that the current model may be too simplistic and not adequately account for complex interactions such as organism detachment. This finding underscores the need for further development of our biofouling models to better understand and mitigate the impacts on maritime structures. The insights from our study provide valuable information for improving the design, operation, and maintenance of future floating infrastructure in rapidly developing coastal areas. This research not only contributes to our understanding of biofouling impacts but also aids in the sustainable development of marine infrastructure.

T. Tu (✉)
Graduate School of Engineering, The University of Tokyo, Chiba, Japan
e-mail: tetu@iis.u-tokyo.ac.jp

J. Zhou · D. Kitazawa
Institute of Industrial Science, The University of Tokyo, Chiba, Japan

A. Takahashi · Y. Yoneyama · I. Yoshida
Ocean Program Department, Emerging Frontiers Div, Shimizu Corporation, Tokyo, Japan

X. L. Zhao et al. (eds.), *Proceedings of the Fourth World Conference on Floating Solutions: WCFS 2024; 2–4 December; Hong Kong*, Lecture Notes in Civil Engineering 597, https://doi.org/10.1007/978-981-96-4569-5_22

Keywords Biofouling · Floating structures · MEC ocean model · Numerical simulation · Tokyo Bay · Water quality

1 Introduction

Rapid urbanization along coastlines worldwide has necessitated the development of maritime structures to support this growth. Among these, floating infrastructures have become increasingly significant due to their potential to expand usable space without extensive land reclamation, thus preserving natural coastal environments [1]. However, these structures face a unique challenge: biofouling. Biofouling refers to the undesirable accumulation of aquatic organisms on submerged surfaces, which can severely affect the structural integrity, functionality, and longevity of these installations. This phenomenon not only increases maintenance costs but also impacts the environmental dynamics of the surrounding marine areas [2].

The focus of this paper is a comprehensive study conducted on the biofouling dynamics on a large floating platform, part of the Mega-Float project in Yokosuka City, within the confines of Tokyo Bay. This research utilized advanced numerical simulations to replicate the water quality conditions of the year 2000 and examine the effects of biofouling under various scenarios. By integrating a specialized biofouling sub-model within a Three-Dimensional Hydrodynamic Ecosystem model—recently named Marine Environmental Committee (MEC) ocean model, this study aimed to provide a detailed analysis of the interactions between biofouling and environmental factors such as water temperature, dissolved oxygen (DO), and nutrient concentration.

The results of study, which revealed minimal differences across various scenarios, suggest that the current model may be too simplistic, primarily replicating the dynamics observed in macrobenthos models without adequately accounting for complex interactions such as the detachment effects of biofouling organisms from the submerged surfaces. This paper will discuss the methodology employed in the simulations, present the results of the scenarios tested, and explore the implications of these findings in the broader context of maritime engineering and environmental science. The significance of this study lies in its potential to enhance the design, operation, and maintenance strategies for future floating infrastructures. By understanding the dynamics of biofouling and its impact on marine ecosystems, engineers and environmental scientists can develop more effective methods to mitigate these effects, thereby promoting the sustainable development of coastal urban areas.

2 Methods and Materials

2.1 Overview of the Three-Dimensional Hydrodynamic Ecosystem Model

This study applied a Three-Dimensional Hydrodynamic Ecosystem model named MEC ocean model, designed to simulate complex interactions within marine environments. This model integrates hydrodynamic and biofouling sub-models to provide a comprehensive understanding of the environmental impacts associated with floating structures. The hydrodynamic sub-model component is based on the fundamental principles of fluid dynamics and is tailored to marine settings. It incorporates the momentum equations, state equation, and the advection and diffusion equations for water temperature and salinity. Key processes have been considered, including surface wind friction, heat and salt fluxes from the surface, and bottom friction. The biofouling sub-model is specifically developed to simulate the growth and accumulation of biofouling organisms on submerged surfaces of the floating structures. It considers various factors such as water temperature, food availability, and dissolved oxygen concentration, which influence the growth rates of biofouling organisms. We tuned the biofouling organism's mortality rate, which is a significant variable due to its direct impact on the extent of biofouling, for model validation (Sect. 2.3), but for the sensitivity test, we set four scenarios to explore the effects on biofouling dynamics and the associated environmental effects (Sect. 2.4).

2.2 Computational Conditions

The model ran from 1 September to 31 October 2000, with a time step of 2 s to capture the transitional environmental conditions of early autumn. The entire Tokyo Bay area with public open data of the water depth was included in the simulation. The model utilizes a nested grid system to enhance regional resolution while reducing the overall computational load. Specifically, the grid resolutions were set at 1350, 450, 150, and 50 m for Ranks 1 through 4, respectively, with the finest resolution near the mega-float area. Other boundary conditions—land boundary, river boundary, meteorological boundary, open boundary, and sea bottom—followed the previous work [3].

2.3 Model Validation

To ensure the accuracy and reliability of the model, it was essential to validate it against observed data. This process involved comparing the simulated results with historical data collected from the Mega-Float project regarding water temperature, salinity, and the horizontal distribution of the flow field at Station E located near the floating structures. The field survey has been summarized in our previous work [4].

2.4 Simulation Scenarios

Recognizing the uncertainty in the mortality rates of attached organisms, the value ($1/(mgC/m^2)/s$) was adjusted at four different levels to assess their impact: (1) Very Low Mortality Rate Scenario (1.16×10^{-12}) represents an extremely rare scenario where biofouling organisms have exceptionally high survival rates, potentially leading to unprecedented levels of biomass accumulation. (2) Low Mortality Rate Scenario (1.16×10^{-8}) represents a scenario where biofouling organisms have higher survival rates, leading to increased biomass accumulation. (3) Medium Mortality Rate Scenario (1.16×10^{-7}) is a balanced scenario reflecting average conditions typically observed in marine environments. (4) High Mortality Rate Scenario (1.16×10^{-6}) depicts a situation where higher mortality rates lead to reduced biofouling. The effects of each scenario were analyzed on the wet weight of biofouling organisms and the DO concentrations around the floating structures.

3 Results and Discussion

3.1 Model Validation

The model was validated mainly by comparing the simulated water temperature and salinity with the observed one. Generally, the model successfully replicated the daily variations in water temperature observed during the monitoring period (Fig. 1b). The simulated temperatures closely matched the measured data, with minor discrepancies likely due to transient weather conditions not fully captured by the model. Similar to temperature, the salinity concentrations predicted by the model were in good agreement with the observed values (Fig. 1a). This indicates that the model accurately captures the salinity dynamics, which are influenced by factors such as freshwater inflow and tidal exchanges.

These results confirm the reliability of the hydrodynamic sub-model, establishing a solid foundation for further studies on biofouling dynamics and their environmental impacts.

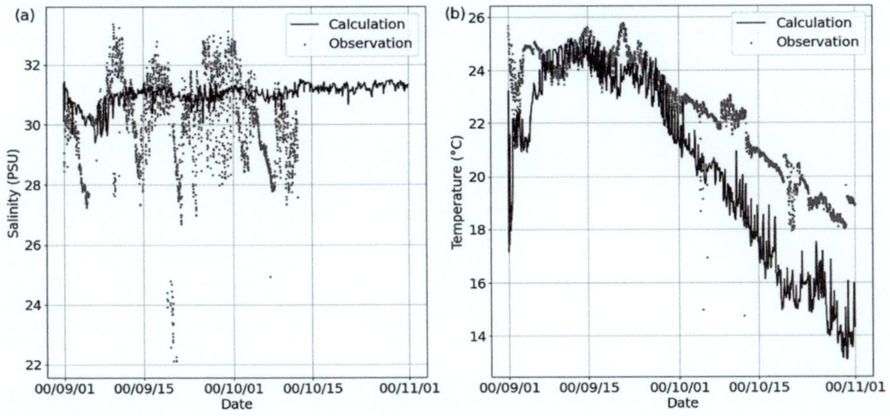

Fig. 1 Comparison of salinity (**a**) and water temperature (**b**) between observation and prediction during the period

3.2 Scenario Analysis

Figure 2 illustrates the change in the wet weight (g) of sessile organisms over the course of September 2000, categorized by four different mortality rates: high, medium, low, and very low. The data shows that organisms with the high mortality experienced a significant decrease in weight, depicted by the blue line. In contrast, organisms with the medium and low mortality rates, represented by the green and orange lines respectively, show a more gradual increase in weight. This suggests that sessile organisms can hardly grow with the high mortality.

The four mortality rate scenarios revealed minimal differences in the DO concentrations (Fig. 3) indicating that the variations in sessile organism biomass did not substantially affect the surrounding water quality. This suggests that the current model may be overly simplistic, primarily replicating the dynamics observed in macrobenthos models without adequately accounting for complex interactions such as the detachment effects of biofouling organisms from the submerged surfaces.

To enhance the model's accuracy and applicability, the following four improvements are recommended. First, integrating a sub-model that specifically addresses the detachment of biofouling organisms could provide a more realistic depiction of biofouling dynamics. This model should consider variables such as shear stress from water currents and the mechanical stability of the biofilm. Second, expanding the model to include more complex interactions between biofouling organisms and their environment, such as predation, competition, and symbiosis, could help in understanding the ecological balance and its impact on biofouling growth. Third, further research is needed to refine the estimates of mortality rates based on empirical data. This could involve detailed field studies and experiments to observe the life cycles and death rates of key biofouling species under various environmental conditions. Fourth, continuous validation of the model with up-to-date and comprehensive field data will

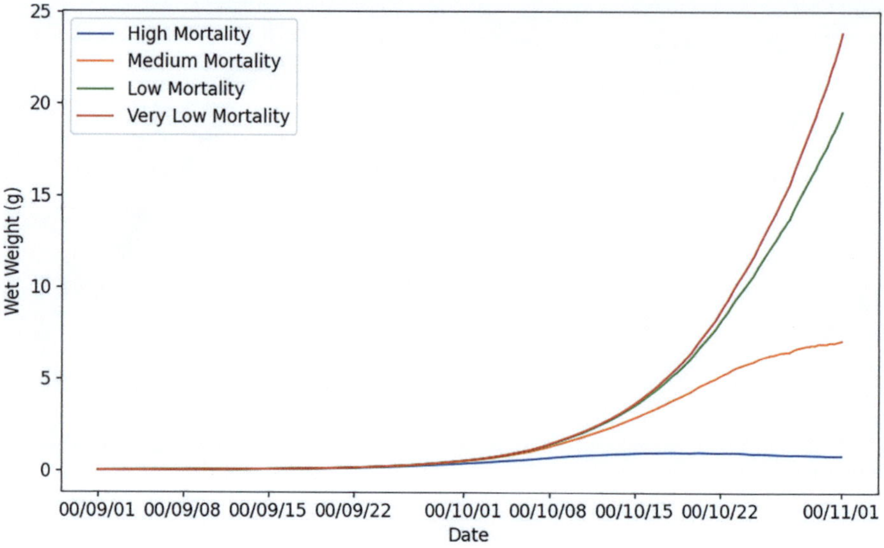

Fig. 2 Wet weight of sessile organisms under four different mortality rates

be crucial. This should include data on biofouling thickness, species composition, and environmental parameters at different depths and locations around the floating structures.

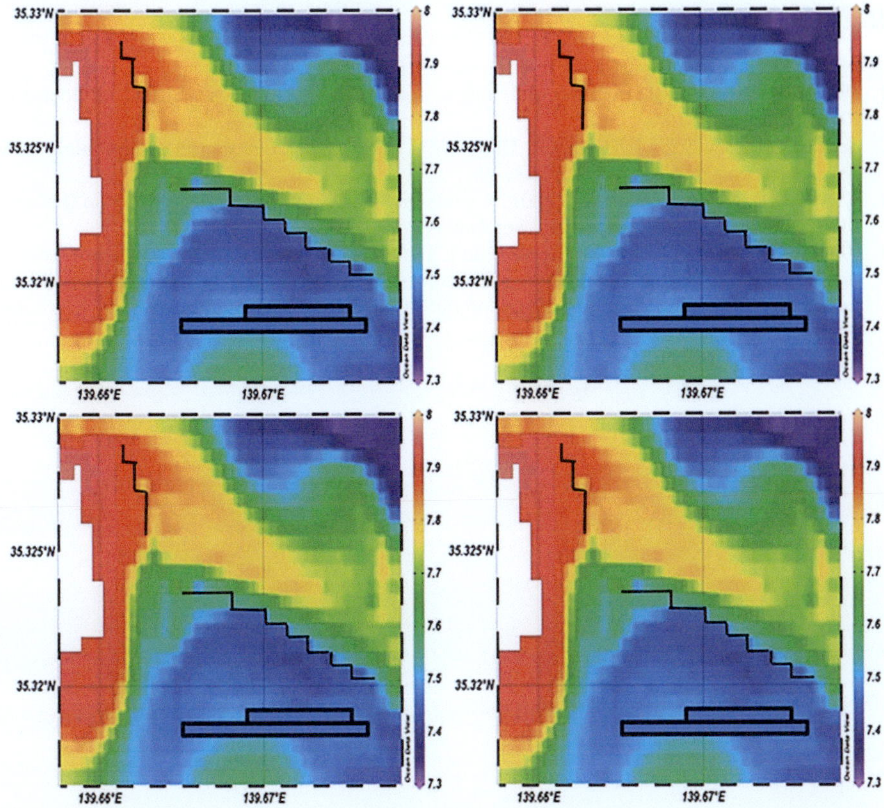

Fig. 3 Horizontal distribution of dissolved oxygen around mega-float area with high mortality rate (upper-left), medium mortality rate (upper-right), low mortality rate (lower-left), and very low mortality rate (lower-right)

4 Conclusion

This study conducted a detailed analysis of biofouling dynamics on a floating platform in Tokyo Bay using the MEC ocean model, aiming to understand the influence of environmental factors on biofouling and its impacts on maritime structures. The findings revealed that the current model might be too simplistic. This suggests a lack of complex biological interaction representation, particularly in the detachment of organisms from surfaces. Future work will focus on enhancing model accuracy by integrating a sub-model for organism detachment, influenced by factors like water currents. These efforts will help improve management strategies for biofouling on floating infrastructures, promoting sustainable coastal development.

References

1. Wang CM, Tay ZY (2011) Very large floating structures: applications, research and development. Procedia Eng 14:62–72
2. Schultz MP, Bendick JA, Holm ER, Hertel WM (2011) Economic impact of biofouling on a naval surface ship. Biofouling 27(1):87–98
3. Kitazawa D, Fujino M, Tabeta S (2001) A study on the environmental effects of Phase II Mega-Float model on marine ecosystem. J Soc Nav Arch Jpn 2001(190):361–371
4. Zhou J, Kitazawa D, Hayakawa Y, Yoshida I, Tabata K, Okumura K (2024) Impacts of sessile organisms attached to artificial floating structures on the surrounding environment: review and update in Tokyo Bay. In: Proceedings of the third world conference on floating solutions, Springer, Singapore

Exploring Ancient Oasis Wisdom for Sustainable Floating Cities in the Dry Tropics

Daniela A. Ottmann

Abstract Coastal regions confront rising sea levels and rapid urbanisation, prompting the quest for sustainable solutions. This study investigates traditional oasis systems as models for self-sustaining urban environments, especially in hot, arid climates. Analysing their principles yields key strategies applicable to sustainable floating architecture and cities. These strategies encompass water management, passive cooling, material use, sustainable agriculture, and ecological integration. Repurposing urban elements into self-sustaining systems enhances connectivity with natural resources, revitalising resource utilisation. A multi-dimensional approach integrates urban structures as active contributors to resource production, meeting escalating demands on limited resources. Utilising locally sourced abiotic resources promotes environmental sustainability while fostering local employment. Incorporating biotic matter and leveraging biomass for nutrients, fuel, and construction strengthen sustainability efforts. Strategic water resource management ensures dependable water distribution, including solar thermal desalination and efficient land use policies. Harnessing wind for passive cooling and other energy conversion contributes to energy efficiency and sustainability. Solar energy technologies offer diverse opportunities for energy generation and reducing reliance on fossil fuels. Furthermore, the modular, organic growth patterns of oasis landscapes might inspire a flexible, adaptable approach to floating city design. This will allow for scalable, resilient development that minimises environmental impact and promotes socio-cultural integration, aligning with the ecological principles embodied in traditional oasis systems. By drawing on these holistic, sustainable strategies, floating architecture and cities can be designed to thrive in hot, arid regions, offering a viable solution to the challenges posed by rising sea levels and coastal urbanisation. Traditional oasis systems' lessons are explored to provide a solid foundation for creating self-sufficient, climate-resilient floating communities.

D. A. Ottmann (✉)
Faculty of Society and Design, Bond University, Gold Coast, Australia
e-mail: dottmann@bond.edu.au

Cities Research Institute, Griffith University, Gold Coast, Australia

X. L. Zhao et al. (eds.), *Proceedings of the Fourth World Conference on Floating Solutions: WCFS 2024; 2–4 December; Hong Kong*, Lecture Notes in Civil Engineering 597, https://doi.org/10.1007/978-981-96-4569-5_23

Keywords Oasis systems · Floating cities · Climate design · Architecture · Sustainable urban planning

1 Introduction

Coastal regions across the globe are increasingly vulnerable to rising sea levels and rapid urbanization, necessitating the development of innovative, sustainable urban environments. Floating cities have emerged as a potential solution to the challenges posed by climate change, especially in hot and arid regions where traditional land-based settlements may no longer be viable.

Coastal cities in hot arid climates are increasingly vulnerable to the impacts of:

- Climate Change: Increasing temperatures, changing rainfall patterns, and frequent extreme weather events pose significant risks to coastal cities. Sea level rise threatens low-lying areas, with cities like Bahrain at risk of substantial land loss.
- Biodiversity Loss: Coastal development and climate change are degrading vital habitats like coral reefs and mangroves, which are crucial for fisheries and local livelihoods [1].
- Water Scarcity: The region faces acute water shortages due to minimal rainfall and groundwater depletion. Seawater intrusion further reduces freshwater availability, while reliance on desalination, which is energy-intensive, increases greenhouse gas emissions.
- Food Security: Limited agricultural capacity and heavy dependence on food imports heighten the region's vulnerability to global supply chain disruptions [2].
- Energy Demand: Rising temperatures will significantly increase the demand for cooling, with projections indicating a 302% rise by the end of the century. The need for more desalinated water will also escalate, demanding substantial energy resources [3].

These challenges threaten critical infrastructure, exacerbate water scarcity, and strain the region's reliance on energy-intensive desalination and air conditioning.

2 Understanding Traditional Oasis Systems

Oasis systems have long been celebrated for their resilience and sustainability in arid environments. These settlements have thrived by developing intricate systems for managing water resources, utilizing local materials, and creating microclimates that enhance thermal comfort. The primary components of these systems include water management through Aflaj irrigation, passive cooling, and the integration of agriculture with urban development. These principles have enabled oasis communities to maintain a delicate balance between human habitation and the natural environment, ensuring their survival in some of the harshest climates on Earth [4–6].

2.1 Water Management

Water management is the cornerstone of any oasis settlement. Traditional oasis communities relied on sophisticated systems such as Aflaj, Qanats, and other forms of irrigation to distribute water efficiently. These systems often involved underground channels to transport water from distant sources to the surface, where it could be used for agriculture, drinking, and other purposes. The careful management of water resources ensured that even in times of drought, oasis communities could sustain themselves and their crops.

2.2 Passive Cooling and Climate Adaptation

Oasis settlements also use passive cooling techniques to create more comfortable living environments. These techniques include strategically placing buildings (Urban Morphology), Urban Canyons and Courtyards to maximize natural ventilation whilst providing shading and incorporating water features of the palm grove irrigation system that cool the surrounding air through evaporation. The architecture of these settlements was designed to minimize heat gain during the day while maximizing the cooling effect of the night-time air.

2.3 Integration of Biotic and Abiotic Resources

Integrating biotic and abiotic resources is another crucial characteristic of oasis settlements. In these communities, natural resources such as palm groves, water, and animals were carefully managed to create a sustainable cycle of production and consumption. Agriculture was often integrated with urban development, with crops grown close to residential areas, ensuring that food production was efficient and sustainable. The use of locally sourced materials, such as loam and stone, further contributed to the sustainability of these settlements by reducing the need for imported resources.

3 Application to Floating Cities

The principles derived from traditional oasis systems can be directly applied to the design of floating cities, providing a potential framework for creating self-sustaining urban environments in the dry tropics.

3.1 Water Management in Floating Cities

Water management in floating cities is critical to ensuring a sustainable and reliable supply of fresh water, particularly given the reliance on seawater as the primary water source. Traditional oasis systems, which have long demonstrated efficient water use in arid environments, provide valuable lessons that can be adapted to modern water management technologies in floating cities.

One of the critical technologies is solar thermal desalination, which offers a sustainable method of converting seawater into potable water. This desalinated water can be distributed throughout the city using a floating channel network, similar to the traditional Aflaj systems in oasis settlements. These channels facilitate the efficient distribution of water and serve as integral design elements that enhance the city's aesthetic appeal and liveability. [7].

Greywater recycling systems are another essential component of water management in floating cities. These systems can significantly reduce overall water demand by reusing greywater for irrigation and other non-potable applications. [8]

Capturing and retaining rainfall through stormwater retention cisterns, rain gardens, and bioswales can help manage water resources and mitigate flooding. [9]

3.2 Passive Cooling and Thermal Comfort

Maintaining thermal comfort is a significant challenge in floating cities, particularly those in hot and arid regions. To address this, passive cooling strategies—similar to those used in traditional oasis settlements—can be effectively integrated into the urban fabric, reducing reliance on energy-intensive air conditioning systems.

One of the key approaches is the strategic orientation of buildings to take full advantage of prevailing winds. By optimizing building positioning, cities can minimize direct solar gain and enhance natural ventilation, thereby reducing indoor temperatures. Additionally, materials with high thermal mass can be employed in building construction to moderate indoor temperatures, as these materials absorb heat during the day and release it at night, maintaining a more stable indoor climate.

The incorporation of shading techniques is another crucial strategy. Structures like breezeways, green (algae) facades, and double-skin systems with air gaps can effectively reduce heat absorption. Green facades and compact mass design—where the surface-to-volume ratio is optimized—further minimize heat gain, creating a more comfortable living environment. Reflective materials on roofs and facades also contribute to lowering ambient temperatures by reducing solar absorption [10].

Radiant cooling techniques using materials optimized for high emissivity can be used to emit radiant heat from surfaces facing cooler environments, such as the clear sky, effectively cooling buildings [11].

Innovative cooling solutions utilizing saline water offer additional methods to enhance thermal comfort. Evaporative cooling facades, where saline water circulates over porous surfaces, provide a cooling effect through evaporation. Saline water can also be integrated into building materials with high thermal mass, enhancing their ability to moderate indoor temperatures without the need for active cooling systems [9].

3.3 Integration of Green and Blue Spaces

Integrating green and blue spaces, with aquaculture as a central component, is essential for creating sustainable and liveable floating cities. These spaces enhance biodiversity, improve air quality, and mitigate the urban heat island effect. Aquaculture systems, strategically placed within the city, contribute to natural cooling, food production, and environmental resilience.

Green and blue spaces also serve practical purposes like stormwater management and irrigation. For instance, rainwater harvesting systems on green roofs can provide water for both irrigation and aquaculture, reducing runoff and flood risks [12]. Using biomass from aquaculture for energy and building materials further enhances sustainability by minimizing environmental impact and supporting local food production.

Incorporating saline water for non-potable uses, such as irrigating salt-tolerant plants and supporting aquaculture, reduces reliance on freshwater resources and integrates seamlessly with the city's blue infrastructure.

3.4 Modular, Organic Growth Patterns

One of the defining characteristics of oasis landscapes is their modular, organic growth patterns [13]. These landscapes evolve naturally, with settlements expanding and contracting in response to environmental conditions and population pressures.

Floating cities can be designed as modular units, allowing for sections of the city to be added or removed based on factors such as population growth, environmental conditions, and resource availability. This modular approach ensures scalable development, enabling the city to adapt to changing circumstances without compromising sustainability. Furthermore, the design can incorporate reconfigurable modules that respond to impact factors such as wind, currents, swells, diurnal patterns, and solar exposure, which can continuously optimize the urban climate through constant adaptation.

By embracing modular, organic growth patterns, floating cities can create more resilient urban environments capable of withstanding the challenges posed by climate change.

3.5 Urban Morphology

Urban morphology in floating cities is crucial for achieving thermal comfort and sustainability. The design of grey spaces—such as buildings, infrastructure, and public areas—must work harmoniously with green and blue spaces to create a balanced urban environment. Key strategies include:

- Optimizing building orientation to capture natural ventilation.
- Designing narrow streets and courtyards to provide shade.
- Using reflective materials on roofs and facades to reduce heat absorption.

Urban canyons and courtyards act as microclimates, enhancing cooling and reducing the urban heat island effect. Using ecological building materials with low embodied energy and proper insulation and thermal mass helps stabilize indoor temperatures and improve energy efficiency. An Urban Microclimate Design Matrix will be proposed accordingly.

4 Case Studies, Simulations and Future Research

Initial testing on traditional settlements in Dubai (UAE) and Jeddah (KSA) has been carried out to investigate how integrating water bodies within floating cities could affect local temperatures and humidity and how building orientation relative to prevailing winds could improve natural ventilation and reduce reliance on artificial cooling. The findings, including potential cooling demand savings, will be presented at the conference.

The author will also introduce an Urban Microclimate Design Matrix for floating cities based on insights from traditional oasis settlements. This matrix expands to cover physical aspects (Urban Morphology), environmental factors (Blue and Green Spaces), and social considerations, each with specific design parameters. By incorporating state-of-the-art climate design strategies, this matrix offers a comprehensive framework for creating resilient floating 'oasis' cities in hot climate zones.

Simulating the microclimates of these environments under various conditions—such as different building orientations, modular and organic growth patterns, materials, and blue/green space configurations—will allow urban planners to assess and adapt these principles for modern floating cities in hot, arid climates, providing valuable insights into their effectiveness.

5 Conclusion

Exploring traditional oasis wisdom offers a promising pathway toward developing sustainable floating cities in the dry tropics. By applying the principles of water management, passive cooling, and the integration of biotic and abiotic resources that have sustained oasis communities for centuries, the promise to create urban environments that are resilient, self-sufficient, and capable of adapting to the challenges posed by climate change stands strong.

However, further research is needed to refine and adapt these principles specifically for floating cities to realise this potential fully. Future studies should focus on:

- Climatic Form-Finding: Developing processes integrating design parameters with climatic analyses to optimize urban design through performance simulations.
- Validation and Benchmarking: Establishing quantitative benchmarks for urban microclimates, focusing on techniques for retrofitting existing spaces and developing scalable strategies for new floating contexts.
- Advanced Simulation Techniques: Utilizing high-resolution climate simulations, Computational Fluid Dynamics (CFD), thermal comfort models, and radiative transfer models to test and refine design strategies of all relevant climate and water data.

These efforts will inform the design of sustainable floating cities that are resilient to climate change and capable of providing comfortable living environments for their inhabitants.

References

1. Al-Olaimy T (2024) Climate change impacts in the GCC|EcoMENA 2024. https://www.ecomena.org/climate-change-gcc/. Accessed 28 May 2024
2. Sherif M, Liaqat MU, Baig F, Al-Rashed M (2023) Water resources availability, sustainability and challenges in the GCC countries: an overview. Heliyon 9:e20543
3. Khourchid AM, Ajjur SB, Al-Ghamdi SG (2022) Building cooling requirements under climate change scenarios: impact, mitigation strategies, and future directions. Buildings 12:1519
4. Potchter O, Goldman D, Kadish D, Iluz D (2008) The oasis effect in an extremely hot and arid climate: the case of southern Israel. J Arid Environ 72:1721–33
5. Ottmann PS 2011–12 by DA. Traditional building knowledge Oman. ArcGIS StoryMaps 2021. https://storymaps.arcgis.com/stories/4cbb6650aa6342758941ec785f04b40d. Accessed 13 May 2023
6. Ottmann D (2016) Urban correlator: strategies for an ecologically adaptable urban and architectural development. Sudwestdeutscher Verlag fur Hochschulschriften AG

7. Baumeister J, Ottmann DA (2016) URBAN: ORGAN–Models for Gulf Cities as Integrated Ecological Systems. Gulf Cities Interfaces, 107, Gulf Research Centre, Cambridge
8. Wang S, Sun M, Wang G, Yao X, Wang M, Li J, Duan H, Xie Z, Fan R, Yang Y (2023) Simulation and reconstruction of runoff in the high-cold mountains area based on multiple machine learning models. Water 15:3222
9. Tahiri AZ et al (2019) Promising water management strategies for arid and semiarid environments|IntechOpen 2019. https://www.intechopen.com/chapters/72574. Accessed 31 May 2024
10. Santamouris M (2005) Passive cooling of buildings. 'Advances Sol Energy' ISES James James Sci Publisher, London 34
11. IAE, Holzer P, Abhishek G, Arens E, Attia S, Corrado V, Gupta R, Hamdy M, Homaei S, Laouadi A, Levinson R, Selkowitz S, Sengupta A, Sodagar B, Wang L (Leon), Zinzi M, Stern P, Czarnecki P (2024) International energy agency—resilient cooling of buildings—key performance indicators report (Annex 80). Institute of Building Research & Innovation
12. Alsaeed BS, Hunt DVL, Sharifi S (2024) A sustainable water resources management assessment framework (SWRM-AF) for arid and semi-arid regions—part 1: developing the conceptual framework. Sustainability 16:2634
13. Luedeling E, Buerkert A (2008) Typology of oases in northern Oman based on Landsat and SRTM imagery and geological survey data. Remote Sens Environ 112:1181–95

Challenges for Characterizing Biofouling on Coastal Floating Infrastructure: A Pilot Investigation of Biofouling on Floating Substrata Installed in Tokyo Bay

Jinxin Zhou, Akane Takahashi, Yoshinobu Yoneyama, Ikuo Yoshida, Kotaro Tabata, Kuniaki Okumura, Teng Tu, and Daisuke Kitazawa

Abstract The increasing coastal urbanization and need for resilience against natural disasters have driven the development of floating structures, such as a three-layer floating ground system proposed by Shimizu Corporation. However, a persistent challenge for virtually all floating structures in water has been biofouling, posing uncertainty for floating project succeed. To evaluate potential impacts on aquatic environment, a pilot study utilizing floating substrata was conducted in Tokyo Bay from August 2022 to March 2023. Monitoring of water quality revealed seasonal variations in the aquatic habitat. Typically, hypoxic conditions and low salinity concentrations characterized the water during the summer and fall seasons, whereas the winter seasons witnessed a decline in water temperatures and chlorophyll *a* concentration. However, monitoring of sessile organism attachment elucidated a generally linear increase in organism growth. Moreover, the relationship between habitat characteristics and sessile organism biomass was assessed and compared to data from the 2000s. Surprisingly, the dominant species of this investigation, comprising nearly 95% of the total individuals and wet biomass, was the small brown mussel (*Xenostrobus securis*). However, two decades ago, the Tokyo Bay shoreline was inhabited predominantly by the blue mussel (*Mytilus galloprovincialis*). Deviations in this relationship are now attributed to biotic factors, such as parasite outbreaks, and abiotic

J. Zhou (✉) · D. Kitazawa
Institute of Industrial Science, The University of Tokyo, Chiba, Japan
e-mail: jxzhou@iis.u-tokyo.ac.jp

A. Takahashi · Y. Yoneyama · I. Yoshida
Ocean Program Department, Emerging Frontiers Division, Shimizu Corporation, Tokyo, Japan

K. Tabata · K. Okumura
Sanyo Techno Marine Inc., Tokyo, Japan

T. Tu
Graduate School of Engineering, The University of Tokyo, Chiba, Japan

X. L. Zhao et al. (eds.), *Proceedings of the Fourth World Conference on Floating Solutions: WCFS 2024; 2–4 December; Hong Kong*, Lecture Notes in Civil Engineering 597, https://doi.org/10.1007/978-981-96-4569-5_24

factors, including climate change and anthropogenic impacts. Notably, this pilot experiment has triggered an ongoing, comprehensive investigation in Tokyo Bay.

Keywords Biofouling investigation · Environmental evaluation · Floating cities · Tokyo Bay

1 Introduction

Riverbeds and floodplains must be taken into account when cities expand quickly, particularly megacities where land that is flood-free has already been exploited. The most dangerous flood zones are seeing far quicker growth than less hazardous areas [1]. However, the extreme weather accompanied by climate change has even put these flood zones in danger [2]. More unfortunately, more nations are continuously increasing their vulnerability to these frequent climate disasters because they lack a more effective urban development plan [3].

Sustainable floating cities have the potential to promote future urbanization, and several significant sustainability factors have been well-researched [4]. One stable platform among them is the artificially created floating land on the sea by very large floating structures (VLFS) [5]. Japan initiated the Mega Float project [6] in order to understand the hydroelastic behavior of VLFS, including the mooring, connecting, and anticorrosion systems, as well as their influence on marine physical environment and ecosystem. The Mega Float project's findings have inspired a three-layer structure system for increased stability, with a flexible pit layer and a load distribution layer positioned on top of the floating structure [7].

However, biofouling has been demonstrated to be a persistent issue for nearly all floating structures in water, casting doubt on the sustainability of floating cities. In the previous Mega Float experiment, for example, blue mussels were found attached to the float and altered the surroundings as they grew [8]. A multi-layered structure has been numerically proposed and experimentally validated to evaluate the effects of their development [9], but the consequences of their detachment are yet unknown.

A pilot experiment was conducted in Tokyo Bay starting in August 2022, originally aiming to investigate the detachment of sessile organisms from the floating structure. A summary of the investigations and some of the results, especially regarding the weight changes of sessile organisms, have already been published [8]. Due to the limited observation period, the attachment process was recorded instead of the detachment process. As a result, the purpose of the pilot experiment has been diverted to provide information on whether the present monitoring site is suitable for the planned detachment investigations, by comparing the surrounding environments and sessile organisms between 20 years ago during the Mega Float project and now. This work presents more detailed monitoring data and, for discussion purposes, a numerical reproduction of the sessile organism weights.

2 Methods and Materials

2.1 Field Survey

The pilot experiment was conducted from August 2022 to March 2023. The field site was selected to reproduce the conditions of the Sumida River estuary, which served as the monitoring site of the Mega Float project [8]. The monitored environmental conditions include water temperature and salinity (Infinity-CTW; JFE Advantech Co., Ltd., Japan), chlorophyll *a* concentration (Infinity CLW, JFE Advantech Co. Ltd.), dissolved oxygen concentration (RINKO-W, JFE Advantech Co., Ltd., Japan), and a dedicated cage for biofouling evaluation. All monitoring instruments and equipment were situated 1 m below the sea surface, and six substrata—four on each of the cage's four sides and two at the bottom, are present on the cage.

2.2 Numerical Model

The possible reasons affecting the weight changes in the sessile organisms were numerically examined. A biomass-based model, originally developed for blue mussel during the Mega Float project [8, 10], was adopted for the purpose. This model considers the total weight following the energy gain and loss due to the mussels' activities, such as food ingestion and mortality. In this work, we applied this model with a strong assumption that blue mussel is the dominant sessile organism during the survey period.

The basic computational conditions followed the survey conditions, including the initial weight and the simulation period. The monitored environmental conditions—water temperature, salinity, and dissolved oxygen concentration—were directly input as the boundary condition, but phytoplankton that is reflected by the chlorophyll *a* concentration is only one of its food resources. Therefore, the monthly Chemical Oxygen Demand (COD) data near the field site was referred using the official dataset [11]. In this work, we assume that the COD value may reflect all food resources in a constant ratio to itself, including phyto- and zooplankton, and particulate and dissolved organic carbon. For the parameterization, the previous values were referred [10] but the parameters of mortality rate and ingestion ability were tuned to match the monitoring data herein. Finally, the time step for this simulation was set at 600 s, following the interval of the monitoring data.

The calculated weight is based on the dry meat of the organism and has been transformed to the carbon weight. As a result, the unit transformation of the sessile organism's calculated weight is the main step of the post-analysis, following the previous practice [10]. In brief, the ratio of carbon to dry meat weight is approximately 40%, and the dry meat weight is approximately 0.088 times the total wet weight. In addition, the area of each substrate in this investigation is 0.04 m^2, the total weight of each substrate can also be calculated. The numerical model was conducted in the Fortran environment, and the post-analysis was conducted using R software.

3 Results and Discussions

3.1 Field Survey

This work focuses on the report of four environmental conditions (Fig. 1). For water temperature, it varied from 8 to 30 °C during the period, with a clear seasonal difference. The maximum temperatures occurred in mid-August, and then gradually dropped until late February. In addition, there were no tidal impacts on the variations in water temperature. For salinity, a significant daily and seasonal variation existed. The daily changes in salinity were largely related to tide because the field site is an estuary area and already affected by river and sea waters. Moreover, salinity remained relatively low, with a maximum value of 25 psu, during the high precipitation period from early August to mid-October. In contrast, salinity remained high, with a value of 29–30 psu, during the low precipitation period from mid-October to mid-March. These seasonal changes indicate the large effects by precipitation and subsequent flooding-out river water. For chlorophyll *a* concentration, it remained in the range of 0–66 ppb throughout the period. It showed a clear daily cycle with its concentration increasing during the daytime. However, after mid-October both the daily fluctuation amplitude and frequency decreased. The fluctuation of dissolved oxygen concentration was tightly coincided with that of chlorophyll *a* concentration. Its concentration tends to fluctuate on a daily cycle with a peak during the daytime. Generally, it ranged between 0 and 18 mg/L, with significant daily variations of up to 10 mg/L from mid-August to mid-October. However, after mid-October, the greatest daily fluctuation was only around 2 mg/L, suggesting a decrease in its daily cycle. The onset of its decrease in the fluctuation amplitude coincided with that of chlorophyll *a* concentration. However, the chlorophyll *a* concentration fluctuated at a low level after October, while the dissolved oxygen concentration showed a gradually rising tendency from mid-October to early March. This difference indicates a decrease in the oxygen consumption, plausibly due to the weakened activities of aquatic organisms.

3.2 Numerical Simulation

Generally, the growth of the sessile organism has been reproduced although there are some variations in each substrate (Fig. 2). A slow increase in weight existed from August to late September possibly due to the initial weight, and it was followed by a rapid growth until early December owing to adequate food and suitable water

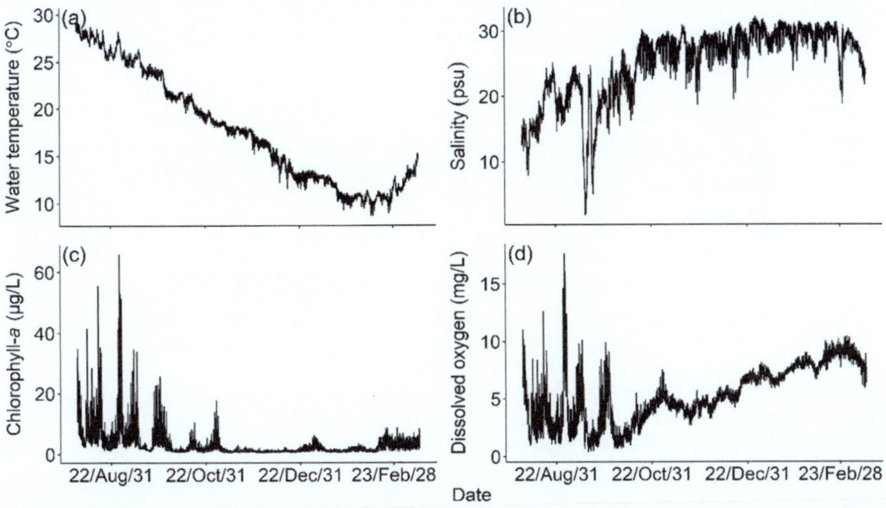

Fig. 1 Monitoring results on the environment. **a** water temperature, **b** salinity, **c** chlorophyll *a* concentration, and **d** dissolved oxygen concentration

temperature and salinity (Fig. 1). After the sharp increase the monitored weight of the sessile organisms tended to exhibit a stable level or slow growth, while the numerical one exhibited a peak and subsequent sharp drop. The numerical results were mainly driven by the food conditions, where a slight increase of chlorophyll *a* concentration promoted the weight increase of the sessile organism. It is reasonable because the current framework is a one-way coupling, and the food availability is the one limitation at the time (Fig. 1). Nonetheless, the survey data, particularly the dissolved oxygen concentration, which showed a general trend toward increase, suggest that the numerical results may have overestimated the fluctuation in January. Because the sessile organism grew slowly in December, phytoplankton was less susceptible to ingestion pressure and quickly multiplied in an environment with constant water temperature (Fig. 1a) and perhaps a sufficient supply of nutrients. However, this hypothesis can not be validated in the current one-way coupling, indicating the necessity of further two-way coupling simulation.

The observed sessile organism challenges the model assumption, although the simulation that targeted blue mussels reached an agreement via parameterization. Instead of blue mussels, small brown mussel (*Xenostrobus securis*) dominated the sessile organisms in the survey. Compared to the Mega Float project, the difference was possibly due to the water salinity (Fig. 1b). From the beginning of the observation until mid-October, more than half of the days there had a salinity below 20 psu and a water temperature over 26 °C. However, the blue mussel become inactive when the salinity falls below 20 psu and the water temperature rises over 26 °C. These abiotic factors together with biotic factors, including parasite outbreaks, caused the disadvantage of its attachment.

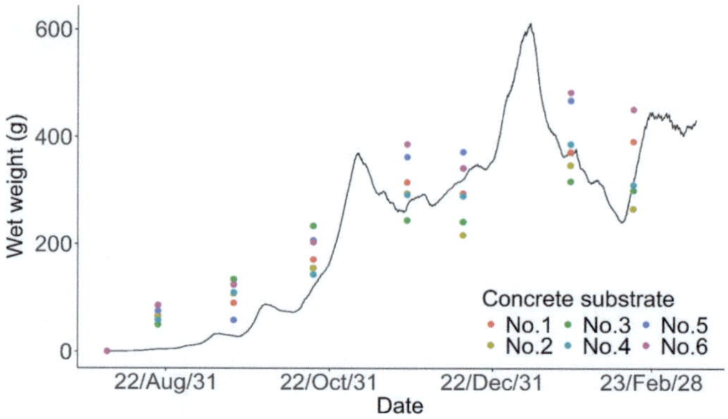

Fig. 2 Comparison of numerical and experimental results on the wet weights of the sessile organisms during the period. Note that blue mussel was assumed to be the dominant organism in the simulation

As a conclusion, a pilot survey was conducted in Tokyo Bay from August 2022 to March 2023, aiming to investigate the environmental effects of sessile organisms. However, the current field site was found unsuitable through the detailed examination on the surrounding environment, monitored sessile organisms, and numerical reproduction.

References

1. Rentschler J, Avner P, Marconcini M, Su R, Strano E, Vousdoukas M, Hallegatte S (2023) Global evidence of rapid urban growth in flood zones since 1985. Nature 622:87–92
2. Otto F, van Oldenborgh G, Eden J, Stott PA, Karoly DJ, Allen MR (2016) The attribution question. Nat Clim Change 6:813–816
3. Hallegatte S (2011) How economic growth and rational decisions can make disaster losses grow faster than wealth. Policy research working paper 5617. World Bank Group. https://doi.org/10.1596/1813-9450-5617
4. Umar T (2020) Making future floating cities sustainable: a way forward. In: Proceedings of the institution of civil engineers-urban design and planning, vol 173, no 6. Emerald Publishing Ltd., England, pp 214–237
5. Wang CM, Tay ZY (2011) Very large floating structures: applications, research and development. Procedia Eng 14:62–72
6. Wang CM, Watanabe E, Utsunomiya T (2008) Very large floating structures. Taylor and Francis, New York
7. Yoshida I, Ohba N, Tamamushi N, Nakashima H, Inoue S, Kobayashi S (2024) Proposal of floating city concept instead of landfill with artificial ground that has a free rebuilding system. In: Proceedings of the third world conference on floating solutions. Springer, Singapore
8. Zhou J, Kitazawa D, Hayakawa Y, Yoshida I, Tabata K, Okumura K (2024) Impacts of sessile organisms attached to artificial floating structures on the surrounding environment: review and update in Tokyo Bay. In: Proceedings of the third world conference on floating solutions. Springer, Singapore
9. Kitazawa D (2011) Competition for space and food among blue mussels. In: Mussels: anatomy, habitat and environmental impact. Nova Science Pub Inc., New York

10. Kitazawa D, Tabeta S, Kato T, Ruardij P (2008) A comparative study of the biomass-based and individual-based models of blue mussels. Ecol Model 215(1–3):93–104
11. Water environment comprehensive information site (Suikankyō sōgō jōhō saito). https://www.env.go.jp/water/mizu_site/. Last accessed 27 Jun 2024

Harvest Simulations of Macroalgae in Japanese EEZ

Zengjie Xia and Shigeru Tabeta

Abstract The greenhouse effect is continuing to threaten our living environment. In response to this situation, the methods of sinking cultured macroalgae to achieve carbon sequestration have been proposed. Macroalgae are highly efficient carbon sequestrators with high carbon to nitrogen ratios and have been observed to have a high net primary productivity. Scientific evidence suggests that wild seaweeds can sequester large amounts of carbon in the ocean by exporting dissolved and particulate organic matters to the deep ocean. In this study, we simulated the potential of mariculture of macroalgae as a carbon dioxide removal method in Japan's exclusive economic zone using algae growth models. We also compared the effects of different growth models on macroalgal production.

Keywords Ocean-based negative emission · Offshore farming · Growth model

1 Introduction

As the second largest inorganic carbon reservoir on Earth, the ocean plays a pivotal role in naturally regulating atmospheric CO_2 concentrations. Therefore, a number of ocean-based carbon dioxide removal (CDR) approaches have been proposed to capitalize on this potential carbon storage capacity. Ocean-based CDR approaches aim to increase the rate of ocean CO_2 uptake and storage by enhancing abiotic processes (i.e., chemical or physical processes such as ocean alkalization [1–4]) or biotic processes (e.g., ocean fertilization [1, 5, 6]). Thus, new topics are emerging that may have great CDR potential to help improve the ability to combat CO_2 emissions, the most promising of which is aquaculture.

Macroalgae (also known as "seaweeds" or "kelp") are highly efficient carbon sequestrants with high carbon to nitrogen ratios [7, 8], and have been observed to have a Net Primary Production (NPP) rate of 91–522 gCm^2yr^{-1}. Scientific evidence

Z. Xia · S. Tabeta (✉)
The University of Tokyo, Kashiwa, Japan
e-mail: tabeta@k.u-tokyo.ac.jp

X. L. Zhao et al. (eds.), *Proceedings of the Fourth World Conference on Floating Solutions: WCFS 2024; 2–4 December; Hong Kong*, Lecture Notes in Civil Engineering 597, https://doi.org/10.1007/978-981-96-4569-5_25

suggests that wild seaweeds may sequester large amounts of carbon in the oceans through the export of organic matter (dissolved and particulate) to the deep ocean (>1000 m), where it is largely buried [8, 9]. While actual burial may occur in more nearshore sediments, the larger algal carbon sink (~90%) is exported through the ocean's "biological pump" [10, 11].

However, as current macroalgae culture facilities are mainly located in coastal areas, the expansion of macroalgae culture is limited by the lack of suitable coastal areas due to variations in nutrient availability and temperature regimes [12, 13]. To address these issues, several offshore macroalgae culture facilities have been designed and evaluated (e.g., Sherman et al.'s seaweed paddocks [14], Buck Buchholz's offshore ring [15]).

Most of the research on large-scale macroalgae aquaculture and carbon sequestration is still in the modeling stage. For example, Wu et al. [16] developed the model "Macroalgae Open Ocean Seawater Farming and Sinking (MOS)" which discusses the CDR potential of MOS and the possible side effects of large-scale aquaculture by combining it with different climate models and geo-biochemical models. They also conducted a sensitivity experiment combining artificial upwelling with MOS to mitigate nutrient deficiencies on the sea surface, and in doing so, expanded the area potentially suitable for mariculture of seaweeds. The "Global MacroAlgae Cultivation MODeling System (G-MACMODS)" established by Arzeno-Solterol et al. [17] optimized seaweed species into four categories and five species on a global scale and identified suitable areas for cultivation through multiple growth simulations. In contrast to the direct access of MOS to external nutrients, G-MACMODS assumed that nitrogen was the limiting macronutrient, included nitrogen cell quota and biomass as state variables. However there are few studies to estimate CDR potential due to macroalgae culture around Japan except for coastal areas [26, 27].

The objective of this study is to estimate the potential of "macroalgae open-sea mariculture and sinking" as CDR method within Japan's EEZ area by using a macroalgae growth model. In the estimation, we compared the effects of using different nutrient uptake models on macroalgal biomass yields and carbon sequestration to discuss their features.

2 Materials and Methods

2.1 Macroalgae Model

The change of seaweed biomass B is calculated by;

$$\frac{dB}{dt} = \mu B - d_M B \tag{1}$$

where μ stands for growth rate per day (1/day), and d_M stands for mortality rate per day (1/day).

The growth rate (μ) of macroalgae is limited by both extrinsic and intrinsic limiting factors:

$$\mu = \mu g(k) g(N) g(T) g(E) \tag{2}$$

$$\mu = \mu g(k) g(Q) g(T) g(E) \tag{3}$$

where $\mu g(k)$ (1/day) is the maximum growth rate of the seaweed species k. The maximum growth rate is constrained by the nitrogen concentration $g(N)$ or nitrogen intracellular quota $g(Q)$, water temperature $g(T)$, and light $g(E)$, all of which are unitless coefficients that vary between 0 and 1.

The temperature limiting term is analogous to the Gaussian probability curve:

$$\begin{aligned} g(T) &= \exp\left(-\beta_1 (T - T_{opt})^2\right), \quad T < T_{opt} \\ g(T) &= \exp\left(-\beta_2 (T - T_{opt})^2\right), \quad T > T_{opt} \\ g(T) &= 1, \quad T = T_{opt} \end{aligned} \tag{4}$$

where T_{opt} is the optimal temperature range for each group of seaweeds, and T is the daily temperature, which was brought to zero by adjusting the β_1 and β_2 coefficients around the lower and upper temperature limits, respectively.

The light limitations were derived to a large extent from phytoplankton studies [18]:

$$g(E) = f \frac{I - I_c}{I_s - I_c} \exp\left(-\frac{I - I_c}{I_s - I_c} + 1\right) \tag{5}$$

where I_s and I_c are daily averages of saturated and compensated irradiance (W/m^2); f is the percentage of daylight used to calculate the dark period, and I is the irradiance reaching a depth of 5 m underwater.

The mortality rate d_M in Eq. 1 is the sum of constant daily mortality rates, i.e., taking into account grazing, senescence, and disease. In this study, different species of macroalgae had the same mortality rate and were set as constant.

2.2 Nitrogen Uptake Models

When the cell quota is not considered, the nitrogen limitation term is as follows,

$$g(N) = \frac{NO_3}{K_N + NO_3} \tag{6}$$

where K_N represents the half-saturation constant for nitrate uptake by macroalgae.

For the cell quota model, the nitrogen quota limit $g(Q)$ is consistent with the Droop model [28]:

$$g(Q) = \frac{Q - Q_{\min}}{Q} \frac{Q_{\max}}{Q_{\max} - Q_{\min}} \tag{7}$$

The changes in cell quota Q is calculated as follows, where Q_{min} is set for each macroalgae species.

$$\frac{dQ}{dt} = V - \mu Q - E(Q - Q_{\min}) \tag{8}$$

where V stands for nitrogen uptake rate [μmol-N/(g-DW h)], E stands for fractional exudation rate per day (1/day). Nitrogen uptake rate V depends on two main aspects: environmental factors and their own intrinsic factors:

$$V = V_{\max} f_1(Q) f_2\left(\left|\vec{v}\right|, T_w, C\right) \tag{9}$$

where $V_{\max}$ represents the maximum rate of nitrogen uptake by seaweed, $f_1(Q)$ represents the dynamic nutrient cell quota, and $f_2\left(\left|\vec{v}\right|, T_w, C\right)$ represents the kinetic and mass transfer limitations of nitrogen uptake. Linear nutrient cell quotas were used in the model [17]:

$$f_1(Q) = \frac{Q_{\max} - Q}{Q_{\max} - Q_{\min}} \tag{10}$$

where $Q_{\min}$ is the minimum nitrogen content (structural nitrogen) in the seaweed cell, $Q_{\max}$ is the maximum nitrogen storage inside the seaweed cell, so that uptake decreases accordingly as the nitrogen concentration in the cell increases, and $f_1(Q)$ is a unitless coefficient between 0 and 1. The function $f_2\left(\left|\vec{v}\right|, T_w, C\right)$ is the limit of uptake based on Michaelis–Menten dynamics and mass-transfer limitations regulated by the surrounding waves and currents [18–20], is unitless and varies between 0 and 1.

2.3 Macroalgae Species

We followed G-MACMODS for seaweed species selection (Fig. 1). The seaweeds simulated globally were divided into four major groups, which contained seaweed species that are among the top ten most cultivated seaweeds in the world by weight [21]: tropical red seaweeds (e.g., *Eucheuma, Gracilaria, Kappaphycus*), tropical brown seaweeds (e.g. *Sargassum*), temperate red seaweeds (e.g. *Porphyra*) and

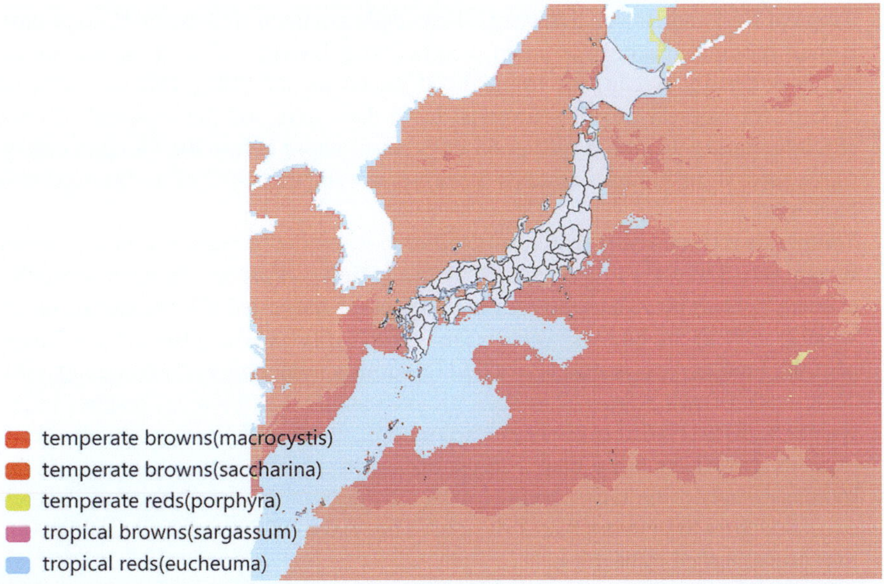

Fig. 1 Seaweed species cultured in each grid. The data comes from G-MACMODS simulation results [17]

temperate brown seaweeds (e.g. *Saccharina*, *Laminaria*, *Macrocystis*). The baseline simulation had a standard run period of one year and was optimized for seeding by selecting a run start date that produced the maximum annual biomass harvest for each grid point.

2.4 Harvest Strategy

Harvest strategies for macroalgae in the baseline simulations were optimized by G-MACMODS. Harvest schedules in G-MACMODS were developed based on available information on existing farming practices and optimized for each seaweed based on standard running trials of three harvest schedules to achieve maximum biomass per harvest: regular harvests, regular harvests with biomass maxima, and conditional harvests (with a double criterion of target weights or mortality over growth). According to the literature, the number of harvests per year is limited; globally temperate brown and red algae are usually harvested twice [22] and six times per year [23], respectively, while tropical brown and red algae are harvested up to eight times per year [24, 25]. Temperate brown algae are allowed to grow without regard to harvest for at least 60 days after seeding. Thus, the specific harvesting scheme for each type of seaweed is as follows:

Tropical Red and Brown Seaweeds: Seaweeds are harvested every 45 days only if their biomass reaches the target weight 800 g-DW/m^2 (27% of the maximum biomass density B_cap) for tropical red seaweeds and 400 g-DW/m^2 (50% of B_cap) for tropical brown seaweeds. If 45 days pass and the seaweeds do not reach the target weight, a further 45 days must elapse before the biomass can be reassessed. If the biomass reached the target weight, then 80% of the biomass was harvested.

Temperate red seaweeds: Harvest is initiated when the biomass reaches its target weight 80 g-DW/m^2 (40% of B_cap) within 150 days of seeding or when mortality exceeds growth by 7 days. 80% of the biomass is harvested if it reaches its target weight; 99% of the biomass is harvested if mortality exceeds growth by > 7 days or if the final harvest period is reached (assuming a final harvest loss rate of 1%).

Temperate Brown Algae: Seaweeds are harvested when biomass reaches target weight 1350 g-DW/m^2 (68% of B_cap) within 220 days of seeding or when death exceeds growth by more than 7 days. 80% of the biomass was harvested if the biomass reached the target weight; 99% of the biomass was harvested if death exceeded growth by more than 7 days or at the end of 220 days (final total harvest loss rate assumed to be 1%).

2.5 *Environmental Data*

Environmental inputs (water temperature, solar irradiance, current, wave, and nutrient concentration) applied to macroalgal growth models are derived from a combined database of satellite measurements and global ocean model outputs spanning many years. To assess the carbon sequestration capacity of macroalgae, in this paper we use a set of data averaged over decades to input models for simulations. The model was run in daily time steps; however, because the data came from different databases and the resolution of the data varied (both 1/4 and 1/12 degree horizontal resolution were included), we used interpolation when entering the data into the model to interpolate all of the environmental inputs into a global grid at 1/12 degree horizontal resolution.

3 Results and Discussions

We simulated all grids within the Japanese EEZ. For the simulations, macroalgae were seeded and cultured below the surface of the open ocean (5 m depth) to avoid storm damage. The total operating cycle is one year, and at the end of the cycle of macroalgae cultivation, a portion of the mature macroalgae rapidly sinks to the seafloor where it offloads its biomass, while the rest of the biomass continues to the next cultivation cycle. This can be thought of as a short-circuiting of the biological pump, bringing marine biomass directly to the seafloor without remineralization along the way [16].

For the model without nitrogen quota (Fig. 2a), the highest biomass production was observed in the northeastern part of the EEZ due to favorable nutrient and water temperature conditions. The grids with annual biomass production above 1000 g-DW/m^2/yr accounted for only about 9% of the total, and those with annual production above 500 g-DW/m^2/yr accounted for about 14.5%. Among them, temperate brown algae dominated the production, accounting for 98.9% of the data greater than 14.5%. Temperate brown algae (*Saccharina* and *Macrocystis*) accounted for 99.9% of the grids with annual yields greater than 500 g-DW/m^2/yr, and almost all of the grids with high yields were contributed by temperate brown algae.

On the other hand, as can be seen in Fig. 2b, by the nutrient uptake model with nitrogen cell quota, the yields of all four macroalgae were more optimistic. Overall,

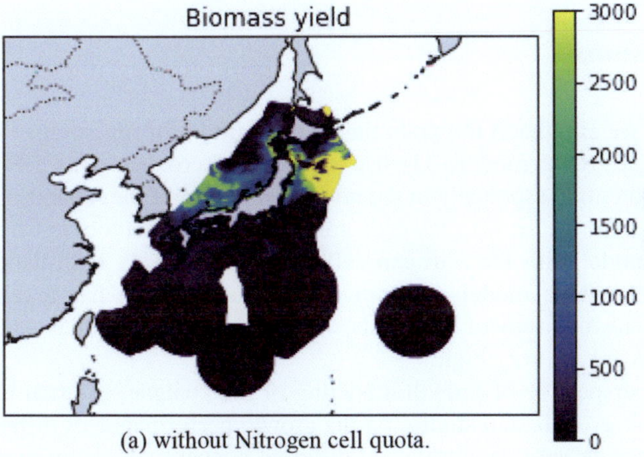

(a) without Nitrogen cell quota.

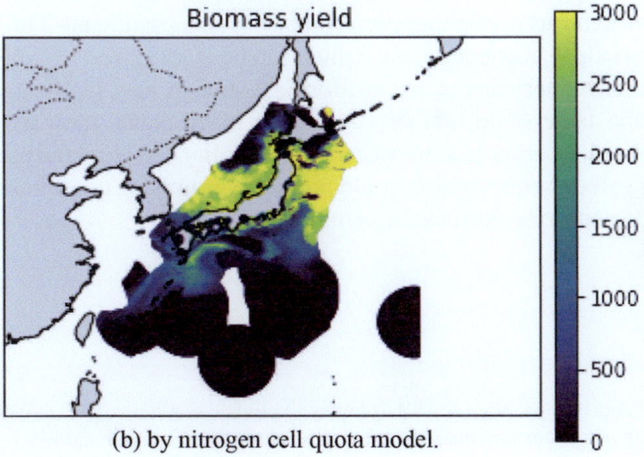

(b) by nitrogen cell quota model.

Fig. 2 Distribution of biomass yield

grids with final annual production greater than 1000 g-DW/m^2/yr accounted for about 23% of the total, and grids with annual production in the range of 1000–500 g-DW/m^2/yr accounted for about 11.5% of the total. Among them, temperate brown algae represented by *Saccharina* and *Macrocystis* were the most widely distributed macroalgae with the highest biomass production accounting for 74% of all grids with annual production greater than 500 g-DW/m^2/yr, which also indicates that temperate brown algae have an important position in carbon sequestration projects through the cultivation of macroalgae. And, thanks to favorable environmental conditions (especially water temperature and nutrients), macroalgae production was highest in the north-central (east coast) and south-central (west coast) regions of the entire EEZ; macroalgae growth was initially very rapid in grids from these and other regions, and most of the very productive grids were able to be harvested multiple times a year, which greatly increased the upper limit of production.

4 Conclusions

In this study, we simulated the potential of mariculture of macroalgae in Japanese EEZ using algae growth models. The use of sinking macroalgae as a CDR method has a promising potential, especially in the northeastern part of the Japanese EEZ. Tropical red and temperate brown seaweeds are the main contributors to carbon sequestration. The model with the nitrogen cell quota is relatively optimistic compared to the nutrient uptake models without cell quota in that it has a relatively low external nutrient requirement, so yields will be higher in the areas where nutrient concentrations are not very high.

From the perspective of growth modeling of macroalgae, external nutrients are crucial for their growth, and studies on the growth performance of different species of macroalgae with various nutrients at different concentrations are important. The design of seaweed farms in offshore environments is also significant. The farm design will affect macroalgae nutrient uptake, which can be investigated by considering the hydrodynamic characteristics of the platform. Although new platforms have been designed for the large-scale offshore culture of macroalgae, they are still in the experimental stage. In order to achieve carbon neutrality, the potential of macroalgae cultivation for carbon sequestration could provide a promising contribution, so more efforts are expected to develop the system.

References

1. Keller DP, Feng EY, Oschlies A (2014) Potential climate engineering effectiveness and side effects during a high carbon dioxide-emission scenario. Nat Commun 5:3304

2. Taylor LL, Quirk J, Thorley R, Kharecha PA, Hansen J, Ridgwell A, Beerling DJ (2016) Enhanced weathering strategies for stabilizing climate and averting ocean acidification. Nat Clim Chang 6:402–406
3. Köhler P, Abrams JF, Völker C, Hauck J, Wolf Gladrow DA (2013) Geoengineering impact of open ocean dissolution of olivine on atmospheric CO_2, surface ocean pH and marine biology. Environ Res Lett 8:014009
4. Albright R, Caldeira L, Hosfelt J, Kwiatkowski L, Maclaren JK, Mason BM, Caldeira K (2016) Reversal of ocean acidification enhances net coral reef calcification. Nature 531:362–365
5. Smetacek V, Klaas C, Strass VH, Assmy P, Montresor M, Cisewski B, Wolf-Gladrow D (2012) Deep carbon export from a southern ocean iron-fertilized diatom bloom. Nature 487:313–319
6. FAO (2018) The state of world fisheries and aquaculture: meeting the sustainable development goals (Food and Agriculture Organization of the United Nations)
7. Buschmann AH, Camus C, Infante J, Neori A, Israel Á, Hernández-González MC, Critchley AT (2017) Seaweed production: overview of the global state of exploitation, farming and emerging research activity. Eur J Phycol 52:391–406
8. Krause-Jensen D, Duarte CM (2016) Substantial role of macroalgae in marine carbon sequestration. Nat Geosci 9:737–742
9. Wernberg T, Filbee-Dexter K (2018) Grazers extend blue carbon transfer by slowing sinking speeds of kelp detritus. Sci Rep 8:17180
10. Raven JA (2017) The possible roles of algae in restricting the increase in Atmospheric CO_2 and global temperature. Eur J Phycol 52:506–522
11. Krause-Jensen D, Lavery P, Serrano O, Marbà N, Masque P, Duarte CM (2018) Sequestration of macroalgal carbon: the elephant in the Blue Carbon room. Biol Lett 14:20180236
12. Duarte CM, Wu J, Xiao X, Bruhn A, Krause-Jensen D (2017) Can seaweed farming play a role in climate change mitigation and adaptation? Front Mar Sci 4:100
13. Oyinlola MA, Reygondeau G, Wabnitz CC, Cheung WW (2020) Projecting global mariculture diversity under climate change. Glob Chang Biol 26:2134–2148
14. Sherman MT et al (2019) Initial modeling and design for a Sargassum Ranch. In: Oceans 2018MTS/IEEE Charleston, SC, USA, 22–25 October 2018, pp 1–6
15. Buck BH, Buchholz CM (2004) The offshore-ring: a new system design for the open ocean aquaculture of macroalgae. J Appl Phycol 16:355–368
16. Wu J, Keller DP (2023) Andreas Oschlies Carbon dioxide removal via macroalgae open-ocean mariculture and sinking: an earth system modeling study. Earth Syst Dynam 14:185–221
17. Arzeno-Soltero, IB, Saenz, BT, Frieder, CA, et al. (2023) Large global variations in the carbon dioxide removal potential of seaweed farming due to biophysical constraints. Commun Earth Environ 4, 185
18. Stevens CL, Hurd CL (1997) Boundary-layers around bladed aquatic macrophytes. Hydrobiologia 346:119–128
19. Fram JP, Stewart HL, Brzezinski MA, Gaylord B, Reed DC, Williams SL, MacIntyre S (2008) Physical pathways and utilization of nitrate supply to the giant kelp, Macrocystis pyrifera. Limnol Oceanogr 53:1589–1603
20. Sanford LP, Crawford SM (2000) Mass transfer versus kinetic control of uptake across solid-water boundaries. Limnol Oceanogr 45:1180–1186
21. Ferdouse F, Holdt SL, Smith R, Murúa P, Yang Z (2018) The global status of seaweed production, trade and utilization. Globefish Research Programme 124, I
22. Bak UG, Mols-Mortensen A, Gregersen O (2018) Production method and cost of commercial-scale offshore cultivation of kelp in the Faroe Islands using multiple partial harvesting. Algal Res 33:36–47
23. Wang X, He L, Ma Y, Huan L, Wang Y, Xia B, Wang G (2020) Economically important red algae resources along the Chinese coast: history, status, and prospects for their utilization. Algal Res 46:101817
24. Hurtado AQ (2013) Social and economic dimensions of carrageenan seaweed farming in the Philippines. Fisheries and aquaculture technical paper no 580

25. Msuya FE, Shalli MS, Sullivan K, Crawford B, Tobey J, Mmochi AJ (2007) A comparative economic analysis of two seaweed farming methods in Tanzania. In: Sustainable coastal communities and ecosystems program. Coastal Resource Center, University of Rhode Island West, Indian Ocean Marine Science Association
26. Fukuzaki K, Tabeta S (2011) Evaluation of the carbon fixation potential by restoring seaweed beds in Japanese coastal waters. Conference Proceedings the Japan Society of Naval Architects and Ocean Engineers 13:251–252
27. Kuwae T, Yoshida G, Hori M, Watanabe K, Tanaya T, Okada T, Umezawa Y, Sasaki J (2019) Nationwide Estimate of the Annual Uptake of Atmospheric Carbon Dioxide by Shallow Coastal Ecosystems in Japan. Journal of Japan Society of Civil Engineers, Ser. B2, 75(1):10–20
28. Droop MR (1973) Some thoughts on nutrient limitation in algae. J Phycology 9:264–272

Floating Tunnel and Mooring System

A Novel Submersible Mooring System Suitable for Honeycomb Modular Floating Structure

Yufei Wu, Longting Qiu, and Xiaoxu Huang

Abstract Improving the survivability of floating structures in harsh marine environments is the key to the large-scale engineering application of floating structures. The authors and their research team have innovatively proposed a new submersible floating mooring system to deal with extreme sea conditions. Being able to switch between floating and diving states according to the environmental conditions, the survivability and safety of floating structures can be guaranteed. Take a cellular modular floating structure as an example, the effectiveness of the proposed submersible floating mooring system was verified in this paper, by conducting the motion response analysis and mooring time domain analysis based on the ocean engineering software SESAM. The results show that the motion response and mooring line tension of the floating structure decrease with the increase of the diving depth. The vertical acceleration of the floating structure was reduced and the comfort was improved correspondingly through diving. The proposed submersible floating mooring system provides promising insights for the application of marine floating structures in extreme sea conditions.

Keywords Modular floating structure · Mooring system · Submersible

1 Introduction

Since the last century, floating structures have been extensively studied as a viable solution to alleviate the shortage of land resources. The advantages of floating structures are obvious, including independence from water depth, excellent seismic resistance, ease of expansion and relocation, and minimal impact on marine ecosystems [1]. Recently, the authors' research team has developed a novel type of modular floating structure- Honeycomb Modular Floating Structure (HMFS). As illustrated in Fig. 1, the HMFS exhibits several prominent advantages. (a) The floating modules are

---

Y. Wu · L. Qiu · X. Huang (✉)
School of Civil and Transportation Engineering, Shenzhen University, Shenzhen, China
e-mail: xxhruby@szu.edu.cn

X. L. Zhao et al. (eds.), *Proceedings of the Fourth World Conference on Floating Solutions: WCFS 2024; 2–4 December; Hong Kong*, Lecture Notes in Civil Engineering 597, https://doi.org/10.1007/978-981-96-4569-5_26

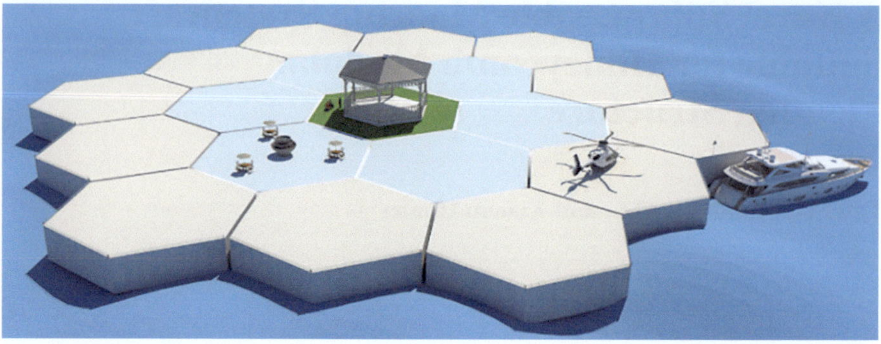

Fig. 1 Honeycomb modular floating structure (HMFS)

constructed with concrete, which is more economical in manufacturing and maintenance compared to traditional steel structures [2]; (b) Different from square modules that are confined to linear expansion, hexagonal floating units allow for dynamic and flexible expansion according to specific requirements [3]; (c) The HMFS is designed as a prefabricated structure, which significantly reduces construction and installation time as well as associated costs; (d) The HMFS features superior sealing performance, enabling the structure to submerge and resurface as needed. The HMFS provides a novel and sustainable approach to the development of floating structures in practical engineering.

The survivability of floating structures under harsh sea conditions is a critical factor influencing their safety, stability, and large-scale application. The ability of floating structures to withstand extreme sea conditions is highly related to factors such as the material and shape of the modules, connectors between modules, and mooring systems [4]. Among them, the mooring system, a key component that provides positioning and restoring force, is an important factor affecting the survival of floating structures in extreme sea conditions. However, in harsh marine environments, the strength of mooring lines decreases as the service life increases, which increases the failure risk of the mooring system. Once the failure of the mooring system occurs, the drift of the floating unit increases significantly, leading to the increasing risk of collision between floating units [5]. Aiming at the mooring design of modular floating structures in extreme sea conditions, a submersible floating mooring system (SFMS) is proposed in this study based on the single floating body module of HMFS. By increasing or decreasing ballast water and retracting or releasing mooring ropes, the mooring system can autonomously switch between floating and submersible states according to changes in environmental conditions. This function quickly realizes the spatial transfer of the module floating structure, improves the wind and wave resistance of the floating structure, and reduces the operation and maintenance costs of the structure. This new submersible and floating mooring system provides an efficient, economical, and feasible solution for the safe service of modular marine floating structures in harsh marine environments.

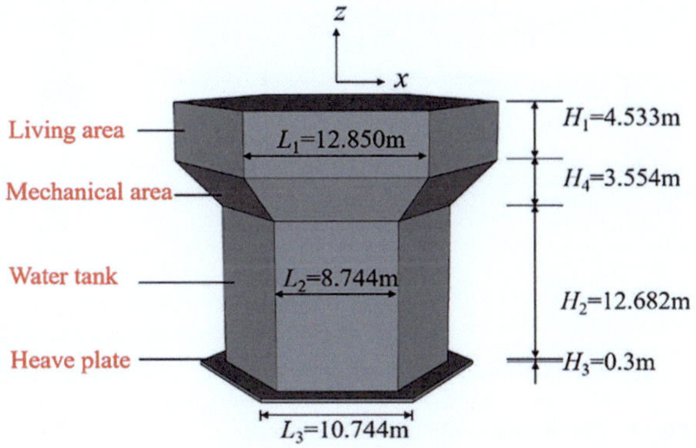

Fig. 2 Configuration and dimension of HMFS unit

2 Description of Submersible Mooring System

2.1 Honeycomb Modular Floating Structure

The honeycomb modular floating structure (HMFS) adopts the design of upper and lower regular hexagons. The floating unit consists of a living layer, a mechanical layer, and a water tank. The geometrical configuration and dimension of the HMFS module unit are depicted in Fig. 2. The HMFS is made of compressed cast concrete with a density of 2500 kg/m^3 and a thickness of 0.3 m. To identify the effect of diving depth on the effectiveness of the submersible mooring system, six states (i.e., S0, SI, SII, SIII, SIV, and SV) are considered here, representing the position of the floating unit. Among them, S0 is a floating status on the sea surface with a freeboard of 4 m; SI–SV represents a submerged status with corresponding diving depths (the distance between the top of the module and the sea surface) of 5 m, 10 m, 15 m, 20 m, and 25 m, respectively.

2.2 Submersible Mooring System

Based on HMFS, a submerged floating mooring system has been proposed in this study, as shown in Fig. 3. This submersible mooring system includes a mooring system and a buoyancy regulating device, in which the mooring system ensures that the floating body remains in a predetermined position, while the ballast tank controls the inflow and discharge of seawater through valves. Additionally, the tension of the mooring line is controlled within a reasonable range by retracting and releasing the line through the winch. The principle of the submerged floating system is shown in Fig. 4 with three stages described as follows:

(a) Working Stage: For the working sea state, the floating unit works normally at sea level;

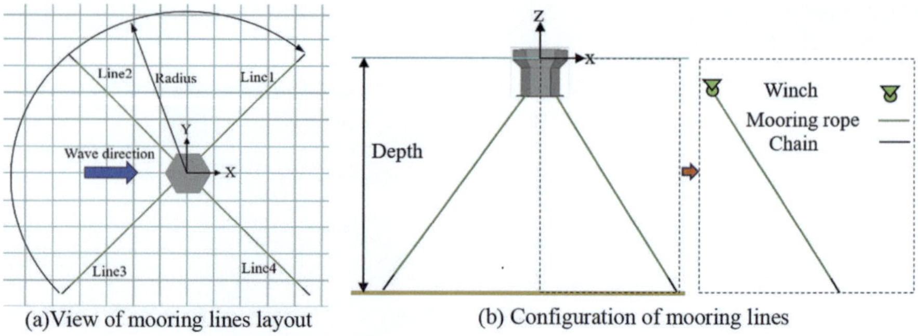

Fig. 3 The submersible mooring system

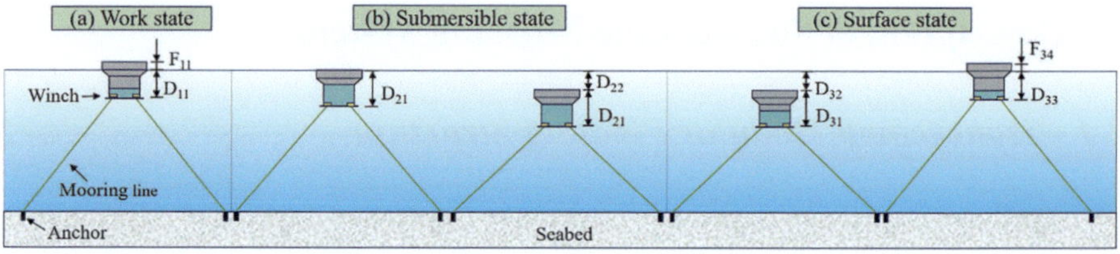

Fig. 4 Submersible mooring system workflow

(b) Submerging Stage: For the self-survival sea state, the floating unit is set to dive to the pre-defined position by filling the ballast tank with water and tightening the line with a winch;
(c) Floating Stage: When the sea state changes to the working condition, the floating body is set to float to the pre-defined position by draining the ballast tank and releasing the line with a winch.

3 Results and Discussion

To verify the effectiveness of the proposed submersible mooring system, several key indicators such as motion response (surge, heave, and pitch), line tension, and vertical acceleration are compared before and after submersion under the survival condition. In addition, the effect of diving depth on the key indicators is identified to find an optimal depth for the floating unit. In this study, the offshore area of the South China Sea with a water depth of 100 m is selected. The JONSWAP spectrum is considered with a peak enhancement factor of 3.3. The environmental action direction is set as 0°. The total simulation duration is 12,000 s, excluding the first 1200 s to avoid transient effects.

Figure 5 presents the results for motion response, line tension, and vertical acceleration of the floating unit. It can be found that:

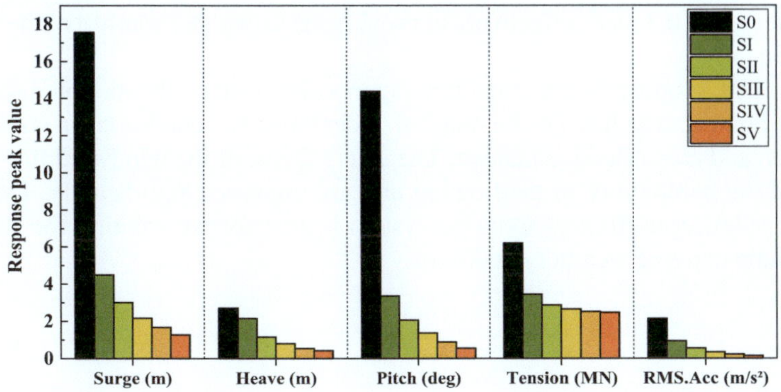

Fig. 5 Comparison of response under surface and submersible status

(1) The floating unit maintains excellent safety and stability in the process of submersion. Compared to the S0 status, the peak values of surge, heave, pitch, and line tension for the SIV status are reduced by 11.34, 2.78, 13.95, and 1.38 times, respectively.
(2) As the diving depth increases, the motion response of the floating unit, line tension, and vertical acceleration decrease progressively. This indicates that deeper submersion enhances the stability and reduces the loads on the floating unit.
(3) When the floating unit is submerged to 20 m (i.e., SIV), the root mean square (RMS) amplitude of vertical acceleration is less than 0.3 m/s^2, meeting the human comfort requirement in the specified standards [6, 7]. This demonstrates that submersion can effectively ensure the habitability and comfort of the floating unit.

4 Conclusion

In this paper, a new submersible floating mooring system (SFMS) has been proposed to address extreme sea conditions. This system allows floating structures to switch between floating and diving states according to environmental conditions, thereby enhancing their survivability and safety. Taking a Honeycomb Modular Floating Structure (HMFS) as an example, the effectiveness of the proposed mooring system was verified based on the SESAM software. The following conclusions can be drawn:

(1) The motion response of the floating unit, line tension, and RMS acceleration all show a decreasing trend as the diving depth increases;
(2) Compared to the floating state on the sea surface (i.e., S0 state), the maximum values of surge, pitch, heave, and the maximum tension of the windward line for the SIV state (i.e., a diving depth of 20 m) decreased by factors of 11.34, 2.78, 13.95, and 1.38, respectively. Additionally, the RMS acceleration was reduced

to less than 0.3 m/s^2, effectively improving the safety and comfort of the floating unit;

(3) With the proposed submerged floating mooring system, the motion response of the floating structure can be reduced under harsh sea conditions in an efficient, safe, and cost-effective manner. The safety factor of the windward line tension and the habitability of the floating unit are improved significantly. Therefore, the submerged floating mooring system is an effective solution for HMFS to handle extreme weather conditions.

References

1. Wang G, Drimer N, Goldfeld Y (2020) Modular floating structures (MFS) for offshore dwelling a hydrodynamic analysis in the frequency domain. Ocean Eng 216:107996
2. Dneg L, Wang B, Xiao Z et al (2016) Conceptual design and performance analysis of a reinforced concrete platform for floating wind turbines. J Huazhong Univ Sci Technol (Nat Sci Ed) 44:11–15
3. Stanković J, Krasić S, Mitković P et al (2021) Floating modular houses as solution for rising sea levels. Conference: eCAADe 2021: towards a new, configurable architecture
4. Drummen I, Olbert G (2021) Conceptual design of a modular floating multi-purpose island. Front Mar Sci 8:615222
5. Wang H, Ke S, Wang T et al (2020) Typhoon-induced vibration response and the working mechanism of large wind turbine considering multi-stage effects. Renewable Energy 153:740–758
6. Wang G, Rosenfeld Y, Drimer N et al (2020) Occupant comfort analysis for rigid floating structures—methodology and design assessment for offshore dwelling module. Ships Offshore Struct 16:184–199
7. Jiang C (2021) Mathematical modelling of wave-induced motions and loads on moored offshore structures. In: Dissertation, Universität Duisburg-Essen, Duisburg, Essen

A New Type of Constant Tension Pulley Mooring System for Hexagonal Floating Units in Shallow Water

Haohan Chen, Yufei Wu, and Xiaoxu Huang

Abstract Floating structure is the basis of marine resource exploitation and marine space utilization. The performance of the mooring system directly affects the overall cost of the floating structure and its survivability in extreme conditions. For a conventional taut mooring system, significant fluctuations in tension occur in mooring lines due to the cyclic loading of waves, which will amplify the fatigue effect and reduce the cable's service life. To overcome these issues, the authors have proposed a constant tension pulley mooring system for floating structures. The proposed mooring system consists of clump weight, pulley, and wires. Based on the time-domain potential flow theory, the motion of the floating structure and the tension in the cables were identified through the software ANSYS-AQWA. By comparing with traditional taut mooring systems, the maximum tension and fluctuation of cable tension were significantly reduced for the proposed mooring system, thereby reducing mooring system costs effectively. This study provides a promising pathway toward sustainable and reliable mooring systems.

Keywords Mooring system · Floating structure · Time-domain analysis

1 Introduction

In the past few decades, offshore floating structures have been widely utilized in marine exploration and production activities, in which mooring systems play a crucial role in ensuring the normal operation of floating structures. Depending on the material, configuration, operational requirements, and layout characteristics of mooring lines, mooring systems can be categorized into several types: catenary mooring, taut mooring, and tension leg mooring systems. Among them, the weight of catenary mooring systems increases with the water depth, leading to higher costs for the floating structures. Taut mooring systems exhibit low dynamic stiffness at low tension

H. Chen · Y. Wu · X. Huang (✉)
School of Civil and Transportation Engineering, Shenzhen University, Shenzhen, China
e-mail: xxhruby@szu.edu.cn

X. L. Zhao et al. (eds.), *Proceedings of the Fourth World Conference on Floating Solutions: WCFS 2024; 2–4 December; Hong Kong*, Lecture Notes in Civil Engineering 597, https://doi.org/10.1007/978-981-96-4569-5_27

levels, making them susceptible to fatigue failure and material aging, especially for high-strength synthetic materials such as nylon and polyester ropes [1, 2].

Recently, installing clump weights in the middle of mooring lines has garnered significant attention as an economical and effective mooring system for floating platforms. Fan et al. [3] innovatively proposed a flexible mooring system with load buffering characteristics by spacing buoys and clump weights along the mooring lines, imparting horizontal stiffness to the system. This design shifts the system's natural frequency away from the wave frequency, significantly reducing dynamic loads and mooring system costs. For semi-submersible wind turbines, Wang et al. [4] introduced a constant tension mooring system suitable for extreme shallow water conditions. This system consists of clump weights and winch wires to release kinetic energy by increasing platform movement, ensuring the instantaneous maximum mooring force does not exceed the chain's breaking strength.

Considering the advantages of taut mooring systems and those with clump weights, this paper proposes a new pulley mooring system (PMS) based on nylon ropes. Unlike existing clump weight arrangements, the PMS places clump weights at the rope's end, connecting the anchor, floating platform, and clump weights through pulleys. This system has significant advantages: the tension is directly provided by the clump weights, avoiding the need to re-tension the ropes due to the relaxation of traditional taut mooring cables during use. Additionally, the setup of the clump weights enables the cable tension to be maintained within a constant range, significantly reducing tension fluctuations and improving the cable's service life.

2 Description of the Pulley Mooring System (PMS)

2.1 PMS Configuration

Traditional taut mooring systems usually require multiple anchors to connect the floating body to the seabed, resulting in high installation and maintenance costs. The randomness of waves leads to uneven tension distribution in mooring ropes, exacerbating fatigue damage and reducing mooring effectiveness. To address these shortcomings of traditional taut mooring systems, in this study a new pulley mooring system (PMS) is proposed based on hexagonal floating units (see Fig. 1 and Table 1). As depicted in Fig. 2, in the PMS configuration, a single anchor and rope connect the floating body to the seabed, with a clump weight at one end of the rope. Fixed pulleys are installed at the bottom of the floating body to alter the force direction on the rope. The rope tension is adjusted by the clump weight's gravity, damping force, and inertia.

Moreover, the PMS effectively enhances stability by controlling the heave and pitch motions of the floating body through rope tension and the system's buoyancy and clump weight gravity. By relaxing constraints on sway, roll, and yaw, the floating body adapts to environmental loads, aligning the rope direction with the load

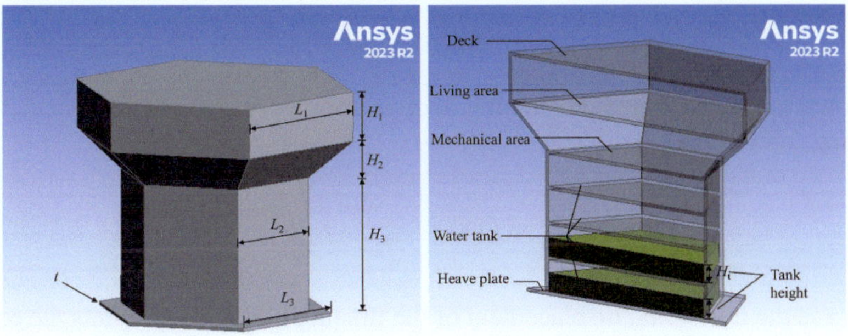

Fig. 1 Schematic of the hexagon-shaped floating unit

Table 1 Main geometric parameters of the unit

Parameter	Value
H_1 (m)	4.531
H_2 (m)	3.554
H_3 (m)	12.384
L_1 (m)	12.85
L_2 (m)	8.744
L_3 (m)	10.744
t (m)	0.3
Freeboard height (m)	4.769
Weight (t)	2359.7
Volumetric displacement (m^3)	3867

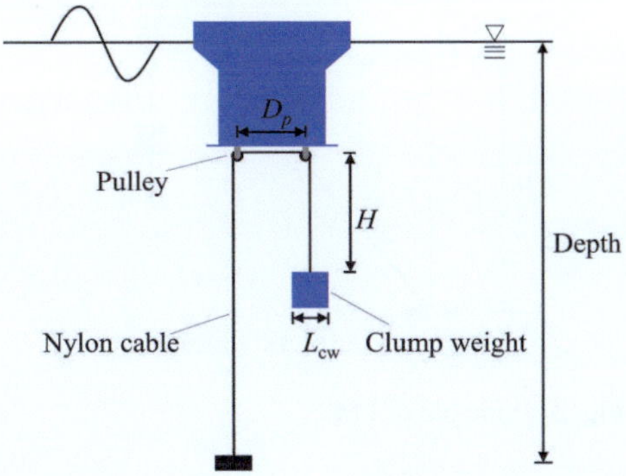

Fig. 2 Schematic diagram of PMS

direction. This approach reduces the need for multiple ropes, significantly lowering mooring system costs and operational expenses.

2.2 Principles of PMS

As shown in Fig. 3a, when the floating body undergoes horizontal displacement, the restoring force F_x is calculated as follows:

$$F_x = f(x) = \frac{G \cdot x}{\sqrt{x^2 + H_s^2}} \tag{1}$$

where x is the horizontal displacement of the floating body, G represents the equivalent gravity of the clump weight, and H represents the depth from the bottom of the floating body to the seabed.

As shown in Fig. 3b, When the floating body rotates, the cable provides a restoring moment to suppress the rotation of the floating body, which is calculated as follows:

$$M(\alpha) = G\left(\frac{|D(h\cos\alpha - d\sin\alpha) - H_s(d\cos\alpha + h\sin\alpha)|}{\sqrt{(D - d\cos\alpha - h\sin\alpha)^2 + (H_s + d\sin\alpha - h\cos\alpha)^2}} - d\cos\alpha + h\sin\alpha\right) \tag{2}$$

where D represents the horizontal distance between the anchor and the center of gravity of the floating body, α denotes the clockwise rotation angle, h represents the vertical distance between the pulley and the center of gravity of the floating body, and d represents the horizontal distance between the pulley and the center of gravity of the floating body.

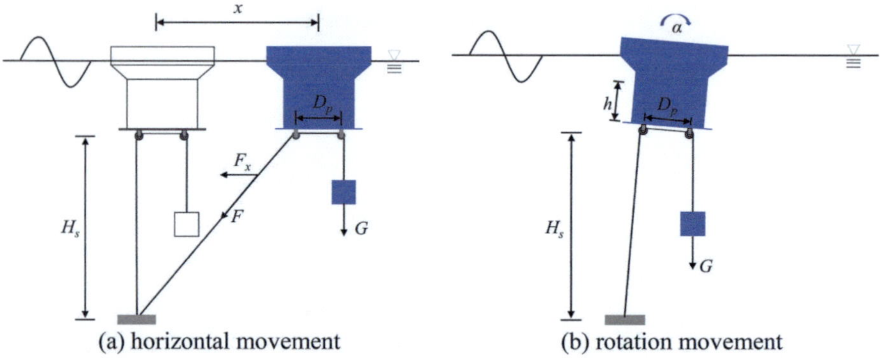

Fig. 3 Principle of PMS

3 Optimization and Comparison

To verify the effectiveness of the proposed pulley mooring system (PMS), this paper carries out a simulation analysis using ANSYS-AQWA software. The working sea conditions of the South China Sea are considered for the analysis, based on the summary of the South China Sea conditions by Sun et al. [5]. The JONSWAP spectrum is used as the wave spectrum, with a significant wave height of 3 m, a peak period of 5.5 s, a wind speed of 15 m/s, and a current speed of 1 m/s.

3.1 *Optimized Design of PMS*

This study aims to enhance the economy of the mooring system by optimizing five key parameters: the filling height of the water tank H_t, the minimum breaking strength of the cable *MBS*, the length of the cable *L*, the side length of the cubic clump weight L_{cw}, and the distance between the pulleys D_p. The optimization problem can be expressed as:

$$Find: \quad H_t, MBS, L, L_{cw}, D_p$$

$$minimize: \quad \text{Cost} = (0.012 \times MBS + 12.116) \times L + 41 \times V_{CW} + 10.198 \times T_{\max} \times SF$$

$$subjected\ to: \begin{cases} 0 \le H_t \le 2.5\text{m} \\ 6000\ \text{kN} \le MBS \le 12000\ \text{kN}, \quad MBS/T_{\max} \ge 2.25 \\ 2.75\ \text{m} \le L_{cw} \le 7.25\ \text{m}, \quad 6\ \text{m} \le D_p \le 17.488\ \text{m} \\ 104\ \text{m} \le L - D_p \le 154\ \text{m}, \quad GM \ge 1\text{m}, \quad RY < 15^\circ \end{cases}$$

where V_{cw} is the volume of the clump weight, $T_{\max}$ is the maximum tension in the cable, *SF* is the safety factor of the anchor (1.5), *GM* is the initial stability height of the floating body, and *RY* is the maximum pitch motion of the floating body.

To improve optimization efficiency, a BP neural network is aopted to establish a proxy model for each relevant parameter (GM, Cost, Cable tension), and combines it with the PSO algorithm for optimization. The iteration process of the mooring cost is shown in Fig. 4. It can be seen that after 300 iterations, the results have essentially converged, and the optimal cost is $36,750. The corresponding optimal values of the design parameters are shown in Table 2. Simulation analysis shows that, compared with the initial design, the cost is reduced by 22.67%. In addition, the maximum pitch is reduced by 56.1%, the fluctuation range of cable tension is reduced by 50.30%, and the maximum cable tension is reduced by 21.99%.

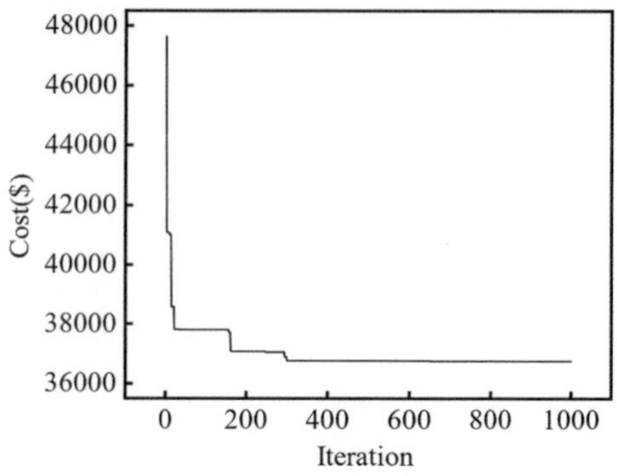

Fig. 4 The iterative process of the cost for the pulley mooring system

Table 2 Design parameters for different moorings

Parameter	Pulley mooring system		Taut mooring system
	Initial design	Optimal design	Initial design
Tank height (m)	0.439	1.222	1.222
MBS (kN)	9804.104	6030.7362	6030.7362
Cable length (m)	148.94	166.953	88.87
Distance between pulleys (m)	9.754	13.000	/
Side length of clump weight (m)	3.989	4.027	/

3.2 Comparison with the Taut Mooring System

To verify the advantages of the proposed pulley mooring system, this section compares it with a conventional taut mooring system, as shown in Fig. 5, The comparison parameters include the floating structure motion response, cable tension, and mooring cost.

Simulations were carried out for 12,000 s with a time step of 0.01 s. Based on the results of the time-domain calculations, the statistical results of the motion response and the mooring line tension for the mooring system were plotted, as shown in Fig. 6. Finally, an economic comparison of the various mooring systems was conducted based on the maximum value of the mooring line tension and the mooring line parameters, as shown in Fig. 6.

4 Discussion and Conclusion

Based on the optimized pulley mooring system (PMS) obtained in Sect. 3, and compared with the traditional tensioned mooring system under the same *MBS* and H_t conditions (Table 2), the following conclusions can be drawn:

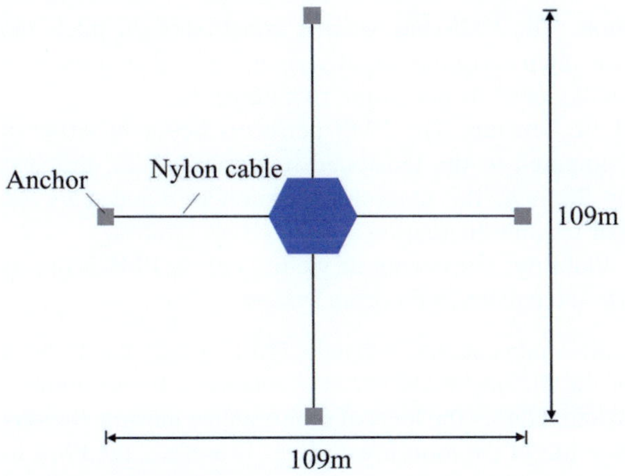

Fig. 5 Schematic diagram of the taut mooring system

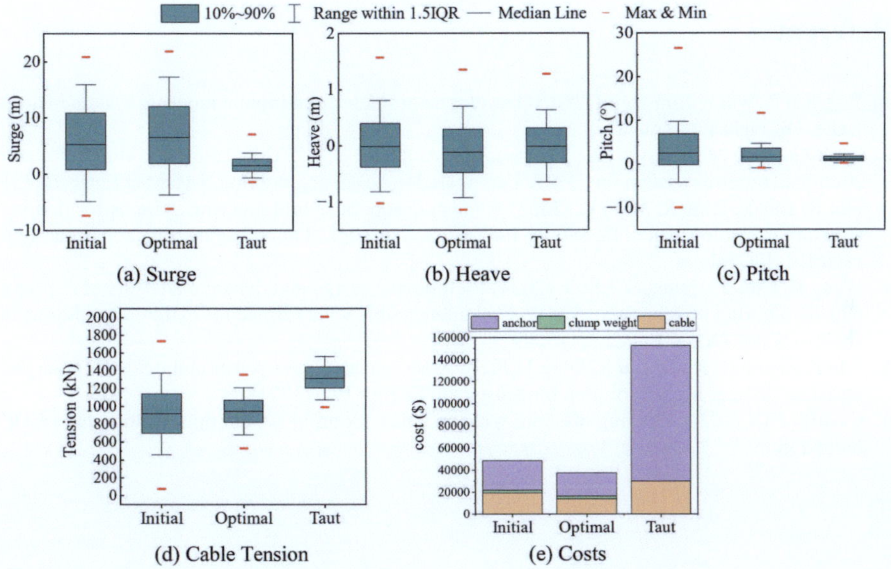

Fig. 6 Comparison of different mooring systems

(1) Surge Motion: The PMS has weaker constraints on surge motion, making its performance in surge significantly better than that of traditional tensioned mooring systems.
(2) Heave Motion: The heave response of the PMS is not significantly different compared to the tensioned mooring system.

(3) Pitch Motion: The PMS has weaker constraints on pitch motion response. Although the pitch motion is slightly greater than that of the tensioned mooring system, it still meets the constraint requirements.
(4) Mooring Line Tension: The PMS performs better in terms of mooring line tension. Compared to the tensioned mooring system, the average tension is reduced by 28.59%, the maximum tension is reduced by 32.57%, and the amplitude of tension fluctuation is reduced by 18.88%.
(5) Economic Viability: The economic viability of the PMS is improved by 75.48% compared to the tensioned mooring system.

In summary, this study concludes that the PMS not only meets the motion response requirements of the floating body but also enhances the economic viability of the mooring system and reduces the level of mooring line tension, thereby improving the safety and service life of the mooring system. Therefore, the PMS has the potential to become a promising mooring method in the field of offshore floating platforms.

References

1. Jing-Rui Z, Wei F, Xun-Ke LI (2013) Development status of multipoint mooring system in deep water. Oil Field Equipment
2. Bach-Gansmo MT, Garvik SK, Thomsen JB, Andersen MT (2020) Parametric study of a taut compliant mooring system for a FOWT compared to a catenary mooring. J Mar Sci Eng 8(6):431
3. Fan H, Liu X, Ding S, Miao Q (2022) A novel design of hybrid mooring systems for floating structure in shallow water. In: ISOPE International ocean and polar engineering conference, pp ISOPE-I. ISOPE
4. Wang K, Chu Y, Huang S, Liu Y (2023) Preliminary design and dynamic analysis of constant tension mooring system on a 15 MW semi-submersible wind turbine for extreme conditions in shallow water. Ocean Eng 283:115089
5. Sun Z, Zhang H, Xu D, Liu X, Ding J (2020) Assessment of wave power in the South China Sea based on 26-year high-resolution hindcast data. Energy 197:117218
6. Castillo FTS (2020) Floating offshore wind turbines: mooring system optimization for LCOE reduction

Theoretical Tools for Linear Analysis of Anomalous Waves in Mooring Systems Combining Data in Time and Frequency Domain

Ulises Torres-Herrera, Alireza Keramat, and Huan-Feng Duan

Abstract In the study of floating structures under the influence of realistic environmental conditions, we have obtained theoretical results of interest for the analysis of anomalous waves. In this work, we present relevant analytical results to study specific short-time waves, both as an isolated event and as a part of a complex long event measured from experiments. This is possible because wave dynamics data acquired by the National Data Buoy System of NOAA are processed via a short-time Fourier Transform. When such data are incorporated in our theoretical results, closed-form expressions of dynamic mooring line displacement and tension are computed, both in frequency and time domain. Our theoretical tools for the analysis of anomalous waves rely on a linearized model of a mooring line, under the assumption of small values of mooring length compared to the characteristic length of elastic longitudinal waves. We show a proof of concept of these tools, by the comparison of three ocean wave events. These results provide a preliminary framework for qualitative analysis of waves acting on mooring systems with minimal computational cost.

Keywords Wave dynamics · Floating structures · Mathematical methods in continuum mechanics · Fourier analysis

1 Introduction

In the development of marine structures under the fast-evolving environmental conditions associated to climate change, there are situations and challenges yet to be solved. Such challenges require interdisciplinary efforts, since the better design of resilient structures is intimately related with the capability to understand and incorporate the actual environmental conditions of the intended place for installment or improvement. For this reason, the progress in this field is achievable only by the

U. Torres-Herrera · A. Keramat · H.-F. Duan (✉)
The Hong Kong Polytechnic University, Hong Kong SAR, People's Republic of China
e-mail: hf.duan@polyu.edu.hk

X. L. Zhao et al. (eds.), *Proceedings of the Fourth World Conference on Floating Solutions: WCFS 2024; 2–4 December; Hong Kong*, Lecture Notes in Civil Engineering 597, https://doi.org/10.1007/978-981-96-4569-5_28

close interaction between experts in computational fluid dynamics, climate change modelling, ocean dynamics sensors and structural mechanics among others [1, 2].

Recent studies on climate change demonstrate that different perspectives are required to re-define what we regard as gold standards for the risk assessment of structures [1]. The current understanding of ocean dynamics demands better tools to integrate the novel, high-quality in-situ measurements of ocean dynamics [3, 4], to consider their implications in the monitoring and design of floating structures for offshore applications.

Recent efforts have been carried out to obtain better tools for simulation of mooring line systems, by incorporation of actual measurements taken from in-situ ocean dynamics. Such tools allow one for simulations adapted to realistic ocean currents and waves, without the need to re-process such data by computational fluid dynamics simulations to adjust their resolution to the minimal requirements of numerical simulations of mooring lines and floating structures. This is possible because analytical tools developed in [5] can be used without convergence or numerical instability problems, since the closed-form expressions developed by the authors allow for a robust scheme to explore the dynamics of mooring systems. Because of the analytical nature of the solution provided by [5], it is possible to develop further qualitative studies that give a clear understanding of features present in such a result.

In this work, we explore a consequence of the results presented in [5] to demonstrate the capability of such a framework to partition a complex ocean dynamic measurement for its analysis and further understanding. Particularly, we develop a framework to extract an isolated wave within a long measurement of sea surface changes, to establish the contribution of such a wave to the global response of a mooring line. Our results are presented both in time and frequency domain, to demonstrate their complementary role for the assessment and study of specific waves of interest for the stability and dynamic properties of floating structures.

2 Methodology

2.1 Equation of Motion

This work starts from a linearized model that describe dynamic displacement from an initially bent mooring line (see [6]), when small values of length compared with the characteristic length of tensile elastic response occur. In such conditions, the equations of motion are given by

$$m\frac{\partial^2 u}{\partial t^2} = \frac{\partial T_D}{\partial s} - Mg\varphi, \tag{1}$$

$$(m + m_a)\frac{\partial^2 v}{\partial t^2} = -k_d\frac{\partial v}{\partial t} + \frac{Mg}{H}T_D + H\frac{\partial \varphi}{\partial s}, \tag{2}$$

$$T_D = EA\left(\frac{\partial u}{\partial s} - \frac{Mg}{H}v\right), \tag{3}$$

$$\varphi = \frac{\partial v}{\partial s} + \frac{Mg}{H}u. \tag{4}$$

where u and v are the dynamic longitudinal and transversal displacement, respectively, whereas s is the position coordinate along the static configuration, t stands for time, m stands for the mooring line mass-per-unit-length, M is the mass-per-unit length of mooring line after reduction by buoyant static force, k_d is the drag constant, m_a is the added mass-per-unit-length, g is the gravitational acceleration, H is the horizontal component of static tension, E is the Young's modulus, A is the cross-sectional area, T_D is the dynamic tension, φ is the dynamic angle with the vertical line.

The mooring line is subject to the following boundary conditions for the attached point to the seabed,

$$u(s = 0, t) = 0, \tag{5}$$

$$v(s = 0, t) = 0, \tag{6}$$

while the boundary conditions for the attachment point with the floating structure are influenced by a balance between the buoyant forces affected by the dynamic change of sea level caused by waves, along with the fluctuation of mooring line tension at the attachment point. Such condition is given as follows in Fourier domain:

$$\begin{aligned}&\left(-K_b + i\omega K_{db} + \omega^2 m_b\right)\hat{u}(s = L, \omega)\\&= (-K_b + i\omega K_{db}) \sin\theta_0 \hat{y}_w(\omega)\\&\quad + \frac{EA}{L}\sin^2\theta_0\left(\frac{d\hat{u}}{ds}(s = L, \omega) - \frac{mgL}{H}\hat{v}(s = L, \omega)\right),\end{aligned} \tag{7}$$

$$\begin{aligned}&\left(-K_b + i\omega K_{db} + \omega^2 m_b\right)\hat{v}(s = L, \omega)\\&= (-K_b + i\omega K_{db}) \cos\theta_0 \hat{y}_w(\omega)\\&\quad + \frac{EA}{L}\sin\theta_0\cos\theta_0\left(\frac{d\hat{u}}{ds}(s = L, \omega) - \frac{MgL}{H}\hat{v}(s = L, \omega)\right),\end{aligned} \tag{8}$$

where θ_0 stands for the mooring line angle at the attachment point with the floating structure, L is the mooring line length, ω is the frequency, whereas K_b and K_{db} are the coefficient for buoyant and drag force of the floating structure. In particular, K_b is given by

$$K_b = (\rho_w - \rho_b)A_b g, \tag{9}$$

The system is depicted in Fig. 1.

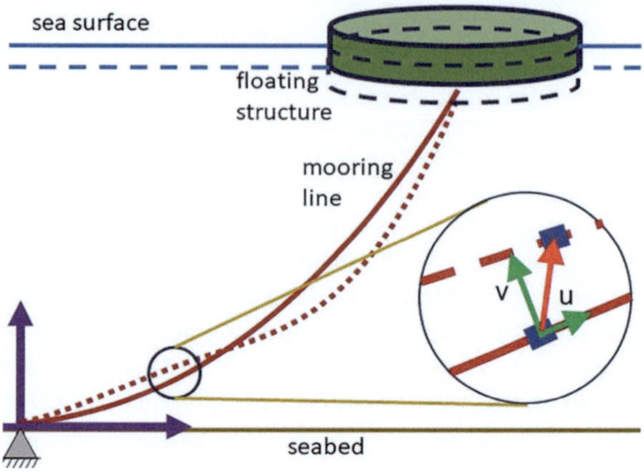

Fig. 1 System of study employed in this work. It consists of a mooring line attached to the seabed at the bottom edge, and to the floating structure at the top edge. It is subject to an arbitrary time-dependent sea surface creating fluctuating buoyant forces

2.2 Analytical Solution in Frequency Domain

The analytical solution is given by substituting the dynamic parameters T_D from Eq. (3) and φ from Eq. (4) into Eq. (2), leading to

$$m\frac{\partial^2 u}{\partial t^2} = EA\frac{\partial^2 u}{\partial s^2} - H\left(\frac{Mg}{H}\right)^2 u - Mg\left(1+\frac{EA}{H}\right)\frac{\partial v}{\partial s}, \tag{10}$$

And, analogously, when substituting the dynamic parameters T_D from Eq. (3) and φ from Eq. (4) into Eq. (2), the following expression is obtained:

$$(m+m_a)\frac{\partial^2 v}{\partial t^2} = Mg\left(1+\frac{EA}{H}\right)\frac{\partial u}{\partial s} + H\frac{\partial^2 v}{\partial s^2} - k_d\frac{\partial v}{\partial t} - EA\left(\frac{Mg}{H}\right)^2 v, \tag{11}$$

Then, Fourier transform in time of Eqs. (10) and (11) leads to

$$EA\frac{d^2\hat{u}}{ds^2} + \left(m\omega^2 - H\left(\frac{Mg}{H}\right)^2\right)\hat{u} - Mg\left(1+\frac{EA}{H}\right)\frac{d\hat{v}}{ds} = 0, \tag{12}$$

$$Mg\left(1+\frac{EA}{H}\right)\frac{d\hat{u}}{ds} + H\frac{d^2\hat{v}}{ds^2} + \left(\omega^2(m+m_a) + i\omega k_d - EA\left(\frac{Mg}{H}\right)^2\right)\hat{v} = 0, \tag{13}$$

Then, Eqs. (12) and (13) define a 2×2 linear system of second-order differential equations for the axial and tangential displacement of mooring line. To solve this system, one method consists of the matrix treatment of differential operators. In such terms, Eqs. (12) and (13) are rewritten as operator equations in matrix form, as

$$\begin{bmatrix} EAD^2 + \left(m\omega^2 - H\left(\frac{Mg}{H}\right)^2\right)D^0 & -Mg\left(1 + \frac{EA}{H}\right)D \\ Mg\left(1 + \frac{EA}{H}\right)D & HD^2 + \left(\omega^2(m + m_a) + i\omega k_d - EA\left(\frac{Mg}{H}\right)^2\right)D^0 \end{bmatrix} \times \begin{bmatrix} \hat{u} \\ \hat{v} \end{bmatrix} = \begin{bmatrix} 0 \\ 0 \end{bmatrix}, \tag{14}$$

From the theory of differential operators, the decoupled solution of this system of differential equations is computed by diagonalization of the operator matrix, as

$$\begin{bmatrix} aD^4 + bD^2 + cD^0 & 0 \\ 0 & aD^4 + bD^2 + cD^0 \end{bmatrix} \begin{bmatrix} \hat{u} \\ \hat{v} \end{bmatrix} = \begin{bmatrix} 0 \\ 0 \end{bmatrix}, \tag{15}$$

where the coefficients a, b and c are defined as

$$a = EAH, \tag{16}$$

$$b = Hm\omega^2 + EA\left(\omega^2(m + m_a) + i\omega k_d\right) + \frac{2EA}{H}(Mg)^2, \tag{17}$$

$$c = \left(m\omega^2 - H\left(\frac{Mg}{H}\right)^2\right)\left(\omega^2(m + m_a) + i\omega k_d - EA\left(\frac{Mg}{H}\right)^2\right), \tag{18}$$

Each decoupled differential equation in Eq. (15) is solved as a fourth-order linear differential equation. The general solution is shown below:

$$\begin{aligned} \hat{u}(s) = {} & \left(\frac{r_p Mg\left(1 + \frac{EA}{H}\right)}{EAr_p^2 - m\omega^2 + H\left(\frac{Mg}{H}\right)^2}\right)(C_2 \sin r_p s - C_1 \cos r_p s) \\ & + \left(\frac{r_m Mg\left(1 + \frac{EA}{H}\right)}{EAr_m^2 - m\omega^2 + H\left(\frac{Mg}{H}\right)^2}\right)(C_4 \sin r_m s - C_3 \cos r_m s), \end{aligned} \tag{19}$$

$$\hat{v}(s) = C_1 \sin r_p s + C_2 \cos r_p s + C_3 \sin r_m s + C_4 \cos r_m s, \tag{20}$$

Finally, the coefficients C_1, C_2, C_3 and C_4 are computed by substituting solution in (19, 20) into the boundary conditions in (5–8), Thus, the particular solution for $\hat{u}(s, \omega)$ and $\hat{v}(s, \omega)$ is of the form

$$\hat{u}(s, \omega) = K_u(s, \omega)\hat{y}_w(\omega), \tag{21}$$

$$\hat{v}(s, \omega) = K_v(s, \omega)\hat{y}_w(\omega). \tag{22}$$

Thus, Eqs. (21, 22) constitute the analytical solution of mooring system dynamics in Fourier domain. Besides from the assumptions on mooring line length, this solution can be carried considering ocean waves of arbitrary time dependence and complexity, and the first step always involves the Fourier Transform of such ocean wave signal. In this work we explore some theoretical tools useful to find analytical expressions of relevance.

3 Generalized Linear Analysis and Schemes of Partition of Ocean Waves

The analytical solution of Eqs. (21–24) exhibits one qualitative property of interest: both transversal and longitudinal displacement of mooring line are proportional to the changing sea level produced by waves. Such a linear response allows us for a high level of freedom to define and study different schemes of partition of waves. Moreover, since Fourier transform is also a linear transformation, it implies that the partition of waves can be carried out whether in time or frequency domain, as desired.

In this work, it is intended to explore a partition of a full wave dynamics that allows us to isolate a single wave located in a specific time location within an arbitrary time-dependent signal. In mathematical terms, this means that, from a function $y_w(t)$ defined over a long interval $0 \leq t \leq T$, we define an auxiliary function $y_{iso}(t)$ such as

$$y_{iso}(t) = \begin{cases} y_w(t) & \text{if } t_{start} \leq t \leq t_{end} \\ 0 & \text{if } 0 \leq t < t_{start} \text{or } t_{end} < t \leq T \end{cases}, \tag{23}$$

and a complementary function $y_{surr}(t)$ as follows:

$$y_{surr}(t) = \begin{cases} 0 & \text{if } t_{start} \leq t \leq t_{end} \\ y_w(t) & \text{if } 0 \leq t < t_{start} \text{or } t_{end} < t \leq T \end{cases}, \tag{24}$$

such that functions $y_{iso}(t)$ and $y_{surr}(t)$ accomplish the following property

$$y_{iso}(t) + y_{surr}(t) = y_w(t), \tag{25}$$

In consequence, Eqs. (23–25) define a valid linear scheme of partition that allows us to isolate a specific wave located in the range $t_1 \leq t \leq t_2$ to study its contribution to the overall response of mooring line dynamics.

3.1 Solution in Time Domain for Mooring Line Dynamics Caused by a Complete Wave

According to the data processing technique used by the National Data Buoy System of the NOAA (see [4]), the acquired ocean waves data for sampling times of one hour are fitted to following function:

$$y_w(t) = \sum_{k=1}^{N_B} d_k A_k e^{i(2\pi \nu_k t + \varphi_k)}, \tag{26}$$

where the parameters A_k correspond to the amplitude of k-th band, ν_k is the frequency of k-th band in Hz, d_k is the bandwidth of k-th band, φ_k is the phase of k-th band and N_B is the number of bands.

By assuming that the isolated wave is smaller than the sampling time employed in Eq. (26), it is possible to study the non-transient dynamics produced by an infinite wave given by Eq. (26) over an infinite period of time. Thus, the Fourier transform in time of sea level $y_w(t)$ in Eq. (26) leads to

$$\hat{y}_w(\omega) = \int_{-\infty}^{\infty} y_w(t) e^{i\omega t} dt = 2\pi \sum_{k=1}^{N_B} d_k A_k e^{i\varphi_k} \delta(\omega + 2\pi \nu_k), \tag{27}$$

and, by substitution into the solution for $\hat{u}(s, \omega)$ and $\hat{v}(s, \omega)$, the analytical solution in time domain is computed as follows:

$$u(s, t) = \sum_{k=1}^{N_B} e^{i\varphi_k} d_k A_k K_u(s, -2\pi \nu_k) e^{2\pi i \nu_k t}, \tag{28}$$

$$v(s, t) = \sum_{k=1}^{N_B} e^{i\varphi_k} d_k A_k K_v(s, -2\pi \nu_k) e^{2\pi i \nu_k t}, \tag{29}$$

Expressions for $u(s, t)$ and $v\left(s, t_q\right)$ in Eqs. (28, 29) allow us for an efficient computation of the response in time domain of a mooring line subject to a complete wave.

3.2 Solution in Time Domain for Mooring Line Dynamics Caused by an Arbitrary Isolated Wave

When the expression for waves $y_w(t)$ in Eq. (26) is substituted into the definition of isolated waves, namely $y_{iso}(t)$ in Eq. (23), the following expression is obtained:

$$y_{iso}(t) = \begin{cases} \sum_{k=1}^{N_B} d_k A_k e^{i(2\pi \nu_k t + \varphi_k)}, & \text{if } t_{start} \le t \le t_{end} \\ 0, & \text{if } 0 \le t < t_{start} \text{ or } t_{end} < t \le T \end{cases}, \tag{30}$$

And its Fourier transform is computed as

$$\begin{aligned} \hat{y}_{iso}(\omega) &= \int_{t_{start}}^{t_{end}} \sum_{k=1}^{N_B} d_k A_k e^{i(2\pi \nu_k t + \varphi_k)} e^{i\omega t} dt \\ &= \sum_{k=1}^{N_B} \frac{i d_k A_k}{\omega + 2\pi \nu_k} e^{i\varphi_k} \left(e^{i t_{start}(\omega + 2\pi \nu_k)} - e^{i t_{end}(\omega + 2\pi \nu_k)} \right) \end{aligned} \tag{31}$$

Finally, the solution in time domain is possible by Fast Fourier Transform, as exemplified below:

$$\begin{aligned} u_{iso}(s, t_q) = {} & \frac{1}{N \Delta t} \sum_{r=1}^{N} \left(\sum_{k=1}^{N_B} \frac{i d_k A_k}{\omega_r + 2\pi \nu_k} e^{i\varphi_k} \left(e^{i t_{start}(\omega_r + 2\pi \nu_k)} - e^{i t_{end}(\omega_r + 2\pi \nu_k)} \right) \right) \\ & \times K_u(s, \omega_r) e^{\frac{-2\pi i (r-1)(q-1)}{N}} \end{aligned} \tag{32}$$

where N is the number of sampling points in fast Fourier transform, Δt is the time step, ω_r is the r-th sampled frequency, given by

$$\omega_r = \begin{cases} (r-1)\left(\frac{2\pi}{N \Delta t}\right), & \text{if } 1 \le r \le 1 + \frac{N-1}{2}, r \in \mathbb{N} \\ (r-1-N)\left(\frac{2\pi}{N \Delta t}\right), & \text{if } 2 + \frac{N-1}{2} \le r \le N, r \in \mathbb{N} \end{cases} \tag{33}$$

Therefore, expressions for $u_{iso}(s, t_q)$ and $v_{iso}(s, t_q)$ in Eqs. (32, 33) provide a framework to study the response of a mooring line to an arbitrary isolated wave in time domain, within a larger measurement.

4 Proof of Concept

As a demonstration of the theoretical tools developed in this work, we carried out simulations using data collected by the National Data Buoy System from the NOAA in station 44,005, on January 1st, 2022, starting at 00:00 h, local time. The main objective of these simulations is to exemplify the capabilities to define and explore features of three specific isolated waves. Such plots are presented in Fig. 2.

The combination of data in time and frequency domain allows us for a complementary study of each isolated event. Analysis of Fig. 2c shows that event 1 does not dominate the time evolution of dynamic tension, since the full response obeys a completely different trend when compared with the full response. In contrast, Fig. 2e

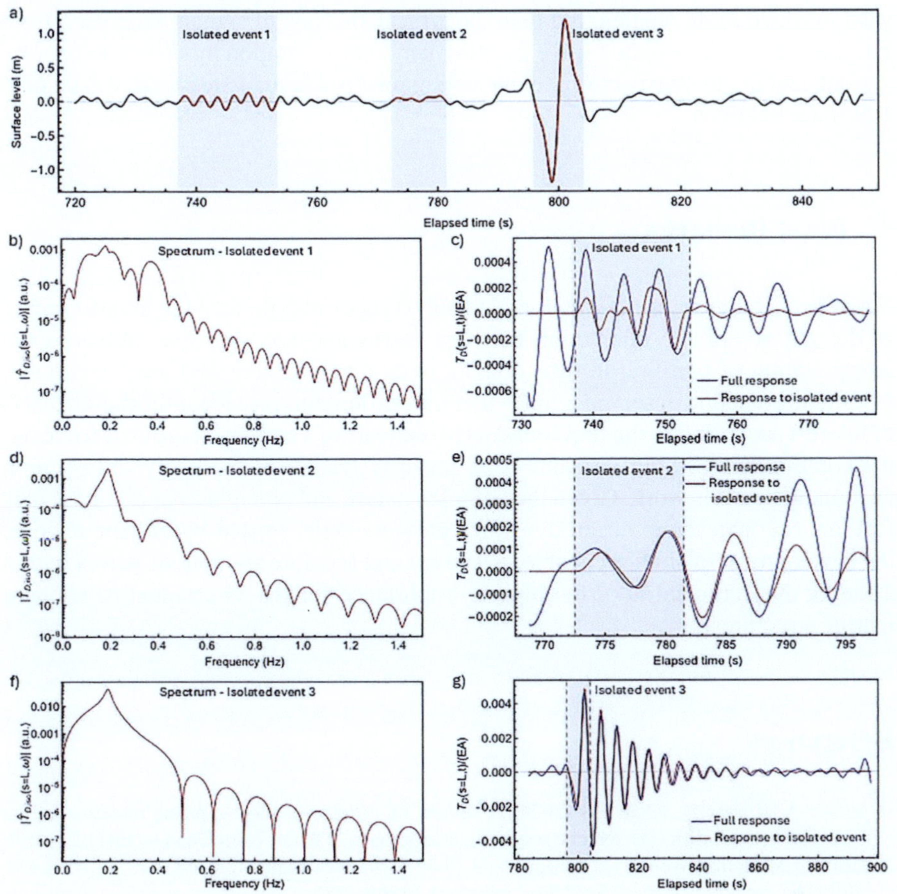

Fig. 2 Combined analysis in time and frequency domain of isolated events within a complete in-situ measurement of ocean waves acting on a mooring system. **a** General overview of the measurement and three exemplified isolated events. The spectrum of dynamic tension at the upper edge is plotted for **b** event 1, **d** event 2 and **f** event 3, whereas dynamic tension at the upper edge in time domain is plotted for **c** event 1, **e** event 2 and **g** event 3

shows that event 2 dominates the whole response during the lapse of the event, but right after the event the dynamics differs from the full response, due to the subsequently immediate event that takes the leading role on the response. Finally, Fig. 2g exhibits that event 3 dominates the dynamics during and after the event, since the isolated response and the full response follow a very similar trend for many seconds after the event 3 ended.

A comprehensive partition and analysis of the complete measurement of ocean waves, in terms of the contribution of each isolated event, is useful to identify the critical events that pose a threat to the stability of mooring systems. This is a task that requires further analysis, since understanding the Fourier transform of an isolated wave is much more challenging than the typical analysis of signals processed from full time series. An isolated wave excites many more vibration modes on a mooring system, leading to unexpected and interesting results where many characteristic time scales are involved.

5 Final Remarks

The role of frequency and time-domain analysis is to complement the understanding of the physics of the interaction between waves and mooring line. Whereas the interpretation of partition in time domain is relatively simple, the mechanism that justifies the reaction observed is not evident. Both domains provide information that is of interest according to the requirements of analysis of a certain structure. Moreover, the combination of both domains is an intrinsic feature of the analytical solution encountered in this work. Given the complex nature and emergent nature of the new features encountered in ocean dynamics measurements caused by climate change, the development of studies capable to analyze and separate anomalous waves within a longer measurement is of the biggest importance for the assessment of resilient marine structures.

References

1. Rawat A, Kumar D, Khati BS (2024) A review on climate change impacts, models, and its consequences on different sectors: a systematic approach. J Water Clim Change 15(1):104–126
2. Climate Modeling Website at Geophysical Fluid Dynamics Laboratory-NOAA, https://www.gfdl.noaa.gov/climate-modeling/. Last accessed 30 Jun 2024
3. Tuckcr MJ (1991) Waves in ocean engineering: measurement analysis and interpretation. Ellis Horwood, New York
4. National Data Buoy Center Homepage, https://www.ndbc.noaa.gov/. Last accessed 30 Jun 2024
5. Torres-Herrera U, Keramat A, Duan H-F (2024) A theoretical framework for robust implementation of in-situ measurements of ocean currents and waves in dynamics of mooring systems. Phys Fluids 36:096604
6. Aranha JAP, Pinto MO (2001) Dynamic tension in risers and mooring lines: an algebraic approximation for harmonic excitation. Appl Ocean Res 23:63–81

Dynamic Mooring Analysis of a Submerged Floating Tunnel

Marcel 't Hart, Dirk Jan Peters, Lisa Koopman-Meijer, and Coen Eggermont

Abstract The paper presents a simulation of loads on the hull of a segmented submerged floating tunnel (SFT) caused by surface waves. The SFT is a cylindrical tunnel, typically 20–30 m below the surface. The tunnel is connected to the seabed by vertical tethers and optionally by taut lines. The method of a Dynamic Mooring Analysis of a moored ship was applied to the SFT. The design wave loads, defined by a wave height, wave period and direction, were derived from monitoring data and used as input for the analysis. The analysis consists effectively out of two parts. In the first part a diffraction analysis of the hull of the tunnel modeled with discrete panels is used to compute pressures caused by waves from all directions in the frequency domain. With the results of this analysis, the effect of a unified wave load is known for different angles as well for different wave lengths. As a second step a time-history wave load scheme was generated using a JONSWAP spectrum converter, including phase-shifts. With this a time-history analysis of the loads on the SFT was performed computing the support forces, and the internal force and deformations in the joints, as well as the bending moments in the hull of the SFT. The methods used are a demonstration of techniques which could be applied in a design process for SFT's.

Keywords Submerged floating tunnel · Diffraction · Dynamic mooring analysis

1 Introduction

An SFT is a structure that "floats" below the water surface and in which traffic can cross the water at large depths. An SFT construction is also known as "Archimedes Bridge", literally a bridge underwater. At locations where immersed or bored tunnels would be too deep and therefore would be too long or where fixed links like bridges,

M. 't Hart (✉) · D. J. Peters
Tunnel Engineering Consultants, Amersfoort, The Netherlands
e-mail: m.thart@tec-tunnel.com

M. 't Hart · D. J. Peters · L. Koopman-Meijer · C. Eggermont
Royal HaskoningDHV, Utrecht, The Netherlands

X. L. Zhao et al. (eds.), *Proceedings of the Fourth World Conference on Floating Solutions: WCFS 2024; 2–4 December; Hong Kong*, Lecture Notes in Civil Engineering 597, https://doi.org/10.1007/978-981-96-4569-5_29

bored tunnels or immersed tunnels cannot be constructed, an SFT is an alternative (see Fig. 1). An SFT is a buoyant structure with a resulting force of buoyancy, gravity, and other loads. In case the resulting force is upward, the SFT needs to be tethered to fix the vertical position below the water surface. In case the SFT has a resulting downward force, the structure is suspended by floating pontoons to fix the vertical position. Current feasibility designs like Bjornafjorden 2016 [1] show both options (see Fig. 2).

In these types of structures, the offshore, (sub)marine, and infrastructural requirements meet both the load impacts and the structural responses. Under marine circumstances, the loads acting on SFTs due to environmental conditions and their corresponding structural responses vary considerably. The responses are influenced by a variety of circumstances that cannot all be analyzed. To estimate the probability of exceeding certain limits of interest for the mechanical responses of the SFT, such as forces, displacements, and accelerations, probabilistic methods are often used in combination with numerical simulations. In this paper, a *dynamic mooring analysis* (DMA) is used for a hypothetical SFT supported by tethers. In these analyses, a diffraction analysis is conducted, sequentially followed by a time-history analysis. To describe environmental loads, real monitored data of wave height, period, and angle of impact are used from the China Strait. The structural responses of interest correspond to the maximum values of forces, displacements, and accelerations at the points of interest along the hypothetical SFT.

Fig. 1 Schematic of different types of crossings

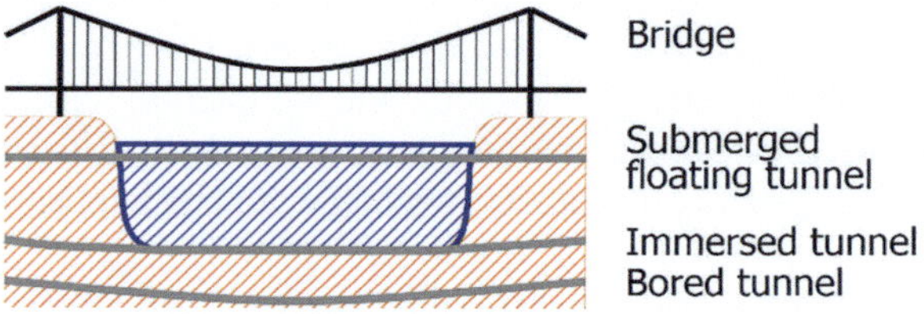

Fig. 2 Schematic of pontoon (left) and tether (right) supported SFT. *Source* NPRA

2 Dynamic Model Analysis

To calculate the dynamic response of the SFT, commercial software is used for the DMA. This software is developed by MARIN and the analysis consists of two sequential parts, a diffraction analysis and a time history simulation of the model. DIFFRAC is a wave diffraction program that is used to calculate the loads and motion responses of floating or moored multibody structures. aNySIM computes the motions of vessels resulting from nonlinear hydrodynamic and mechanical loading. Initially, this software was developed for DMA on ships to find their responses and interactions. In this paper, the SFT is divided into different elements and the interaction between the elements as well as the response of the elements. Both packages are used in different publications on wind turbines [2–6]. More detailed information on both tools can be found at the website of MARIN [7].

3 Dynamic Response Model

The hypothetical SFT is represented by a model presented in Fig. 3, the SFT is made up of 20 elements of 100 m length and is vertically supported by 5 tethers, and in the center of the structure, the model is laterally supported by 2 mooring lines. The tethers and mooring lines are represented by spring elements and because the structure is positively buoyant, the tethers have an initial pretension. At both ends, the structure is supported by hinged connections.

The analysis is set up as a staggered analysis in which a diffraction analysis (DIFFRAC) is combined with a time history analysis (aNySIM). The results of the DIFFRAC calculation are stored in a so-called hydrodynamic definition containing the first-order and second-order wave exciting forces and moments together with the added mass and damping coefficients. Using this hydrodynamic definition, the

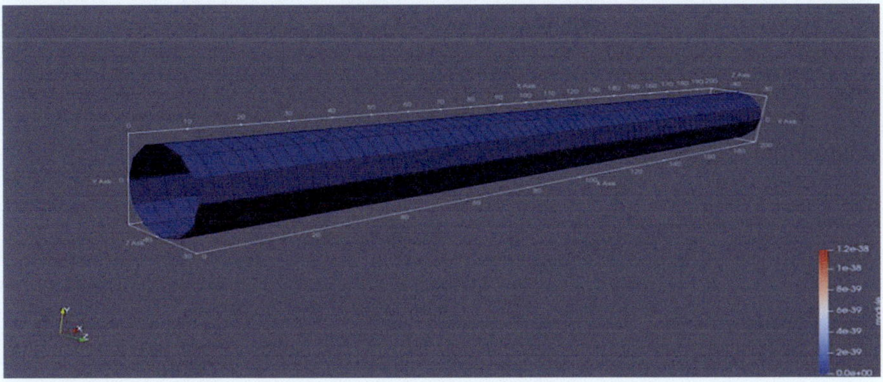

Fig. 3 Diffraction model

wave loads and motion response of a tunnel element for one wave component with certain wave height and frequency can be calculated. Based on linear superposition, the outcoming (ir)regular first and second order wave forces are used in the aNySIM model to calculate the response (motion) of the tunnel. aNySIM allows to generate simple regular waves or the spectral (irregular) waves at any location in the domain of the simulation and determine the direction of propagation, thus permitting to address the crossing of different wave systems (like two crossing swell systems in combination with a locally generated wind wave system). Wave probes can be placed at any location in the domain to monitor the variation of the water elevation. For each wave case, the spatial positions of the tunnel elements and the wave forces together with the water elevations are calculated and recorded in time to investigate the tunnel behavior in such specific wave case.

The interaction between the elements (bodies) is represented by a spring/damper which relates the cross-sectional results of forces and moments to the 6 degrees of freedom (3 displacements and 3 rotations), their derivatives (velocity) and second derivatives (accelerations) in time.

For the time history analysis (aNySIM), a wave load as a function of time is generated based on a JONSWAP analysis [8], given a wave height and a wave period. Combining the JONSWAP wave signal and the diffraction result (velocity potential) to a time-dependent input on the structure gives a wave load over time. Using this wave load over time enables a time history analysis and the collection of a time history response of the structure (tether forces, mooring forces, displacements, rotations, and accelerations) based on a specified wave height, wave period, and angle of impact. For representative points in the structure and the tethers, the response over time is stored as output. Initially the structure is encountering start-up distortion and, therefore, only results after 150 s were considered.

The total DMA sequence is schematically visualized in Fig. 4.

For this research, as input wave period (T_p [s]), wave height (H_s [m]) and angle of impact (φ [°]) were used and based on monitoring data. The monitoring data was retrieved from ERA5 [9]. Torres et al. [10] give a description of how to obtain the input data set. ERA5 provides climate reanalysis data sets. For monitoring data, observations in the time span 2000 to 2019 have been used in the China Strait. The wave data set was checked for outliers; an outlier was considered if it deviates more

Fig. 4 Hypothetical SFT layout

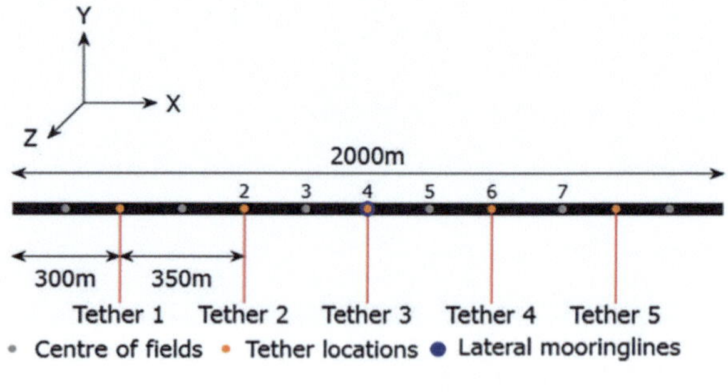

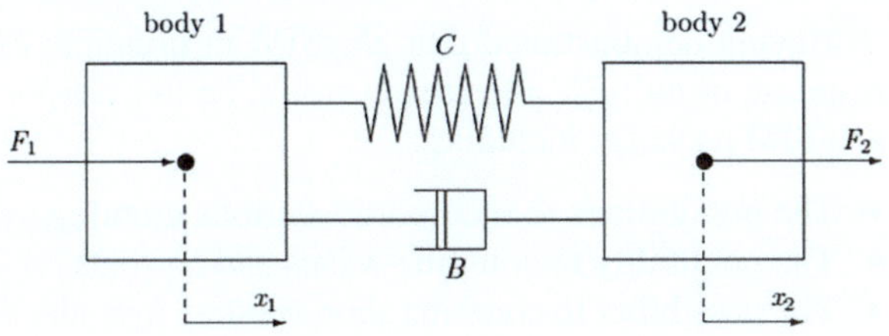

Fig. 5 Connection between 2 elements

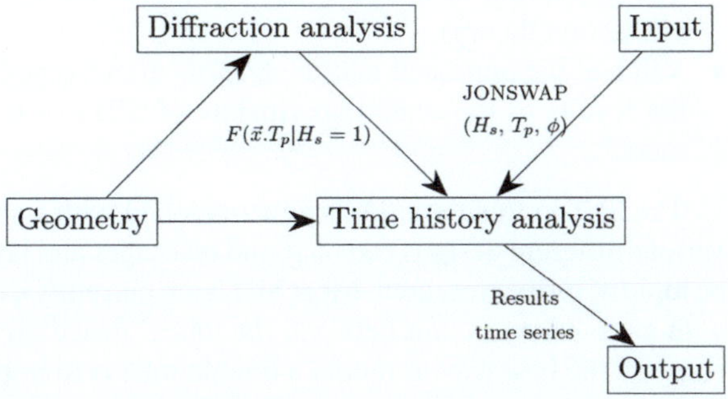

Fig. 6 Dynamic mooring analysis sequence

than 7 times the standard deviation of the monthly data from its mean. As SFTs are vulnerable to swell waves, a conditional draw from the monitoring data was performed by considering only wave periods greater than 7.8 s (Figs. 5 and 6).

As the total time history analyses were intense in terms of computational effort and, within a limited time frame, 497 different analyses were performed.

4 Conclusions and Recommendations

To establish the wave forces on and the hydrodynamic properties (added mass and damping) of the SFT, a well proven method from vessel hydrodynamics has been used: 3D potential diffraction theory. Starting with a mesh of the tunnel element, the (ship motion frequency dependent) added mass, the (ship motion frequency dependent) damping coefficient and the (ship motion frequency and direction dependent) wave forces are calculated and stored in a so-called hydrodynamic definition of the SFT element. By using the monitored wave conditions (height, period and direction) as input, aNySIM, using the parameters in the hydrodynamic file, calculates the forces acting on the SFT and subsequently the motion response of the SFT. In this research nearly 500 situations have been calculated covering all sorts of situations. The results and animation of the aNySIM calculation show that the combination of DIFFRAC and aNySIM is a good way to simulate the dynamic response of SFT structure. The calculation time is relatively low and therefore this approach appears very suitable to fit in a design process of the development of an SFT.

Having demonstrated that aNySIM in principle can be applied to calculate the response of an SFT, a next step would be to incorporate the full capabilities of the aNySIM package, including:

- The possibilities to incorporate various mooring systems and arrangements.
- The possibility to combine waves and currents.
- The possibility to combine short crested, irregular waves (including direction and frequency spectra).
- The possibility to included more complex tunnel shapes (through 3D potential diffraction theory).
- Validate the approach and to show its effectiveness and accuracy by comparing the results to the scale experiments of SFTs or to a more advanced numerical model.

The SFT in this research is now modelled with a single circular tube. In practice various different designs can be found on shapes and layouts. Different shapes should be topic of further research. Most SFT's are currently designed in a double tube layout with an interconnection between the tubes. Based on the experience gained in this research the best way to model a double tube is to implement two tubes in aNySIM and not a single equivalent beam. The option to use a single equivalent beam in aNySIM would require a more sophisticated DIFFRAC model whereas adding a second beam is easy and not increasing computational time significantly. Adding a second tube will also not complicate the derivation of the coupling matrix.

Acknowledgements This research was supported by the Submerged Floating Tunnel (SFT) Team. This research project is commissioned by the Chinese engineering and construction company China Communications Construction Co., Ltd. (CCCC) and is jointly carried out by eight universities, scientific research institutes, engineering consulting firms, design, and construction companies.

References

1. Norconsult (2016) K3/K4 Technical report—Bjørnafjord submerged floating tunnel
2. Jonkman J, Robertson A, Gueydon S, Wuillaume P, Platt A (2015) Comparison of second-order loads on a tension-leg platform for wind turbines. In: Twenty-fifth international offshore and polar engineering conference. Conference date: 21 June 2015 through 26 June 2015. American English
3. Bayati I, Gueydon S, Belloli M (2015) Study of the effect of water depth on potential flow solution of the oc4 semisubmersible floating offshore wind turbine [12th Deep sea offshore wind R&D conference, EERA Deep-Wind'2015]. Energy Procedia 80:168–176. https://doi.org/10.1016/j.egypro.2015.11.419
4. Gueydon S, Duarte T, Jonkman J (2014) Comparison of second-order loads on a semisubmersible floating wind turbine. https://doi.org/10.1115/OMAE2014-23398
5. Gueydon S, Lindenburg K, Savenije F (2013) Coupling of two tools for the simulation of floating wind turbines. https://doi.org/10.1115/OMAE2013-11174
6. Koop A, Leverette S, Bangs A, Withall T, Castillon Valverde F, Van der Zan-den J, Heerink R, de Wilde J (2023) Investigating the effect of damping originating from mooring and risers on

the VIM response of a semi-submersible floater using coupled CFD-time-domain simulations, 04 May 2023. https://doi.org/10.4043/32285-MS
7. Marin (n.d.) Maritime Research Institute Netherlands. https://www.marin.nl/en
8. Hasselmann K, Barnett T, Bouws E, Carlson H, Cartwright D, Enke K, Ewing J, Gienapp H, Hasselmann D, Kruseman P, Meerburg A, Muller P, Olbers D, Richter K, Sell W, Walden H (1973) Measurements of wind-wave growth and swell decay during the joint north sea wave project (jonswap). Deut Hydrogr Z 8:1–95
9. Hersbach H, Bell B, Berrisford P, Biavati G, Horányi A, Muñoz Sabater J, Nicolas J, Peubey C, Radu R, Rozum I, Schepers D, Simmons A, Soci C, Dee D, Thépaut J-N (2019) Era5 hourly data on single levels from 1940 to present. LNCS Homepage. http://www.springer.com/lncs. Last accessed 21 Nov 2016
10. Torres et al (n.d.) A vine-copula model for simulation of extreme metocean loads and estimation of hydrodynamic forces for a submerged floating tunnel at the Qiongzhou strait—under review

Dynamic Analysis of a Concrete TLP for Offshore Wind Turbine in Different Wave Conditions

Zuntao Feng and Jian-Fei Chen

Abstract In recent years, a number of semi-submersible floating offshore wind turbine systems have been deployed in transitional water regions with depths ranging from 40 to 150 m. However, these designs are large and expensive. They were constructed with steel or steel reinforced concrete, susceptible to steel corrosion. This study proposes a submerged TLP for offshore wind turbines in transitional water regions. The dynamic responses of the wind turbine during the operation are investigated, with emphasis on the effect of wave height. Based on fully coupled time domain simulations, it is found that the standard deviations of the pitch are dependent on the wave height, while those of the surge and yaw motions are largely controlled by wind loads. In addition, the heave motion is mainly caused by the sinking effect of surge.

Keywords Floating wind turbines · Tension leg platforms (TLP) · Dynamic responses

1 Introduction

In recent years, research and engineering practices in offshore wind power have witnessed rapid global advancements, leading to a continuous increase in the capacity of individual wind turbines. While this enhances power generation efficiency, the growing weight and size of turbines also place greater demands on foundations [1]. Moreover, as water depth increases, particularly in areas exceeding 50 m, floating foundations offer a more economically advantageous solution compared to fixed foundations, making them the inevitable choice for deep-sea wind energy development. Therefore, designing large-scale floating wind turbine platforms suitable for transitional water depths has significant practical implications.

Z. Feng · J.-F. Chen (✉)
Department of Ocean Science and Engineering, Southern University of Science and Technology, Shenzhen, China
e-mail: chenjf3@sustech.edu.cn

X. L. Zhao et al. (eds.), *Proceedings of the Fourth World Conference on Floating Solutions: WCFS 2024; 2–4 December; Hong Kong*, Lecture Notes in Civil Engineering 597, https://doi.org/10.1007/978-981-96-4569-5_30

Floating wind turbines in the ocean are subject to various environmental loads such as wind, waves and currents. Due to the larger dimensions of their foundation platform compared to fixed wind turbines, advanced methods such as potential flow theory [2] and numerical programs including those using computational fluid dynamics [3] are required to conduct proper analyses. Wayman et al. [4] conducted early research on the motion coupling characteristics of floating wind turbines. The Morison formula was used to conduct simplified initial hydrodynamic analyses for the slender single-column Hywind support platform. Li et al. [5] used the SESAM software to investigate the impact of second-order difference frequency wave forces on floating wind turbines. Different from traditional marine platform load analysis, wind load is the dominant load in the analysis of floating wind turbine motion performance. For TLP floating wind turbines, model tests had been conducted [6, 7], showing that the coupling effect of the wind load and wave load is very important. It was also highlighted that the predictions of common software may have noticeable differences with test results at some conditions.

This study proposes a concrete tension-leg platform (TLP) to take advantages of the low cost and good durability of concrete. TLPs have an advantage in that their mooring system occupies only a very small sea floor area, making much more efficient use of sea areas. In this paper, the dynamic responses of the concrete TLP floating wind turbine were studied. A fully coupled numerical model of hydrodynamic-aerodynamic of the TLP floating wind turbine with mooring system is established using the FAST software.

2 Parameters of the TLP Wind Turbine

As depicted in Fig. 1, the platform comprises a single column, floating tanks, and pontoons. During operation of the wind turbine, the pontoons and floating tanks are submerged below the mean sea level (MSL), with the upper surface of the pontoons positioned 8 m below the sea level. This submerged configuration notably serves as the primary source of buoyancy for the wind turbine. A column is centrally located within the arrangement of four pontoons. The wind turbine tower is connected to its upper section and is integrated with the floating tanks through bracings. The diameter of the column is designed to match the tower's base diameter, thereby minimizing the waterplane area of the platform and optimizing its hydrodynamic performance during operation.

The platform is anchored to the seabed by four sets of tensioned mooring lines. The key parameters are presented in Table 1. The platform design offers several advantages:

(1) Due to its semi-submersible nature and inherent stability during transport, the platform can be wet-towed from the quayside to the offshore installation site, with the wind turbine installed in place;

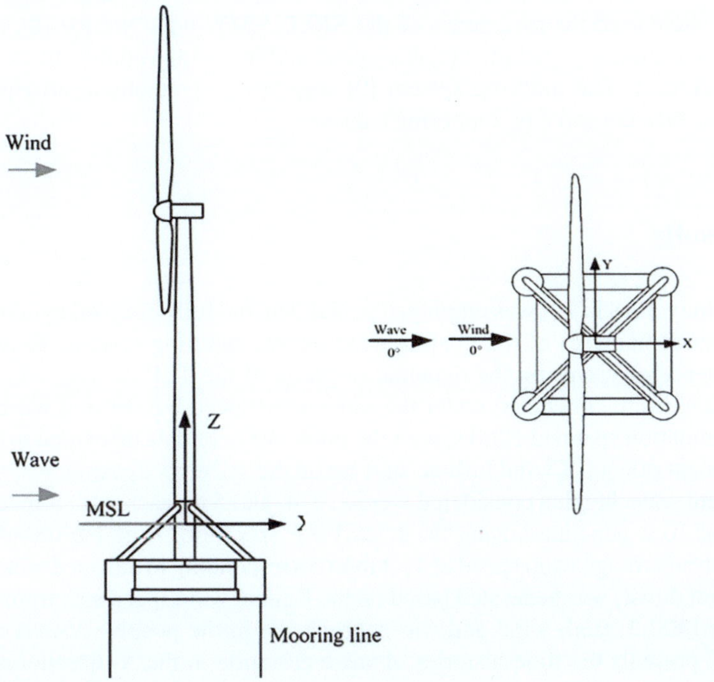

Fig. 1 Schematic layout of the concrete TLP wind turbine

Table 1 Properties of the platform

Characteristics	Value
Draft (m)	20.0
Column diameter (m)	6.5
Diameter of vertical pontoon (m)	10.0
Height of vertical pontoon (m)	12.0
Distance between vertical pontoons (m)	40.0
Width and height of floating tanks (m)	6.0, 4.0
Platform mass (kg)	5.8×10^6
Volume of displacement (m^3)	10.2×10^6
Water plane area during operation phase (m^2)	51.5

(2) During operation, the platform, anchored to the seabed through tensioned mooring lines, maintains a submerged configuration with a relatively small waterplane area. This design minimizes the wave loads and enhances the hydrodynamic performance;

(3) The above arrangement means that its size is relatively small compared with other arrangements such as semi-submissible platforms and thus more economical.

This study used the parameters of the NREL 5 MW wind turbine [8], a conventional three-bladed, upwind, variable-speed turbine with variable blade-pitch-to-feather control. The mooring system [9] employs a symmetrical arrangement of 4 groups, each comprising 2 mooring cables.

3 Results

In this study, FAST [10] was employed to simulate the fully coupled hydrodynamic-aerodynamic-analyses of the floating wind turbine-mooring system. To investigate the influence of waves on the dynamic response of the TLP floating wind turbine, simulations were conducted under the same wind speed but varying wave heights. Each simulation spanned 1200 s, with the initial 400 s of data discarded to eliminate the transient effects of wind turbine start-up on the system's dynamic response. The significant wave heights considered were 2, 3, 4, and 5 m, each with a spectral peak period of 10 s, generated using the JONSWAP spectrum. A global turbulent wind field with an average wind speed of 11.4 m/s (corresponding to the rated wind speed at maximum thrust) was generated based on the Kaimal wind speed spectrum provided in IEC 61400-3. Both wind and waves propagate in the positive X-axis direction. Figure 2 presents the time histories of wave elevation in the X-direction under the different wave conditions.

Figure 3 shows the power spectral density (PSD) curves of the motion response under different wave conditions. It shows in Fig. 3a that the surge motion is mainly coupled by the low-frequency wind load and by wave. With the increase of wave height, the wave frequency response increases, while the surge resonance motion induced by the low-frequency wind load decreases. In Fig. 3b, the variation of the heave motion of the platform at low frequency and wave frequency with wave height is similar to that of surge, that is, with the increase of wave height, the wave frequency response increases, while the heave motion response caused by the surge motion decreases. Figure 3c shows the pitch motion of the platform is mainly generated by

Fig. 2 Time history of wave

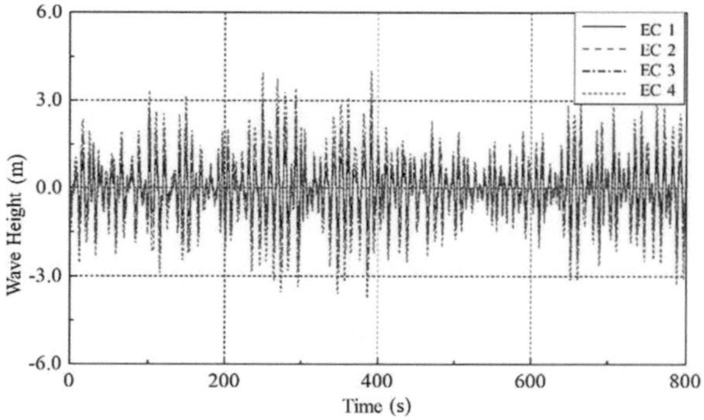

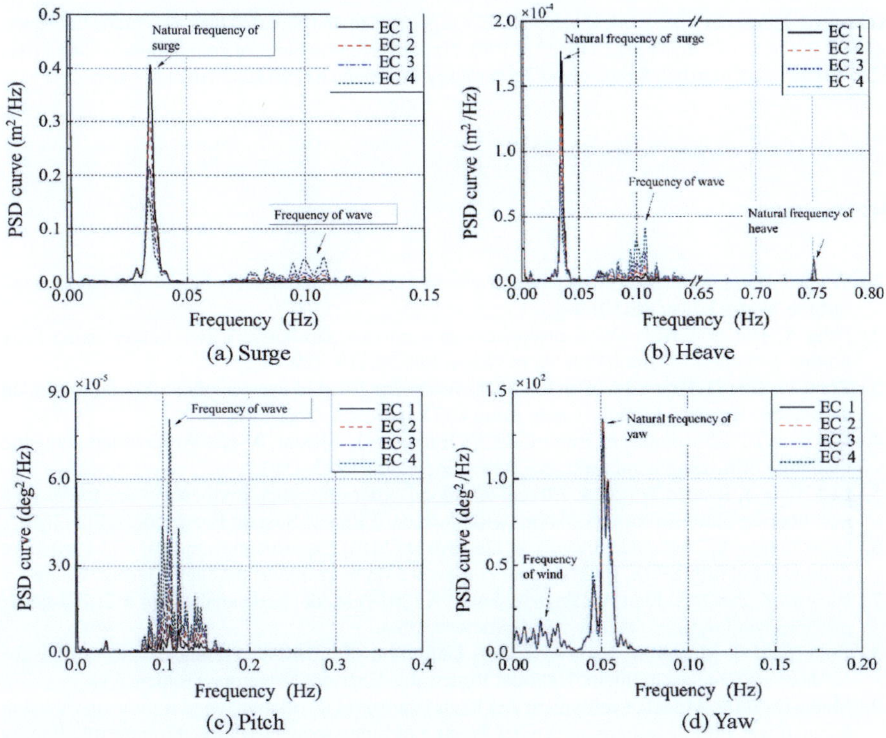

Fig. 3 PSD curves of the motion response under different wave conditions

wave excitation. Figure 3d shows the yaw motion of the platform is mainly caused by wind load and does not change significantly with wave height.

4 Conclusions

This paper has presented an initial design of a novel concrete TLP floating wind turbine. A coupled hydrodynamic-aerodynamic-control-mooring system model has been employed to analyse the dynamic characteristics of both the floating platform and the supported NREL-5 MW wind turbine. The analysis revealed that the platform's surge motion is primarily governed by wind loads. The heave motion is predominantly induced by the set-down effect associated with surge. Due to the high natural frequency of heave, no significant resonant response with waves is observed. The amplitude of pitch motion remains small, primarily driven by wave loads. Yaw motion is largely controlled by wind loads and exhibit minimal variation under different wave conditions.

Acknowledgements The work was financially supported by the National Key Research and Development Program of China (2023YFB3711500), National Natural Science Foundation of China (No. 52178218), and Shenzhen Science and Technology Program (KCXST20221021111409023).

References

1. Wang F, Hao J, Li S (2022) Key technologies and development trends of floating offshore wind turbine. Water Power 48(10):9–12
2. Feng X, Bai W (2015) Wave resonances in a narrow gap between two barges using fully nonlinear numerical simulation. Appl Ocean Res 50:119–129
3. Zhou Y, Xiao Q, Peyrard C, Pan G (2021) Assessing focused wave applicability on a coupled aero-hydro-mooring FOWT system using CFD approach. Ocean Eng 240
4. Wayman E, Sclavounos P, Butterfield S, Jonkman J, Musial W (2006) Coupled dynamic modeling of floating wind turbine systems. Wear 302:1583–1591
5. Li J, Tang Y, Ronald W (2014) Effects of second-order difference-frequency wave forces on a new floating platform for an offshore wind turbine. J Renew Sustain Energy 6(3):332–361
6. Oguz E, Day A, Clelland D, Incecik A, Gustavo D (2016) Experimental study of a TLP offshore floating wind turbine. In: International conference on maritime technology. Harbin, China
7. Ricardo Z, Pablo G, Juan A, Diego V, Souto A (2014) Model scale analysis of a TLP floating offshore wind turbine. OMEA, San Francisco, USA
8. Butterfield S, Musial W, Scott G (2009) Definition of a 5-MW reference wind turbine for offshore system development. National Renewable Energy Laboratory, Golden, CO
9. Matha D (2009) Model development and loads analysis of an offshore wind turbine on a tension leg platform with a comparison to other floating turbine concepts. National Renewable Energy Laboratory (NREL), Golden, CO
10. Han Y, Le C, Ding H, Cheng Z, Zhang P (2017) Stability and dynamic response analysis of a submerged tension leg platform for offshore wind turbines. Ocean Eng 129:68–82

Pull-Out Capacity of Finned Suction Anchors in Sand Considering Torsional Effect

Min-Hao Zhang, Zhen-Yu Yin, and Yong Fu

Abstract The suction anchor is an increasingly used offshore foundation for the floating structures, while the relevant studies on the scenarios with torsional load and anti-torsion solutions appear quite meagre. Therefore, this paper reports numerical investigations into the pull-out capacity of novel finned suction anchors in sand considering torsional effect. The advanced SIMSAND soil model is adopted for the sand constitutive behavior. A comprehensive parametric study is conducted to examine the influence of padeye position, fin number and fin width. These findings may provide insights in research and engineering practice for suction anchors encountering torsional load issues.

Keywords Finned suction anchor · Pull-out capacity · Torsion

1 Introduction

The suction anchor is gaining popularity as a foundation for mooring systems due to its easy installation and removal, low cost, dependable pull-out capacity, and reusability. It connects to floating structures through mooring lines and provides the necessary resistance to environmental loads such as wind and waves. Suction anchors are commonly used for spar platforms in deep water and floating wind turbines. As a result, the pull-out capacity of suction anchors is a central concern in scientific research and engineering practice [12].

In the past, scholars have conducted a lot of research on the pull-out capacity of suction anchors using different methods including field tests [7], centrifuge tests

M.-H. Zhang · Z.-Y. Yin (✉)
Department of Civil and Environmental Engineering, The Hong Kong Polytechnic University, Hong Kong, China
e-mail: zhenyu.yin@polyu.edu.hk

M.-H. Zhang · Y. Fu
Department of Ocean Science and Engineering, Southern University of Science and Technology, Shenzhen, China

X. L. Zhao et al. (eds.), *Proceedings of the Fourth World Conference on Floating Solutions: WCFS 2024; 2–4 December; Hong Kong*, Lecture Notes in Civil Engineering 597, https://doi.org/10.1007/978-981-96-4569-5_31

[2], 1 g physical model tests [5] and numerical simulations [12]. However, most of these studies did not consider the torsional loading due to eccentricity of the mooring lines. According to Suroor and Hossain [9], the torsion can be caused by improper positioning during suction anchor installation and deviation due to current forces. Only a small number of scholars have investigated the torsional effects of suction anchors in clay through numerical modelling [8, 9], and there is a paucity of relevant studies in sandy soils.

This paper presents a numerical study on the pull-out performance of new finned suction anchors in sand, with a focus on torsional effects. The study uses the advanced SIMSAND model to simulate soil behavior. Initially, it evaluates the optimal padeye position for finned anchors compared to traditional finless anchors. Following this, the paper examines how varying the fin number and width affects anchor pull-out capacities.

2 Finite Element Model

Three-dimensional finite element model is established via ABAQUS/EXPLICT. The finned suction anchor is assumed "wished-in-place" and it is modelled as a rigid body. The shaft has a diameter D, a skirt length L, and a skirt thickness t and the fins have a width w. The dimensions of the anchor shaft, as specified by Bang et al. [2] and Ahmed and Hawlader [1], are adopted for this study: $D = 3$ m, $L = 6$ m, and $t = 0.1$ m. Additionally, the fin length and thickness are assumed to be the same as the shaft dimensions. The soil domain has a diameter of $7D + 14w_f$ and a height of $3L$, ensuring it is sufficiently large to minimize boundary effects. Steel is chosen for the anchor, with a unit weight $\gamma_a = 78$ kN/m^3 and a Young's modulus $E_a = 2.1 \times 108$ kPa. The velocity boundaries perpendicular to the bottom and side surfaces of soil domain are restricted. Based on the mesh convergence analyses, the minimum mesh size of soil adjacent to the suction anchor is set as $1.5t$ to ensure computation efficiency and accuracy. Regarding the soil-anchor interface, the Coulomb friction model is employed, i.e., $\tau = k\phi_c$, where k is a parameter typically ranging within 0.5–0.7 in sand [10] and in this study $k = 0.5$ is used. A load-control method is used to simulate the pull-out of the suction anchor. A concentrated pull-out load P is applied at the padeye, with a loading angle θ and a torsion angle β. The general model sketch and definitions of the angles are illustrated in Fig. 1.

The advanced SIMSAND model proposed by Yin et al. [11] is employed for the soil constitutive behavior in the present study. This model is a critical state-based nonlinear Mohr–Coulomb model enhanced with nonlinear elasticity, plastic hardening, and stress–dilatancy effects. It is implemented using the VUMAT subroutine in ABAQUS/Explicit. Parameters for this study, as provided in Table 1 and based on Jin [6], have been calibrated using experimental results from drained triaxial tests on Baskarp sand (Aalborg University Sand No. 0) by Houlsby and Byrne [4].

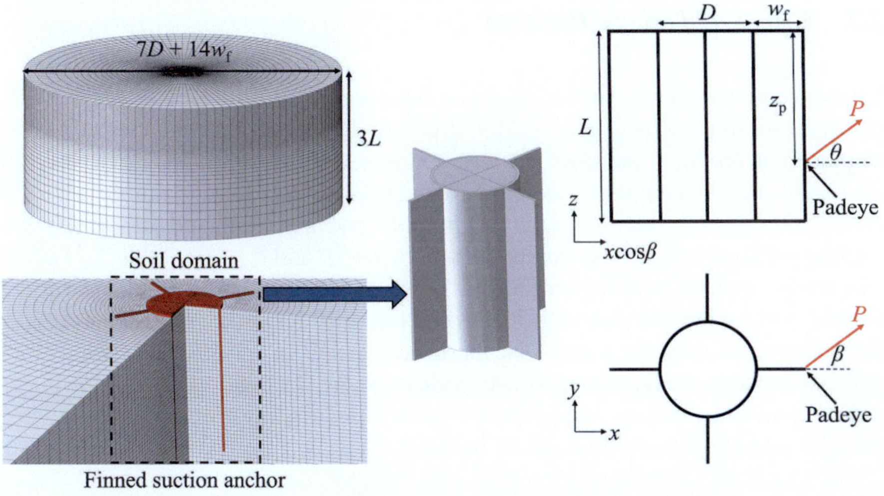

Fig. 1 3D finite element model and definitions of loading angle and torsion angle

Table 1 Parameters used in the SIMSAND model in this study (Jin, 2019)

Parameters	Symbols	Value
Referential bulk modulus (dimensionless)	K_0	344
Poisson's ratio	υ	0.25
Elastic constant controlling the nonlinear stiffness	n	0.58
Critical state friction angle	ϕ_c	35.1 (°)
Initial critical state void ratio	e_{ref}	0.9
First constant controlling the nonlinearity of CSL	λ	0.38
Second constant controlling the nonlinearity of CSL	ξ	0.11
Constant of magnitude of the stress-dilatancy	A_d	0.45
Plastic modulus related constant	k_p	0.0034
Peak friction angle related constant	n_p	1.6
Phase transformation angle related constant	n_d	4.4

3 Results and Analyses

A series of cases are performed to investigate pull-out capacity of finned suction anchor in sand considering torsional effect. Three influencing factors are examined: padeye position, fin number and fin width. The pull-out capacity of suction anchor is determined using the pull-out force when the anchor padeye displacement reaches ~0.1D following other researchers [1].

3.1 *Effect of Padeye Position*

For conventional finless suction anchors, the optimal padeye depth is around 0.7–0.75L according to Zhang et al. [12]. Differently for finned anchors, the padeye can be placed on the fin or the skirt. Therefore in this study, the optimal padeye position for finned suction anchors is examined for anchors with various fin numbers (Fig. 2, $w = 0.5D$). Normal conditions without torsional load are considered and two extreme loading paths are adopted for comparison (i.e., $\theta = 0°$ and $\theta = 90°$, Table 2). Two-fin and six-fin anchors hold higher pull-out capacities for both horizontal and vertical loading paths when the padeye is located on the skirt. Regarding four-fin and eight-fin anchors, locating the padeye on the fin can result in ~3–8% higher horizontal pull-out capacity, while scenarios with padeye on the skirt will have ~6–11% higher vertical pull-out capacity. Based on aforementioned findings, in the further studies the optimal padeye position is set on the skirt.

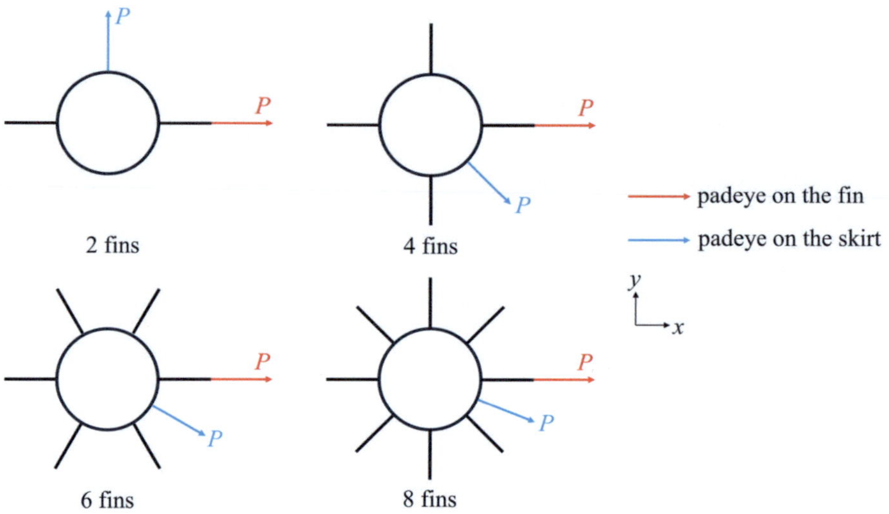

Fig. 2 Illustration of padeye on the fin and skirt for anchors with varying fin numbers

Table 2 Comparisons of horizontal and vertical pull-out capacities for different padeye positions

Fin number	Padeye on the fin		Padeye on the skirt	
	$\theta = 0°$ (kN)	$\theta = 90°$ (kN)	$\theta = 0°$ (kN)	$\theta = 90°$ (kN)
2	5700	1632	8550	1980
4	9000	2231	8300	2376
6	9400	2548	9680	2790
8	10,120	2813	9840	3135

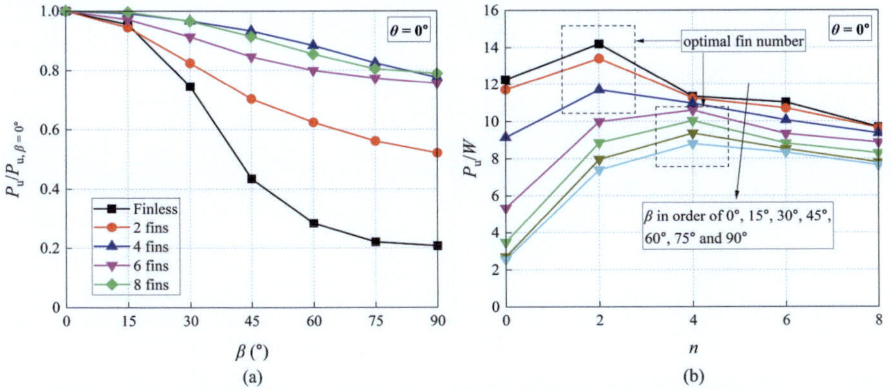

Fig. 3 Effect of fin number on the pull-out capacity: **a** reduction ratio, **b** normalized pull-out capacity considering material cost

3.2 *Effect of Fin Number*

Figure 3a shows the reduction ratio of the pull-out capacity ($P_u/P_{u,\beta=0°}$) varying with the torsion angle and fin number for loading angle $\theta = 0°$ ($w = 0.5D$). The data reveal a consistent "slow-rapid-slow" pattern across anchors with different fin numbers except for 4-fin anchor, representing translational, combined translational and torsional, and torsional failure modes, respectively [12]. Increasing the number of fins from 0 to 4 notably improves anti-torsion performance. However, anchors with 6 or 8 fins do not show further enhancement in anti-torsion capabilities. Figure 3b displays the results for the normalized pull-out capacity considering material cost (P_u/W, W is the anchor weight). The data clearly show that the 2-fin and 4-fin anchors are the most cost-effective options for torsion angle lower and higher than 45°, respectively. According to Eltaher et al. [3], the typical allowable torsion angle of suction anchor is around 7.5°, meaning that the 2-fin anchor emerges as the optimal choice for small-scale torsion angles in normal conditions.

3.3 *Effect of Fin Width*

Figure 4a illustrates the reduction ratio of the pull-out capacity ($P_u/P_{u,\beta=0°}$) varying with the torsion angle and fin width for loading angle $\theta = 0°$. Here 2-fin anchor is adopted based on the findings in the previous section. It is evident that increasing the fin width from 0 to $0.3D$ significantly enhances the anchor's torsional resistance. However, further widening the fin from $0.3D$ to $0.5D$ yields only a modest improvement in torsional resistance. Figure 4b summarizes the normalized pull-out capacity in relation to material cost. For small torsion angles (i.e., $0° \leq \beta \leq 30°$), the normalized pull-out capacity increases nearly linearly with fin width. However, for larger

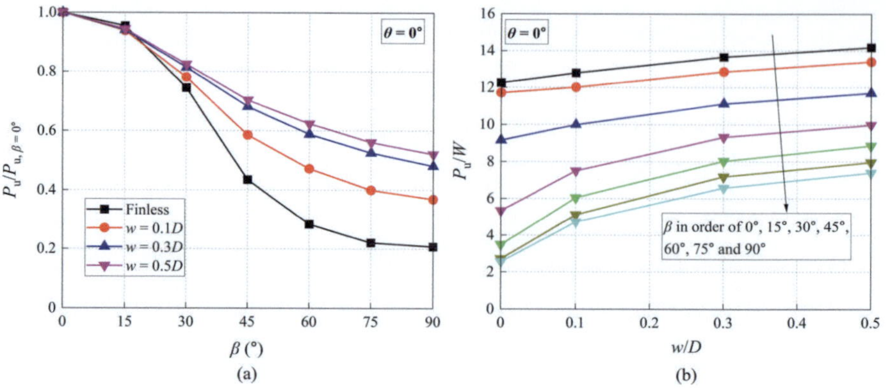

Fig. 4 Effect of fin width on the pull-out capacity: **a** reduction ratio, **b** normalized pull-out capacity considering material cost

torsion angles (i.e., $\beta \geq 45°$), the increase in normalized pull-out capacity begins to level off.

4 Conclusions

This paper employs three-dimensional finite element analyses to examine the anti-torsion pull-out performance of suction anchors in sand, using the advanced SIMSAND model. Initially, the study evaluates the optimal padeye position for anchors with different fin numbers, recommending placement on the anchor skirt. Following this, the impact of fin number is analyzed, concluding that the 4-fin anchor provides the best anti-torsion performance and 2-fin anchor is the most cost-effective choice. Finally, it is found that increasing fin width can enhance both anti-torsion performance and normalized pull-out capacity considering material cost.

References

1. Ahmed SS, Hawlader BC (2015) Numerical analysis of inclined uplift capacity of suction caisson in sand. Int J Offshore Polar Eng 25:145–155
2. Bang S, Jones KD, Kim KO, Kim YS, Cho Y (2011) Inclined loading capacity of suction piles in sand. Ocean Eng 38:915–924
3. Eltaher A, Rajapaksa Y, Chang KT (2003) Industry trends for design of anchoring systems for deepwater offshore structures. In: Offshore technology conference (OTC), p 15265
4. Houlsby GT, Byrne BW (2005) Design procedures for installation of suction caissons in sand. Proc Inst Civil Eng Geotech Eng 158:135–144
5. Gao Y, Qiu Y, Li B, Li D, Sha C, Zheng X (2013) Experimental studies on the anti-uplift behavior of the suction caissons in sand. Appl Ocean Res 43:37–45

6. Jin Z (2019) Numerical investigation of caisson foundations in sand under combined monotonic loadings for offshore wind turbines. PhD Thesis, École centrale de Nantes
7. Ravichandran V, Maji VB, Gandhi SR (2015) Field testing of suction anchors for mooring applications. Indian Geotech J 45:267–277
8. Saviano A, Pisanò F (2017) Effects of misalignment on the undrained HV capacity of suction anchors in clay. Ocean Eng 133:89–106
9. Suroor H, Hossain J (2015) Effect of torsion on suction piles for subsea and mooring applications. In: Proceedings of the 3rd International symposium on frontiers in offshore geotechnics. ISFOG III, Oslo, p 325
10. Tiwari B, Al-Adhadh AR (2014) Influence of relative density on static soil-structure frictional resistance of dry and saturated sand. Geotech Geol Eng 32:411–427
11. Yin Z-Y, Hicher P-Y, Dano C, Jin Y-F (2017) Modeling mechanical behavior of very coarse granular materials. J Eng Mech 143:C4016006
12. Zhang M-H, Yin Z-Y, Fu Y (2023) Numerical investigation on the pull-out capacity of suction anchors in sand considering torsional and mooring line effects. Comput Geotech 160:105524

Numerical Study of Lateral Responses of Monopiles with Local Scour Holes in Non-homogenous Clay

Hang Feng and Zhen-Yu Yin

Abstract Monopiles are widely used as foundations for offshore wind turbines (OWTs), with local scour significantly affecting their lateral responses. This study numerically investigates the lateral behavior of monopiles with local scour holes in non-homogenous clayed soils. A finite element model is first established and validated against existing field tests. The model is then extended to evaluate lateral responses under varying soil properties, foundation geometries, and scour geometries. The findings offer valuable insights for monopile design in offshore engineering.

Keywords Monopile · Lateral response · Finite element analysis · Local scour

1 Introduction

Monopiles with large diameters are widely employed in offshore wind industry. During the service period of monopiles, this type of foundation is required to withstand lateral loads caused by horizontal winds, waves, and current loads [1, 5, 7, 8]. Furthermore, waves and currents can eliminate the soils around the monopile to generate local scour holes which has a substantial impact on the lateral responses of monopiles. Consequently, it is significant to investigate the lateral behavior of the monopile subjected to combined lateral loads and local scour holes.

In the existing literature, the influence of local scour on the lateral responses of monopiles has been extensively studied through numerical simulations, centrifuge tests, and theoretical analyses [3, 6, 9–11]. However, most of these studies have

H. Feng · Z.-Y. Yin (✉)
Department of Civil and Environmental Engineering, The Hong Kong Polytechnic University, Hong Kong, China
e-mail: zhenyu.yin@polyu.edu.hk

H. Feng
e-mail: 23039257r@connect.polyu.hk

X. L. Zhao et al. (eds.), *Proceedings of the Fourth World Conference on Floating Solutions: WCFS 2024; 2–4 December; Hong Kong*, Lecture Notes in Civil Engineering 597, https://doi.org/10.1007/978-981-96-4569-5_32

primarily focused on the ultimate bearing capacity, with limited attention given to the lateral response, particularly the load–displacement deformation relationships.

To address this gap, this study investigates the lateral response of monopiles with local scour holes in non-homogeneous clay. The paper is organized as follows: Sect. 2 presents the three-dimensional finite element model, which is validated against existing field tests. Section 3 discusses the lateral responses obtained from the validated model, considering the effects of monopile geometries, soil properties, and local scour characteristics.

2 Finite Element Model and Validation

2.1 Model Setup

Figure 1 shows the 3D finite element model established using Abaqus. Due to the symmetry of the loading condition, only half of the soil and monopile are modeled. The monopile has a diameter (D) of 4 m and a length (L) ranging from 12 to 48 m (i.e., $L/D = 4$–12). To omit boundary effects, the soil domain extends 20D horizontally and 2L vertically. The bottom boundary is fully fixed, and the horizontal displacement of the lateral boundary is constrained. Following Feng et al. [5], the horizontal element size varies from $D/20$ inside the caisson to $D/4$ at half the model diameter and remains constant at $D/4$ outward. The arc in the radial direction is uniformly meshed with 16 points, and the vertical element size is $D/2$. The soil is modeled with C3D8 elements and the monopile with C3D8R elements.

The monopile is modeled as a linear elastic material with a Poisson's ratio of 0.3. Its stiffness is calculated via $E_{equ} = E_{steel}(D^4 - d^4)/D^4$ (where d is the inside

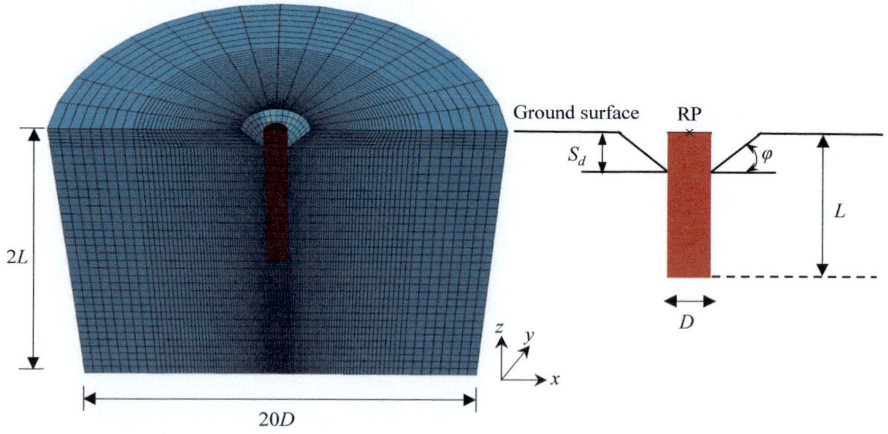

Fig. 1 The mesh in the 3D finite element model for the monopile with local scour holes

diameter of the pipe pile, defined as $d = D - t$; t is the wall thickness of the pipe pile, which is taken as $D/20$; E_{steel} is the stiffness of the pipe pile, which is 2.1×10^5 MPa). The undrained clay is simulated using the Tresca criterion, with Poisson's ratio of 0.49 to keep the volume constant. The non-homogenous strength profile $S_u = kz$ (where k is taken as 2 kPa/m) is employed. The soil stiffnesses normalized by undrained shear strength (E_u/S_u) from 50 to 400 are considered. A submerged soil unit weight of 6 kN/m^3 is employed. Under undrained conditions, the pile-soil interface at the bottom is bonded. A surface-to-surface interface is used for lateral monopile-soil interaction, with no-separation hard contact for the normal contact, and a Coulomb friction model with a friction coefficient $\mu = 1$ for tangential contact.

Based on DNV's [4] recommendation of a maximum scour depth ratio $S/D = 1.3$, scour depth ratios of 0, 1, and 1.3 are considered. A constant scour angle φ of 45° is adopted considering the scour angle has a minimal influence on the lateral responses.

In summary, the established finite element model investigates the monopile's lateral H-y responses with the following effects: (i) the soil stiffnesses E_u/S_u with values of 50, 100, 200, 400; (ii) the aspect ratio L/D with values of 3, 4, 6, 8, 10, 12, and (iii) the local scour depth ratio S/D with value of 0, 1, 1.3.

2.2 Validation

Field test results by He [7] are used to validate the effectiveness of the 3D finite element model. He [7] conducted the field test on a monopile subjected to horizontal loading in undrained clays. The prototype pile had a diameter of 0.8 m and a length-to-diameter ratio (L/D) of 36.25. The undrained strength profile of clays has been reported as $S_u = 20$ kPa [2]. A stiffness ratio E_u/S_u of 1000 was used for the undrained clay. The stiffness and Poisson's ratio of the monopile take the value of 3 GPa and 0.2, respectively. Figure 2 compares the lateral H-y responses from the field test and the finite element analysis (FEA). This comparison confirms that the established FE model can effectively capture the nonlinear lateral behavior of the monopile in undrained clays.

3 Results and Analysis

First, it is necessary to illustrate the effect of the aspect ratio L/D on the behavior of the monopile with local scour holes. Figure 3a presents lateral load–displacement H-y curves for monopiles under varying aspect ratio L/D. It demonstrates that the lateral ultimate bearing capacity rises with an increase in the aspect ratio L/D, evidenced by the ascending trend of H-y curves with rising L/D values.

Figure 3b presents lateral load–displacement H-y curves for monopiles under varying soil stiffness E_u/S_u. It is observed that, despite differing E_u/S_u values, these curves arrive at the same ultimate bearing capacity. However, an increase in E_u/S_u is

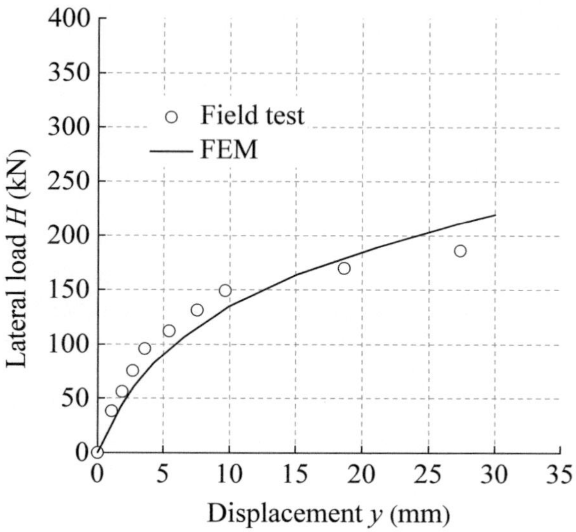

Fig. 2 Comparison between lateral H-y curves from FEM and the field tests by He [7]

associated with a steeper slope of the H-y curve, indicating a higher initial stiffness in the lateral H-y response as E_u/S_u increases.

Figure 3c depicts the lateral response for monopiles with different local scour holes. The ultimate lateral bearing capacities decrease with an increase in the local scour depth ratio. This highlights the significance of investigating the impact of local scour on the lateral bearing characteristics of monopiles.

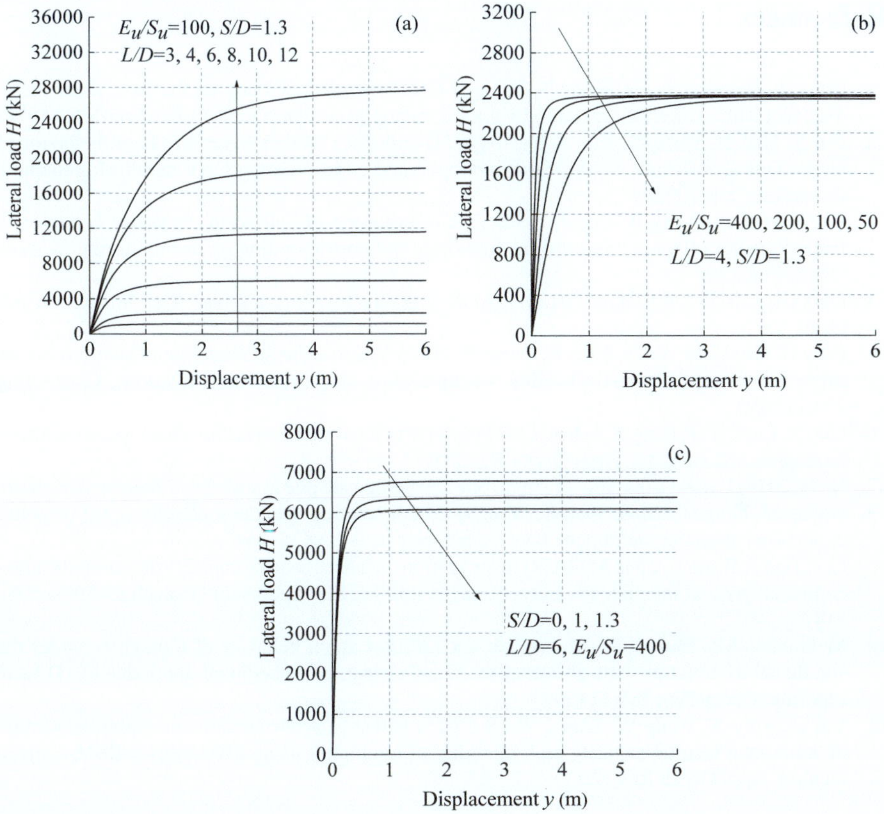

Fig. 3 Lateral H-y responses for monopiles with influence aspect ratio L/D, soil stiffness E_u/S_u, and scour depth ratio S/D

4 Conclusion

This study has investigated the lateral responses of monopiles with local scour hole in non-homogenous clay via the 3D Finite Element Analysis (FEA). The effects of soil properties, foundation geometries, and local scour geometries on lateral responses are investigated. It is found that the increase of the aspect ratio L/D and decrease of scour depth ratio S/D can reduce the lateral ultimate bearing capacity of monopiles. While the soil stiffnesses E_u/S_u does not influence the capacity, the increase of E_u/S_u increases macro stiffness of H-y responses.

References

1. Chen Z, Zhang Z, Li W, Wang T, Zhang J (2023) Numerical study on the base shear force-displacement relationship for laterally loaded monopiles in dense sand. Ocean Eng 286:115527
2. Dai S, Han B, Wang B, Luo J, He B (2021) Influence of soil scour on lateral behavior of large-diameter offshore wind-turbine monopile and corresponding scour monitoring method. Ocean Eng 239:109809
3. Dai S, Han B, Wang B, Luo J, He B (2021) Influence of soil scour on lateral behavior of large-diameter offshore wind-turbine monopile and corresponding scour monitoring method. Ocean Eng 239
4. DNVGL (2018) DNVGL-ST-0126-support structure for wind turbines. Det Norske Veritas, Oslo
5. Feng H, Huang M, Shi Z, Shen K, Wang B (2024) Macro-element modeling of suction caisson subjected to vertical tensile loading via up-scaling soil stress-strain relations. Ocean Eng 304:117850
6. Guo X, Liu J, Yi P, Feng X, Han C (2022) Effects of local scour on failure envelopes of offshore monopiles and caissons. Appl Ocean Res 118
7. He B (2016) Lateral behaviour of single pile and composite pile in soft clay. Zhejiang University
8. Huang M, Feng H, Shi Z, Shen K, Wang B (2024) Inferred Winkler model for uplift response of suction caisson in undrained clays. Acta Geotech 19(1):437–454
9. Li F, Tian P, Wang L, Chen M (2020) Investigation on lateral bearing capacity of monopile under combined vertical-lateral loads and scouring condition. Mar Georesour Geotechnol 39(4):505–514
10. Mehrabadi AR, Moradi M, Malakshah RR (2023) Lateral behavior of monopiles under the simultaneous effect of vertical-lateral loads and change of embedment depth due to 1D local scouring. Ocean Eng 280:114509
11. Yin Q, Xing X, Wang W, Zhai J, Xu S (2023) Effect of scour erosion and riprap protection on horizontal bearing capacity and reliability of monopiles using FEM-BPNN-RSM coupled method. Appl Ocean Res 140

Effect of End Boundary Conditions on the Structural Response of the Submerged Floating Tunnel

Kiwon Jeong, Hyunwoo Shin, Jinwoo Kim, and Seungjun Kim

Abstract This study focuses on the practical application of submerged floating tunnels (SFTs), emphasizing the need to effectively resist environmental loads while maintaining serviceability and safety. A comprehensive analysis involving various boundary conditions was conducted to understand the behavior of SFTs. The structural responses of SFTs were analyzed through hydrodynamics-based numerical methods, specifically using ABAQUS-AQUA. The study investigated the dynamic motions and moment distributions under different mooring types and boundary conditions. It was found that double-inclined mooring (Type3) displayed the most stable displacement and the least moment in the free condition, providing valuable insights for the fundamental design of SFTs.

Keywords Submerged floating tunnel · Tether · Boundary conditions · Hydrodynamics · ABAQUS

1 Introduction

The submerged floating tunnel (SFT) is an innovative underwater transportation infrastructure, suitable for use in deep waters. The SFT, constructed on land and then installed underwater, simplifies construction and reduces costs and time. Structural responses are mainly induced by wave loads, affecting both serviceability and stability. Therefore, appropriate mooring lines and cross-sectional designs are required to sufficiently suppress wave loads and ensure stability. Studies have been actively conducted to rigorously calculate the structural response of SFTs to environmental loads and to develop simple yet reasonable methods to estimate this response.

K. Jeong · H. Shin · J. Kim · S. Kim (✉)
School of Civil, Environmental and Architectural Engineering, Korea University, Seoul, South Korea
e-mail: rocksmell@korea.ac.kr

X. L. Zhao et al. (eds.), *Proceedings of the Fourth World Conference on Floating Solutions: WCFS 2024; 2–4 December; Hong Kong*, Lecture Notes in Civil Engineering 597, https://doi.org/10.1007/978-981-96-4569-5_33

Studies by Jeong et al. [1], Won et al. [2], and Østlid [3] have primarily focused on the impact of external loads on the tunnel, examining various mooring methods and tunnel cross-sections. However, most research has been limited to a single combination of boundary conditions and mooring types. The dynamic behavior of SFTs varies complexly depending on combination of boundary conditions at both ends and mooring methods. More comprehensive research that considers various combination of boundary conditions and mooring methods is required for more rational design.

2 Analysis Method to Obtain Structural Responses of SFTs

2.1 Mathematical Formulation

A coordinate system and variables are established as shown in Fig. 1, and the tunnel is divided into a segments according to the spacing of the mooring lines. Using Hamilton's principle, it's possible to obtain the equations of motion without considering force balance in the free-body diagram, through repetitive calculations.

According to Hamilton's principle, the following Eq. (1) is established.

$$\int_{t_1}^{t_2} (\delta T - \delta V + \delta W_{nc})dt = 0 \tag{1}$$

where, T, V and W_{nc} is the kinetic energy, potential energy, and energy due to non-conservative forces. Therefore, by solving the above variational equation, the following Eqs. (2–4) of motion can be obtained.

$$EI_z \frac{\partial^4 y_n}{\partial x^4} + \frac{m_n}{L_n} \frac{\partial^2 y_n}{\partial t^2} = F_y \tag{2}$$

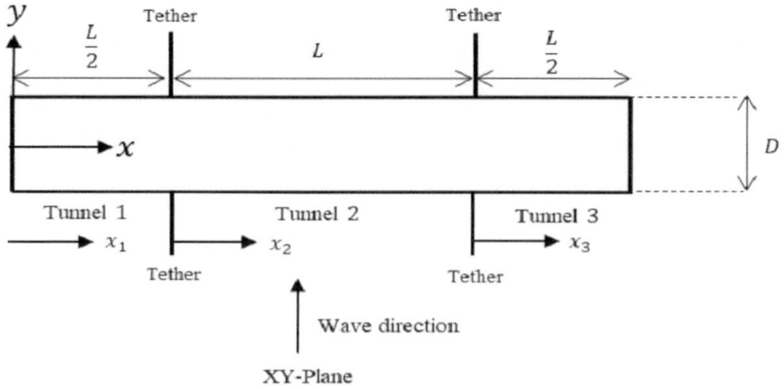

Fig. 1 Coordinate system and variables

$$EI_y \frac{\partial^4 z_n}{\partial x^4} + \frac{m_n}{L_n} \frac{\partial^2 z_n}{\partial t^2} = F_z + w + B \tag{3}$$

$$GJ \frac{\partial^2 \theta_n}{\partial x^2} = \frac{I_n}{L_n} \frac{\partial^2 \theta_n}{\partial t^2} \tag{4}$$

Here, m_n and I_n mean the self-weight and moment of inertia of the nth tunnel. Additionally, E is the modulus of elasticity, G is the shear modulus, I_z and I_y are the second moments of area about the z and y axes, respectively, J is the polar moment of inertia of the section, m_0 is the mass per unit length, I_0 is the moment of inertia per unit length, L_n is the length of the nth tunnel, w is the self-weight per unit length, and B is the buoyancy force per unit length, F_y, F_z is the horizontal and vertical force components.

The objective of this study is to determine the structural response of tunnels under irregular wave conditions in actual sea areas. Irregular waves are simulated by defining the energy spectrum using the Joint North Sea Wave Project (JONSWAP) wave spectrum. In ocean engineering, potential theory is typically used to calculate wave forces on floating bodies. These potential calculations generally require numerical solutions based on integral equations. However, it has been shown in numerous studies [4–8, 12–15] that for cylindrical objects like SFTs and moorings, the Morison equation can yield results almost like potential theory in many cases. Therefore, in this study, F_y, F_z are calculated by using Morison equation.

2.2 *Hydrodynamics-Based Numerical Analysis Method*

The equations of motion for the SFTs given by Eqs. (2–4) are nonlinear differential equations, thus necessitating numerical methods for their solution. Commercial finite element analysis programs like ABAQUS or Orcaflex have been utilized by many researchers. Particularly, ABAQUS has been used by various research groups globally, as evidenced in studies like Min et al. [6] and Xiang et al. [9], and is recognized as a reliable analysis program. Previous research by Min et al. [6], which utilized ABAQUS-AQUA to obtain the structural response of SFTs and compared it with the experimental results from Oh et al. [10], demonstrated considerable reliability.

The tunnel is defined as a beam element and modeled as B31 elements, while mooring lines are set as truss elements, using T3D2 elements. The analysis of the SFTs is conducted in two steps: the initial buoyancy-induced static bending deformation in Step 1, and Step 2 performs dynamic analysis, obtaining the dynamic response to waves over 3800 s. This dynamic analysis employs the direct time integration method.

3 Structural Responses Under Various Mooring and Boundary Conditions

3.1 Environmental Conditions and Model Specifications

In this study, the Messina Strait in Italy was selected as the target area. To analyze the structural responses of the SFTs under extreme environmental conditions, 100-year return period irregular waves, representative of extreme conditions in this area, was considered [11]. Figure 2. illustrates the time series of wave heights impacting the SFTs over 3800 s. The overall environmental conditions and specifications of the tunnel and mooring line are shown in Table 1. Also, Fig. 3 shows the side view of numerical model and front view of the SFT.

The three mooring types and four boundary conditions are considered to investigate the structural behavior under different mooring types and boundary conditions. The selected mooring type is shown in Fig. 4. The boundary conditions include free ends on both sides, fixed ends on both sides, hinged ends on both sides, and a combination of hinged and fixed ends. Transportation infrastructure, such as bridges, typically relies on fixed and hinged supports as boundary conditions. Fixed supports offer high stability and resistance, though they also induce significant bending moments. Hinged supports, on the other hand, provide structural flexibility and efficiently distribute loads but can result in substantial bending deformations. In this study, we also considered free boundary conditions to account for the specific characteristics of submerged floating tunnels (SFTs), which are heavily influenced by mooring lines. This approach effectively reduces the bending moments on the tunnel but may increase the load exerted on the mooring lines.

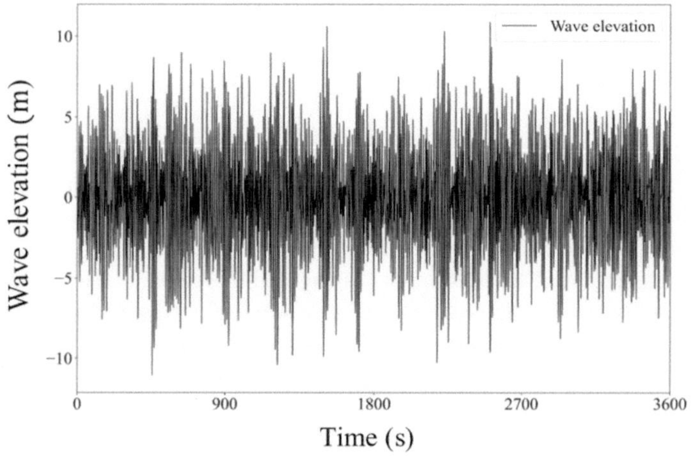

Fig. 2 Wave elevation

Table 1 Tunnel specifications and environmental conditions

Parameters	Values
1. *Environmental condition*	
Water depth (m)	80.0
Significant wave height (m)	13.5
peak wave period (s)	11.5
Draft (m)	40.0
2. *For tunnel*	
Diameter (m)	23.0
Thickness (m)	1
Elastic modulus (GPa)	25.0
Buoyancy–weight ratio	1.3
Drag/added mass coefficient	1.0/1.0
3. *Tether (Mooring)*	
Diameter (m)	0.95
Thickncss (m)	0.04
Elastic modulus (GPa)	210.0
Outside tether inclination angle (°)	45.0
Tether spacing (m)	40.0

(a) Side view of numerical model

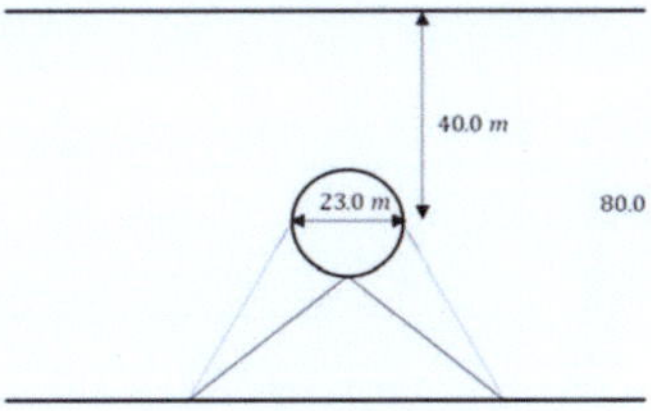

(b) Front view

Fig. 3 Numerical analysis model and environment condition

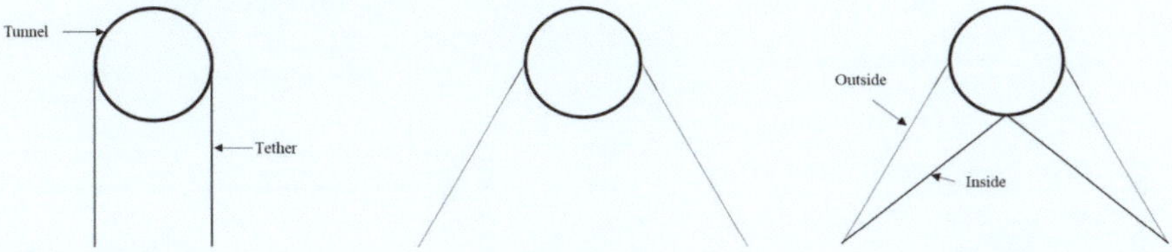

Type 1 Type 2 Type 3

Fig. 4 Various mooring types

3.2 Structural Responses of the SFTs Based on Various Boundary Conditions

In Fig. 5, the sway and heave of the SFT at peak wave height were examined. Under free boundary conditions, the sway and heave appeared constant for all types, while under hinge and fixed boundary conditions, they exhibited curved forms. This indicates that rigid body motion dominates the total displacement under free conditions, whereas bending deformation dominates under the other conditions. Additionally, Fig. 5d shows the setdown effect of heave due to large bending sway in the middle position.

In Fig. 6, Types 1 and 2, the maximum sway and heave were smaller under conditions where the tunnel end's rotation was restricted, such as fixed or hinge-fix. Type 3 exhibited the smallest absolute displacement magnitude compared to the other types, and the displacements under free boundary conditions were like those under other boundary conditions. This is attributed to the tether arrangement suppressing the roll of the tunnel.

The maximum moment of the SFT tunnel was examined through Fig. 7. For all types, the free boundary condition results in a significantly smaller bending moment compared to the other constrained boundary conditions due to the dominance of

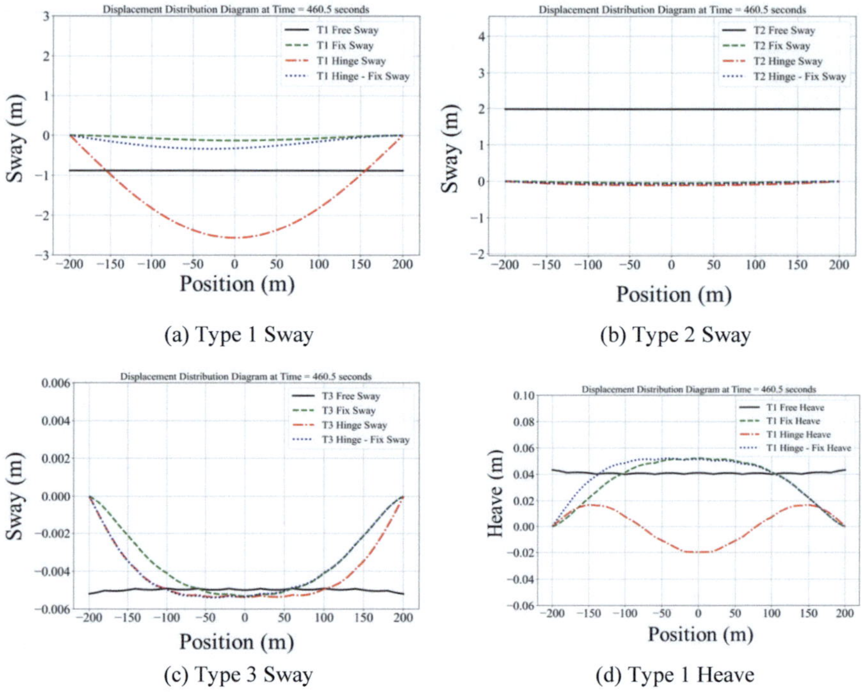

Fig. 5 Displacement distribution at peak wave elevation (460.5 s)

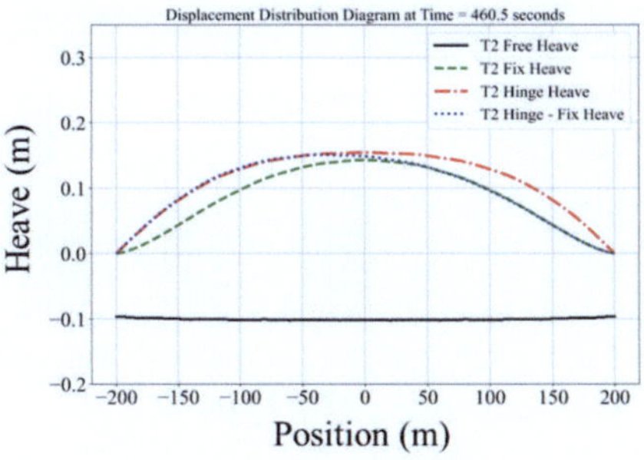

(e) Type 2 Heave

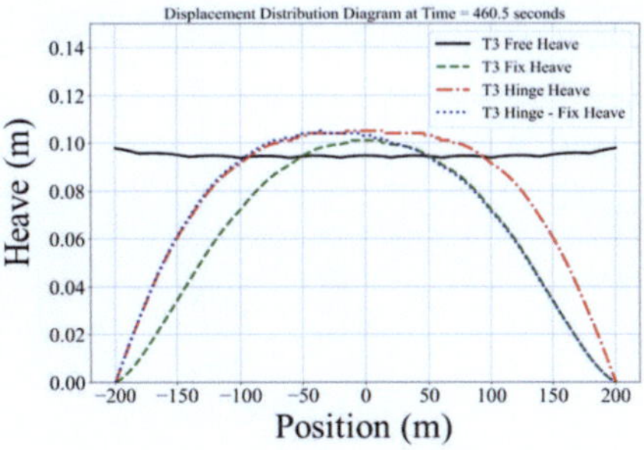

(f) Type 3 Heave

Fig. 5 (continued)

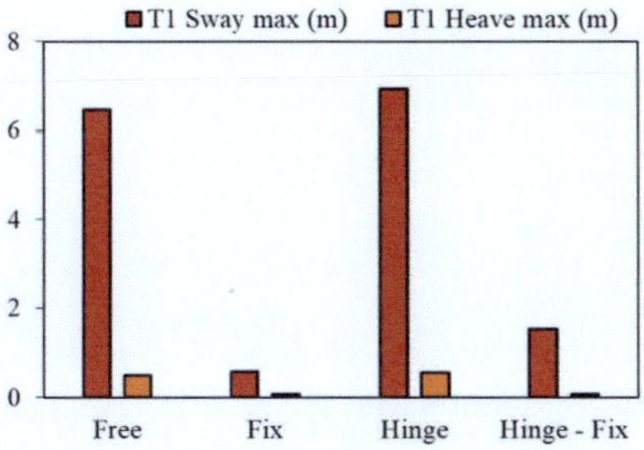

(a) Type 1

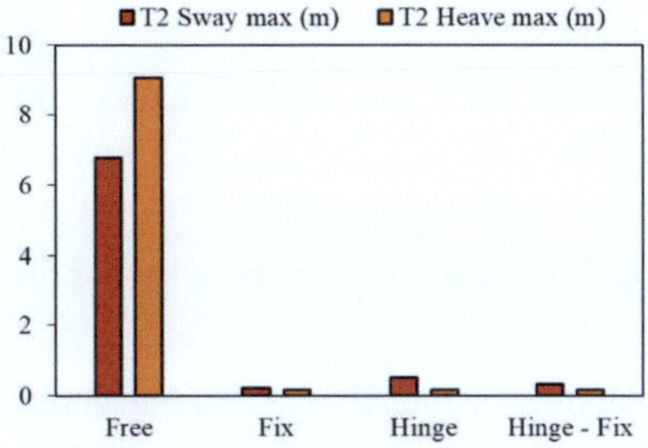

(b) Type 2

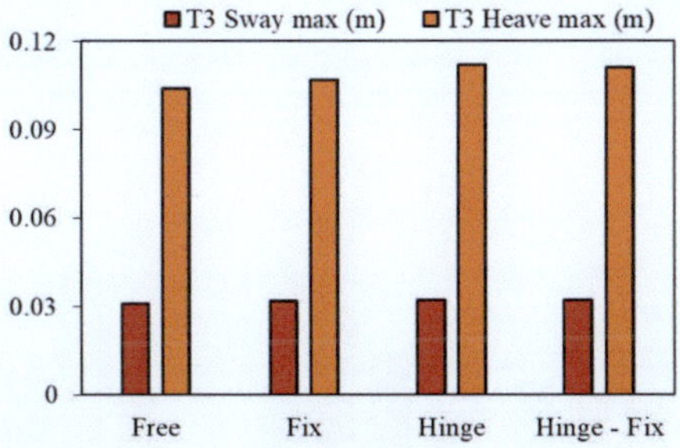

(c) Type 3

Fig. 6 Comparison of maximum displacement

rigid body motion. Considering both Figs. 6 and 7, Type 3 with free boundary condition appears to be the most reasonable combination because it best suppresses displacement and generates minimal moments.

In addition to the displacement and moment of the tunnel, it is also important to examine the stress occurring in the mooring lines. Figure 8 compares the maximum stress for each type and boundary condition against the mooring line design strength

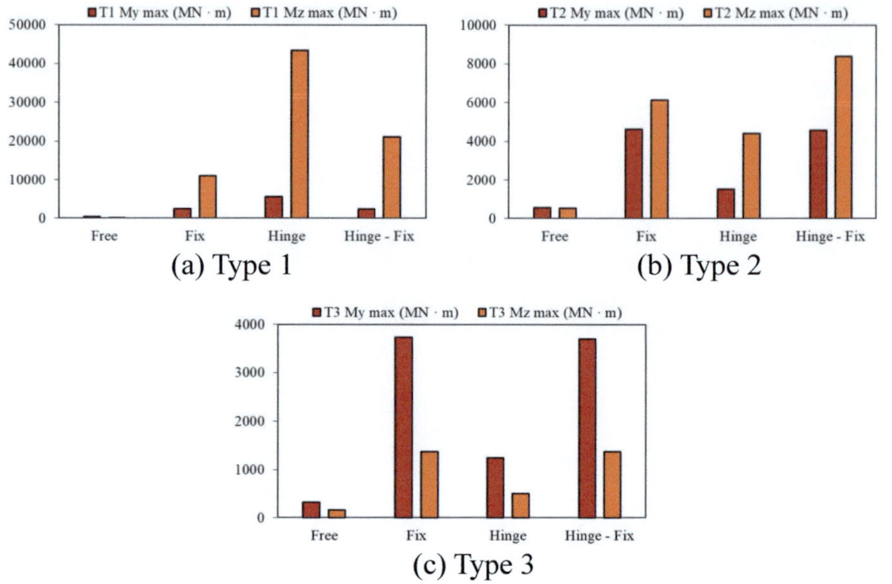

Fig. 7 Comparison of maximum bending moment

specified by the API for extreme condition. Consequently, the tether stress for Type 3 under the free boundary condition, which minimizes displacement and moment, meets the API strength criteria.

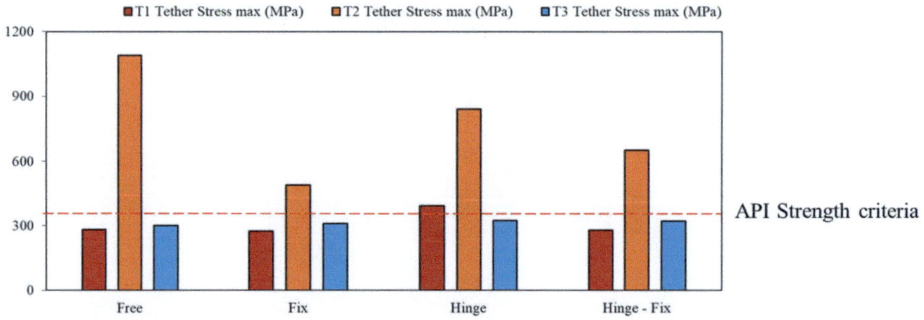

Fig. 8 Maximum tether stress comparison

4 Conclusion

In this study, the structural behavior of a SFT was investigated under various combinations of boundary conditions and mooring types to achieve a rational design. The focus was on comparing the displacement, bending moment of the tunnel and the maximum stress of the tethers, which are directly related to serviceability and safety. The key results are as follows:

1. For Types 1 and 2, the smallest displacement was observed under boundary conditions with rotational constraints, whereas Type 3 exhibited the smallest displacement regardless of the boundary condition. This is due to the tether arrangement effectively restraining roll motion.
2. In terms of bending moment, the free condition resulted in the smallest values for all types. This is because the free condition induces rigid body motion.
3. From the perspective of displacement and moment, the Type 3 free condition is the most rational choice. Additionally, under this condition, the stress in the mooring lines was very low, satisfying the API strength criteria.

References

1. Jeong K, Min S, Jang M, Won D, Kim S (2022) Feasibility study of submerged floating tunnels with vertical and inclined combined tethers. Ocean Eng 265:112587
2. Won D, Seo J, Kim S (2021) Dynamic response of submerged floating tunnels with dual sections under irregular waves. Ocean Eng 241:110025
3. Østlid H (2010) When is SFT competitive? Procedia engineering, first international symposium on Archimedes Bridge (ISAB-2010), vol 4. Elsevier, Amsterdam, pp 3–11
4. Cifuentes S, Kim S, Kim MH, Park WS (2015) Numerical simulation of the coupled dynamic response of a submerged floating tunnel with mooring lines in regular waves. Ocean Syst Eng 5(2):109–123
5. Seo S, Mun H, Lee J, Kim J (2015) Simplified analysis for estimation of the behavior of a submerged floating tunnel in waves and experimental verification. Mar Struct 44:142–158
6. Min S, Jeong K, Noh Y, Won D, Kim S (2022) Damage detection for tethers of submerged floating tunnels based on convolutional neural networks. Ocean Eng 250:111048
7. Min S, Jeong K, Lee Y, Kim S (2023) Estimation of unmeasured structural responses of submerged floating tunnels using pattern model trained via long short-term memory. Ocean Eng 277:114284
8. Jeong K, Kim S (2024) Structural response of submerged floating tunnels with free-end boundary condition based on an analytical approach. Appl Ocean Res 143:103861
9. Xiang Y, Chen Z, Yang Y, Lin H, Zhu S (2018) Dynamic response analysis for submerged floating tunnel with anchorcables subjected to sudden cable breakage. Mar Struct 59:179–191
10. Oh SH, Park WS, Jang SC, Kim DH (2013) Investigation on the behavioral and hydrodynamic characteristics of submerged floating tunnel based on regular wave experiments. KSCE J Civ Eng 33(5):1887–1895 (in Korean)
11. Faggiano B, Panduro J, Mendoza Rosas MY, Mazzolani FM (2016) The conceptual design of a roadway SFT in Baja California, Mexico. Procedia Eng 166:3–12
12. Won D, Kang YJ, Park WS, Kim S (2021) Dynamic behavior of the submerged floating tunnel moored by inclined tethers attached to fixed towers, Ocean Eng 237:109663. https://doi.org/10.1016/j.oceaneng.2021.109663

13. Min S, Jeong K, Noh Y, Won D, Kim S (2022) Convolutional neural network-based damage detection of the tethers of submerged floating tunnels using structural response data under various incident waves, In: Bridge Safety, Maintenance, Management, Life-Cycle, Resilience and Sustainability, CRC Press, pp. 1170–1177. https://www.taylorfrancis.com/chapters/edit/10.1201/9781003322641-141/convolutional-neural-network-based-damage-detection-tethers-submerged-floating-tunnels-using-structural-response-data-various-incident-waves-min-jeong-noh-kim-won
14. Ryu S, Min S, Kang YJ, Kim S (2024). LSTM-based structural pattern recognition for floating bridges. In: Proceedings of international conferences on bridge maintenance, safety, and management. IAMBAS. https://doi.org/10.1201/9781003483755-84
15. Jeong K, Kim J, Shin H, Kim S (2025) Design strategy for submerged floating tunnels: Structural response analysis and rational mooring-boundary condition combinations under wave loads, Ocean Eng 317:120097. https://doi.org/10.1016/j.oceaneng.2024.120097

Submerged Floating Tunnel High Potential Energy System

Wei Lin, Mark Cassidy, Yinghui Tian, and Ming Lin

Abstract Although proposals to build submerged floating tunnels (SFTs) have existed for over a century, their history is littered with well-meaning suggestions of form and type, but very few have been built. More recently, however, their potential as a highly cost-effective and sustainable crossing has rejuvenated industry interest. The main types of SFTs are pier-, tether-, pontoon-supported and free-spanned. With the introduction of a novel high-potential energy system, the benefit of providing axial tension along a tether-supported SFT is studied. The system imposes permanent axial tension in the tube by pre-tensioning the tube from the shore during construction. The system enables a positive adjustment of the stiffness and natural vibration frequency. This novel energy system has three main advantages: (i) expensive and challenging construction in the offshore and submerged environment can be dramatically reduced; (ii) introducing design flexibility can ensure a globally optimised design; and (iii) the risks of system failure due to changing environments or uncertain system behaviours can be eliminated. Design comparisons are made of the SFT's effectiveness against wave loading, with and without the high potential energy system. An optimisation approach allows for a desk-top study that minimises system cost while satisfying multiple design criteria, such as tube deflection and acceleration, tether breaking strength, no slack-and-snap in the tethers, and SFT weight balance equilibrium.

Keywords Submerged floating tunnel · Design · Construction · Optimisation

W. Lin (✉) · M. Cassidy · Y. Tian
Faculty of Engineering and Information Technology, University of Melbourne, Melbourne, VIC, Australia
e-mail: wei.lin1@student.unimelb.edu.au

M. Lin
China Communications Construction Company Ltd., Beijing, China

X. L. Zhao et al. (eds.), *Proceedings of the Fourth World Conference on Floating Solutions: WCFS 2024; 2–4 December; Hong Kong*, Lecture Notes in Civil Engineering 597, https://doi.org/10.1007/978-981-96-4569-5_34

1 Introduction

Submerged floating tunnels (SFT) have the potential to cross broader and deeper waterbodies than the longest cable bridges in the world because they use natural buoyancy to balance their own weight. Historically, ambitious efforts have been put forth to cross some of the world's most famous waterways [1–12]. Many extreme concepts have been proposed, including a futuristic TransAtlantic tunnel with a nuclear-powered thruster system to keep the tunnel at the desired location [13].

An SFT generally consists of three subsystems: a tube, a support (for the tube's main body) and a shore connection (for the tube's ends). Based on the different approaches for support, SFTs can be classified into four types: tether-supported, pontoon-supported, pier (column)-supported and unsupported free-spanned. To date, only two pier-supported SFTs have been constructed [6]. Challenges in building other types of SFTs include uncertainties of hydrodynamic loadings and the highly complex system behaviours between the loading, tube, supports and soils.

To address the challenges, and to allow an SFT to positively adapt to the environment and improve its structural behaviour, this paper investigates a novel SFT structure [14]. It allows the SFT to self-tune the structural stiffness (and thus natural vibration frequency) before or in the operational stage after construction. This also shifts some high-risk and expensive offshore construction to relatively cost-effective and more accessible land works. This paper studies its effectiveness when subjected to wave loading (response versus design criteria) using a recently published SFT design optimisation framework [15].

2 SFT High Potential Energy System

2.1 Concept

Potential (internal) energy can be stored on an SFT by imposing permanent axial tension on the tube's body through its ends. This tension increases the tube's geometric bending and torsional stiffness. Any change in stiffness changes the system's natural vibration frequency, and a more optimised natural vibration frequency can be obtained by adjusting the magnitude of axial tension.

Axial tension can be applied into the SFT in several ways. For instance, even a simple change in temperature from the installed conditions and the operating conditions can change the axial tension. If the SFT is installed in summer and both ends fixed axially, by winter the temperature drop will induce an axial tension [9, 16] and the SFT will keep some potential energy.

However, a more active system may be required. Figure 1 shows two scaled SFT models with axial tension imposed by hydraulic cylinders. Figure 1a is a scaled tethered support SFT model in a wave basin. The SFT tube was tensioned by 0, 20,000, 40,000, 80,000 newtons of force and then excited by knocking, waves or currents,

(a) Tethered SFT test in basin [14,30,31] (b) SFT tube free vibration decay test [14]

Fig. 1 SFT scaled models subjected to axial tension [14, 17, 18]

during which the tube's strains, deflections, accelerations and tethers' tension were recorded. As expected, and it was even observable by eye that the more tension imposed the less response of the model. Figure 1b is a free-spanning SFT tube model axially tensioned by a maximum of 100,000 newtons of force [14]. The free-decay vibration was produced by enforcing an initial displacement at the mid-point of the tube and then releasing it suddenly, both in air and in still water. It was observed that as the axial tension increases the model's vibration frequency increases. Further results from these tests are not shown in this paper for brevity but will be published in an upcoming PhD thesis.

2.2 Design Considerations

Tension and Anchor System of the Shore Connections. Figure 2 shows the details of a possible shore connection to tension the SFT tube to obtain permanent potential energy. One possible option is tensioning the tube by both ends and anchoring it to the shore. Specifically, tension rods or cables are connected to the tube on one end and to rock or precast concrete blocks on the other, as shown in Fig. 2a. Tension works can reference proven technologies, such as tendon cables used in cable bridges or external prestressing used in girder bridges. If the tube is very long, or the temperature fluctuation is large, the length of the tube may be subject to cyclic elongation and shortening. An expansion joint, as shown in Fig. 2a, can be set to accommodate the large axial movement of the tube. The large-diameter expansion joint is a proven technology used in, for instances, the pipelines of China's Three Gorges project [19], and the connection between bored tunnel and immersed tunnel of the Bosphorus strait [4]. Figure 2b shows a cross-sectional view where the tension rod or cables are evenly set along the circumferential outline of the tube. Another possible form tensions the tube from one end, with the other set as a passive end, as shown in Fig. 2c. The tube is anchored into rock or concrete blocks that bear the axial tension. Figure 2d shows

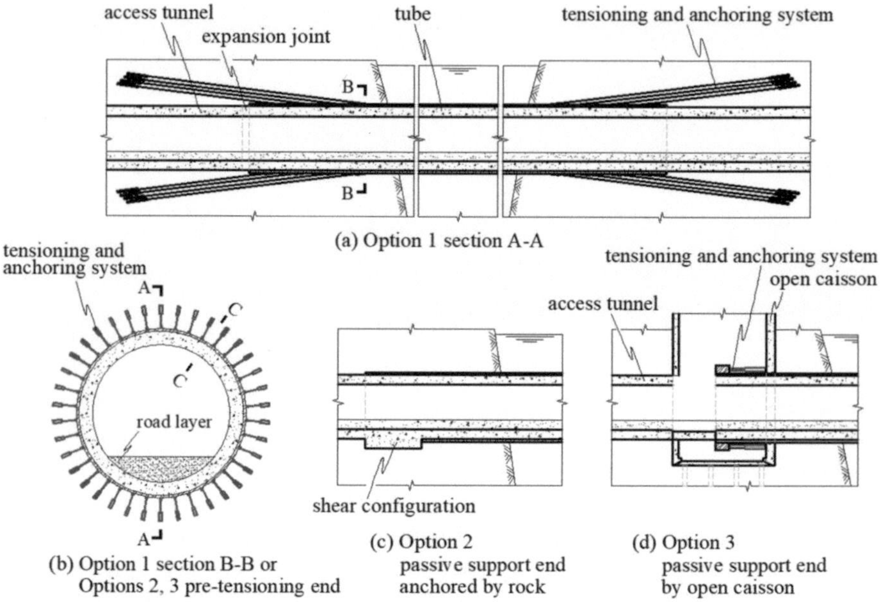

Fig. 2 Shore connection details

an alternative for soft ground, where the potential energy for an SFT can be obtained by building an open caisson in advance [16]. The caisson ensures a strong anchorage for the tube and facilitates tube installation works to the shoreline.

Tube Wall Configurations. The tube wall can adopt composite structures, e.g., concrete-steel with shear studs, concrete-steel allow-sliding, steel–concrete-steel for double waterproof, or concrete-rubber-steel for anti-collision.

3 Optimisation

3.1 Method

The SFT tube was modelled as Euler–Bernoulli beam in Eq. 1. The modelling of the tether accounted for geometry and elastic stiffness using [20] but disregarded mass and fluid drag damping (this ensures that the response prediction is conservative). The tether material assumed steel wires and their elastic stiffness, submerged weight and Maximum-Breaking-Strength (MBS) was estimated as a function of tether's nominal diameter using [21]. With these assumptions the system's stiffness, mass and damping matrices were assembled in Rayleigh–Ritz form [22]. The random waves used a JONSWAP spectrum [23] with the wave load estimated by the linearised Morison equation. The response computation was in the frequency domain [22, 24] with three hours peak responses derived [23]. The response calculation was verified by the free vibration decay model test (i.e., the test shown in Fig. 1b), the analytical solution [25], and the finite element time domain programming [26].

For the system verification, the SFT with and without High Potential Energy (HPE) system was optimised with the objective function of system cost of tube, tether, and anchor. The SFT response was constrained to design criteria when subject to wave loading, an extension of the methodology and work of [15]. The constraint functions are (i) that the tube's deflections and accelerations remain less than 0.3 m and 0.5 m/s^2 respectively under wave load [27], (ii) the tube's cross-section structural capacity [28] derived from concrete and rebar steel's stress–strain relation [29] must be larger than the detrimental combination of extreme value of the moment and axial tension of the tube, (iii) all tethers have no overstressing or slack (or in other words the sign-convention of the computed tether angle at the anchor must remain the same), and (iv) the SFT is buoyant (i.e. the system buoyancy is greater than or equal to the weight. Further, the interaction between the tube's residual buoyancy and the tethers' pre-tension was considered in the optimisation framework. The open-sourced software OpenMDAO was used as a highly efficient, easy-coding, and supporting gradient-based design optimisation tool [30].

3.2 Results

Results are presented for the following design case. The SFT must be 200 m long, crossing 50m water depth, submerged under water at 5 m depth and having a circular (inner diameter) cross-section of 5 m and a composite wall configuration, as shown in Fig. 3a. Three pair of vertical tethers connect the tube on one end and the gravity anchor on the other end. The significant wave height is 2m and the wave mean period is 10s. The design variables being optimised for were (i) the SFT tube concrete wall thickness and (ii) steel shell thickness, (iii) axial tension (zero, or value if HPE activated) and (iv) nominal diameters for each pair of tethers.

The results for this base-case design are provided in the first two columns of Table 1. The first column is the optimised design with no HPE and the second column the optimised design with HPE. The first row is inner diameter of the tube; default setting is 5.00 m as the HPE design because the stiffness can be ensured via axial tension. However, in the no HPE design, the stiffness can only be ensured by increasing second moment of area and thus inner diameter of tube needs to increase to 6.72 m. The second and third rows are the wall thicknesses of inner concrete layer and outer steel shell. No HPE design chooses thicker concrete wall and thinner steel shell while HPE design chooses the opposite; these show different design strategy: (i) the no HPE design uses as much concrete as possible because the unit cost of concrete is cheaper than that of steel in this model setting, and secondly, large volume of concrete is used to compensate the large buoyancy resulted from the enlarged tube, so as to suppress the tether's pretension and the anchor design cost. (ii) the HPE design uses as little concrete wall as possible because the tube's outer diameter can be further

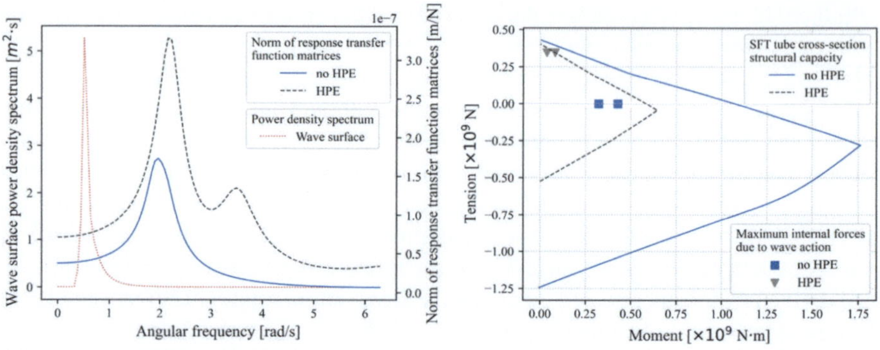

(a) Norm of response transfer function matrix of SFT system versus power density spectrum of wave

(b) SFT tube cross-section structural capacity versus maximum internal forces

Fig. 3 Structural characteristics of the optimised design of vertically tethered SFT the base cases

reduced to decrease wave load, and at the same time, the lightweight structure with more steel and less concrete yields a high Buoyancy-Weight-Ratio (BWR) design that allows the tethers and anchors contributing more to the system stiffness from two sources, the tether's elastic stiffness of larger nominal diameter tether, and the tether's geometric stiffness of the high pretension resulted from high residual buoyancy. The fifth row is axial tension only concerned with HPE model. This value is optimised to 0.34×10^9 N. The sixth to eleventh rows are design constraints relevant indexes. As seen, both designs reach tube deflection limit but neither reach acceleration limit; HPE design reaches the limits of tube cross-section structural capacity and tethers' MBS while no HPE design has redundancy safety for both. And both designs have a positive BWR: it means a residual buoyancy is reserved for pretension the tethers, and thus implies that the tether is not abandoned by the optimiser, in another word, the free-spanning design is not the optimum option for the two base cases. The last three rows are costs: as seen the no HPE design has higher cost and most cost came from tube and shore, on the contrary the HPE design has higher cost in tether an anchor; these cost distribution patterns are in line with the design strategy discussed.

The results for the optimised case are shown in Fig. 3 for both the no HPE and HPE design. In the frequency domain Fig. 3a shows the norm of response transfer function matrices versus wave surface power density spectrum. Both designs avoid the peak frequency of waves by setting the SFT's natural vibration frequency several times higher. However, no HPE design has a smaller value of the norm of response transfer matrices than that of the HPE design, because it endures larger wave forces due to a larger tube. On the other hand, HPE design has twin spectral peaks, indicating a more flexible design, but at the same time the primary vibration mode has a higher frequency than that of the no HPE design. In summary, the optimised HPE design shows the potential advantage of higher flexibility in re-distributing internal forces and higher 'rigidness' in avoiding resonance.

Table 1 Results summary of six optimised design of different tethers arrangement and with or without HPE

	Base cases		Sensitivity study of tethers arrangement			
	Tether A—0, 0, 0 (°)		Tether B—45, 45, 45 (°)		Tether-C—0, 45, 0 (°)	
	No HPE*	HPE	No HPE	HPE	No HPE	HPE
Inner diameter (m)	6.71	5.00	5.00	5.00	5.00	5.00
Concrete wall thickness (m)	0.90	0.20	0.63	0.58	0.20	0.20
Steel shell thickness (m)	0.02	0.59	0.02	0.02	0.02	0.02
Tether diameters (mm)	27, 29, 27	55, 52, 55	34, 74, 34	10, 20, 10	69, 90, 69	69, 86, 69
Tube axial tension ($\times 10^9$ N)	0.00	0.34	0.00	0.30	0.00	0.13
Maximum tube deflection (m)	0.30	0.30	0.30	0.30	0.27	0.23
Maximum tube acceleration (m/s^2)	0.25	0.25	0.26	0.26	0.22	0.19
Tube cross-section safety factor	2.51	1.00	3.74	1.00	6.15	1.00
Tether MBS safety factor	1.02	1.00	1.00	1.00	1.00	1.00
Minimum force of tether's anchor end (slack-and-snap check) (N)	0.00	0.00	0.00	0.00	0.00	0.00
Buoyancy-Weight-Ratio (BWR)	1.03	1.51	1.05	1.00	2.20	2.20
Tube and shore cost ($\times 10^6$ USD)	6.55	3.34	3.66	4.38	1.64	1.64
Tether and anchor cost ($\times 10^6$ USD)	0.61	1.00	1.51	0.93	2.87	2.71
Total (system) cost ($\times 10^6$ USD)	7.15	4.34	5.17	4.51	4.51	4.35

Figure 3b shows the tube cross-section capacity envelop versus the maximum internal forces. The envelop for the no HPE design is much larger particularly in compression, which is typical for a reinforced concrete tube. HPE design has a much smaller capacity envelop in compression, but the envelop is nearly identical in combined compression and tension. This is because the steel's yielding stress and stress–strain relation is equal in both compression and tension [29]. In terms of internal forces, in HPE design, though the pre-imposed larger axial force drives the scatters of internal forces towards the envelop, the moment decreases significantly. The reason for the significant decrease in moment internal forces for HPE design are not only the smaller tube size (that attracts less wave load) and stiffer tethers (that convey more load to mooring system then to the shore end), but also the system 'flexibility' and higher natural vibration frequency (as shown in Fig. 3a).

Two design sensitivities were also studied, with results presented in the third to sixth columns of Table 1 and in Fig. 4. In the first the layout of the tethers was changed and in the second the severity of the design sea-state was increased.

The base-case design (Tether A) assumed all tethers for the mooring system were vertical. This was compared to tethers inclined at 45° (Tether B) and a design where the mid-span tether was inclined with the rest vertical (Tether C), with the design

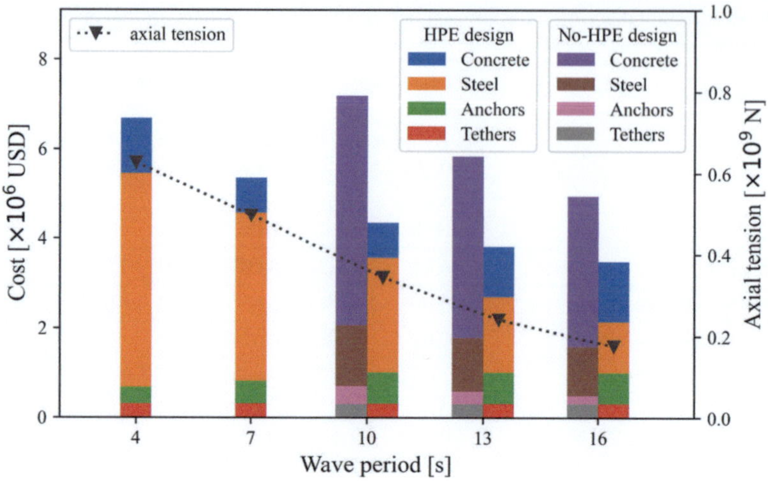

Fig. 4 Sensitivity study against wave period

variables set the same as the base case. As shown in Table 1, HPE always yielded lower costs than those without HPE, irrespective of the tether arrangements. However, Tether B, the design with inclined tethers, has a lower system cost than design with vertical tethers. This is because the tethers can advantageously restrain the tube's lateral response.

As a second sensitivity case the wave period was varied and 4, 7, 10, 13, 16 s were studied. All other base case conditions were maintained (including Tether A and HPE, as in the second column of Table 1) with design variables being the tube's inner diameter, wall thicknesses of concrete and steel shell, and the axial tension (when using HPE). Figure 4 shows the results. SFT designs with HPE still yield lower system cost than those without HPE. For wave periods of 4 and 7s the SFT without HPE was unable to find a valid design. Designs with HPE could be optimised because the axial tension could suppress the response to be within the allowed criteria. From engineering design optimisation view [31], adding axial tension as a design variable indicates that a new dimension for design space exploration is added. Therefore, a better design can be achieved.

4 Conclusions

In this paper use of a new high potential energy system to optimise the design of a Submerged Floating Tunnel (SFT) was presented. Further, how optimisation analysis could be used to select an appropriate design under different wave loading conditions and different tether designs was presented. It was shown that axial tension helps the system better distribute structural inner member forces, increase system stiffness and alter natural vibration frequency to prevent resonance again wave action. Those benefits reduce the outer dimension of the tube and thus attract less wave load—a virtuous cycle. Design optimisation results show that the SFT design with HPE has a higher chance of yielding a possible design to survive in unfavourable wave conditions, and always yields a more cost-effective scheme.

The optimisation framework does not represent a detailed design, but as a desk-top study provides confidence to construction of a very long SFT in deep and open sea environments. The approach has potential for engineering contractors to quickly and reliably develop a potential design concept, on which detailed assessment. The results also show the potential of adding a HPE system to unlocking the economics of SFTs.

References

1. Ingerslev C (2010) Immersed and floating tunnels. Procedia Eng 4:51–59
2. Andrew CE (1950) Floating tunnel for long wave crossing. ASCE 76(8):1–13
3. Grant AB (1972) Floating submerged tunnel for the strait of Messina. TBILISI, pp 53–60
4. Gursoy A (1996) Seismic joints and transition tunnels for the Messina straits fixed-link crossing. In: Ozdemir L (eds) North American tunnelling '96. Balkema, Rotterdam, pp 203–209
5. Ahrens D (1997) SFT—a concept whose time has arrived. Tunn Undergr Space Technol 12(2):317–336
6. Minoretti A, Peiretti HC, Fjeld A et al (2020) Guidelines for submerged floating tube bridge. In: fib Bulletin 96
7. Hiroshi S et al (1995) Underwater tunnel research and investigation submarine tunnel manual vol 2 case studies. Underwater tunnel research investigation. Legal Person (1995)
8. Kanie S (2010) Feasibility studies on various SFT in Japan and their technological evaluation. Procedia Eng 4:13–20
9. Tveit P (2000) Ideas on downward arched and other underwater concrete tunnels. Tunn Undergr Space Technol 15(1):69–78
10. Mazzolani FM, Landolfo R, Faggiano B et al (2008) Structural analyses of the submerged floating tunnel prototype in Qiandao lake (PR of China). Adv Struct Eng 11(4):439–454
11. Fjeld A (2012) Feasibility study for crossing sognefjorden submerged floating tunnel. Dr.techn. Olav Olsen, Reinertsen
12. Project Team, Soreide TH, Haugerud SA (2016) Bjørnafjord submerged floating tube bridge K3/K4 technical report. Dr.techn. Olav Olsen & Norconsult, Reinertsen
13. Østlid H (2010) When is SFT competitive. Procedia Eng 4:3–11
14. CCCC Submerged Floating Tunnel Engineering Technical Joint Research Team (2021) SFT build methodology, novel structure, mechanism test, and engineer's dialogues. Science Press, Beijing (in Chinese)
15. Lin W, Tian Y, Cassidy MJ (2022) A preliminary study on design optimisation of submerged floating tunnel. In: 42nd International conference on ocean, offshore & arctic engineering. ISOPE, Melbourne, pp 1–9
16. Lin W, Liu L, Tian Y, Cassidy MJ (2023) Building submerged floating tunnel in Kanas Lake. In: 43rd International conference on ocean, offshore & arctic engineering. ISOPE, Rhodes, pp 1–9
17. Pan W, Zhou Z, Gu Q et al (2024) Research on the manufacturing method of 3D hydroelastic submerged floating tunnel model. Dalian Marine Time University Press, Dalian (in Chinese)
18. Zhang H, Yang Z, Li J et al (2021) A global review for the hydrodynamic response…. Ocean Eng 225:108825

19. Cui J (2019) Private conversation
20. Thai H-T, Kim S-E (2011) Nonlinear static and dynamic analysis of cable structures. Finite Elem Anal Des 47(3):237–246
21. Ma K-T, Luo Y, Kwan T, Wu Y (2019) Mooring system engineering for offshore structures. Elsevier
22. Clough RW, Penzien J (1993) Dynamics of structures, 2nd edn. McGraw-Hill College, New York
23. Faltinsen OM (1998) Sea loads on ships and offshore structures. Cambridge University Press
24. Remseth S, Leira BJ, Okstad KM et al (1999) Dynamic response and fluid/structure interaction of submerged floating tunnels. Comput Struct 72:659–685
25. Duan W, Li Z, Zhang J et al (2020) Study on mathematical model and exact solution of overall structural response and natural frequency of SFT. In: 30th International conference on ocean, offshore & arctic engineering. ISOPE, Shanghai
26. Newmark NM (1959) A method of computation for structural dynamics. J Eng Mech Div Proc ASCE EM3:67–94
27. Utsunomiya T, Watanabe E (2003) Analysis and design of floating bridges. Prog Struct Eng Mater 5:127–144
28. Zhao G (2012) Advanced reinforced concrete structure. Mechanical Industry Press (2012) (in Chinese)
29. MHURDPRC (2015) Code for design of concrete structures. GB 50010-2015 (in Chinese)
30. Gray JS, Hwang JT, Martins JRRA et al (2019) OpenMDAO: an open-source framework for multidisciplinary design, analysis, and optimisation. Struct Multidiscipl Optim 59(4):1075–1104
31. Martin JRRA, Ning A (2022) Engineering design optimisation. Cambridge University Press, Michigan

Effect of Mooring Pretension Force of a Soft-Connected Floating Photovoltaic System Under Regular Waves

Jenish Baniya, Jian Dai, Zhiyu Jiang, Chi Zhang, and Arvind Keprate

Abstract The rise in global temperature has accelerated the demand for replacing fossil fuels with sustainable renewable energy. Alongside wind, waves and current energies in the oceans, there has been huge interest in developing large floating photovoltaic (FPV) plants in offshore environments. Applying FPV structures in offshore environments remains challenging due to harsh environmental conditions. This paper numerically studies a recently developed soft-connected FPV concept for harsh offshore conditions. The case of a small array consisting of 6 modules interconnected by soft ropes that were experimentally examined is considered. A numerical model for the time domain analysis is built using the commercial solver SIMO. The numerical model is calibrated and validated according to the available experimental results. Next, a parametric study is conducted to examine the effect of different mooring line pretensions on the global responses of the FPV array under regular waves. The study will provide a better understanding of the effect of mooring pretension on the global FPV design. Moreover, recommendations are presented for further optimisation and development of such FPV systems.

Keywords Floating structure · Offshore floating photovoltaic · Hydrodynamic analysis · Numerical calibration

J. Baniya (✉) · J. Dai
Department of Built Environment, Oslo Metropolitan University, Oslo, Norway
e-mail: jeban4151@oslomet.no

Z. Jiang
Department of Engineering Sciences, University of Agder, Grimstad, Norway

C. Zhang
Technology Centre for Offshore and Marine Singapore, Singapore, Singapore

A. Keprate
Green Energy Laboratory, Department of Mechanical, Electrical and Chemical Engineering, Oslo Metropolitan University, Oslo, Norway

X. L. Zhao et al. (eds.), *Proceedings of the Fourth World Conference on Floating Solutions: WCFS 2024; 2–4 December; Hong Kong*, Lecture Notes in Civil Engineering 597, https://doi.org/10.1007/978-981-96-4569-5_35

1 Introduction

There has been an increased interest in harvesting sustainable and efficient energy sources to replace fossil fuels. In the past decades, efficient alternatives to fossil fuels have been substantially developed, such as hydropower, solar and wind. Among these alternatives, solar energy generates electricity by directly converting solar radiation through photovoltaic (PV) cells, and energy production is sustainable and stable because of the abundant amount of solar radiation. The advancement in solar technology has also significantly reduced the cost of developing solar farms, a large cluster of PV panels. Many large solar farms have been installed on land. Installation of such solar farms on land requires a large amount of open space as land space is scarce to sustain the growing population. Large-scale solar project such as Sarnia in Canada occupies a land area of 365 hectares [1]. Therefore, there has been a huge interest in developing floating photovoltaic (FPV) for land-scarce countries and regions with high land prices. Installing such plants over water bodies frees spaces for other purposes such as agriculture and housing.

A typical FPV system consists of a pontoon with mounted solar panels, connection systems, mooring and anchoring, spreading over a huge waterplane area. Developing FPV farms has numerous benefits compared to land-based PV farms, e.g., the natural cooling effect which increases power generation, reduces water evaporation, reduces eutrophication, and improves water quality. Most FPV farm projects are installed in lacustrine environments in lakes and reservoirs with calm and self-contained water. Compared to the in-land water body, the ocean provides significantly more space for FPV farms. However, there are many challenges in developing FPV plants for offshore conditions, where the structures are exposed to saltwater, marine growth, and UV radiation. These can reduce the life span due to material degradation. Moreover, FPVs in marine environments are constantly exposed to complex excitations from wind, waves, tides, and currents. For these reasons, despite a few studies on FPV, the application is still limited to calm [2] and relatively mild conditions [3]. However, these studies were conducted for FPV developed for calm and benign water conditions. There is still a need for research on FPV farms in harsh environments. There are two challenges for developing FPV farms in harsh environments. First, the dynamic motions of large FPV structures induced by strong wind and waves can stress the mechanical structures and fatigue compromising the structural integrity. Second, the FPV farm is a complex multi-body system, and the multi-body interactions with waves introduce complexities to the system's behaviour. This study will investigate the FPV systems in harsh marine conditions for safe installation and operation of the FPV.

In this study, we investigate the hydrodynamic response of a novel FPV system through numerical studies, focusing on the effect of mooring systems. The experimental results are used to calibrate the numerical model which are presented previously in the study [4]. The numerical model is built in the time domain. Based on the calibrated numerical model, a parametric study is conducted to examine the impact of different mooring pretension forces on the global structural response of an FPV array under regular waves. The results from the numerical study provide a better understanding of the effect of mooring pretension on the global design of such FPV systems.

2 Conceptual Design of the FPV System

The FPV system studied in this study consists of multiple interconnected single FPV modules proposed in the article [4]. The single FPV module consists of a lightweight porous pontoon to support four PV panels and the buoyancy for the module is provided with six cylinders with a diameter of 0.6 m. Both the porous pontoon and the cylinders are hollow inside to reduce the overall weight of the system. The single modules are interconnected with soft connecting ropes to form a large array which is anchored with a mooring system. The conceptual design for this single FPV module can be found in Fig. 1a and an array of interconnected single modules can be found in Fig. 1b. The concept has been extensively tested through a laboratory experiment in the towing tank at the Canal de Ensayos Hidrodinamicos of the Universidad Politecnica de Madrid [4]. The experimental study was conducted for six pontoons arranged in a small array of 3×2. For the model test, the single float is idealised to reduce the complexity of the proposed FPV system. In addition, the model test was conducted at a 1:60 Froude scale in the towing tank whereas the numerical model was established at full scale. The full-scale numerical model is based on the model test for a valid comparison between experimental results and the numerical model. Figure 2 presents the model test setup for the numerical study. The numerical model has a spacing of 1 m between pontoons connected with connection ropes and a mooring system. The full-scale single module has a dimension of 4.7 m in length, 2.9 m in width and 0.6 m in height. In addition, the floating modules are submerged with a draft of 0.2 m. The surge direction for both the numerical and experimental model is along the wave direction which is propagated in the longitudinal direction of the pontoon. The wave height and period for the experiment were conducted for nine wave cases typical of wind-generated ocean waves for coastal and nearshore areas. A detailed summary of the experimental study is presented in [4].

3 Methodology

3.1 Dynamics of Global Model

As explained in Sect. 2, the FPV array comprises 6 pontoons interconnected by connection ropes. The corner pontoons are restrained by mooring lines. In the numerical modelling, the pontoons are treated as rigid bodies interacting with the fluid. The connection ropes and mooring lines are modelled as bilinear springs that can

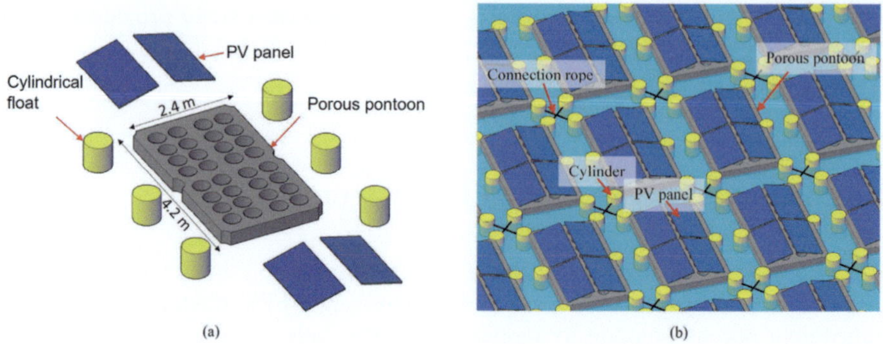

Fig. 1 Conceptual design of FPV module (**a**) and an array of interconnected single modules with connection ropes (**b**) [4]

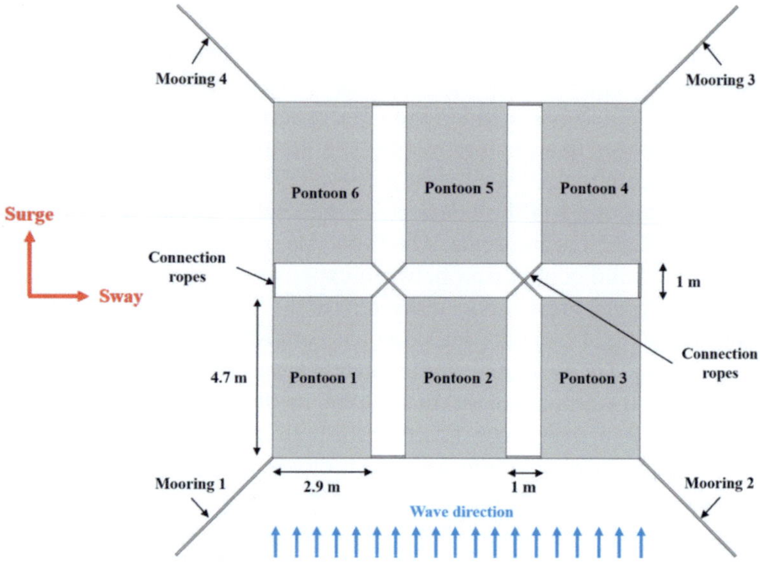

Fig. 2 Idealised full-scale numerical model with pontoons, connection ropes and mooring lines

only withstand tensile forces. For the pontoons, the dynamics of fluid motions are governed by the opposing actions of different forces and moments exerted on the structures [5]. The forces on floating structures are mainly exerted by hydrostatic forces, hydrodynamic forces, and mooring forces. Such forces are influenced by fluid inertia, weight, viscous stresses, and secondary effects such as surface tension. According to Newton's second law of motion, the governing equations of motion of the floating structure are expressed by:

$$M\ddot{x} + C\dot{x} + D_1\dot{x} + D_2 f(\dot{x}) + K(x)x = q_{WA} + q_{pre} \tag{1}$$

where M is body mass and the frequency-dependent added mass matrix, C is the frequency-dependent potential damping matrix, D_1 is the linear damping matrix, D_2 is the quadratic damping matrix, and K is the stiffness matrix due to mooring stiffness, connection stiffness and hydrostatic restoring stiffness. The q_{pre} denotes the pretension force applied on the mooring system. Additionally, the wave excitation forces q_{WA} are expressed as,

$$q_{WA} = q_{WA}^1 + q_{WA}^2 \tag{2}$$

where q_{WA}^1 and q_{WA}^2 are the first and second-order wave excitation forces respectively.

3.2 Potential Flow Theory

The hydrodynamic analysis model was created based on the Panel model. The frequency domain analysis was conducted with Wadam to calculate the wave-structure interaction of the single FPV panels. The method based on Potential flow theory is used to calculate the system's response. The potential flow theory assumes the fluid is ideal, namely, inviscid, irrotational and incompressible. The velocity field of an irrotational fluid can be represented as a scalar function $\phi(x, z, t)$ or as a velocity potential. Thus, the potential is expressed by [5],

$$V = \overline{V}\phi \tag{3}$$

The vector field needs to satisfy the conservation of mass and momentum for the Laplace equation. Additionally, the velocity potential of the harmonic wave must fulfil the criteria for the seabed boundary condition, free surface dynamic boundary condition and free surface kinematic boundary condition. In addition to wave velocity potential, the flow must incorporate the reflected waves. Thus, the wave potential can be expressed with a more general velocity potential [6],

$$\phi(x, y, z, t) = \phi_r + \phi_w + \phi_d \tag{4}$$

following, ϕ_r is the radiation potential from the motion of the body, ϕ_w is the undisturbed wave potential and ϕ_d is the diffracted wave potential.

3.3 Morison Element

The potential flow theory simplifies the laws of fluid mechanics. As previously mentioned, the fluid is assumed to be inviscid where the effect of the viscosity is not considered by the potential flow theory. However, the viscosity has a significant influence on the fluid behaviour of small and slender structures. The calculation based on the potential flow method underestimates the wave drift forces on the structures [7]. Similar outcomes were also observed in other studies conducted by [8, 9]. The discrepancies provided by idealised flows will be accommodated by introducing the Morison element in the numerical model by applying the viscous drag force [10]. The quadratic viscous drag force $F(t)$ is given by

$$F(t) = \frac{1}{2}\varrho_w C_D D(\dot{u}_w - \dot{u}_m)|\dot{u}_w - \dot{u}_m| + \varrho_w C_M V(\ddot{u}_w - \ddot{u}_m) + \varrho_w V\ddot{u}_w \quad (5)$$

where C_D denotes the dimensionless quadratic drag coefficient, ϱ_w is the fluid density, V is the volume and D is the diameter of the slender elements. $\dot{u}_w$ and $\dot{u}_m$ is the flow velocity of the fluid and slender elements respectively. While $\ddot{u}_w$ and $\ddot{u}_m$ is the acceleration of the fluid and slender elements.

3.4 Numerical Procedure

The numerical simulation of the FPV system is conducted with a combination of different software packages. The geometry of the FPV modules and discretised meshes of the wet surface were generated in the GeniE, a preprocessing and FE analysis software package. Then, a frequency-domain boundary element-based hydrodynamic analysis was conducted in Wadam, a wave diffraction and radiation analysis solver based on 3D linear potential flow theory. The hydrodynamic analysis was conducted for a single pontoon. The effect of hydrodynamic coupling between pontoons is neglected in this study. Frequency-dependent hydrodynamic coefficients such as added mass, potential damping, and wave excitation forces are obtained and imported into SIMO, which is a time-domain analysis of the entire array with connection ropes and the mooring system. This model is used to evaluate the global response, especially the interactions among the structural components and incident waves.

4 Results and Discussion

4.1 Calibration and Validation of Numerical Model

The numerical model is calibrated with the responses obtained from an experimental model test [4]. The calibration is made according to the measure characteristics of the mooring system, free decay test, and the heave and surge response amplitude operators (RAOs) of all six pontoons under regular waves. The detailed calibration and validation process for the numerical model is presented below.

Mooring and Connection Systems. The mooring and connection systems significantly affect the dynamic response of the FPV system. A four-point above-water-level mooring line connected with a spring to a load cell was used for the experimental setup. During the experimental study, the spring stiffness of 3.43 kN/m was determined to obtain restoring properties of the mooring lines. The spring length for the mooring line was determined to avoid slacking of mooring lines during the experimental study. Therefore, the mooring system for the numerical model in SIMO is defined as a fixed elongation coupling with a spring element. The spring element is defined as a specified length–force relation with the stiffness obtained from the numerical study. The defined mooring system act as a coupled element between the floats and anchored points. In addition, the mooring system is applied with a pretension of 53 kN as a station-keeping force, consistent with the model test setup. However, the damping of the mooring lines was neglected for the numerical study due to the limitation of experimental study. The damping of the mooring lines was not measured during the model test. The mooring system in the numerical model was verified by examining the applied tension forces during the static and dynamic simulation. The connection system between the FPV panels for the numerical model was defined as simple wire coupling. However, the stiffness of the connection ropes was not measured during the model test. For the numerical model, the axial stiffness is reasonably set to 3.9 GPa, which is in accordance with a new generation of high-performance fibre ropes for marine and offshore applications [11].

Decay Test. The decay tests were first conducted using the time-domain coupled model. The decay test was conducted for a free-floating single pontoon in calm water. The natural periods and damping ratios of the single pontoon in heave, roll, and pitch were compared to the experimental results. Note that the damping was calculated according to the logarithmic damping decrement method [4]. According to the decay test results, the numerical model based on the potential flow theory does not require additional damping in pitch and heave. However, a linear damping of 700 Nsm in the roll was required for the numerical model to match the model test.

Motion Responses Under Regular Waves. The RAO of pontoons for the numerical model in SIMA and the experimental model test were generated for the nine regular wave cases presented in the article [4]. The RAOs in surge and heave were obtained from the model test. Figure 3 presents the RAOs in heave and surge generated for nine regular wave cases reported in the model test and the numerical simulations. Note that in the numerical model, the motion RAOs were generated for motion response after a simulation time of 400 s when the transient response is damped out. Also note that the RAOs presented are the average values of all six pontoons, which is identical to the experimental study. The average response amplitude of individual six pontoons depicts the response of the whole 3×2 array. It can be found from Fig. 3a that the heave RAOs from the numerical model have good agreement with the model test results with a discrepancy between 0.3 and 7.2%. The surge motions were calibrated by introducing additional viscous drag via the employment of slender elements in SIMO. The drag coefficient for the slender elements was calibrated iteratively until reasonable surge motion was obtained compared to experimental results. As can be seen in Fig. 3b, a good agreement is also observed for surge RAOs between the numerical and experimental results.

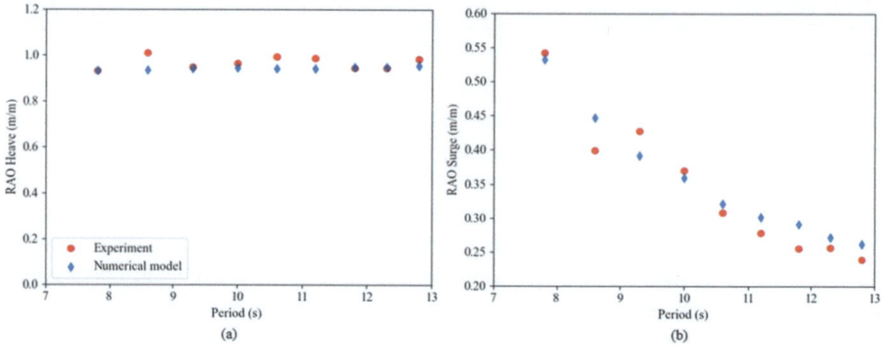

Fig. 3 Comparison of average RAO of six pontoons between the experimental and numerical results: **a** heave and **b** surge

4.2 Effect of Mooring Pretension on Motion Responses

The mooring system is a critical component of the FPV farm to withstand wave-induced forces. During the experimental model test, the mooring lines in the experiment were pretensioned to simulate the behaviour of a realistic catenary mooring system. The pretension force of 53 kN for each mooring line was determined after a numerical screening study carried out in OrcaFlex [4]. In this study, the effect of mooring pretension force on the motion responses of the FPV farm is based on a parametric study based on the calibrated numerical model in SIMA-Riflex. The parametric study is conducted with two cases. In the first case, the pretension force is reduced by 37.8% to a value of 30 kN and in the second case, the pretension force is increased by 37.8% to a value of 70 kN. The wave cases for the parametric study are conducted for periods between 3 and 12 s with a unit amplitude of 1 m for all cases.

Figure 4 shows the heave and surge RAOs of FPV farms from the parametric study. According, to the parametric study the pretension force influences the motion responses of the FPV system. Generally, the motion response in both heave and surge decreases with a higher pretension force. The surge RAOs are increased by 40% when the pretension force is reduced by 37.8% to 30 kN at a period of 6 s. But when the pretension force increased by 37.8% to a value of 70 kN, the surge RAOs were reduced by 24% at a period of 6 s. When the pretension force was increased by 37.8% at a period of 6 s the RAOs increased by 3% in heave. However, the RAOs are reduced by 2% when the pretension force decreases by 37.8%. In general, the surge RAOs are more sensitive to the change of the pretension forces as compared to heave. This is because heave motion is mainly dominated by hydrostatic stiffness for long waves, while surge motion relies on the mooring system to provide restoring forces. When the pretension force increases, the mooring lines experience high stresses which makes the FPV system more rigid to withstand wave-induced forces. Therefore, the mooring pretension force should be carefully determined to ensure safe design and

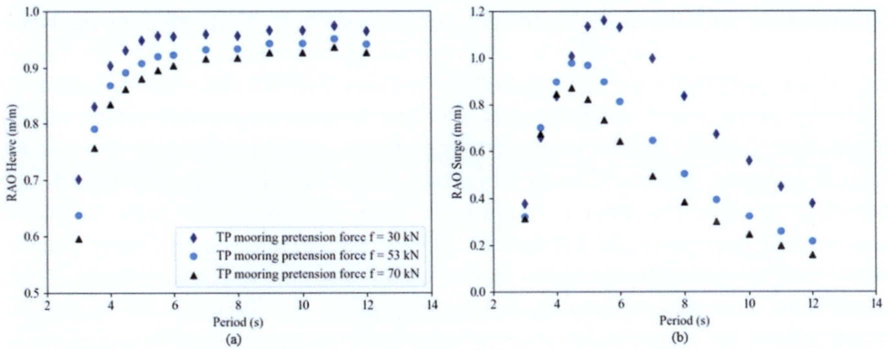

Fig. 4 The Response Amplitude Operators (RAO) of the calibrated numerical model with unit wave amplitude. **a** shows the RAO in heave and **b** in the surge

operation. Additionally, it is found that the effect of the pretension force is greater for shorter waves compared to longer waves on the heave motion.

In addition to the pretension force, the motion responses of the FPV system are significantly influenced by the wave period. In the heave direction, the minimum RAOs were observed for waves with the shortest and maximum RAOs for the longest waves. The short wave exhibits high-frequency forces on the FPV system compared to longer waves. The numerical response shows that the FPV system can withstand rapid and small-scale oscillations from short waves. The heave motions under high-frequency forces from the short wave are reduced due to the inertia of the FPV system and increased pretension in the mooring lines. However, the maximum RAOs in the surge are observed around the resonance period due to wave excitation. The surge motion of the FPV system is amplified due to constructive interference between the wave-induced force and the natural oscillation of the structure. The RAOs in surge increase drastically with a maximum horizontal displacement observed at the resonance period. The motion response of the FPV system is reduced gradually after the resonance period according to the numerical simulation. The maximum RAOs in the surge direction is at a wave period of 4.5 s for a model with a pretension force of 53 and 70 kN. The maximum surge motion for the model with 30 kN pretension force is observed around the wave period of 5.5 s. The natural period of the FPV system is influenced by the pretension force. The resonance period of the FPV system is found to increase when the pretension force in the mooring lines is reduced.

5 Concluding Remarks

In this paper, a numerical study of a floating photovoltaic (FPV) system in regular waves is conducted. The numerical model is developed based on frequency domain analysis in HydroD and subsequent time domain analysis in SIMO. The study is benchmarked with the available results from a model test. Through numerical calibration, the numerical model is found to provide accurate predictions of the motion response in both the heave and surge of the pontoons in a moored array. A parametric study is next conducted, and the results show that the mooring pretension force affects the motion response in both heave and surge directions. The difference in the heave motion response due to different pretension forces is small when compared with the surge motion. The latter is found to be more sensitive to the wave frequency and mooring pretension. In addition, the FPV system exhibits significantly reduced heave motions under shorter waves that are commonly seen in coastal regions, which is desirable from the engineering design perspective. For the surge, the large RAO amplitudes of the pontoons are observed near the resonance period. This needs to be taken into account in the design practice.

The numerical model for this study was based on the simplified model without the effect of hydrodynamic coupling between pontoons. Additionally, the parametric study is limited to the effect of mooring pretension force under regular waves. In the future, a more detailed study is fundamental to obtain insight into the effect of different parameters, including the configuration and stiffness of connection ropes as well as the gap between neighbouring pontoons, on the response of the FPV system. Besides, the numerical study is based on a small array comprising 6 pontoons. There is also a necessity to develop an efficient numerical model for larger arrays.

References

1. de Castro C et al (2013) Global solar electric potential: a review of their technical and sustainable limits. Renew Sustain Energy Rev 28:824–835
2. Dai J et al (2020) Design and construction of a floating modular photovoltaic system for water reservoirs. Energy 191:116549
3. Zhang C et al (2024) Development of compliant modular floating photovoltaic farm for coastal conditions. Renew Sustain Energy Rev 190:114084
4. Jiang Z et al (2023) Design and model test of a soft-connected lattice-structured floating solar photovoltaic concept for harsh offshore conditions. Mar Struct 90:103426
5. Newman JN (1977) Marine hydrodynamics. The MIT Press
6. Journée JMJ et al (2000) Offshore hydromechanics. TU Delft
7. Fonseca N, Stansberg CT (2018) Calibration of a time-domain numerical hydrodynamic model for mooring analysis of a semi-submersible. In: ASME 2016 37th International conference on ocean, offshore and arctic engineering (OMAE2018). Spain
8. Ma S et al (2020) The numerical study of viscous drag force influence on low-frequency surge motion of a semi-submersible in storm sea states. Ocean Eng 213:107511
9. Stansberg CT et al (2015) Challenges in wave force modelling for mooring design in high seas. In: Offshore technology conference
10. Morison JR, Johnson JW, Schaaf SA (1950) The force exerted by surface waves on piles. J Petrol Technol 2(05):149–154
11. Leech CM et al (1993) Modelling tension and torque properties of fibre ropes and splices. In: The third international offshore and polar engineering conference

Experimental Study on Dynamic Response of Submerged Floating Tunnel Moored with Different Materials Cables

Chunyan Li, Zhiwen Wu, Canrong Xie, Xianlong Xu, Xiao Lin Zhao, and Yonghui An

Abstract Submerged floating tunnel (SFT) is an innovative transport structure across straits, large rivers, and waterways. Mooring cables occupy a critical role within the SFT system, necessitating their ability to endure the rigors of complex marine environmental forces, including waves, currents, and the impact of vehicular traffic. However, the high density and susceptibility to corrosion of traditional steel mooring cables have limited their application in SFT. In contrast, synthetic fiber materials, provide a new option for the mooring system of SFT due to their lightweight, high strength, and excellent corrosion resistance. In this study, a physical scale segment model with five different cable materials of SFT is designed and investigated. The dynamic response and tension performance of the SFT moored with traditional steel wire cable and chain is compared with the SFT moored with synthetic fiber (CFRP, nylon and aramid) cables under irregular wave in the wave-flow flume. The results show that the dynamic response and tension performance are related to the stiffness and structural damping of the cables. Among the cables made of five different materials, the model motion response is greatest under nylon mooring and smallest under chain mooring, the CFRP and steel slightly less than chain. While an inverse result is found in the tension responses, chain and steel systems exhibit a more significant tension response, then the CFRP system. This research holds important engineering implications for the design of structural materials for SFT located in complex marine environments.

C. Li · Z. Wu (✉) · C. Xie · Y. An
Key Laboratory of Disaster Prevention and Structural Safety of Ministry of Education, Guangxi Key Laboratory of Disaster Prevention and Structural Safety, College of Civil Engineering and Architecture, Guangxi University, Nanning, China
e-mail: wuzhiwen20031985@163.com

X. Xu
Guangxi City Vocational University, Chongzuo, China

X. L. Zhao
Research Institute for Land and Space, The Hong Kong Polytechnic University, Hong Kong, China

X. L. Zhao et al. (eds.), *Proceedings of the Fourth World Conference on Floating Solutions: WCFS 2024; 2–4 December; Hong Kong*, Lecture Notes in Civil Engineering 597, https://doi.org/10.1007/978-981-96-4569-5_36

Keywords Submerged floating tunnel · Hydrodynamic behavior · Different materials · Mooring system

1 Introduction

Existing transport structures such as bridges and immersed tunnels are unable to meet the requirements of crossing long and deep-water areas due to various scientific and technical difficulties in construction. Therefore, the submerged floating tunnel (SFT) is proposed [1]. It is mainly composed of a tunnel tube suspended at a certain depth underwater by buoyancy force, the mooring system limiting the excessive displacement of the tube, the deep-water foundation, and the revetment section connecting both sides [2]. The SFT is suspended in water through the balance of gravity, buoyancy, and the tension of mooring lines. The mooring cables hold a pivotal position in the entire SFT, it must be able to withstand all the complex load. Therefore, as a key component, mooring cables have attracted the attention of researchers. At present, the analysis related to the structure of mooring cables mainly focuses on their load-bearing capacity and stability under the action of various environmental loads (waves, currents, earthquakes, etc.) [3, 4]

However, the traditional steel mooring cables, characterized by their high density and vulnerability to corrosion, have faced limitations in their use for SFT applications. On the other hand, synthetic fiber materials, like CFRP, nylon and aramid polyester cables, emerge as a promising alternative for SFT mooring systems. This is attributed to their combination of low weight, superior strength, and outstanding resistance to corrosion [5–7]. These materials can not only reduce the weight of the mooring cables and lower the costs of installation and maintenance, but also, to some extent, enhance the dynamic response performance of the SFT. The hydrodynamic performance of SFT under the anchoring effect of synthetic fiber ropes is still unclear. To advance the construction of suspended floating tunnels, it is urgent to conduct research on the dynamic response of SFT with different materials cables mooring. In this research, we explore the dynamic behaviors of SFT when anchored by mooring cables made from five distinct materials, with the goal of identifying the optimal cable material for SFT anchoring, thereby offering valuable insights for the practical application of SFT projects.

Table 1 Major parameters of the physical model of SFT

Component	Parameter	Unit	Model scale
Tunnel tube	Length	m	1.4
	Material type	–	PVC
	Outer diameter	m	0.23
	Elastic modulus	Gpa	2.33
	Density	$kg \cdot m^{-3}$	1.18
	Buoyant weight ratio	–	1.5
	Immersion depth of tube	m	0.3
	Section type	–	Circular
Anchor cables	Length	m	0.8
	Material	–	CFRP, Steel wire, Chain, Nylon, Aramid
	Inclination angle	°	45
	Cable interval	m	0.7
	Cable quantity	–	6

2 Experimental Model Design

2.1 Experimental Model Setup

In this experiment, a scheme of a SFT for crossing Funk Bay is taken as the prototype. Since this study focuses on the hydrodynamic performance of the SFT under the wave loadings, the geometry similarity law, and the Froude scaling requirement should be satisfied to simulate the inertial, gravitational and restoring forces in the experiment [8, 9]. In this study, the scale ratio is adopted as 100:1. Based on the dimensions of the wave-current flume, the scaled segment model of the SFT is designed, and the scales reaction for various physical quantities are summarized in Table 1. A detailed sketch of experimental model set up is shown in Fig. 1a. The experimental model consists of wave-current flume, scaled-model tube, anchoring cables, boundary restrain device and data acquisition devices. The numbers of the cables are named C1–C6, and C1–C3 and C4–C6 arranged at the downstream and upstream sides of the flume respectively.

2.2 Experimental Facilities and Instruments

The experiment is carried out in the wave-flow flume that belongs to Laboratory of Disaster Prevention and Mitigation and Engineering Safety, Guangxi University in China. The size length/width/height is 30 m × 1.5 m × 1.5 m. The photograph of

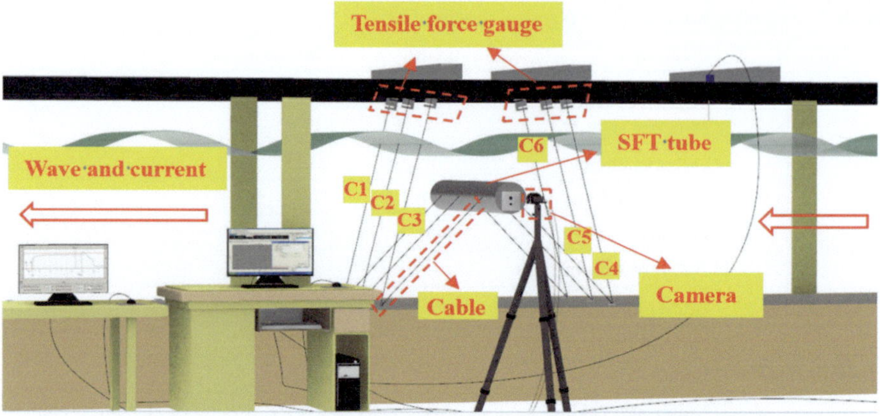

(a) Sketch of the experimental setup for SFT model. SFT test setup.

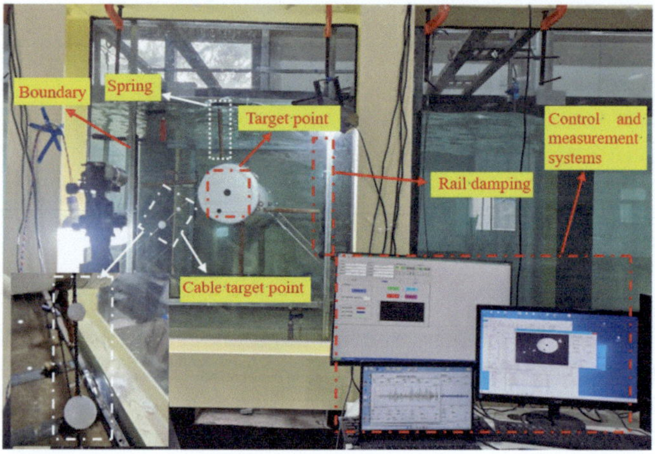

(b) Photograph of

Fig. 1 Schematic diagram of physical model experiment device for SFT

the experimental setup is shown in Fig. 1b. The movements of the SFT model are measured by image acquisition processing.

3 Results and Discussions

3.1 Dynamic Response of SFT

The two main motion responses of SFT are studied, and the results of different materials under independent irregular wave conditions are compared in Fig. 2. To show the result clearly, only 100 s data are displayed for clarity. In Figs. 2a, c, e, it is seen that the sway, heave and roll motion shows obvious wave frequency characteristic. The motion amplitude of the chain system is slightly smaller than that of the other four systems. The possible reason is that the mass of the chains is the largest, which makes the center of gravity of the whole structure drop, and the damping of the SFT with chains is the largest. The spectral analysis is then performed, and the results are shown in Fig. 2b, d, f. Two notable spectral peaks are observed in sway, heave and roll spectral density curves, one appears near the natural period of the structure and the other one occurs near the wave peak period. It is seen that the sway, heave and roll spectral peaks of the nylon system are the largest, indicating that the nylon system shows the most significant three main motion responses. Presumably because nylon rope has the least stiffness, soft rope will cause a lot of movement. The spectral peak values in ascending order are nylon, aramid, CFRP, steel wire and chain system. This result is caused by the stiffness of the rope and the damping of the SFT structure.

3.2 Tension of Typical Cable

Figure 3 shows the tension behavior of typical cables for five SFT models subjected to the same irregular wave. Based on classical mooring theory, it is noted that the motion amplitude of the SFT and the cable tension burden are a pair of contradictions. Figure 3a shows the tension amplitude of the chain system is slightly larger than that of the other four systems, which meets the basic mooring theory. The spectral analysis is then performed, and the results are shown in Fig. 3b, it is seen that a notable spectral peak is observed in tension spectra near the natural period of the structure. The spectral peak values in ascending order are chain, steel wire, CFRP, aramid and nylon system, which is the opposite of the motion response. The reason for this phenomenon may be related to the fact that the wave energy of water is converted into the kinetic energy of SFT and the elastic potential energy of the anchor cable. When more kinetic energy is converted into SFT, the elastic potential energy of the anchor cable becomes smaller. In other words, the greater the motion response of SFT is, the smaller the tension of the anchor cable is.

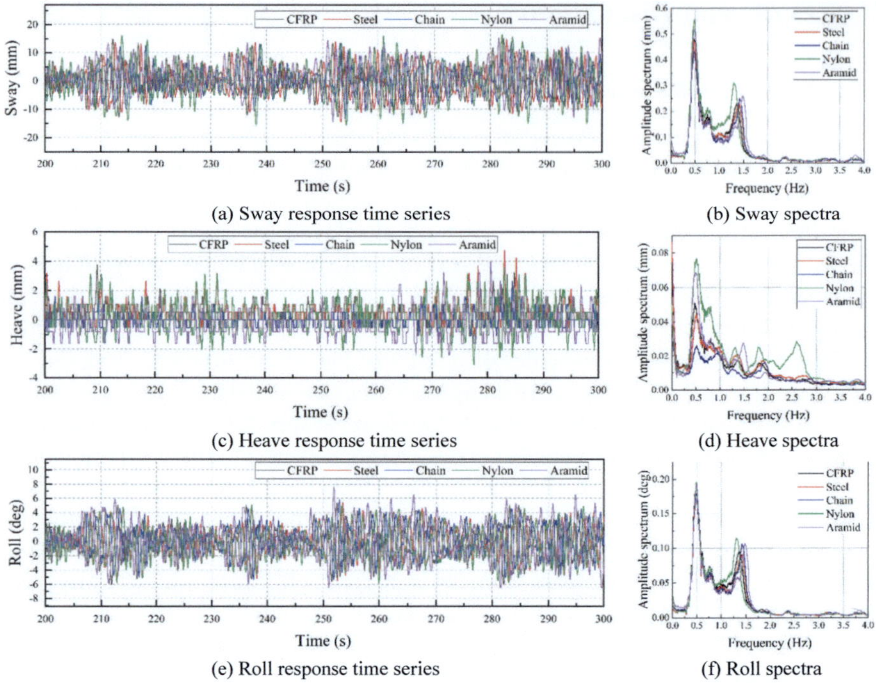

Fig. 2 Dynamic response of tube for five SFT models subjected to independent irregular wave

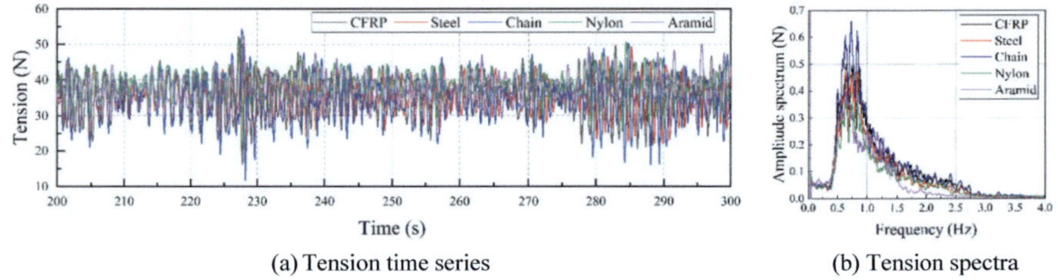

Fig. 3 Tensions of typical cable for five SFT models subjected to independent irregular wave

4 Conclusions

Five SFT models were subjected to independent irregular wave tests with significant wave height (Hs) of 10 cm and wave period (Ts) of 1.5 s, each employing different mooring systems: CFRP, steel wire, chain, nylon and aramid. Among the cables made of five different materials, the model motion response is greatest under nylon mooring and smallest under chain mooring, CFRP and steel slightly less than chain. While an inverse result is found in the tension responses, the chain and steel wire systems exhibit a more significant tension response, then is CFRP system. The motion of the chain and steel wire is slightly less than that of the other models, but their amplitude of the cable tension is higher. The finding matches the stiffness and structural damping of the cables.

More work is being carried out by the authors on sensitivity analysis for the key parameters of SFT moored with different materials cables conducted using hydrodynamic parameters (such as wave height, wave period, and current velocity), mechanical properties of cable materials (such as elastic modulus and damping), cable failure (such as failure position).

References

1. Jiang B, Liang B, Wu S (2018) Feasibility study on the submerged floating tunnel in Qiongzhou strait, China. Polish Maritime Res 25(s2):4–11
2. Minoretti A, Xiang X, Johansen I, Eidem M (2020) The future of the tunnel crossing: the submerged floating tube bridge. Struct Eng Int 30(4):493–497
3. Lu W, Ge F, Wu X, Hong Y (2013) Nonlinear dynamics of a submerged floating moored structure by incremental harmonic balance method with FFT. Mar Struct 31:63–81
4. Seo S, Mun H, Lee J, Kim J (2015) Simplified analysis for estimation of the behavior of a submerged floating tunnel in waves and experimental verification. Mar Struct 44:142–158
5. Hao Z, Yan N, Bai X, Cui L, Yin J, Diao H, Bi J, Liu J (2024) Mechanical characteristics and application of CFRP cable anchor support structures in coastal area. Ocean Eng 299:117366
6. Wang X, Wu Z, Wu G, Zhu H, Zen F (2013) Enhancement of basalt FRP by hybridization for long-span cable-stayed bridge. Compos B Eng 44(1):184–192
7. Wang X, Wu Z (2010) Integrated high-performance thousand-meter scale cable-stayed bridge with hybrid FRP cables. Compos B Eng 41(2):166–175
8. Liu C, Zhang S, Hao E (2017) Joint earthquake, wave and current action on the pile group cable-stayed bridge tower foundation: an experimental study. Appl Ocean Res 63:157–169
9. Ding Y, Ma R, Shi Y, Li Z (2018) Underwater shaking table tests on bridge pier under combined earthquake and wave-current action. Mar Struct 58:301–320

Food, Energy, and Architecture

Multiple Cold Water Pipes Concept of a Floating OTEC Plant

Ryoya Hisamatsu and Tomoaki Utsunomiya

Abstract Ocean Thermal Energy Conversion (OTEC) is one of the ocean renewable energies that utilizes the temperature differences in the ocean. Since its power production is very stable, OTEC can configure a redundant and clean power network combining with other natural energy sources. In addition, this energy source would provide an attractive integration with other industries such as an aquaculture with ocean nutrient enhancement, desalination, deep seawater cooling, and hydrogen production. A commercial-scale of OTEC requires a floating plant and an ultra-large Cold Water Pipe (CWP), e.g., 12 m in diameter for a 100 MW plant. Such a CWP is still the most challenging component for floating OTEC. This paper addresses a preliminary design using multiple medium diameter CWPs instead of a single ultra-large CWP. The floating structure is first assumed to be used a pre-owned oil tanker conversion. The multidisciplinary design process takes into account: design trade-off of number of pipes, internal flow velocity and diameter, pressure loss, the risk of contact between the CWPs, structural design, and eventual cost of energy. This preliminary study may be useful in a future feasibility study of floating OTEC.

Keywords OTEC · Cold water pipe · Plantship · Design

1 Introduction

Tropical oceans are constantly divided into the surface layer, thermocline and deep layer. The surface layer reserves heat from sunlight. While the deep layer is hardly affected by sunlight and the seawater temperature is constantly low value. Ocean Thermal Energy Conversion (OTEC) produces electricity by driving a heat engine with the temperature difference between the surface and deep layers [1]. Notably, this energy source is superior to the stability, i.e., it can supply power 24/7. Thus, it is possible to configure a redundant and clean power grid combining with other natural

R. Hisamatsu (✉) · T. Utsunomiya
Kyushu University, Fukuoka, Japan
e-mail: hisamatsu@nams.kyushu-u.ac.jp

X. L. Zhao et al. (eds.), *Proceedings of the Fourth World Conference on Floating Solutions: WCFS 2024; 2–4 December; Hong Kong*, Lecture Notes in Civil Engineering 597, https://doi.org/10.1007/978-981-96-4569-5_37

energies. In addition, this system is expected to be integrated with other industries: an aquaculture with ocean nutrient enhancement, desalination, seawater cooling, and hydrogen production.

Currently, several onshore plants integrated with deep seawater utilization industries are going to be deployed for tropical islands and Small Island Developing States (SIDS). While the feasibility of a larger-scale plant has been also examined worldwide. For example, the IEA-OES has recently reported that the levelized cost of electricity for a 10 MW-scale floating plant is $0.37–0.46/kWh [2]. Note that this figure can significantly reduce by scaling up the plant capacity. Basically, although the main technologies appear to be at the commercial level, reducing costs and improving the technical reliability of floating plants would remain challenges.

The authors have been working on technical development of a 100 MW-scale floating plant. The system configuration of a floating plant has been described in detail in the literature [3]. In particular, the Cold Water Pipe (CWP) has been historically identified as one of the challenging components due to its size, e.g., a 100 MW-scale plant requires a CWP with 12 m in diameter to transport 235 m^3/s of seawater [4]. For this reason, several literatures recently discuss concepts with multiple CWPs in currently available diameter to avoid initial costs, technical difficulty and uncertainty of the cost estimation in the development [2, 5]. Although this approach is inferior in terms of hydraulic head loss, it has been suggested that the economic impact is substantially smaller. Therefore, understanding the technical feasibility and structural performance of such multiple-CWPs concepts would be helpful in improving the accuracy of cost estimates in short to medium term.

This paper discusses the technical feasibility of a multiple CWPs concept of a Floating OTEC Plant. Specifically, this study uses the preliminary design of a 100 MW-net scale plantship.

2 Design Conditions

2.1 Baseline Design

This paper first assumes a baseline design with a single CWP. This concept is a 100 MW-net scale plantship as shown in Fig. 1. This subsection outlines the definition of the baseline design. Subsequently, we discuss the analytical impact of making change the number of pipes, without changing the plant configuration.

The seawater temperature is 29 °C at the surface and 8 °C at 800 m deep. Assuming a plant capacity of 100 MW-net, the required deep water intake flow rate is estimated to be 235 m^3/s [4]. The floating platform is a ship-shaped body. This shape is expected to reduce the capital cost by reusing a pre-owned ship. The hull geometry is set to KVLCC2M which is scaled down to 275 m long, 50 m wide, and 17.9 m in draft. The plant configuration and the general arrangement can be found in [4]. The hydrodynamic characteristics is summarized in [6].

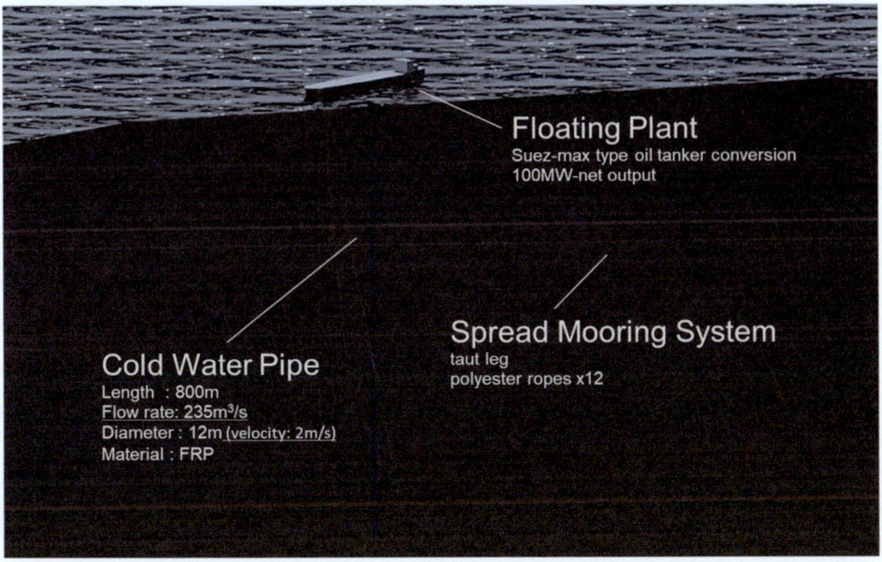

Fig. 1 Schematic view of a baseline design of 100 MW-net OTEC plantship [3]

The mooring system is designed as a spread mooring system which has 12 taut legs with three mooring lines in each of the four directions. The design water depth is 1000 m. Each line consists of 1750 m long of polyester rope, which has an axial stiffness of 294 MN and a weight in water of 7.52 kg/m. The hydrodynamic coefficients of the section of the rope are a drag coefficient $C_d = 1.6$ and an added mass coefficient $C_a = 1.0$.

The CWP is 800 m long and transports deep seawater in 235 m^3/s. The design flow velocity is 2.0 m/s, thus the inner diameter was first decided to 12 m. The CWP consists of GFRP wall with a thickness of 19 mm. The material properties of GFRP are assumed a density of 1520 kg/m^3 and the elastic modulus of 13,776 MPa. The CWP is attached to the bottom of the hull with pinned joint.

The extreme load condition is considered to validate the preliminary design. The extreme environmental conditions are considered that waves, wind, and currents come from 90° transverse of the hull. The waves are modeled as a JONSWAP spectrum with a significant wave height of 3.65 m, peak period of 11.1 s and a peak factor of $\gamma = 2.79$. These values reflect the feature of the spread mooring system with the bow heading in the most powerful direction. The wind is modeled as an NPD Spectrum with a 1 h mean wind speed of 28.6 m/s. The currents are represented as a steady flow and the profile is assumes the 1/7 law with a surface speed of 0.85 m/s. This environmental condition is representative of South China Sea without typhoon [7].

2.2 Methodology

This section describes methodology for designing multiple CWPs. First, a flow rate and length are constrained so as not to significantly change the plant configuration and the gross output. The diameter, flow velocity, and number of pipes are considered to be design variables. However, hydraulic head losses of the CWPs should be taken into account since it plays a crucial role on the net power output. The power loss is evaluated as [4],

$$\Delta P = \left(k + f\frac{L}{D}\right)\frac{\rho A v^3}{2\eta} \tag{1}$$

where, ΔP is the power loss, ρ is fluid density, L is the length, D is the hydraulic diameter, A is the hydraulic area, v is flow velocity, η is the pumping efficiency. Although the pump efficiency varies strictly with flow rate and water head, a constant value of $\eta = 0.9$ is assumed here. The k is the entrance loss coefficient caused at the inlet, here, we assume $k = 1.0$. The f is the friction loss coefficient that is evaluated by the Colebrook formula. The heat loss of the CWPs may be effectiveness; however, we decided to neglect the heat loss of the CWPs. Unless the diameter is small (< 1 m) and the flow velocity is very low, the heat transfer between the inner and outer water of the CWPs are small as in [8].

The structure of the CWP is GFRP with the same constitution as the baseline. The wall thickness t is constrained to the ratio of $D/t = 60$. A dynamic analysis with OrcaFlex is performed under extreme environment conditions to assess stresses. Furthermore, a contact between the CWPs is considered to present a risk, therefore, the analysis also evaluates the pipe clearance.

We roughly estimate the economic impact due to the change in the material cost and the head loss. The evaluation method is the theoretical LCOE, and we assume a discount rate of 10%, a plant life of 20 years, a capacity factor of 95%, and OPEX costs percentage of CAPEX is 5% [9]. The capital cost of CWP with the baseline design is assumed as 5% of total CAPEX [9]. For the multiple CWPs, the cost is approximated as varying with the ratio of the volume of material against the baseline CWP. It is notably that this estimation does not include the manufacturing, transportation and installation, ship conversion, and learning effect.

3 Results and Discussion

3.1 Design Space Exploration

To discuss the design trade-offs, we conduct the design space exploration: the inner diameter, velocity and number of CWPs. Figure 2 illustrates the set of possible designs and the results of cost estimation. Figure 2a shows the power loss for a

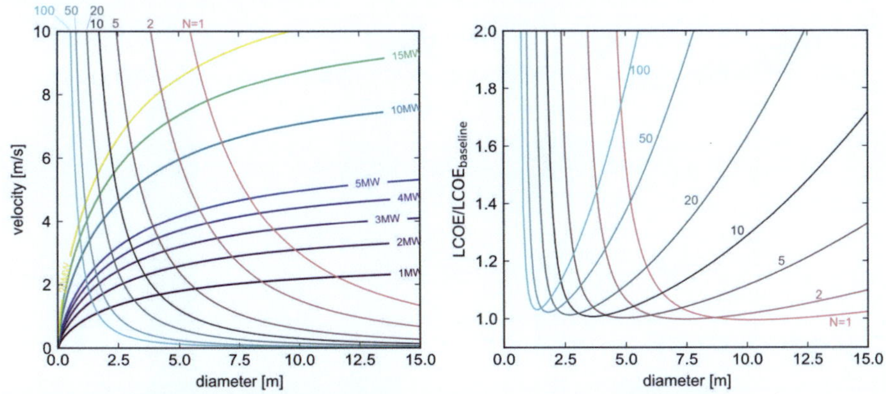

(a) power loss and Q=235m³/s curves

(b) LOCE curves

Fig. 2 Design space of multiple CWPs: N is the number of CWPs; Q is the flow rate; $LCOE_{baseline}$ is the cost of baseline design, \$0.2/kWh

combination of the diameter and velocity. In general, an increase in flow velocity allows for a reduction in pipe diameter, although this results in a significant increase in losses. As a result, the cost increases rapidly as indicated in Fig. 2a. By contrast, an increase in diameter by decreasing the velocity results in an increase in the cost due to the significant rise in material volume. Consequently, there exists a theoretical optimal combination of diameter and velocity for each number of pipes.

Importantly, while the optimal cost increases as the number of pipes increases, the increase is very slight. For example, in the case of $N = 100$ with a diameter of 1.34 m, the increase is only 3% from the baseline design. This conclusion has been also mentioned by Langer et al. [5]. The cost of manufacturing and installation might be rather more important.

We assess the structural performance by extracting several design examples.

- $D = 5$ m, $v = 2.39$ m/s, $N = 5$, $LCOE/LCOE_{baseline} = 1.002$.
- $D = 3$ m, $v = 2.22$ m/s, $N = 15$, $LCOE/LCOE_{baseline} = 1.010$.
- $D = 2$ m, $v = 1.92$ m/s, $N = 39$, $LCOE/LCOE_{baseline} = 1.019$.

The first example represents a scenario that is needed to develop in near future. The next 3 m in diameter is currently considered to be the maximum feasible diameter [2]. The 2 m diameter is assumed to be a more readily available option. Smaller diameters may be possible; however, considering the necessity of the discharge pipes and hull reinforcement, we decided that the minimum diameter of the CWP for a 100 MW-scale is approximately 2 m.

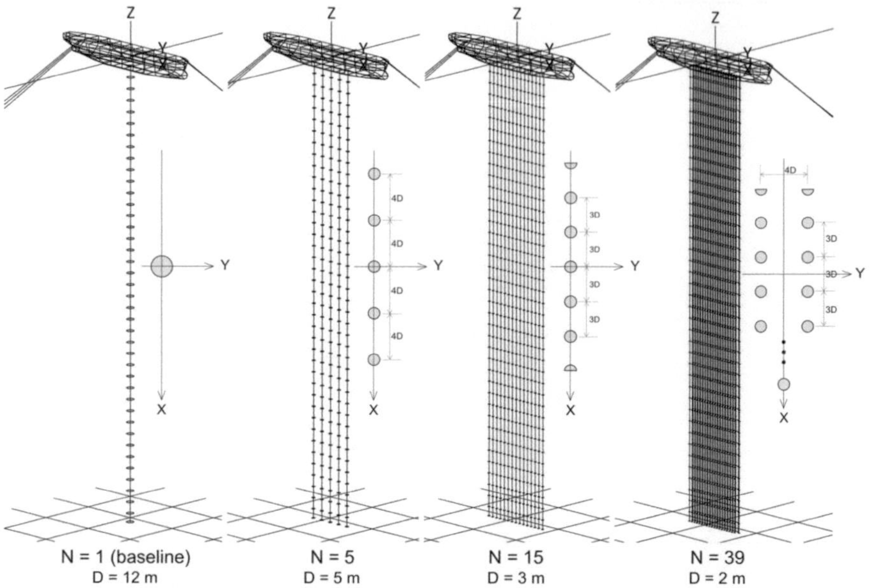

Fig. 3 Design examples of CWP arrangement: *N* is the number of CWPs; *D* is the inner diameter

3.2 Structural Analysis

The arrangement of the CWPs is a significant point on the risk of contact, yet in practice it would be determined by the plant layout and hull structure. Thus, we first conducted a preliminary study by using the CWP arrangement shown in Fig. 3. The hydrodynamic coefficients of the CWPs are estimated from the DNV-RP-C205 [10]. While currents are possible to lead wake, flow-induced vibration and change in hydrodynamic characteristics. In order to first clarify the feasibility, we did not investigate this point in detail in this paper.

A coupled analysis has been performed by OrcaFlex 11.4c. The detail of the coupled analysis and its rationality can be seen in [6]. The extreme values are evaluated by using 10 random seeds and each simulation durations are 3 h (10,800 s).

Figure 4 shows the results of the maximum sway of the hull and the tension of the mooring. The significant increase in sway and mooring tension by increasing the number of pipes as shown in Fig. 4 is attributed by current forces acting on the pipe array. Another important aspect is the displacement of the CWPs themselves due to current forces. It is worth noting that in the case of the lightest and most flexible 2 m pipe, a displacement of 266 m occurred at the lower end. The results highlight the need for consideration of these points in sites with strong currents.

The maximum von Mises stress and the minimum pipe's clearance are shown in Fig. 5. Basically, the elastic vibration of the pipes is slight. Also, since the natural

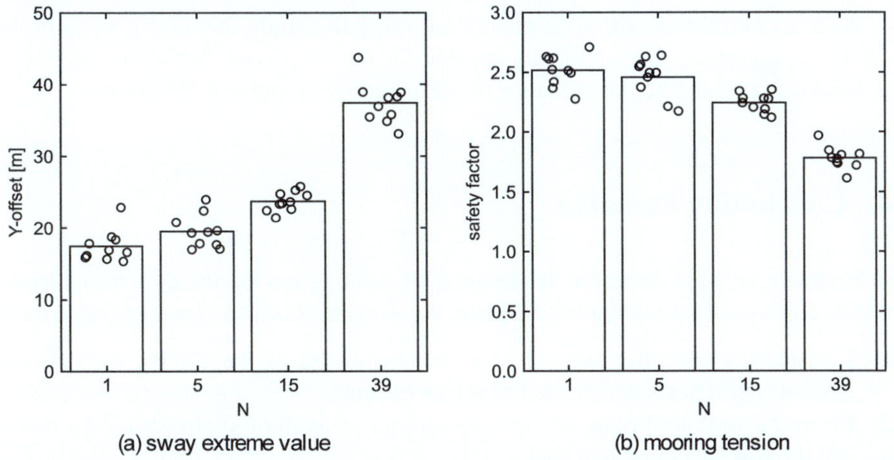

Fig. 4 Extreme value of sway and mooring tension: the bar is the mean value of 10 times simulations; the scatter plot is each maximum values

period of the first mode is even longer than the sway of the mooring body, no significant oscillation occurred; therefore, the risk of contact between them is extremely small here as shown in Fig. 5b. Besides, as the diameter of the CWP decreases, its modal properties change significantly. The smaller diameter pipes would eliminate resonances in the lower-order modes that is possible to reduce the response. As a result, the case of $N = 15$, 39 has been evaluated with smaller value of stress than larger diameter CWP(s) as presented in Fig. 5a.

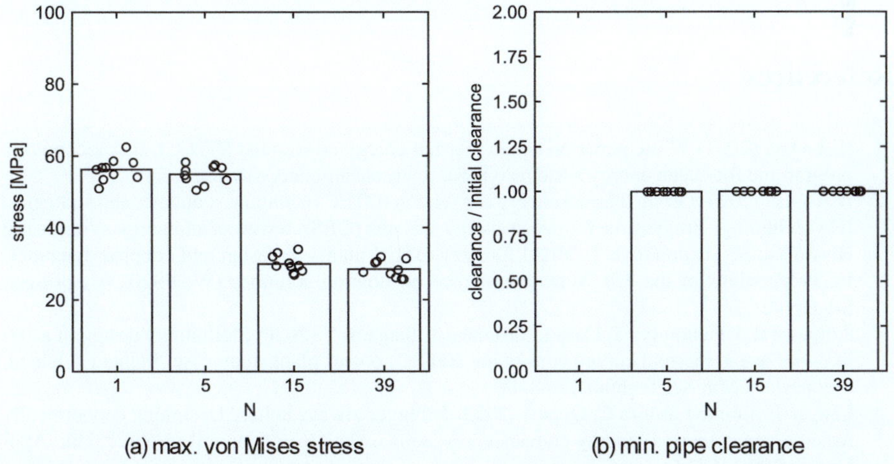

Fig. 5 Extreme value of von Mises stress and pipe's clearance along the CWP: the bar is the mean value of 10 times simulations; the scatter plot is each maximum values

As described above, within the range analyzed in this study, the concept of multiple CWPs is suggested to be sufficiently feasible. On the other hand, further investigation is needed to investigate the influence of wake and other incident directions.

4 Concluding Remarks

In this study, we have discussed the design methodology and feasibility of the multiple CWPs concept of a Floating OTEC plant. As a result, this study has suggested that:

1. Considering only the material cost and head loss of the CWPs, the optimal configuration has a negligible impact on the cost.
2. From the structural point of view, the design of medium-diameter CWPs made of GFRP seems to be feasible.
3. The effect of current force acting on array of CWPs would be considerable impact on the mooring system.

This study is limited by lack of information of the cost estimation of manufacturing, installation and assembly of the CWPs. We hope this study will contribute to future detailed estimates and feasibility studies.

In addition, GFRP were used in this study, while a high-density polyethylene (HDPE) CWP in diameter of 3 m has been positioned as a currently available pipe as in [2]. It would be a surge of importance to discuss lighter weight and more flexible HDPE pipe. Besides, it will need to gain a better understanding of the effect of wake, the hydrodynamic interaction between pipes and internal flow effects [11] from engineering point of view.

References

1. IEA-OES (2021) White paper on ocean thermal energy conversion (OTEC). IEA technology programme for ocean energy systems (OES). www.ocean-energy-systems.org
2. IEA-OES (2024) Ocean thermal energy conversion (OTEC) economics: updates and strategies. IEA technology programme for ocean energy systems (OES). www.ocean-energy-systems.org
3. Hisamatsu R, Utsunomiya T (2024) Floating OTEC plant—a design and coupled dynamics. In: Proceedings of the 3rd World conference on floating solutions (WCFS2023). Springer, Singapore
4. Adiputra R, Utsunomiya T, Koto J, Yasunaga T, Ikegami Y (2020) Preliminary design of a 100 MW-net ocean thermal energy conversion (OTEC) power plant study case: Mentawai Island, Indonesia. J Mar Sci Technol 25:48–68
5. Langer J, Infante Ferreira C, Quist J (2022) Is bigger always better? Designing economically feasible ocean thermal energy conversion systems using spatiotemporal resource data. Appl Energy 309:118414
6. Hisamatsu R, Utsunomiya T (2022) Coupled response characteristics of cold water pipe and moored ship for floating OTEC plant. Appl Ocean Res 123:103151
7. DNVGL (2018) DNV-OS-E301 position mooring. Det Norske Veritas (DNV), Norway

8. Mao L, Wei C, Zeng S, Cai M (2023) Heat transfer mechanism of cold-water pipe in ocean thermal energy conversion system. Energy 269:126857
9. IEA-OES (2015) International levelised cost of energy for ocean energy technologies. IEA technology programme for ocean energy systems (OES). www.ocean-energy-systems.org
10. DNV (2019) DNV-RP-C205 environmental conditions and environmental loads. Det Norske Veritas (DNV), Norway
11. Hisamatsu R, Utsunomiya T (2023) Free vibration and stability of a fully submerged pipe aspirating water: an experiment and new physical insights. J Fluids Struct 116:103789

Design and Operation of Ocean Aquaculture Installations

P. A. Åland

Abstract This paper address some of the most significant technical trends in the development of ocean space technology. It concerns possibilities and challenges related to utilization of ocean space to supple healthy and sustainable sea food proteins to a growing world population. As a result, several inventive solutions to increase capacity, sustainability and safety in ocean fish farming are being engineered and piloted. An outline of background for this industrial expansion followed by a description of the most common concepts being explored follows.

Keywords Ocean space · Aquaculture · Offshore fish farming · Exposed operations · Offshore installations

1 Introduction

All aquaculture market projections uncover a dramatic supply–demand imbalance in the next two to three decades. Focusing on fin fish supply specifically uncovers that output level is close to a saturation point for traditional areas with sheltered aquaculture [1].

Fish farming is growing out of sheltered, near-shore areas and moved to locations further offshore together with several new land-based Recirculation Aquaculture System (RAS)/flow-through plants being established around the world. A number of innovative solutions is being tried out in offshore fish farming in the process of industrialization of the aquaculture segment. The objective is a higher output volume of fin-fish to supply the global market with marine proteins combined with improved fish welfare. As the world's population grows, the options for producing enough healthy, sustainable food are becoming limited. Several interesting solutions for efficient, large-scaled fish farming in offshore/exposed environment are in the pipeline [3].

P. A. Åland (✉)
DNV-Det Norske Veritas, Høvik, Norway
e-mail: per.arild.aland@dnv.com

X. L. Zhao et al. (eds.), *Proceedings of the Fourth World Conference on Floating Solutions: WCFS 2024; 2–4 December; Hong Kong*, Lecture Notes in Civil Engineering 597, https://doi.org/10.1007/978-981-96-4569-5_38

2 Novel Concepts and Solutions

The numerous concepts planned and implemented in the ocean fish farming segment may be split into four main categories: Open trusswork with traditional net or grating, semi-closed units, submerged fish farms and at last closed aquaculture installation. All these concepts may be operated with permanent manning or a combination of remote operation and daily visits to carry out fish farming duties together with inspection and maintenance. A more specific description of the main solutions applied in the development of ocean farming is presented in this section.

2.1 Open Trusswork

Further development of the traditional sheltered net pen usually with rigid floating collar in steel or concrete. Additional structural arrangements for added buoyancy, ballasting/de-ballasting and distention of net panels are the main features of open trusswork installations. This category may be split in two subsets. One includes operating at the ocean surface only while an increasing number of designs have the option of submerged position in order to avoid splash-zone dynamics at rough sea states. Havfarm 1 has outer shape and outline of a ship but is actually a framework structure in combination with semisubmersible platform when it comes structural capacity—see Fig. 1.

Fig. 1 Havfarm 1—installation farming Atlantic salmon in Norway

Fig. 2 Aquatraz G4 being operated by Salmar in Norway

2.2 *Semi-closed Units*

Main purpose of this solution is to separate fish environment from sea for the first few meters of the water column. This separation result in less exposure of parasites that lived in the upper part of the water column. Less environmental impact to the fish stock is also considered an advantage due to solid barrier between fish and surrounding water in the splash zone. Aquatraz G4 is one of many new semi-closed concepts. Four units are in-operation with promising results—see Fig. 2.

2.3 *Submerged*

Submerged fish farming installations are becoming a common solution. Environmental impact from rough sea is avoided for structure and net as splash-zone dynamics at rough sea states are prevented. The fish is exposed to less harmful parasites as the upper part of the water column is evaded. Several of these concepts are permanently submerged during the complete cycle of fish farming while others only during periods of high sea states. Hence, a wide variety of operational modes are put to use depending on species, sea conditions and level of technology applied.

2.4 *Closed Fish Farming Units*

This concept is based on total separation between fish farming environment and marine environment at sea. This solution gives good control of what comes into the fish pen and what is discharged. There is a possibility to collect fresh water

Fig. 3 Conceptual model for floating closed fish farm, hexacage from Westcon

with favorable temperature containing a minimum of parasites at relevant depth to improve wish welfare. Closed fish farming units enables operator to filter and clean return water from fish environment to reduce, or even totally remove, environmental impact to marine environment.

A wide variety of solutions are currently being explored in this category. Several closed floating fish farming concepts are being designed and engineered, very few are put into operation on the date of publishing this article. The modular solution Hexacage illustrates well the general direction of this development—see Fig. 3.

3 Essential Functions and Operations

There are several advantages utilizing advanced, high volume fish farming installations to increase supply of marine proteins. The most relevant items include better fish welfare, less conflicts with interests in the ocean space as well as possibility to scale up output volume without compromising environmental sustainability.

The main categories of solutions being planned and tested listed in Sect. 2 have several advantages and weaknesses that have to be considered and evaluated. The various solutions ability to perform with high efficiency and sufficient safety depends on a number of aspects [4]. The most essential parameters to be taken into consideration assessing total performance of fish farming installation are addressed below.

3.1 Oxygen Level

O_2-level in sea water is essential for fish welfare is the ability to keep continuous O_2-level above lower limit in all parts of the fish farming installation. In-depth analysis

Fig. 4 Sliding bulkhead for fish crowding on ocean farm 1

must be performed to ensure no "dead pockets" where the fish density at periods of low water circulation may be too high compared to required oxygen level for good fish welfare and optimal growth.

3.2 Crowding and Live Transfer of Fish

These are processes that cause significant stress and physical strain. Again, resulting in reduced resistance towards diseases, etc. and may lead to great losses of fish combined with low fish welfare. The industrialized solutions applied for offshore fish farming often utilized advanced mechanical systems to facilitate these processes, failure may lead to operational interruption and fish escape. An example of an advanced crowding solution is the sliding bulkhead installed in the offshore fish farming installation Ocean Farm 1—see Fig. 4.

3.3 Dead Fish Handling

Dead fish collection and ensilage processing is an essential process operating an aquaculture facility. Typically, automated processes that involves much less manual handling compared to traditional sheltered methodology where these tasks are not carried out at the pen, but rather on the feed barge or similar.

3.4 Logistics

Logistics of supplies and goods going both on and off the fish farming installation result in several and more advanced marine operations compared with traditional fish farming. Higher sea states with more dynamics in both supply vessel and offshore fish farming installation is a major risk contributor to be mitigated when emergency preparedness is planned.

3.5 Personnel Safety

Personnel safety is mainly addressing arrangement for emergency escape and fire safety on board. This included lifesaving appliances, launching equipment and similar as well as fire detection and extinguishing. In addition, other common safety items such as falling objects, working at heights and pinching/crushing must be considered [2].

Other crucial operational issues include ability to perform cleaning to improve hygiene as well as feed handling and control may also be added to the list above considering the total operational performance of a high-volume fish farming installation.

To address and mitigate issues mentioned above there is a significant potential to utilize competence from traditional offshore- and maritime industry to support operators of exposed fish farming units in design and tuning operational modes for efficient production. The industry is currently exploring possibilities and collecting experience from the various concepts being planned and put in operations. This learning process is an important phase in the development of offshore fish farming to supply future generations with sustainable and healthy marine proteins.

References

1. DNV (2023) Aquaculture forecast 2050
2. Åland PA (2024) Overall risk perspective for ocean aquaculture. In: WAS–EAS AQUA2024 conference, 26–30 Aug 2024, Copenhagen, Denmark
3. Åland PA (2024) Industrialization of ocean fish farming contributing to boost supply of marine proteins—solutions and challenges. In: 43rd International conference on ocean, offshore and arctic engineering (OMAE 2024), 9–14 June 2024
4. Åland PA (2024) Reducing risk in design and operation of ocean aquaculture. In: 43rd International conference on ocean, offshore and arctic engineering (OMAE 2024), 9–14 June 2024

Digital Solutions in Ocean Aquaculture—Possibilities and Challenges

P. A. Åland

Abstract Digital solutions implemented on aquaculture installations is an advanced approach that utilize sensor technology, data analytics and machine learning to provide real-time insights into the condition and performance of these advanced assets in ocean space. This paper address some of the major possibilities and challenges associated with implementing digital condition monitoring systems in maritime environments. The potential gains include improved safety and reliability, predictive maintenance, operational efficiency, and streamlined regulatory compliance. However, the integration of these systems also brings up challenges such as data management complexities, cybersecurity hazards and the continuously shifting environmental and operational conditions. This paper aims to provide an overview of the possibilities and challenges related to digital solutions such as data driven condition monitoring and offers insights into future directions for further development in this field.

Keywords Condition based monitoring · Risk based inspection · Digital asset management · Digital solutions · Ocean fish farming

1 Introduction

Smart aqua operations are developing to be a common solution in modern aquaculture installations. Real time monitoring of fish condition combined with surveying the integrity of fish farming installation itself are considered a competitive advantage to promote premium product with high sustainability score. The aquaculture industry is in the middle of a major transition as vessels and fish farms are transformed into sensor hubs, generating data, and linked in an expanding interconnected digital web. The concept often referred to as the "smart game" is just and valid for aquaculture as

P. A. Åland (✉)
DNV—Det Norske Veritas, Høvik, Norway
e-mail: per.arild.aland@dnv.com

X. L. Zhao et al. (eds.), *Proceedings of the Fourth World Conference on Floating Solutions: WCFS 2024; 2–4 December; Hong Kong*, Lecture Notes in Civil Engineering 597, https://doi.org/10.1007/978-981-96-4569-5_39

for other industries. The core idea is that both sustainability and profit may only be achieved through implementation of robust digital solutions—as illustrated in Fig. 1.

While digital driven condition monitoring offers numerous possibilities for enhancing safety, reliability, and operational efficiency, it also results in major challenges that must be addressed. By utilizing advanced technologies and developing robust strategies to overcome these challenges, the aqua culture industry may achieve significant improvements in the management and maintenance of both its assets as well as fish stock [1].

As fish farming installations and -operations become increasingly complex and situated in more challenging environments, the need for reliable digital monitoring systems is increasing. Digital condition monitoring is powered by advanced technologies such as smart sensors, big data analytics and machine learning to provide real-time insights into the state and performance of the aquaculture facility. All stored and processed in a cloud-based scheme, as demonstrated in Fig. 2.

Fig. 1 Digital solutions result in both sustainability and profit

Fig. 2 Digital monitoring of fish farming site based on decentralized cloud solutions

2 Main Approach

Smart hull monitoring of ocean fish farming installations has several advantages. This approach enables the operator to get full insight into utilization status at any time and henceforth establish improved trust in the design. Along with this goes the ability to apply data stream from current operation into analytical models to improve designs of future aquaculture installations being scheduled in the future. Smart hull monitoring is also an efficient way to document and demonstrate control of asset integrity towards essential stakeholders such as authorities, insurance, finance institutions and similar [2]. The stakeholder confidence scheme together with the fact that it may be applied to determine parameters for design re-calculations with significantly reduced uncertainties explains why digitally based structural monitoring is an attractive and efficient approach.

Digital driven monitoring is a method often put to use to gather input data for dynamic Risk-Based-Inspection (RBI) regime for optimized and cost-effective inspection program.

Digital asset management is a term often used in this context where inspection as well as operational decisions are based on continuous data streams supplied by sensors on board. The dynamic RBI concept enables the operator to manage his inspection and maintenance campaigns according to actual conditions of the installation based on input from the digital condition monitoring system. The result is that less time is lost on non-essential inspection items which is the case if a standard periodic maintenance list is followed.

Digitally driven condition-based maintenance result in improved reliability by continuously assessing the condition of fish farming installation together with monitoring of the fish in it. The objective is to identify potential issues before they become major problems and take proactive measures to prevent failures and unplanned downtime.

Examples of typical data streams harvested on board a fish farming installation covers a wide extent of critical operations. This may range from fish health monitoring through structural (hull) surveillance to emission monitoring [3]. See Fig. 3 for more details.

Digital asset management will help improve efficiency and reduce operational costs. For example, by identifying equipment that needs maintenance or replacement, fish farmers can avoid costly breakdowns and reduce downtime.

The digitally driven operational monitoring is based on the process of capturing information, analyze data collected, learn combined with synthesizing, and finally act accordingly. The integrity of these solutions is mainly based on the reliability of sensors and their location in the fish farm. Operational onboard servers together with data transfers solutions and intermediate storage are also important aspects in the utilization of digital asset management infrastructure. Stability and performance of software applied to process/analyze collected data represents a significant risk and needs to undergo a robust assurance scheme to verified integrity.

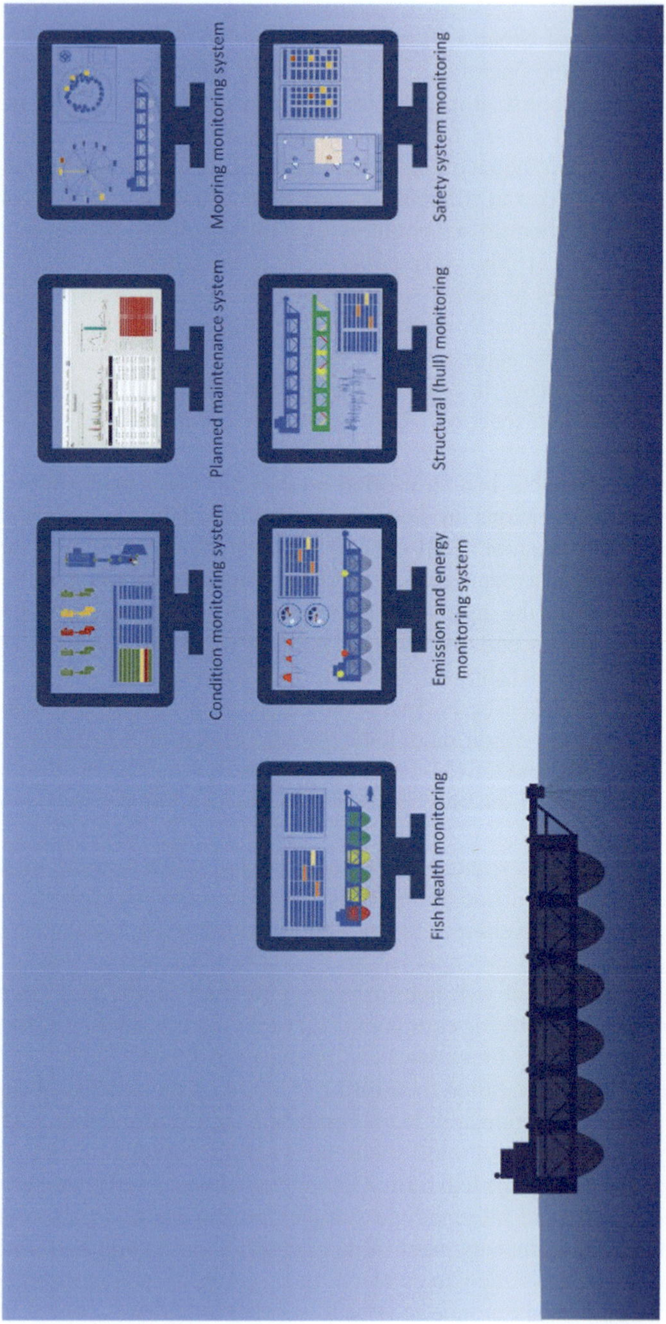

Fig. 3 Aqua-digital connections and data streams

Finally safe data transfer and cyber security are vital items in the overall digital performance. While digital asset management can offer several benefits, there are also a number of potential risks and challenges associated with relying on this technology. Together with data security risks and technical issues, challenges related to data quality is a major concern. The monitoring system relies on high-quality, accurate data to provide meaningful insights into asset condition and performance.

3 Conclusion

Smart aqua is an efficient operational concept that more and more operators are chasing, but to what extent are we able trust the status and green lights on our operational dashboard? We need to fully rely on the data stream provided and the algorithms processing the date stream supplied before the result ends up on the operator's panel. The industry at large is expected to put significant effort into assuring the integrity of the data streams continuously feeding information to our operational decision-making tools.

There are several efficient methods to mitigate the dramatic risks introduced as advanced digital technology is applied in operations of fish farms. One of the most rational approaches includes focus on sensor reliability, integrity of operational onboard servers together with robustness of data transfers solutions. Overall perspective for digital assurance model is illustrated in Fig. 4.

One of the more resent concepts introduced to mitigate digital risk includes integrating a standardized data collection and tampering-proof routine into the date processing chain to present proof of non-tampering of data in a way that can easily be examined by stakeholders [4]. Hence, this enables the stakeholder to be able to

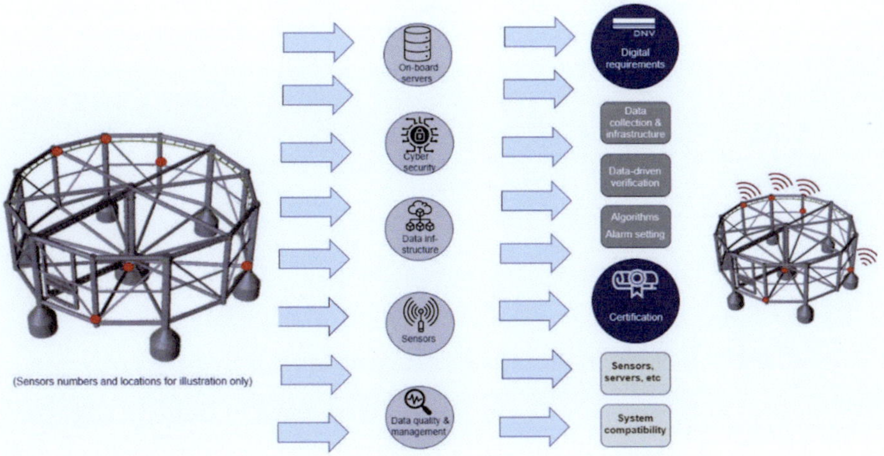

Fig. 4 Model of overall scheme for digital assurance

trust the data set the operational decisions regarding inspection and maintenance is based on. It needs to be investigated further how these concepts may be implemented effectively.

References

1. Åland PA (2024) Reducing risk in design and operation of ocean aquaculture. In: 43rd International conference on ocean, offshore and arctic engineering (OMAE 2024), 9–14 June 2024
2. Åland, PA (2024) Unlocking the power of digital asset management: navigating the possibilities and overcoming the challenges. In: 43rd International conference on ocean, offshore and arctic engineering (OMAE 2024), 9–14 June 2024
3. Åland PA (2024) Overall risk perspective for ocean aquaculture. WAS—EAS AQUA2024—conference, 26–30 Aug 2024. Copenhagen, Denmark
4. DNV Homepage. https://www.dnv.com/research/maritime/data-driven-services/nfts-and-oracles-beyond-the-hype/. Last accessed 07 Aug 2024

Floating Architecture Strategies: A Case Study of an Adaptive Pavilion in the Venice Lagoon

Diego Antonio Bonilla-Ávila, Andrés Medardo Calero-Trávez, Gabriele Porporato, Roberto Giordano, Alice Rosiello, and Sergej Antonello Sirigu

Abstract This paper examines the potential of sustainable floating architecture as an adaptive strategy to address coastal risk and sea level rise challenges associated with climate change in the Italian context. Using a research-by-design approach, through a case study focusing on a temporary floating pavilion proposed for the city of Venice, the paper establishes a theoretical framework for floating architecture and sea level rise research, to guide the conceptualization and schematic design of the pavilion. It emphasizes the incorporation of sustainable principles into the design process, focusing on practices aimed at minimizing environmental impact, offering insights into climate adaptation measures through passive strategies. By the testing of the SEAform solution for floating platforms, this research contributes to the advancement of best practices in floating architecture, fostering reflection on future scenarios. It aims to inform decision-making processes for upcoming projects by providing insights and recommendations.

Keywords Floating architecture · Research-by-design · Adaptive design · Climate resilience

D. A. Bonilla-Ávila (✉) · S. A. Sirigu
Department of Mechanical and Aerospace Engineering, Politecnico di Torino, Turin, Italy
e-mail: diego.bonilla@polito.it

A. M. Calero-Trávez · G. Porporato · R. Giordano
Department of Architecture and Desing, Politecnico di Torino, Turin, Italy

A. Rosiello
SEAform, Turin, Italy

X. L. Zhao et al. (eds.), *Proceedings of the Fourth World Conference on Floating Solutions: WCFS 2024; 2–4 December; Hong Kong*, Lecture Notes in Civil Engineering 597, https://doi.org/10.1007/978-981-96-4569-5_40

1 Introduction

The climate crisis presents a significant global challenge with diverse and far-reaching consequences. While its impacts are universal, it is essential to comprehend the localized implications to formulate effective mitigation and adaptation strategies [1].

According to the Intergovernmental Panel on Climate Change (IPCC) [2], projections for sea level rise vary significantly depending on the greenhouse gas emission scenarios. Under the RCP 8.5 scenario, which represents a pessimistic future with high emissions, sea levels are projected to rise by 0.63–1.01 m by the end of this century. Extending these projections to 2300 under RCP 8.5, sea levels could rise significantly, with some studies suggesting an increase of up to 5 m, posing catastrophic risks to coastal regions worldwide.

The Italian context, characterized by its extensive coastline, is particularly susceptible to these impacts. The Veneto, Emilia-Romagna, and Friuli-Venezia Giulia regions, adjacent to the northern Adriatic Sea, are the most vulnerable to sea level rise [3, 4]. Even under moderate scenarios, significant areas of these regions could face flooding. As a peninsula, Italy has a high concentration of urban centers and population along its coastline, exacerbating the nation's vulnerability and necessitating comprehensive coastal protection and adaptation measures.

Venice, situated in one of Italy's most vulnerable regions, faces not only the threat of sea level rise but also significant subsidence issues [5]. These unique environmental challenges, alongside the continuous need to maintain its historical infrastructure, substantially increase real estate costs [6]. Consequently, Venice has consistently ranked as one of the most expensive cities for real estate in Italy [7], a reflection of high demand and limited housing supply, and a highlight of the substantial stakes involved in its preservation.

This combination of factors makes Venice an exemplary case study for understanding the challenges and potential solutions for cities facing rising sea levels. Despite the widespread use of floating infrastructure solutions in Northern Europe, such strategies have not been adequately explored in Mediterranean countries like Italy. Given the urgency of the climate crisis, it is imperative to consider and implement these innovative solutions to ensure the resilience and sustainability of coastal urban centers.

2 Research Goal and Method

The primary objective of this research is to establish fundamental parameters for the conceptualization and schematic design of future floating structures, particularly aimed at a broader audience, including designers and planners who may not have specialized knowledge in marine contexts. This objective is driven by the urgent need to develop adaptive strategies to address coastal risks and the challenges posed by sea level rise due to climate change, especially within the Italian context.

To achieve this objective, a research-by-design method was employed. This approach involved synthesizing findings from existing literature and state-of-the-art practices in floating architecture. The insights from these studies were transformed into actionable guidelines that can be integrated into the iterative cycle of design processes.

A case study was conducted involving the design of a pavilion for the Venice lagoon. This design utilized technological innovations developed by the MOREnergy Lab at Politecnico di Torino, within the Floating Solution research project [8], which will be engineered and commercialized by SEAform s.r.l. (Co. Ltd). These innovations include modular platforms, an innovative connection system between platforms, a ballast system, and taut mooring. This case study exemplifies the research-by-design method, demonstrating the practical application and testing of theoretical insights and technological features in a real-world context.

3 Case Study: A Pavilion for the Venice Lagoon

The case study involves the preliminary design of a floating pavilion consisting of three interconnected hexagonal concrete bouyant platforms, each with an 18 m diameter, as shown in Fig. 1. This pavilion serves as a prototype for floating architecture, showcasing sustainable practices and technologies aimed at raising public awareness through spatial experiences. To meet adaptation goals, the pavilion incorporates features such as self-sufficiency services and equipment, renewable energy systems, and a modular design suitable for off-site production. These design elements ensure that the pavilion not only serves its educational purpose but also aligns with principles of sustainability and resilience.

Detailed design specifics of the pavilion are not extensively covered herein. Interested readers are encouraged to consult References [9, 10], which provide a broader description of the architectural project.

4 Design Guidelines

The following guidelines integrate findings from literature reviews on existing floating structures. These guidelines were developed to guide the initial design phase, applying insights from research to practical design considerations. They cover site selection, blueprint configurations, architectural design on floating platforms, and sustainable practices, exemplified by the case study area, the Venice Lagoon.

Fig. 1 The floating pavilion architectural project. On the left is a plan view of the three platforms comprising the floating pavilion. On the right is a view of the pavilion situated in the Venice lagoon in front of the 'Arsenale' area. These drawings were produced by Andrés Calero and Gabriele Porporato in 2023

4.1 Aquatic Site Selection

- **Prioritize deeper and wider bodies of water**. Deeper and wider water bodies are preferable for floating structures due to their ability to provide better lighting, more resilient ecosystems, and enhanced water mixing [11, 12]. Such conditions are less likely to suffer from hypoxia or accumulate harmful substances. In contrast, smaller and enclosed water bodies are prone to prohibitive conditions detrimental to aquatic life, such as low oxygen content and high concentrations of pollutants. Although the pavilion is initially proposed for the Venice Lagoon, it could be relocated to a deeper body of water to better adhere to this guideline. Additionally, the mooring and anchoring system developed by SEAform in Italy exhibits significant advantages over alternative solutions in deeper waters.
- **Choose orientations that have a limited influence on natural water currents, wave patterns, and tides**. Floating structures should minimally alter natural water movements. This consideration is particularly relevant in the Venetian Lagoon, which benefits from constant water exchange due to its connection with both rivers and the sea. The brackish nature of the lagoon's waters also poses less of a threat to materials compared to marine environments.

4.2 *Blueprint Configuration*

- **Favor long narrow platforms over compact and wide ones**. Smaller, narrower platforms minimize the shadow cast on the seabed, which is beneficial for the survival of seagrass and benthic algae [11].The pavilion's design, with its elongated hexagonal platforms arranged in a C-shape, optimizes light penetration and reduces the environmental footprint compared to more compact shapes.
- **Favor N-S to W-E orientation for the main axis**. A north–south orientation for the main axis of floating buildings is recommended to enhance light levels beneath the platform [11]. The pavilion, originally planned with a SW-NE orientation, was adjusted to a more effective N-S direction to align with solar paths, minimizing the darkened water areas.
- **Ensure a small local coverage percentage**. Maintaining a low coverage percentage of the water surface is vital to ensure adequate air–water interface, which facilitates gas exchange essential for aquatic life [11, 14]. The pavilion, isolated and with a relatively low coverage area, adheres to this guideline, thereby minimizing its environmental impact.
- **Avoid secluded water areas**. Designs should avoid creating enclosed water areas that restrict natural water flow and mixing. The pavilion's layout, with its platforms not generating significant secluded areas, ensures better water circulation and reduces negative impacts.

4.3 *Upper-Structure Design*

- **Use light weight materials**. To ensure platform stability, it is crucial to select lightweight materials. Prioritizing materials with low densities from the conceptual phase is essential. For the pavilion, Cross-Laminated Timber (CLT) was chosen as the main structural material due to its low density and structural efficiency.
- **Privilege materials without polluting components and finishes**. Materials used in floating structures should be resistant to marine environments and free from harmful emissions. Concrete used should have minimal additives to reduce the risk of leaching heavy metals [15, 16]. External components of the pavilion, including glass and polycarbonate, are chosen for their inert properties, and aluminium components rely on their natural patina for protection.
- **Avoid rainwater runoff into the sea**. Efficient drainage systems are necessary to prevent rainwater from carrying pollutants into the water body. The pavilion utilizes perimetral gutters and a rainwater tank to manage runoff, capturing the first and most contaminated portion of rainwater for proper disposal.
- **Integrate bioclimatic design strategies**. Incorporate elements such as natural ventilation, passive heat gain, adaptive facades, photovoltaic systems, water precooling and pre-heating systems, renewable energy sources, and efficient water

and waste management. The goal is to enhance the upper-structure's energy efficiency, environmental resilience, and user comfort. As shown in Fig. 2, some of the bioclimatic strategies considered in the design process of the pavilion were:

1. *Natural Ventilation*: While external air and water can damage structural elements in a marine environment, controlled natural ventilation is beneficial for user comfort and reduces risks such as fungi growth, molding, and rotting, especially for wooden structures. The pavilion features two natural ventilation systems: one for external spaces and one for the central hall. External spaces use metallic gratings for a chimney effect, pulling in air from the bottom and expelling it from the top. The central hall utilizes heated air from greenhouses, enriched with oxygen, improving air quality and preheating the space in winter.
2. *Passive Heat and Adaptive Facades*: Maximize natural light in greenhouses with transparent facades, supplemented by artificial lighting for consistency. Control solar radiation with automated horizontal solar screens to reduce energy demand in winter and prevent overheating in summer. Lightweight, water-resistant, and non-reflective aluminum screens are used.
3. *Adaptive and Responsive Control*: An automated system adjusts the solar screens and ventilation gratings based on real-time data from sensors monitoring air temperature, humidity, glass surface temperature, and weather forecasts. This adaptive facade responds dynamically to environmental changes, optimizing thermal performance and preventing excessive temperature fluctuations.

- **Ensure modular and reusable upper structures**. One of the primary advantages of floating platforms is their inherent stability, coupled with the potential for easy relocation—an attribute not typically found in land-based structures. It is imperative that the architecture built on these platforms also embodies flexibility and is designed with its entire life cycle in mind, including plans for reuse. To achieve these goals, the pavilion employs practices such as modularity, dry assemblage and off-site construction.

4.4 Under-Structure Design

- **Design for shallow buoyant platforms with minimal submersion**. Local water depth is critical. A higher platform above the bottom ensures greater lighting under the platform, crucial for maintaining healthy benthic environments [11]. Shallow water bodies are more impacted by floating structures, as changes in water currents and mixing patterns are usually confined to the top 3–4 m of water [13].
- **Prevent the formation of filter feeder colonies or provide the possibility to remove them**. Floating structures can inadvertently support the growth of filter

feeders, which can disrupt local ecosystems [12, 17, 18]. Design features should discourage their formation or include mechanisms for their removal.
- **Favour the growth of photosynthetic organisms**. Encouraging the growth of algae and other photosynthetic organisms helps maintain ecological balance. This can be achieved by integrating dedicated substrates and anchor points in well-lit areas [12, 14, 19, 20].
- **Favour the growth of local flora and fauna**. Designs should enhance local biodiversity by incorporating features that promote native species and discourage invasive ones, preserving the lagoon's ecological balance and resilience.

For the pavilion, concrete anchors were identified as key opportunities for habitat creation. Due to solar rotation, these anchors experience minimal shading, except for the northern anchors, which might have longer shaded periods depending on the season. Being distant from the area directly beneath the floating structure, the risks of excessive organic concentration and hypoxia are negligible. The anchors are located in nutrient-rich zones, benefiting from colonies of filter feeders expected to develop on the submerged surfaces, fostering aquatic life.

The anchors, with a compact core for weight and resistance, are designed with holes, niches, crevices, and bumps to provide multiple anchor points and hiding areas for aquatic plants and macrofauna. This design encourages the proliferation of underwater life, which can expand to the surrounding area. When the pavilion is removed, the concrete anchors can remain as integral parts of the aquatic ecosystem.

5 Conclusions

This research underscores the potential of sustainable floating architecture, attuned to the marine environment, as a robust adaptive strategy to address the challenges posed by rising sea levels in the Italian context. By focusing on Venice, a city of significant heritage value and one of the regions most susceptible to climate change, the study stimulates discourse on future urban development strategies. This paper encourages stakeholders to delve deeper into the exploration of floating solutions, offering preliminary guidelines that equip developers and designers with a foundational understanding of how to approach floating technologies. These guidelines also highlight the comparative advantages of floating structures over traditional urban development techniques specially regarding the environmental footprint.

One significant barrier to the implementation and funding of floating projects is the lack of social acceptance. By expanding and synthesizing information for a broader audience, this research aims to engage the general public and influence decision-makers to consider these innovative solutions. The fifteen guidelines presented in Chap. 4 encapsulate the critical parameters that facilitated the design process for the pavilion case study.

Ultimately, this study advocates for the inclusion of floating architecture in urban planning, emphasizing its viability and benefits in adapting to the impacts of climate

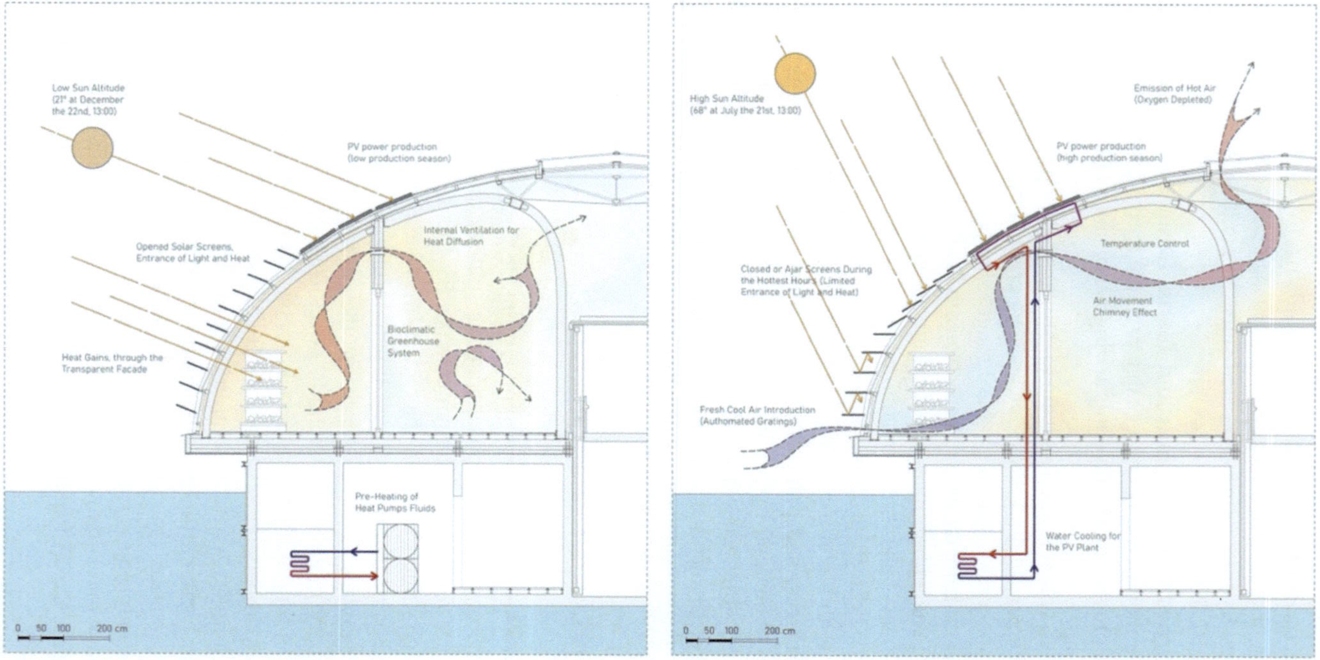

Fig. 2 Section of the floating pavilion illustrating bioclimatic design strategies. The left panel depicts the winter season scenario, while the right panel represents the summer season scenario. These drawings were produced by Andrés Calero and Gabriele Porporato in 2023

change. The research-by-design approach not only advances best practices in the field but also fosters a deeper understanding of the potential for floating structures to contribute to resilient and sustainable urban development.

References

1. Santos FD, Ferreira PL, Pedersen JST (2022) The climate change challenge: a review of the barriers and solutions to deliver a Paris solution. Climate 10:75
2. Oppenheimer M, Glavovic BC, Hinkel J, van de Wal R, Magnan AK, Abd-Elgawad A, Cai R, Cifuentes-Jara M, DeConto RM, Ghosh T, Hay J, Isla F, Marzeion B, Meyssignac B, Sebesvari Z (2019) Sea level rise and implications for low-lying islands, coasts and communities. In: Pörtner HO, Roberts DC, Masson-Delmotte V, Zhai P, Tignor M, Poloczanska E, Mintenbeck K, Alegría A, Nicolai M, Okem A, Petzold J, Rama B, Weyer NM (eds) Special report on the ocean and cryosphere in a changing climate. IPCC, Switzerland, pp 321–445
3. Floodmap. https://www.floodmap.net. Accessed 15 June 2024
4. Barausse A, Meulenberg C, Occhipinti I, Abordi M, Endrizzi L, Guadagnin G, Piron M, Visintin F, Vižintin L, Manzardo A (2022) A methodological proposal for the climate change risk assessment of coastal habitats based on the evaluation of ecosystem services: lessons learnt from the INTERREG project ECO-SMART. Sustainability 14:7567
5. Tosi L, Teatini P, Carbognin L, Brancolini G (2009) Using high resolution data to reveal depth-dependent mechanisms that drive land subsidence: the Venice coast, Italy. Tectonophysics 474:271–284
6. Broxterman D, Gatzlaff D, Letdin M, Sirmans GS, Zhou T (2023) Introduction to special issue: topics related to real estate market efficiency. J R Estate Financ Econ 66:197–202
7. Idealista. https://www.idealista.it/en/news/property-sale-italy/2019/11/04/2595-most-expensive-areas-italy-buy-property. Accessed 15 July 2024
8. Sirigu SA, Boccanfuso V, Carapellese F (2024) Multi-body analysis of modular floating islands: optimisation of connector stiffness. In: Ikoma T, Tabeta S, Lim SH, Wang CM (eds) The third world conference on floating solutions (WCFS 2023). Springer, Singapore, pp 313–328
9. Porporato G (2023) FloatScapes: a floating adaptive pavilion, a proposal to lessen the environmental impact of floating architecture on aquatic ecosystems. A study in the Venice Lagoon. Master thesis, Politecnico di Torino, Italy
10. Calero-Travez AM (2023) FloatScapes: a floating adaptive pavilion, a proposal for a performant prefabricated modular building in the maritime context. A study in the Venice Lagoon. Master thesis, Politecnico di Torino, Italy
11. Burdick DM, Short FT (1999) The effects of boat docks on eelgrass beds in coastal waters of Massachusetts. Environ Manage 23:231–240
12. de Lima RLP, de Graaf-van Dinther RE, Boogaard FC (2022) Impacts of floating urbanization on water quality and aquatic ecosystems: a study based on in situ data and observations. J Water Clim Chang 13:1185–1203
13. Kyozuka Y, Kato S, Nakagawa H (2001) A numerical study on environmental impact assessment of Mega-Float of Japan. Mar Struct 14:147–161
14. Foka E, Rutten M, Boogaard F, de Graaf R, Lima R, van de N (2015) The effect of floating houses on water quality. In: Amsterdam international water week (IWW). The Netherlands
15. Vollpracht A, Brameshuber W (2016) Binding and leaching of trace elements in Portland cement pastes. Cem Concr Res 79
16. Lu H, Wei F, Tang J, Giesy JP (2016) Leaching of metals from cement under simulated environmental conditions. J Environ Manage 169:319–327
17. Giles H, Pilditch CA (2006) Effects of mussel (Perna canaliculus) biodeposit decomposition on benthic respiration and nutrient fluxes. Mar Biol 150:261–271

18. Callier MD, Richard M, McKindsey CW, Archambault P, Desrosiers G (2009) Responses of benthic macrofauna and biogeochemical fluxes to various levels of mussel biodeposition: an in situ "benthocosm" experiment. Mar Pollut Bull 58:1544–1553
19. Exley G, Page T, Thackeray SJ, Folkard AM, Couture R-M, Hernandez RR, Cagle AE, Salk KR, Clous L, Whittaker P, Chipps M, Armstrong A (2022) Floating solar panels on reservoirs impact phytoplankton populations: a modelling experiment. J Environ Manage 324:116410
20. Alexander C, Robinson M (2006) Assessing the impacts of floating docks on bottom character and benthic productivity in coastal Georgia. Coastal Zone Management Program Coastal Resources Division, Georgia Department of Natural Resources, Brunswick, GA, United States of America

Some Ways to Enhance the Sustainability on Floating Architecture

Changho Moon

Abstract The aim of this study is to propose some ways to enhance the sustainability on floating architecture. Sustainability can be defined as the potential for long-term maintenance of well-being, which may have environmental, economic, and social dimensions. Through literature review of sustainability and empirical study of floating architecture cases, not only enough budget securement for renewable energy system and prefabrication and passive design system for the sustainability in environmental and economic dimension but also residents' psychology, community spirit and social life in social dimension are considered as enhancing the comprehensive sustainability.

Keywords Floating architecture · Sustainability · Way of enhancement

1 Introduction

Owing to the increase of economic income level, people want to live around or on the water. And due to climate change, there have been unprecedentedly severe and frequent natural disasters. Proper countermeasures in terms of architecture are needed and diverse floating architectures are emerging as realizable alternatives. Floating architecture also requires sustainability according to the global architecture trends.

This paper aims to discuss some ways to enhance the sustainability of floating architectures, to investigate the sustainable characteristics of floating buildings from the view point of environmental, economic and social dimensions, and suggest some ideas for the reference of design and construction in new floating architecture projects.

C. Moon (✉)
Kunsan National University, Gunsan, Korea
e-mail: mchangho55@gmail.com

Balance Architecture Studio, Gunsan, Korea

X. L. Zhao et al. (eds.), *Proceedings of the Fourth World Conference on Floating Solutions: WCFS 2024; 2–4 December; Hong Kong*, Lecture Notes in Civil Engineering 597, https://doi.org/10.1007/978-981-96-4569-5_41

Research methods include the site-visits experience of some floating buildings in Europe and USA, the review of related literatures, and the navigation of sustainability and floating building homepages.

2 Concept of Sustainability and Floating Architecture

2.1 Concept of Sustainability

According to Wikipedia, sustainability can be defined as the capacity to maintain or improve the state and availability of desirable materials or conditions over the long term. Sustainability is improving the quality of human life while living within the carrying capacity of supporting eco-systems. Sustainability of architecture has been discussed mainly in terms of physical issues like energy and ecology, but the concept of sustainability needs to be extended to the non-physical aspects like residents' psychology, community spirit and social life.

For humans, sustainability is the potential for long-term maintenance of well-being, which can have environmental, economic, and social dimensions. In this paper the sustainable elements of floating buildings on water will be analyzed and investigated based on the 3 dimensions as references [1].

2.2 Concept of Floating Architecture

Floating architecture can be defined as a building for living/working space that floats on water via a floatation system, is moored in a permanent location, does not include a water craft designed or intended for navigation, and has a premises service (electricity, water/sewage, gas) system served through connection line by a permanent supply/return system between the floating building and adjacent land, or has self-supporting service facilities for itself.

3 Sample Cases of Floating Architecture

Well known cases of floating architecture with sustainable elements are going to be reviewed in terms of environmental, economic, and social dimensions. Actually, floating architecture itself is already sustainable to sea/river water level change.

Fig. 1 Concept picture of Four seasons hotel (left, Popular Mechanics January 1988 p. 86 [2]) Hotel Haegumgang (right, Maeil Business Newspaper [3])

3.1 Four Seasons Hotel

Four Seasons (Great Barrier Reef) Hotel with 140 double rooms and 34 luxury suites was designed by Swedish naval architect and built in Singapore in 1988, moved to and operated in Great Barrier Reef area in Queensland, Australia for a while, moved to Ho Chi Mihn City on the Mekong River, Vietnam in August 1989 to mitigate the shortage of quality hotel and operated until 1997, returned and remodeled in Singapore in 1997, and bought by Hyundai Asan and moved to North Korea in 1999, operated as Mt. Kumgang hotel for south Korean tourists until 2010, and unfortunately dismantled in 2022 for political reason (Fig. 1).

This floating hotel had been moved four times and operated in three different countries for more than 20 years and could have a possibility to use for more years. Furthermore, the floating hotel has a self-supporting facility with desalination plant, sewage and waste treatment system, and a power plant with at almost noiseless level for environmental consideration.

As long-term usage of this hotel in different places by different people due to mobility and some environmental considerations of the facility can contribute to the conservation of environment, this floating hotel can be considered as having the sustainability in environmental dimension.

3.2 IBA Dock

IBA Dock in Hamburg, Germany, the international building exhibition IBA had a slogan 'City in a Climate Change' with a goal of CO_2-neutral city development and multiple possibilities were tried to provide energy supply from the water temperature of the Elbe river combined with solar energy (Fig. 2).

This floating building employs a prefabrication and modular system in the design and construction process. Most parts are manufactured in a factory, transported, and

Fig. 2 IBA Dock (left, photo by the author), modular construction concept (right, Archdaily [4])

assembled and installed on site. Maximum savings in construction waste can be made and the floating building can be easy and economical to maintain.

While energy systems such as hydrothermal use and solar energy can be regarded as having the sustainability of environmental dimension, the prefabrication and modular system can contribute to the sustainability in economic dimension.

3.3 Salt & Sill Hotel

In October 2008, the first floating hotel in Sweden, opened alongside the famous restaurant "Salt & Sill" in Klädesholmen. The hotel consists of six two-story buildings on floating pontoon. The hotel has 23 rooms with 46 beds, and all the rooms have their own entrance and access to an outdoor seating area. People can feel a cozy & personal atmosphere and a style characterized by modern Scandinavian simplicity (Fig. 3).

In the floating hotel, heating energy is actually generated from the warm sea water underneath the floating pontoon in winter. The hotel was built using local raw

Fig. 3 Salt & Sill Hotel (left, WiKipedia [5]), (right, photo by the author),

materials such as pine wood from Swedish forests, and only environmentally friendly paint. Even the left-over quarrying stone was used to build a new lobster reef under the concrete pontoon in consideration of the environment [6].

While comfortable atmosphere and modern style of the hotel can be considered as having the sustainability in social dimension, usage of hydrothermal energy, local materials, and environmentally friendly material, and reuse of left-over stone can be regarded as enhancing the sustainability in environmental dimension.

3.4 Oregon Yacht Club

Oregon Yacht Club (OYC, Fig. 4) is a floating home community located on the Willamette River in Portland, Oregon, USA. The close proximity of downtown Portland while still in a pastoral setting has contributed to OYC's being one of the premier floating home moorages in the area. The 38 floating homes adjoin eight acres of natural riverfront property [7].

The residents enjoy the peaceful and comfortable atmosphere on water within the natural setting. They have great interest in conserving the natural environment like wild birds and watershed vegetation, have to cooperate in managing natural disasters like flooding and typhoons, and should negotiate legal regulations with the city officers. Solid social sustainability is essential and easily achieved there.

The floating home community is usually more secure against crime than housing sites on land. Because the residents know each other, the entrance is controlled, and the community is surrounded by water, unwanted guests cannot easily access to the floating home community.

Peaceful atmosphere, interesting in conserving nature, cooperation in disaster, negotiation with city officers, residents' good relationship and secure environment can be considered as having the sustainability in social dimension.

Fig. 4 Oregon Yacht Club (left & right, photos by the author)

Fig. 5 Schoonschip (left & right, Architizer Homepage [8])

3.5 Schoonschip

Fully completed in 2021, the floating community in Amsterdam, Netherlands consists of 46 water plots and is home to more than 100 residents. The residents live in highly eco-efficient buildings that were remotely manufactured (Fig. 5).

Built on a circular community model, solar panels on the roofs are connected to a smart grid where residents can trade energy with the help of smart technology. It contains decentralized and renewable solutions to water, energy, and waste systems, as well as submersed heat exchangers for heating and cooling [8].

Heat comes from pumps that use hydrothermal technology to extract warmth stored naturally in the canal below; electricity comes from solar panels connected to a shared grid; and roofs are partially covered by plants that absorb water, reflect sunlight and capture carbon dioxide from the atmosphere. There are no gas connections, nor private cars, but rather, a system for sharing electric vehicles and bikes with all residents [9].

Renewable energy system (hydrothermal, solar energy) and capturing carbon dioxide can be considered as having the sustainability in environmental dimension, remotely manufacturing system can contribute to the sustainability in economic dimension, and circular community model and sharing mobilities can be considered as having the sustainability in social dimension.

4 Conclusions

As seen from the sample cases, floating architectures have various sustainable elements in environmental, economic and social dimension according to the building functions and purposes. Floating architecture itself is already sustainable to sea/river water level change and especially hydrothermal use of sea/river water beneath the floating building might be a great advantage.

To enhance the sustainability in floating architecture, enough budget securement for the renewable energy system application like hydrothermal and solar energy would be necessary from the planning stage as renewable energy system may be a core factor of the sustainability in environmental dimension.

And introduction of prefabrication system, passive design system, and local and environmentally friendly material usage in the process of design and specification in the floating architecture can enhance the sustainability in economic and environmental dimension.

Sustainability of floating architecture has been discussed mainly on the physical issues like renewable energy and ecological measures in economic and environmental dimension, but the concept and application of sustainability need to be expanded to the non-physical aspects like residents' psychology, community spirit and social life in social dimension. Therefore, comprehensive sustainability on floating architecture can be enhanced.

References

1. Wikipedia Homepage, Sustainability, http://en.wikipedia.org/wiki/Sustainability. Accessed 03 July 2024
2. Popular Mechanics January 1988 p 86, Floating Hotel Near Australia's Great Barrier Reef, https://books.google.co.kr/books?id=IuQDAAAAMBAJ&pg=PA86&lpg=PA86&dq=Floating+Hotel+Near+Australia%27s+Great+Barrier+Reef&source=bl&ots=wg2uloj_Z2&sig=-lSOUajeG9rJ9yCZ33-h6vNFTOE&hl=en&sa=X&redir_esc=y#v=onepa. Accessed 05 July 2024
3. Maeil Business Newspaper Homepage. https://www.mk.co.kr/news/politics/10249637. Accessed 05 July 2024
4. Iba Dock/Architech—Architecture and Technology, Homepage of Archdaily. https://www.archdaily.com/288198/iba-dock-architech. Accessed 03 July 2024
5. Image of Salt & Sill. https://sv.wikipedia.org/wiki/Fil:Salt_%26_Sill_ifr%C3%A5n_bron.jpg. Accessed 05 July 2024
6. Salt & Sill Homepage. https://www.saltosill.se/en/. Accessed 03 July 2024
7. Oregon Yacht Club Homepage. https://oregonyachtclub.com/. Accessed 03 July 2024
8. Schoonschip, Architizer Homepage. https://architizer.com/projects/schoonschip/. Accessed 05 July 2024
9. Rebecca Redelmeier, Sea level rise is coming for cities. But who can afford to 'float' their way out? THE GLOBE AND MAIL. https://www.theglobeandmail.com/business/article-sea-level-rise-is-coming-for-cities-but-who-can-afford-to-float-their/. Accessed 05 July 2024

Feasibility Research on the World's Largest Floating Infinity Pool Edge

Kimmo Saharinen

Abstract Bluet is a Finnish company pioneering innovative floating constructions, such as their solution for the Mandarin Oriental, Lago di Como hotel. This project is the world's largest floating infinity pool with a 40 m infinity edge which extends 7 m along both short sides of the pool. Special analysis was required to ensure the feasibility of the structure and to maximize the infinity edge effect. Ensuring the continuous water flow to maintain the infinity edge effect while concealing the edge is more challenging on a floating structure due to the movement. The research began by capturing data from Lake Como and using that information for computer modelling of possible solutions. Bluet built a 1:10 scale model of the shoreline and proposed pool for analysis in a research center wave pool. The test concentrated on optimizing the functionality of the infinity edge and the pool's conformity [1] with the wave patterns of Lake Como. The experiment subjected the pool to a range of waves, including simulations of storm conditions. The analysis focused on determining the necessary water volume in the pool to maintain the desired effect and quantifying the spillage over the edge into the gutter. Throughout the testing phase, adjustments were made to the angles of the infinity edge gutter bottom and the gutter size, the pool weight, anchoring methods and shape of the structure. The process allowed the optimal design for the gutter, necessary water volume, pumping speed, and overflow tank size to be determined.

Keywords Floating pool · Floating infrastructure · Floating foundation · Floating infinity pool

K. Saharinen (✉)
Bluet Floating Solutions, Vantaa, Finland
e-mail: kimmo.saharinen@bluet.fi

X. L. Zhao et al. (eds.), *Proceedings of the Fourth World Conference on Floating Solutions: WCFS 2024; 2–4 December; Hong Kong*, Lecture Notes in Civil Engineering 597, https://doi.org/10.1007/978-981-96-4569-5_42

1 Introduction

Mandarin Oriental Lago di Como is a 5-star luxury resort located in Blevio, a small commune on the shores of Lake Como in northern Italy. The resort had a floating swimming pool in the lake but were unsatisfied with its stability and how the sundeck limited the lake views. The client wanted a new floating pool which met the following goals:

- Strict safety for guests and boaters.
- Less movement of the structure to improve comfort for guests.
- "Wow" factor; unprecedented size and scope.
- Panoramic infinity effect to improve view of lake.
- Adequate size of pool and sundeck for normal number of guests.
- Design should complement the architectural style of property.
- Budget considerations.
- Deadline considerations.

1.1 Main Partners

The consultant on the project was Bluet, a Finnish company specializing in floating construction solutions [2]. The client was Mandarin Oriental Hotel Group, a Hong Kong-based company dating back to 1963 [3]. The architect was Herzog & de Meuron, a Swiss architectural firm famous for projects like the Tate Museum of Modern Art in London and FC Barcelona's stadium Camp Nou [4]. The hydrological consultant was VTT Technical Research Centre of Finland, a state-owned research and technology center.

2 Background

The shores of Italy's Lake Como, where the Mandarin Oriental Lago di Como is located, is a densely built-up area [5]. As shoreside real estate is at a premium, the hotel built a floating swimming pool for their guests. However, they were unsatisfied because of the uncomfortable movement of the pool. Also, the pool was a square surrounded by a sundeck, which limited one of the most desirable aspects of the property: the view of Lake Como. For these reasons the client decided to build a new floating pool.

The client desired an infinity pool. These are particularly popular at luxury resorts and residences with the ocean or lakes in the background. The client worked with the architect and floating construction expert to develop a solution.

3 Challenges

An infinity edge pool is more challenging to develop, construct, and maintain than a traditional pool. It requires additional structural, mechanical, and architectural solutions which are exacerbated when the pool is floating [6].

A common practice to minimize challenges in infinity pools is to limit the infinity edge, such as using the infinity edge on one short edge of a rectangle-shaped pool. However, the client and architect wanted a much more impressive pool. As it would be floating on the lake there was a unique opportunity to expand the infinity effect, not only on one long edge of the pool, but along both short edges as well. The pool would offer a panoramic infinity edge along three sides.

3.1 Engineering Challenges

The new floating pool was required to be much more stable than the previous pool. An infinity pool is designed so the water flows over one edge, producing a visual effect of water with no boundary. The overflowing water is collected in a gutter which is pumped back into the tank which supplies the pool. The movement of the floating platform could cause sloshing and ruin the infinity effect. Also, it was important to avoid waves splashing into the gutter and water escaping from the gutter into the lake, which would cause cross-contamination.

3.2 Architectural Challenges

The hotel and waterfront are historic sites, and the new pool had to fit into the overall aesthetic. The architect is world-famous for giant, unique projects, and they brought the same philosophy to the infinity pool concept. Some of the initial brainstorming ideas included circular and oval pools, which would have been extremely challenging to achieve with current technologies and techniques. Bluet invented some extraordinary solutions which, in theory, could have achieved these shapes but they were ultimately rejected [6].

The parties have long experience with working with diverse partners in challenging projects. While architects can't overrun technical requirements, the parties agreed to push the boundaries as much as possible while accepting engineering limits.

4 Data Gathering

In December 2021, the team visited the site on Lake Como. Sensors were placed on the old pool to measure:

- Longitudinal acceleration (milli-g)
- Lateral acceleration (milli-g)
- Vertical acceleration (milli-g)
- Pitch motion (deg)
- Roll motion (deg).

A buoy was installed at the site to measure the direction and strength of waves while video recordings were made of the motions of the existing pool.

Detailed scans were also made of the lake bottom and waterfront where the pool would be installed. The water depdth next to the shoreline is 5m. From there the lake bottom deepens in a 60 degree angle to 15m. The waterlevel variation is 3 meters.

Most waves were generated by passing boats. Wave winds and swell were also recorded but were not a serious concern. Wave heights up to 0.32 m were recorded.

Most the existing pool's motions were on a narrow frequency range from about 0.1 to 0.4 Hz. The largest amplitudes were typically between 0.25 and 0.30 Hz. Pitch motion was observed to be more significant than roll motion.

The data gathered on site was complemented by general lake data collected and published by the Italian authorities as a public service. The data-gathering period turned out to have more waves and adverse weather conditions than normal, but this was judged to be good for analysis and design, as any proposed solution would be measured against unusually harsh waves.

5 Design

The new floating infinity pool was designed with CAD. Three possible new designs were created.

The existing pool design was compared against the three possible new designs using theoretical methods. The three possible new designs were analysed and one was picked as most likely to meet the client's goals. The first choice was a rectangular shape, much longer than the existing pool (Table 1).

Typically, leisure floating structures have a simple flat-bottomed monohull shape. However, a catamaran structure was chosen instead. A catamaran-like design can add stability due to having less contact with the water and hence less wave impact.

Table 1 Main dimensions of the floating pools

	Existing pool	New pool
Floating platform		
Length (m)	22.8	40
Width (m)	14.6	15
Draught (m)	1.4	1.72
Estimated weight (t)	445	457
Swimming pool		
Length (m)	15.9	39.3
Width (m)	8.4	6.3
Water depth (m)	1.35	0.23 – 1.20
Volume (m^3)	160	173

6 Testing [7]

The testing partner was VTT Technical Research Centre of Finland. One of their specialties is the maritime sector, so they maintain modern software, a wave pool and hydrology experts. The tank VTT uses in Espoo, Finland, is 130 m long, 11 m wide and 5.5 m deep. Ansys Aqwa, commercial software for hydrodynamic assessments, was used in the analysis [8].

A 1:10 scale model of the proposed new design was constructed of stainless steel and the lakeshore model was built of wood. The shoreline was approximately three times the width of the new infinity pool.

In total, 100 tests were performed with regular beam waves (90 degree), regular oblique waves (134 degree), and irregular waves (134 degree). The waves were based on the most demanding boat and wind waves as observed with the Lake Como buoy as well as computer modelling of potential weather. In addition, decay tests were performed to determine the natural period of roll and the roll damping coefficient. Variations included:

- Dampening plates
- Infinity pool gutters
- Elastic-type moorings
- Hydraulic-type moorings
- Wave dampers in the swimming pool
- Buoyancy.

The tests specific to the infinity effect included variations in the infinity overflow gutter and how they reacted to waves. These variations included gutter size and angle, as well as experiments with the edge. One possibility was to make the infinity edge a comb-like structure, with the water flowing through the grooves, which would break the surface tension and maintain the infinity edge effect. This would have lowered the power usage and made irregular sloshing less likely, but would have negatively impacted the infinity illusion effect. Ultimately, the comb structure was found not to be necessary.

6.1 Testing Conclusions

The roll damping plates, adapted from their use on boating piers, were found to be particularly effective. They decreased the roll about 40% when three plates were used on each corner [9].

7 System Description

The pool area is 39.3 m by 6.3 m, with a water depth of 0.23–1.20 m. The total volume is 173 m^3.

The infinity effect runs along the entire 39.3 m of the long edge of the pool, around the corners, and back along the two 6.3 m-short edges to the sundeck. The result is a pool with a panoramic infinity edge along three of the four sides [10].

To maintain the infinity effect the volume of water spilling over the infinity edge was increased by boosting the pumping to about three times that of a regular infinity pool. If the infinity effect was ever lost due to the movement of the structure, it would quickly be regained with the higher pumping volume.

The architect wanted the overflow gutter as low as possible for aesthetic reasons, but the greater water flow over the infinity edge and the possibility of high waves caused concerns about cross-contamination between pool and lake, which was unacceptable due to water treatment rules.

It was decided to place the gutter high enough to avoid a 1-m-high wave, which was nearly three times the height of any observed during data gathering at Lake Como. The gutter angle is slightly changed and the gutter is also slightly larger than a similar-sized infinity pool because of the increased water volume.

8 Use

The world's largest floating infinity pool opened early in spring 2023. It has turned into a tourist attraction, with tour guides pointing out the pool to guests on tour boats.

9 Conclusion

A major reason for the success of the project is that Bluet was involved early in the project, when it was still in the conceptual phase. Leisure floating solutions remain a specialised, niche industry, and it is common for ideas to be impractical. An infinity pool design may work perfectly on land but fail in floating conditions. Many floating solutions have been designed but not executed, most likely because during the design phase the challenges have not been fully addressed. Transportation and installation must also be considered early in the process. If floating solution experts are not involved at an early stage, significant time and resources can be wasted on solutions which will not succeed [11].

It was determined that the best way to affect the stability of a floating platform were the main dimensions (length, width, draft) and mass distribution. As such, traditional naval architecture tools and techniques can also be appropriate for floating, stationary structures. Damping solutions, such as the damping plates and moorings, were also important. Attempts to affect the waves, such as with breakwaters, were not feasible.

The benefits of preliminary, onsite preparations were confirmed. The depth to the lake floor and angles from the stone arches created unusual wave patterns which did not appear in general lake data collected and published by Italian authorities. The lakefront where the floating structure was to be placed created a unique "microclimate" which was not accurately reflected in public data.

Even more importantly, extensive, long-term hydrologic data has been collected for Lake Como, but this was found to be of limited use. Lake Como hydrological data is a statistical system built for wind-generated waves. Typical lake wave data removed outliers, which can be normal statistical procedures. This process treated abnormal waves—particularly those caused by boats—as measurement or processing errors. However, these were precisely the waves the project needed to consider. It was critical for the project's success to collect bespoke data and use that alongside traditional public hydrologic data [12].

References

1. Mansfield N (2004) Human response to vibration. CRC Press
2. Bluet. https://bluet.fi/en. Accessed 05 July 2024
3. Mandarin Oriental, https://www.mandarinoriental.com/en. Accessed 05 July 2024
4. Herzog, de Meuron, https://www.herzogdemeuron.com/. Accessed 05 July 2024
5. Blevio, Villa Pasta e Villa Roccabruna, https://www.lombardiabeniculturali.it/percorsi/ville-como/4.1/. Accessed 05 July 2024
6. Saukko D, interviews
7. Happonen K, Ville V, Martio J (2022) VTT como lake floating pool—assessment of motions in waves
8. Happonen K, interviews
9. Bluet: VTT Tank Test Summary. Bluet (2022)

10. Infinity pool by Herzog & de Meuron aims to "blend into the waters" at Lake Como, https://www.dezeen.com/2024/02/02/infinity-pool-herzog-de-meuron-lake-como/. Accessed 05 July 2024
11. Sirola T, interviews
12. Bertoli L, Balzarolo D, Todini E (2022) On the benefits of collaboration between decision makers and scientists: the case of lake Como. Hydrology 9(11):187

Modularity and Feasible Installation

Kimmo Saharinen

Abstract Modularity is advantageous in construction. The creation of prefabricated modules in controlled factory environments improves quality, accelerates installation, and enables scalability. These benefits are particularly amplified in floating construction, where building on water presents unique technical requirements. The importance of modularity has been proven in projects such as in the floating Seal Basin, Bluet's Multiuse platform (product), Vök Baths, Lake Como floating infinity pool and Jubileumsparken, and the Allas Pool. The Korkeasaari Zoo "seal basin" is a unique multi-purpose floating pool, originally designed for the Saimaa ringed seal. The design of the floating basin had to consider the needs of the animal and visitors, the construction site, special requirements of floating construction, and landscaping. Various options were studied with crane and cargo companies and finally the best solution was to use heavy-duty barge modules, which could be transported to the site by truck and lifted by crane to be assembled on water. When developing the Bluet Multiuse Platform in 2019, similar challenges were encountered. This platform comprises modules which can be connected end-to-end, side-to-side, or end-to-side. The versatility of these modules allows for the creation of floating event platforms, music stages, pop-up restaurants, temporary walkways, and construction platforms. The modular design allows transportation via trucks, even to remote inland locations like lakes. Furthermore, assembly and disassembly are quick and simple. The Allas Sea Pool is part of an urban spa and event center with a large floating foundation. The complicated structure was built with modules. Recognizing the importance of modularity in floating construction, Bluet consistently prioritizes modularity in its solutions, ensuring that products are optimized for transportation and feasible, time efficient installation.

Keywords Floating pool · Floating platform · Floating event stage · Floating infrastructure · Floating foundation · Floating modular construction

K. Saharinen (✉)
Bluet Floating Solutions, Vantaa, Finland
e-mail: kimmo.saharinen@bluet.fi

X. L. Zhao et al. (eds.), *Proceedings of the Fourth World Conference on Floating Solutions: WCFS 2024; 2–4 December; Hong Kong*, Lecture Notes in Civil Engineering 597, https://doi.org/10.1007/978-981-96-4569-5_43

1 Introduction

The benefits of modular construction have been recognized for decades, if not longer. Some of the main advantages include:

- Cost savings
- Ecological
- Flexibility
- Improved quality control
- Less waste
- Reduced disruption
- Reliability
- Safety
- Scalability
- Speed

Bluet, a Finnish company specializing in floating construction solutions, has found the benefits of modularity extend to floating constructions. Illustrative examples include the Bluet's Multiuse platform (product), Vök Baths, Lake Como floating infinity pool and Jubileumsparken [1].

1.1 Seal Basin

The Seal Basin was designed for Korkeasaari Zoo in Helsinki, Finland. It was originally planned for the endangered Saimaa ringed seal and consists of a 20 × 20 m pool which is 3.2 m deep. Surrounding the pool area is a viewing platform and landscaping to imitate the seal's natural environment.

1.2 Bluet Multiuse Platform

The Bluet Multiuse Platform was designed to be versatile, such as for concerts, theatre performances, pop-up art exhibitions, fashion shows, product launches and also private events, for example, weddings. There is also use as an additional space as a grandstand, terrace area, fair booth, sports field, café or bar area. Each module is 6.6 × 3.3 × 0.75 m and can be connected to others to fulfill a variety of needs.

1.3 Allas Sea Pool

The Allas Sea Pool in downtown Helsinki is part of an urban spa and event center. A restaurant is onshore, while the floating structure consists of two heated pools, one seawater pool, and a 2700 square meter sundeck. The Allas Sea Pool was opened in 2016 and floating saunas were added in 2022.

2 Background

Historically, most evolution in floating constructions has been through major industrial projects, particularly due to the offshore industry. These include floating structures for the drilling, production, storage, and offloading of hydrocarbons located in marine environments. Floating wind turbines are also maturing.

More recently, new technologies, materials, and techniques have made floating constructions viable for many more uses, such as leisure facilities or temporary structures. An emerging trend is floating megastructures [2].

Coastal areas are extremely desirable for development, but economic and sustainability factors make shoreline construction sometimes difficult. A logical solution is floating platforms, which can be developed faster, cheaper, and more sustainably than similar-sized projects on the shore. They are also less disruptive, as they can be manufactured and assembled elsewhere and towed to their eventual location [3].

Smaller floating leisure structures are better suited for more sheltered waters, typically 1–10 m deep with waves less than 0.3 m in height. However, the real issue for floating structures is the wave length, not the wave height [4].

Whenever possible, Bluet tries to use modularity in floating construction. Otherwise it is necessary to use a shipyard or submergible solution.

3 Challenges

Floating construction has most of the structural, mechanical, and architectural challenges of land-based construction, as well as additional ones related specifically to floating structures. Two of the most important of these are stability and mooring.

3.1 Modularity Challenges

Modularity in floating solutions brings additional challenges. Some of the most prominent are:

- Incorrect connections between modules

- Too complicated connections between modules
- Risk to workers during assembly
- Dimensions not being accurate enough
- Different buoyancies of different modules.

4 Solutions for Modularity's Challenges

The particular challenges caused by modularity in floating solutions can be solved with proper care and attention.

4.1 Higher Precision

The modules must be designed and manufactured to a high degree of precision as there is a lower margin of error. The modules must be crafted to precise dimensions for transportation and assembly. The connections must be easy to use and properly aligned, because the assembly environment over open water is almost always challenging.

4.2 Make the Components Modular

In addition to making the platform modular, Bluet makes different components modular as well. These include technical, heating, sewage, water treatment, and container modules. A floating pool will need a water treatment module, for instance, while a floating concert stage will not. Making these components modular allows different floating structures to be built quickly, efficiently, and easily.

4.3 Design for Single Modules as Well as Groups

A floating structure might have a variety of modules, with different types each having a different buoyancy and weight load. For example, a heavy technical module filled with equipment will behave differently in the water compared to a buoyant end module which bears the load, and when they are connected the pair will behave unalike. Bluet's goal is that all floating pieces will float individually, even with different loading capacities.

On the platform with various structures on deck, there are multiple weight loads positioned in different locations on the platform. The challenge is how to make the overall platform work as a whole, which can be solved with proper forward-thinking design.

4.4 *Plan for Installation*

Work on the water's surface or under the surface is more hazardous and should be handled carefully. Connecting modules underwater requires divers and is expensive and dangerous. As much as possible, the design for floating structures should be planned for assembly on top of water to avoid the use of divers.

Preassemble as much as possible to minimize assembly time on site. This can lower costs and improve safety, too [5].

5 Cases

Bluet endeavors to use modularity in nearly all floating solutions. Some relevant examples are the Seal Basin, Bluet Multiuse Platform, and the Allas Sea Pool.

5.1 *Seal Basin*

The Seal Basin was conceived for Korkeasaari Zoo in Helsinki, Finland for amphibious animals such as seals. The complicated structure contains three separate pools (main, feeding, and nesting), natural landscaping, a service area, and a viewing platform. Equipment is concealed below deck.

It was originally designed for a single endangered Saimaa ringed seal and consists of a 20 × 20 m pool which is 3.2 m deep. However, zoos and wildlife rehabilitation centers often have changing space needs for animals. Therefore, the Seal Basin was made to be modular so it could be expanded if they needed more space for more animals.

The area where the zoo wanted the floating exhibit was in a small inlet connected to the sea with a depth of 4.0 to 4.5 m. Installation was challenging due to the location and limited space for heavy equipment. In this location, the waterways were not suitable for assembly elsewhere and towing the structure to the zoo. Instead, the modules had to be delivered by truck and assembled over the water. Preassembly was completed to verify the accuracy of the connecting bolts and holes.

5.2 *Bluet Multiuse Platform*

The Bluet Multiuse Platform was designed to be versatile, for a variety of temporary or permanent uses. Each platform is 6.6 × 3.3 × 0.75 m and can be connected end-to-end, side-to-side, or end-to-side. Common platform sizes are small with three modules, medium with eight modules, and large with ten modules. A rigid platform can be built to about 20 × 20 m; anything larger normally needs movement joints to allow some flexing.

Each steel pontoon module weighs about 3200 kg and has a max point load of 2000 kg/m^2 and continuous load of 8000 kg. The modules contain a watertight compartment that can be filled with water or sand to help balance the final platform. Accessories include anchoring systems, superstructure fixing implements, side-to-side connections, and lifting fittings.

The platform was designed to be suitable for temporary solutions, such as short summer festivals. The modules can be quickly assembled and disassembled in one or two days without special tools.

5.3 *Allas Sea Pool*

The Allas Sea Pool in downtown Helsinki is part of an urban spa and event center. The floating structure consists of three swimming pools and a 2700 square meter sundeck. It was opened in 2016 and new floating saunas were added in 2022.

While the simpler Bluet Multiuse Platform uses steel structures, the Allas Sea Pool uses steel, concrete, and plastic modules. Steel was found to be best for complicated modules which need space under the deck, but maintenance can be challenging. Concrete has a long lifetime and low maintenance. For example, the technology in steel modules might need maintenance or replacing in 20–30 years, but a concrete module could last 100 years, far outliving typical permits for long-term floating structures.

Installation of the Allas Sea Pool was challenging, as conventional tools such as laser levelers were useless with the flexing platform. It was shaped like an inverted U before weight was put on its center. It was critical to take these changing loads into consideration in the design and planning stages of the project.

6 Modularity and Floating Megastructures

An emerging trend is floating megastructures, sometimes called Very Large Floating Structures (VLFSs). These floating solutions can be more environmentally friendly alternatives than the traditional land reclamation method, which is an important concern for development projects. A variety of VLFSs have been built, such as the Mega-Float runway in Tokyo and the 2350-m-long Evergreen Point Floating Bridge in Seattle. Other proposed VLFSs are for civil engineering, industrial, military, residential, and even agricultural purposes [6].

A VLFS is usually constructed in modules. Sometimes the modules are welded together at sea or joined with rigid connectors. More recently, research has suggested hinged or semi-rigid connectors be used instead because they better reduce hydroelastic response. Some of the most vibrant research on modularity and VLFSs is occurring in East Asia universities [7]. The idea of using flexible connectors for larger platforms matches Bluet experience [8].

7 Conclusion

Modularity brings a variety of economic, safety, and environmental benefits to all types of floating structures. These include cost savings, improved quality control, more reliability, and faster deployment. Environmental benefits are less disruption to the ecosystem and less waste. Modularity also improves safety, as it makes assembly less dangerous for workers and minimizes the use of divers. Finally, modularity brings flexibility and scalability.

References

1. Bluet. https://bluet.fi/en. Accessed 05 July 2024
2. Wang CM, Tay ZY (2011) Very large floating structures: applications, research and development. In: Lam HF (ed) The twelfth east Asia-pacific conference on structural engineering and construction (EASEC12). Elsevier, Netherland, pp 62–72
3. Sirola T, interviews
4. Bluet: VTT Tank Test Summary, Bluet (2022)
5. Saukko D, interviews
6. Lamas-Pardo M et al (2015) A review of very large floating structures (VLFS) for coastal and offshore uses. Ocean Eng 109:677–690
7. Zhang HC et al (2015) Nonlinear network modeling of multi-module floating structures with arbitrary flexible connections. J Fluids Struct 59:270–284
8. Happonen K, Ville V, Martio J (2022) VTT Como lake floating pool—assessment of motions in waves

Studies on Integrating a Carbon Capture System with Hydrogen Fuel Production in FLNG Power Generation

We Lin Chan, Arun Dev, and Ivan C. K. Tam

Abstract The burning of hydrocarbon fuels releases greenhouse gases (GHGs) that are linked to climate change and pose a health risk to people. To offset the marine industry's transition to renewable energy fuel, reduced emissions are going to be required (IMO, "Greenhouse Gas Emissions. Retrieved from IMO", 2018). LNG fossil gas was seen as the most environmentally friendly option due to its lower carbon emissions compared to other fuels. The increasing cost and demand for LNG have facilitated the exploration of inland gas deposits (GIIGNL, "RETAIL LNG HANDBOOK", 2015). However, most gas reservoirs are situated in offshore areas, which encourages the growing number of FLNG, which stands for Floating Liquefied Natural Gas, widely recognised and utilized in both nearshore and offshore. The initial phase of the clean energy trial initiative for FLNG involves meeting the power generation needs aboard through the utilization of clean energy sources. Adopting a clean energy mindset is crucial for maximizing the sustainable use of hydrocarbon fuel. The main reason for choosing hydrogens is preferred over other energy sources is because it has a better energy-to-mass ratio (H/C ratio). The preferred method for hydrogen production is the methane reforming process, which is highly efficient in generating large quantities of hydrogen. The reforming method is determined by the choice of either steam or oxygen as the reactant for natural gas. This choice leads to three possible processes: steam methane reforming (SMR), partial oxidation (POX), or autothermal reforming (ATR) (Alicia Boyano in Steam Methane Reforming System for Hydrogen Production, 2012). Moreover, restricted space availability which design SMR or ATR will be implemented to enhance the small-scale plant with CCS as an onboard FLNG. In conjunction with CCS (Carbon Capturing System), research on the integration of extra Clean Conversion modules to produce green hydrogen fuel is presented in this study. The goal is to help provide clean fuel for FLNG's power generation, which has been engineered for offshore conditions and emits no carbon.

Keywords Carbon capture system (CCS) · Hydrogen · Reforming; FLNG

W. L. Chan (✉) · A. Dev · I. C. K. Tam
Newcastle University in Singapore, Newcastle University, Newcastle Upon Tyne, UK
e-mail: W.L.Chan3@newcastle.ac.uk

X. L. Zhao et al. (eds.), *Proceedings of the Fourth World Conference on Floating Solutions: WCFS 2024; 2–4 December; Hong Kong*, Lecture Notes in Civil Engineering 597, https://doi.org/10.1007/978-981-96-4569-5_44

1 Introduction

1.1 FLNG Power Generation with Green Fuel

To satisfy the FLNG process energy demand, the FLNG completely depends on hydrocarbon fuels. This leads to growing carbon emissions. To cope with the trend of increasing carbon emissions, it is even more important to use fossil hydrocarbon fuels sustainably. Research idea of using NG/LNG fuel which has a higher hydrogen/carbon ratio than the other fossil fuels to produce hydrogen onboard FLNG. Using hydrogen as a fuel onboard FLNG can improve power generation efficiency and lower carbon emissions. This is positive for the environment because of growing concern of carbon emissions. Hydrogen fuel is likely to be favored in the future. Additionally, hydrogen is a very environmentally friendly energy carrier. There are numerous ways to produce hydrogen, and it has potential applications across all end-use sectors [4].

Hydrogen as Marine Fuel. Hydrogen (H_2) is well-known for its adaptability as a clean energy source and for its potential to completely transform as fuel energy for electricity production. Now, fossil fuels account for 95% of H2 production due to its affordability and technological maturity, with the SMR process is a fossil-based H_2 production technique that is additionally cost-effective. As a result, it accounts for 75% of global H_2 production. The lower heating value per mass of this substance is significantly higher than that of methane, gasoline, and coal, with values that are 2.4, 2.8, and 4 times greater, respectively. The combustion of hydrogen does not result in the production of CO_2 or, in the case of low temperature fuel cells, nitrous oxide, sulphur dioxide, or carbon monoxide [5, 6].

2 Methodology

2.1 Reforming Method

The most significant approach for producing hydrogen at a high level is through the process of methane reforming. The reforming process takes place when natural gas undergoes a reaction with either steam or oxygen. This chemical process results in the production of hydrogen gas. The selection of reaction types dictates the appropriate process model to apply. There are several choices available to enhance research. The methods used for the above reason are Steam Methane Reforming (SMR), Partial Oxidation (POX), and Autothermal reforming (ATR). The distinction between the SMR and ATR processes lies in the method by which heat is introduced into the system boundary.

The SMR reforming process is a reaction that produces heat. The system relies on external combustion heat for its operation (exothermic reaction). The process of ART reforming is a reaction that absorbs heat. The system boundary necessitates the presence of internal combustion heat (endothermic reaction). The research analysis will provide a concise explanation of the advantageous methodologies for hydrogen production with different carbon capture systems through the utilization of natural gas. Moreover, the main challenges for FLNG are the constraints of space and economic factors. To effectively minimize carbon emissions, it is imperative to choose either the SMR or ATR design for the small-scale plant on board FLNG. The research scope was defined as the Clean Conversion module. It is employed to produce Hydrogen for the purpose of generating power in FLNG [7, 8].

2.2 Steam Methane Reforming Method

Steam methane reforming (SMR) method of hydrogen production from fossil fuel is one of the most exciting processes in the marine industry. SMR is used in around 40% of the world's hydrogen production It includes various oxygen transfer materials or oxidants. This reforming technique is a catalytic process with a catalyst-packed reactor. Methane is reacted with water vapor at temperatures of 700–850 °C. This is the primary process of steam reforming [3].

The reaction product equation of Reform 1 and Meth 1 is syngas, which is a mixture of hydrogen and carbon monoxide. Syngas contains approximately 12% carbon monoxide. This carbon monoxide can be converted to more hydrogen by the Water–Gas Shift (WGS) as shown in equation Shift 1 as Eq. (3). In this step, carbon monoxide reacts with water vapour over a shift catalyst to produce. Once more, WGS improves hydrogen yields.

In summary, the SMR process is endothermic in thermodynamic terms. It requires or absorbs thermal energy from its surroundings, usually in the form of heat. The energy input is the limiting factor to maintain the required reaction temperature in the SMR process [8].

$$\text{Reform 1} \quad CH_4 + H_2O = CO + 3H_2 \text{(Conversion)} \tag{1}$$

$$\text{Meth 1} \quad CH_4 + CO_2 = 2CO + 2H_2 \text{(Conversion)} \tag{2}$$

$$\text{Shift 1} \quad CO + H_2O = CO_2 + H_2 \text{(Equilibrium)} \tag{3}$$

All cases contain High Temperature Shift (HTS) and Low Temperature Shift (LTS) processes. Using both processes increased the hydrogen yield. This ensures a higher hydrogen concentration in the gas. More carbon monoxide reacted with steam, creating hydrogen. Carbon monoxide is very harmful to the atmosphere. Carbon monoxide reacts with oxygen. This process prevents methane gas from being emitted directly into the atmosphere. It also reduces the risk of carbon emissions. However, both HTS and LTS ensure a higher hydrogen concentration in the gas. This may lead to a smaller hydrogen purification unit. It may also help establish a carbon capturing system. After partial combustion of methane, the gas enters the water–gas shift step. In this stage, carbon monoxide reacts with steam to create carbon dioxide and hydrogen. The last step will be hydrogen purification.

2.3 Autothermal Reforming (ATR) Process

The autothermal reforming process is a combination of the Steam Methane Reforming process (SMR). In thermodynamic terms, exothermic is a process or reaction that releases energy from the system to its surroundings, usually in the form of heat. The total reaction is exothermic. Inside the ATR, natural gas reacts with steam to provide syngas, as in the SMR process, additionally, parts of the fuel react with oxygen.

There are the pre-reformer and main reformer, Steam supplies the process before the gas enters the reformer stage, there will be production of syngas occurs in the reformer, where hydrocarbons react with steam and oxygen and create syngas with help from catalysts [8].

The chemical equilibriums in the reformer stage are the following reaction equations in the ATR process.

$$\text{Partial 1} \quad CH_4 + {}^{1}/_{2}\, O_2 = CO + 2H_2 (\text{Equilibrium}) \tag{4}$$

$$\text{Reform 1} \quad CH_4 + H_2O = CO + 3H_2 (\text{Equilibrium}) \tag{5}$$

$$\text{Shift 1} \quad CO + H_2O = CO_2 + H_2 (\text{Equilibrium}) \tag{6}$$

3 Theory and Equations

3.1 Steam to Carbon (S/C) Ratio

Steam to carbon S/C ratio is usually used to analyse the steam reforming in both SMR and ATR processes to optimise the system. The S/C ratio is based on the ratio between the mole of steam and the carbon at the inlet of the system to evaluate the feeding amount of steam to the reformer. If the S/C ratio is lower, will produce the coke slug formation, the feeding of the steam is important to avoid the thermal cracking of hydrocarbon. If the S/C ratio increases, the temperature of the reform outlet will be lower, indirectly, reformer consumption energy will be lower. The design steam to carbon is typically 2.5 with a range between 1.0 and 5.0. To prevent the carbon deposit and coke deposit formation, the ratio is to be kept equal or above 2.5, or higher [5, 9].

3.2 Other Thermodynamic Variables

Product Yields. The chemical equilibrium analysis of the methane reforming reaction is analysed to predict the product yields during the reaction. The chemical equilibrium of reactions can be calculated by the CEA Program which calculates the product concentrations from any set of reactants based on the minimization of Gibbs free energy and the equation as below [10, 11],

$$i, yield = \frac{n_{i,,Out}}{n_{j,in}} \tag{7}$$

where i is the product and j is the feed gas, n is the total moles of species

The product yields of carbon monoxide, carbon dioxide and hydrogen as per the equation

$$\mathrm{CO_2\ yield} = \frac{\mathrm{n_{CO_2,\ Out}}}{\mathrm{n_{CH_4,\ in}}} \tag{8}$$

4 Simulation and Model Analysis

Aspen HYSYS simulation software is utilized to run simulations of the hydrogen manufacturing and CCS process. These include the Steam Methane Reforming (SMR) and the Autothermal Reforming (ATR). When performing process modeling, create the LNG systems as steady-state models. Use the "Peng-Robinson" fluid package. This will facilitate the evaluation and improvement of the underlying design concepts of LNG systems. This study focused mostly on optimizing and modeling the Steam Methane Reforming (SMR) and Autothermal Reforming Process (ATR) with the CCS system. The specified flow rate of natural gas is 214.2 kgmole/h for ratio 30% H_2 fuel usage for 50 MW gas turbine. The pressure decrease in the heat exchanger is expected to be 10 Kpa in the cooler and 35 Kpa in the reactor. The SMR and ATR process is set up with an air-to-fuel ratio of 2.5. The temperature of the reforming product depends on the amount of oxygen injected. The simulated processes will prioritize the production of hydrogen entirely. For example, in specific models, the boundary will align with the discharge of the LT_WGS reactor. There will be a specific level of carbon dioxide (CO_2) emissions, and the primary emphasis of this study on the Carbon Capture System [7, 11–13].

4.1 Model Development

The primary determining factors in the reforming process are the steam-to-carbon ratio (S/C ratio) and the oxygen-to-carbon ratio. In the SMR and ATR process, an excessive intake of oxygen will lead to a decrease in hydrogen production and an increase in temperature. To regulate the temperature ranges of the feed gas stream, a pre-reformer, four reactors, a heater, Amine and a chiller will be utilized. The research report will primarily focus on two models: ATR, and SMR Amine process.

During the simulation, the various stages of heating and cooling are carried out to achieve specific temperature ranges. These stages include preheating the feed natural gas from a temperature more than 400 °C to a temperature greater than 700 °C, the water gas shift process from a temperature greater than 900 °C to a temperature greater than 300 °C, and the conversion process at about a temperature greater than 250 °C. These processes result in the production of the desired product [6].

4.2 SMR Process with MEA Amine CCS

This study applies the control methodology to examine two significant processes and determine the optimal controlled variables for maximizing hydrogen production and minimizing carbon emissions. The techniques under analysis involve the transformation of natural gas into hydrogen through the SMR process, followed by the capture of CO_2 after undergoing treatment in both HT WGS (High-Temperature Water Gas Shift) and LT WGS (Low-Temperature Water Gas Shift) reactors. The capture of CO_2 is an essential component in the generation of green hydrogen from hydrocarbons. Monoethanolamine (MEA) Amine absorption/stripping processes have been used to eliminate CO_2. Figure 3 illustrates a carbon dioxide (CO_2) capture system stationed downstream of the conversion of fossil fuels to hydrogen.

The MEA CCS methods consist of two columns. In the initial column, referred to as the absorber, carbon dioxide (CO_2) is taken in by a solution of low-concentration amine solvent through rapid chemical reactions. The end-product solution, referred to as rich amine, is next sent to the second column, often known as the stripper, where the CO_2 component is eliminated. Stripping necessitates a substantial quantity of energy, which is supplied through the transfer of heat from a recuperative heat exchanger via the cross flow of rich amine and lean amine (see Fig. 1). This energy conservation represents approximately 15–30% of the whole electricity produced by the facility. The power challenge is essential for efficiently operating this process on board the FLNG to maximize energy savings in a cost-effective manner. Therefore, the primary goal of this thesis is to create a simple control system that can efficiently handle on-board operations and operate the process as close to the optimal level as possible. The study concentrated on the dynamic modeling of the stripper and examined two separate configurations to enhance energy efficiency on board FLNG with a significant electrical requirement [14].

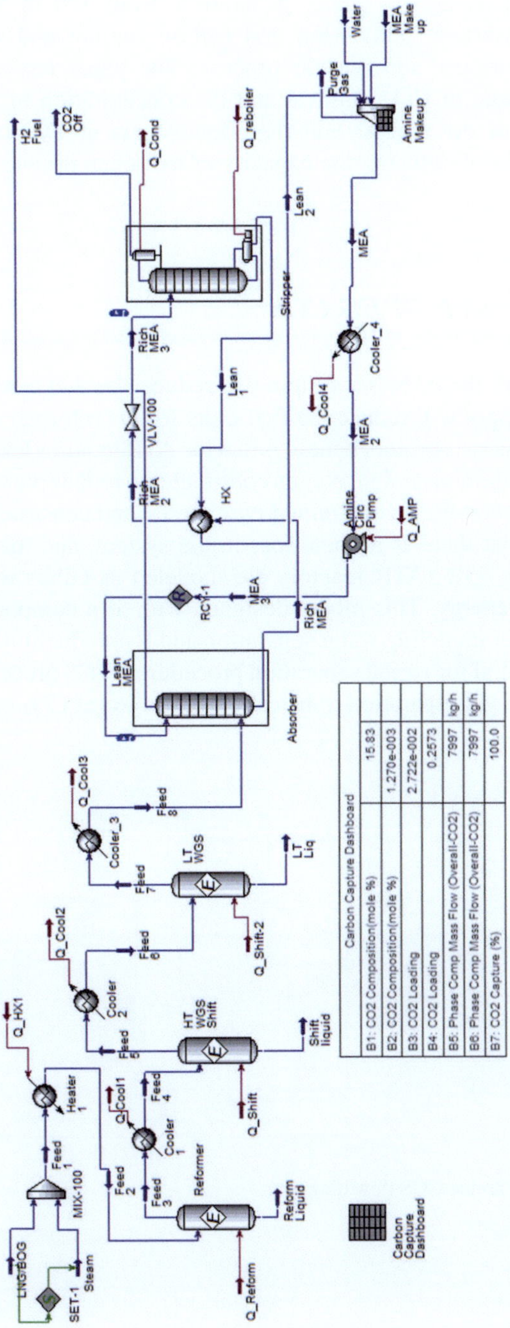

Fig. 1 SMR process with amine CCS flow diagram

The authors developed a dynamic model specifically for the absorber and analyzed the effects of a load change in power generation, from 100 to 50%. An integrated system for hydrogen production and carbon capture and storage (CCS) that incorporates a stripper and recycle process. The paper has examined how changing the percentage of CO_2 removal and the concentration of amine (MEA) affect the efficiency of the process and the effectiveness of CO_2 capture. Moreover, the consequences of reducing the capacity of hydrogen production have been investigated.

4.3 ATR Process with DPEG CCS

In comparison to SMR, the ATR is a system that reduce size and complexity. There is no external heat supply will reduce the fuel costs for the reformer and result in a decrease in CO_2 emissions. Another benefit is that the ATR requires less water than an SMR, increased methane conversion oxygen contributes to reduce the soot formation. Nonetheless, the temperature is constrained by material and construction expenses. It is also presumed that there is no heat loss in the system, and the compressor's adiabatic efficiency is 75%. ATR reactors are modeled as Gibbs reactors, which minimize Gibbs free energy. This model designed with as a component splitter is furthermore utilized for modeling water separation and H_2 production Refinement to demonstrate the potential for varied separation procedures. CCS process used DEPG and used separators to get separation to be easy step out to get CO_2 off (see Fig. 2).

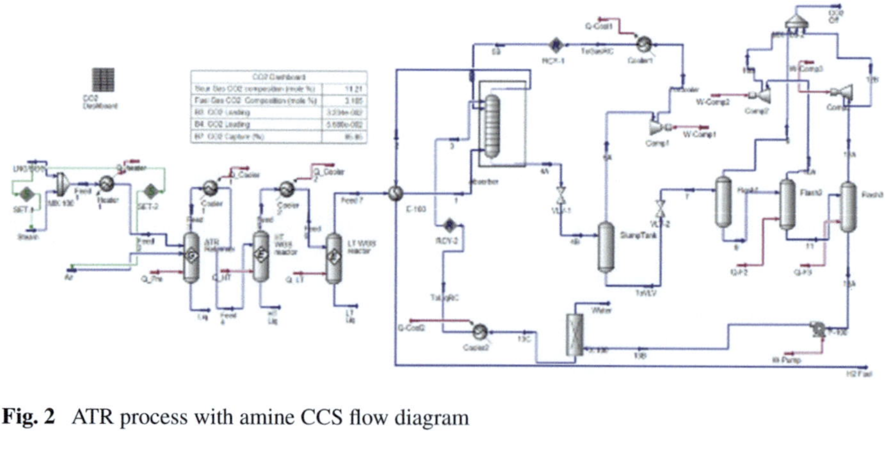

Fig. 2 ATR process with amine CCS flow diagram

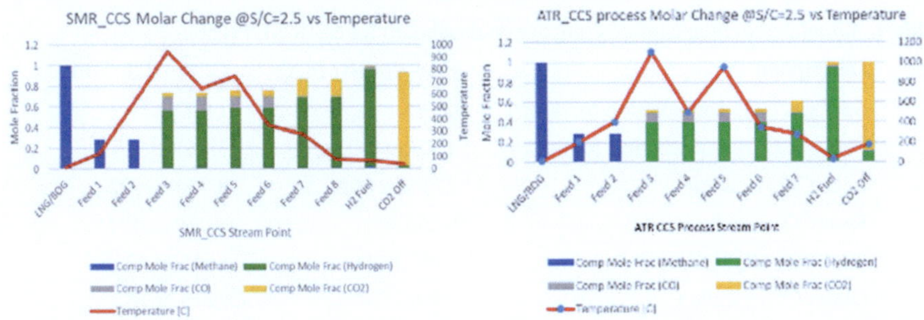

Fig. 3 Molar change of S/C = 2.5 veruss temperature for SMR_CCS and ATR_CCS process

5 Result Discussion and Conclusion

5.1 Simulation Result Analysis

Molar Change of Components vs Temperature. The effect of temperature on reforming leads to enhancements in the process due to changes in the mole fraction of the components. As a result, based on the S/C ratio of 2.5, temperature change at the process stream point and mole change of mixing components, are plotted as per graph, (see in Fig. 3), Molar changes in terms of temperature variation are presented for all process models. The optimal operating temperature is to be evaluated to maximise based on the generating of hydrogen.

The results are dependent on the thermodynamic analysis to compare the SMR and ATR Amine processes, the molar composition change for hydrogen across the outlet stream points are monitored with graph (see in Fig. 3). Both systems are well performing except to consider for floating FPSO space limit that ATR is more considerable than SMR. Furthermore, simplifications and assumptions made throughout the simulation process in the way of four reforming processes will be compared under optimum conditions for each process.

This should all be incorporated with the overall evaluation based on the carbon capture system analysis. The main process design to see the H_2 and CO_2 molar flow rate variable directly as carbon capture dashboard which shown the sour gas and fuel gas mole percentage composition, lean and rich amine loading, inlet and outlet CO_2 loading, CO_2 capture percentage, Net GHG emission and carbon fee as Table 1. The most important output of both simulation is the CO_2 capture and carbon fee which affects the overall system performance across all processes. To get an efficient design, have to evaluate the temperature which influences the mole fraction and exergy of the reformer, and also separation process in absorber and stripper. Related to the flow is minimise to meet the hydrogen fuel requirement for the power generation turbine only, in this approach throughout the system will determine to get a smaller and efficient design for FLNG to meet the concern of economy and unit sizing, to suit the FLNG's space constraint. The review of the various analyses of hydrogen production

Table 1 Simulated carbon capture comparison dashboard for SMR SE Process

Sr	Description	CCS_SMR(MEA)	CCS_ATR (DEPG)
1	Sour gas CO_2 composition (mole %)	15.83	11.21
2	Fuel gas CO_2 composition (mole %)	1.270e−003	3.161
3	Lean CO_2 Loading	2.722e−002	3.234e−002
4	Rich CO_2 Loading	0.2573	5.479e−002
5	Inlet CO_2 (Kg/hr)	7997.1472	8441.4143
6	Outlet CO_2 (Kg/hr)	7996.7551	7216.6676
7	CO_2 capture (%)	100	85.49
8	Heat utility emissions (Kg/hr)		1208
9	Power utility emissions (Kg/hr)	2.934e+004	424.5
10	Inlet emissions (Kg/hr)	7.216e+004	7.216e+004
11	Outlet emissions (Kg/hr)	1.308e+004	8777
12	Net GHG emission (Kg/hr)	1.004e+006	7.233e+005
13	Carbon fee ($)	5.119e+007	8.680e+008

* According to US SCC regulatory uses an SCC of around $51 per ton

from all simulation models is required to identify the performance characteristics of FLNG applications which include energy and exergy efficiency, space constraints, and carbon emissions. The result conclusion is based on the model of the process both in same the environment with all conditions and assumptions. Further extending the investigation to incorporate more variables can bring one closer to identifying an optimization.

Regarding CO_2 capture, SMR with MEA Amine are the best alternatives. SMR Amine is the most efficient process for producing hydrogen based on hydrogen yields, followed by ATR with DEPG amine. As per the simulation result, and the comparison graph (see Fig. 3), SMR is slightly efficient, however ATR is one feasible solution for FLNG applications in terms of further research to reduce the equipment. The ATR process is the combination of SMR, which requires to control of both oxygen and steam. The oxygen-to-fuel ratio drives carbon emission, hence, oxygen to fuel ratio used in the simulation will affect to reduce the carbon emission. The reformer temperature is substantially greater in the ATR reactor than in the SMR reformer because of the absence of steam and soot or coke formation which can be controlled by the scrubber after the reformer outlet [2].

6 Conclusion

The key objective of this study is to explore the green energy potential of hydrogen fuel as a substitute for conventional fuels through a reforming process, addressing the energy requirements of floating platforms. The comprehensive examination of the variance in molar change, efficiency, and carbon capture ratio will determine which system is preferable. The ATR system is better in terms of carbon fee savings and equipment size reduction; however, the optimal choice is SMR MEA Amine, as determined by the molar change analysis conducted for hydrogen production at each stage. To obtain acceptable study findings, it is essential to understand all assumptions and simplifications applied in the interpretation of the data. It is recommended to widen the system boundaries to provide a more detailed assessment. Future system optimization studies may involve reconfiguring the reformer with an enhanced stream subsequent to the reactor, so increasing the volumetric flow rate within the reforming reactor; nevertheless, the dimensions of the reformer must be evaluated concerning the space and cost constraints of the FLNG. The expense of hydrogen production will be affected by numerous energy-related variables. This indicates that, following assessment, the SMR and ATR methods for producing hydrogen are more efficient for FLNG.

References

1. IMO, "Greenhouse Gas Emissions. Retrieved from IMO" (2018). [Online]. Available: http://www.imo.org/en/OurWork/Environment/PollutionPrevention/AirPollution/Pages/GHG-Emissions.aspx
2. GIIGNL, "RETAIL LNG HANDBOOK" (2015). [Online]. Available: https://giignl.org/sites/default/files/PUBLIC_AREA/Publications/giignl_retail_lng_handbook_0.pdf
3. Alicia Boyano AMBMTMGT (2012) Steam methane reforming system for hydrogen production. [Online]. Available: https://dergipark.org.tr/en/download/article-file/65813
4. Angus Campbell A (2016) Interviewee, LNG fuel is not a solution for reducing GHGS (Marine & Industrial_Magazine Report). [Interview]
5. Gupta R (2008) Hydrogen fuel: production, transport, and storage. CRC Press, Boca Raton
6. Europe, "Hydrogen Production. Retrieved from Hydrogen Europe" [Online]. Available: https://hydrogeneurope.eu/index.php/hydrogen-production-0
7. A. T. HYSYS®2004.2; Operations Guide, @aspentech (2004)
8. Antonio Vita LP (2018) Steam reforming, partial oxidation, and autothermal reforming of ethanol for hydrogen production in conventional reactors. Science Direct, Hydrogen Production. [Online]. Available: https://www.sciencedirect.com/topics/chem
9. Gillespie M (2018) Hyper for hydrogen: our world first for carbon-free fuel. Retrieved from CSIROscope. [Online]. Available: https://blog.csiro.au/hyper-for-hydrogen-our-world-first-carbon-free-fuel/
10. N. Glenn Research Center, Chemical Equilibrium with Applications, [Online]. Available: https://www1.grc.nasa.gov/research-and-engineering/ceaweb/
11. Tam ICK AB (2020) Thermal systems—an overview. Energies 14(1):175
12. Silberova B (2005) Production of hydrogen by short contact time partial oxidation and oxidative steam reforming of propane. J Catalysis 99(1–2):69–76
13. Chan WL (2023) Clean energy hypothesis for future marine and offshore. Research Gate 378684136
14. Liu CSAVSK (2010) Hydrogen and syngas production and purification. Wiley, USA

Comparison Between the Practical Floating PV Generation and Simulation Models

Haojie Luo, Suijie Liu, Zhenyu Dou, Qi Wu, Sunliang Cao, and Hongxing Yang

Abstract With the limited availability of land resources in urban areas and the increasing energy demands, there is a growing interest in exploring alternative renewable energy solutions in coastal regions. This study focused on the potential of utilising nearshore space by designing a micro-floating PV panel prototype and evaluating its performance under actual ocean conditions in Hong Kong. The primary objectives were to monitor photovoltaic generation and investigate the characteristics of the generation data while also assessing the environmental impact on the performance. For deck-on PVs, PV temperature significantly affects power generation. On clear days, peak PV temperature at noon can reduce generation by 5.8% despite a 39.9% increase in solar radiation. Ambient temperature and wind speed impact PV temperature, with a 6.8 °C increase reducing generation by 35% even with a 31.3% increase in radiation. The enclosed floater temperature was heated heavily due to the strong greenhouse effect under intense solar radiation. The open-circuit voltage showed a general correlation with PV generation. The simulation-experiment comparison identifies key factors impacting PV performance. Solar radiation and shade effects bring experimental uncertainty. Surroundings influence PV temperature, with barge radiative heating and floater conductive heating causing differences between experiments and simulations. Barge radiative heating is more significant, resulting in a 16.5% PV generation difference during the afternoon on a clear sky for deck-on PVs.

H. Luo · S. Liu · Z. Dou · Q. Wu · S. Cao (✉) · H. Yang
Renewable Energy Research Group (RERG), Department of Building Environment and Energy Engineering, Faculty of Construction and Environment, The Hong Kong Polytechnic University, Kowloon, Hong Kong SAR, China
e-mail: sunliang.cao@polyu.edu.hk

S. Cao
Research Institute for Smart Energy (RISE), The Hong Kong Polytechnic University, Kowloon, Hong Kong SAR, China

S. Cao · H. Yang
Research Institute for Sustainable Urban Development (RISUD), The Hong Kong Polytechnic University, Kowloon, Hong Kong SAR, China

X. L. Zhao et al. (eds.), *Proceedings of the Fourth World Conference on Floating Solutions: WCFS 2024; 2–4 December; Hong Kong*, Lecture Notes in Civil Engineering 597, https://doi.org/10.1007/978-981-96-4569-5_45

Keywords Floating solar system · Nearshore space · Sustainable energy solution

1 Introduction

1.1 Background

Floating photovoltaic (PV) generation is gaining attention as an alternative to land-based solar installations, especially in land-constrained regions. Deploying floating PV systems on water bodies like reservoirs and lakes offers space efficiency, increased energy generation, and reduced environmental impact. Hong Kong's extensive coastline and islands provide a unique opportunity for floating PV systems [1]. These systems can overcome land constraints, utilizing seawater for solar energy generation. Positioned on the sea surface, they maximize sunlight exposure, achieving higher efficiency than land installations. In Hong Kong's push to reduce its carbon footprint and embrace cleaner energy sources, integrating floating PV systems into the energy mix is crucial. These systems play a vital role in meeting renewable energy goals, cutting greenhouse gas emissions, and enhancing energy security by diversifying electricity sources [2, 3]. Compared to land-based setups, floating PV systems offer environmental benefits by reducing land conversion and preserving ecosystems. They also help maintain the cityscape while bolstering the local renewable energy mix. However, successful implementation demands meticulous planning, design, and assessment. Recent global floating PV projects highlight the technology's potential, with the global market expected to exceed 6 GW by 2031. By then, over 15 countries are projected to surpass 500 MW of cumulative installations, with Indonesia, India, and China leading the trend [4]. Real-world projects offer valuable insights into performance, efficiency, and challenges, while simulation models aid in assessing feasibility and predicting system performance under various conditions [5, 6].

1.2 Objective

This study aims to compare practical floating PV generation with simulation models to assess their effectiveness and identify any discrepancies. By analysing real-world experiments and simulation results, the research aims to understand floating PV's performance and simulation model accuracy. The study will use data from existing floating PV installations, focusing on deck-on and micro floating PV systems in seawater. It will evaluate energy generation and efficiency under diverse weather conditions to identify key factors influencing system effectiveness and assess real-world potential. Furthermore, the study will review and analyse simulation models used to predict the performance of floating PV systems. By comparing simulation outcomes with practical data, it aims to evaluate model accuracy and reliability, identifying any discrepancies for improvement. The research will explore factors influencing performance differences, such as environmental conditions, system design, and energy losses. Understanding these elements will refine models for more accurate future predictions.

2 Outline of the Project

2.1 Deck-on Photovoltaic System

As shown in Fig. 1, the deck-on PV system is set on a boat staying at the To Kwa Wan Typhoon Shelter. The system consists of 5 PV panels with 450 W capacity each. All these panels are connected to a maximum power point tracking (MPPT) and a charge controller. A set of battery packs is connected to the PV and an inverter through the controller to utilize the energy for supporting the external load. A remote data uploader is installed to collect and monitor the system data including voltage, current and power of PV panels and inverter, together with the back surface temperature of PV panels. The time-resolution of the online database is two minutes.

To record the actual maximum power generation of the system. Three LED lights of 0.06 kW are installed to consume extra energy from the battery packs. The impacts of the back surface temperature of PV panels influenced by ambient temperature and wind speed on energy generation under different weather conditions (clear sky,

Fig. 1 The location of deck-on photovoltaic system

scattered cloudy sky and overcast sky) are studied. The comparisons of the energy generation data of PV panels obtained through testing and simulation under different weather conditions are also conducted.

2.2 Micro Floating PV System

The micro floating PV system has two test models, as shown in Fig. 2. The first is a closed-circuit model, where one or two floating PV panels charge a portable battery (Battery 1), with Battery 2 acting as a load connected via a DC cable. Battery 1's input power indicates PV generation, and both charging and discharging powers are monitored every 15 min, along with FSOC, floater angle, and various temperatures. The second model is open-circuit, where each PV panel is connected to a sensor recording voltage every minute. All panels, each with a 40 W capacity, are placed on the sea surface in Tsing Yi harbour, Hong Kong (Figs. 3 and 4).

Using the recorded data, the relationship between PV panel temperature, floater temperature, seawater temperature, floater angle, and PV generation, as well as open-circuit voltage, is examined. Differences between the test data and computer simulations under three weather conditions (clear sky, scattered clouds, and overcast) are analyzed, with results presented in Sect. 3.

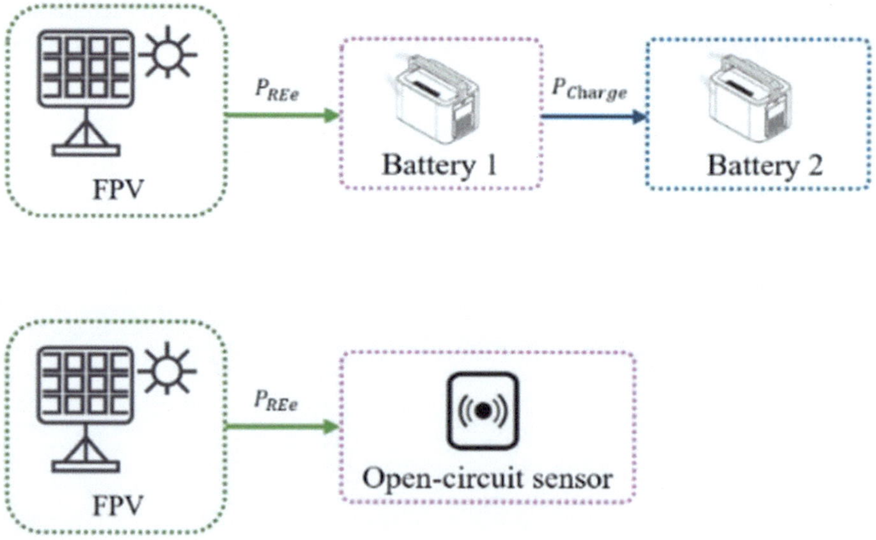

Fig. 2 The two test model of schematic design of floating PV system

Fig. 3 The location of floating PV panel system

Fig. 4 Site scene photo

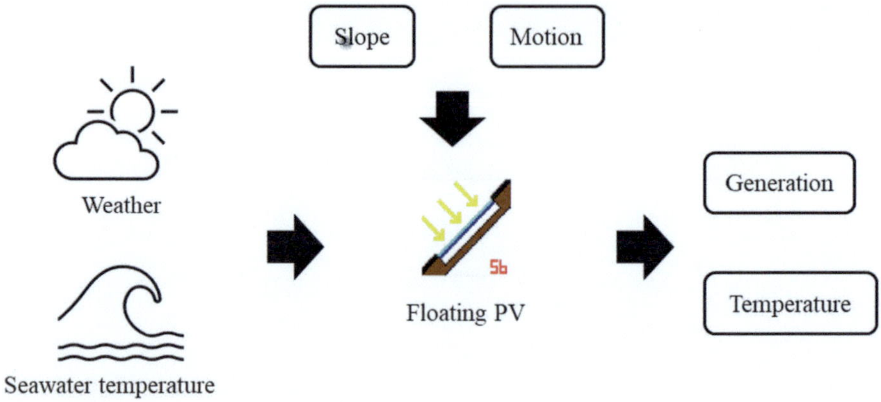

Fig. 5 Schematic of the simulation model

2.3 Simulated Photovoltaic System

Simulations of the floating PV systems are conducted to compare the simulation model result and real-world PV generation for uncertainty characterisation. The simulation is performed in TRNSYS 18, a commercial tool to simulate the behaviour of transient systems. Type 567 is chosen to simulate the floating PV dynamic generation. Type 99 is adopted to provide the weather information. The weather information is collected from the Hong Kong Observatory database. Weather information includes the wind velocity, wind direction, solar irradiance, and ambient temperature. The seawater temperature is collected from the experiment data by Infrared imaging. The system diagram is depicted in Fig. 5. Based on the data of weather and seawater temperature, the simulation model exports the generation and temperature. The real experiment is affected by the motion and slope of the floating PV which are not considered in the simulation.

3 Result and Discussion

3.1 Performance of Deck-On Photovoltaic

The analysis is based on the data under three typical weathers. Due to the orientation of the deck and shading effect by other objects, only the data of afternoon is utilized for analysis. 29th May, 2nd June and 30th May are considered as typical clear, scattered cloudy and overcast sky respectively, the data of these days are shown in Figs. 6, 7 and 8 separately. On a clear day, after the temperature of PV panels reaches the peak around 44.7 °C at 13:40, the power generation would decreased by 5.8% to 1430.3 W while the global radiation increased by 39.9% to 1087.0 W/

m^2 at 13:50. This phenomenon can also be noticed in a scattered cloudy day, after the PV temperature reaches the peak of that day around 52.5 °C at 13:20, the power generation of PV panels remains almost the same even when the global radiation increases by 13.0%.

The impact of air temperature and wind speed on the power generation of the deck-on PV system is also investigated. Compared the period from 13:00 to 14:00, the average global radiation in Fig. 7 is 940.7 W/m^2 and is 31.3% higher than that in Fig. 6. However, the corresponding average power generation in Fig. 7 is 35.0% lower than that in Fig. 6. This may because the air temperature and wind speed differences. The corresponding average air temperature of Fig. 7 is 31.7 °C and the average wind

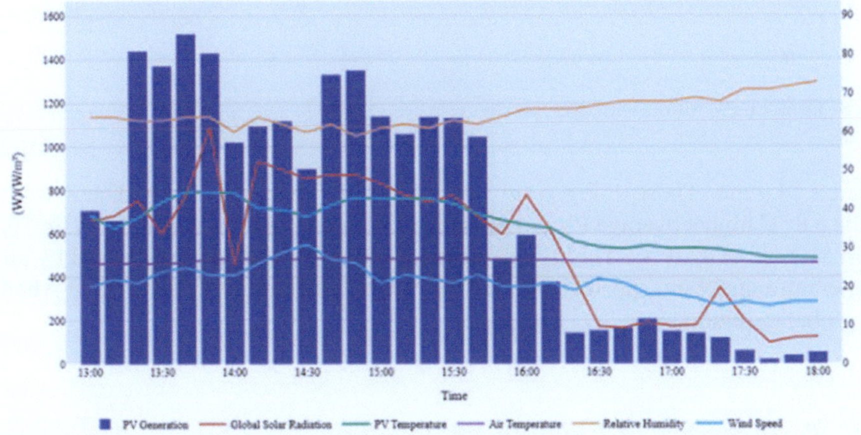

Fig. 6 Deck-on PV results with clear sky

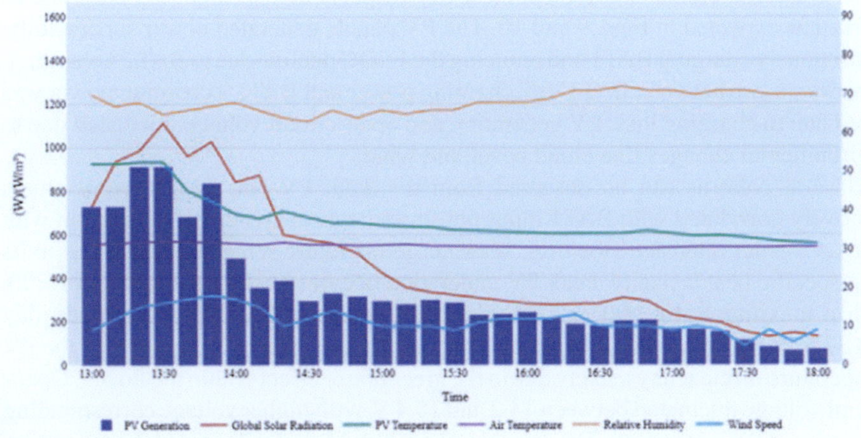

Fig. 7 Deck-on PV results with scattered cloudy sky

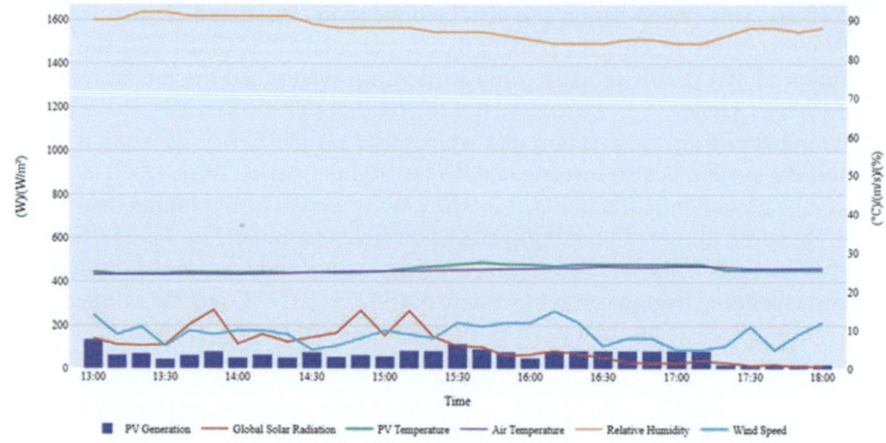

Fig. 8 Deck-on PV results with overcast sky

speed is 14.9 m/s. Those of Fig. 6 are 26.4 °C and 22.6 m/s respectively. These result in the 6.8 °C higher average PV temperature of the noon in Fig. 7 than that in Fig. 6, from 41.1 °C to 47.9 °C. This 16.5% increase of the PV temperature caused by air temperature and wind speed constrains the power generation by 35.0% even when the global radiation has a 31.3% increase.

3.2 Performance of Floating Photovoltaic

The floating PV tests, conducted in Tsing Yi harbour on 8th May and 29th May, under scattered clouds and typical clear sky respectively, showed that the systems worked as expected in Figs. 9 and 10. The PV panels generated power successfully, continuously charging BAT1 and reducing the FSOC decline due to BAT2's charging. However, a gap between BAT1's discharging power and BAT2's charging power was noted due to charging loss. PV generation and open-circuit voltage fluctuated due to environmental changes like cloud cover and wind.

Several patterns can be observed from the data: PV and floater temperatures positively correlated with BAT1 input power, indicating higher PV generation with increased solar intensity. However, seawater temperature remained stable due to its high specific heat capacity. Peak PV generation occurred between 13:00 and 15:00, with a maximum at 14:00 on both test days, reaching 28.3 W on a cloudy day and 37.5 W on a clear day. The floater temperature was generally higher than the PV temperature on clear days, likely due to the greenhouse effect within the floater. Open-circuit voltage fluctuated between 14.2 and 15.4 V, with higher voltage corresponding to increased PV generation. The second test on 29th May also recorded floater angles, showing that waves influenced PV generation, a topic for further research.

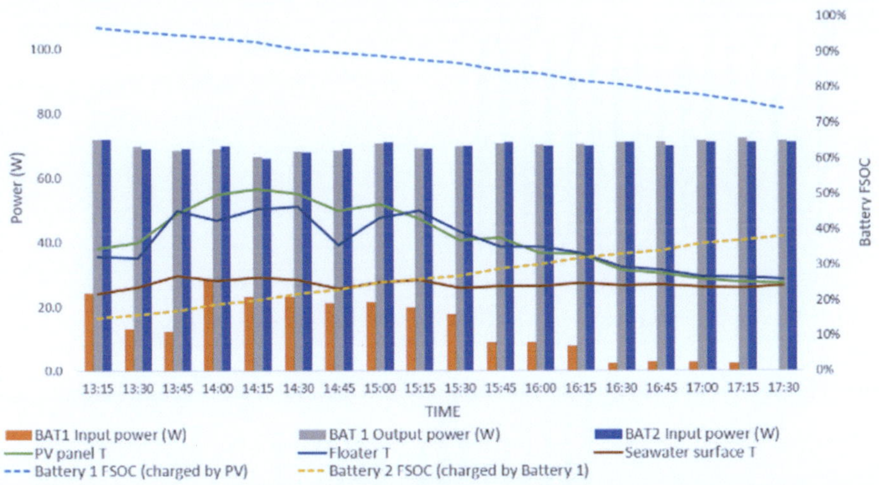

(a)

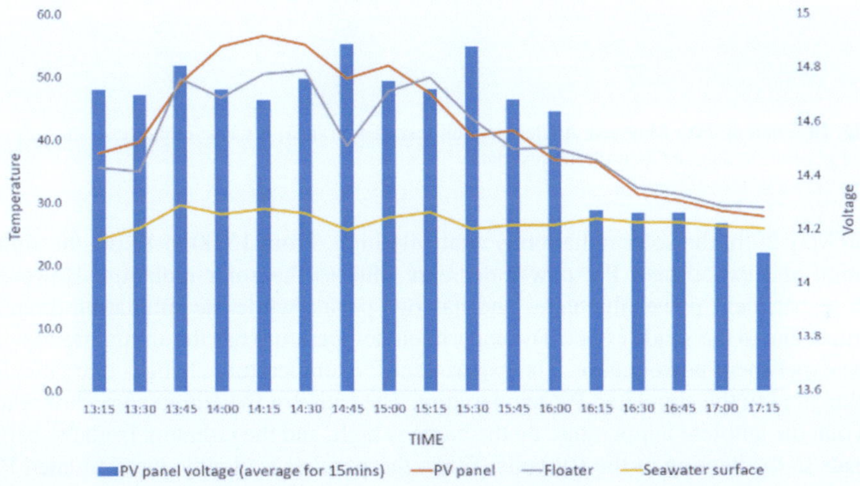

(b)

Fig. 9 Data of 8th May test. **a** Close-circuit test data; **b** open-circuit test data

3.3 Comparison Between the Actual Experiment and Simulation

The experiment and simulation results are investigated in this section. Figure 11 depicts the results for deck-on PVs. The power and temperature of PVs are presented.

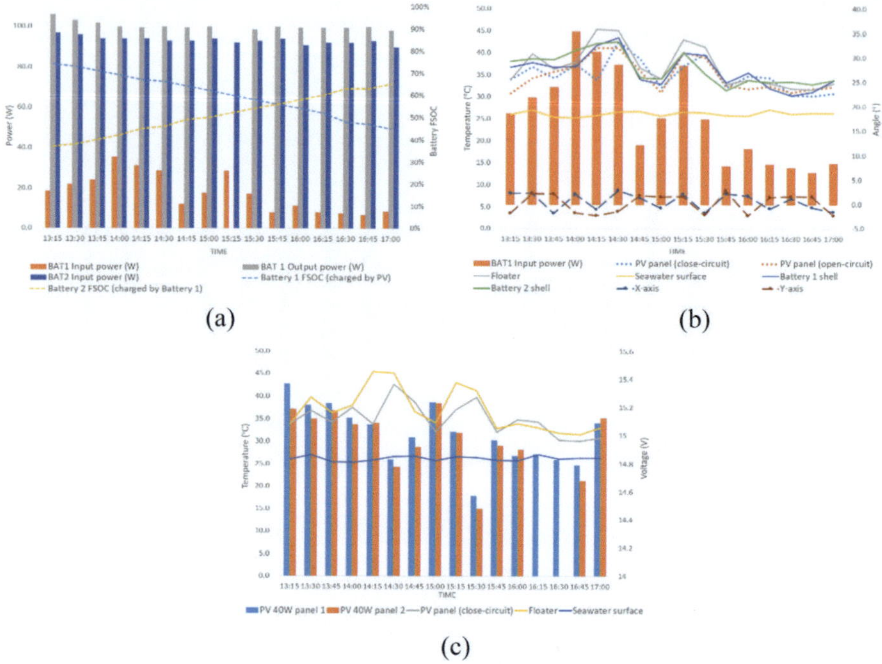

Fig. 10 Data of 29th May test. **a** Close-circuit test system data; **b** Close-circuit test temperature and generation data; **c** Open-circuit test data

On May 29th, the solar radiation is relatively high. From 13:00 to 18:00, the simulation and experiment PV power decrease, due to the solar radiation. However, the experiment power fluctuates and has two peaks, while the simulation directly drops, due to the shade effect. The experiment temperature has the similar trend with the experiment power curve. The experiment PV temperature changes dramatically compared to the simulated PV temperature. The cause of the temperature difference is that the ambient temperature on the barge is high, and the radiation from the barge leads to the heating of the PV cells. When the weather is cloudy, the simulated PV power is higher than the experiment PV power. The cause of this power difference is the shade from the cloud. The experiment PV temperature is higher before 14:00, and then the experiment and simulation PV temperatures are similar. These results indicate the radiative heating of barge is significant when the solar radiation is high.

Figure 12 depicts the simulation and experiment results for floating PV. On May 8th, the solar radiation is high. The simulated and experiment power have similar value after 14:00. The sudden drop of PV power between 13:00 to 14:00 is attributed to the shade. The simulation and experiment temperatures have the difference reaching 20 °C. The potential cause is the floater temperature rise due to solar radiation. The floater temperature reaches over 50 °C and hinders the heat dissipation of PV cells. On May 29th, the weather is cloudy. The experiment PV power is similar to the simulation but fluctuates after 14:00. The cause is the temperature change of

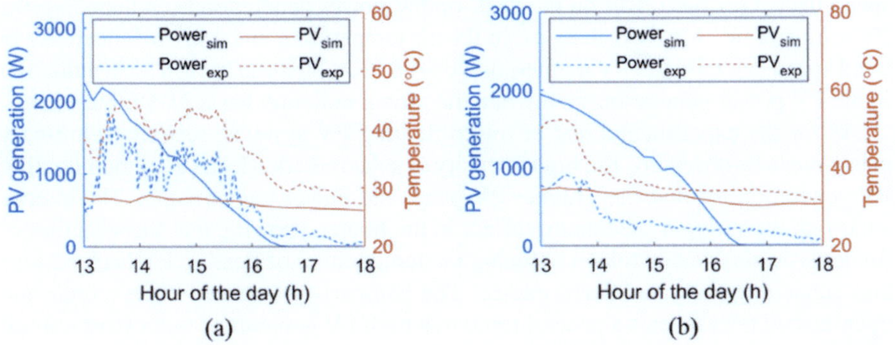

Fig. 11 Comparison of deck-on PV experiment and simulation results on **a** May 29th, and **b** June 2nd

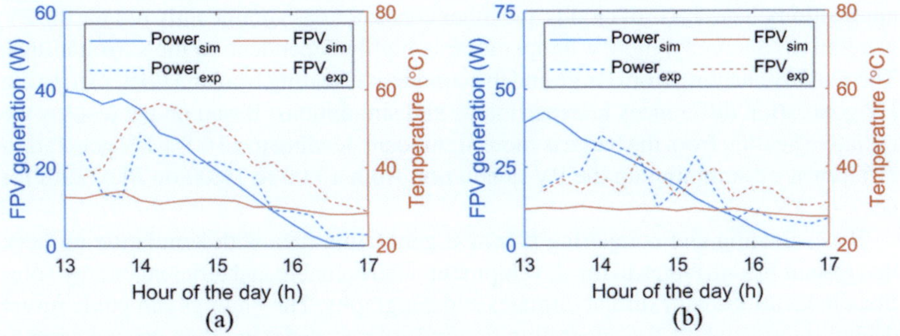

Fig. 12 Comparison of floating PV experiment and simulation results at **a** May 8th, and **b** May 29th

PV floater., which affects the PV cell temperature and the efficiency. The experiment temperature is higher than the simulation, caused by the floater heating.

4 Conclusions

Through a thorough comparison between real-world floating PV generation data and simulation models, this research aims to bridge the existing gap between theoretical understanding and practical application, enabling stakeholders to make informed decisions and maximise the benefits of floating PV technology.

For the power generation of the deck-on PV system, the impact of the PV temperature is significant. Once the PV temperature reaches the peak at noon on a clear day, the power generation of the system may even have a 5.8% reduction when the global solar radiation has a 39.9% increase. Furthermore, the ambient temperature and wind speed have a serious effect on the deck-on PV power generation by influencing the PV temperature. A 5.3 °C increase in the air temperature and a 7.7 m/s decrease in wind speed can raise the PV temperature by 6.8 °C, which can bring a 35% reduction in the PV power generation even when the global radiation has a 31.3% increase.

As for the experimental test of micro floating PV systems, several preliminary patterns can be observed. The high intensity of solar radiation leads to a larger floating PV generation, while brings higher PV panel and floater temperatures. The intense solar radiation trigger greenhouse effect in the floater, showing that the selection of floater type plays a key role in reducing the temperature of floating PV back surface and enhancing the panel performance. The comparison between close-circuit and open-circuit tests reveals a general trend that high PV generation under close-circuit conditions corresponds with high PV voltage under open-circuit test in the same working environment.

Comparison of simulation and experiment results reveal the potential key impact factors affecting the PV performance. The first is the uncertain solar radiation in real applications. The shade of clouds and other constructions significantly reduce the PV generation. Secondly, the PV temperature is highly dependent on the surroundings. The radiative heating from barge and the conductive heating from the floater led to the PV generation differences in experiments and simulations. Based on the results, the radiative heating from the barge is more significant, leading to a 16.5% PV generation difference by summing up the PV generation from 13:00 to 18:00 on May 29th for deck-on PVs.

By evaluating and comparing practical generation data with simulation models, this research contributes to the development of sustainable and efficient energy solutions in locations with similar climates and geography. The findings can guide future design improvements for integrating floating solar systems in urban coastal areas.

References

1. World Bank Group. Where Sun Meets Water: Floating Solar Market Report. https://documents1.worldbank.org/curated/en/579941540407455831/pdf/Floating-Solar-Market-Report-Executive-Summary.pdf. Accessed 01 July 2024
2. Environment Bureau. Hong Kong Climate Action Plan 2030+. https://www.hkgbc.org.hk/eng/engagement/file/ClimateActionPlanEng.pdf. Accessed 01 Aug 2024
3. Environment Bureau (2015) Hong Kong Climate Change Report 2015. https://www.eeb.gov.hk/sites/default/files/pdf/ClimateChangeEng.pdf. Accessed 01 July 2024
4. Global floating solar to top 6GW threshold by 2031. https://www.woodmac.com/press-releases/global-floating-solar-to-top-6gw-threshold-by-2031/. Accessed 01 July 2023
5. Elminshawy NAS, Osama A, El-Damhogi DG, Oterkus E, Mohamed AMI (2021) Simulation and experimental performance analysis of partially floating PV system in windy conditions. Solar Energy 230:1106–1121
6. Zhou S, Cao S, Wang S (2022) Realisation of a coastal zero-emission office building with the support of hybrid ocean thermal, floating photovoltaics, and tidal stream generators. Energy Convers Manag 253:115135

Hydrodynamic Analysis

Impact of Different Tendon Configurations on the Hydrodynamic Performance of a TLP Floating Offshore Wind Turbine

Zhe Tan and Ronghua Zhu

Abstract This study investigates the impact of different tendon configurations on the hydrodynamic performance of a Tension Leg Platform (TLP) floating offshore wind turbine. Utilizing the NREL 5 MW turbine and the MIT TLP design, three new mooring system configurations were developed, including a mooring layout with two tendons arranged in an X shape between adjacent vertical tendons. The configurations were tested under various load cases using OpenFAST. Results indicate that the X layout configurations reduce the surge and yaw motions compared to the original layout under operating conditions. Eight tendon failure scenarios with different failed tendons were also analyzed.

Keywords TLP · Floating offshore wind turbine · Mooring system · Hydrodynamic performance · Tendon failure

1 Introduction

The increasing demand for renewable energy has driven significant interest in offshore wind energy, particularly floating offshore wind turbines (FOWTs). Among various FOWT floating foundation designs, the tension leg platform (TLP) has an advantage of stability, including low motion response in pitch [1]. However, the performance of TLPs is highly dependent on the configuration of their mooring systems. Recent studies have explored various aspects of TLP FOWT design and performance, including hydrodynamic behavior, structural integrity, and mooring system dynamics. Wang et al. [2] proposed a 5 MW typhoon-resistant fully submerged TLP FOWT, which adopts a hybrid mooring scheme with tendons and

Z. Tan · R. Zhu (✉)
Ocean College, Zhejiang University, Zhoushan, China
e-mail: Zhu.richard@zju.edu.cn

Yangjiang Offshore Wind Power Laboratory, Yangjiang, China

X. L. Zhao et al. (eds.), *Proceedings of the Fourth World Conference on Floating Solutions: WCFS 2024; 2–4 December; Hong Kong*, Lecture Notes in Civil Engineering 597, https://doi.org/10.1007/978-981-96-4569-5_46

catenary mooring lines. Ren et al. [3] studied the dynamic characteristics of a tension-leg floating wind turbine under tendon failures through model experiments. Zhang et al. [4] investigated the dynamic response of a TLP floating offshore wind turbine using vertical carbon fiber composite tendons and circumferential catenary mooring lines for combined mooring. Rony et al. [5] studied the coupled dynamic characteristics of a tension leg floating wind turbine-wave energy hybrid platform under different tendon configurations, including four mooring configurations: four-line, five-line, eight-line, and nine-line.

Existing studies on TLP floating wind turbine mooring systems mostly use vertical tendons with different numbers of tendons. In this paper, a mooring layout with two tendons arranged in an X shape between adjacent vertical tendons is tested. The study employs a 5 MW wind turbine model and an MIT TLP design as the basis for analysis. Three mooring configurations are developed, including an original vertical mooring layout and an X layout. The performance of these configurations is evaluated through numerical simulations, and six degrees of freedom platform motions are analyzed. The simulations are under operational conditions, extreme load conditions, and tendon failure scenarios to provide a comprehensive understanding of the mooring system's impact on TLP FOWT performance.

2 Model Description

The wind turbine and the tension leg platform are based on the research of NREL, including a 5 MW turbine and an MIT TLP design [6–9]. The main properties of the wind turbine, TLP platform, and original mooring system are listed in Table 1. The original mooring system consists of 8 tendons, with two tendons per fairlead. Based on the original mooring system, three new mooring system configurations were generated, with the total tension on each fairlead remaining unchanged. The three mooring system configurations have the same total mooring line volume as the original configuration. This setup is intended to ensure that different mooring configurations have the same material cost. Under this assumption, if the cross-sectional area of some mooring lines increases, enhancing their extensional stiffness, the cross-sectional area of other lines in this mooring system will decrease to keep the total volume of the mooring lines constant. This makes different mooring system configurations more comparable.

As shown in the Fig. 1, the three mooring configurations include two layouts. The left layout is the same as the original layout of the MIT TLP, but each fairlead has three vertically downward tendons, increasing the total number of tendons to 12. These three tendons are equivalent to the original two tendons, with the same connection positions, total extensional stiffness, and unstretched length. In the right layout (X layout), each fairlead has one vertically downward tendon numbered from 1 to 4, and additionally, there are two thinner tendons arranged in an X layout between adjacent vertical tendons. This layout is feasible because the solver does not account

Table 1 Main properties of the wind turbine, TLP platform, and original mooring system

Properties	Value
Rated power	5 MW
Cut-in, rated, and cut-out wind speed	3, 11.4, and 25 m/s
Platform diameter	18 m
Platform draft	47.89 m
CM location below still water level	40.61 m
Water displacement	12,180 m^3
Water depth	200 m
Radius to fairleads	27 m
Depth to fairleads	47.89 m
Number of mooring lines	8
Line diameter	0.127 m
Line mass per unit length	116.03 kg/m
Line extensional stiffness	1,500,000,000 N
Upstretched line length	151.73 m

for the interactions between mooring lines. However, in a real-world scenario, the positions of the fairleads need to be adjusted to avoid line collisions and crossings.

The properties of the three mooring configurations can be seen in Table 2. The twelve tendons of MC1 have the same properties. MC2 and MC3 using an X layout have two kinds of tendons, including the thicker tendons numbered from 1 to 4 and the thinner tendons numbered from 5 to 12 arranged in an X layout. The detailed parameters of the load cases are shown in Table 3. The wave is generated using

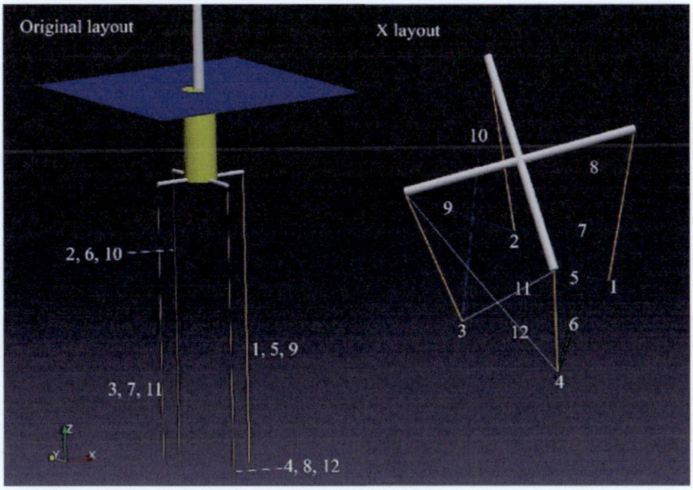

Fig. 1 Diagram of different layouts with numbered tendons

the JONSWAP spectrum and the wind is steady. LC1 and LC2 represent operating conditions and LC3 represents parked conditions. In LC2, the wind and waves are misaligned. Tendon failure simulations were also performed. Failure scenarios are shown in Table 4. The tendons in the table were selected to break because they are the ones under the highest tension. And load case, LC3, was applied. The simulations were run by OPENFAST with a duration of 1800s, and the data from the last 1000 s was used for analysis.

Table 2 Properties of three mooring configurations

Mooring configurations	Line diameter (m)	Line mass per unit length (m)	Line extensional stiffness (N)
MC1 (Original layout)	0.1037	77.3513	1.00E+09
MC2 (X layout)	0.127	116.027	1.50E+09
	0.0884	56.2643	7.27E+08
MC3 (X layout)	0.147	154.703	2.00E+09
	0.072	37.5064	4.85E+08

Table 3 Parameters of load cases

Case name	Wave height (m)	Peak-spectral period (s)	Wave direction (°)	Wind speed (m/s)	Wind direction (°)
LC1	7	13	0	11.4	0
LC2	7	13	15	11.4	0
LC3	14	19	0	42	0

Table 4 Failure scenarios of tendon failure simulations

Failure scenarios	Mooring configurations	Number of failed tendons	Abbreviated as
1	MC1	3	MC1 F3
2	MC1	3 and 7	MC1 F3&7
3	MC1	3, 7, and 11	MC1 F3&7&11
4	MC3	3	MC3 F3
5	MC3	10	MC3 F10
6	MC3	3 and 10	MC3 F3&10
7	MC3	10 and 11	MC3 F10&11
8	MC3	3, 10, and 11	MC3 F3&10&11

3 Results and Discussion

3.1 Response Analysis of Different Mooring Configurations

The motion data of the TLP platform under different load cases is shown sequentially in Fig. 2 (LC1), Fig. 3 (LC2), Fig. 4 (LC3). The data metrics used include the mean, maximum, minimum, standard deviation (Std), and root mean square (Rms). In LC1, the X layout (MC2, MC3) shows a better effect in suppressing surge motion. Their mean, maximum, and root mean square values of surge motion are smaller than those of the original layout (MC1). In terms of sway, heave, and roll motions, MC1, MC2, and MC3 show little difference. For pitch motion, the X layout has a slightly larger amplitude. It can be seen that the vertical tendons play a major role in suppressing pitch motion. In the X layout, each fairlead has only one vertical tendon, resulting in an increased pitch amplitude, but the maximum value is still within 1°. Both use the X layout, but the vertical tendons of MC3 have a larger extensional stiffness, resulting in a smaller pitch amplitude than MC2. For yaw motion, the X layout can significantly reduce the yaw amplitude, which is only about half of that of MC1. Moreover, under LC2 where the wind and waves are in different directions, the X layout has a more obvious effect in suppressing yaw motion. In LC3, due to the large wave height and wind speed, the amplitude of all degrees of freedom increases. The X layout no longer has an advantage in surge, sway, and heave amplitudes, and the roll and pitch amplitudes are larger than those of the original layout, but the yaw amplitude is still smaller than that of MC1. Due to the stronger vertical tendons of MC3, its overall motion performance is better than that of MC2.

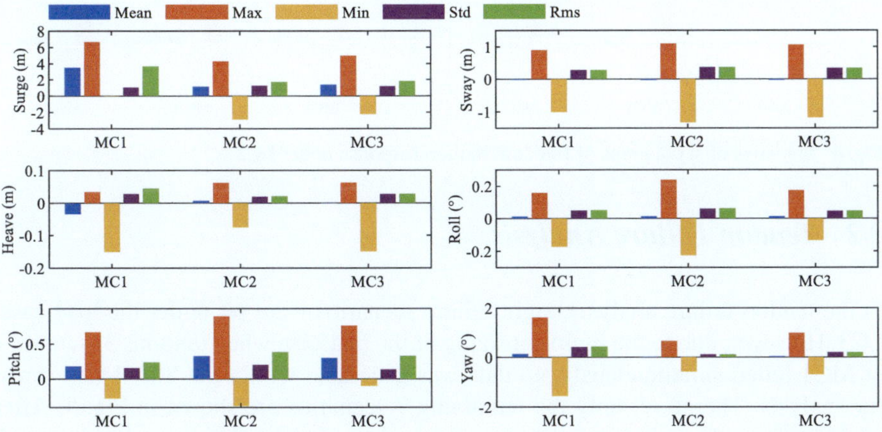

Fig. 2 Statistics of six degrees of freedom motion response under LC1

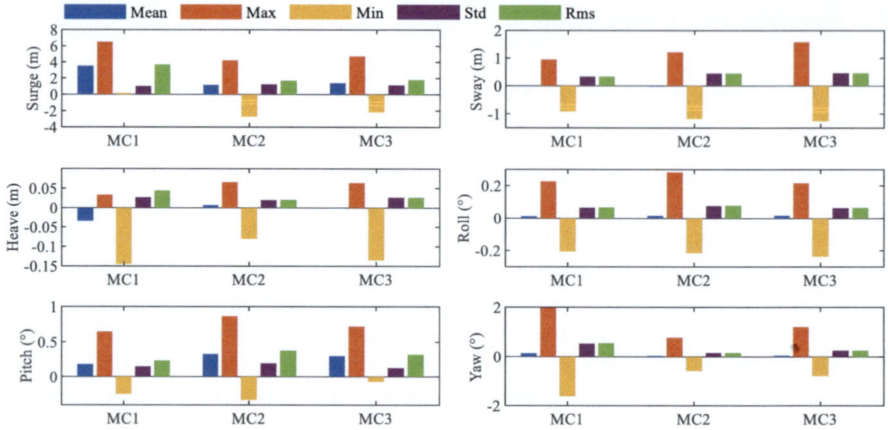

Fig. 3 Statistics of six degrees of freedom motion response under LC2

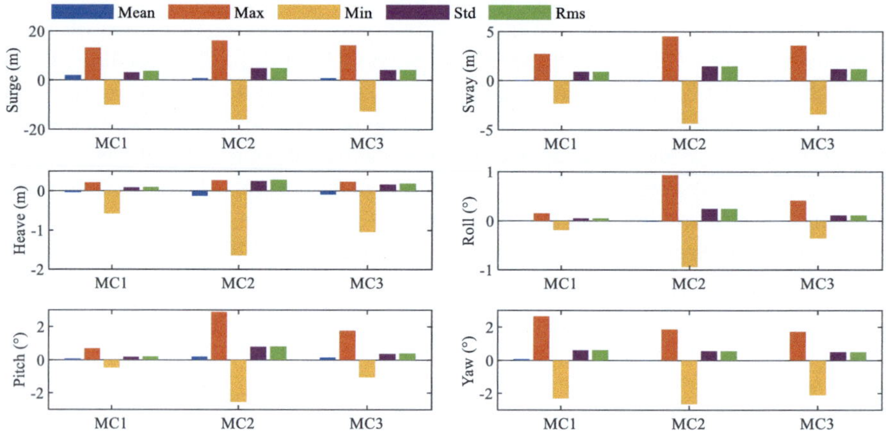

Fig. 4 Statistics of six degrees of freedom motion response under LC3

3.2 *Tendon Failure Analysis*

In the tendon failure analysis, eight failure scenarios were set under the load case, LC3. However, due to the rapid capsizing of the platform when tendons 3, 7, and 11 of MC1 failed simultaneously, no data was generated within the 800–1800s range for analysis. Therefore, only the remaining 7 scenarios are shown in Fig. 5. After tendon failure, the motion amplitudes of sway, heave, and roll did not change significantly compared to the intact LC3 condition, but the amplitudes of surge, pitch, and yaw motions increased significantly. The maximum surge values in all scenarios exceeded 10 m, and the surge amplitude was greater under MC3 tendon failure, with a maximum value approaching 20 m. As for pitch, the scenarios with smaller amplitudes included MC1 F3, MC1 F3&7, MC3 F10, and MC3 F10&11. However, the maximum pitch value in scenarios where tendon 3 failed within MC3 reached about 5°, indicating that losing vertical tendons in the X layout has a highly adverse effect on pitch response. MC3 F10 and MC3 F3&10 had significantly larger yaw amplitudes due to the failure of the tendons disrupting the symmetry of the mooring system, with minimum values approaching − 10°. Overall, the two tendon failure scenarios under MC1 performed better in six degrees of freedom motion. However, the platform quickly capsized when tendons 3, 7, and 11 of MC1 failed simultaneously, whereas the platform remained stable within a certain range of motion when tendons 3, 10, and 11 of MC3 failed simultaneously.

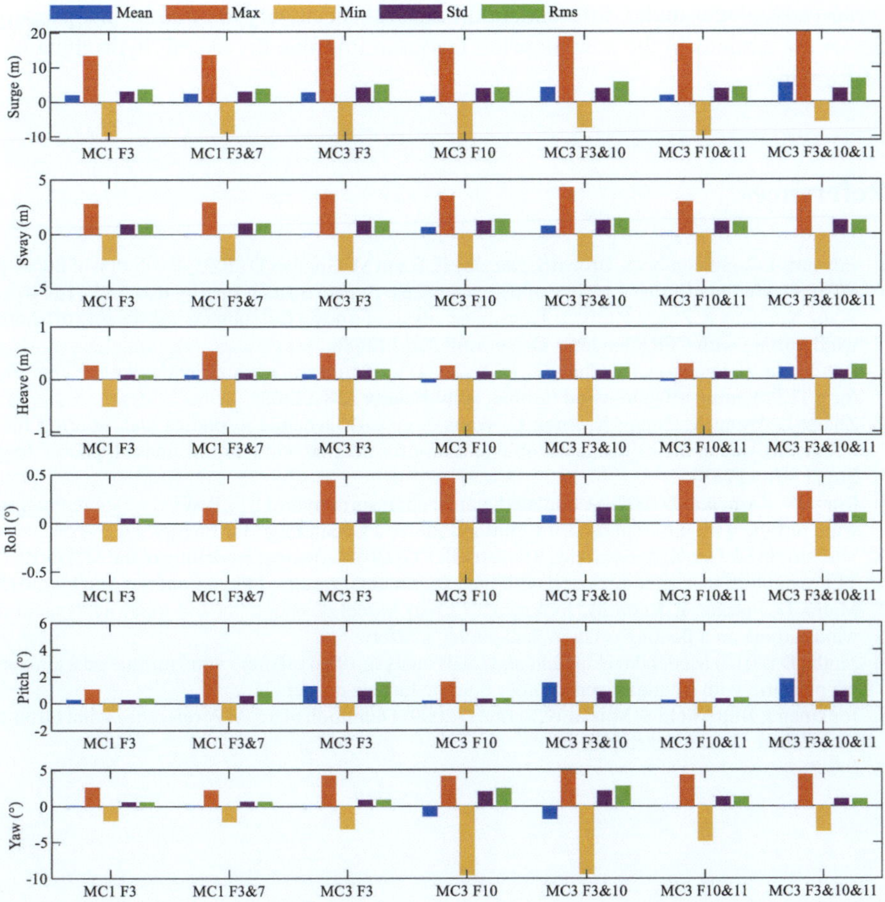

Fig. 5 Statistics of six degrees of freedom motion response under tendon failure scenarios

4 Conclusions

This study investigates the impact of different tendon configurations on the hydrodynamic performance of a TLP floating offshore wind turbine. The analysis reveals that the X layout mooring configurations (MC2, MC3) effectively suppress surge and yaw motions compared to the original layout (MC1) under operating conditions. In the extreme condition (LC3), the X layout's advantage diminishes, but MC3's stronger vertical tendons provide better overall motion performance. Tendon failure scenarios indicate that the platform's stability is significantly affected by the loss of vertical tendons. MC3 is better at maintaining stability compared to MC1 under multiple tendon failures. Overall, the tendon configuration in the X layout has certain advantages under different conditions. It is necessary to optimize the properties of the tendons in the new mooring layout to improve the overall hydrodynamic performance.

References

1. Edwards EC, Holcombe A, Brown S, Ransley E, Hann M, Greaves D (2023) Evolution of floating offshore wind platforms: a review of at-sea devices. Renew Sustain Energy Rev 183:113416
2. Wang X, Zhang Z, Zhuge P (2022) Transient analysis of a novel full submersible floating offshore wind turbine with CFRP tendons. Ocean Eng 266:112686
3. Ren Y, Shi W, Venugopal V, Zhang L, Li X (2024) Experimental study of tendon failure analysis for a TLP floating offshore wind turbine. Appl Energy 358:122633
4. Zhang Z, Wang X, Zhang X, Zhou C, Wang X (2024) Dynamic responses and mooring line failure analysis of the fully submersible platform for floating wind turbine under typhoon. Eng Struct 301:117334
5. Rony JS, Karmakar D (2023) Coupled dynamic analysis of hybrid STLP-WEC offshore floating wind turbine with different mooring configurations. J Ocean Eng Mar Energy 9:623–651
6. Vardaroglu M, Gao Z, Avossa AM, Ricciardelli F (2020) Numerical modelling of the MIT/NREL TLP wind turbine and comparison with the experimental results. J Phys Conf Ser 1669:012015
7. Matha D, Fischer T, Kuhn M, Jonkman J (2010) Model development and loads analysis of a wind turbine on a floating offshore tension leg platform
8. Matha D (2010) Model development and loads analysis of an offshore wind turbine on a tension leg platform with a comparison to other floating turbine concepts
9. Jonkman J, Butterfield S, Musial W, Scott G (2009) Definition of a 5-MW reference wind turbine for offshore system development

Hydroelastic Responses of Submersible Cages Under Extreme Waves

Mingyuan Ma, Hong Zhang, Xinyu Zhang, and Chien Ming Wang

Abstract The accurate prediction of the dynamic responses of fish cages under extreme sea states is of vital importance for the structural reliability of marine aquaculture infrastructure. The present study investigates the hydroelastic behaviour of flexible fish pens under different extreme wave conditions with specified recurrence periods. Initially, environmental contours for extreme wave conditions were derived using the direct Inverse First Order Reliability Method (IFORM), by fitting the generalised extreme value (GEV) distribution to the declustered storm peaks and extrapolating to conditions with specified recurrence periods. The wave condition contours are then mapped to the response space of the fish cage according to the Norwegian standard NS9415 using finite element method (FEM) numerical simulations. The research results indicate that significant differences in the wave responses of the fish cage occur under severe non-severe extreme wave conditions. The study also examines the reliability and serviceability of the fish cage under these extreme sea states. It is found that while the fish cage exhibits a substantial hydroelastic response at low frequencies and high energy, this response is reduced under severe sea conditions. The techniques and findings provide valuable insights for the reliability analysis of marine aquaculture infrastructure.

Keywords Extreme sea states · Marine aquaculture infrastructure · Finite element analysis

M. Ma · H. Zhang (✉) · X. Zhang
School of Engineering and Built Environment, Griffith University, Gold Coast, QLD, Australia
e-mail: hong.zhang@griffith.edu.au

C. M. Wang
School of Civil Engineering, The University of Queensland, St Lucia, QLD, Australia

Blue Economy Cooperative Research Centre, Launceston, TAS, Australia

X. L. Zhao et al. (eds.), *Proceedings of the Fourth World Conference on Floating Solutions: WCFS 2024; 2–4 December; Hong Kong*, Lecture Notes in Civil Engineering 597, https://doi.org/10.1007/978-981-96-4569-5_47

1 Introduction

Marine aquaculture is an important component of the global food supply chain. According to DNV [1], global production of marine aquaculture is expected to increase by about 1.5 times by 2050. In Australia, aquaculture is a rapidly growing industry known for providing high-quality, safe, and sustainable seafood, with an output value of A$3.3 billion in 2020 [2]. Storm Bay in Tasmania has been identified by both industry and the Tasmanian Government as a suitable location for sustainable development of the industry, particularly for offshore fish farming [3]. However, unlike traditional fish cages located in nearshore, sheltered waters, fish cages exposed to high-energy environments offshore are at great risk of structural failure, which could result in loss of personnel and property. Therefore, it is necessary to analyze the dynamic response of such structures, especially under extreme sea conditions.

Numerical modelling is an effective and flexible tool, as physical tests and field monitoring are limited by conditions and external factors. Currently, there are extensive models for simulating the dynamic response of fish cages. Due to the slender and flexible characteristics of fish cages, the hydrodynamic loads can be estimated using Morison equations [4] or screen models [5], and the dynamic responses are analyzed using the lumped-mass method [4] or finite element method (FEM) [5].

In addition, forecasting the extreme sea states is an open topic. The common approach is to establish a joint probability distribution based on buoy measurement data or hindcast data and extrapolate to the sea state under a specified probability of exceedance. Various extreme modelling methods have been tested and compared in [6].

Based on the aforementioned background and literature review, the present study hopes to investigate the hydroelastic response of fish cages under the sea states of a given extreme recurrence period.

2 Stude Area and Employed Data

The selected study area is located at a potential fish farming site (Longitude: 147.47°, Latitude: – 43.40°, water depth: 75.5 m) in Storm Bay, Tasmania, Australia. Storm Bay has been identified as a priority area for the potential expansion of salmonid marine farming in the Government's Sustainable Industry Growth Plan for the Salmonid Industry (refer to Fig. 1).

In the present study, the potential fish farming site analysed is located in the offshore waters on the western side of Bruny Island. The Collaboration for Australian Weather and Climate Research (CAWCR) wave hindcast dataset [7] from 1990 to 2023, which is simulated using the third-generation spectral wave model Wave Watch III [8], is employed to extrapolate to the extreme sea states reaching up to 50 years.

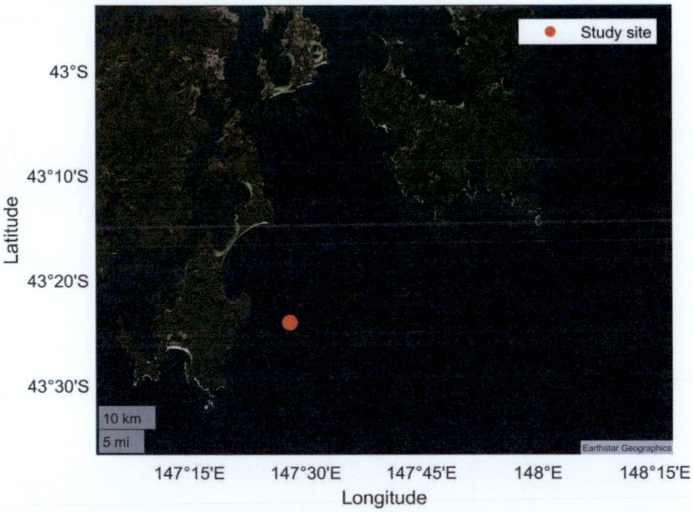

Fig. 1 Study site of fish farms in Storm Bay, Tasmania, Australia

3 Methodologies

To predict extreme sea conditions with a specific recurrence period, a common practice is to use the entire dataset for parameter estimation to ensure sufficient samples and thus reduce the variance of the estimated values. However, probability distribution estimation usually requires the sample data to be independent and identically distributed. Mackay [9] have pointed out that time series of meteorological and ocean variables typically exhibit significant short-term serial correlation. Therefore, resampling the sample data is necessary. We employ an algorithm to decluster the sea state series into a series of storm peaks (Fig. 2), ensuring that the interval between adjacent storm extremes is not less than five days, according to [13], to eliminate the autocorrelation in the series.

The declustered storm peak variables, X^{peak} can be fitted by the generalised extreme value (GEV) distribution:

$$f_{\text{GEV}}(x;\mu,\sigma,\xi)=\begin{cases}\frac{1}{\sigma}\exp\left[-\left(1+\xi\frac{x-\mu}{\sigma}\right)^{-1/\xi}\right]\left(1+\xi\frac{x-\mu}{\sigma}\right)^{-1-1/\xi},\ \xi\neq 0\\ \frac{1}{\sigma}\exp\left[-\exp\left(-\frac{x-\mu}{\sigma}\right)-\frac{x-\mu}{\sigma}\right],\ \xi=0\end{cases},\quad (1)$$

where μ, σ and ξ are the location, scale, and shape parameters, respectively. To describe the tail of the distribution more accurately, we suppose a two-part model to fit the distribution of the storm peaks (Fig. 3a). The probability of exceedance $P(X^{\text{peak}} > x)$ is derived as

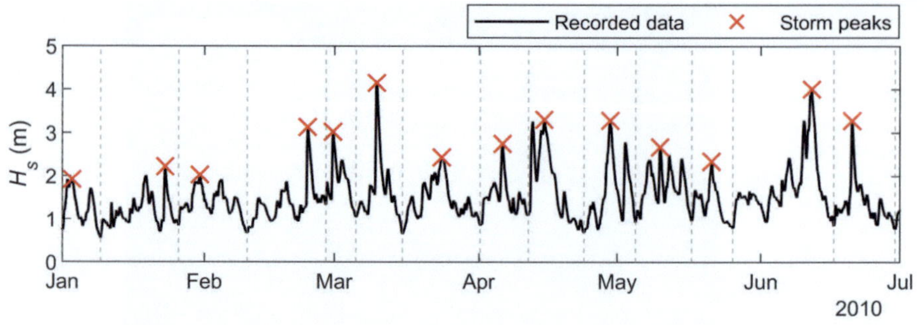

Fig. 2 Declustering the wave series into storm peaks; the dashed line represents the minima of the adjacent peaks

$$P\left(X^{peak} > x\right) = \begin{cases} 1 - \int\limits_{-\infty}^{x} f_{\mathrm{GEV}}(x)dx, \; x \leq \theta \\ P\left(X^{peak} > \theta\right)\left[1 + \xi\left(\frac{x-\theta}{\sigma}\right)\right]^{-1/\xi}, \; x > \theta \end{cases}. \tag{2}$$

In Eq. (2), the part of samples over the threshold θ is fitted by the generalised Pareto (GP) distribution:

$$f_{\mathrm{GP}}(x; k, \varepsilon, \theta|x\rangle\theta) = \frac{1}{\varepsilon}\left[1 + k\frac{(x-\theta)}{\varepsilon}\right]^{-1-1/k}, \tag{3}$$

where $k \neq 0$ and ε are the shape and scale parameters, respectively. The threshold θ is determined from the empirical estimates of recurrence values of variables from 3 to 6 months, and the one with the minimum root mean square error for the empirical estimates of recurrence values of one year and above is chosen. Therefore, the recurrence period T_R corresponds to a specified probability of exceedance of P ($X^{\text{peak}} > x$) is calculated as

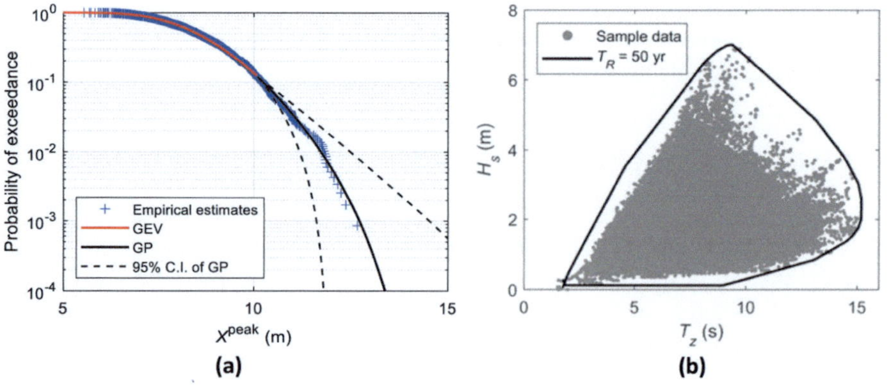

Fig. 3 Estimating extreme sea states: **a** fitting of the univariate distribution, and **b** constructed EC

$$T_R = \frac{1}{MP\left(X_s^{peak} > x\right)}. \tag{4}$$

where M is the average number of storm peaks per year. In order to determine the joint extreme sea states of the two variables, namely the significant wave height H_s and the mean zero-up wave period T_z, we used a direct Inverse First-Order Reliability Method to describe the environmental contours (EC) of $T_R = 50$ years. The two variables H_s and T_z are projected on a straight line through the origin at an angle φ, so that the exceedance bound can be constructed using the univariate estimate of the variable $X = T_z\cos\varphi + H_s\sin\varphi$ (as seen in Fig. 3b).

On the other hand, the commercial software AquaSim [10] based on the FEM was used to simulate the dynamic response of the fish cage, where the floating collar was modelled by beam elements, the fish cage net was modelled as membrane elements, and the mooring ropes were modelled as truss elements. The hydrodynamic forces dF on the structural element are estimated by the Morison equation:

$$dF = \frac{1}{2}\rho C_d DdL(u - \dot{x})|u - \dot{x}| + \rho(1 + C_m)\frac{\pi D^2}{4}dL\ddot{u} - \rho C_m \frac{\pi D^2}{4}dL\ddot{x}. \tag{5}$$

where ρ is the water density, u is the wave water velocity, x is the displacement of the structural element, D is the element cross-section diameter, dL is the length of the element, $C_d = 1.2$ is the drag coefficient, and $C_m = 1$ is the added mass coefficient. In addition, the maximum wave height for a sea state shall be assumed around twice H_s when using a regular sea for the analysis according to the Norwegian standard NS9415 [11]. In addition, NS9415 suggests that the sea state under a recurrence period of $T_R = 50$ years is considered as the design condition.

4 Results Discussions

For fish cages exposed to high-energy environments, submerging them underwater is beneficial in avoiding strong wave loads, especially during extreme sea conditions. As illustrated in Fig. 4, we consider a fish cage submerged 20 m below the mean water level and analyse its hydroelastic response under all possible sea states under $T_R = 50$ years. The dimensions and material parameters of all structural components of this cage can be found in Table 1. Additionally, as described in the methodology, twelve representative sea states are selected from the EC in Fig. 3b for analysis. The specific wave conditions are shown in Table 2. For each wave condition, the wave steepness is defined as $S_z = 2\pi H_s/(gT_z^2)$.

Currently, only unidirectional monochromatic waves, as shown in Fig. 4, are considered. Based on the FEM simulation results, the most severe failure of the floating collar occurs at the connection point with the mooring lines, as shown by the distribution of von-Mises stress in Fig. 5a. The maximum vertical displacement occurs at the mid-span of the mooring constraints from Fig. 5b. In the direction of the incoming waves, the mooring lines on the windward side experience greater

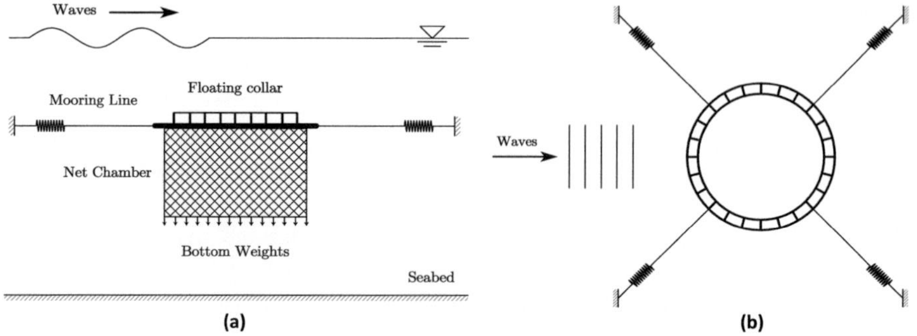

Fig. 4 Sketch of a submersible fish cage below the water surface: **a** isometric view, and **b** plan view

Table 1 Dimensions and material parameters of structural components of the fish cage

Structural components	Parameter	Value
Floating collar	Outer/Inner diameter (m)	42/40
	Cross-section outer diameter/Pipe thickness (m)	0.45/2.65E−2
	Mass in water (kg/m)	21.06
	Bending stiffness (N·m^2)	2.83E5
	Poisson's ratio	0.4
Net chamber	Mask width/Thread diameter (m)	0.03/0.0025
	Mass in water (kg/m)	3.51E−4
	Line stiffness (kN/m)	80.6
Mooring line	Diameter/Length (m)	0.064/56.56
	Mass in water (kg/m)	0.37
	Line stiffness (kN/m)	153.57
Bottom weight	Mass (kg/m)	50

Table 2 Analysed wave conditions from the sea state with recurrence period $T_R = 50$ years

No.	H_s (m)	T_z (s)	Sz	No	H_s (m)	T_z (s)	S_z
#1	7.01	9.27	0.0522	#7	0.25	9.68	0.0017
#2	5.92	11.30	0.0297	#8	0.13	5.92	0.0023
#3	4.82	13.17	0.0178	#9	0.13	1.89	0.0233
#4	3.52	14.65	0.0105	#10	2.01	3.37	0.1134
#5	2	15.19	0.0056	#11	3.52	4.58	0.1075
#6	0.84	13.01	0.0032	#12	5.44	6.86	0.0740

mooring forces. Additionally, the volume change of the cage is also a crucial indicator to monitor, as it is related to fish welfare. Therefore, these four indicators will be used in the subsequent analysis. An example of the time series for these four indicators is plotted in Fig. 6, where wave condition #1 in Table 2 is applied, corresponding to the peak point of the EC contour and can be considered as a severe condition counted as exceedances.

The maximum responses of the four indicators for the fish cage under all possible extreme sea states with a $T_R = 50$ years are plotted in Fig. 7, associated with the

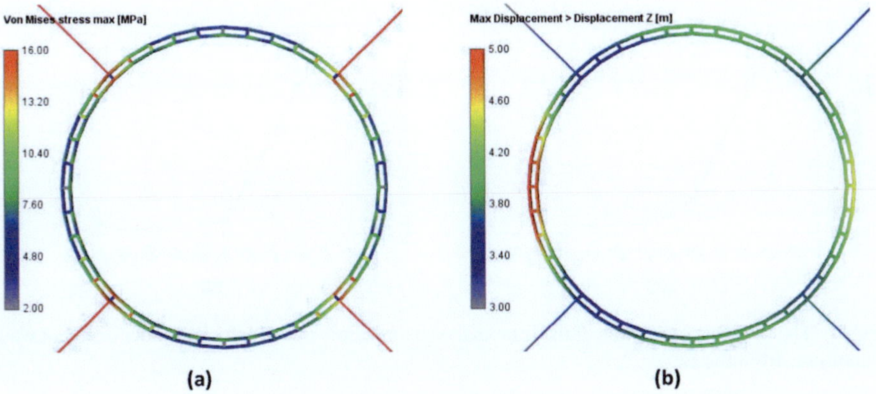

Fig. 5 The part of the fish cage floating collar with the most critical response based on FEM simulation: **a** von-Mises stress, and **b** vertical displacement

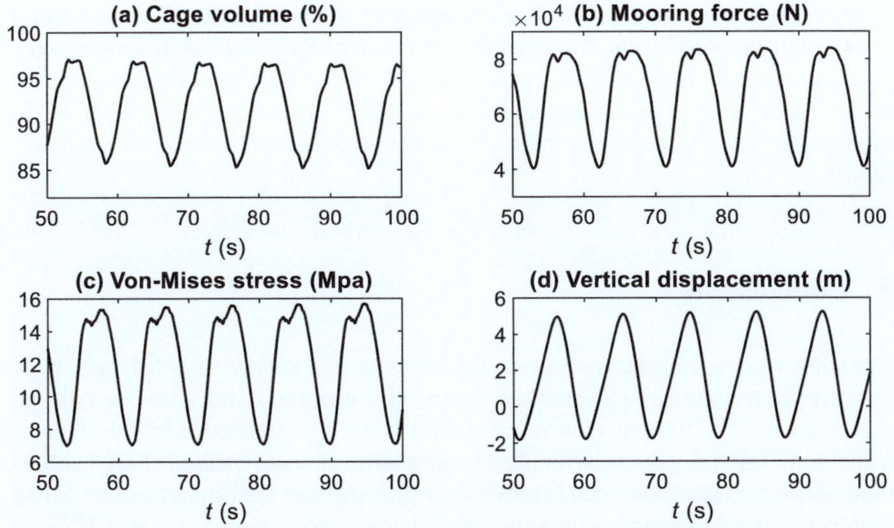

Fig. 6 Time series plots of the four analysed indicators of the fish cage based on FEM simulation

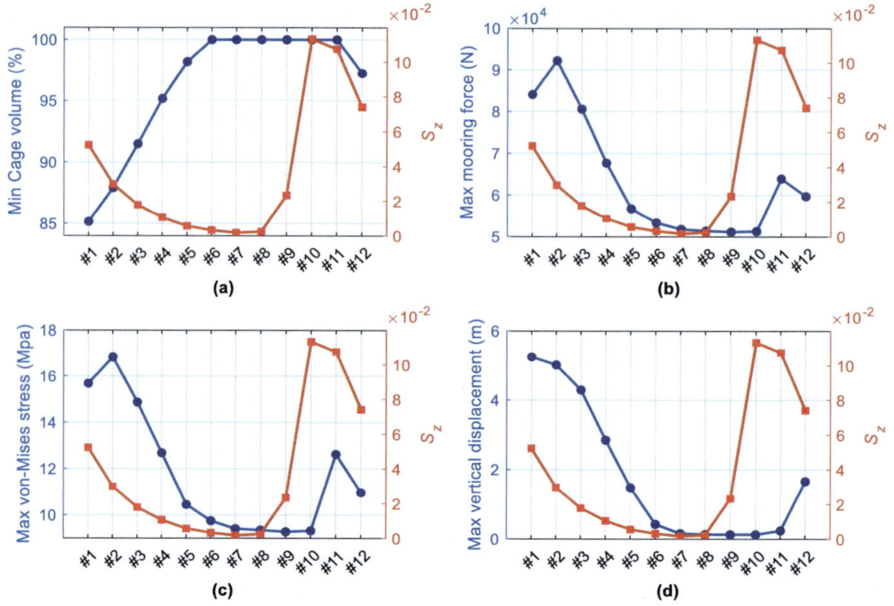

Fig. 7 The max response plots of the four analysed indicators of the fish cage under different wave conditions from the EC

varying wave steepness S_z. For wave conditions #1 to #4, the fish cage exhibits a significant dynamic response, and the response intensity gradually decreases as the steepness decreases, although the mooring force and von-Mises stress slightly increase for wave condition #2. For sea states #5 to 8, the period T_z varies significantly, but the significant wave height H_s is relatively small, indicating a minor response of the fish cage. The steepness for sea states #9 to #12 increases noticeably, but the dynamic response of the cage is not as significant as for those of #1 to 4. This implies that low-frequency but high-energy sea states contribute more to structural failure.

5 Conclusions

The study presents a case analysis on the response of a submersible fish cage under extreme sea conditions in Storm Bay, Tasmania, Australia. Using a hindcast dataset, sea states with a 50-year recurrence period were extrapolated, and the dynamic response of the fish cage was simulated using finite element method (FEM) simulations. Representative wave conditions were extracted from the Environmental Conditions (EC), and design wave conditions were determined in accordance with NS9415. The dynamic response of the fish cage was analyzed to identify the most critical conditions. The study reveals that significant stress concentrations develop at the mooring connections, accompanied by notable vertical displacement between these connections. The wave conditions in the EC can be categorized into three regimes, with the fish cage exhibiting substantial dynamic response under low-frequency, high-energy sea conditions. These findings provide valuable insights for engineers analyzing the extreme response behavior of offshore aquaculture structures. Future studies will focus on incorporating realistic irregular waves to further refine the analysis.

Acknowledgements The authors acknowledge the financial support of the Blue Economy Cooperative Research Centre, established, and supported under the Australian Government's Cooperative Research Centres Program, grant number CRC-20180101. The authors also would like to thank Aquastructures AS for granting access to their software AquaSim (https://aquasim.no).

References

1. DNV: Marine Aquaculture Forecast to 2050. Det Norske Veritas Group, Hovik (2021)
2. Chen D, Wang CM, Zhang H (2021) Examination of net volume reduction of gravity type open-net fish cages under sea currents. Aquacult Eng 92:102128
3. Peter & Una Rockliff: ENVIRONMENTAL IMPACT STATEMENT—Draft Storm Bay North Marine farming development plan. Department of Natural Resources and Environment Tasmania, Hobart (2017)
4. Zhao YP, Li YC, Dong GH, Gui FK, Wu H (2008) An experimental and numerical study of hydrodynamic characteristics of submerged flexible plane nets in waves. Aquacult Eng 38(1):16–25
5. Kristiansen T, Faltinsen OM (2012) Modelling of current loads on aquaculture net cages. J Fluids Struct 34:218–235
6. Haselsteiner AF, Coe RG, Manuel L, Chai W, Leira B, Clarindo G, Guedes Soares C, Hannesdótti, Á, Dimitrov N, Sander A, Ohlendorf JH, Thoben KD, Hauteclocque GD, Mackay E, Jonathan P, Qiao C, Myers A, Rode A, Hildebrandt A et al (2021) A benchmarking exercise for environmental contours. Ocean Eng 236:109504
7. Durrant T, Hemer M, Smith G, Trenham C, Greenslade D (2019) CAWCR wave Hindcast—aggregated collection. v5. CSIRO. Service Collection. http://hdl.handle.net/102.100.100/137152?index=1. Accessed 26 Aug 2024
8. Tolman HL (1991) A third-generation model for wind waves on slowly varying, unsteady, and inhomogeneous depths and currents. J Phys Oceanogr 21:782–97
9. Mackay E, de Hauteclocque G, Vanem E, Jonathan P (2021) The effect of serial correlation in environmental conditions on estimates of extreme events. Ocean Eng 242:110092
10. Aquastructures Homepage. https://aquasim.no/resources/documentation.html. Accessed 26 Aug 2024
11. Norge S (2021) NS9415 Marine fish farms. Requirements for site survey, risk analyses, design dimensioning, production, installation and operation. Standards Norway, Lysaker

Hydroelastic Response of Large Floating Offshore Base for Offshore Wind Farm

Saika Iwamatsu, Mako Yamamoto, Ananda Raiz Pambela, and Kazuhiro Iijima

Abstract The offshore wind industry is a rapidly growing sector, and it is estimated that the number of offshore Wind Farms (WF) will increase in the future. At present, offshore WF is based at close-by onshore substations or port facilities where power transmission, construction, and operation and maintenance (O&M) are performed. This is reasonable for offshore area close to coastal area. However, it is no more justified for far offshore areas such as EEZ. In the case of such offshore WF, the construction of offshore base is required since it can reduce the trip time between the base port and the offshore WF site and increase the workability. Therefore, this study examines a very large floating structure with substation and O&M functions by numerical simulations. Similar projects such as the North Sea Wind Power Hub are reported, but they are assumed to be deployed at a water depth of less than 50 m. In our companion paper, we propose a new concept of an offshore base which is a kind of a very large floating structure installed at a water depth more than 50 m in Japan's EEZ. The proposed offshore base has a complex geometry and properties. It is semi-submersible and rectangular shape, but with a few cut-outs for its functionalities such as the port function as a base, and with heavy concentrated-mass objects mounted on its upper deck. In this study, the elastic response of a very large floating structure with non-uniform mass and with complex geometry, is analyzed by using numerical calculations.

Keywords Hydroelasticity · Very large floating structure · Offshore wind farm

S. Iwamatsu (✉)
Joint Research Chair for Offshore Wind System Integration, The University of Osaka, Osaka, Japan
e-mail: iwamatsu.saika.b4k@osaka-u.ac.jp

M. Yamamoto · A. R. Pambela · K. Iijima
Department of Naval Architecture and Ocean Engineering, The University of Osaka, Osaka, Japan

X. L. Zhao et al. (eds.), *Proceedings of the Fourth World Conference on Floating Solutions: WCFS 2024; 2–4 December; Hong Kong*, Lecture Notes in Civil Engineering 597, https://doi.org/10.1007/978-981-96-4569-5_48

1 Introduction

In the North Sea, plans are progressing for the construction of artificial islands within the offshore wind farm (WF) development areas. A notable project is the North Sea Wind Power Hub ([1], see Fig. 1). This artificial island features a substation function and is designed to interconnect with multiple planned offshore WFs in the vicinity. Additionally, it incorporates storage via P2X and operational and maintenance (O&M) functionalities, aiming to enhance its utility as a central hub for offshore WF operations in the North Sea, thereby potentially reducing costs relative to other options. Generally, for offshore WFs located beyond a 60 km distance from shore, it is suggested from an O&M perspective that using a Service Operation Vessel (SOV) for several weeks of offshore anchorage may lead to cost savings compared to daily Crew Transfer Vessel (CTV) operations [2].

In Japan, the majority of waters, including its Exclusive Economic Zone (EEZ), have depths far exceeding 50 m, necessitating the essential development of floating wind turbines suitable for deep waters. Similarly, artificial islands created through land reclamation, serving as bases for offshore WFs, are not well-suited for Japan. Therefore, Yamamoto [3] has proposed and designed a Semi-Submersible Offshore Base tailored for Japan's marine environment (see Fig. 2). This concept includes not only substation and O&M functions but also assembly capabilities. It is semi-submersible and rectangular shape, but with a few cut-outs for its functionalities such as the port function as a base, and with heavy concentrated-mass objects mounted on its upper deck. Verification is required using a non-rectangular, non-uniform mass model to ensure its functionalities and safety. Hence, this study focuses on examining its response by using non-uniform mass massive floating structures with openings. Refer to the companion paper for detailed concept design of the Semi-Submersible Offshore Base [4].

During assembly and O&M operations, it is concerned that heavy items such as cranes and wind turbine components will move on the deck, necessitating consideration of their impact on the structural response of the floater due to their movement.

Fig. 1 North Sea wind power hub program [1]

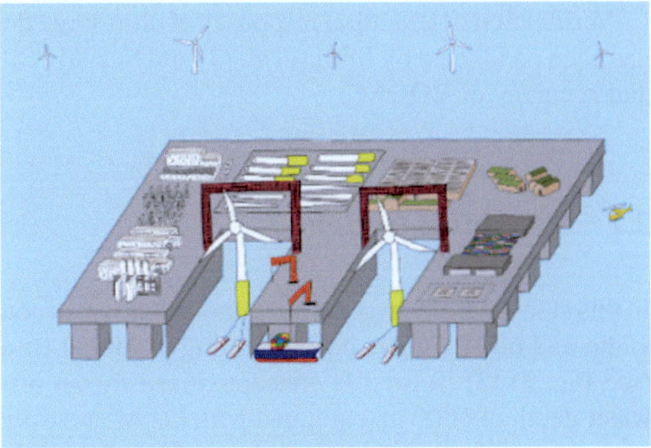

Fig. 2 Offshore wind firm float base concept [3]

The cut-out part of the structure may exhibit complex hydroelastic response. Therefore, the objective of this study is to elucidate the elastic response of a mega floating structure with openings and loaded with heavy items to non-uniform loads.

2 Numerical Method

The numerical analysis theory utilized in this paper employs VODAC (Very Large Floating Structure Oriented Dynamic Analysis Code), developed by Iijima et al. [5]. Here, an overview of its theory is briefly given.

In VODAC, for the fluid domain, the method of singular point distribution and Kagemoto's hydrodynamic interference theory [6] are combined, while in the structural domain, the concept of Group body is incorporated, and for the structural domain, the substructure method is applied, enabling handling of even larger structures. As given in Fig. 3, it is a hierarchical hydroelasticity method both in fluid and structural domains.

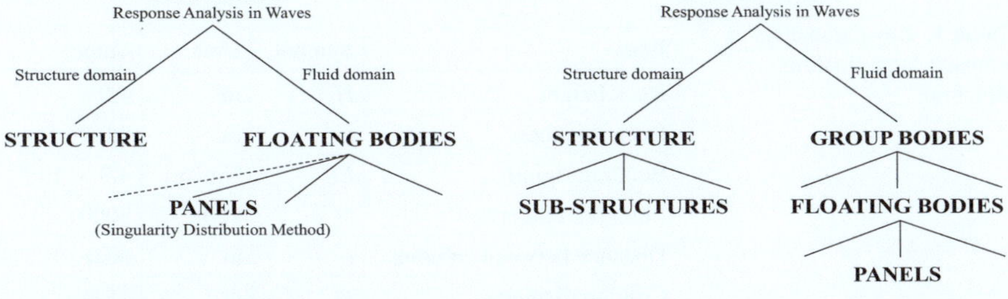

Fig. 3 Hierarchy of the present method [5]

Verification of structural responses of ultra-large floating structures using VODAC has been conducted by Iijima et al. [7] and Suzuki et al. [8], confirming the capability and accuracy of VODAC.

2.1 Structural Design

In this study, we envisage an ultra-large floating structure with dimensions of 1200 m width and 600 m depth, corresponding to Model Base 3 as proposed by Yamamoto (see Fig. 2) [3]. Base 3 is conceptualized for an offshore distance of 300 km and water depth of 5000 m, equipped with O&M and assembly functions. The structural analysis of this model utilized the analytical model of the mega floating structure proposed by Suzuki et al. [9]. This approach clarifies the characteristics of structural responses and serves as an effective tool for guiding decisions in selecting structural forms. For detailed calculation methods, refer to our companion paper [4]. In this paper, the parameters outlined in the companion paper are presented in Table 1.

The Dogger Bank Wind Farm substation [10] is referenced for planning the offshore base. It is assumed that it accommodates an offshore substation capable of 1.2 GW (total weight 7000 tons).

Wind turbine components stored on the deck for assembly and O&M are based on the IEA 15 MW wind turbine [11]. The masses of these components are 65.25 tons for the blades, 860 tons for the tower, and 820 tons for the nacelle. The mass and moment of inertia of the blades are calculated assuming three stacked blades, using a simplified assumption of a rectangular prism for the moment of inertia.

The crane used for assembly is assumed based on existing cranes on catalogues. While Yamamoto [3] concept includes a gantry crane, ring lift cranes and crawler cranes are also potential options. Referring to the weight and dimensions of a 3200-ton lifting capacity crawler crane [12] capable of assembling 15 MW wind turbines, the assumed weight including the load and main body is 4000 tons. This assumption exceeds the weight of existing crawler cranes because there is potential to lift floating foundations or wind turbines larger than 15 MW in the future.

Table 1 Key parameters of semisub-large floating structure

Item	Symbol	Unit	Value
Deck height	H	m	4.0
Deck thickness	t	m	0.025
Bending rigidity	EI	Nm^2/m	4.2×10^{10}
Mass per unit area	M	kg/m^2	3000
Distance between columns	L	m	60.0
Column diameter	D	m	25.0
Draught	d	m	22.0

Each port function as a harbor opening is assumed to measure 240 m in width and 300 m in depth. This assumption is based on the 240 m rotor diameter of the IEA 15 MW wind turbine [11], and the approximate length of hydrogen transport ships [13].

2.2 Model Case

Model-1 is depicted in Fig. 4, serving as the rectangular base model for this study. This is a reference model.

Model-2 and 3 are shown in Figs. 5 and 6, respectively. Compared to Model-1, Model-2 and 3 maintain the width and depth while incorporating harbor and assembly functions through an opening.

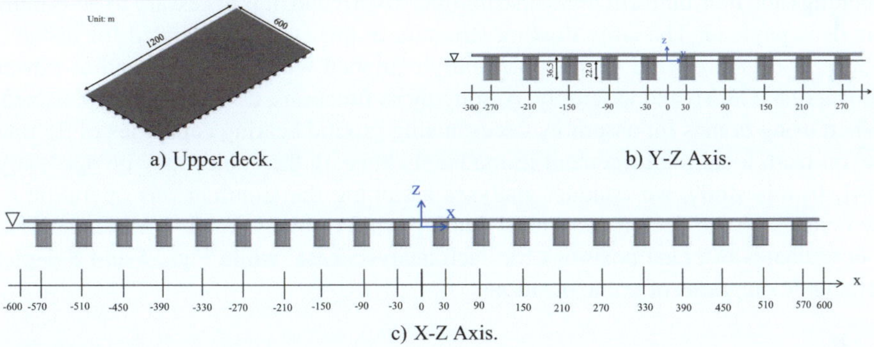

Fig. 4 Model-1 of large floating offshore base

Fig. 5 Model-2

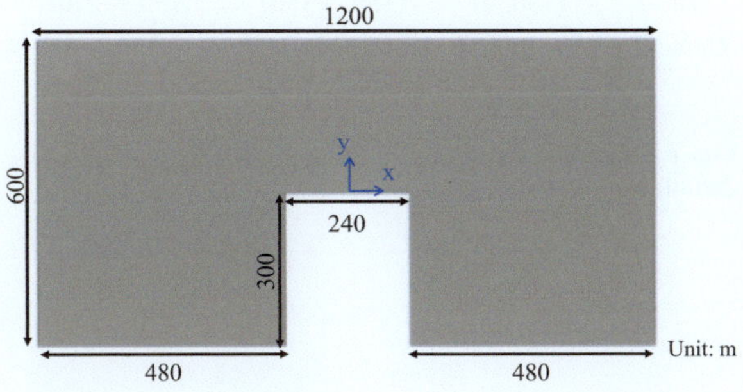

Fig. 6 Model-3

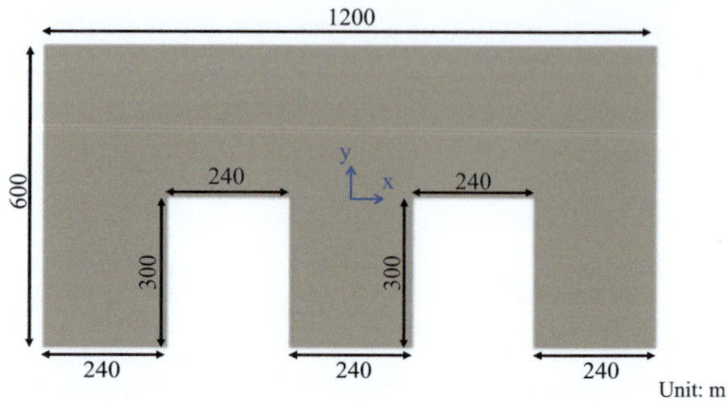

2.3 Analysis Case

In this study, we investigate the structural response of a large floating structure with openings and non-uniform mass distribution. To this end, it is necessary to determine the deck payload. The large floating structure in this study is intended for installation near offshore wind farms, potentially equipped with functions such as power conversion, O&M, and assembly. Among these functions, assembly is emphasized. When using cranes for assembly necessitating ground bearing capacities of 35 ton/m^2 on land, local reinforcement immediately beneath the cranes may be necessary [14]. In this study, we conduct analyses assuming the simultaneous operation of two cranes on the deck to assemble floating wind turbines. Table 2 presents the x, y coordinates of crane positions for each analysis case, while Figs. 7 and 8 depict schematic diagrams of crane positions.

Table 2 Crane placement conditions in Model-2

Crane No.	Case-1	Case-2	Case-3	Case-4	Case-5
Crane-1	(150, 0)	(150, −150)	(150, −270)	(150, −270)	(150, 0)
Crane-2	(−150, 0)	(−150, −150)	(−150, −270)	(−150, 0)	(−150, −270)

Fig. 7 Case-1, 2, 3 definitions

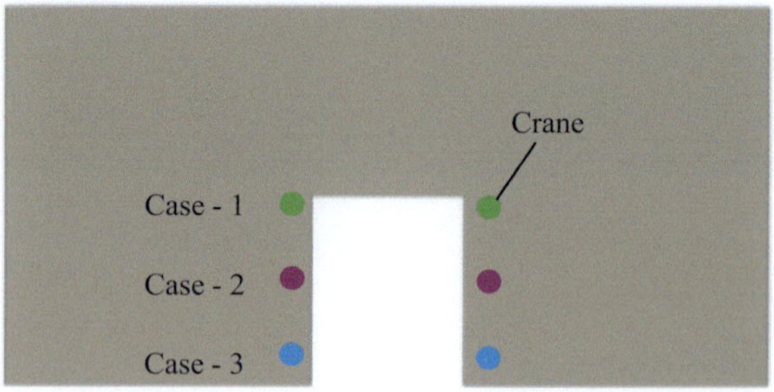

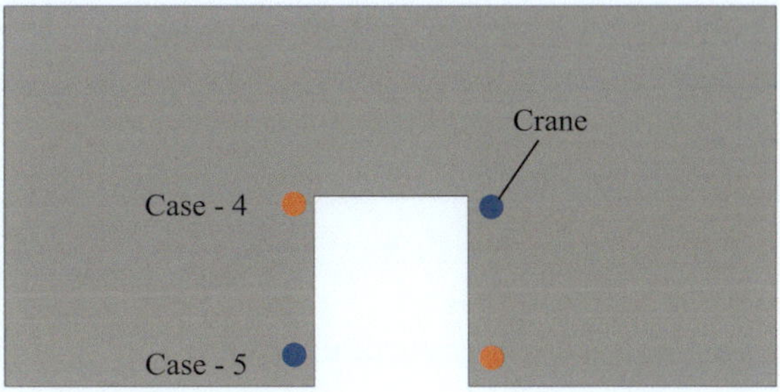

Fig. 8 Case-4, 5 definitions

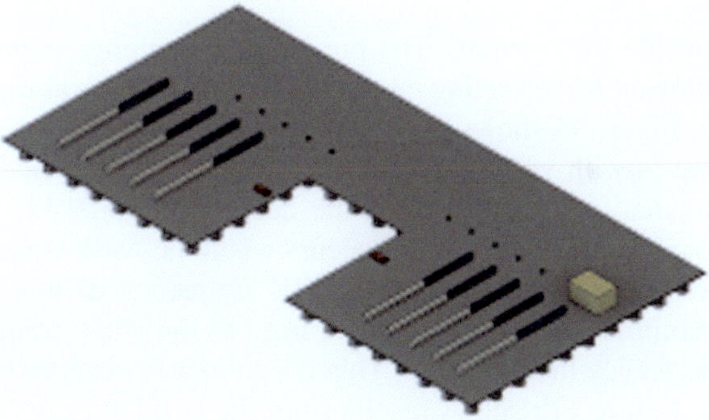

Fig. 9 Fully loaded

Furthermore, we load all assumed payloads and conduct structural assessments. The payloads assumed include a substation, two cranes, and ten IEA15MW wind turbines. Figure 9 illustrates a schematic diagram depicting all payloads loaded.

3 Results and Discussion

3.1 The Effect of Openings on the Floating Structure

Figure 10 shows the results of static analysis of vertical displacement distribution and angular displacement at Y = 0 m. Compared to Models-1, 2 and 3 exhibit significant variations at the location of the openings. From Fig. 10a, it is observed that in Model-1, except at the ends, the deck vertical displacement distributes within a range from 0.039 to 0.080 m. In Models-2 and 3, the vertical displacement at the areas with openings are −0.045 m and −0.035 m, respectively.

From Fig. 10b, it is evident that while the openings in Model-2 are at X = −120 to 120 m, rapid changes in angular displacement around Y-axis occur at X = −150 and 150 m, indicating occurrence at the edge of the opening. This phenomenon is similarly observed in Model-3, where results of angular displacement are obtained

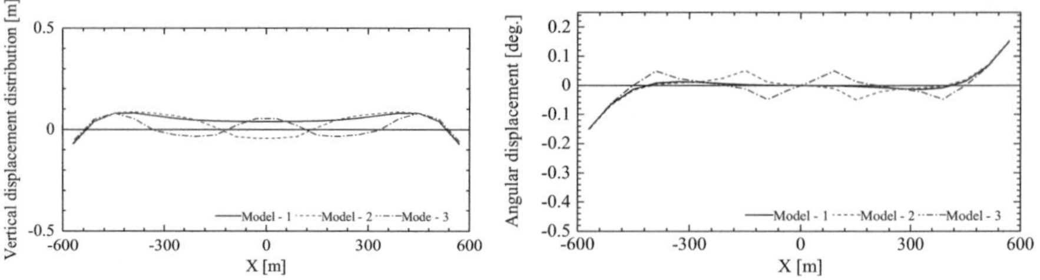

a) Vertical displacement distribution. b) Angular displacement.

Fig. 10 Displacements due to openings in the structure (Y = 0)

outside the opening. The presence of openings is presumed to cause non-uniform buoyant forces acting on the floaters, leading to abrupt changes near the openings.

Figure 11 shows the RAO (response amplitude operators) of bending stress under waves with varying frequencies. Figure 11a indicates the point at the center coordinates (x, y) = (0, 0) of the decks for Models-1–3. In large floating structures, maximum bending stress occurs when the wave frequency is near the characteristic frequency [9]. The characteristic frequency of this model is 0.362 rad/s [3], and maximum bending stress occurred at the wave frequencies near the characteristic frequency in all models. This is commonly observed for all models.

Figure 11b shows results from the point at (x, y) = (−150, 0) for Models-1 and 2. This point is at the base of the opening in Model-2, which is presumed to be a vulnerable area prone to damage. From Fig. 11b, it is observed that similar to the deck center coordinates, maximum bending stress occurred near the characteristic frequency of 0.362 rad/s. Furthermore, while Fig. 11a showed no significant difference in bending stress among all models, the bending stress at this point is 25 MPa for Model 1 and 37 MPa for Model-2, indicating that the stress in Model-2 is 1.48 times higher. This suggests that thorough validation at the base of the openings is essential.

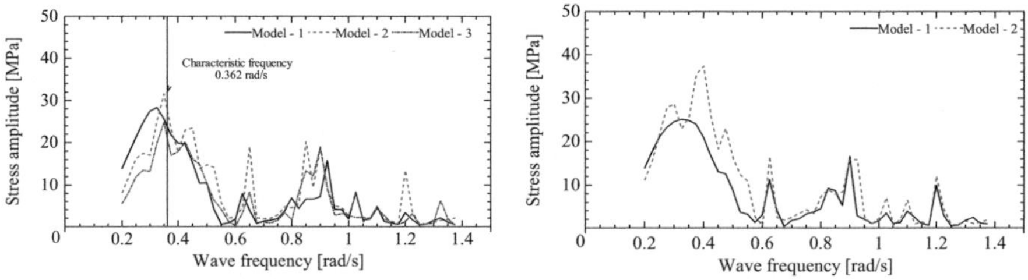

a) Stress amplitude (X=0, Y=30). b) Stress amplitude (X=-120, Y=30).

Fig. 11 Stress due to openings in the structure under wave conditions of a significant wave height of 13.0 m and an average wave period of 14.0 s (Y = 0)

Comparing Models-2 and 3 based on Figs. 10 and 11, similar trends in structural deformation and bending stress can be inferred from the results of Model-2 to Model-3.

3.2 Structural Response to Crane Movement on the Deck

Figure 12 shows the results of static analysis for vertical displacement distribution and angular displacement around Y-axis along the line at Y = 0 m, while Fig. 13 depicts the same but at X = 30 m.

From Fig. 12a, it is observed that in Case 4 and 5 where two cranes are positioned near the center of the deck and at the tip of the protrusion, the vertical displacement reaches approximately −0.8 m. The displacement is smallest in Case 1, where both cranes are placed at the center, with a displacement of about −0.05 m. Figure 13a indicates that in the vertical direction along the Y-axis where the crane is installed, Case 3, positioned at the tip of the protrusion, shows a displacement of −2.5 m. This is influenced by the buoyancy supported by surrounding columns; in the central part of the deck like Case 1, surrounded by columns, vertical displacement is minimized.

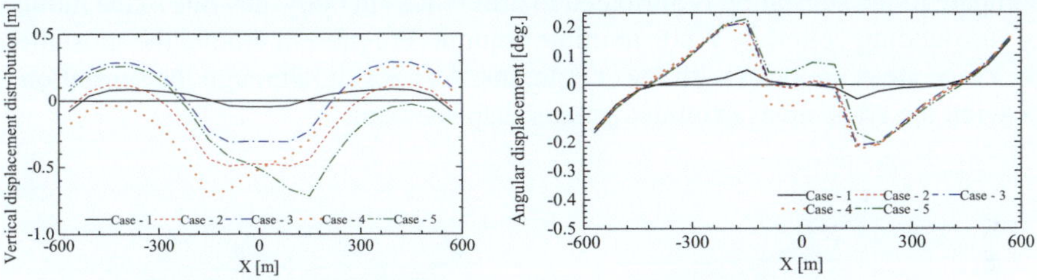

a) Vertical displacement distribution. b) Angular displacement.

Fig. 12 Displacements due to changing the positions of two cranes installed around the openings in the structure (Y = 0)

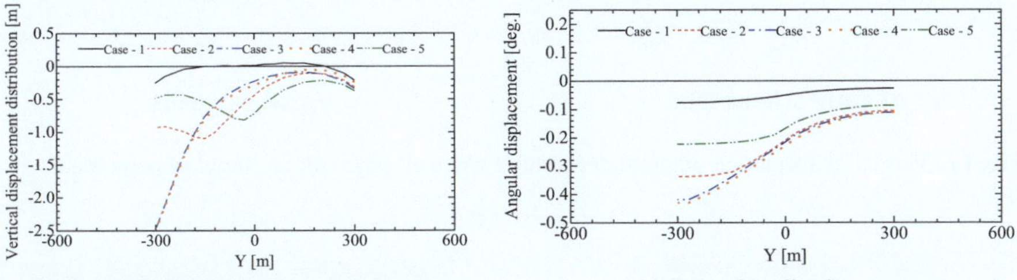

a) Vertical displacement distribution. b) Angular displacement.

Fig. 13 Displacements due to changing the positions of two cranes installed around the openings in the structure (X = 30)

However, when the crane is positioned at the tip of a protrusion as in Case 3, fewer columns can support the crane's mass, resulting in a displacement of −2.5 m.

From Fig. 12b, it is noted that angular displacement around X-axis shows a similar trend across all Cases. Only Cases 4 and 5 do not exhibit angles near 0° within the range of X = −120 to 120 m, attributed to the diagonally offset crane positions. Figure 13b illustrates that the installation of cranes causes the floater to tilt. Even in Case 3 and 4 where the crane is positioned at the tip of an opening, the inclination angle is approximately −0.4°, suggesting minimal overall structural inclination.

3.3 *Fully Loaded*

Figure 14 shows the vertical displacement distribution from the static analysis. In this numerical simulation, point loads representing a substation, two cranes, and ten 15 MW wind turbines (nacelles, and blades) are applied. At the location of the heaviest substation, the displacement is −1.0 m. In contrast, the greatest displacement occurred at the point near the crane installation location, specifically at X = −150 m, Y = −90 m, where the displacement is −1.2 m. The displacements at each crane position are both about −1.0 m. The larger displacement around the cranes compare to the substation is attributed to differences in buoyancy due to the number of surrounding columns. While multiple columns are present around the substation, the crane areas are open with fewer columns. As a result, although the substation is heavier, the crane areas exhibited greater displacements.

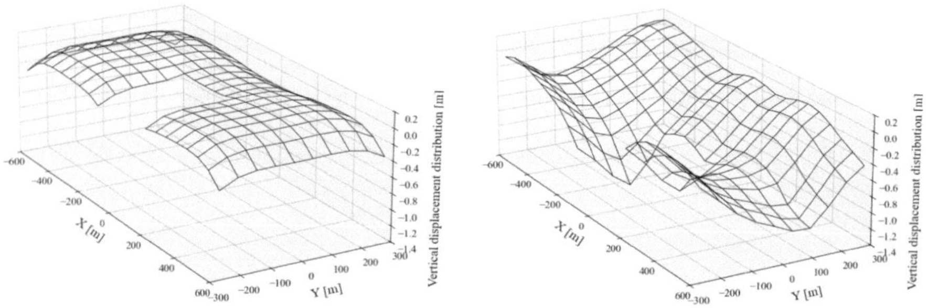

a) Only self-weight. b) All payloads.

Fig. 14 Vertical deformation diagram of Model-2 when all payloads are input as point loads

4 Conclusion

This study investigates the structural response of a mega floating structure with non-uniform mass distribution and with complex geometry, designed for a floating offshore base. The key findings are summarized as follows:

- Using models without openings (Model-1), with one opening (Model-2), and with two openings (Model-3), the effect of openings on the hydroelastic response was compared. The influence of the openings was evident in the vertical displacement and inclination angles. When focusing on bending stress, there was no significant difference in trend among the three models at the center of the deck, but at the vicinity of the openings, bending stress was approximately 1.8 times higher in Model-1 compared to the others.
- In scenarios where assembly operations involve crane movements on the deck, the structural response was investigated by varying the crane positions along the openings in Model-2. Comparing vertical displacements at different crane positions—near the center of the deck (Case-1) and at the tip of the openings (Case-3)—revealed displacements of −0.8 m and −2.5 m, respectively. This disparity is attributed to the fewer columns surrounding the crane position at the tip of the openings compared to the center of the deck. The maximum inclination angle near the crane installation was −0.4°, indicating minimal overall structural inclination due to the crane's position at the opening's tip.
- Using Model-2, the structural response was examined while loading equipment equivalent to a 1.2 GW substation, components for 10 units of 15 MW wind turbines, and 2 cranes. The results confirmed the feasibility of accommodating all planned equipment on Model-2.

Countermeasures against the effect of the openings, especially strengthening of local structures may have to be considered in the actual design.

Acknowledgements This work was supported by the Sasakawa Scientific Research Grant from The Japan Science Society.

References

1. North Sea Wind Power Hub. https://www.tennet.eu/north-sea-wind-power-hub. Last accessed 19 Jul 2024
2. CATAPULT Offshore Renewable Energy. https://guidetoanoffshorewindfarm.com/wind-farm-costs. Last accessed 19 Jul 2024
3. Yamamoto M (2024) Study on the basic concept of offshore bases for EEZ wind farms. Bachelor Thesis, Osaka University
4. Yamamoto M, Iwamatsu S, Iijima K, Shibata M, Tatsuhara H (2024) A proposal of floating offshore base for wind farm deployed in EEZ, to appear in Proceedings of WCFS2024, Hong Kong

5. Iijima K, Suzuki H, Yoshida K (1997) Structural analysis of very large semi-submersibles in waves. J Soc Naval Archit Jpn 181:281–288 (in Japanese)
6. Kagemoto H, Dick KPY (1986) Interaction among multiple three-dimensional bodies in water waves: an exact algebraic method. J Fluid Mech 166:189–209
7. Iijima K, Yoshida K, Suzuki H (1999) Structural design methodology of VLFS from the viewpoint of dynamic response characteristics. In: Proceedings of the third international workshop on very large floating structures (VLFS' 99), vol. 1. Honolulu, pp. 249–258
8. Suzuki H, Yoshida K, Iijima K, Kobayashi K (2009) Response characteristics of semisubmersible-type-megafloat in waves and accuracy of hydroelastic response analysis program VODAC. OMAE2002, vol. 2. Oslo, Norway, pp 773–781
9. Suzuki H, Yoshida K (1995) A consideration on the dynamic behavior and the structural design of large scale floating structure. J Soc Naval Archit Jpn 178:473–483 (in Japanese)
10. Aibel. https://aibel.com/news/three-giants-at-the-yard-in-haugesund. Last accessed 19 Jul 2024
11. Evan G, Jennifer R, Latha S, Frederik Z, Benjamin A, Garrett B, Nikhar A, Fanzhong M, Pietro B, Witold S, George S, Roland F, Henrik B, Katherine D, Matt S, Christopher A, Anthony V (2020) Definition of the IEA 15-Megawatt offshore reference wind. National Renewable Energy Laboratory, Golden, CO
12. MIC Co., Ltd. https://www.micjp.com/wp-content/uploads/2020/07/1ca2f8d06b3984d3f5a40d90250c0e80.pdf. Last accessed 19 Jul 2024
13. Kawasaki Heavy Industries Ltd. https://www.khi.co.jp/pressrelease/detail/20220422_1.html. Last accessed 19 Jul 2024
14. IHI Transport Machinery Co., Ltd. https://www.iuk.co.jp/crane/offshorewind.html. Last accessed 19 Jul 2024

A Transient Analysis of the Tension Leg Platform for Floating Wind Turbine

Wei Huang

Abstract Tension leg platform is one of classical supporting structures for floating wind turbine. The motion and tension response of the tendon is crucial to be estimated accurately in the detailed design stage. In this study, the floating structure model has been established considering several sources related to wind turbine. Thus, the aerodynamic analysis model, hydrodynamic model and numerical model for mooring tendons have been set up and the coupled effects have been taken into account. The aerodynamic model is based on the blade element momentum theory. The panel model of the platform has been established based on the classical potential flow software i.e. Hydrostar, and the mooring tendons has been modelled based on the lumped mass method. Several cases are designed considering the wave loads extracted from the metocean data report. One tendon sudden failure case is the focus of the present work, the enlargement of the tension of the adjacent tendon under the severe environment has been discussed thoroughly. The detailed effect of the failure of the tendon can be seen in the discussion, which would be interested in the engineering application for the tension leg floating wind turbine.

Keywords Tension leg platform · Floating wind turbine · Transient effect

1 Introduction

The development of offshore wind energy is advancing rapidly, with technology becoming a key focus for researchers. Over the next decade, offshore wind capacity is expected to increase by over 380 GW [1]. As wind farms move into deeper waters, floating structures are needed to support the turbines. These floating platforms, which are more cost-effective than fixed foundations, come in four main types: Tension Leg Platform (TLP), Semisubmersible, Spar platform, and barge. Mooring systems for these platforms include catenary, taut mooring and tendon supported mooring types.

W. Huang (✉)
School of Civil Engineering, Guang Zhou, China
e-mail: huangw288@mail.sysu.edu.cn

X. L. Zhao et al. (eds.), *Proceedings of the Fourth World Conference on Floating Solutions: WCFS 2024; 2–4 December; Hong Kong*, Lecture Notes in Civil Engineering 597, https://doi.org/10.1007/978-981-96-4569-5_49

TLP offers a smaller footprint and better vertical stability, with horizontal motion being more flexible. While TLP is not widely used for wind turbines yet, they offer significant potential due to their balance of performance and cost. Yang and Kim [2] studied an extended tension leg platform (ETLP) in the Gulf of Mexico, finding that the failure of one or two tendons alters system stiffness and natural periods. Wu et al. [3] analyzed the WindStar TLP at 160 m depth with one tendon failure, revealing limited effectiveness of shutdown strategies in reducing platform responses. Yang et al. [4] investigated a 10 MW FOWT at 110 m depth, showing that tendon failures significantly increased roll and yaw motions and peak tension. Tabeshpour et al. [5] examined the ISSC TLP at 230 m depth under tendon damage, noting similar response amplitude operators (RAOs) for tension, pitch, and roll, and critical wave heading effects. Ren et al. [6] developed a 5 MW Tri-Floater with a heave plate for 60 m depth, finding that tendon failures doubled adjacent tendon tension and affected heave, pitch, and roll motions. They also conducted an experimental study on a scaled model [7]. Besides, the transient effects when mooring failures for other types like semi-submersible or Spar platforms are also studied [8, 9]. Previous research indicates that TLP are well-established for oil and gas fields but show significant potential for wind turbines. However, there is limited study on the transient response of TLP floating offshore wind turbines (FOWTs) to sudden tendon failures or damage, especially under varying environmental conditions and for different factors such as maximum tension, yaw angle, or offset. This paper investigates these issues by designing a nine-tendon system for the OC4 FOWT. Hydrodynamic analysis using Hydrostar provides key coefficients and forces, while Orcaflex models the mooring tendons. The study includes transient analysis for single tendon failure. The paper is structured as follows: Sect. 2 details the methodology, including hydrodynamic methods, time-domain motion equations, and lumped mass methods. Section 3 covers parameters for the OC4 and mooring tendons, environmental conditions, and design cases. Section 4 presents numerical simulation results on TLP motions and tendon tensions. Section 5 concludes the study.

2 Methodology

2.1 Hydrodynamics

Normally the floating structure is assumed to be a rigid body without forward speed on the perfect fluid in some specific position, under the presence of the incidence propagative wave. So the boundary value problem for the velocity potential ϕ can be written as,

$$\nabla \phi^2 = 0 \quad \textit{in the fluid domain} \tag{1}$$

$$\phi_z = 0 \quad \text{at the sea floor} z = -h \tag{2}$$

$$g\phi_z - \omega^2\phi = 0 \quad \text{at the mean free surface } z = 0 \tag{3}$$

$$\lim_{n\to\infty} \sqrt{R}\left(\frac{\partial\phi}{\partial R} - ik\phi\right) = 0 \quad \text{in the far field} \tag{4}$$

where h denotes the water depth, R is the radial distance $\sqrt{x^2 + y^2}$. The motion equation in the frequency domain can be expressed as [10],

$$\left[-(\mathbf{M} + \mathbf{M_a}(\omega))\omega^2 - \mathbf{iB}(\omega)\omega + \mathbf{C} + \mathbf{K_A}\right]\vec{x} = i\omega\rho \iint\limits_{S_{B_0}} (\phi_I + \phi_D)\vec{N}_0 \mathrm{d}S \tag{5}$$

where $\mathbf{M}$ is the mass matrix, $\mathbf{M_a}$ is the added mass matrix, $\mathbf{B}$ is the radiational damping matrix, $\mathbf{C}$ is the hydrostatic stiffness matrix, $\mathbf{K_A}$ is the stiffness matrix, ρ is the sea water density, ω is the circular frequency, ϕ_I denotes the velocity potential of the incoming wave, and ϕ_D denotes the velocity potential of the wave system diffracted by the body.

2.2 Motion Equation in Time Domain

Based on the Newton's second law, the motion equation in the time domain can be obtained as [11],

$$[\mathbf{M} + \mathbf{M_a}(\infty)]\ddot{\mathbf{x}}(t) + \int_0^t \mathbf{K}(t-\tau)\dot{x}(\tau)\mathrm{d}\tau + Cx(t) = F(t) \tag{6}$$

where $\mathbf{K}$ is the retardation function, which can be expressed as,

$$K(t) = \frac{2}{\pi}\int_0^\infty \mathbf{B}(\omega)\cos(\omega t)d\omega \tag{7}$$

And $\mathbf{M_a}(\infty)$ can be obtained by,

$$\mathbf{M_a}(\infty) = \mathbf{M_a}(\omega) + \frac{1}{\omega}\int_0^\infty K(t)\sin(\omega t)\mathrm{d}t \tag{8}$$

2.3 Lumped Mass Method

The mooring line is simulated based on the Software Orcaflex, indeed, the lumped mass method is employed to establish the line model. One of the main assumptions of this method is that the line is divided into a finite number of elements constituting of a linear spring with the mass of the element concentrated at each node, there are two nodes for one element. The motion equation of the element can be expressed as,

$$\mathbf{M}_{\mathrm{L}}\ddot{\mathbf{S}} = T + \mathbf{F}_{\mathbf{D}} + \mathbf{F}_{\mathbf{I}} - \mathbf{F}_{\mathbf{W}} + Q \quad (9)$$

where $\mathbf{M}_{\mathrm{L}}$ denotes the mass of the element, $\ddot{\mathbf{S}}$ denotes the acceleration, $\mathbf{T}$ is the axial tension, $\mathbf{F}_{\mathbf{D}}$ and $\mathbf{F}_{\mathbf{I}}$ are the fluid induced drag and inertial force, $\mathbf{F}_{\mathbf{W}}$ denotes the weight of the element in the fluid, $\mathbf{Q}$ is the shear force.

3 TLP Wind Turbine Parameters and Designed Cases

3.1 OC4 and Tendon Parameters

This paper analyzes the OC4 semisubmersible floating offshore wind turbine using Hydrostar for hydrodynamic coefficients, including added mass and radiation damping. The setup features a 200 m water depth, 24 m diameter columns spaced 50 m apart, and a 20 m draft. The platform weighs 1.3473×10^7 kg with its center of gravity at 13.46 m. Nine tendons are designed for mooring, and panel models are illustrated in Fig. 1.

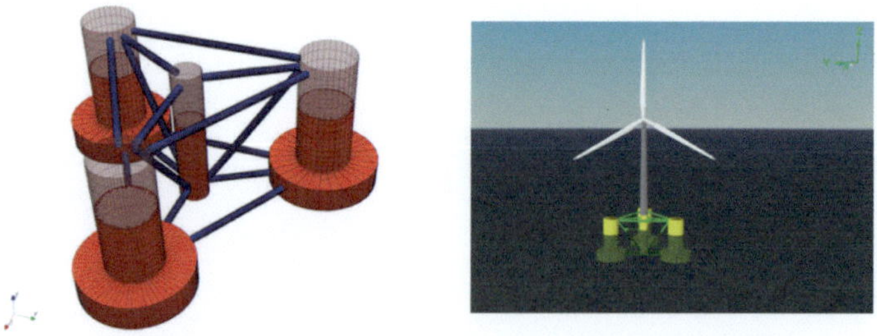

(a) Panel model (b) TLP FOWT

Fig. 1 The panel model and TLP model for OC4 wind turbine

Table 1 Environmental conditions

Case	H_S (m)	T_p (s)	γ	Heading (°)
C1	1.71	5.84	1	45
C2	3.61	8.5	1	45
C3	4.9	10.3	1	45

3.2 Designed Cases

Three designed wave loading cases are selected as the input condition for the transient analysis. Different wave significant height and peak periods are taken into account, and one wave heading 45° is chosen, and γ equals 1, the detailed parameters are all listed in Table. 1.

4 Results and Discussions

4.1 Tensions of the Tendon

This section presents and discusses the tension and motion results for the OC4 wind turbine under both intact and transient conditions. Typically, transient analysis for a single tendon breakage is essential during the design phase. Here, Tendon 7 is assumed to break at 2700 s, and the tension time series for the top of the porch over 500 s is analyzed. Figure 2a shows the tension comparisons for Tendons 2, 5, 7, and 8 between the intact case and the transient case with Tendon 7 broken. The wave significant height H_S is 1.71 m, and the peak period T_p is 5.84 s. The tension response varies significantly when a tendon fails suddenly. Before the break, tensions for all tendons are nearly identical to the intact case. Tendon 8, adjacent to the broken Tendon 7, shows the largest increase: the mean tension rises by 48.8% from 2478.24 to 3687.04 kN, and the maximum tension increases by 55.6% from 2792.96 to 4346.36 kN. Tendon 2 experiences a mean increase of 5.9% and a maximum increase of 6.6%. Conversely, Tendon 5 shows a mean decrease of 7.8% and a maximum decrease of 4.8%.

Figure 2b shows the tension time history for tendons with a significant wave height H_S of 3.61 m and peak period T_p of 8.5 s. Before Tendon 7 breaks, the tension is similar to the intact case. After Tendon 7 fails, tension changes significantly. The adjacent Tendon 8 sees a major increase, with mean tension rising by 48.7% and maximum tension by 52.6%. The mean tension of Tendon 2 increases by 5.9%, and maximum tension by 8.6%. In contrast, the mean tension of Tendon 5 decreases by 7.8% and maximum tension by 2.8%. Figure 2c illustrates the tension time history for four tendons with a significant wave height of 4.90 m and a peak period of 10.3 s. After the rupture, Tendon 8, adjacent to the broken tendon, shows the largest

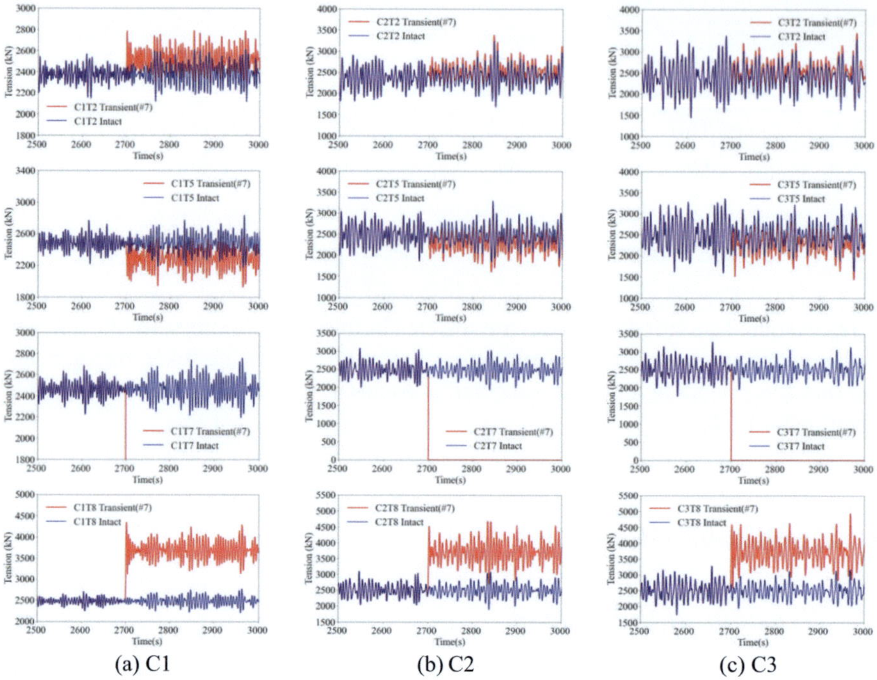

Fig. 2 Tendon tensions

increase: mean tension rises by 48.7% and maximum tension by 40.2%. The mean tension of Tendon 2 increases by 5.9%, while its maximum decreases by 6.0%. Tendon 5 experiences a 7.9% drop in mean tension and a 6.0% drop in maximum tension. This indicates the adjacent tendon undergoes the most significant increase in tension (about 50%).

4.2 Motions of FOWT

Figure 3 compares pitch responses in the time domain for intact and transient cases when Tendon 7 broke suddenly at 2700 s. The mean pitch angle increased over three times for all three cases (3.56 for C1, 3.39 for C2, and 3.48 for C3) compared to the intact state. Maximum pitch errors are 67.8% for C1, 37.0% for C2, and 18.3% for C3. Sudden tendon failure leads to significant changes in platform rotation, emphasizing the need for careful monitoring of motion and tension based on operational criteria.

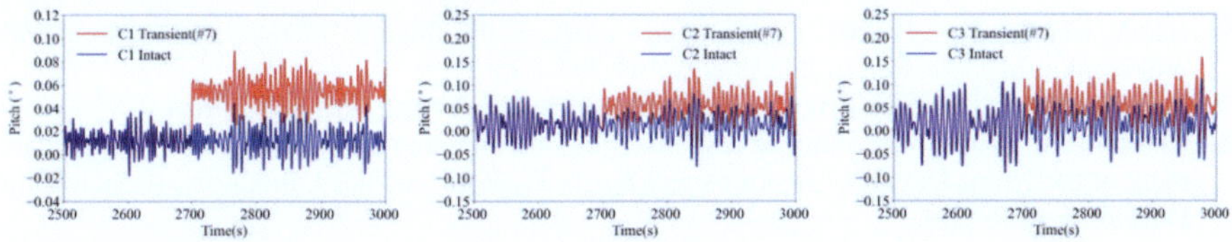

a. C1 b. C2 c. C3

Fig. 3 Pitch motion

5 Conclusions

A TLP supporting the OC4 floating offshore wind turbine features nine tendons grouped at three corners. A panel model was created, and hydrodynamic analysis was performed using Hydrostar to obtain added mass coefficients, radiation damping coefficients, and second mean drift load for the transient time-domain analysis. Tendon modeling was done with Orcaflex, using the lumped mass method for line structures. The study focused on the transient effects of a sudden break in Tendon 7, using three wave parameter sets from a 1-year return period. Key findings include, the adjacent tendon experiences a significant tension increase (40–50%) after one tendon fails, and the mean pitch motion triples, with maximum pitch changes ranging from 18 to 67%, depending on environmental conditions and sea loads. These results underscore the need for further studies to build a comprehensive database on dynamic properties, including effects of multiple tendon failures and contributions from different loading frequencies.

References

1. Global Wind Energy Council (2023) Global offshore wind report 2023, vol 19. GWEC, Brussels, Belgium, pp 10–12.
2. Yang CK, Kim MH (2010) Transient effects of tendon disconnection of a TLP by hull–tendon–riser coupled dynamic analysis. Ocean Eng 37:667–677
3. Wu HY, Zhao YS, He YP, Shao YL, Mao WG, Han ZL et al (2021) Transient response of a TLP-type floating offshore wind turbine under tendon failure conditions. Ocean Eng 220:108486
4. Yang Y, Bashir M, Michailides C, Mei X, Wang J, Li C (2021) Coupled analysis of a 10 MW multi-body floating offshore wind turbine subjected to tendon failures. Renewable Energy 176:89–105
5. Tabeshpour MR, Ahmadi A, Malayjerdi E (2018) Investigation of TLP behavior under tendon damage. Ocean Eng 156:580–595
6. Ren YJ, Venugopal V, Shi W (2022) Dynamic analysis of a multi-column TLP floating offshore wind turbine with tendon failure scenarios. Ocean Eng 245:110472
7. Ren Y, Shi W, Venugopal V, Zhang L, Li X (2024) Experimental study of tendon failure analysis for a TLP floating offshore wind turbine. Appl Energy 358:122633
8. Li Y, Zhu Q, Liu L, Tang Y (2018) Transient response of a SPAR-type floating offshore wind turbine with fractured mooring lines. Renewable Energy 122:576–588

9. Bae Y, Kim M, Kim H (2017) Performance changes of a floating offshore wind turbine with broken mooring line. Renewable Energy 101:364–375
10. Molin B (2023) Offshore structure hydrodynamics. Cambridge University Press
11. Cummins W (1962) The impulse response function and ship motion. Report 1661, Department of the Navy, David W Taylor Model Basin, Hydromechanics Laboratory, Research and Development Report

Dynamic Response Analysis of Pontoon-Type Floating Bridges Subjected to Wave and Traffic Loads

Dongqi Jiang, Bo Wu, Bin Peng, and Yuhang Shao

Abstract Floating bridges have attracted great interests in coastal countries for sea-crossing transportation infrastructures, such as Norway, Japan, Singapore, etc. Wave and traffic loads are two critical actions to be taken into consideration in the dynamic analysis of floating bridges, which involves both fluid–structure and vehicle-bridge interactions. This paper presents a fully coupled computational fluid dynamics (CFD)—finite element method (FEM) approach to predict dynamic responses of pontoon-type floating bridges when subjected to combined wave and traffic loads. The sophisticated numerical model is developed using ABAQUS and Star-CCM + co-simulation software platform, where a two-way data communication scheme is applied between CFD and FEM solvers. CFD simulation exports the fluid pressure for the determination of structural responses in the FEM solver, while pontoon motions are fed back to deform the mesh in the CFD solver. Moving vehicles are idealized as mass-spring-damper mechanical systems and the floating bridge is modeled with beam and plate elements. A simple traffic model composing of different numbers of trucks is adopted to assess the vehicle-induced dynamic response of the floating bridge. Research outcome indicates that the traffic load causes downward effects on the dynamic response of pontoon-type floating bridges. The traffic-induced dynamic impact effect varies with different spans of the floating bridge. Dynamic responses caused by the simple multi-truck load are relatively small as compared with wave action effects, and the random traffic load model is suggested to be considered in the engineering practice.

Keywords Floating bridge · Moving load · Vehicle-bridge interaction · Dynamic responses · Co-simulation

---

D. Jiang (✉) · B. Peng · Y. Shao
Nanjing University of Science and Technology, Nanjing, China
e-mail: jiangdongqi@njust.edu.cn

B. Wu
China Ship Scientific Research Center, Wuxi, China

X. L. Zhao et al. (eds.), *Proceedings of the Fourth World Conference on Floating Solutions: WCFS 2024; 2–4 December; Hong Kong*, Lecture Notes in Civil Engineering 597, https://doi.org/10.1007/978-981-96-4569-5_50

1 Introduction

Ever-increasing population and economic activities in coastal areas stimulate greater interests and demands for essential transportation infrastructures to link main cities and islands [1, 2]. During the past decade, several world-famous sea-crossing bridges have been built in China, such as Quanzhou Bay Bridge (2015), Hong Kong-Zhuhai-Macao Bridge (2018) and Lingdingyang Bridge (2023), etc. In the construction site with deep water and soft seabed, the floating bridge is considered to be a preferable engineering solution as compared to conventional bottom-founded bridge structures because it utilizes the natural buoyancy force to balance the vertical loads [3]. Under such circumstance, the floating solution is able to reduce the construction difficulties and costs significantly by omitting the need for building huge-size foundations [4, 5]. Moreover, floating bridges constructed with modular units can be relocated, expanded and removed whenever necessary, which can further decrease the costs of maintenance and reconstruction [6, 7]. Ancient floating bridges date back to 480 BC for military purposes, and several modern floating bridges were built in clam waters in the past century, such as Evergreen Point Bridge, Bergsoysund Bridge, Yumemai Bridge, etc., while no permanent floating bridges have been constructed in severe sea conditions until now [8].

In-service floating bridges are subjected to various live load actions, including wave load, wind load, traffic load, etc. Several numerical analysis approaches have been developed to study the dynamic response of discrete pontoon floating bridges. Løken et al. [9] analyzed a scaled model of Bergsoysund bridge under wave actions using the potential flow theory, while Petersen and Øiseth [10] proposed a sensitivity-based finite element method (FEM) to analyze the structural response of Bergsoysund Bridge. Xu et al. [11] developed a time-domain method to study the dynamic behavior of a suspension bridge with two floating pylons. Computational fluid dynamics (CFD) methodology provides another time-domain analysis option to simulate flow processes in general and wave actions in particular. The moving load is one of critical live loads in the design of bridge superstructures, and the vehicle-induced dynamic effect has attracted great research interests. The vehicle-bridge interaction system consists of two sub-systems: the vehicle and the bridge, which can be idealized with mass-spring-damper units and FEM models [12]. It is quite challenging to properly analyze the discrete pontoon floating bridge when subjected to the combined wave actions and moving vehicle loads because the whole system involves both fluid–structure and vehicle-bridge interaction problems. The effects of aforementioned two live loads on dynamic responses of floating bridge superstructures need further thorough investigations.

This paper presents a co-simulation method using CFD solver Star-CCM + and FEM solver ABAQUS to predict the dynamic response of discrete pontoon floating bridges when subjected to combined wave and moving vehicle loads. A concrete pontoon-type floating bridge case is taken as an example to illustrate the numerical model development of the vehicle-bridge-wave system. The traffic-induced dynamic amplification effect is further evaluated through parametric studies, which provides the guidance for a reliable and economical design of floating bridges.

2 Numerical Methodology

A two-way data communication scheme is applied between CFD solver Star-CCM+ and FEM solver ABAQUS to develop numerical models of discrete pontoon floating bridges. Figure 1 depicts CFD-FEM co-simulation flow chart for the analysis work. The external fluid pressure exported from the CFD simulation is used to derive the structural response in the FEM solver, and structural deformations are fed back into the CFD solver to deform the mesh.

Figure 2a–c show the bridge and vehicle models developed in the ABAQUS software. The bridge superstructure is modeled with S4R shell elements or B31 generalized beam elements based on different simplification considerations. The pontoon is modeled with S4R shell elements. The mass and rotational inertia values are assigned to the reference point attached at the centroid location of the discrete pontoon. Multi-point constraints (MPC) method is used to connect the bridge superstructure and discrete pontoons. The floating bridge case considered herein owns a uniform span length of 50 m and an overall length of 300 m with the prestressed concrete twin-box single cross section, whose boundary conditions are fixed at both ends. Table 1 lists key parameters in the modelling of the beam and pontoon.

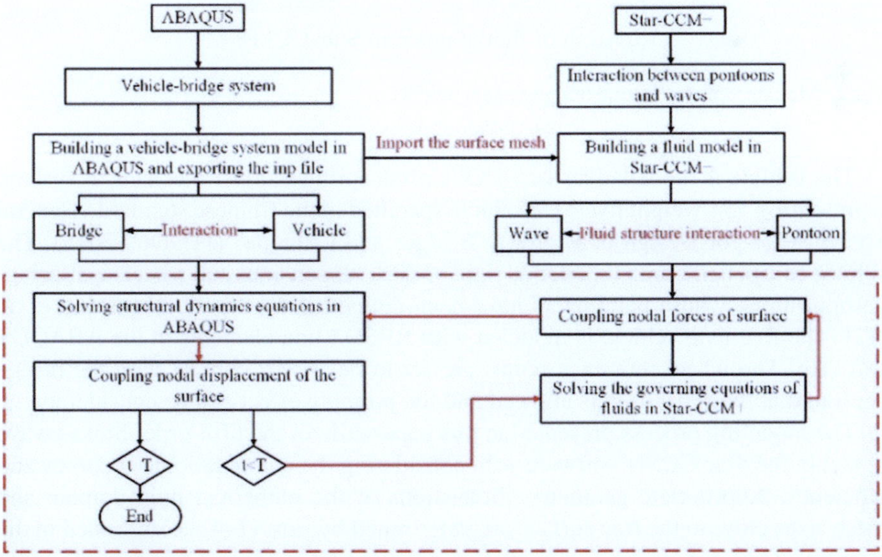

Fig. 1 CFD-FEM co-simulation flow chart

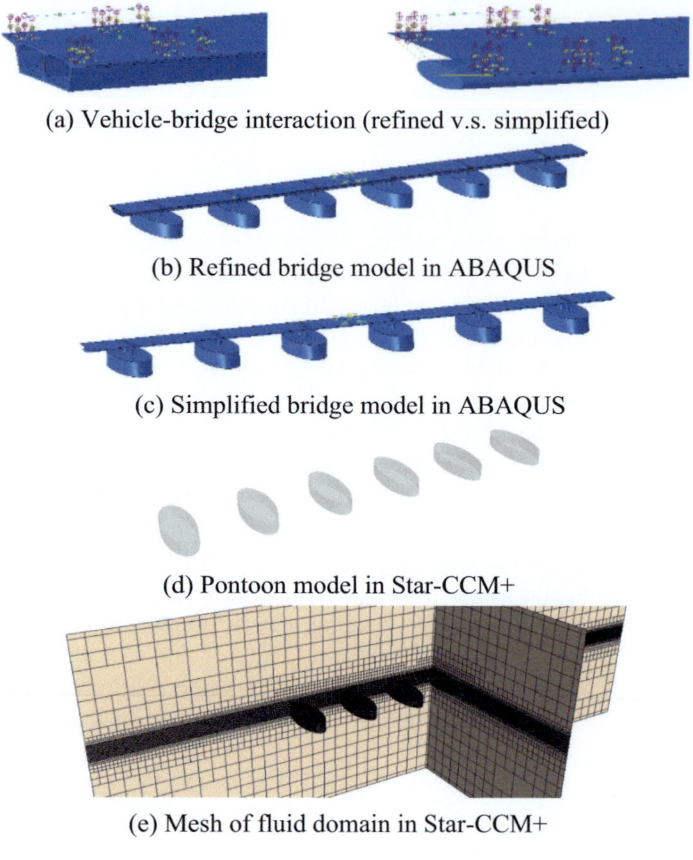

Fig. 2 Moving vehicle-floating bridge system model

The vehicle is modelled as a 15-DOFs mass-spring-damper mechanical system, representing 55 t-weight five-axle vehicle specified in the Chinese standard "General Specification for Design of Highway Bridges and Culverts" (JTGD60-2015). The vehicle components are modelled as rigid bodies, suspensions and wheels via spring-dashpot units, whose parameters have been defined and calibrated by Deng et al. [12]. The five-axle vehicle is modeled with RB3D2 line elements in the ABAQUS software. The wheel reference points are set to be "hard" contact with the bridge deck so that the separation is allowed and the jumping effect can be considered.

The modeling process presented in this paper utilizes the fifth-order Stokes wave model in the Star-CCM+ software with considering the computational accuracy and efficiency. Appropriate geometric dimensions of the numerical fluid domain and mesh sizes close to the free surface are determined by Eqs. (1–4) as suggested in the reference [13]. Figure 2d, e show the pontoon model in Star-CCM+ and the mesh arrangement of the fluid domain. Geometric sizes of the fluid domain are 600 m (length) × 300 m (width) × 200 m (height) and the basic mesh size is 12 m. For the

Table 1 Parameters of beams and pontoons

Components	Parameter	Unit	Value
Beam	Mass, m	t/m	36.23
	Area, S	m^2	14.68
	Moment of inertia, I_{xx}	m^4	72.3
	Moment of inertia, I_{yy}	m^4	425
	Moment of inertia, I_{zz}	m^4	23.1
Pontoon	Major axis, a	m	50
	Minor axis, b	m	20
	Depth, h	m	9
	Draft, d	m	5.75
	Mass, M	t	2350
	Centroid of gravity, COG	m	4
	Roll inertia, I_{tr}	t·m^2	5.02×10^5
	Pitch inertia, I_{tp}	t·m^2	1.105×10^5
	Yaw inertia, I_{ty}	t·m^2	5.48×10^5

range of 1 m in height close to the free surface, horizontal and vertical mesh sizes are 2.64 m and 0.1 m, respectively.

$$L \geq 2\lambda + l \tag{1}$$

$$D \geq 2\lambda + d \tag{2}$$

$$40 \leq \lambda / m \leq 100 \tag{3}$$

$$h / n \geq 20 \tag{4}$$

where L and D are the length and width of the fluid domain. λ is the wavelength, l and d are the length and width of the pontoon, m and n are mesh sizes in the horizontal and vertical directions, h is the wave height.

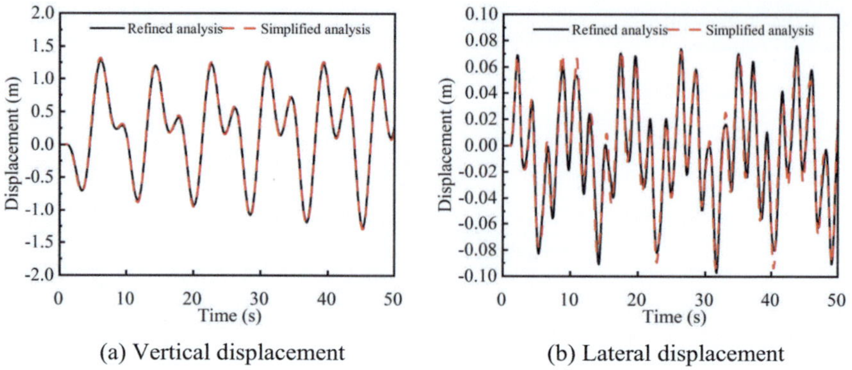

Fig. 3 Time-history responses of interior pontoons using two different modeling approaches

3 Results and Discussion

3.1 Comparison of Different Modeling Techniques

The use of shell elements in bridge superstructures increases the computation cost, especially for large-scale bridge models, which is not practical for the numerical parametric study. Therefore, the simplified model (Fig. 2c) developed with generalized beam section elements is proposed and compared with the refined analysis model (Fig. 2b). Note that equivalent beam section properties representing the original bridge section configuration are determined by carefully comparing static responses of both models, such as flexural rigidity EI_y and EI_z, torsional rigidity GJ, etc. Analysis results indicate that mode shapes from the first mode to tenth mode are the same for both refined and simplified analysis models and the corresponding natural frequency values are very close to each other. Figure 3 depicts the vertical and lateral displacement time-history responses of interior pontoons in floating bridges under only wave actions, which is obtained by both refined and simplified analysis models. It can be seen that the displacement magnitude and time-history curve trends are almost identical for two modeling techniques. Therefore, the simplified analysis approach with user-defined beam section is reliable and it will be used to perform the numerical parametric study on discrete pontoon floating bridges when subjected to combined wave actions and moving vehicle loads.

3.2 Vehicle-Induced Dynamic Amplification Effect

In order to explore the dynamic effect of moving vehicles on the response of discrete pontoon floating bridges, two-truck and six-truck loaded cases are analyzed without considering the road surface roughness. Note that the six trucks are arranged in three rows with the spacing of 7.2 m between two adjacent vehicles in the longitudinal direction. The vehicles are symmetrically arranged across the transverse direction and run along the path at a speed of 10 m/s. Figures 4 and 5 show vertical displacement time history curves at each span of the discrete pontoon floating bridge when subjected to two and six moving vehicles. The regular wave action with period of 8.5 s is also considered in the analysis. The magnitude of static displacements is labelled in the right vertical axis. It is seen that the vehicle load generally causes downward action effects as compared with the curves without vehicle loads. The downward effect in the six-truck load case is more obvious due to a heavier load action, but the curve trend does not change significantly. As compared with the wave action, dynamic responses caused by two or six trucks are quite small. It is essential to consider the random traffic load in the future analysis to represent the real live load situation.

4 Conclusion

A CFD-FEM co-simulation method is developed to analyze dynamic responses of discrete pontoon floating bridges subjected to combined wave and moving vehicle loads. By comparing with the refined shell element model, the simplified analysis model using beam elements is capable of assessing the dynamic response of floating bridges. Analysis results indicate that the traffic load causes downward effects on the dynamic response of pontoon-type floating bridges. Dynamic responses caused by the simple multi-truck load are relatively small as compared with wave action effects, and the random traffic load model is suggested to be considered in the future analysis to represent the real live load situation.

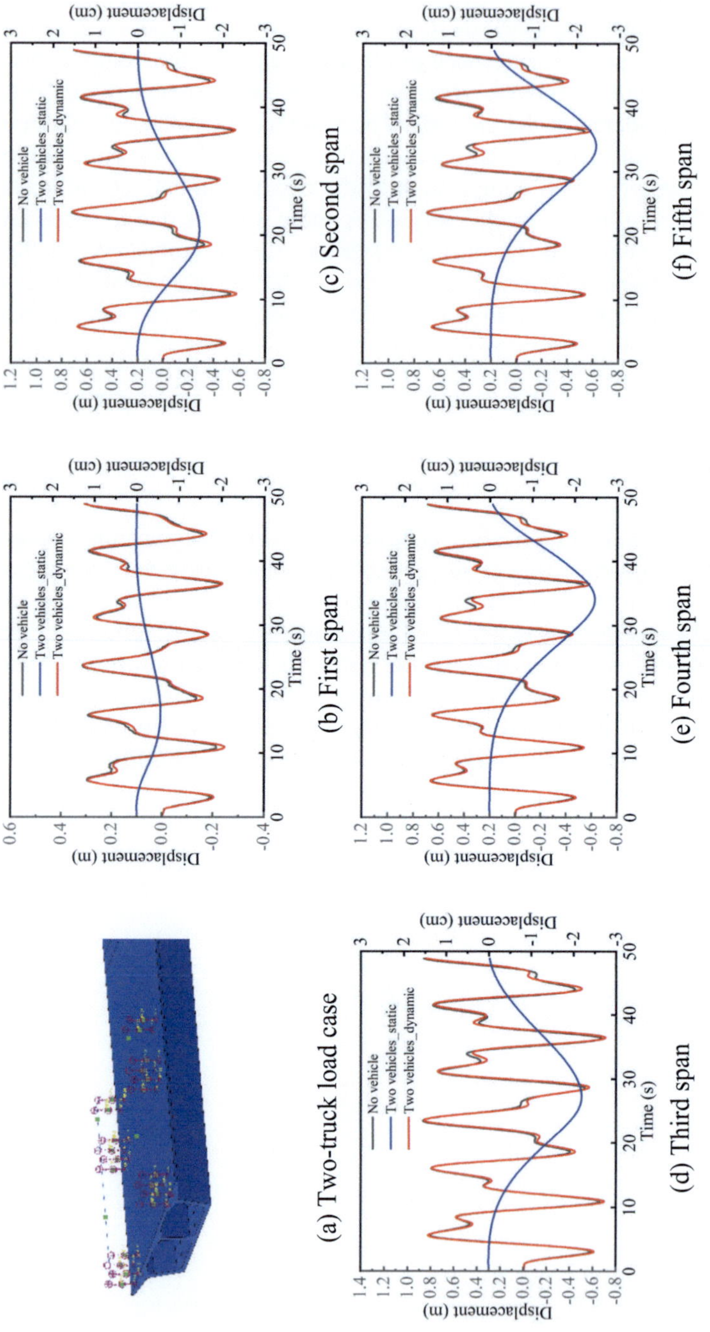

Fig. 4 Vertical displacement responses at each span of the floating bridge subjected to two moving vehicles and the wave action

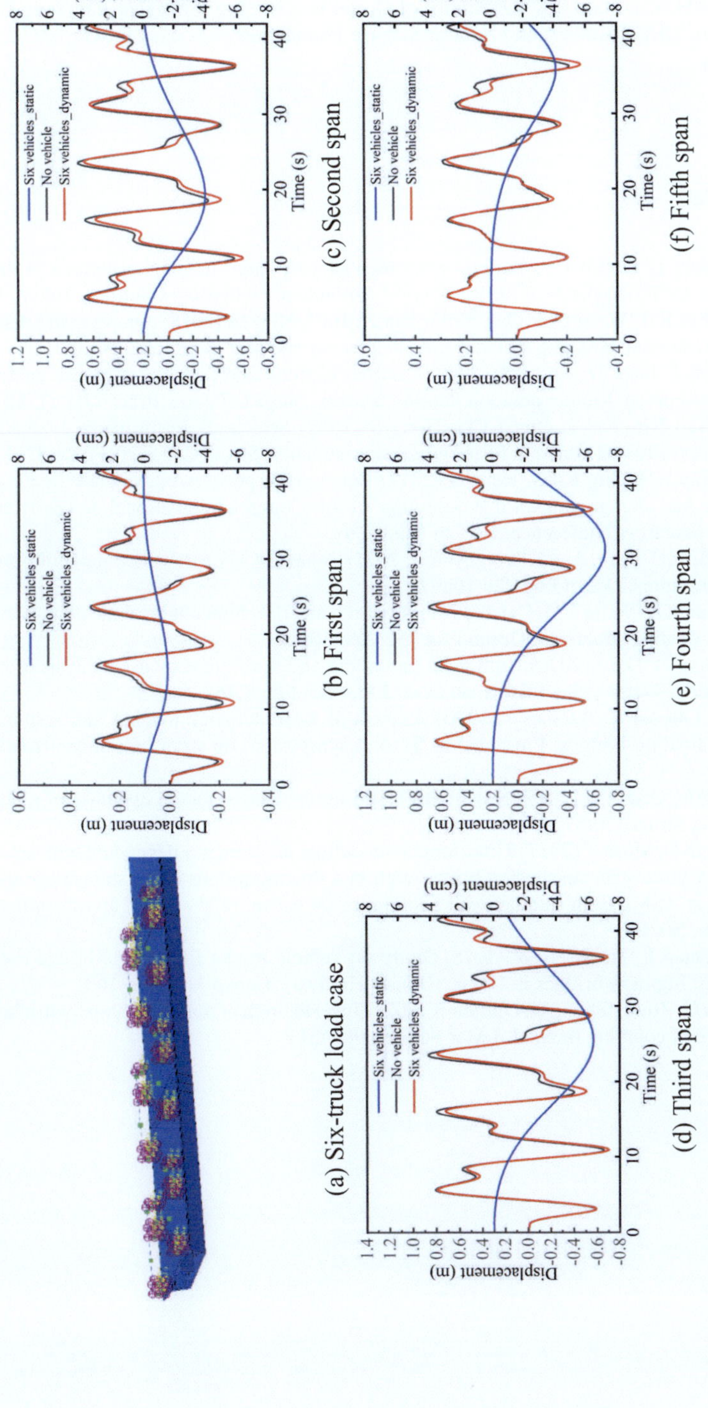

Fig. 5 Vertical displacement responses at each span of the floating bridge subjected to six moving vehicles and the wave action

Acknowledgements This work is supported by Fundamental Research Funds for the Central Universities (Grant No. 30922010907), National Natural Science Foundation of China (Grant No. 52478180, 51808292), State Key Laboratory of Ocean Engineering (Shanghai Jiao Tong University) (Grant No. GKZD010089) and Natural Science Foundation of Jiangsu Province (Grant No. BK20180487).

References

1. Luo C, Jiang D (2023) Hydroelastic responses of very large floating structures in damage conditions. In: Proceedings of the third world conference on floating solutions. Tokyo, Japan
2. Jiang D, Tan KH, Wang CM, Ong KCG, Bra H, Jin J, Kim M (2018) Analysis and design of floating prestressed concrete structures in shallow waters. Mar Struct 59:301–320
3. Wan L, Dai J, Jiang D, Ang KK (2022) Parametric study and dynamic response analysis of three single curved discrete pontoon floating bridges. Ships Offshore Struct 17(8):1788–1801
4. Jiang D, Tan KH, Dai J, Ong KCG, Heng S (2019) Structural performance evaluation of innovative prestressed concrete floating fuel storage tanks. Struct Concr 20(1):15–31
5. Jiang D, Tan KH, Ong KCG, Heng S (2017) Behavior of prestressed concrete floating self-stabilizing fuel storage tanks. In: Proceedings of 4th Congrès International de Géotechnique Ouvrages structures conference. Ho Chi Minh City
6. Jiang D, Tan KH, Dai J, Ang KK, Nguyen H (2021) Behavior of concrete modular multi-purpose floating structures. Ocean Eng 229:108971
7. Jiang D, Tan KH, Wang CM, Dai J (2021) Research and development in connector systems for very large floating structures. Ocean Eng 232:109150
8. Wan L, Jiang D, Dai J (2020) Numerical modelling and dynamic response analysis of curved floating bridges with a small rise-span ratio. J Mar Sci Eng 8:467
9. Løken A, Oftedal R, Aarsnes J (1990) Aspects of hydrodynamic loading and responses in design of floating bridges. Presented at Second symposium on strait crossings, Trondheim, Norway
10. Petersen ØW, Øiseth O (2017) Sensitivity-based finite element model updating of a pontoon bridge. Eng Struct 150:573–584
11. Xu Y, Øiseth O, Moan T (2017) Time domain modelling of frequency dependent wind and wave forces on a three-span suspension bridge with two floating pylons using state space models. Presented at ASME 36th International conference on ocean, offshore and arctic engineering. Trondheim, Norway
12. Deng L, Duan L, He W, Ji W (2018) Study on vehicle model for vehicle-bridge coupling vibration of highway bridges in China. China J Highway Transp 31(7):92–100
13. Sun Z, Liu G, Zou L, Zheng H, Djidjeli K (2021) Investigation of nonlinear ship hydroelasticity by CFD-FEM coupling method. J Mar Sci Eng 9(5):511

Development and Verification of a Simulation Tool for Floating Offshore Wind Turbines Using MBDyn

Tomoaki Utsunomiya, Ryoya Hisamatsu, Takayuki Morifuji, Kodai Uematsu, Riku Wakasa, Hisashi Okubo, Hayami Sato, and Takuya Kamijo

Abstract For design of Floating Offshore Wind Turbines (FOWTs), time domain simulation of the FOWT model for many Design Load Cases (DLCs) is essential. For this purpose, several commercial codes for aero-hydro-elastic-mooring dynamics coupled analysis are under development. In this paper, we attempt to develop such a code using opensource multibody dynamics solver MBDyn. Particular attention is made to use an elastic beam element (beam3) in MBDyn, instead of using linear springs to connect rigid bodies as before. Then, verification of the simulation code and models using MBDyn has been made by comparing with OrcaFlex. The Volturn semi-submersible model is the model used here for comparisons. This tool would be valuable for checking and verifying their FOWT models to be used with commercial codes.

Keywords Floating offshore wind turbine · FOWT · Simulation · MBDyn

1 Introduction

In order to make floating offshore wind power generation practical and the main power source, there has been a noticeable trend in recent years toward larger wind turbines to reduce power generation costs. In 2020, IEA Wind TCP Task 37 released a 15 MW Reference Wind Turbine [1], and the community is expected to develop technology for large wind turbines.

T. Utsunomiya (✉) · R. Hisamatsu
Department of Marine Systems Engineering, Kyushu University, Fukuoka, Japan
e-mail: utsunomiya@nams.kyushu-u.ac.jp

T. Morifuji · K. Uematsu · R. Wakasa
Department of Naval Architecture and Ocean Engineering, Kyushu University, Fukuoka, Japan

H. Okubo · H. Sato · T. Kamijo
Engineering Department-Offshore Wind, Eurus Energy Holdings Corporation, Tokyo, Japan

X. L. Zhao et al. (eds.), *Proceedings of the Fourth World Conference on Floating Solutions: WCFS 2024; 2–4 December; Hong Kong*, Lecture Notes in Civil Engineering 597, https://doi.org/10.1007/978-981-96-4569-5_51

The design of floating offshore wind power generation facilities considers complex natural environmental conditions such as wind, waves, and currents, as well as structures such as floats, moorings, nacelles, and blades under these conditions. In the load analysis, coupled analysis is required after modeling fluid forces and structural elements under many assumptions in order to handle a large number of design load cases. For this reason, the reliability of modeling and analysis tools is extremely important. In response to this, Utsunomiya et al. [2] considered that numerical analysis using multibody dynamics (MBD) would be useful to consider the finite amplitude motion of floats and structures, and developed a design tool using a general-purpose analysis software MSC Adams. The validity of this tool has been demonstrated for a 2 MW spar-type offshore wind turbine in actual sea areas [2, 3].

On the other hand, one of the practical challenges in dealing with a large number of design load cases is the limitation imposed by the number of licenses for existing commercial codes for simultaneous parallel analysis. For this reason, the authors' research group has developed a design tool for offshore wind turbines using the open source general-purpose MBD analysis software MBDyn [4]. It has been confirmed that this tool is capable of performing analysis equivalent to MSC Adams for spar-type offshore wind turbines [5].

For large wind turbines above 15 MW scale, the applicable water depth becomes deeper and installation of wind turbine onto the spar-type floaters becomes more difficult, so it is considered that semi-submersible floaters are more practicable. Therefore, the purpose of this study is to establish a design tool for semi-submersible offshore wind turbines using MBDyn. In this paper, we report on the modeling of the aforementioned IEA 15 MW offshore wind turbine and the effectiveness of the analysis by comparing with the commercial code OrcaFlex [6].

2 Target Model

The analysis target of this study is the IEA 15 MW Offshore Reference Wind Turbine [1] and its floating platform VolturnUS-S [7], which are open source and available for the purpose of promoting technological development in the community. Figure 1 shows an outline of the design.

VolturnUS-S is a semi-submersible floating body consisting of four cylindrical columns and three rectangular cross-section lower hulls. When the effects of diffraction/radiation due to short-period wave conditions cannot be ignored, it is necessary to evaluate the hydrodynamic forces by treating the floating body as a rigid body using potential theory. On the other hand, in wave conditions with long periods and large floating body motion, which are the range of application of the Morison's formula, analysis using a three-dimensional elastic frame model that can consider nonlinearities such as finite amplitude motion and drag, and can directly obtain the cross-sectional forces of the members, is more faithful and efficient. In this paper, we first construct a case that can be considered as a three-dimensional frame model using Morison's formula for wave loading.

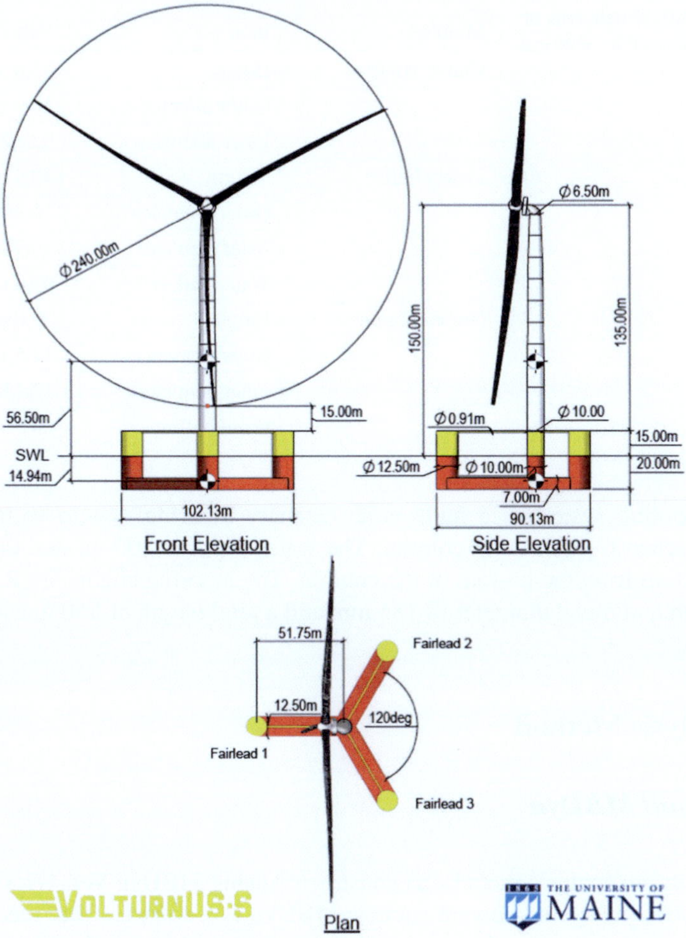

Fig. 1 General arrangement of the VolturnUS-S

The lower hull of the float is a horizontal member with a rectangular cross section, and the viscous effects of vortex shedding from the bilge are particularly significant. In addition, careful discussion of the effects on the angle of attack is thought to be necessary. In this paper, the lower hull was first changed from a rectangular member to a cylindrical member with equivalent mass and cross-sectional area, and an analysis was performed. The specifications of each member of the semi-submersible float are shown in Table 1. The density of the structural material of each member is 8.5 t/m^3, and the Young's modulus is 210 GPa. In addition, the hydrodynamic coefficients are assumed to be the drag coefficient Cd of 0.6 and the added mass coefficient Ca of 1.0 for each member. In addition, the drag coefficient Cdv = 2.0 for the bottom of the column member, and the added mass coefficient Cav = 1.0 based on a hemisphere.

Table 1 Main dimensions of the semisubmersible platform

Member	Item	Value
Center column	Height	35.0 m
	Outer dimeter	10.0 m
	Inner diameter	9.889 m
Lower hull	Length	40.5 m
	Outer diameter	10.869 m
	Inner diameter	10.744 m
	Water ballast	93.0 t/m
Outer column	Height	35.0 m
	Outer diameter	12.5 m
	Inner diameter	12.431 m
	Fixed ballast	847 t

The mooring system is a multi-point catenary mooring system with mooring chains attached to each outer column. The water depth is 200 m, and the fairlead is located 6 m from the bottom of the column. The mooring chains are R3 studless chains with a nominal diameter of 185 mm and a total length of 850 m each.

3 Analysis Method

3.1 About MBDyn

In this study, an analysis model was constructed using MBDyn Ver. 1.7.3. General-purpose MBD analysis software such as MBDyn is a software that mechanically derives equations of motion for any coupled system by inputting information on interactions between rigid bodies (kinematic constraints, elastic forces, etc.) and external forces, and calculates the motion and internal forces by solving them numerically.

MBDyn is an open source software, but has a high degree of completeness and capability of extension. In our previous research, it has been confirmed that analysis of spar-type floating bodies is practically possible by adding buoyancy, fluid forces, and mooring reaction forces as extension modules [5]. In addition, MBDyn has an aeroelastic module implemented because it was developed for rotorcraft in the field of aeronautics, and it has been suggested that practical analysis is also possible for wind turbines. In addition, it has been confirmed that the aerodynamic calculation module AeroDyn, which is released by the National Renewable Energy Laboratory (NREL), can also be introduced as an extension module.

In our previous study, the elastic bodies of the floating body, tower, and blades were modeled as multiple rigid bodies and springs. However, it has been reported that the analytical results of this model differed from those of analytical models whose validity has been confirmed [5]. Therefore, in this study, we have tried to use the three-node beam (beam3) element implemented in MBDyn to analyze the elastic bodies.

Aerodynamic Element. An aerodynamic load calculation element implemented in MBDyn. Aerodynamic loads are evaluated using blade element theory. On the other hand, when dealing with a rotor, the effect of induced velocity must be considered using momentum theory. In this module, the induced velocity is calculated by converging the following equation at each time step.

$$T - 2\rho Au\sqrt{(u+v)^2 + w^2} = 0 \tag{1}$$

Here, u is the induced velocity, v is the opposing component of the relative velocity of the rotor plane, and w is the parallel component. T is the rotor thrust evaluated by the blade element theory considering the induced velocity, ρ is the air density, and A is the area of the rotor. T in Eq. (1) refers to the aerodynamic component of the axial force on the rotor shaft. In other words, it should be noted that Eq. (1) calculates the induced velocity under the assumption that the inflow velocity to the rotor is uniform and steady.

User Defined Element. An extension element that users can add to MBDyn. In this study, we introduce a fluid force element (Spardyn Element) and a mooring reaction force element (Catenary Element) for slender members. The effect of hydrostatic pressure is applied to the submerged nodes as a nodal load based on the hydrostatic pressure distribution, not as a vertical buoyancy force. The effect of dynamic hydrodynamic pressure is evaluated using the Morison's formula, which considers the relative velocity between the node and the water particles. The mooring reaction force is evaluated quasi-statically by catenary theory that does not consider elongation and dynamic effect, based on the positional relationship between any nodes and a spatially fixed anchor point.

3.2 Floating Body Model

Figure 2 shows an overview of the modeling of the floating body. The floating body was constructed as a 3D elastic frame model using beam3 element. The number of elements for each member is five (e.g. 11 nodes) for the lower hull and four (e.g. 9 nodes) for the column. However, as shown in Fig. 2, the connection part of each column member with the lower hull is modeled as a rigid body, and the relative displacement of the connection point is restrained. The fluid force is directly applied to each node by the Spardyn Element. For each end face of the member, the end face of the lower hull is actually the internal face connected to the column, but a load due to hydrostatic pressure is applied to ensure consistency with the hydrostatic pressure distribution on the column side. On the other hand, the effect of dynamic

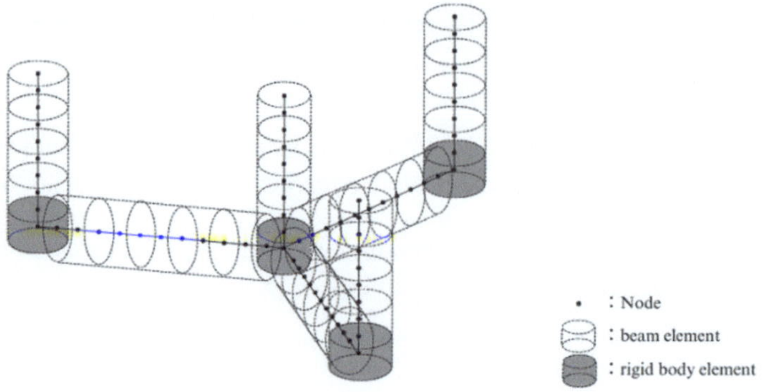

Fig. 2 Model configuration of the platform

water pressure is not applied as a fluid force here. This point may need to be discussed in future when considering actual phenomena. The mooring reaction force is set so that the fairlead's position in the global coordinate system can be referenced at each time step by constraining the node representing the fairlead to the nearby node.

3.3 Wind Turbine Model

The tower is modeled using beam3 element. However, the aerodynamic load acting on the tower is not considered here. The wind turbine to be analyzed has a direct drive mechanism. Here, the generator is not modeled in detail, and only the rigid body and constraints are modeled. The drive-train system is represented by the nacelle as a rigid body, and the rotor shaft is represented by constraining the axial position relative to the nacelle. The blades are modeled with 16 elements using beam3 element and aerodynamic element. There are four integration points for the aerodynamic load for each element, and the aerodynamic coefficients and chord length at the integration points are evaluated by interpolation. The inertia matrix, aerodynamic coefficients, stiffness matrix, Rayleigh damping coefficients, etc. of the beam required for modeling were input to MBDyn by interpolating and coordinate conversion from the publicly available OpenFAST input file [8].

3.4 Control System

In this paper, we first focus on confirming the validity of MBDyn for each structural element and load evaluation module, and only introduce a simple control system. In the wind turbine analyzed, in the reference wind speed range of 6.98–10.59 m/s, control is performed by a torque controller with the optimal value of Tip Speed Ratio

(TSR) as the target value when the blade pitch angle is 0 degree. In the constructed model, a simple PI controller for this range was implemented using only the MBDyn input file.

3.5 Comparison Model Using OrcaFlex

OrcaFlex is a commercial software for dynamic analysis of marine structures, and has been especially trusted by the offshore petroleum industry. In recent years, wind turbine functions have been developed, and analysis of floating wind turbines has become practical. In this study, we verify the effectiveness of MBDyn by comparing the analysis results for an equivalent design object between MBDyn and OrcaFlex. An example modeling of the IEA 15 MW wind turbine, which is the subject of our research, has been published by OrcaFlex. However, since the floating body is a rigid body and the fluid forces are calculated using potential theory, we modeled it as an elastic frame structure by combining line elements, as in the case of MBDyn.

4 Analysis Results

4.1 Regular Waves

Regular waves with the wave height of $H = 4$ m and the period $T = 10$ s are applied. The results are shown in Fig. 3 for platform motions, sectional forces at the tower base, and the mooring reaction forces. Here, the wind turbine is modeled only as a point mass. As we can see in Fig. 3, good agreement of the results between MBDyn and OrcaFlex can be observed.

4.2 Regular Waves and Constant Wind

Regular waves with the wave height of $H = 1.4$ m and the period $T = 10$ s together with a constant wind speed of 7 m/s are applied. The results are shown in Fig. 4 for platform motions, sectional forces at the tower base, the rotor speed, the generator torque and the generator power. Here, the wind turbine is modeled as a whole model using the IEA 15 MW wind turbine specifications. As we can see in Fig. 4, basically a good agreement of the results between MBDyn and OrcaFlex can be observed. However, some discrepancy is also seen, probably because of the incomplete aerodynamic modelling of the wind turbine in MBDyn when using aerodynamic element. This point will be further considered in future research.

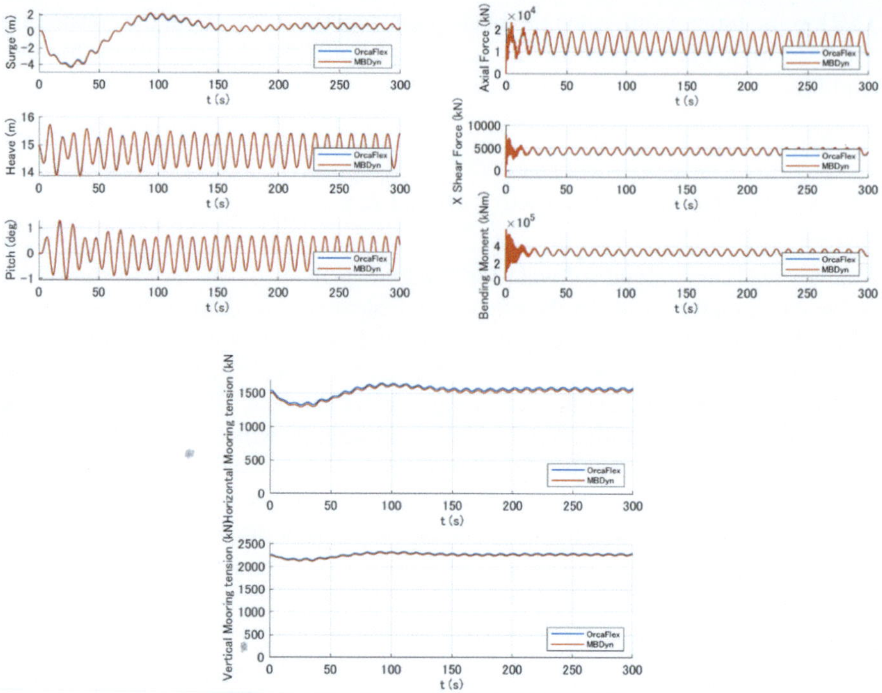

Fig. 3 Numerical results for VolturnUS-S model under regular waves of $H = 4$ m and $T = 10$ s

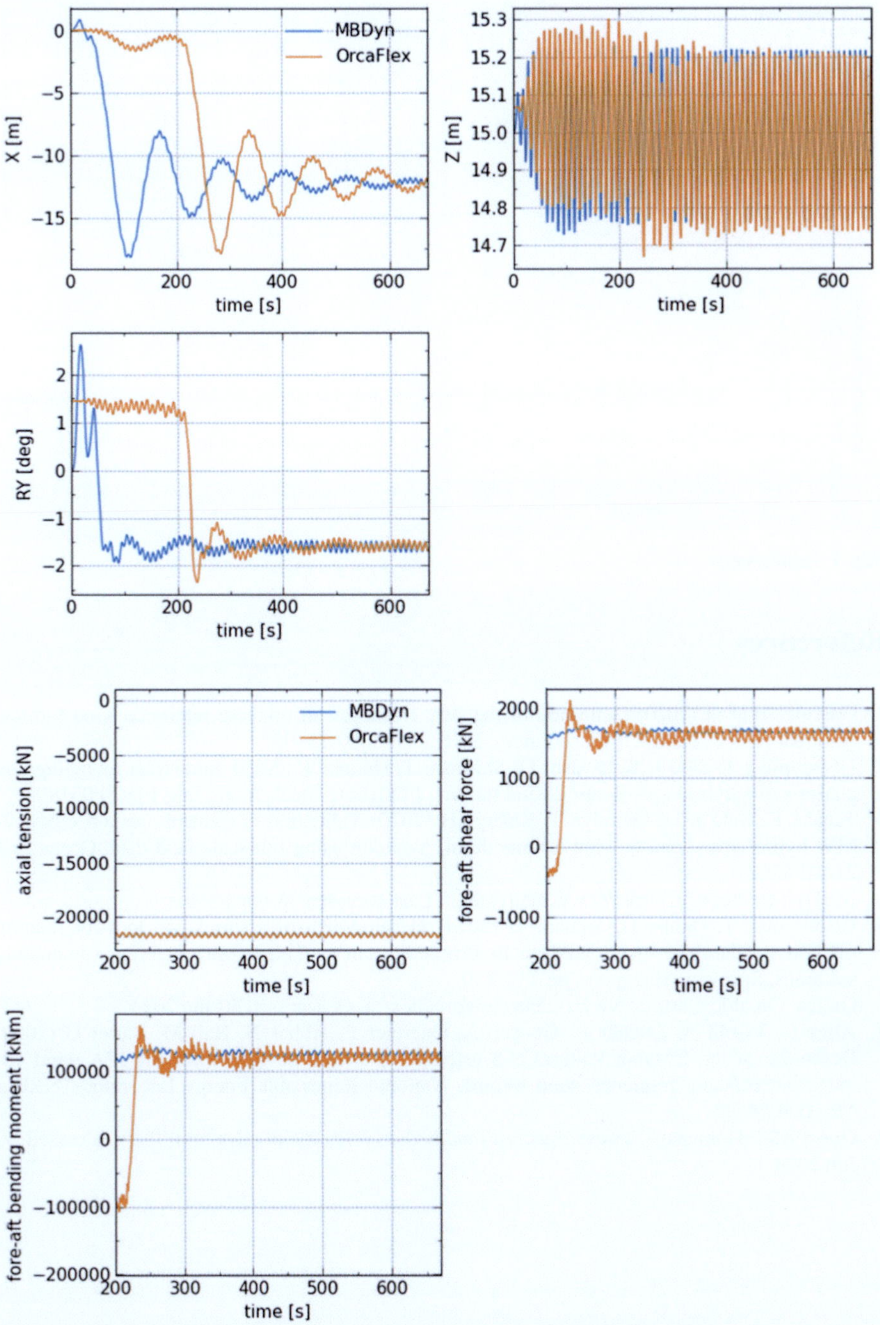

Fig. 4 Numerical results for VolturnUS-S model under regular waves of $H = 1.4$ m and $T = 10$ s with a constant wind speed of 7 m/s

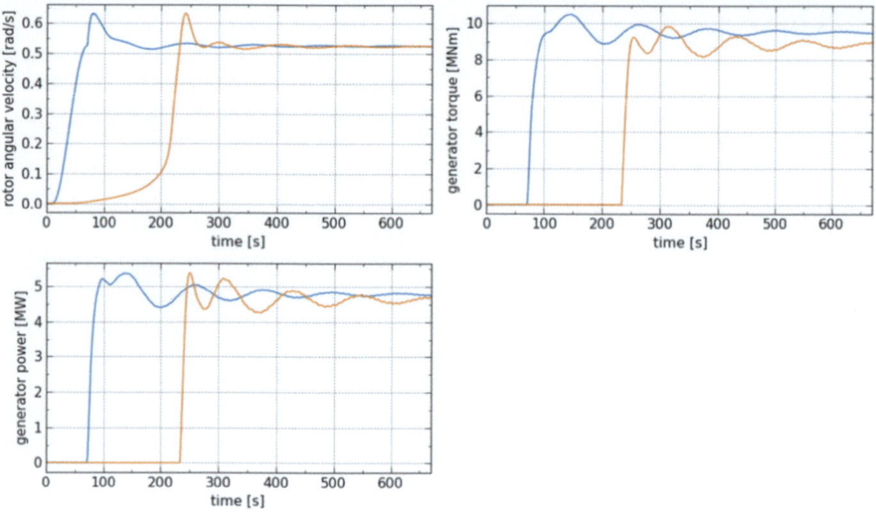

Fig. 4 (continued)

References

1. Gaertner E et al (2020) Definition of the IEA 15-megawatt offshore reference wind turbine. National Renewable Energy Laboratory. NREL/TP-5000-75698
2. Utsunomiya T, Sato I, Kobayashi O, Shiraishi T, Harada T (2019) Numerical modeling and analysis of a hybrid-spar floating wind turbine. J Offshore Mech Arctic Eng 141(3):031903
3. Tanaka K, Sato I, Utsunomiya T, Kakuya H (2020) Validation of dynamic response of a 2-MW hybrid-spar floating wind turbine during typhoon using full-scale field data. Ocean Eng 218:108262
4. MBDyn Homepage. https://www.mbdyn.org/. Last accessed 30 Jun 2024
5. Utsunomiya T, Okubo H, Yamada H (2024) Development of a simulation tool for floating offshore wind turbines using MBDyn. In: Proceedings of the Third world conference on floating solutions. Springer, Singapore, pp 737–750
6. Orcica. OrcaFlex, https://www.orcina.com/orcaflex/. Last accessed 30 Jun 2024
7. Allen C, Viselli A, Dagher H, Goupee A, Gaertner E, Abbas N, Hall M, Barter G (2000) Definition of the UMaine VolturnUS-S reference platform developed for the IEA wind 15-megawatt offshore reference wind turbine. National Renewable Energy Laboratory. NREL/TP-5000-76773
8. OpenFAST Homepage. https://openfast.readthedocs.io/en/main/index.html. Last accessed 30 Jun 2024

Wave Attenuation Effect of Various Breakwater for Tropical Climate

Zhi Yung Tay and Nyan Lin Htoo

Abstract This study investigates the vital role of floating breakwaters in mitigating wave energy in intermediate water environments, crucial for shoreline protection and ecosystem preservation. Various floating breakwater types are analyzed for their effectiveness in attenuating the wave force. Through numerical simulations of computational fluid dynamics in ANSYS Fluent, different breakwater cross sections are evaluated in time-series domain for wave approaching at varying wave periods. The comparative analysis encompasses a range of breakwater designs, including box-type, stepped-box type, pontoon-type, box-with-plates type, dual-box type and T-shaped breakwater. The transmission of waves when passing through these breakwaters are studied to assess breakwater efficiency in mitigating the wave forces. The incident and transmitted wave elevations in each breakwater type observed at a selected time among the upstream and downstream locations are tracked and compared. The most efficient breakwater type for wave attenuation is then chosen to demonstrate its impact over time. The study aims to inform decision-makers and coastal engineers on selecting optimal floating breakwater solutions suitable for tropical climate. This research enhances our understanding of wave attenuation, aiding coastal protection and sustainable development while also exploring its impacts on ecosystem dynamics.

Keywords Floating breakwater · Wave attenuation · Intermediate water · Computational fluid dynamics · Tropical climate

1 Problem Statement

A floating breakwater (FB) is a type of structure used to protect coastal areas, harbors, and marinas from the impact of waves. Unlike traditional fixed breakwaters, FBs are buoyant and anchored to the seabed, allowing them to absorb and dissipate wave

Z. Y. Tay (✉) · N. L. Htoo
Engineering Cluster, Singapore Institute of Technology, Singapore, Singapore
e-mail: zhiyung.tay@singaporetech.edu.sg

X. L. Zhao et al. (eds.), *Proceedings of the Fourth World Conference on Floating Solutions: WCFS 2024; 2–4 December; Hong Kong*, Lecture Notes in Civil Engineering 597, https://doi.org/10.1007/978-981-96-4569-5_52

energy while floating on the water's surface. This design is particularly advantageous in locations where deep water or soft seabeds make the construction of traditional breakwaters impractical. FBs are used to enhance coastal protection, improve ecological resilience, and support sustainable development practices by mitigating wave energy and reducing shoreline erosion.

Numerous research on FB types have been assessed both numerically and experimentally. Lokesha et al. [4] conducted experimental research on the stepped-box breakwater design. Huang et al. [3] and He et al. [2] explored the box-with-plates breakwater type. Wang et al. [7] focused on the dual breakwater design. The studies by Sun et al. [6] included the state-of-the-art reviews of wave attenuation capabilities of different breakwater designs including pontoon and box types. Building upon the relevant prior separate research studies on FBs, this paper investigates six types of FB section types altogether, such as box-type, stepped-box type, pontoon-type, box-with-plates type, dual-box type, and T-shaped FB, to analyze and compare their wave attenuation effects in the mild seastate of the tropical climate. All FBs share the same two-dimensional footprint boundary, with width and height combining draft and freeboard. The characteristic width of the FB along the wave propagation direction is 5 m, the draft is 3 m, and the freeboard is 2 m. Detailed geometries of the FBs are shown in Table 1 and wave properties considered are given in Table 2.

The water depth D is assumed to be 10 m, with incident wave periods T of 4, 5, and 6 s. These wave periods are typical of the relatively benign waters of the Singapore sea which also involves primarily higher frequency waves shorter than 4 s. Since the FB is fully effective against high-frequency waves, $T = 4$ s is chosen as the lower bound and $T = 5$, and 6 s are considered to study the varying FB' effects on the wave attenuation. The incident wave height H was set at 2 m, i.e., $H = 2A$, where A is the wave amplitude. The water depth category and wavelengths were determined

Table 1 Various floating breakwater sections

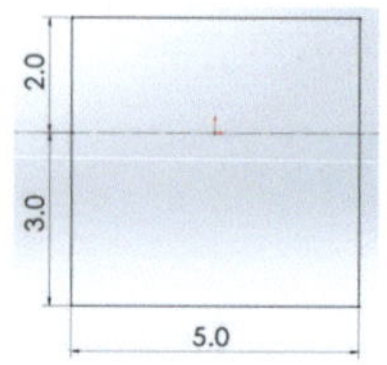

Box-type

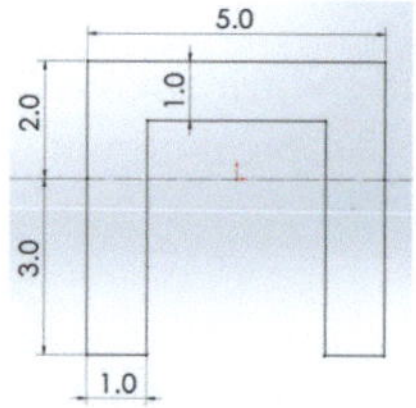

Pontoon type

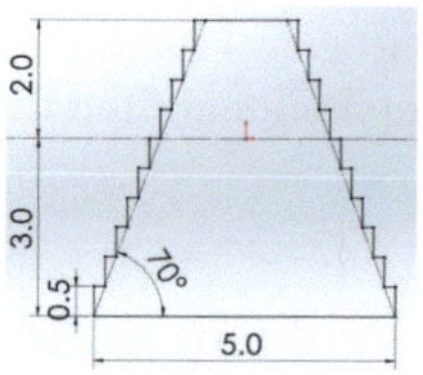

Stepped box-type

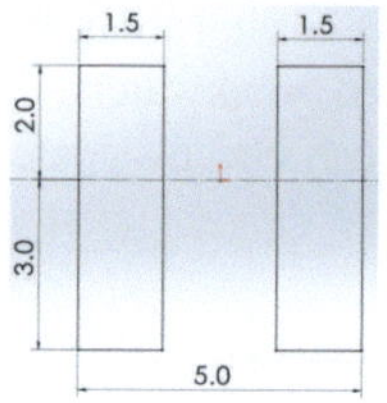

Dual box-type

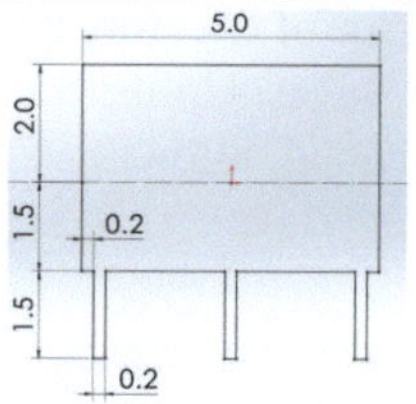

Box with plates-type

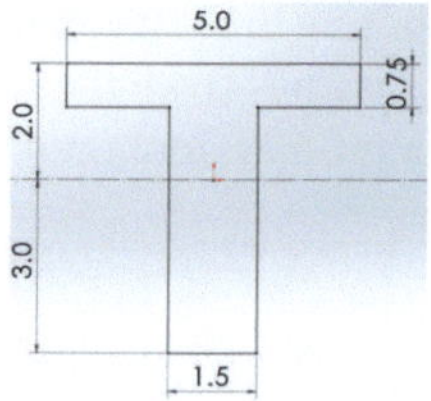

T-shaped-type

Table 2 Wave properties

T (s)—wave periods	4	5	6
D (m)—water depth	10	10	10
λ (m)	24.68	36.59	48.41
D/λ	0.41	0.27	0.21
Water depth category*	Intermediate		

* **Shallow water depth**: $\frac{D}{\lambda} < 0.05$, **Intermediate water depth**: $0.05 \leq \frac{D}{\lambda} \leq 0.5$, **Deepwater depth**: $\frac{D}{\lambda} > 0.5$

based on the frequency dispersion relation [5]. According to the wave dispersion equation given in Eqs. (1–3) and the corresponding depth-to-wavelength ratio D/λ to categorize the wave characteristic, the 10 m water depth falls into the intermediate water depth category. The corresponding wavelengths for wave periods of 4 s, 5 s, and 6 s are 24.68 m, 36.59 m, and 48.41 m, respectively.

$$\left(\frac{2\pi}{T}\right)^2 = \frac{2\pi g}{\lambda}\tanh\left(\frac{2\pi h}{\lambda}\right) \quad \text{(for intermediate water depth)} \tag{1}$$

$$\lambda = \frac{gT^2}{2\pi} \quad \text{(for deep water)} \tag{2}$$

$$\lambda = T\sqrt{gh} \quad \text{(for shallow water)} \tag{3}$$

Higher frequency waves in relatively low water depth and with high wave amplitudes can induce wave breaking, especially when analyzed using linear Airy wave theory. To ensure wave breaking and stability limit, mandatory checks are performed, including relative wave height, wave steepness, Ursell number, and wave regime checks. The 3rd order Stokes wave theory of regular periodic waves is employed, upon consistently satisfying the established checks across wave periods of 4, 5, and 6 s.

A typical numerical fluid domain incorporating a FB is established using the Reynolds-Averaged Navier–Stokes (RANS) equations. The simulations are conducted with the computational fluid dynamics (CFD) software, ANSYS® Fluent [1]. A two-dimensional numerical wave tank is considered with a clearance of 150 m set on both upstream and downstream of the FB. This setup also meets the flow regime requirements and allows for sufficient wave propagation through the domain while being computationally feasible for multiple runs. The study analyzes six FB types given in Table 1 for three wave periods.

The fluid domain with the FB modelled is meshed with refinements to ensure the convergence of the viscous turbulent model. A refined mesh size of 0.15 m is used around the free surface water level and wall regions. The polyhedral meshing is applied, achieving a minimum orthogonal quality standard of 0.2. The Volume of Fluid (VOF) model is employed for the multiphase simulation of immiscible water and air phases, with an emphasis on tracking the free surface of the VOF model. The numerical wave tank is set up with wave boundary conditions using the aforementioned wave parameters. A numerical beach is configured with an effective length of two wavelengths from the outlet of the numerical wave tank. This setup aids in absorbing wave energy, simulating the effect of natural beaches or wave-absorbing structures, thereby preventing artificial wave reflections at the outlet of the fluid domain. The SST k-omega viscous turbulent model is utilized, and a flat free surface solution of numerical wave tank is initiated.

The required time step size is calculated to be 0.01 s. This is based on the longest wavelength of 48.41 m over 6 s (fastest velocity of 8.07 m/s) and the smallest mesh size of 0.15 m, giving a required time of 0.0185 s. The chosen time step size of 0.01 s is smaller than the required time, ensuring sufficient node coverage during wave propagation. The simulation runs for 18,000-time steps with a 0.01 s step size, covering 180 s of real time, with solutions exported every second.

The FB is assumed to be stationary by assuming that all three degree of freedoms (i.e., heave, pitch and surge) is constrained from moving by the taut mooring or pile mooring systems. These mooring methods are particularly suitable for FBs in intermediate water depth due to their minimal motion which optimizes wave transmission reduction. Although taut and pile mooring systems generate greater mooring and foundation forces compared to conventional catenary mooring systems, the tropical sea conditions, such as those around Singapore, are relatively benign. This makes the design of these mooring systems more feasible in this context. For this study, assuming the FB is stationary allows for a clearer and more efficient investigation of various FB shapes and their performance in wave attenuation, based on the specified environmental parameters.

2 Results and Discussion

2.1 Incident and Transmitted Waves

The CFD simulation is conducted for a duration of 180 s for each FB type. All cases are initiated with the same incident wave height of 2 m (unit amplitude), allowing for a direct comparison of the transmitted wave elevations across different FBs to determine the superior attenuation performance. Figures 1, 2 and 3 highlight the wave attenuations for each FB type over the fluid domain with the incident wave periods of 4 s, 5 s, and 6 s, respectively. These figures illustrate the incident and transmitted waves at the time of 180 s after the commencement of wave propagation in the numerical wave tank.

The wave attenuation for $T = 4$ s is similar for all considered FBs due to the shorter wavelengths. This indicates that FBs will remain effective for wave periods shorter than 4 s. The transmitted wave coefficients η are around 0.1 in all cases, up to 0.15. The stepped and dual box FBs have the best performance with an almost

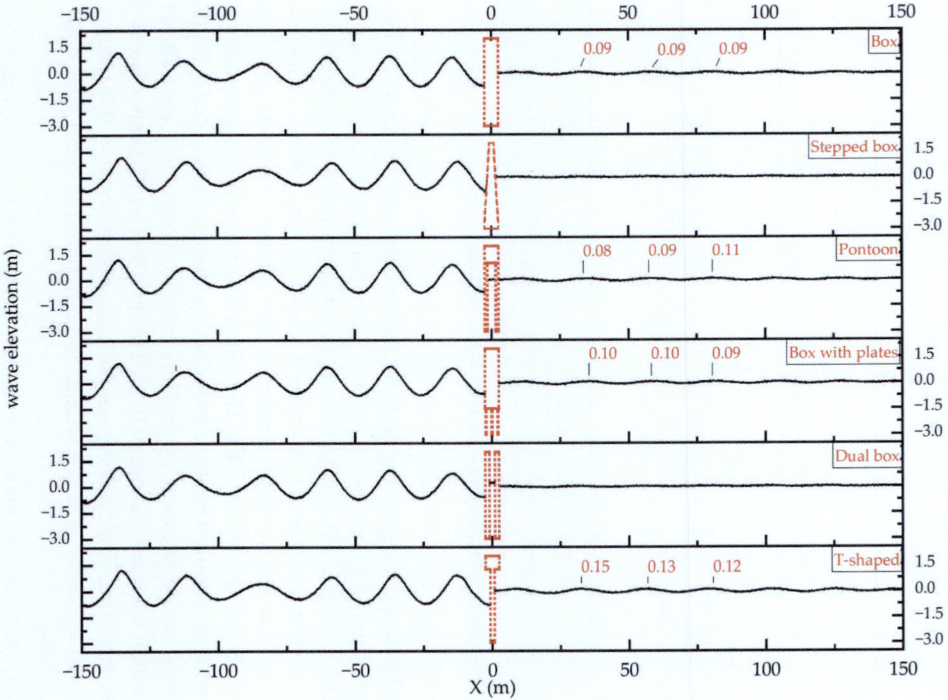

Fig. 1 Wave elevation along horizontal X-axis of numerical wave tank for different FBs at $t = 180$ s. $T = 4$ s. *Note* Floating breakwater geometry is not to scale

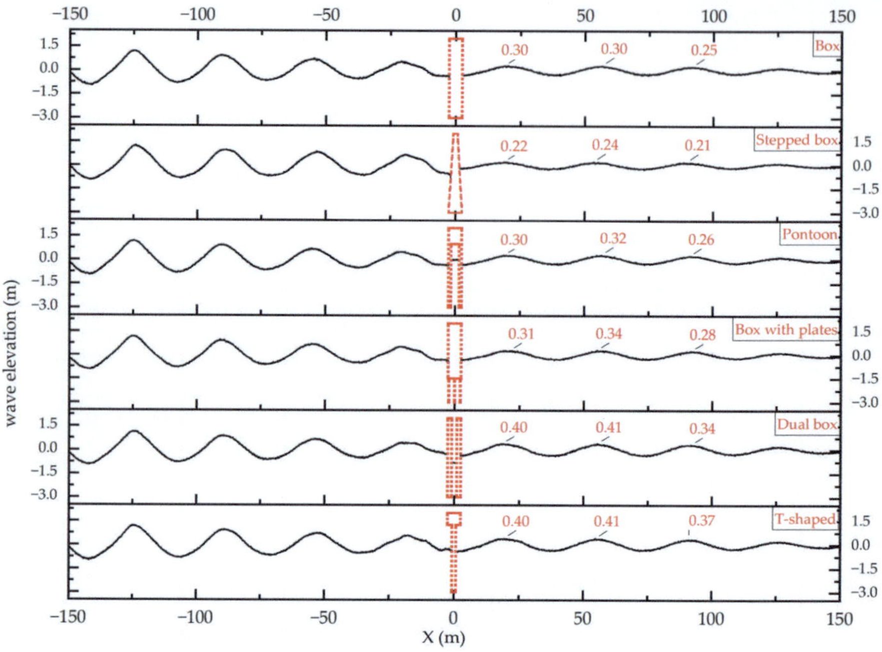

Fig. 2 Wave elevation along horizontal x-axis of numerical wave tank for different FBs at $t =$ 180 s. $T = 5$ s. *Note* Floating breakwater geometry is not to scale

flat transmitted free surface showing no wave peaks. The box, pontoon and box with plates FBs shows similarly consistent and low transmitted wave coefficient peaks of around 0.1. The T-shaped FB produces a higher transmitted wave coefficient peak of 0.15, indicating less efficiency compared to other types. When T increases to 5 s, Fig. 2 shows that the wave attenuation is still effective but with more varying performance observed in different FBs. The stepped box type once again outperforms the others, having the smallest transmitted wave coefficient peak at 0.21. Ranking in order of efficiency are the pontoon, typical box, and box with plates, each showing an average transmitted wave coefficient of 0.30. The dual box and T-shaped breakwaters generate the highest transmitted wave coefficients, approximately 0.40, suggesting that they are least effective in wave attenuation compared to other FBs with same footprint boundaries.

At $T = 6$ s, there is less wave attenuation compared to $T = 4$ and 5 s, due to the FB being less effective when encountering longer incident wavelength. The stepped breakwater shows the lowest transmitted wave coefficeint peaks, averaging 0.38, making it consistently the best performer. This is followed by the pontoon and box with plates types, with a difference of around 0.1 in transmitted wave elevation compared to the stepped breakwater. The box breakwater is observed to be moderately effective, preceding the dual box and T-shaped designs, which are the least effective. The T-shaped type produces the highest transmitted wave coefficient at

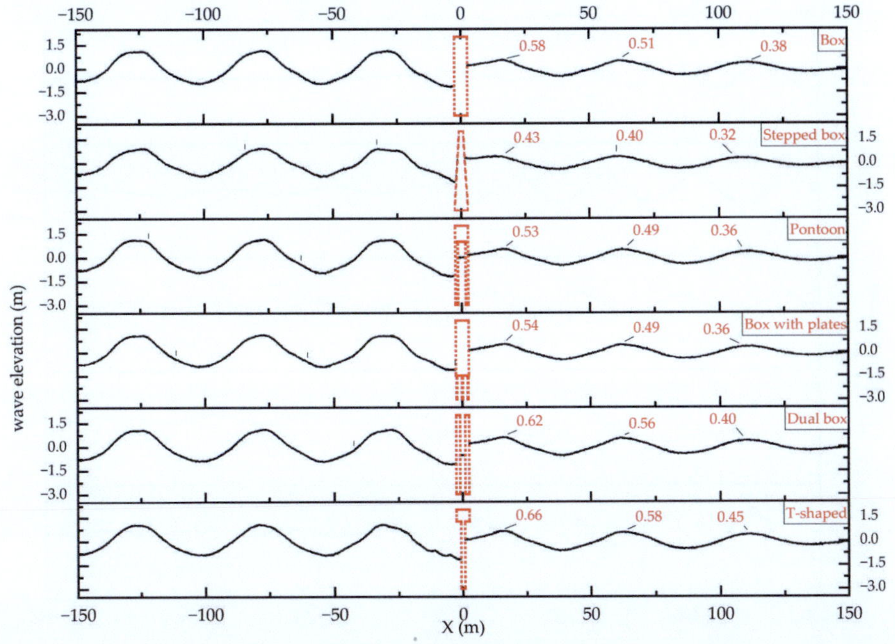

Fig. 3 Wave elevation along horizontal X-axis of numerical wave tank for different FBs at $t = 180$ s. $T = 6$ s. *Note* Floating breakwater geometry is not to scale

0.66, indicating the poorest performance for wave attenuation. In all cases, the last wavelength near the outlet of the numerical wave tank is damped due to the numerical beach effect, which acts as a wave absorber.

2.2 Wave Attenuation of the Best Performing FB

The evolution of wave propagation and attenuation of the best performed stepped box FB is further studied for various time intervals (i.e., $t = 22, 45, 90, 135$, and 180 s) in Fig. 4. The incident wave period selected in this case is 5 s. This also serves to validate the CFD time series simulation of 180 s, demonstrating that computations sufficiently reach steady state by $t = 180$ s. At $t = 45$ s, the response of the transmitted waves remains in a transient state, where it begins to stabilize at $t = 135$ s and continues to display persistent peaks in transmitted waves, thereby confirming achievement of the steady-state solution.

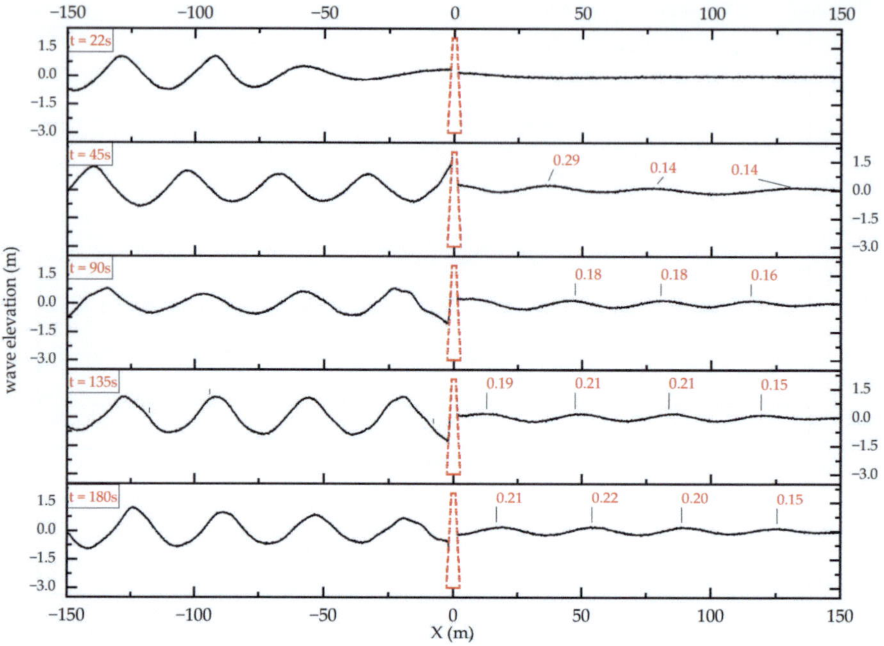

Fig. 4 Wave attenuation of the stepped box breakwater type in wave period 5 s at different time intervals: $t = 22, 45, 90, 135$, and 180 s. *Note* Floating breakwater geometry is not to scale

3 Conclusion

Based on the analysis outcome, it can be summarized that the stepped type breakwater has the best performance across all the wave periods considered. Pontoon type performs well, ranking below the stepped breakwater. While not as effective as the stepped or pontoon types, the box with plates design demonstrates similar effectiveness to the standard box, despite having only vertical plate covering half of the draft underwater, rather than a full box-shaped structure. To enhance the assumptions and limitations of the current study, future research should include more detailed design parameters including the motion response of FB, reflecting the design of slack mooring in addition to stationary, three dimensional CFD approach, hydro-elasticity and validating the findings through experiments.

Acknowledgements The authors are grateful to the National Research Foundation of Singapore, PUB (Singapore's National Water Agency) and the Coastal Protection and Flood Resilience Institute (CFI) Singapore for supporting this work done under the Coastal Protection and Flood Management Research Programme (CFRP). The conclusions put forward in this publication reflect the views of the authors alone and not necessarily CFI Singapore or any of the supporting entities.

References

1. Ansys® Fluent (2023) Release 2023 R2, User's guide. ANSYS Inc., p 5552. http://ansyshelp.ansys.com
2. He F, Li J, Pan J, Yuan Z (2023) An experimental study of a rectangular floating breakwater with vertical plates as wave-dissipating components. Appl Ocean Res 133. https://doi.org/10.1016/j.apor.2023.103497
3. Huang Z, He F, Zhang W (2014) A floating box-type breakwater with slotted barriers. J Hydraul Res 52(5):720–727. https://doi.org/10.1080/00221686.2014.888690
4. Lokesha KNB, Sannasiraj SA, Sundar V, Schlurmann T (2015) Experimental investigations on wave transmission at submerged breakwater with smooth and stepped slopes. Procedia Eng 116(1):713–719. https://doi.org/10.1016/j.proeng.2015.08.356
5. Phillips OM (1977) The dynamics of the upper ocean. Cambridge University Press, 7319331
6. Sun B, Zhang H, Li C, Li Z (2023) Research progress and prospect of wave attenuation performance and integration of wave-energy converter of floating breakwater. J Environ Eng 149(4):1–17. https://doi.org/10.1061/joeedu.eeeng-7118
7. Wang G, Xie S, Yuan H, Wang R, Zhang T, Liu X, Tu J (2024) Numerical investigation of hydrodynamic characteristics of a dual floating breakwater. Ocean Eng 294. https://doi.org/10.1016/j.oceaneng.2024.116728

Hydroelastic Analysis of a Large Container Ship with Forward Speed Based on Multi-module Method

Yusong Ye, Shixiao Fu, and Shiyuan Zhang

Abstract In this study, a beam-connected-discrete-modules (BCDM) hydroelastic method is employed to conduct hydroelastic analysis of large container ships with forward speed. The large container ship is discretized into several rigid modules connected by a hull girder. Considering the hydrodynamic interactions between modules, the multi-body hydrodynamic theory is adopted to obtain the hydrodynamic coefficients of each module. Subsequently, the hydroelastic equation is established by coupling the hydrodynamics of modules with the structural stiffness of the hull girder. The BCDM method replaces the boundary value problem solved in the modal space with extensively studied multi-body hydrodynamics. In particular, the derivation of generalized restoring matrices associated with flexible modes could be avoided, which has been ongoing debate for more than two decades. The effect of flexible deformation on a large container ship is investigated. To account for the influence of forward speed, the wave natural frequency is replaced by the encounter frequency. The results show that the flexible deformation increases sectional forces, especially at the structural vibration frequency.

Keywords Beam-connected-discrete-module method · Large container ship · Global hydroelasticity · Forward speed

1 Introduction

With the enhancement of shipping demand, large container ships play an increasingly important role in shipping. However, due to the large dimensions, the structural vibration frequency of the ship decreases. When ships advance in waves, the encounter

---

Y. Ye · S. Fu (✉) · S. Zhang
State Key Laboratory of Ocean Engineering, Shanghai Jiao Tong University, Shanghai, China
e-mail: shixiao.fu@sjtu.edu.cn

Collaborative Innovation Centre for Advanced Ship and Deep-Sea Exploration, Shanghai, China

X. L. Zhao et al. (eds.), *Proceedings of the Fourth World Conference on Floating Solutions: WCFS 2024; 2–4 December; Hong Kong*, Lecture Notes in Civil Engineering 597, https://doi.org/10.1007/978-981-96-4569-5_53

frequency will approach the structural vibration frequency, which excites springing response and increases fatigue damage significantly.

To ensure the structural safety of the container ship, it is necessary to carry out the hydroelastic analysis of large ships with forward speed. The modal superposition method is a popular approach to solve the problem. Wu et al. [1] employed the three-dimensional frequency domain linear hydroelastic method to conduct a series of numerical calculations on a submerged ellipsoid near the wave surface and a semi-ellipsoidal elastic surface hull navigating on the wave surface. Tian [2] established a second-order nonlinear hydroelastic analysis method for sailing ships considering the steady flow, and analyzed the hydroelastic response of the SWATH ship. Duan et al. [3] investigated hydroelastic response of ship with speed in time domain based on the TEBEM method. Chen et al. [4] established a three-dimensional time-domain nonlinear hydroelastic analysis method in irregular oblique waves, and validated the method by comparing the experimental and numerical results of a high-speed trimaran. Zhang [5] integrated the computational advantages of the two-dimensional to extend three-dimensional time-domain hydroelasticity theory, and put forward the determination method of nonlinear design wave loads on ships with large bow flare. Hu et al. [6] used the second-order nonlinear three-dimensional frequency domain hydroelasticity method to calculate the springing response of a large ore ship. Despite the widely use of the modal superposition method, there are still some limitations, such as the controversial generalized hydrostatic coefficient [7, 8].

In addition to the aforementioned superposition method, in recent years, the beam-connected-discrete-module (BCDM) [9] method for analyzing the hydroelastic response of very large floating structures (VLFS) and ships has been developed. This method avoid the problems of the controversial generalized hydrostatic coefficient. Zhu et al. [10] analyzed the hydroelastic response of the VLFS with moving loads by the BCDM method. Bispo et al. [11] investigated a moored VLFS composed by a set of hinged plates, and calculated the effects of the hinged plates on the structural displacement and the mooring cable loads. Wei et al. [12] employed the BCDM method to calculate the hydroelastic response of the pontoon bridge in inhomogeneous sea environment, and analyzed the influence of the inhomogeneous waves with spatial varying wave heights/periods. Li et al. [13] extended the BCDM method to calculate hydroelastic response of a floating wind turbine with a large pontoon, and the effects of elastic deformation on bending moment are analyzed. Based on the second-order hydrodynamics of multi-body, Li et al. [14] considered the effect of the first-order hydroelastic response on the flow field and extended the BCDM method to the second-order hydrodynamic. Zhang et al. [15] developed a two-step stress analysis method to incorporate the global hydroelasticity and local bending effect based on the BCDM method, which extended the method to the stress analysis.

Previous studies of the BCDM method focused on no-speed hydroelastic analysis. For the hydroelastic analysis with forward speed, Bakti et al. [16] and Zhang et al. [17] used Wigley hulls to investigate hydroelastic responses of ship with forward speed based on BCDM method. In this paper, the BCDM method is utilized to analyze the hydroelastic response of a large container ship with forward speed, realizing the engineering application of the BCDM method in ship with forward speed. The large container ship is discretized into several rigid modules, which are connected by a hull girder. The hydroelastic equation is established by coupling the hydrodynamics of modules with the structural stiffness of the hull girder. In order to consider the effect of speed, the wave natural frequency is replaced by the encounter frequency. Subsequently, the hydroelastic response of the container ship is investigated.

2 Theoretical Background

According to the BCDM method, the large container ship is firstly discretized into several rigid modules connected by a hull girder. Then, the multibody hydrodynamic theory is adopted to obtain the wave excitation force, added mass, and damping coefficient acting on each module. After that, the multibody hydrodynamic force is coupled with the structural stiffness of the beam to establish the frequency domain hydroelastic equation (see Fig. 1). Due to the warping phenomenon in large container ships, a thin-walled beam element is adopted [18].

$$\left(-\omega^2[M + A(\omega)]_{7N\times 7N} - i\omega[c + C(\omega)]_{7N\times 7N} + [K + k]_{7N\times 7N}\right)\{u\}_{7N\times 1} = \{f(\omega)\}_{7N\times 1} \tag{1}$$

where N is the number of thin-walled beam nodes with seven DOF; $[k]$ is the stiffness matrix of the thin-walled beam; $[c]$ is the structural damping matrix of the thin-walled beam; $[M]$, $[A]$, $[C]$, $[K]$ are the mass, added mass, damping coefficient and hydrostatic restoring coefficient matrices of the rigid bodies, respectively; $\{u\}$ is the node displacement vector; and $\{f\}$ is the wave excitation force vector.

In order to consider the effect of the forward speed, the encounter frequency ω_e is used to replace the wave natural frequency ω when calculating $[A]$, $[C]$ and $\{f\}$. The relationship between ω and ω_e is

$$\omega_e = \omega\left(1 - \frac{U\omega}{g}\cos\theta\right) \tag{2}$$

where U is the ship speed, g is the gravity acceleration, and θ is the wave direction. Equation (1) changes to

$$\left(-\omega_e^2[M + A(\omega_e)]_{7N\times 7N} - i\omega_e[c + C(\omega_e)]_{7N\times 7N} + [K + k]_{7N\times 7N}\right)\{u\}_{7N\times 1} = \{f(\omega_e)\}_{7N\times 1} \tag{3}$$

Although this approach only considers the encounter frequency and ignores the steady flow caused by ship speed, it is widely used in engineering and can analyze the effect of speed on the flexible deformation [19].

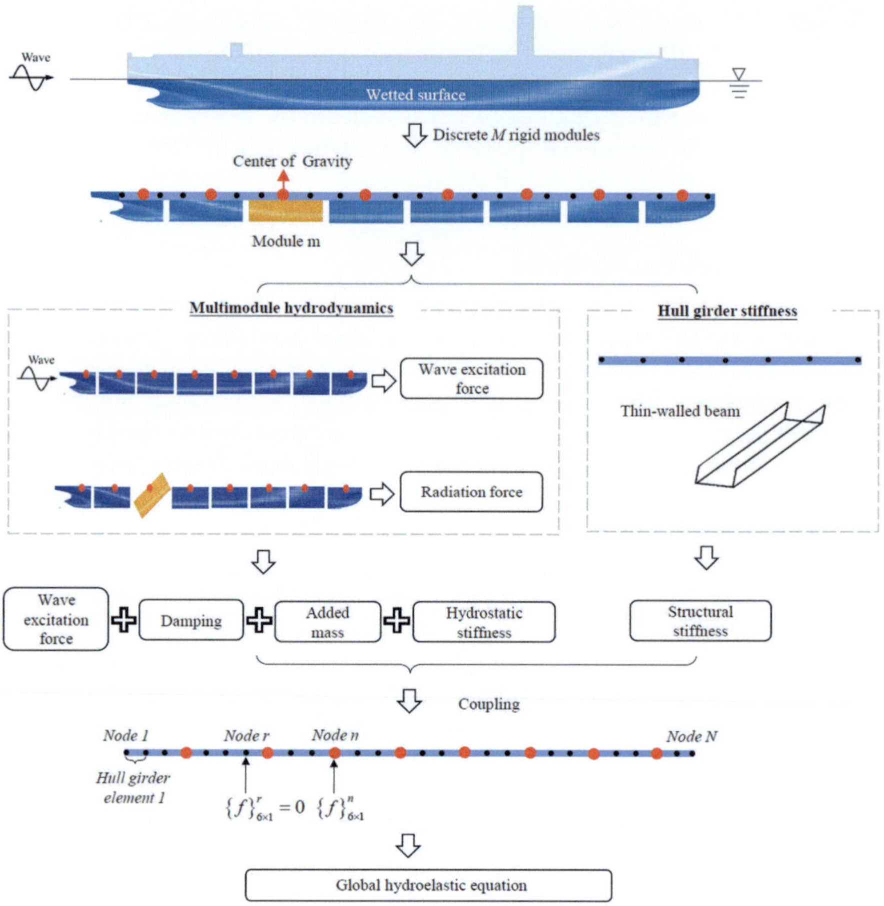

Fig. 1 Flow diagram of the hydroelastic analysis of the container ship

3 Numerical Model

A 14,000 TEU container ship is considered in this study (see Fig. 2). The main particulars of the full loaded container ship are shown in Table 1.

The 1D thin-walled beam model is established according to the mass and stiffness distribution of the container ship. In order to prove that the 1D beam model can reflect the vibration characteristics of the container ship, the structural dry modes based on the one-dimensional beam model are calculated and compared with the 3D FE model, as shown in Table 2 and Fig. 3. The results demonstrate the efficacy of the established 1D beam model.

Fig. 2 Model of the 14,000TEU container ship

Table 1 Main particulars of the full loaded container ship

Designation	Value
Length (m)	335.9
Breath (m)	51.0
Depth (m)	29.9
Draft (m)	14.5
Mass (t)	173,509.4
Center of gravity in X-axis (m)	165.9
Center of gravity in Z-axis (m)	22.7
Roll inertia (ton · m^2)	33,570,000
Pitch inertia (ton · m^2)	1,160,000,000
Yaw inertia (ton · m^2)	1,154,000,000
Products of inertia about xz axis (ton · m^2)	1,608,000

Table 2 Natural frequency based on 1D beam and 3D FE model

Mode number	Mode shape	Natural frequency (Hz)		Error (%)
		3D FE	1D beam	
7	1st Tor. and H.B.	3.073	3.123	−1.601
8	2nd Tor. and H.B.	3.424	3.487	−1.807
9	1st V.B	4.096	4.016	1.993
10	3rd Tor. and H.B.	6.126	6.057	1.139
11	2nd V.B.	7.351	7.684	−4.334

4 Hydroelastic Results

According to the above methodology, the global hydroelastic response of the container ship at service speed (22 kn) and zero speed is considered. To investigate the flexibility effect on sectional forces, the results are also given under the assumption of rigid body.

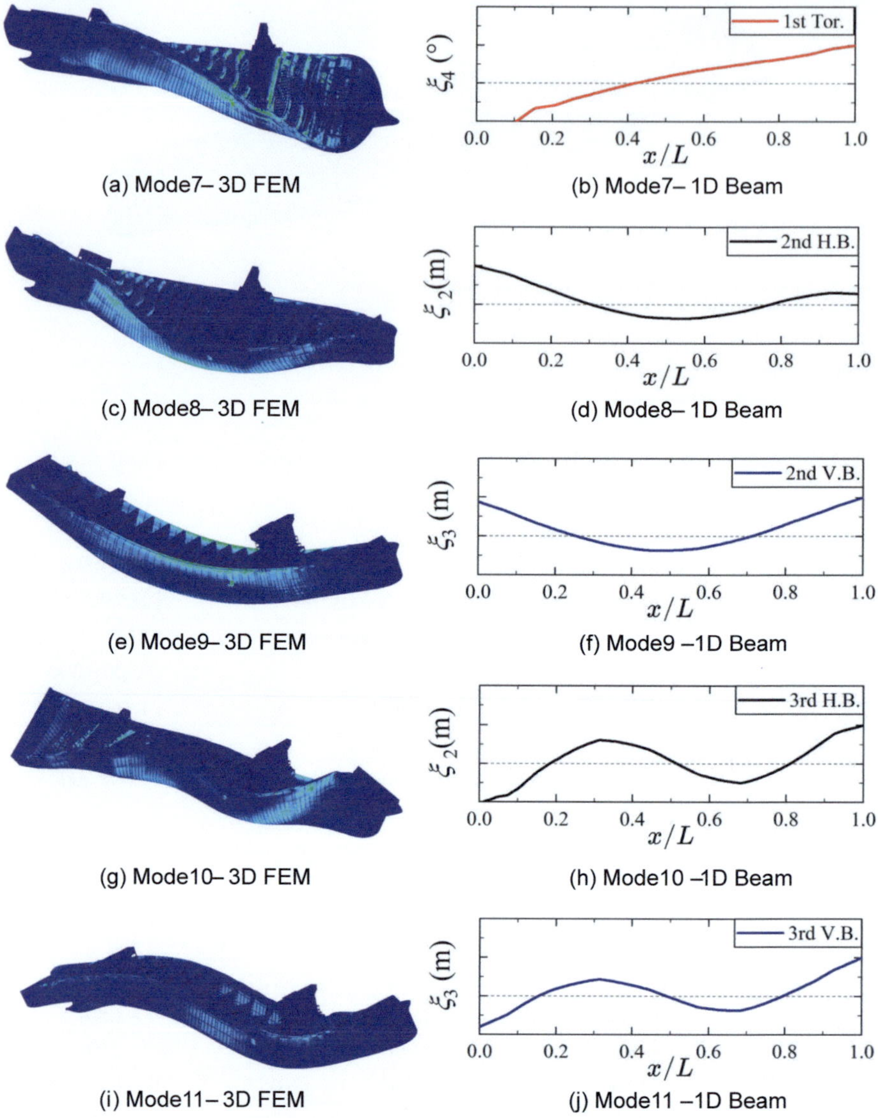

Fig. 3 The first 5 flexible mode shapes of the container ship in dry conditions

Figure 4a, b display the amplitude of vertical bending moment at mid-ship position and vertical shear force at the position of $x/L = 0.25$ under $\theta = 180°$ wave direction respectively. For rigid ship, the results of the sectional forces at 22kn speed are basically the same as those at zero speed, which indicates that the effect of speed is not significant. For flexible ship, when the encounter frequency is close to the structural vibration frequency, springing response is excited, and the sectional forces increase significantly. As shown in Fig. 4a, the response amplitude at the first-order vertical bending resonance frequency is about 1.2 times of the maximum amplitude of the response at the non-resonance frequency. In addition, for the vertical shear force, as shown in Fig. 4b, the peak occurs at both the first-order and second-order vertical bending resonance frequency. However, in the non-resonant frequency interval, the sectional forces of the flexible ship are consistent with the rigid ship.

Figure 5 displays the amplitude of vertical bending moment, horizontal bending moment and torsional moment at midship and vertical shear force at $x/L = 0.25$ under $\theta = 135°$ wave direction respectively. Similar to the case of $\theta = 180°$, when the encounter frequency is close to the structural vibration frequency, the elastic response is excited and the sectional forces increase. It is worth noting that for large container ships, due to the low torsional structural vibration frequency, the torsional modes are excited when sailing in oblique waves, which also leads to higher sectional forces, as shown in Fig. 5c, d. Both the first-order and second-order torsion and horizontal bending modes are excited. The resonance peak is almost equal to the maximum amplitude of the response at the non-resonance frequency. Therefore, extra attention must be paid to the torsional loads of large container ships.

In order to investigate the effect of structural stiffness on the sectional forces along the longitudinal direction of the ship, Fig. 6 illustrates the amplitude of vertical bending moment, horizontal bending moment and torsional moment along the longitudinal direction of the container ship under $\theta = 135°$ wave direction respectively. Two types of wave natural frequency are considered: $\omega = 0.5$ rad/s and wave natural frequency which will excite elastic modes (for vertical bending $\omega = 1.4$ rad/s and for torsion and horizontal bending $\omega = 1.3$ rad/s). When $\omega = 0.5$ rad/s, the encounter frequency is away from the structural vibration frequency. In Fig. 6a, the vertical bending moment of the flexible ship is basically consistent with that of the rigid ship. Furthermore, in Fig. 6b, c, the horizontal bending moment and torsional moment of

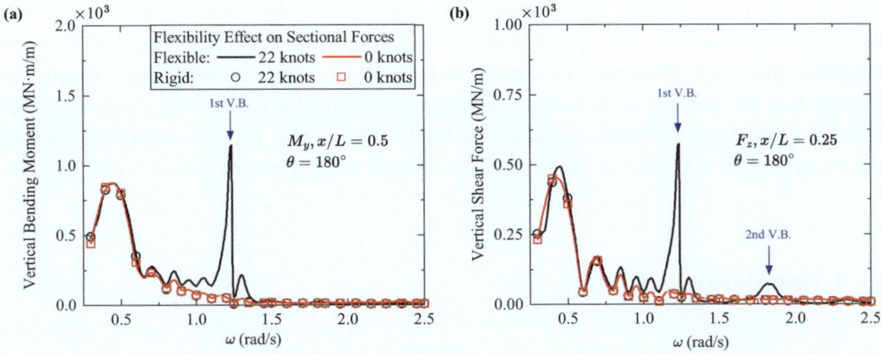

Fig. 4 Effect of structural stiffness on the sectional forces of navigating and freely floating container ships under wave direction of $\theta = 180°$: **a** vertical bending moment at mid-ship position; **b** vertical shear force at the position of $x/L = 0.25$

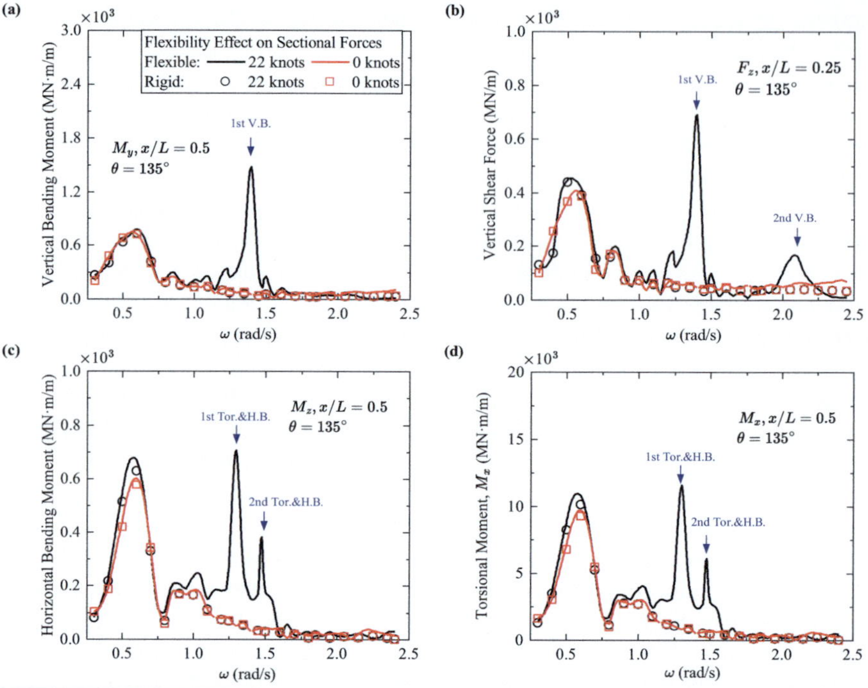

Fig. 5 Effect of structural stiffness on the sectional forces of navigating and freely floating container ships under wave direction of $\theta = 135°$: **a** vertical bending moment at mid-ship position; **b** vertical shear force at the position of $x/L = 0.25$; **c** horizontal bending moment at mid-ship position; **d** torsional moment at mid-ship position

the flexible ship are slightly higher than those of the rigid ship. Therefore, flexible deformations can be ignored in long wave length. Besides, for both rigid and flexible ships, the forward speed will increase the horizontal bending moment and torsional moment. However, as shown in Fig. 6d–f, when the encounter frequency approaches the structural vibration frequency, the elastic modes are excited, which induces significant increase of sectional loads. Therefore, it is necessary to conduct the hydroelastic analysis of large container ships with forward speed.

5 Conclusions

The beam-connected-discrete-module (BCDM) method is applied to the hydroelastic analysis of large container ships with forward speed. In this method, the large container ship is divided into several rigid modules connected by a hull girder. The global hydroelastic equation is established by coupling multibody hydrodynamics and beam structural dynamics equation. The BCDM method avoid the problem of the

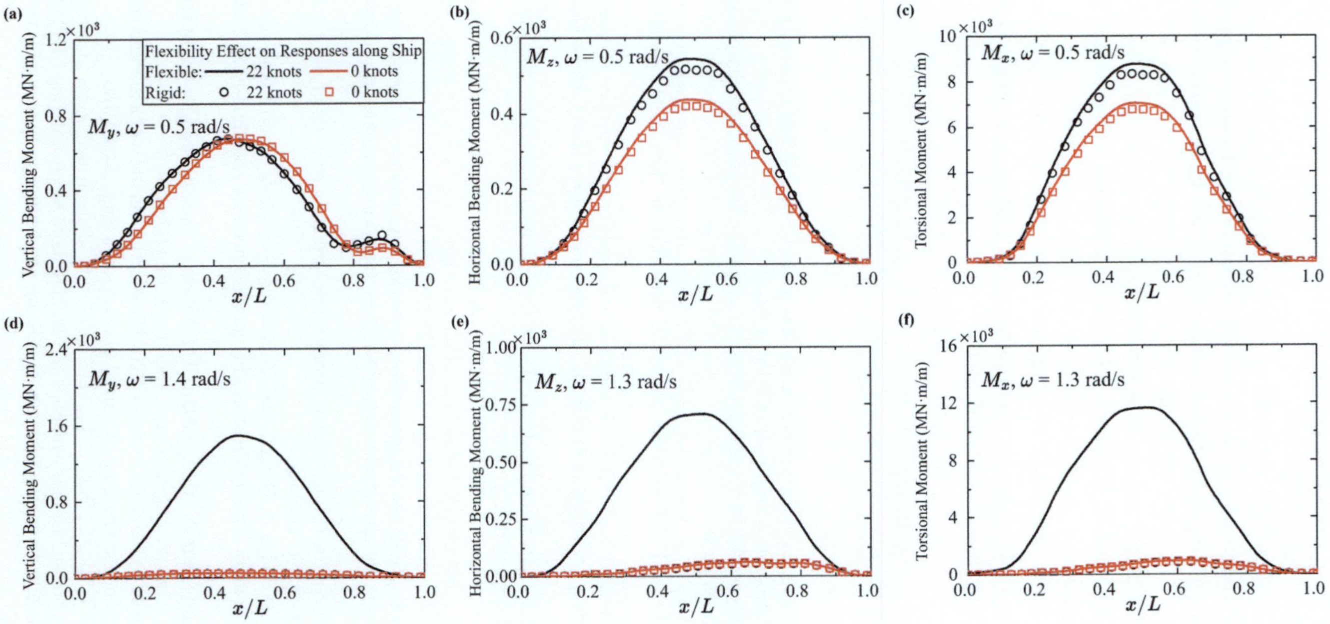

Fig. 6 Effect of structural stiffness on the sectional forces along the longitudinal direction of the container ship under the wave direction of $\theta = 135°$: **a** vertical bending moment when ω = 0.5 rad/s; **b** horizontal bending moment when ω = 0.5 rad/s; **c** torsional moment when ω = 0.5 rad/s; **d** vertical bending moment when ω = 1.4 rad/s; **e** horizontal bending moment when ω = 1.3 rad/s; **f** torsional moment when ω = 1.3 rad/s

controversial generalized restoring matrices by using extensively studied multi-body hydrodynamics. Besides, to account for the speed effects, the wave natural frequency is replaced by the encounter frequency.

Based on the BCDM method, the hydroelastic response of a 14,000 TEU container ship is investigated. The results show that when the encounter frequency approaches the structural vibration frequency, the elastic modes of the large container ship will be excited, resulting in higher sectional forces, especially in the case of oblique waves, which will excite the torsion & horizontal bending vibrations. Therefore, the hydroelastic response must be considered for safety checking in the design process of large container ships.

In the future research, the influence of steady wave caused by forward speed can be considered in the hydrodynamic calculation to extend BCDM method in large ship hydroelastic analysis with forward speed.

References

1. Wu Y, Du S (1998) Influence of speed and steady flow on the hydroelastic mechanical properties of navigating hull structures. Ship Build China S1:90–99
2. Tian C (2007) Study on the theory and applications of nonlinear hydroelasticity of ships with forward speed. Doctor thesis
3. Duan W, Wang H, Chen J, Wang H (2022) TEBEM for springing responses of a container ship with forward speed and nonlinear effects in time domain. Eng Anal Boundary Elem 140:406–420
4. Chen Z, Gui H, Dong P, Yu C (2019) Numerical and experimental analysis of hydroelastic responses of a high-speed trimaran in oblique irregular waves. Int J Naval Archit Ocean Eng 11:409–421
5. Zhang K (2018) Research on nonlinear wave loads' time-domain hydroelastic analysis method and application of ships with large bow flare. Doctor thesis
6. Hu J, Wu Y, Tian C, Wang X, Zhang F (2012) Hydroelastic analysis and model tests on the structural responses and fatigue behaviours of an ultra-large ore carrier in waves. Proc Inst Mech Eng Part M: J Eng Maritime Environ 226(2):135–155
7. Malenica Š, Bigot F (2020) On the generalized motions/deformations of the floating bodies. In: Proceedings of the 35th International workshop on water waves and floating bodies. South Korea
8. Pan Z, Zhang S, Fu S (2024) Integrated hydrodynamic-structural analysis of flexible floating structures. Ocean Eng 310:118644
9. Wei W, Fu S, Moan T, Lu Z, Deng S (2017) A discrete-modules-based frequency domain hydroelasticity method for floating structures in inhomogeneous sea conditions. J Fluids Struct 74:321–339
10. Zhu L, Shao F, Xu Q, Sun Y, Ma Q (2018) The effects of moving load on the hydroelastic response of a very large floating structure. Math Probl Eng 6062586:9
11. Bispo IBS, Mohapatra SC, Guedes Soares C (2022) Numerical analysis of a moored very large floating structure composed by a set of hinged plates. Ocean Eng 253:110785
12. Wei W, Fu S, Moan T, Song C, Deng S, Lie H (2019) A time-domain method for hydroelasticity of a curved floating bridge in inhomogeneous waves. J Offshore Mech Arct Eng 141(1):014501
13. Li L (2022) Full-coupled analysis of offshore floating wind turbine supported by very large floating structure with consideration of hydroelasticity. Renewable Energy 189:790–799
14. Li S, Fu S, Zhang S, Moan T (2022) Second-order hydroelastic analysis of a flexible floating structure under spatially inhomogeneous waves. Mar Struct 86:103306

15. Zhang S, Fu S, Li S, Moan T, Xu Y, Pan Z (2024) Frequency-domain hydroelastic stress analysis considering local bending effect based on a two-step procedure. Mar Struct 95:103580
16. Bakti FP, Jin C, Kim MH (2021) Practical approach of linear hydro-elasticity effect on vessel with forward speed in the frequency domain. J Fluids Struct 101:103204
17. Zhang S, Fu S, Pan Z, Han K, Ye Y, Springing responses of ships with forward speed based on a multi-module hydroelastic method. Mar Struct 101:103733
18. Senjanović I, Tomašević S, Vladimir N (2009) An advanced theory of thin-walled girders with application to ship vibrations. Mar Struct 22:387–437
19. Jiao J, Chen Z, Chen C, Ren H (2020) Time-domain hydroelastic analysis of nonlinear motions and loads on a large bow-flare ship advancing in high irregular seas. J Mar Sci Technol 25(2):426–454

Hydrodynamic Investigation on Floating Offshore Wind Turbine Platform Integrated with Porous Shell and OWC Wave Energy Converter

Dezhi Ning and Sijia Deng

Abstract The motion response of a floating offshore wind turbine (FOWT) platform is closely related to ensuring its safety. It is significant to decrease the motion response in the design of FOWTs. By integrating an external porous shell and internal Oscillating Water Column (OWC) devises with a floating wind turbine platform, a new-formed platform is designed to reduce the motion response. A numerical model, based on the nonlinear higher-order boundary element method (HOBEM), investigating the coupling effect of the platform-porous shell-OWCs was developed and validated by comparison with the physical experiments. The results of the motion response were compared with those of the conventional platform without the integrating parts. The motion response of the two kinds of platforms was simulated in the time domain. The study demonstrates a floating platform integrated with porous shell and OWC to achieve reduced motion response is feasible and can serve as a new design approach for the development of deep-sea offshore floating wind turbines.

Keywords FOWT · Platform · Porous shell · OWC · Motion response

1 Introduction

Offshore wind turbines are extensively deployed along coastlines and in oceans to harness wind power [1]. Compared to fixed offshore wind turbines, floating turbines exhibit more complex responses due to their six degrees of freedom in motion. Their motion is a critical factor in the operation and maintenance of turbines, as reducing the motion response can effectively lower maintenance costs and increase the energy yield of wind projects. Numerous studies have focused on mitigating the

---

D. Ning (✉) · S. Deng
State Key Laboratory of Coastal and Offshore Engineering, Dalian University of Technology, Dalian, China
e-mail: dzning@dlut.edu.cn

Dalian Key Laboratory of Offshore Renewable Energy, Dalian University of Technology, Dalian, China

X. L. Zhao et al. (eds.), *Proceedings of the Fourth World Conference on Floating Solutions: WCFS 2024; 2–4 December; Hong Kong*, Lecture Notes in Civil Engineering 597, https://doi.org/10.1007/978-981-96-4569-5_54

motion response of turbines. Currently, extensive research is aimed at reducing the motion response amplitude. Rao et al. [2] conducted a 1:100 scale model experiment to investigate the heave damping and added mass of a classical spar with a heave plate, comparing experimental results with numerical observations. Hegde et al. [3] explored the impact of heave plate diameter and position relative to draft ratio on damping, identifying an optimal configuration. Yang et al. [4] installed two tuned mass dampers on the spar and nacelle of a floating offshore wind turbine (FOWT) to control vibration response. The stiffness and damping coefficients of the TMD were optimized, and the damping effect under wind and wave loading was investigated.

In recent years, many researchers have combined wind turbines with wave energy converters to reduce turbine motion. An innovative integration of a cylindrical WEC device into a seabed-mounted wind turbine foundation has been proposed for the combined exploitation of wave and ocean wind energy in coastal areas [5]. Perez-Collazo et al. [6] conducted a feasibility study on the multi-purpose integration of an OWC device with a jacket-frame substructure. Perez-Collazo et al. [7], Zhou et al. [8], and Cong et al. [9] systematically studied the hydrodynamic flow around an OWC WEC fitted to a fixed wind turbine monopile. An exterior shell was installed concentrically around the wind turbine monopile to form the OWC chamber, trapping water and air inside.

In addition to studies on reducing the motion response of FOWTs, porous structures are frequently used in the marine industry to dissipate energy and reduce the kinematic response of structures. Porous structures are commonly employed in fixed and floating breakwaters [10, 11] and are also used to enhance the motion damping of marine structures [12–16]. Mackay et al. [17] investigated the use of porous materials for reducing loads on fixed and offshore structures and confirmed that porous structures have significant potential for load reduction and motion damping on floating structures. Saha et al. [18] examined the scattering phenomena in water waves from a bottom porous composite cylinder with an annular porous cap, finding that wave forces on an inner concentric column diminished when the porous effect parameter was low.

Based on the review of the above studies, the concept of platform-porous shell-OWC was developed. This study aims to propose a new method to reduce the amplitude of the motion response of the offshore floating wind turbine spar platform. The remainder of the manuscript is structured as follows: Sect. 2 presents the description of the hybrid system. Section 3 outlines the mathematical formulation and numerical methodology. The reliability of numerical model is verified in Sect. 4. Section 5 illustrates the research results. Finally, the research conclusions are discussed in Sect. 6.

2 Description of the Hybrid System

The hybrid system consists of a FOWT system, the OWC device and a porous shell, which have the rigid connection with each other. The OWC device lies in the middle of the main column. The porous shell is integrated with the vertical semisubmersible platform at the incident wave height region. The configuration of the whole FOWT system consists of a new form floating platform structural and mooring system properties. Details of the hybrid system are provided in Fig. 1.

The semisubmersible platform is employed. This semisubmersible platform consists of two cylinders vertically. The diameter of the top and bottom cylinder are 12 m and 24 m respectively. The draft of the float is 20 m. Three mooring lines are selected for station keeping of FOWT and anchored to the sea bed at a water depth of 200 m. The mooring lines are attached at the bottom of the platform, with 120 apart in each line. Diagram illustrating a mooring system. Fig. 1(a) shows a top view with a central structure surrounded by concentric circles, indicating wave direction. Red and green dots represent anchors and fairleads, labeled L1# to L4#. Fig. 1(b) presents a side view with a vertical structure labeled MC 1, showing water surface level (SWL) and seabed anchors. The depth is marked as 200 meters. Arrows indicate wave direction on both views. The properties of the mooring system are summarized in Ref. [19].

The OWC devices are rigidly integrated into the floating platform, motioning in sync with the platform. The PTO is modeled as linear damping without considering the complexity of the various control strategies. The porous shell outer diameter is 24 m, the inner diameter 23.8 m, the wall thickness 0.2 m and the porosity $\tau = 0.3$. The wall height is 14 m. The openings in the porous cylinder are rectangular and arranged on a regular square grid.

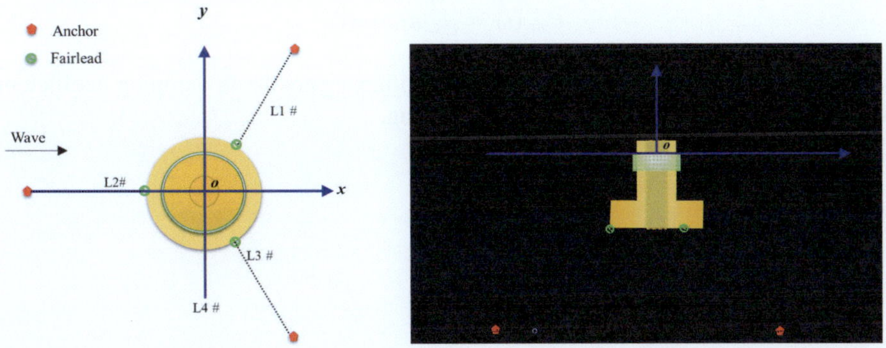

(a) **(b)**

Fig. 1 The hybrid system including floating platform, mooring lines, porous shell, and the mooring lines: **a** top view and **b** side view

3 Mathematical Formulation and Numerical Methodology

For generality, the constrained motion equations of a hybrid system consisting of a new form platform. ϕ is complex velocity potential satisfying the Laplace equation,

$$\frac{\partial^2 \varphi}{\partial x^2} + \frac{\partial^2 \varphi}{\partial z^2} = 0 \tag{1}$$

The corresponding boundary conditions of all velocity potentials can be described as,

$$\frac{\partial \varphi_s^{(m)}}{\partial n} = -\frac{\partial \varphi_i^{(m)}}{\partial n}, \quad \text{on} \quad S_D \quad \text{and} \quad S_B \tag{2}$$

where $\frac{\partial}{\partial n}$ denotes the normal derivative on the solid surface. To analyze the wave motion in a finite domain, a sponge layer is introduced to absorb the reflected waves from the platform. Using Taylor's expansion, the kinematic and dynamic boundary conditions on the free surface S_F can be expressed as

$$\frac{\partial \eta_s^{(m)}}{\partial t} = \frac{\partial \varphi_s^{(m)}}{\partial z} - f_{m'} - \nu_{(r)} \eta_s^{(m)} (m = 1\&2) \tag{3}$$

$$\frac{\partial \varphi_s^{(m)}}{\partial t} + g\eta_s^{(m)} - f_m^{''} + \nu_{(r)} \varphi_s^{(m)} = \begin{cases} -\frac{1}{\rho} P_{air} + \mu_1 \frac{\partial \varphi^{(1)}}{\partial n} & (m = 1) \\ -\mu_1 \left(\eta^{(1)} \frac{\partial \varphi^{(1)}}{\partial n} + \frac{\partial \varphi^{(2)}}{\partial n} \right) & (m = 2) \end{cases} \tag{4}$$

The air pressure P_{air} can be linked to the square of the flow velocity [20]:

$$P_{\text{air}}(t) = \mu_2 |u(t)| u(t), \tag{5}$$

where u is the air flow velocity, μ_2 is the nonlinear pneumatic damping coefficient which characterizes the OWC turbine damping.

4 Validation

4.1 Wave-Porous Shell Interaction

The novel aspect of the design proposed in this study consists of the addition of the concentric porous structure around the floating platform to attenuate the platform motion. Therefore, the performance of the model with the addition of this component should be validated. To achieve the model validation, a numerical model consisting of a solid cylinder with an external diameter of a = 250 mm and a porous cylinder with an external diameter of b = 375 mm is created. In the porous component, the holes have a circular form of radius r, satisfying a porosity $\tau = 0.3$. The wall thickness of the porous structure is 3 mm and the effect of the wall thickness is ignored in the numerical calculation as shown in Fig. 2. The model is located in a water depth of 1 m and the height of the porous structure extends above the maximum amplitude of the incident waves. By comparing the results obtained from present numerical calculations with the results of analytical solutions using the boundary element method [17], the reliability of the shell part in the present model is established as presented in Fig. 3.

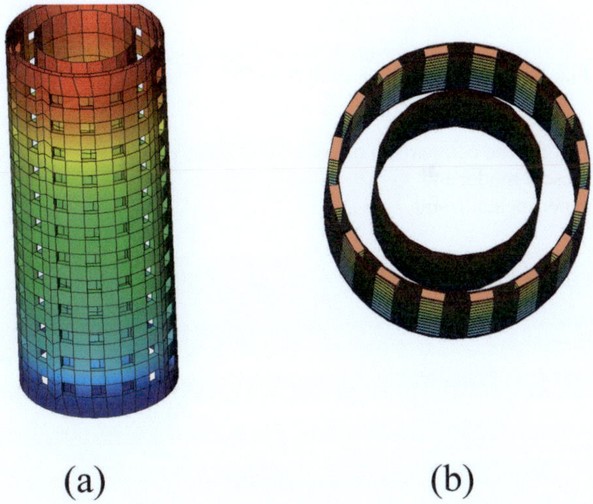

Fig. 2 The validation mesh of the porous shell: **a** side view and **b** top view

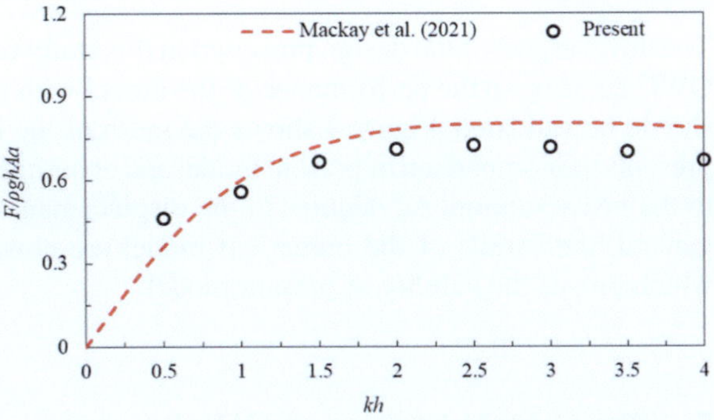

Fig. 3 The comparison of present model and analytical solutions

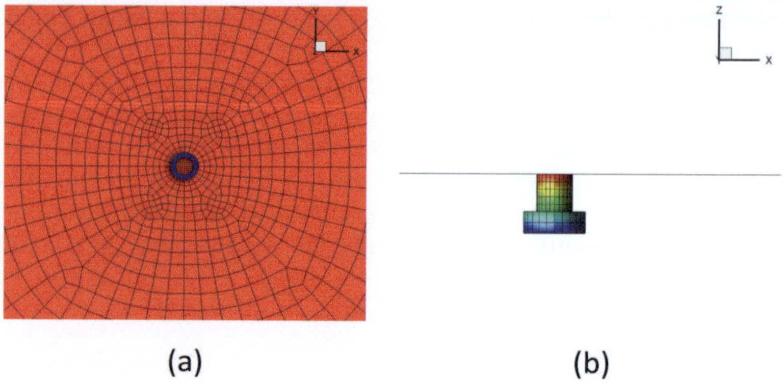

Fig. 4 The validation mesh of the OWC device: **a** side view and **b** top view

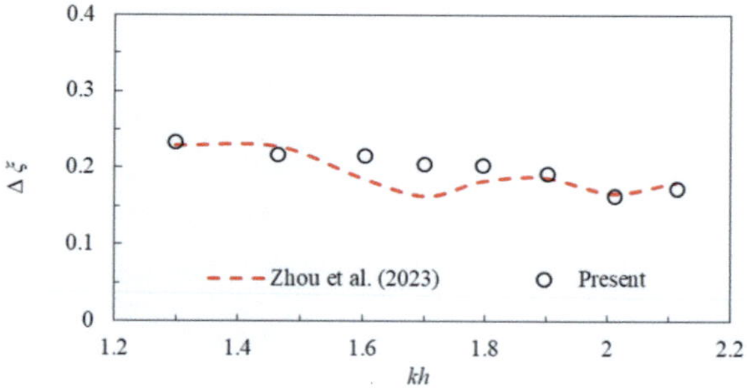

Fig. 5 The comparison of present model and experimental result

4.2 Wave-OWC Device Interaction

The novel aspect of the design proposed in this study consists of the integration of the OWC devices, so the performance of the model with the addition of this component should be validated. Figure 4 shows the mesh of the FOWT-OWC model. Figure 5 presents the comparison of present model and experimental result [21]. The variations in the heave motion, $\Delta\xi$ denotes to the displacement difference in heave modes. In general, the results of the numerical model are close to the experimental results, which proves the validity of present model.

5 The Coupled Effect of OWC and Porous Shell

A regular wave environment case is adopted to test the platform motion response. The wave height (H) and the wave period (T) are assumed to be 1 m and 7 s, respectively. The wave propagates along the x-axis. Because both the structure and the waves are symmetrical along the x-axis, the sway, roll and yaw modes shift too little to be negligible. Hereafter, the surge, heave and pitch mode motion responses are discussed, as shown in Fig. 6. In the surge mode, the motion responses are basically the same, where slight differences are shown in the peaks and troughs. In the heave and pitch modes, the motion responses of hybrid platform are decreased obviously.

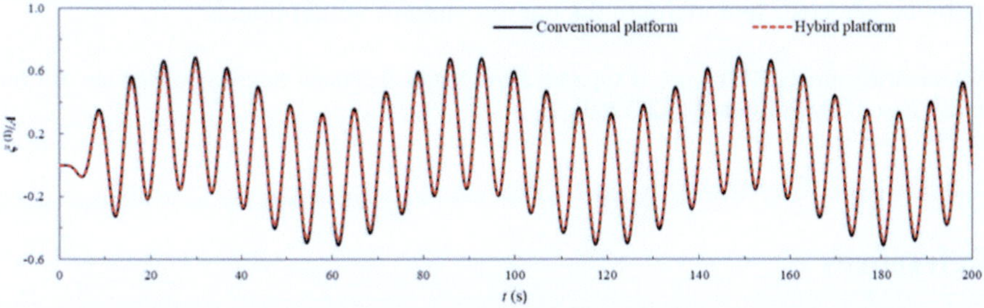

(a)

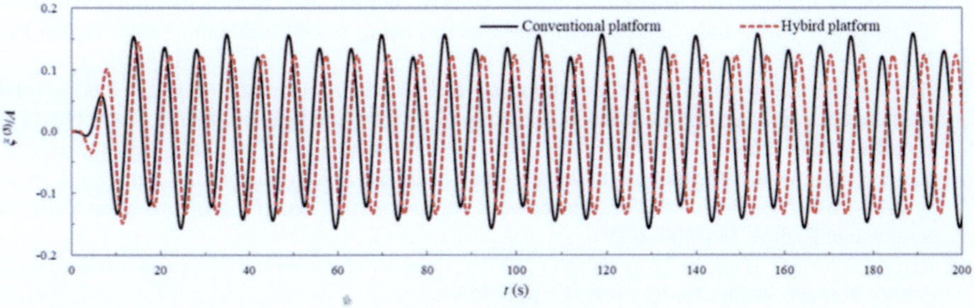

(b)

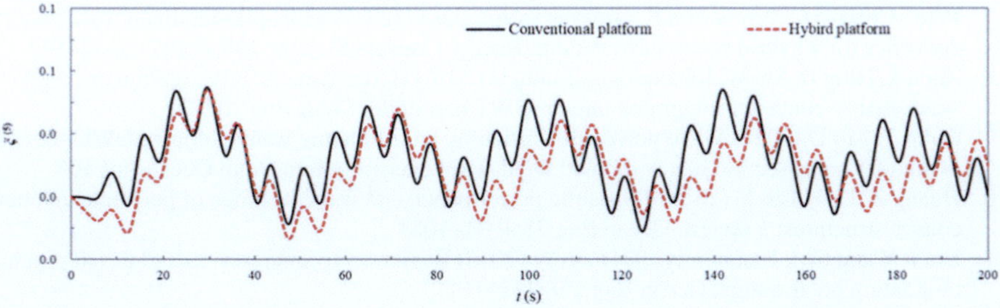

(c)

Fig. 6 The motion response comparison of conventional platform and the hybrid platform: **a** surge, **b** heave, and **c** pitch

6 Conclusions

In this study, a novel hybrid platform with OWC devices and a porous shell are proposed. Based on the HOBEM method, the motion responses of the conventional and the hybrid platform are compared. The result shows that the hybrid platform motion in heave and pitch modes decreased obviously, which is helpful to lower maintenance costs and increase the energy yield of wind projects.

Acknowledgements This work is supported the National Natural Science Foundation of China [Grant Nos. U22A20242 and 52271260].

References

1. GWE Council (2024) GWEC global wind report 2024. Global Wind Energy Council
2. Rao SK, Nallayarasu MJ, Bhattacharyya S (2021) Numerical and experimental studies of heave damping and added mass of spar with heave plates using forced oscillation. Appl Ocean Res 111
3. Pooja H, Nallayarasu S (2023) Investigation of heave damping characteristics of buoy form spar with heave plate near the free surface using CFD validated by experiments. Ships Offshore Struct 18:1650–1667
4. Yang Y, Bashir M, Michailides C, Li C, Wang J (2020) Development and application of an aero-hydro-servo-elastic coupling framework for analysis of floating offshore wind turbines. Renewable Energy 161:606–625
5. Pérez-Collazo C, Greaves D, Iglesias G (2015) A review of combined wave and offshore wind energy. Renew Sustain Energy Rev 42:141–153
6. Perez-Collazo C, Greaves D, Iglesias G (2018) A novel hybrid wind-wave energy converter for jacket-frame substructures. Energies 11
7. Perez-Collazo C, Pemberton R, Greaves D, Iglesias G (2019) Monopile-mounted wave energy converter for a hybrid wind-wave system. Energy Convers Manage 199:111971.1–111971.13
8. Zhou Y, Ning D, Shi W, Johanning L, Liang D (2020) Hydrodynamic investigation on an OWC wave energy converter integrated into an OWT monopile. Coast Eng 103731
9. Cong Y, Liu D (2021) Wave power absorption by an oscillating water column (OWC) device of annular cross-section in a combined wind-wave energy system. Appl Ocean Res 107
10. Huang Z, Li Y, Liu Y (2011) Hydraulic performance and wave loadings of perforated/slotted coastal structures: a review. Ocean Eng 38:1031–1053
11. Dai J, Wang CM, Utsunomiya T, Duan W (2018) Review of recent research and developments on floating breakwaters. Ocean Eng 158:132–151
12. Downie MJ, Graham JMR, Hall C (2000) An experimental investigation of motion control devices for truss spars. Mar Struct 13:75–90
13. Molin B (2001) On the added mass and damping of periodic arrays of fully or partially porous disks. J Fluids Struct 15:275–290
14. Tao L, Dray D (2008) Hydrodynamic performance of solid and porous heave plates. Ocean Eng 35:1006–1014
15. An S, Faltinsen OM (2012) Linear free-surface effects on a horizontally submerged and perforated 2D thin plate in finite and infinite water depths. Appl Ocean Res 37:220–234
16. An S, Faltinsen OM (2013) An experimental and numerical study of heave added mass and damping of horizontally submerged and perforated rectangular plates. J Fluids Struct 39:87–101

17. Mackay E, Shi W, Qiao D, Gabl R, Davey T, Ning D, Johanning L (2021) Numerical and experimental modelling of wave interaction with fixed and floating porous cylinders. Ocean Eng 242
18. Saha S, Bora S, Das S (2023) Time-dependent water wave scattering by a bottom-mounted porous compound cylinder fitted with an annular porous lid. In: Waves in random and complex media, pp 1–22
19. Robertson A, Jonkman J, Masciola M, Song H, Goupee A, Coulling A, Luan C (2014) Definition of the semisubmersible floating system for phase II of OC4
20. Zhou Y, Ning D, Shi W, Johanning L, Liang D (2020) Hydrodynamic investigation on an OWC wave energy converter integrated into an offshore wind turbine monopile. Coast Eng 162:103731
21. Zhou Y, Ning D, Chen L, Mayon R, Zhang C (2023) Experimental investigation on an OWC wave energy converter integrated into a floating offshore wind turbine. Energy Convers Manage 276:116546

A Frequency-Domain Modeling Approach for Hydroelastic Responses of Pontoon-Type Floating Bridges: An Experimental Validation

Minghao Cui, Zhengshun Cheng, Peng Chen, and Torgeir Moan

Abstract Very large floating bridges are distinguished by their capacity to cross wide and deep waters without the need for fixed foundations. Code-based design checks of floating bridges require a great number of simulations to ensure their serviceability and safety under various load conditions. However, time-domain simulation requires tremendous computational efforts. A promising alternative is to utilize a frequency-domain approach. In this study, a frequency-domain hydroelastic modeling approach is developed for a general pontoon-type floating bridge, in which the pontoons are modeled as rigid bodies whereas the superstructure is represented using beam elements. Hydrodynamic interactions between pontoons and viscous loads are considered. This study aims to validate the method by utilizing a public-available dataset of a truncated floating bridge tested in SINTEF Ocean basin. Time-domain simulations based on SIMA are also conducted as comparisons. It is shown that the proposed approach offers reasonable estimates of modal properties and RAOs. Moreover, with linearized second-order loads and hydrodynamic interactions considered, it captures low- and wave-frequency resonant peaks with adequate accuracy.

Keywords Floating bridge · Frequency-domain approach · Experimental validation · Hydroelasticity · Hydrodynamic interaction and viscous loads

M. Cui · Z. Cheng (✉) · P. Chen
State Key Laboratory of Ocean Engineering, Shanghai Jiao Tong University, Shanghai, China
e-mail: zhengshun.cheng@sjtu.edu.cn

Hainan Research Institute, Shanghai Jiao Tong University, Sanya, China

T. Moan
Norwegian University of Science and Technology, Trondheim, Norway

Ningbo University, Ningbo, China

X. L. Zhao et al. (eds.), *Proceedings of the Fourth World Conference on Floating Solutions: WCFS 2024; 2–4 December; Hong Kong*, Lecture Notes in Civil Engineering 597, https://doi.org/10.1007/978-981-96-4569-5_55

1 Introduction

Floating bridges are large-scale infrastructures supported by floating pontoons. Without the need for fixed foundations, these bridges are particularly suitable for spanning wide and deep waters, such as fjords, or areas with very soft seabeds [1].

Modern floating bridges can extend for several kilometers, which significantly increases the overall flexibility, leading to complex hydroelastic responses under diverse loading conditions. In code-based designs, numerous simulations are required to represent their dynamic behaviors and ensure adequate serviceability and safety.

Time-domain simulations are typically employed to capture nonlinear and transient effects. However, due to the large dimensions and coupling between the environment and the flexible structure, each simulation is extremely time-consuming. Consequently, a frequency-domain approach is promising, but ensuring accuracy through appropriate modeling and linearization techniques poses a further challenge.

Therefore, this paper introduces a frequency-domain approach for the hydroelastic responses of floating bridges. The focus is set on its extensive applications and experimental validation regarding Eigen modes, RAO, response spectra, and statistics. A publicly available dataset is utilized, based on the model tests of a truncated straight floating bridge conducted by Rodrigues et al. [2] at the SINTEF Ocean basin.

2 Frequency-Domain Modeling of the Experimental Model

The experimental model applies a truncated segment scaled by 1:33 with the full-scale length exceeding 1000 m. As shown in Fig. 1, the bridge consists of a continuous girder with ballast bars and rigid clamps, boundary brackets and bar elements, nine pontoons (of two types), and taut moorings for lateral stiffness. The model is fixed at the right hand with a bar element and ball joints. Details on the model description are referred to [2].

The FD approach assumes minor deformations of the pontoon-column pairs, such that the motion equations of the girder and the pontoons can be established based on the beam model and the multi-rigid-body hydrodynamics, respectively. With multi-body interaction, wave viscosity, and second-order wave loads incorporated, the ultimate hydroelastic equation of motion can be solved for Eigen values, response statistics and response spectra within only a few minutes.

In addition, another numerical model is established by the commercial software SIMO-RIFLEX (SIMA) [4]. The widely applied simulation tool applies Cummins's theory for coupled time-domain analyses, where structural dynamics and hydrodynamics are considered based on the finite element method and the potential flow theory, respectively. Each seed of a 3h simulation requires a solution time of over ten hours.

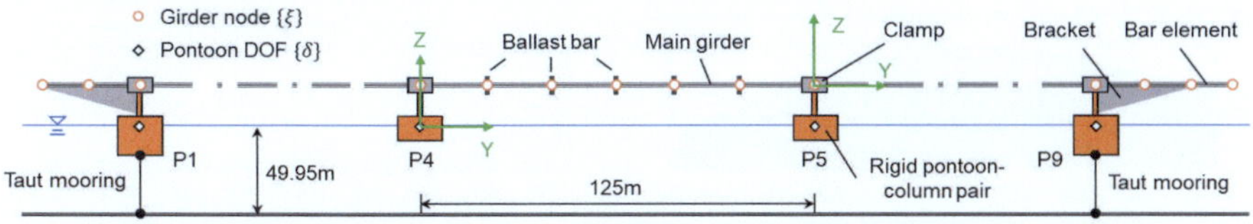

(1) $[M_p + M_a]\{\ddot{\xi}\} + [C_h + C_v]\{\dot{\xi}\} + [K_h]\{\xi\} = \{F_w\} + \{F_{cp}\}$

(2) $[M_s]\{\ddot{\delta}\} + [C_s]\{\dot{\delta}\} + [K_s]\{\delta\} = \{F_{cg}\}$

Frequency-dependent equation of motions for pontoons and for girder nodes

(3) $\{F_{cp}\} = -[T_l]\{F_{cg}\}$

(4) $\{\xi\} = [T_d]\{\delta\}$

Transfer matrices for motions and loads between pontoons and pontoon-axis nodes

$$\left[[M_p] + [M_a] + [M_s]\right]\{\ddot{\delta}(\omega)\} + \left[[C_h] + [C_v] + [C_s]\right]\{\dot{\delta}(\omega)\} + \left[[K_h] + [K_s]\right]\{\delta(\omega)\} = \{F_w(\omega)\}$$

Fig. 1 Frequency-domain (FD) modeling of the tested floating bridge (viewed from incoming wave direction). Details on the modeling approach can be found in our previous study [3]

3 Experimental Validation

Figure 2 shows an overview of the tested bridge and relevant instruments [2]. Due to the existence of equipment and cables, the mass properties of pontoons and clamps at different pontoon-axis vary slightly; however, they are assumed to have a uniform distribution in the TD and FD methods. The stiffness of the bridge girder is calibrated by comparing static pull-out results between TD and tested results.

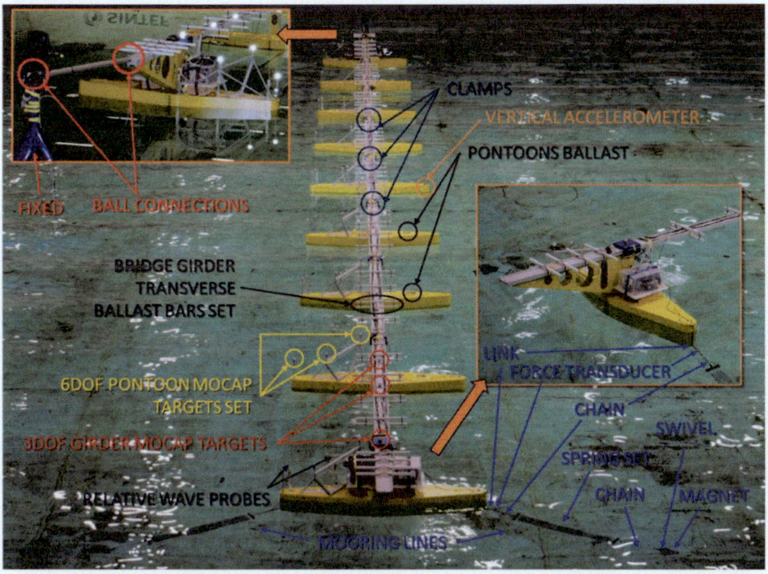

Fig. 2 Experimental model and instrumentations photographed by Rodrigues et al. [2]

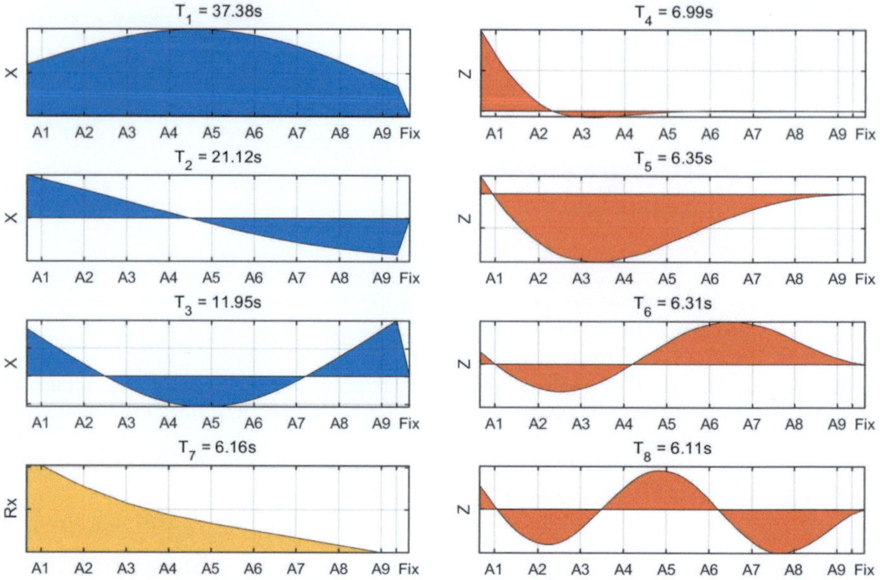

Fig. 3 Eigen-periods and model shapes by the frequency-domain approach considering frequency-dependent added mass of the pontoons

In Fig. 3, Eigen values of the entire system are estimated by the FD approach with frequency-dependent added mass of the pontoons, serving as a basis for model calibration and dynamic analysis. The modal shapes are generally close to those shared in [2], indicating a reasonable representation of the mass and stiffness term. Located in low- and wave-frequency, these modes can be excited by wave loads and increase the responses significantly.

Response amplitude operators (RAOs) of pontoon motions are shown in Fig. 4. During white noise analysis, hydrodynamic interactions and viscosity are not considered for both TD and FD models. Besides, the taut moorings are modeled in SIMA as bar elements with Morison's damping term, while for the FD model a simplified stiffness matrix is used. The differences in damping terms result in the peak differences for surge and pitch RAO, while a remarkable peak in heave is caused by the interaction effect.

Figure 5 shows the standard deviations of pontoon motions subjected to irregular wave actions. The viscosity and multi-body interaction in the experiments are represented in the FD approach. Satisfactory consistency is found between FD and Test results in surge of all pontoons and in heave and pitch of only central pontoons.

The differences are revealed by response spectra in Fig. 6. For surge, the low-frequency resonances are dominating and well captured by the FD approach. A significant peak in heave motion, corresponding to 6.6 s, is caused by hydrodynamic interactions for both FD and Test model, while for P1 the heave PSD is less consistent.

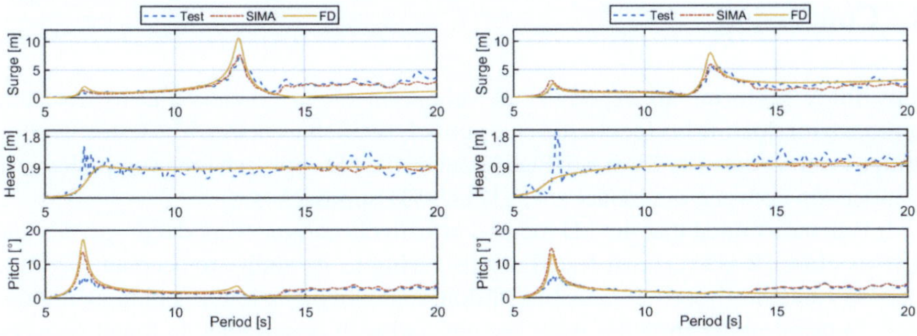

(a) P1 (b) P5

Fig. 4 RAOs of surge, heave, and pitch motions of pontoon **a** P1 and **b** P5 by the model test (Test), the time-domain simulation (SIMA), and the proposed frequency-domain approach (FD)

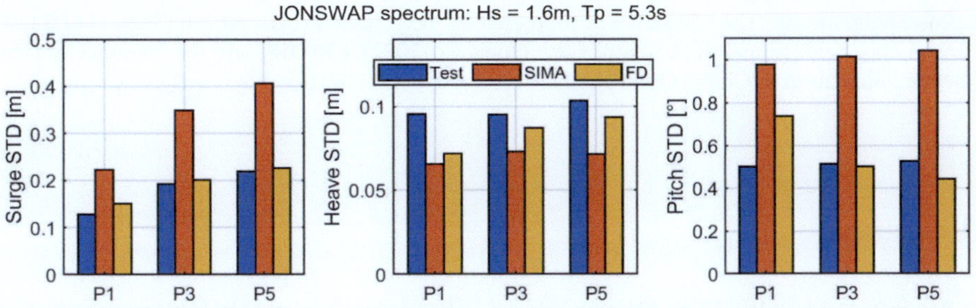

Fig. 5 Standard deviations of surge, heave and pitch motions of pontoon P1, P3 and P5. Hydrodynamic interaction and wave viscous damping are considered in tests and by the FD approach

The pitch PSD is lower for Test of P1, indicating larger damping effects related to the mooring lines.

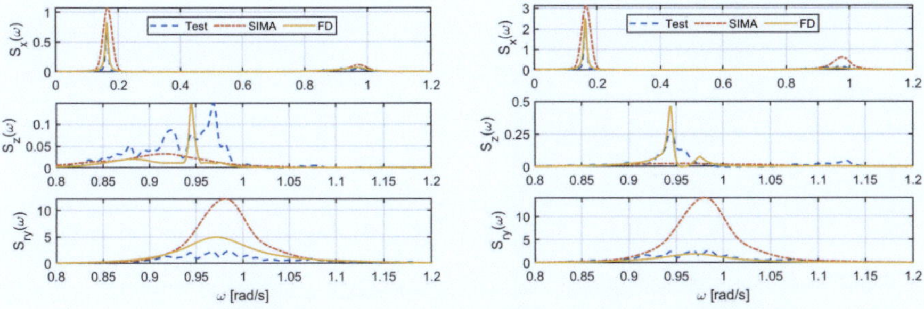

(a) P1 (b) P5

Fig. 6 Power spectral density (PSD) of surge, heave and pitch motions of pontoon P1 and P5

4 Conclusion

This study performs an experimental validation of an efficient FD hydroelastic approach for floating bridges based on a publicly available dataset. The tested floating bridge is represented by SIMO-RIFLEX and by the FD approach for a comprehensive comparison between the model test, TD and FD approaches.

It is demonstrated that the FD method effectively estimates modal properties and derives closer RAOs against tested results owing to its ability to account for wave viscosity and multi-body effects. In addition, standard deviations of pontoon motions are compared under irregular wave actions and well consistency is found between FD and Test results, except for the side pontoon that is affected by the nonlinearities of mooring lines and boundary conditions. Response spectra further reveal that the FD method captures linear wave responses, low-frequency resonances as well as wave-frequency resonances induced by the hydrodynamic interaction of pontoons.

Acknowledgements The research was supported by the Hainan Provincial Joint Project of Sanya Yazhou Bay Science and Technology City, Grant No 2021JJLH0018, and the National Natural Science Foundation of China, Grant No 52201330 & No 42176210.

References

1. Moan T, Eidem ME (2020) Floating bridges and submerged tunnels in Norway—the history and future outlook. In: Proceedings of the world conference on floating solutions. Springer, Singapore, pp 81–111
2. Rodrigues JM, Viuff T, Økland OD (2022) Model tests of a hydroelastic truncated floating bridge. Appl Ocean Res 125:103247
3. Cui M, Cheng Z, Chen P, Moan T (2024) A frequency-domain hydroelastic approach for modeling and stochastic dynamic analysis of pontoon-type floating bridges. Eng Struct 318:118740
4. SINTEF Ocean (2022) SIMA documentation, version 4.2.0

Numerical Investigation and Optimization on the Performance of Wave Energy Converter Array

Mengjie Wang, Ling Wan, Chi Zhang, and Zhiyung Tay

Abstract To reduce the emission of the greenhouse gas, exploration of renewable energy sources with abundant reserves could be viable solutions. Wave energy as renewable resource, has a substantial potential in the ocean. Nevertheless, the advancement of wave energy utilization technologies encounters obstacles such as suboptimal energy conversion efficiencies and high costs. To overcome these obstacles, dynamic properties and designs of single wave energy converter (WEC) and WEC array need to be studied. This work investigated floating WEC array composed of point absorbers (PA), focusing on the coupling effects among WECs in an array configuration. By employing fundamental theories and computational techniques related to ocean structures in the frequency, the study performs numerical simulations using ANSYS-AQWA software. In frequency domain, the study focuses on the effects of array types, float spacings, float parameters, and on the energy capture performance of the WEC array. A comparison was made between the energy capture efficiency of the WEC array and a single WEC, indicating that the WEC array demonstrated a superior capability to harness wave energy resources effectively.

Keywords Wave energy converter · Wave energy converter array · Design parameters · Energy capture performance

---

M. Wang · L. Wan (✉)
Faculty of Maritime and Transportation, Ningbo University, Ningbo, China
e-mail: wanling@nbu.edu.cn

C. Zhang
Technology Centre for Offshore and Marine, TCOMS, Singapore, Singapore

Z. Tay
Engineering Cluster, Singapore Institute of Technology, Singapore, Singapore

X. L. Zhao et al. (eds.), *Proceedings of the Fourth World Conference on Floating Solutions: WCFS 2024; 2–4 December; Hong Kong*, Lecture Notes in Civil Engineering 597, https://doi.org/10.1007/978-981-96-4569-5_56

1 Introduction

With global attention on the climate change due to the rising CO_2 level, generating electricity from renewable sources is once again an important area of research [1]. Wave energy utilization has garnered significant interest from governments and industry due to its high energy density, which surpasses that of solar and wind energy among renewable sources [2]. Currently, three main wave energy converters (WEC) types are used for harnessing wave energy: overtopping (OP) devices, oscillating bodies (OB), and oscillating water column (OWC) [3]. Among the myriad devices, OWC have gained extensive research attention owing to their structural simplicity, robust survivability, and ease of maintenance [4]. However, most existing OWCs utilize a single float, limiting power generation efficiency and increasing costs, hindering commercialization [5]. Consequently, arranging multiple WECs into an array has become a key research focus in wave energy converters [6, 7].

To improve the energy capture efficiency of array WECs, many scholars have studied the hydrodynamic characteristics and optimized the array layout based on their findings. Siddorn et al. [8] studied the hydrodynamic performance of a four-float cylindrical array using numerical analysis, and provided the expressions of wave excitation force, added mass, radiational damping and other parameters acting on each float. Babarit [9] examined how float spacing affects absorbed power, finding that larger spacing reduces interaction impact on energy absorption, with regular waves having a greater influence than irregular waves. Agamloh et al. [10] studied the interaction between two cylindrical floats and waves by using a three-dimensional numerical flume simulator, and found that changes in wave parameters and float spacing would greatly affect the interaction between floats. In addition, there are many works that combines WEC with wind turbines [11], that plays a role to stabilize the floater motion.

In this study, a heave-type WEC array is proposed with flexible designs tailored to power demand, layout area, and wave conditions. To enhance efficiency and stability, it investigates the effects of array form, float parameters, wave incidence direction, and float spacing on energy capture characteristics. The findings will provide practical insights for optimizing WEC arrays and understanding their impact on large-scale wave fields.

2 Mathematical Model

In this study, we only consider WEC with heave degree of freedom in frequency-domain, therefore, viscous force, mooring force and other nonlinear forces are ignored. The kinematic equation of the array float in the frequency domain is established based on the linear potential flow theory.

According to Newton's second law, the kinematic equation of the array float in the heave direction is:

$$M\ddot{Z}(j\omega) = F_h(j\omega) + F_{pto}(j\omega) \tag{1}$$

where, M is the mass matrix of the array float, $\ddot{Z}(j\omega)$ is the heave motion displacement matrix of the array float, $F_{pto}(j\omega)$ is the PTO force matrix of the array float, and $F_h(j\omega)$ is the hydrodynamic matrix of the array float in the heave direction, which is expressed as:

$$F_h = F_e + F_r + F_s \tag{2}$$

Further analysis, F_h includes wave excitation force matrix F_e in the heave direction of the array float, wave radiant force matrix F_r in the heave direction and the hydrostatic restoring force matrix F_s of the array float are obtained. They can respectively be expressed as:

$$F_{\mathrm{r}} = -M_{\mathrm{a}}(\omega)\ddot{Z}(j\omega) - B_{\mathrm{r}}(\omega)\dot{Z}(j\omega) \tag{3}$$

$$F_s = -K_s Z(j\omega) = -\rho g A_{\mathrm{w}} Z(j\omega) \tag{4}$$

In Eqs. (3) and (4): $M_{\mathrm{a}}(\omega)$ is the added mass matrix of the array float in the heave direction. $B_{\mathrm{r}}(\omega)$ is the radiational damping coefficient matrix of the array float in the heave direction. K_s is the hydrostatic restoring force coefficient matrix of the array float; A_{w} is the array float waterplane area matrix.

In this study, only linear PTO damping force is considered, so F_{pto} can be expressed as:

$$F_{pto} = -C_{pto}\dot{Z}(j\omega) \tag{5}$$

By combining the above equations, the kinematic equation of the array float in the frequency domain can be obtained as:

$$-\omega^2[M + M_{\mathrm{a}}(\omega)]Z(\mathrm{j}\omega) + j\omega\left[B_{\mathrm{r}}(\omega) + C_{pto}\right]Z(\mathrm{j}\omega) + \rho g A_{\mathrm{w}} Z(\mathrm{j}\omega) = F_{\mathrm{e}}(\mathrm{j}\omega) \tag{6}$$

The average output power of the array floats can be expressed as:

$$P = \tfrac{1}{2}\omega^2 |Z(\mathrm{j}\omega)|^T C_{pto} |Z(\mathrm{j}\omega)| \tag{7}$$

Array interaction q factor is defined as:

$$q = \frac{\Sigma_{j=1}^{N} P_j}{N P_0} \tag{8}$$

This study assumes that all devices in the array are exactly the same. In the equation, P_0 is the output power of a single device layout.

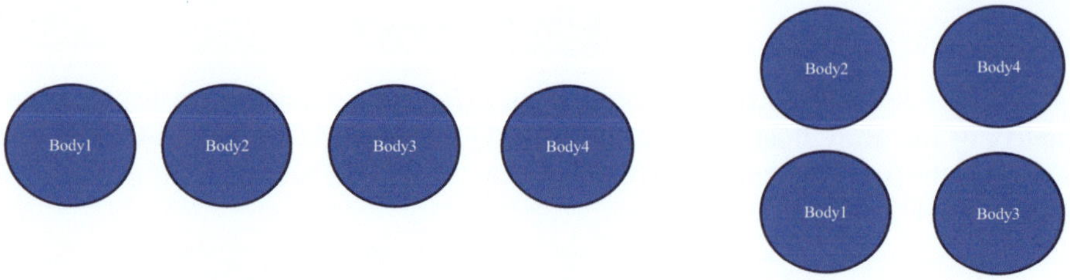

Fig. 1 Linear layout and square layout

3 Numerical Model

For the hydrodynamic analysis of point absorption wave energy converter, it is necessary to establish a reasonable hydrodynamic model. In this study, ANSYS-AQWA is used to establish the hydrodynamic model of floating body and calculate the hydrodynamic parameters of the float, then the calculation results are imported into ORIGIN and programmed to solve the energy capture power of the array float.

In this study, four identical floats were selected to form two array configurations with linear shape and square shape, respectively, as shown in Fig. 1. In the simulation analysis, deep water condition with unit wave height, and incident wave frequency of 0–4 rad/s are assumed. To simulate the PTO system, linear damping is set in the vertical motion direction of the array float.

After establishing the panel model in AQWA, the frequency domain analysis is carried out. The diagonal length of the panel size is small enough (<< λ/6) [12] that can capture the effects from the smallest wave length analyzed.

4 Hydrodynamic Performance of Wave Energy Converter

The WEC design parameters are shown in Table 1, the effects of array layout, WEC spacing, wave incidence direction and float number on the hydrodynamic performance of the array wave energy converter are studied.

4.1 Influence of Array Form on Hydrodynamic Performance

Numerical simulation parameters: the float is a cylindrical float, the float spacing is 4R, the number of floats N is 4, and the rest parameters remain unchanged in Table 1. Considering the symmetry of array form, the incident wave direction of linear array is 0°,15°, 30°, 45°, 60°, 75°, 90° when numerical simulation is carried out. When the square array is numerically simulated, the incident wave direction is 0°, 15°, 30° and 45°. Figure 2 shows the capture power of the two array layouts at the 90°

Table 1 Design parameters of wave energy converter array

Parameter	Value
Float radius R	2.5 m
Float draft B	4 m
Water depth H	1000 m (Deep water condition)
Number of floats N	2–4
Float spacing L	4–10 R
WEC array layout	Linear layout, Square layout
Incident wave frequency ω	0–4 rad/s
PTO Damping coefficient C	10,000 N/(m/s)

wave direction. Tables 2 and 3 show the peak energy capture power of the two array layouts.

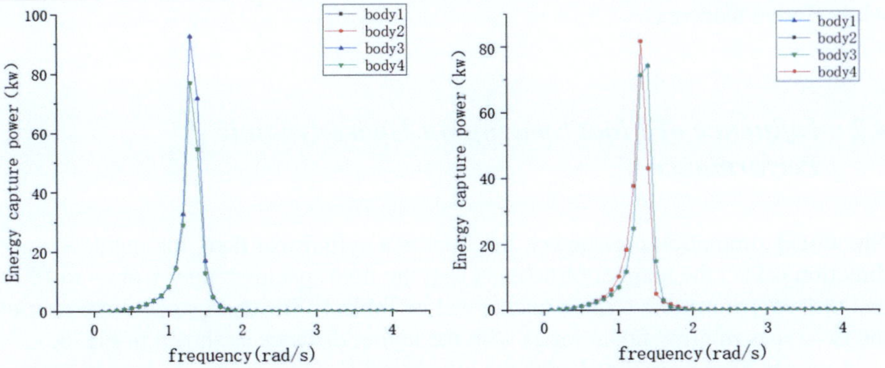

(a) Capture energy power (Linear layout) (b) Capture energy power (Square layout)

Fig. 2 Capture energy power of two array layouts at 90° wave direction (heave mode only)

Table 2 Captured power peak of linear array layout under different wave directions

Wave incident direction	Peak capture power (kW)			
	Body1	Body2	Body3	Body4
0°	75.69	75.463	76.77	70.17
15°	77.61	73.73	76.55	68.35
30°	82.74	69.45	76.60	64.92
45°	87.00	66.03	79.51	65.80
60°	83.68	68.68	87.57	74.16
75°	76.23	79.74	95.43	80.44
90°	76.95	92.57	92.58	76.96

Table 3 Captured power peak of square array layout under different wave directions

Wave incident direction	Peak capture power (kW)			
	Body1	Body2	Body3	Body4
0°	81.60	81.61	74.20	74.21
15°	81.55	85.09	79.86	67.40
30°	83.79	89.38	86.88	60.91
45°	85.09	90.50	90.49	58.23

From Table 2, Body1's peak capture power increases between 0° and 45°, while Body2, Body3, and Body4's decrease. At 60°–75°, Body1's power drops as the others increases. At 90°, Body1 and Body4 have similar peak power, as do Body2 and Body3. Table 3 shows that at 0°, the peak power of Body1 and Body2, as well as Body3 and Body4, is nearly identical due to the same wave force distribution. As the wave incidence angle increases, Body1, Body2, and Body3's power increases, while Body4's decreases.

4.2 *Influence of Float Spacing on Hydrodynamic Performance*

Numerical simulation parameters: the float is a cylindrical float, the incident wave direction is 90°, the number of floats N is 4, the float spacing ranges from 4 to 10 R, and the rest parameters remain unchanged in Table 1. The energy capture power of the two types of array floats varies with the layout distance as shown in Fig. 3.

As can be seen from Fig. 3, with the increase of layout spacing, the energy capture power of the array floats decreases first and then increases, and the energy capture

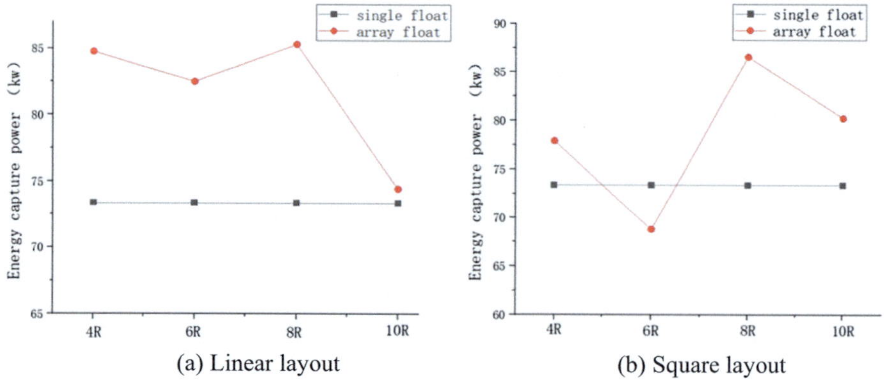

Fig. 3 The energy capture power of the two types of array floats varies with the layout distance

power of any spacing of the linear layout is greater than the generating power when the floats move alone. The square layout reaches its maximum value when the spacing is 8 R.

4.3 Influence of Number of Floats on Hydrodynamic Performance

Numerical simulation parameters: the float is a cylindrical float, the array form is linear layout, the incident wave direction is 90°, the float spacing is 4R, the number of floats N range from 2 to 5, and the rest parameters remain unchanged in Table 1. The average output power of two arrays with different number of floats is obtained by simulation and shown in Fig. 4.

As shown in Fig. 4, increasing the number of floats amplifies q factor fluctuations around 1. When q > 1, adding more floats improves energy capture efficiency;

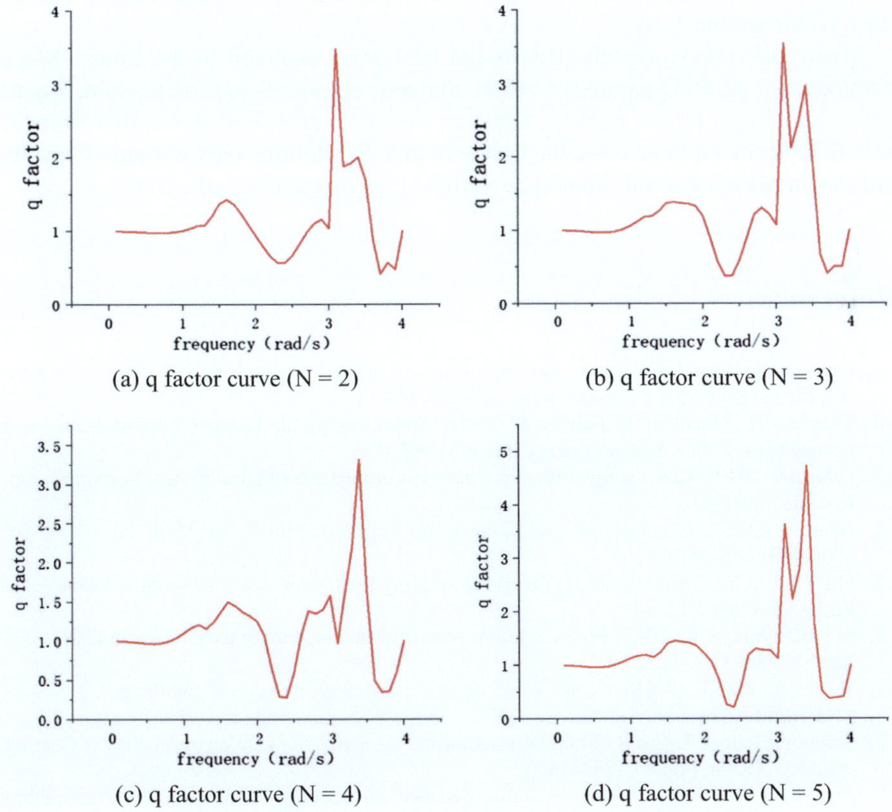

Fig. 4 Array interaction q factor curve when the number of floats N = 2–5

however, if the incident wave frequency falls within the range where $q < 1$, it negatively impacts efficiency. Therefore, selecting the appropriate number of floats based on the sea area's wave information is crucial for optimizing the WEC array's energy capture performance.

5 Conclusion and Future Work

A WEC array is proposed to supply power for moderate ocean operations, such as fish farms and ocean observation. Using linear potential flow theory and ANSYS-AQWA software, the energy capture characteristics of the WEC array are studied.

Different float layout significantly impacts the hydrodynamic performance of the WEC array. Selecting an optimal layout based on wave conditions is crucial for improving performance. As float spacing increases, energy capture power initially decreases, then increases, peaking at 8 R. After this point, the peak frequency of energy capture declines with greater spacing. Additionally, increasing the number of floats enhances hydrodynamic interactions, as indicated by increased fluctuations in the q factor around 1.

There are various aspects that should be further explored in the future. More parameters, e.g., PTO parameter effect, dynamic responses in time domain, where various nonlinear forces are considered, as well as control strategies in time domain should be included to increase the power output. In addition, only numerical results are shown here, the results should be verified experimentally in the future.

References

1. Drew B, Plummer A, Sahinkaya M (2009) A review of wave energy converter technology. Proc Inst Mech Eng Part A J Power Energy 223:887–902
2. Clement A, Mccullen P, Falcao A (2002) Wave energy in Europe: current status and perspectives. Renew Sustain Energy Rev 6(5):405–431
3. Falcao A (2010) Wave energy utilization: a review of the technologies. Renew Sustain Energy Rev 14(3):899–918
4. Heath T (2012) A review of oscillating water columns. Phil Trans Math Phys Eng Sci 370(1959):235–245
5. Shi YT, Liu YJ, Zhao LZ (2021) Overview of array float wave power generation technology. Energy Environ 2:12–14
6. Shi HD, Wang CK (2017) Progress and prospect of marine energy technology in China. Sol Energy 3:30–37
7. Tunde A, Hua L (2018) Ocean wave energy converters: status and challenges. Energies 11(5):1250–1275
8. Siddom P, Eatock Taylor R (2008) Diffraction and independent radiation by an array of floating cylinders. Ocean Eng 35:1289–1303
9. Babarit A (2010) Impact of long separating distances on the energy production of two interacting wave energy converters. Ocean Eng 37(8–9):718–729

10. Agamloh E, Wallace A, Von J (2008) Application of fluid-structure interaction simulation of an ocean wave energy extraction device. Renewable Energy 33(4):748–757
11. Wan L, Moan T, Gao Z, Shi W (2024) A review on the technical development of combined wind and wave energy conversion systems. Energy 294:130885
12. DNV (2019) Recommended practice dnv-rp-f205: global performance analysis of deepwater floating structures

Potential Analysis for Coupling Effect of Ship Motion with Liquid Sloshing

Sheng-Chao Jiang and Qian Gu

Abstract Potential analysis for coupling effect of ship motion and liquid sloshing is considered. A three-dimensional FPSO ship with three partially-filled prismatic tanks are carried out for the global ship response. It is found that the coupling effects are dominant in beam waves. The roll RAOs show typical anti-rolling effect, so that the roll motion decreases dramatically near ship motion resonance and double peaks at sloshing motion resonant frequencies can be observed. The filling mass, that is, the number of filling tank in this work, has significant effect on the ship rolling RAOs, especially at around the ship motion natural frequency.

Keywords Coupling effect · Sloshing · Ship motion · Roll RAO

1 Introduction

Ship motion and inside liquid sloshing react on with each other in ships with liquid cargos, meaning that ship motion and liquid sloshing are not independent. Therefore, the coupled effects between ship motion and internal sloshing should be considered for practical ship response and cargo damage design. Such a coupling problem is important for the global ship responses. In detailed, the coupling effects can lead to the stronger and induces significant influence if the natural frequency of sloshing flow is closer to that of ship motion. In particularly, it is usually happen on LNG and FPSO carriers because the size of liquid tank configurations.

Many numerical simulations on coupling analysis have been introduced, such as Rognebakke and Faltinsen [1], Newman [2] and Gou et al. [3]. In these studies, the linear potential theory is adopted, both on external wave and internal sloshing flow. Based on the physical experimental results by Rognebakke and Faltinsen [1], even if the violent sloshing occurs in the cargo, the steady-state ship motion is still almost

S.-C. Jiang (✉) · Q. Gu
State Key Laboratory of Coastal and Offshore Engineering, Dalian University of Technology, Dalian, China
e-mail: jiangshengchao@foxmail.com

X. L. Zhao et al. (eds.), *Proceedings of the Fourth World Conference on Floating Solutions: WCFS 2024; 2–4 December; Hong Kong*, Lecture Notes in Civil Engineering 597, https://doi.org/10.1007/978-981-96-4569-5_57

linear and sinusoidal with the frequency of the linear incident waves. It implies that the higher-order harmonics of the sloshing force are filtered out by the system. The similar physical experimental results can also be found in Nam et al. [4]. In a word, the existing studies indicate that the assumption of linear ship motion seems adequate in coupled analysis. However, it is speculated that the linear sloshing flow assumption may be not reasonable.

Therefore, a time domain solution is necessary due to the importance of nonlinearities for sloshing flow. A viscous numerical wave flume for direct calculation of ship motion is currently gaining popularity, but the significant computational cost is still required. If coupling with internal sloshing flow effects, it is much more time consuming. Since the linear assumption can satisfy the necessary of ship motion and external wave action, an impulse-response function (IRF) is applied for the time-domain analysis of ship motion problem. Converting the frequency domain solutions to time domain, the IRF method can provide much faster simulations than the direct solvers. This method is suggested by Cummins [5] firstly and a detailed description is given by Ogilvie [6]. Lee and Newman [7] suggested an analytic form of IRF for compensating cut-off frequency error. Based on the IRF method, the nonlinear effects of the internal sloshing flow can be considered for coupled analysis. Lee et al. [8] consider the effects of LNG-tank sloshing on the global motions of LNG carriers. The pattern of coupling effects between vessel motion and liquid sloshing appreciably changes with filling level are investigated. The nonlinear coupled effects of internal sloshing flow is investigated by Kim et al. [9], according to simulate a rectangular barge with partially filled tanks and a ship with rectangular passive-type ART, respectively. It is found that the nonlinearity of sloshing flow is very important in coupled motion response, and the ship motion shows a strong sensitivity to wave slope. A coupled analysis between ship motion response and inside liquid sloshing are carried out by Jiang et al. [10], in which a LNG-FPSO with two partially filled tanks is considered. The numerical study is extended to the observation of pressure field inside the tanks.

In this paper, a coupled model based on frequency domain is adopted for analyzing a FPSO ship with three liquid tanks. Three kinds of floating conditions, corresponding to different filling conditions, are considered for investigating the coupled effects between ship motion response and liquid sloshing.

2 Numerical Setup

In present work, a FPSO with three partially-filled tanks is considered. The molded lines and the main dimensions of the hull and tanks are shown in Table 1, respectively. Table 2 shows the parameters of the tank. In the experiment, only heave, roll and pitch motions are allowed, and surge, sway and yaw motions are restrained. As shown in Table 3, three kinds of draught conditions with two filling conditions are adopted for the simulations. For each draught, the radius of gyration for different filling conditions are kept on constants.

Table 1 Main particulars of FPSO ship

Parameters	Dimension		
	$T = 6.0$ m	$T = 8.0$ m	$T = 10.5$ m
Length, L, (m)		210.00	
Breadth, B, (m)		32.8	
Draught, T, (m)	6.0	8.0	10.5
Displacement, V, (t)	37,794	50,450	67,541
K_{xx}, K_{yy} (m)	(11.11, 73.60)	(9.55, 66.44)	(8.89, 58.65)
GM (m)	12.32	11.93	11.17
KG	109.77	109.72	109.66

Table 2 Main particulars of tanks

Parameters	Tank
l	25.0
b	28.0
h	18.5

Table 3 Filing conditions

Draught (m)	Filing percent (%)	Total mass	Water mass	Mass percent (%)
6.0	20	37,794	8219	21.75
	30	37,794	12,452	32.95
8.0	40	50,450	16,921	33.54
	50	50,450	21,272	42.16
10.5	60	67,541	25,623	37.94
	70	67,541	29,975	44.38

3 Numerical Results

3.1 Coupling Effect of Filling Level

The numerical model established above is adopted for the numerical simulations. The target of this work is to obtain the basic hydrodynamic behavior of coupling sloshing effect, especially under the resonant conditions. Therefore, no artificial damping is adopted in this work.

The ship motions of FPSO of $T = 6.0$ m different filling conditions are simulated, firstly. As shown in Figs. 1 and 2, two kinds of filling conditions, 20 and 30% in three tanks, by three incident wave directions are considered. Little discrepancy of the heave and pitch RAOs among the various filling conditions, including the zero-filling condition, can be observed, indicates that the coupling effect is insignificant on heave and pitch motions.

The ship rolling RAOs at different filling conditions in 45 and 90° incident waves are compared, including that of without sloshing conditions in Fig. 3. A typical anti-rolling effect generated by the internal sloshing flow can be visualized, in which a radical decrease of rolling RAOs near 'without sloshing resonant frequency of

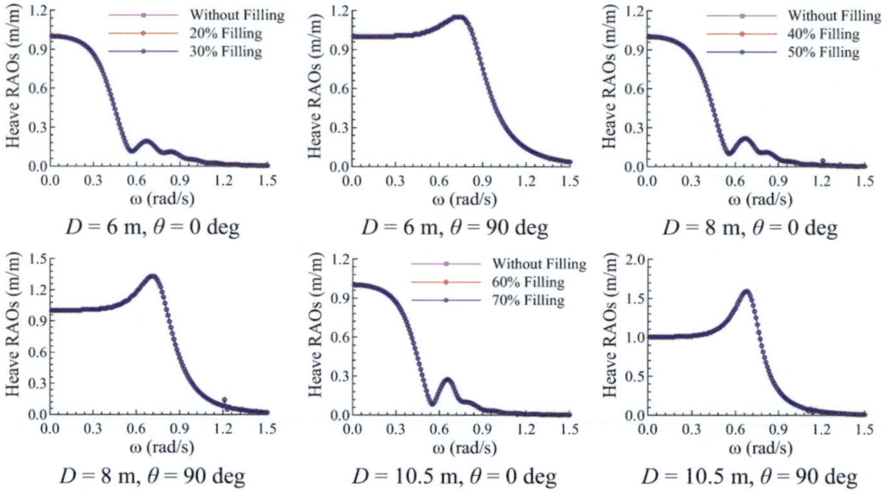

Fig. 1 Comparisons of heave RAOs under wave action

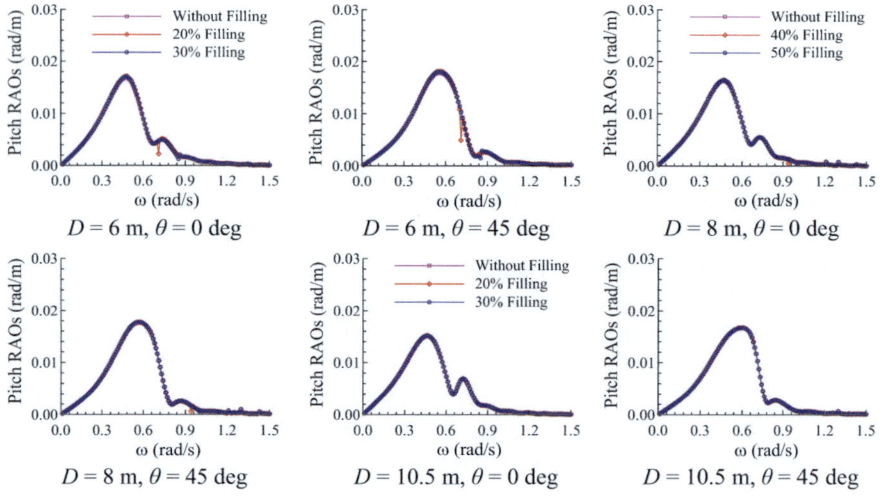

Fig. 2 Comparisons of pitch RAOs under wave action

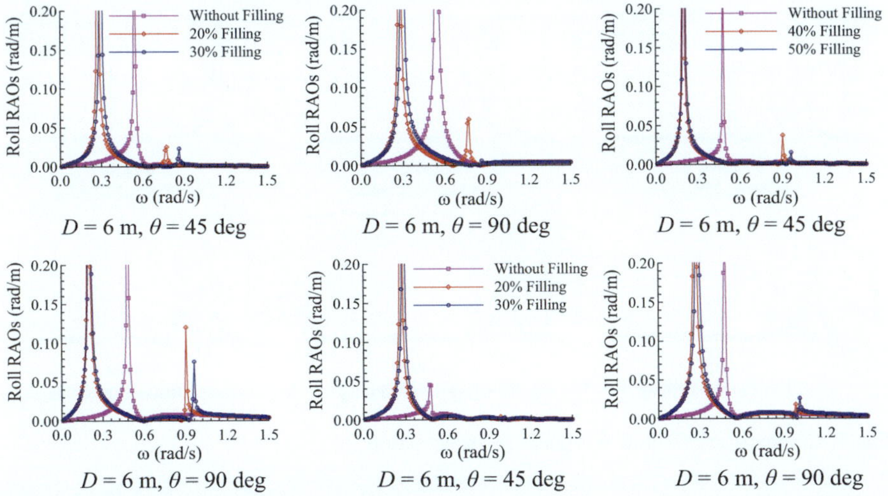

Fig. 3 Comparisons of roll RAOs under wave action

the ship' can be obtained. Besides, a second peak near 'sloshing motion natural frequency' region is also observed, which comes from the violent sloshing flow action due to the resonance happen.

3.2 Coupling Effect of Filling Mass

The coupling effect by the filling mass, that is, the number of filling tank, is considered in this section. As shown in Fig. 4, it can be observed that the filling mass has significant effect on the rolling RAOs under wave action. The larger filling mass can lead to a smaller natural frequency ship motion response and a larger natural frequency of sloshing flow motion, respectively.

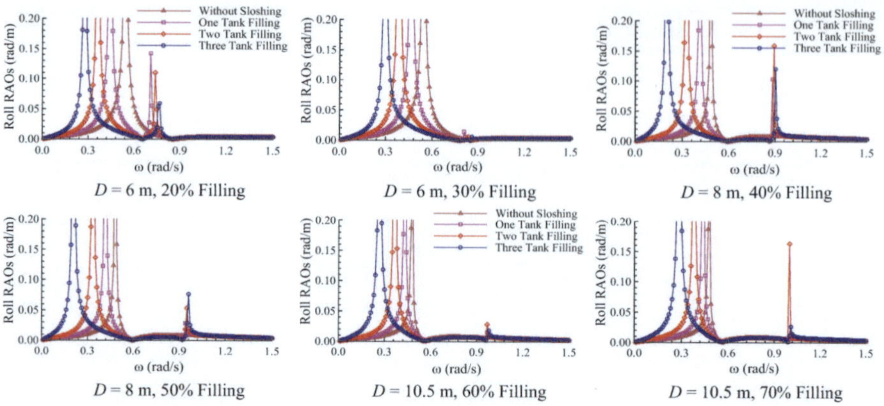

Fig. 4 Comparisons of roll RAOs under 90° wave action

4 Conclusion

Potential analysis for the coupling effect between ship motion response and internal sloshing flow is investigated in this paper. Numerical simulations for a three-dimensional FPSO ship with three partially-filled prismatic tanks are carried out for the global ship response. The sloshing-induced coupling effect is relatively small for heave and pitch motion of the ship in head sea condition. In Beam Sea, the coupling effect on global ship response is dominant and it shows the typical anti-rolling characteristics. The ship motion response shows strong sensitivity to filling mass, especially at around the ship motion natural frequency.

References

1. Rognebakke OF, Faltinsen OM (2003) Coupling of sloshing and ship motions. J Ship Res 47(3):208–221
2. Newman J (2005) Wave effects on vessels with internal tanks. In: Proceedings, 20th Workshop on water waves and floating bodies. Spitsbergen, Norway
3. Gou Y, Kim Y, Kim TY (2011) A numerical study on coupling between ship motions and sloshing in frequency and time domain. In: Proceedings of 21th International offshore and polar engineering conference. Maui, Hawaii, USA, pp 158–164
4. Nam BW, Kim Y, Kim DY, Kim YS (2009) Experimental and numerical studies on ship motion responses coupled with sloshing in waves. J Ship Res 53(2):68–82
5. Cummins WE (1962) The impulse response function and ship motions. In: Proceedings, symposium on ship theory. Schiffstechnik, Hamburg, Germany, p 9
6. Ogilvie TF (1964) Recent progress toward the understanding and prediction of ship motions. In: Proceedings, Fifth symposium on naval hydrodynamics. Bergen, Norway, pp 3–79
7. Lee CH, Newman JN (2005) Computation of wave effects using the panel method. In: Chakrabarti S (ed) Numerical modeling in fluid-structure interaction. WIT Press, Southampton
8. Lee SJ, Kim MH, Kim JW, Shin Y, Kim Y (2007) The effect of LNG-tank sloshing loads on the motions of LNG carriers. Ocean Eng 34:10–20

9. Kim Y, Nam BW, Kim DW, Kim YS (2007) Study on coupling effects of ship motion and sloshing. Ocean Eng 34(16):2176–2187
10. Jiang SC, Teng B, Bai W, Gou Y (2015) Numerical simulation of coupling effect between ship motion and liquid sloshing under wave action. Ocean Eng 108:140–154

Hydrodynamic Analysis of Novel Offshore Floating Concrete Modules and Mooring System

Peicen Wang, Xiao Lin Zhao, and Jinping Ou

Abstract This work investigates the hydrodynamic performance of various configurations of concrete floating modules and mooring systems, which are suited for a shallow water marine environment with a depth of 60 m. A comparative analysis of the hydrodynamic shapes of floating modules was conducted, including modules combining flat plate and semi-submersible design, semi-submersible modules, and column-type modules. This work also compared the catenary chain's performance, including catenary chains and catenary chains complemented with segmented weight clumps. The paper underlines the functional characteristics of floating modules, the performance of distinct hydrodynamic shapes, and both the economic indicators and restoring force of the mooring systems. Findings suggest that flat plates with circular truncated cone modules aptly fulfill the hydrodynamic performance requirements and can provide superior load-bearing efficiency. Furthermore, using catenary chains with segmented weight clumps can improve the motion response performance of the platforms, under the same mooring chain length and safety requirements, thereby bringing greater economic benefits. This work can potentially aid in the investigation of the hydrodynamic performance and design of mooring systems for modular floating structures.

Keywords Floating platform module · Hydrodynamic performance · Mooring system

P. Wang · J. Ou (✉)
School of Civil and Environmental Engineering, Harbin Institute of Technology, Shenzhen, China
e-mail: oujinping163@163.com

P. Wang · X. L. Zhao
Department of Civil and Environmental Engineering, The Hong Kong Polytechnic University, Hung Hom, Kowloon, Hong Kong, China

X. L. Zhao
Research Institute for Land and Space, The Hong Kong Polytechnic University, Hung Hom, Kowloon, Hong Kong, China

X. L. Zhao et al. (eds.), *Proceedings of the Fourth World Conference on Floating Solutions: WCFS 2024; 2–4 December; Hong Kong*, Lecture Notes in Civil Engineering 597, https://doi.org/10.1007/978-981-96-4569-5_58

1 Introduction

The accelerated urbanization, rising sea levels due to climate change, and the scarcity of land and resources have driven global efforts to seek sustainable floating solutions [1]. Concurrently, advancements in marine renewable energy development and marine engineering have highlighted the potential of Very Large Floating Structures (VLFS). Extending hundreds of meters or even kilometers, VLFS offer significant advantages such as positional flexibility, minimal environmental impact, and resilience against rising sea levels. These attributes make VLFS a viable solution for coastal cities facing land shortages and flooding threats, serving various purposes including residential areas, industrial zones, airports, and renewable energy installations [2].

However, due to their large structure, VLFS are highly flexible and significantly affected by wave-induced forces [3]. The hydroelastic effects from the interaction between the structure and fluid are critical. Additionally, the fabrication and transportation of a single, large VLFS present challenges, leading to the proposal of modular VLFS. Modular VLFS alleviates large internal bending moments and forces, offering redundancy and scalability. This study aims to design and research the hydrodynamic shape of individual modules to achieve optimal hydrodynamic performance for a multi-module VLFS system.

This paper compares three different platform hydrodynamic shape modules in shallow water: a flat plate with semi-submersible design platform, a flat plate, a semi-submersible platform, and a column-type platform. It also proposes a new mooring system that significantly improves dynamic performance without increasing mooring line length by adding segmented clumps. Using the ANSYS-AQWA program, the hydrodynamic characteristics of modular platforms under different wave periods were studied, and the hydrodynamic response of the new mooring system was analyzed across various sea conditions.

2 Hydrodynamic Analysis of Floating Platforms

2.1 Numerical Computation Theory

ANSYS AQWA is a component of the ANSYS engineering simulation software suite, specializing in the hydrodynamic analysis of marine and offshore structures. In ANSYS AQWA, the hydrodynamics around marine structures are modeled using potential flow theory, which involves solving potential flow equations under specific boundary conditions. Potential flow theory assumes the fluid is incompressible and irrotational, and the fluid flow field can be described using a velocity potential function. This velocity potential function must satisfy the Laplace equation:

$$\frac{\partial^2\phi}{\partial x^2}+\frac{\partial^2\phi}{\partial y^2}+\frac{\partial^2\phi}{\partial z^2}=0 \tag{1}$$

Additionally, it is assumed that the motion of the floating body in small amplitude waves is minimal. According to linear wave theory, the velocity potential function must meet the seabed boundary condition, the free surface condition, the submerged body surface condition, and the radiation condition:

$$\frac{\partial\phi}{\partial z}=0, \quad at\ z=-h \tag{2}$$

$$\frac{\partial^2\phi}{\partial t^2}+g\frac{\partial\phi}{\partial z}=0, \quad at\ z=0 \tag{3}$$

$$\frac{\partial\phi}{\partial n}=\sum_{j=1}^{6}v_j f_j(x,y,z) \tag{4}$$

$$\lim_{R\to\infty}\phi=0 \tag{5}$$

2.2 Design and Parameters of Floating Platforms

This paper designs three types of floating platforms, suitable for a water depth of 60 m. The side, top, and 3D views of each platform are illustrated in Figs. 1 and 2. Parameters such as weight, displacement, and rotational inertia are provided in Table 1. Scheme 1 is a semi-submersible platform with five columns and a flat bottom pontoon. The central column aims to reduce mid-span deflection and the bending moment of the upper deck. Scheme 2, based on Scheme 1, removes the central column and changes the flat bottom pontoon to a ring-shaped pontoon. Scheme 3 is inspired by some column-shaped and hourglass-shaped FPSOs, with the cylinder contracted at the waterline to reduce the waterline area. For comparison, the displacement of these schemes is set at approximately 18,000 tons.

The three designed floating platform schemes are mainly used as supporting platforms for floating factory facilities. Therefore, two five-story steel modular assembly buildings are placed on top of the floating platform. The total weight of these two buildings is approximately 3248.3 tons. In practical calculations, the two buildings are simplified as rectangular prisms. The schematic diagram of the superstructure and the floating platform is shown in Fig. 3.

The Response Amplitude Operator (RAO) of the platform motion can be calculated using the AQWA-LINE module to obtain the RAO for each degree of freedom of the platform. Figures 4 and 5 show the pitch RAO and heave RAO respectively. Due to the strong symmetry of the platform, the roll RAO is omitted in this paper.

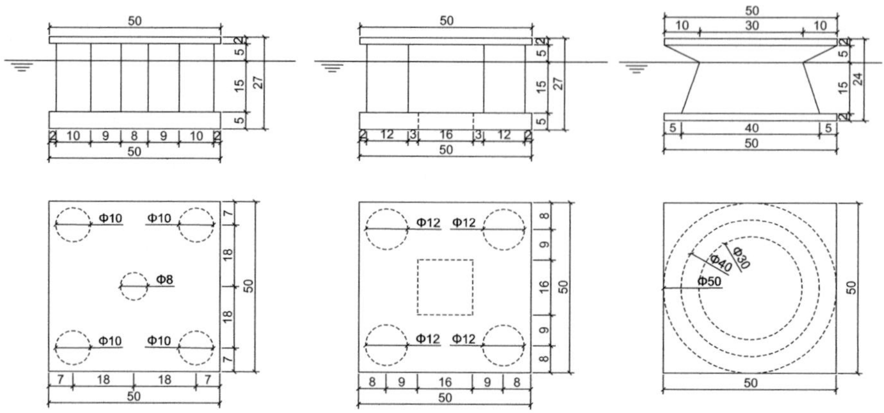

Fig. 1 Side and top views of floating platforms, **a** Scheme 1, **b** Scheme 2, and **c** Scheme 3

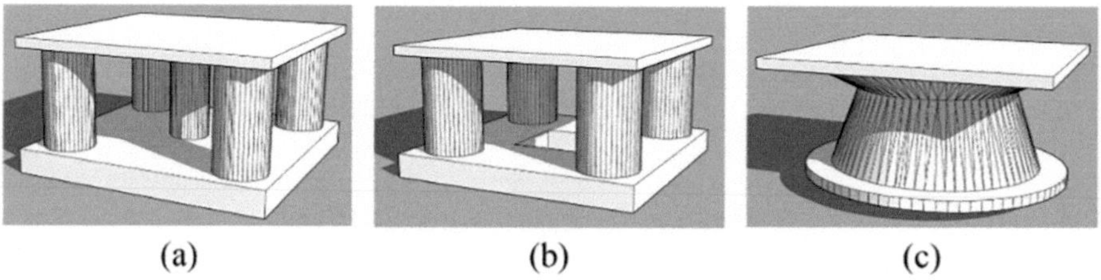

Fig. 2 3D models of floating platforms, **a** Scheme 1, **b** Scheme 2, and **c** Scheme 3

Table 1 Parameters of floating platforms

Scheme number	Scheme 1	Scheme 2	Scheme 3
Waterline area (m^2)	360	449	707
Draft (m)	20	20	17
Displacement (t)	17,900	17,949	18,447
Self-weight (t)	9580	9431	9843
Maximum load weight (t)	8320	8518	8604
Moment of inertia I_{xx} ($kg \cdot m^2$)	3.20×10^9	3.28×10^9	2.40×10^9
Moment of inertia I_{yy} ($kg \cdot m^2$)	3.20×10^9	3.28×10^9	2.40×10^9
Moment of inertia I_{zz} ($kg \cdot m^2$)	4.68×10^9	4.83×10^9	3.16×10^9

For the pitch RAO, different wave directions significantly impact the pitch RAO, with the greatest impact on Scheme 1. Since the undamped natural frequencies of the pitch for all three schemes are less than 0.2 rad/s, and the frequency range shown in Fig. 4 is 0.25–2 rad/s, each scheme only has a small peak in the figure. The pitch

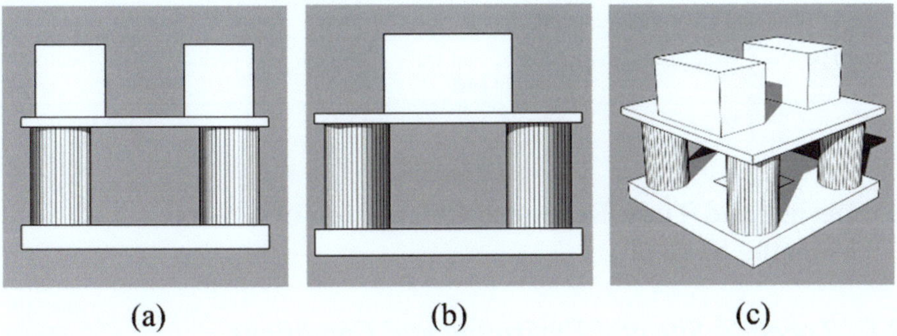

Fig. 3 Schematic diagram of the superstructure and the floating platform, **a** left view, **b** front view, and **c** isometric view

RAO peaks for all three schemes are concentrated around 0.65 rad/s, with Scheme 1 having the largest pitch RAO at 0.62. For the heave RAO, the values for different wave directions are almost identical. Since the undamped natural frequencies of the heave for Schemes 2 and 3 are greater than 0.2 rad/s, a larger peak is observed in the figure, with Scheme 3 having the highest peak, reaching a heave RAO of 2.1.

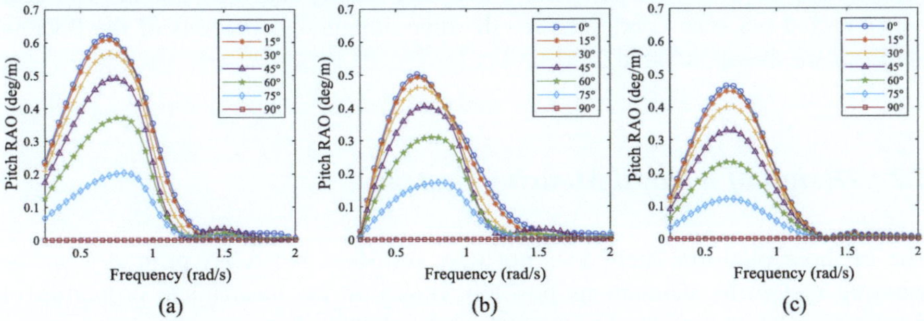

Fig. 4 Pitch RAO, **a** Scheme 1, **b** Scheme 2, and **c** Scheme 3

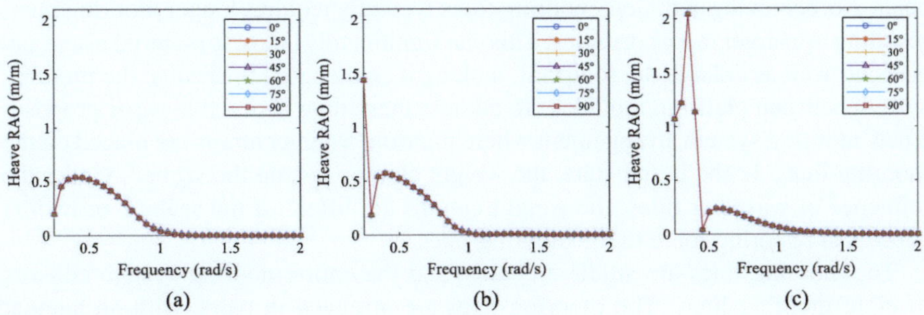

Fig. 5 Heave RAO, **a** Scheme 1, **b** Scheme 2, and **c** Scheme 3

Table 2 Wind and wave conditions [5]

Return period	Wind speed (m/s)	Significant wave height (m)	Spectral peak period (s)	Wave spectrum
100 years	34.22	10.65	14.13	JONSWAP

3 Design and Time-Response Calculation of a Novel Mooring System

3.1 Location Site and Environmental Conditions

The floating platform, which serves as a support structure for floating factory facilities is assumed to be located at 114° 50′ E and 22° 00′ N in the South China Sea, where the water depth is approximately 60 m [4]. The basic data for this location is sourced from the waveclimate.com database provided by BMT ARGOSS. This database utilizes measurements from radar altimeters, radar scatter meters, and imaging radar, capturing average wave parameters within a 50 km radius centered at the specified coordinates. The data has been validated and calibrated with buoy testing data. Samples were collected every three hours from 1992 to 2016, totaling 73,056 samples. The wave parameters were statistically analyzed, and the 100-year return period sea state conditions for the time-domain calculations of the floating platform are shown in Table 2 [5].

3.2 Design of a Novel Mooring System

The offshore platform itself does not have thrusters and relies primarily on the mooring system to maintain its position. Based on the mechanical properties of the mooring lines, the mooring system can be classified into two main categories: catenary mooring and taut mooring. Catenary mooring is often used in shallower water depths, and the mooring force is mainly provided by the self-weight of the chain. So, achieving sufficient mooring force typically requires longer mooring lines, resulting in a larger mooring radius. This can significantly increase material usage and costs, as well as enlarge the footprint, making it challenging to arrange the mooring system between platform modules. To address these drawbacks, this paper proposes a new mooring system arrangement where multiple weight clumps are placed on the mooring lines. In the initial state, the weight clumps rest on the seabed. Under the influence of waves or tides, the weight clumps are lifted off the seabed, providing additional restoring force to the mooring lines.

The mooring lines are studless chains, and the entire mooring system consists of eight mooring lines. The mooring lines are arranged in pairs, with an angular separation of 15°. Each mooring line is equipped with four groups of a total of eight

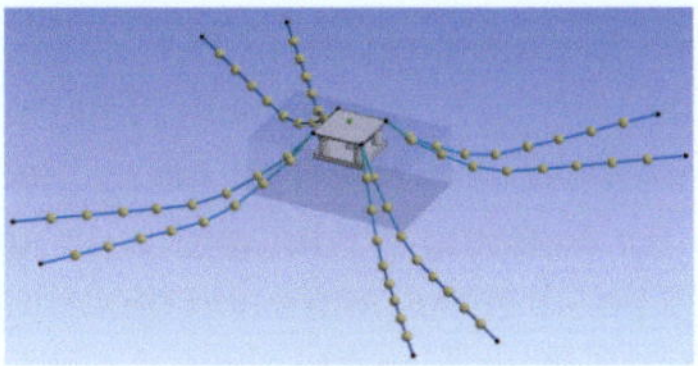

(a)

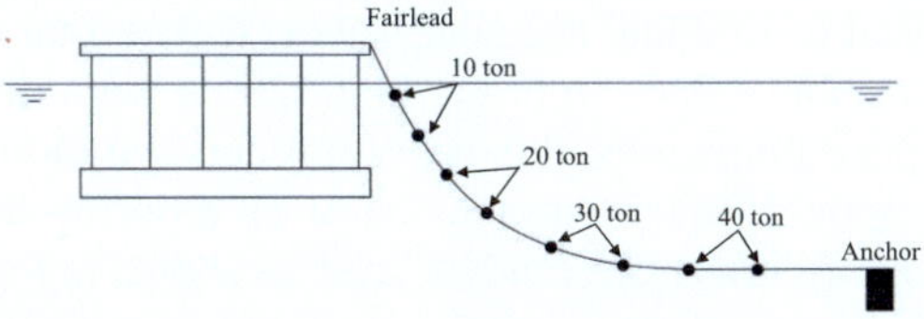

(b)

Fig. 6 Schematic diagram of mooring lines and clumps, **a** mooring system, and **b** single mooring line and clumps

Table 3 Parameters of mooring lines

Mass/unit length (kg/m)	Stiffness (N)	Maximum expected tension (N)	Added mass coefficient (–)	Transverse drag coefficient (–)	Longitudinal drag coefficient (–)	Total length (m)
219.4	9.415×10^8	7.693×10^6	0.8	2.0	1.15	300.0

Table 4 Coordinate information of fairlead points and anchor points

Mooring lines	Fairlead point (m)	Anchor point (m)	Mooring lines	Fairlead point (m)	Anchor point (m)
No. 1	(25, 25, 7)	(250.2, 191.98, −60)	No. 5	(−25, −25, 7)	(−250.2, −191.98, −60)
No. 2	(25, 25, 7)	(191.98, 250.2, −60)	No. 6	(−25, −25, 7)	(−191.98, −250.2, −60)
No. 3	(−25, 25, 7)	(−191.98, 250.2, −60)	No. 7	(25, −25, 7)	(191.98, −250.2, −60)
No. 4	(−25, 25, 7)	(−250.2, 191.98, −60)	No. 8	(25, −25, 7)	(250.2, −191.98, −60)

concrete weight clumps. The submerged weights of each group are 10, 20, 30, and 40 tons, respectively. These eight weight clumps are evenly distributed along the mooring line, with a spacing of approximately 33.3 m. The platform and mooring line arrangement are shown in Fig. 6. The parameters of the mooring lines and the coordinates of the fairlead and anchor points are shown in Tables 3 and 4.

3.3 *Time Response Calculation*

This section simulates the wind and wave parameters outlined in Table 2 by invoking the AQWA-DRIFT module. For each design scheme, 10 different random seeds were used to simulate irregular waves, with a total simulation duration of 12,000 s. The unstable values from the first 1200 s were discarded in every final result. Due to space limitations, this paper presents the motion response and No. 4 mooring line response (maximum tension of all mooring lines) for Scheme 2 with and without clumps under one random seed, as shown in Fig. 7. The surge response and mooring response values for all three schemes are displayed in Tables 5 and 6.

From Fig. 7, and the data in Tables 5 and 6, it is evident that the introduction of clumps significantly reduces the maximum and mean values of the surge response, with the maximum value decreasing by over 20% and the mean value by over 40%. At the same time, the introduction of clumps does not significantly affect the heave

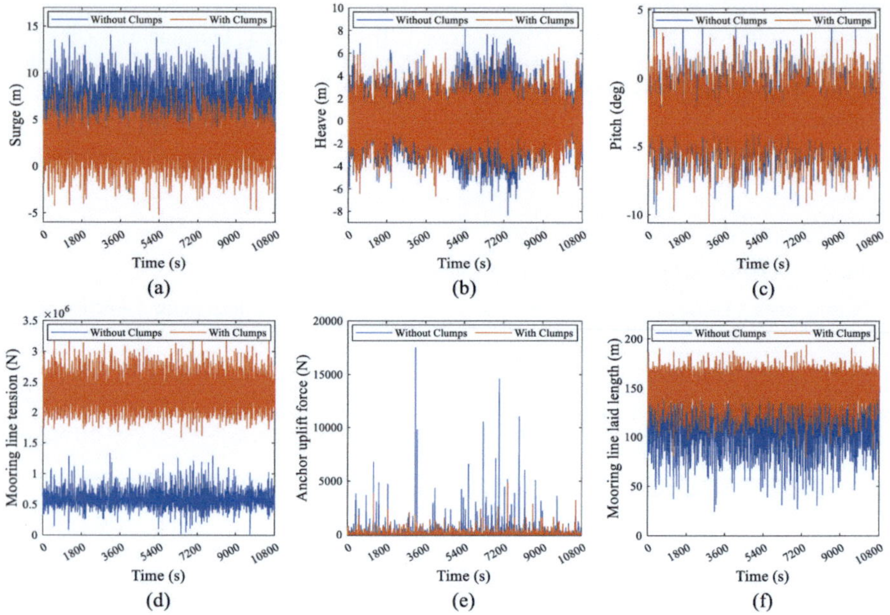

Fig. 7 Motion and mooring response of Scheme 2

Table 5 Surge and mooring response (without clumps)

Scheme	Scheme 1	Scheme 2	Scheme 3
Maximum surge (m)	22.01	15.40	25.33
Mean surge (m)	12.12	6.55	13.68
Maximum mooring line tension (N)	2.07×10^6	1.66×10^6	8.67×10^6
Mean mooring line tension (N)	5.93×10^5	5.81×10^5	6.70×10^5
Mooring safety factor (–)	3.71	4.63	0.89
Maximum anchor uplifting force (N)	1.31×10^5	3.98×10^4	1.34×10^6
Minimum laid length (m)	7.13	13.82	3.24

Table 6 Surge and mooring response (with clumps)

Scheme	Scheme 1	Scheme 2	Scheme 3
Maximum surge (m)	16.71	10.38	19.22
Mean surge (m)	6.84	2.59	6.05
Maximum mooring line tension (N)	3.51×10^6	3.59×10^6	5.68×10^6
Mean mooring line tension (N)	1.89×10^6	2.35×10^6	2.46×10^6
Mooring safety factor (–)	2.19	2.14	1.35
Maximum anchor uplifting force (N)	7.30×10^3	6.78×10^3	1.81×10^5
Minimum laid length (m)	69.08	70.13	10.95

or pitch response. Additionally, adding clumps significantly increases the laid length of the mooring lines and reduces the anchor uplifting force. However, it should be noted that the introduction of clumps increases both the maximum and mean values of mooring line tension. Despite this, the increased tension remains within safety factor requirements. Adding concrete weight clumps to each mooring line results in a relatively small increase in overall material costs, providing good economic value. By comparing the three schemes in Tables 5 and 6, it is evident that the mooring system performance of Schemes 1 and 2 is quite similar and significantly better than that of Scheme 3. However, under the same mooring conditions, the surge response of Scheme 2 is reduced by more than 30% in both maximum and mean values compared to Scheme 1.

4 Conclusion

This study conducted a comprehensive analysis of various floating platform configurations and mooring systems to identify optimal designs for shallow water marine environments with a depth of 60 m. Among the three configurations analyzed, the semi-submersible combined with a ring-shaped pontoon demonstrated better hydrodynamic performance, showing reduced motion response, making it suitable for offshore applications.

The proposed mooring system, incorporating segmented weight clumps, significantly enhanced the dynamic performance of the floating platforms without increasing the mooring line length. This system effectively reduced the surge response. Although the introduction of weight clumps increased mooring line tension, the levels remained within safe limits, ensuring the integrity of the mooring system. The innovative mooring system provides a cost-effective solution for offshore floating structures.

In conclusion, the findings of this study highlight the potential of using a novel floating platform combined with a segmented weight clump mooring system to achieve optimal hydrodynamic performance for offshore floating structures. These insights contribute to the design and development of more efficient and cost-effective marine platforms, supporting the growing need for sustainable solutions in marine engineering.

References

1. World Economic Forum. https://www.weforum.org/agenda/2021/10/floating-buildings-climate-change-rising-sea-levels/. Last accessed 25 Jun 2024
2. Lamas-Pardo M, Iglesias G, Carral L (2015) A review of very large floating structures (VLFS) for coastal and offshore uses. Ocean Eng 109:677–690
3. Wang CM, Watanabe E, Utsunomiya T (2008) Very large floating structures. Taylor & Francis, Abingdon
4. Zheng S, Li C, Wang X, Ou J (2023) Equivalent static wave-induced loads based on long term responses of a 10-MW floating semi-submersible wind turbine system. Available at SSRN 4520907
5. Zheng S, Li C, Wang P, Zhou S, Xiao Y (2023) Wind tunnel and wave flume testing on directionality dynamic responses of a 10 MW Y-shaped semi-submersible wind turbine. J Renew Sustain Energy 15(1)

Numerical Investigation of Hydrodynamic Responses of Straight Floating Bridge with Pontoons

Jihun Song, Chungkuk Jin, and Seungjun Kim

Abstract Floating bridges offer alternative design solutions for sea-crossing where conventional bridges with piers and abutments are not feasible. However, their complex hydrodynamic interactions necessitate detailed modeling and analysis to ensure structural safety, durability, and serviceability. Utilizing OrcaFlex software, this study introduces an advanced computational framework to examine the structural behavior of these bridges under various ocean environmental loading conditions. Based on previous work that highlights the importance of considering inhomogeneous waves, we successfully created a realistic model of a straight pontoon bridge and integrated inhomogeneous wave conditions into the OrcaFlex model. The study mainly focused on the examination of the effects of different mooring configurations on the structural behaviors under extreme conditions. We found that optimized mooring configurations can substantially improve the bridge's resilience to hydrodynamic forces. This research is anticipated to contribute valuable insights into the basic and detailed design of floating pontoon bridges, thereby improving safety and efficiency.

Keywords Floating bridge · Hydrodynamics · Global performance · Mooring

1 Introduction

Floating bridges, which rely on natural buoyancy to support the bridge deck, offer a viable alternative to conventional bridge types in locations with deep water or extremely soft seabeds [1–3]. Globally, the advantages of floating bridges have

J. Song · S. Kim (✉)
Department of Architectural, Civil and Environmental Engineering, Korea University, Seoul, Korea
e-mail: rocksmell@korea.ac.kr

C. Jin
Department of Ocean Engineering and Marine Sciences, Florida Institute of Technology, Melbourne, USA

X. L. Zhao et al. (eds.), *Proceedings of the Fourth World Conference on Floating Solutions: WCFS 2024; 2–4 December; Hong Kong*, Lecture Notes in Civil Engineering 597, https://doi.org/10.1007/978-981-96-4569-5_59

spurred the construction of such infrastructure, including notable examples like the Hood Canal Bridge, the Bergsøysund Bridge, the Nordhordland Bridge, and the Yumemai Bridge. The Norwegian Public Road Authority (NPRA) launched the E39 coastal highway project, aiming to reduce travel time by replacing ferry connections with floating infrastructures. For the Bjørnafjord, which has a width of up to 5 km, various types of floating bridge concepts have been designed and examined, supported by numerous research activities. Based on field measurements, the inhomogeneity within wave parameters in the Bjørnafjord was assessed [4–6], and the impact of inhomogeneous waves on the hydrodynamic responses of a floating curved bridge was explored [7, 8]. Some researchers used the Simulating Waves Nearshore (SWAN) model to simulate wave conditions and calibrate them with field measurement data, followed by increasing the number of data points to perform extreme value analysis for deriving design wave parameters [9, 10]. The same research group also investigated inhomogeneous load effects on a straight and side-anchored floating bridge, which is the primary literature referred on this study [11]. Viuff et al. [12–14] studied numerical analysis method for a long side-anchored floating curved bridge with a stay-cable tower, investigating the uncertainty with a model test. They further studied the uncertainty in various numerical models of floating bridges to improve experimental methodologies and numerical analyses [15, 16]. Despite extensive research, current commercial software does not provide functionalities to implement inhomogeneous wave conditions, which are crucial for accurately assessing the structural responses and limit states under realistic sea states. Previous studies have emphasized the significance of considering inhomogeneous wave characteristics in structural response evaluations and limit state assessments, as they can significantly impact the behavior of floating bridges. Therefore, it is essential to integrate inhomogeneous wave conditions into commercial solvers to improve design effectiveness. Furthermore, the effects of directional inhomogeneity of waves on the structural responses under various mooring configurations are not well understood. Investigating these effects is important for optimizing the design and improving the resilience of floating bridges, thus leading to better design practices and enhanced safety and performance.

In this study, a referred straight floating pontoon bridge was reproduced using OrcaFlex to evaluate wave-induced hydro-elastic responses of structural components. In addition, an advanced framework to integrate inhomogeneous wave conditions into the commercial solver was proposed. Following the approach, various inhomogeneous wave cases were simulated for the straight floating bridge model, focused on the directional inhomogeneity. Finally, parametric studies considering the characteristics of mooring configurations were conducted to evaluate and improve the resilience capacity against hydrodynamic forces.

2 Theoretical Background

2.1 Motion of Equation

Pontoons are the major components that interact with waves, causing complex wave-induced behaviors of structural components above the water. In general, the hydrodynamic coefficients of pontoons are evaluated in the frequency domain analysis based on the potential theory. Then, the pre-evaluated wave load characteristics are mapped on the lumped mass. The governing equations for pontoons of floating bridge in the time domain can be written by Cummin's equation as

$$\left(\tilde{M} + \tilde{M}_a(\infty)\right)\ddot{\tilde{u}}_p(t) + \int_{-\infty}^{t} \tilde{R}(t-\tau)\dot{\tilde{u}}_p(\tau)d\tau + \left(\tilde{K}_H + \tilde{K}_s\right)\tilde{u}_p(t) = \tilde{F}_{ext}(t) \tag{1}$$

where $\tilde{M}$ is the inertia matrix, $\tilde{M}_a(\infty)$ represents the added mass at infinite frequency matrix, $\tilde{R}(t-\tau)$ is the retardation function to involve the energy dissipation corresponding to the velocity by convolutional integral for the time lag, $\tilde{K}_H$ is the hydrostatic stiffness matrix, $\tilde{K}_s$ is the coupled stiffness matrix of the floating bridge, $\tilde{u}_p$ is the displacement vector, $\dot{\tilde{u}}$ is the velocity, $\ddot{\tilde{u}}$ is the acceleration, and $\tilde{F}_{ext}(t)$ is the external force vector including environmental loads.

2.2 Inhomogeneous Wave

In offshore engineering, the ocean wave is described as a function of the power spectrum, directional spreading, wave heading and individual phase angles of wave components. The Joint North Sea Wave Observation Project (JONSWAP) spectrum was utilized for generating irregular waves in this study and is expressed as

$$S_\zeta = \frac{5}{16}(1-0.287)\ln(\gamma)H_s^2\frac{\omega_p^4}{\omega^5}e^{-\frac{5}{4}\left(\frac{\omega}{\omega_p}\right)^{-4}}\gamma^{e^{-\frac{1}{2}\left(\frac{\omega-\omega_p}{\sigma\omega_p}\right)^2}} \tag{2}$$

where ω_p is the peak angular frequency, H_s is the significant wave height, γ is the extra peak enhancement factor, σ is the spectrum width parameter which equals to 0.07 for $\omega \leq \omega_p$ and 0.09 for $\omega > \omega_p$.

The directional spreading refers to the distribution of wave energy across different directions, while maintaining the total energy. The specific form with the normalizing constant used by OrcaFlex is given by

$$D_\zeta = \frac{\Gamma\left(1+\frac{n}{2}\right)}{\sqrt{\pi}\Gamma\left(\frac{1}{2}+\frac{n}{2}\right)}\cos^n(\theta-\theta_p) \tag{3}$$

where θ_p is the principal wave direction, n is the spreading coefficient to adjust energy concentration towards the principal wave heading.

For the wave forces acting on the floating body, the first order wave force and the second order difference frequency forces can be written as

$$F^1_{ext} = \sum_{m=1}^{N_\omega}\sum_{n=1}^{N_\theta} \mathrm{Re}\left(\left|H^1(\omega_m,\theta_n)\right|\sqrt{2S_\zeta(\omega_m)D_\zeta(\theta_n)\Delta\omega\Delta\theta}\, e^{i\left(\omega_m t - k_m x\cos(\theta_n) - k_m y\sin(\theta_n) + \varepsilon_{mn} + \varphi_{H^1_{mn}}\right)}\right) \tag{4}$$

$$F^2_{ext} = \sum_{l=1}^{N_\omega}\sum_{m=1}^{N_\omega}\sum_{n=1}^{N_\theta} \mathrm{Re}\left(\left|H^2(\omega_l,\omega_m,\theta_n)\right|\sqrt{2S_\zeta(\omega_m)D_\zeta(\theta_n)\Delta\omega\Delta\theta}\, e^{i\left((\omega_l-\omega_m)t + \varepsilon_{ln} - \varepsilon_{mn} + \varphi_{H^2_{lmn}}\right)}\right) \tag{5}$$

where $\boldsymbol{H}^1(\omega_m,\theta_n)$ is the linear transfer functions from the diffraction and radiation analysis, $\boldsymbol{H}^2(\omega_l,\omega_m,\theta_n)$ is the quadratic transfer functions, $S_\zeta(\omega_m)$ is the unidirectional wave spectrum, $D_\zeta(\theta_n)$ is the directional spreading, ε_{mn} is the phase angle of wave components, k_m is the wave number calculated by the dispersion relation, φ_H is the phase angle of load transfer functions. If more than one of wave parameters differ by pontoon locations, the wave condition has inhomogeneity and regarded as the inhomogeneous wave field.

3 Target Floating Bridge Model

3.1 System Particulars

The foundational model described by Multiconsult [17] incorporates the stay-cable tower, the seabed terrain characteristics of the target site, and additional intricate elements in the design of the floating bridge at Bjørnafjord. This study adopts a straight and side-anchored floating pontoon bridge model, mirroring the numerical framework proposed by Dai et al. [11] which provides a simplified yet effective representation of the original design's behavior. As shown in Fig. 1, the stay-cable tower has been reduced to a line element with equivalent sectional stiffness, with pontoons, girders, and columns uniformly positioned at the same elevation. Spanning a total length of 4600 m, the bridge features pontoons spaced 125 m apart. The 14.5 m-tall columns are firmly connected to both the upper surfaces of the pontoons and the girders. Mooring lines are connected to four designated pontoons, each located 1000 m apart, with each set comprising eight strands clustered together.

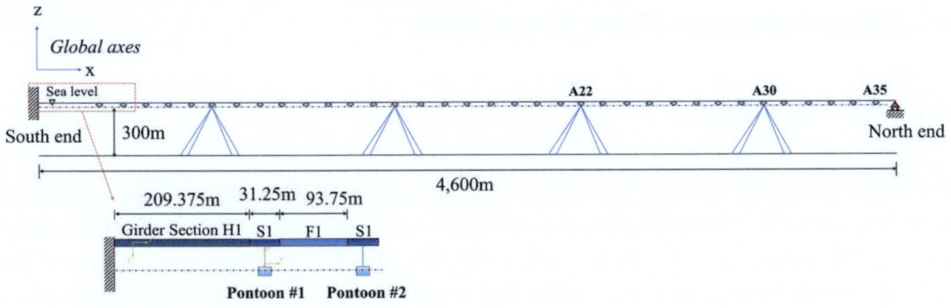

Fig. 1 Idealized straight floating bridge in plan and side view

3.2 Numerical Model of Straight Pontoon Bridge in OrcaFlex

The numerical model for the floating pontoon bridge was constructed using OrcaFlex, as demonstrated in Fig. 2. Within OrcaFlex, the local z-axis of the line element aligns with its length, with the default setting being a shear-flexible beam element. Each line element's beam orientation was configured as illustrated in Fig. 1. To prevent overlapping connections, Girder S1 was placed with free-end connections at a height of 18 m, while Girder F1 connected to Girder S1. All connections were modeled with dummy spring elements which have infinity stiffness. A fixed boundary condition was imposed at the southern end, whereas only axial elongation and transverse rotation were permitted at the northern end. Pretension was applied to the mooring lines, with 1400 kN for the outer clusters and 1522 kN for the inner clusters. The adjusted length of the mooring lines influences the total stiffness and natural periods of the system.

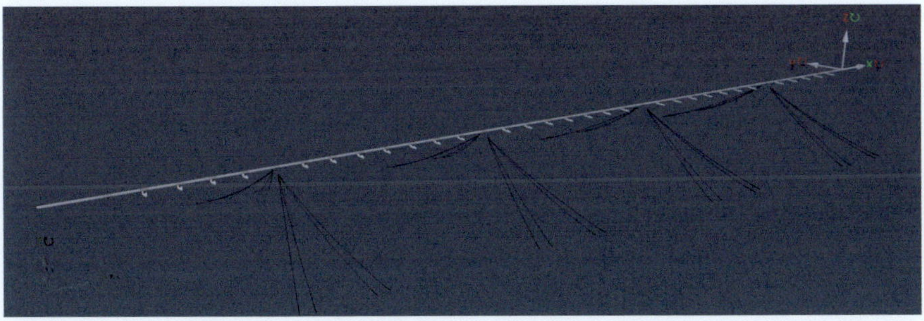

Fig. 2 Numerical model of the straight floating pontoon bridge in OrcaFlex

4 Inhomogeneous Wave Simulation

4.1 Methodology

OrcaFlex generates wave forces in the time domain with given parameters according to the Eqs. (4), (5). However, the built-in feature of the software applies identical wave loads at all locations in the fluid field, indicating that the software only supports the homogeneous wave conditions. Here, we propose a methodology to generate inhomogeneous waves by using the applied load feature in OrcaFlex. Basically, OrcaFlex solves the discretized equation of motion with implicit method after constructing inertial, damping, and stiffness matrix of fully coupled structures. Instead of using the built-in feature of generating waves, users can define external forces with tabular data based on the wave loads at specific nodes. Then, the numerical solver yields identical motion response vectors discretized for each time step. Similarly, users can define different force vectors for each node with sets of pre-arranged wave parameters to simulate inhomogeneous wave conditions. OrcaFlex provides the environment for text-editable input files so that users can add a large amount of prepared data in the appropriate input file format with Python codes. This approach is summarized as switching external force vectors with pre-arranged tabular data, skipping the generation of time-series wave forces with wave parameters. The serial process is illustrated in Fig. 3.

The proposed approach was successfully validated with the simulation using the built-in wave generating feature for homogeneous wave case. The effects of 1st and 2nd difference frequency wave loads, peak period, wave height, wave headings, directional spreading were examined and showed good agreement, while individual phase angles of wave components were set to be fully correlated.

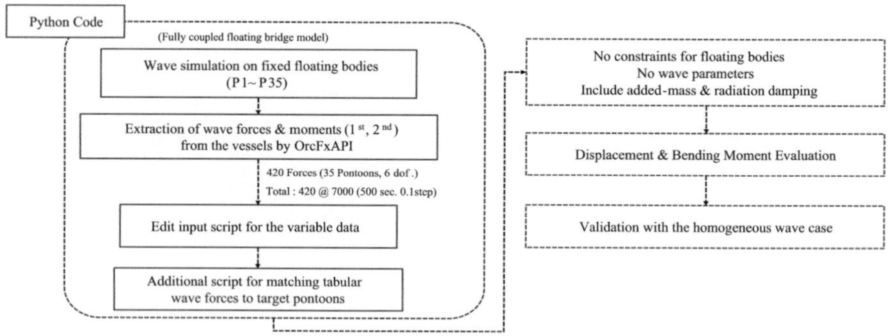

Fig. 3 Flowchart of the proposed methodology

Fig. 4 Arbitrary inhomogeneous wave cases in the conventional representation

4.2 Arbitrary Inhomogeneous Wave Cases

This study aims to introduce an advanced framework for examining the structural behaviors of a floating bridge under inhomogeneous wave conditions using OrcaFlex, rather than considering the appropriate design wave conditions of Bjørnafjord. Referring to the 100-year period inhomogeneous wave conditions given by Dai et al., we created several arbitrary inhomogeneous wave cases as shown in Fig. 4. Similar to the conventional representation of wave conditions, the radial axis indicates the frequency range, the angles indicate wave heading, and the spectral density of the wave is presented as a contour plot. For each wave case, only the distribution of wave headings varied, while the peak period was set to 5.9 s and the significant wave height was 2.4 m. The phase angles of the wave components were random but fully correlated, and non-directional spreading was configured.

Figure 5 shows the standard deviation of the wave-induced moment at each node of the girder, indicating the expected dynamic response during the simulation of three inhomogeneous wave cases (IHWC) introduced in Fig. 4 and two homogeneous wave cases (HWC). Directional inhomogeneity was revealed to have a critical effect on bending moments in the vertical direction. Based on local wave headings, the minor bending moment in the girders near the south end increased 2–4 times in inhomogeneous wave cases. This is due to the pontoon's hydrodynamic shape, which increases pitch moments and minor girder bending as the wave heading shifts from head sea to beam sea. This is supported by the fact that the homogeneous wave cases serve as the upper and lower bounds in the bending moment diagrams. On the other hand, the directional inhomogeneity showed insignificant variance in the major bending moments and torsional moments of the girders.

4.3 Variation in the Mooring Configuration

The current mooring configuration is categorized as a semi-taut type, stretching outwards from the moored pontoon. This system provides additional horizontal and vertical stiffness to the moored pontoons while offering minimal restoring capacity for pitch motion. In the range of wave periods that the floating bridge is examined, the

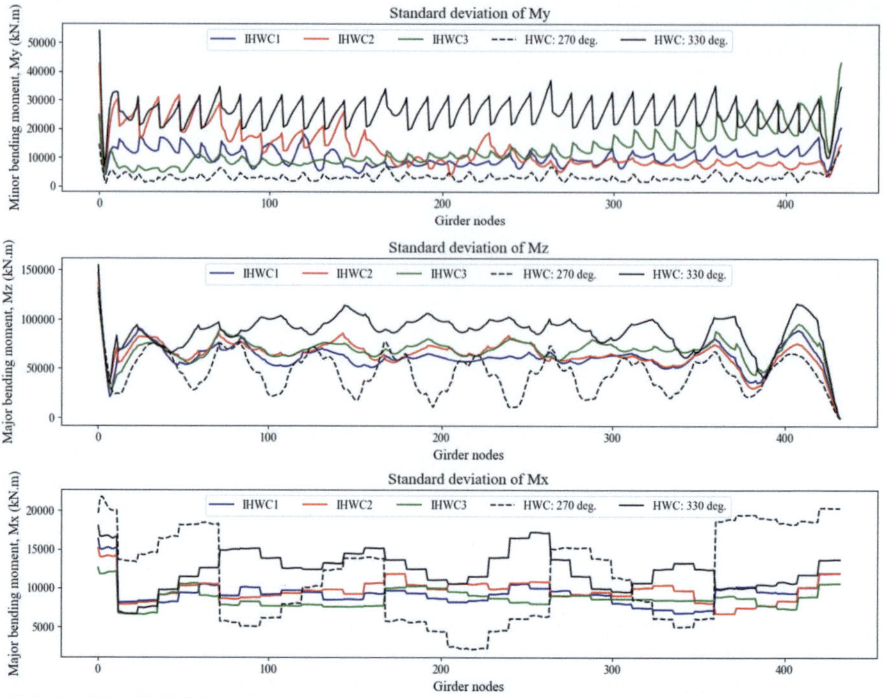

Fig. 5 Wave-induced moments of the girders under the inhomogeneous and homogeneous wave cases

dominant vibration mode is rotation with respect to the longitudinal axis. Considering the practical limit states of joints and structural components, two different mooring configurations were investigated under inhomogeneous wave case 3 as an example. One mooring type consists of a combination of semi-taut and tension legs, while the other consists solely of tension legs. As shown in Fig. 6, the modified mooring types showed improved resilience for the pitch of the moored pontoons, reducing wave-induced torsional moments in the center girders by 45%. However, the use of solely tension leg moorings increased the major bending moments in the girders by 20% overall due to the reduced horizontal mooring stiffness. In contrast, the combination of mooring types improved resilience to torsion while maintaining the capacity for horizontal responses.

5 Concluding Remarks

This study aims to introduce an advanced framework to examine the structural behaviors of floating bridges under inhomogeneous wave conditions using OrcaFlex. A straight floating pontoon bridge model was reproduced by referring to previous works

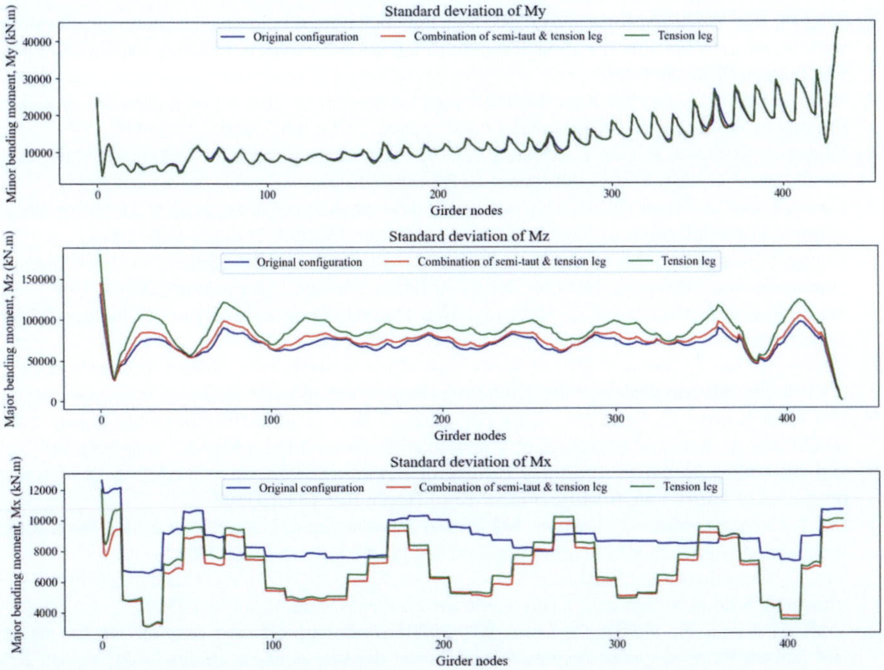

Fig. 6 Wave-induced moments of the girders for different mooring configurations

and validated based on the natural periods to capture the behavioral characteristics of the target model. The methodology for integrating inhomogeneous wave conditions into OrcaFlex by utilizing the script environment was proposed. Using this approach, arbitrary inhomogeneous wave cases were simulated for the reproduced model, revealing that directional inhomogeneity has a critical effect on the vertical responses of the girders due to the hydrodynamic characteristics of the pontoon. Finally, different mooring configurations were simulated under inhomogeneous wave conditions, indicating that the model has a greater capacity to improve resilience to torsion and optimize the design.

Acknowledgements This work was supported by a Korea Agency for Infrastructure Technology Advancement (KAIA) grant funded by the Ministry of Land, Infrastructure, and Transport (Grant RS-2023-00250727) through the Korea Floating Infrastructure Research Center at Seoul National University.

References

1. Jang M, Lee Y, Won D, Kang YJ, Kim S (2020) Static behaviors of a long-span cable-stayed bridge with a floating tower under dead loads. J Mar Sci Eng 8(10):816

2. Jang M, Lee Y, Kim S, Kang YJ (2020) Dynamic behavior of a long-span cable-stayed bridge with floating towers after the sudden failure of tethers and cables under irregular waves. J Mar Sci Technol 28(6):593–601
3. Won D, Lee K, Kang YJ, Kim S (2020) Short-term fatigue damage of tethers of long-span floating cable supported bridges under harsh waves. J Mar Sci Technol 28(6):602–609
4. Cheng Z, Svangstu E, Gao Z, Moan T (2019) Field measurements of inhomogeneous wave conditions. J Waterw Port Coast Ocean Eng 145:05018008
5. Cheng Z, Gao Z, Moan T (2019) Numerical modeling and dynamic analysis of a floating bridge subjected to wind, wave, and current loads. J Offshore Mech Arct Eng 141:011601
6. Cheng Z, Svangstu E, Moan T, Gao Z (2021) Assessment of inhomogeneity in environmental conditions in a Norwegian fjord for design of floating bridges. Ocean Eng 220:108474
7. Cheng Z, Gao Z, Moan T (2018) Hydrodynamic load modeling and analysis of a floating bridge in homogeneous wave conditions. Mar Struct 59:122–141
8. Cheng Z, Gao Z, Moan T (2018) Wave load effect analysis of a floating bridge in a fjord considering inhomogeneous wave conditions. Eng Struct 163:197–214
9. Dai J, Stefanakos C, Leira BJ, Alsos HS (2021) Effect of modelling inhomogeneous wave conditions on structural responses of a very long floating bridge. J Mar Sci Eng 9(5):548
10. Cui M, Cheng Z, Moan T (2023) Effects of inhomogeneous wave modeling on extreme responses of a very long floating bridge. Appl Ocean Res 134:103505
11. Dai J, Leira BJ, Moan T, Kvittem MI (2020) Inhomogeneous wave load effects on a long, straight and side-anchored floating pontoon bridge. Mar Struct 72:102763
12. Viuff T, Ravinthrakumar S, Økland OD, Gryta OA, Xiang X (2023) Model test of a hydroelastic truncated floating bridge with a stay-cable tower. Appl Ocean Res 135:103539
13. Viuff T, Xiang X, Øiseth O, Leira BJ (2020) Model uncertainty assessment for wave- and current-induced global response of a curved floating pontoon bridge. Appl Ocean Res 105:102368
14. Cheng Z, Gao Z, Moan T (2020) Extreme responses and associated uncertainties for a long end-anchored floating bridge. Eng Struct 219:110858
15. Viuff T, Xiang X, Leira BJ, Øiseth O (2020) Software-to-software comparison of end-anchored floating bridge global analysis. J Bridg Eng 25(5):04020022
16. Rodrigues JM, Viuff T, Økland OD (2022) Model tests of a hydroelastic truncated floating bridge. Appl Ocean Res 125:103247
17. Multiconsult (2017) SBJ-31-C3-MUL-22-RE-100-0—analysis and design (base case). Multiconsult AS

Dynamic Behavior of an Operational Floating Offshore Wind Turbine Subjected to Ship Collision

Xudong Ye and Kaiming Bi

Abstract This paper investigates the dynamic responses of a NREL 5 MW spar-type floating offshore wind turbine (FOWT) under ship collisions. Physics-based finite element models (FEMs) are established by LS-DYNA. Aerodynamics of FOWT are introduced using the steady wind field simulation results. While hydrodynamics of floaters are considered by the MCOL method. The collision forces of ship and vibration frequencies of FOWT match well with the reference values, indicating the rationality of the developed model. Four wind velocities are adopted to represent different operating status for the FOWT. The collision process can be divided into primary and secondary collisions. The maximum force of the secondary collision only has 60% of the primary one. The results show the collision forces with 0 and 7 m/s wind speeds are larger than other cases. This is because the wind thrust delays the collision process through changing the pitch angle of the FOWT. Also, it is observed that the indentation of the spar has the same regular pattern since it is related to the collision force. The secondary collision has almost no effect on the indentation.

Keywords Floating offshore wind turbine · Ship collision · Operational condition · Dynamic response

1 Introduction

With the development of the offshore wind power industry and the recovery of the global marine economy, the frequency of ship collisions with offshore wind turbines is increasing. Recently, on April 25, 2023, the Gode Wind 1 wind farm in Germany was struck by a cargo ship, resulting in a large hole on the starboard bow [1]. On September 19, 2024, at the Hornsea One offshore wind farm in the UK, a service operation vessel (SOV) collided with an operating wind turbine, causing

X. Ye · K. Bi (✉)
The Hong Kong Polytechnic University, Hung Hom, Kowloon, Hong Kong, China
e-mail: kaiming.bi@polyu.edu.hk

X. L. Zhao et al. (eds.), *Proceedings of the Fourth World Conference on Floating Solutions: WCFS 2024; 2–4 December; Hong Kong*, Lecture Notes in Civil Engineering 597, https://doi.org/10.1007/978-981-96-4569-5_60

minor damage to the starboard side of the vessel and the turbine foundation [2]. These incidents indicate the necessity of predicting the ship-FOWT collision process and consequence.

Many scholars have conducted research on ship collisions with fixed wind turbines. However, research on collisions involving floating wind turbines remains relatively superficial. Ren [3] studied the influence of different impact velocities and hull rigidity on platform damage, without considering the wind-wave coupling effect on the turbine. Zhang [4] proposed an aero-hydro coupling method to simulate hydrodynamics during ship-FOWT collisions, but his research focused on the rigidity of the colliding object, with fewer scenarios related to the operational state of the turbine.

Existing literature indicates a lack of analysis on the operational floating wind turbines. Therefore, this paper develops a FOWT model that considers both aerodynamic and hydrodynamic characteristics, considering operational states of wind turbines. Section 2 introduces the FE model and validates its effectiveness. Section 3 presents the collision scenarios and discusses the simulation results.

2 Validation of FE Models

2.1 Modeling of Ship and FOWT

This paper adopts a 3000 DWT ship to collide with the NREL 5 MW spar-type floating offshore wind turbine. Figure 1 shows the FE model established in LS-DYNA. For the ship bow, the thickness of the steel plates varies from 6 to 12 mm. The material of the steel is simulated using the keyword MAT_PIECEWISE_LINEAR_PLASTICITY, with parameters identical to [5].

Details of the FOWT model can be found in [6]. The impacted area of the spar has a diameter of 6.5 m and a thickness of 40 mm. The blades are simulated using beam elements and can rotate around the hub. The FOWT uses the same material model as the ship.

Hydrodynamic characteristics of the ship and the FOWT are considered using the MCOL method while the viscous damping and mooring system by custom functions. Rigid body dynamics are investigated to balance the hydrodynamic forces and other forces acting on the ship [7]:

$$(M_S + M_A) \cdot \dot{y} + G \cdot y = F_W + F_H + F_V + F_C \tag{1}$$

where M_S is the mass matrix of ship; M_A is the added mass matrix of ship; y is the velocity of the ship; G is the gyroscopic matrix; F_W is the wave force; F_H is the hydrostatic restoring force; F_V is the viscous force; F_C is the collision force.

Incorporating aerodynamics, a series of wind forces obtained from the steady wind field simulation in FAST are applied to each node of the blades. The contact between

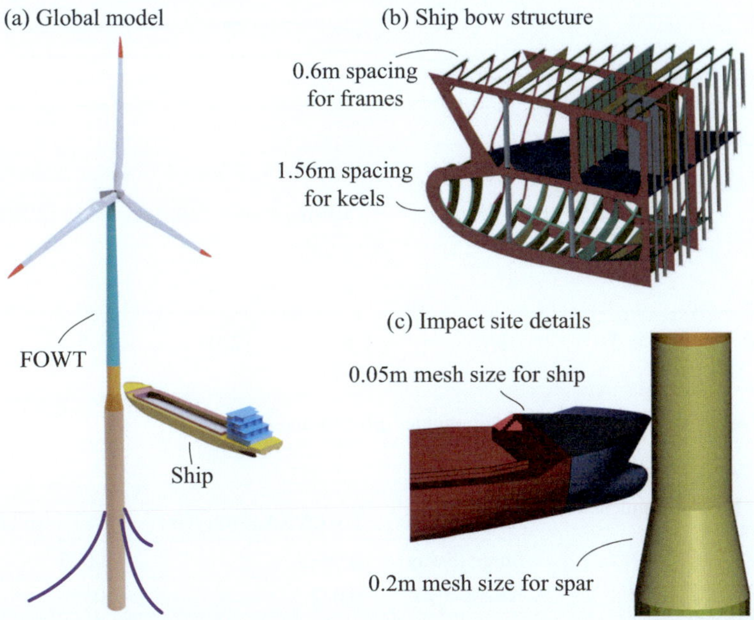

Fig. 1 Ship and FOWT models

the ship and the FOWT is managed by the keyword CONTACT_AUTOMATIC_SURFACE_TO_SURFACE, with a friction coefficient of 0.3.

2.2 *Mesh Convergence Analysis*

According to [8], the ratio of plate thickness to mesh size should be between 1:5 and 1:10. The most suitable mesh size is 40–80 mm. Figure 2 compares the indentation of the bow when it collides with a rigid wall under different mesh sizes. It can be seen that the results are similar for mesh sizes of 25 and 50 mm. Balancing computational accuracy and efficiency, 50 mm is finally selected as the mesh size for the bow. Similarly, the mesh size for the impacted part of the FOWT is determined to be 200 mm.

2.3 *Verification of FOWT's Maneuverability*

To validate the effectiveness of the aerodynamic and hydrodynamic simulation methods, the results of LS-DYNA and FAST models are compared. Table 1 lists the natural frequencies of the FOWT. Good consistency shows the effectiveness of

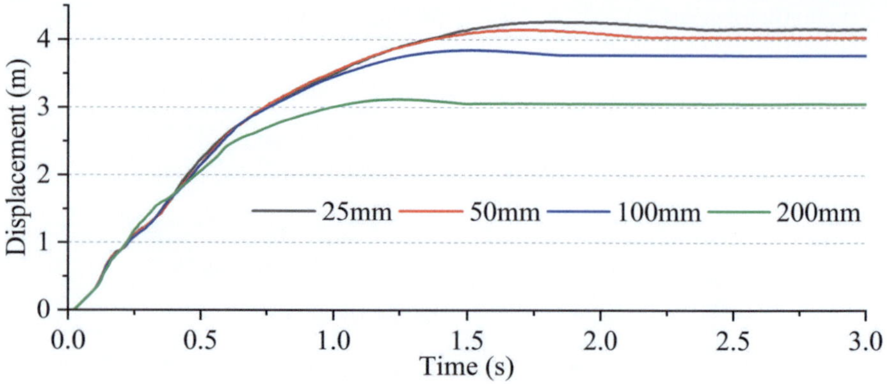

Fig. 2 Displacement of bow indentation under different mesh sizes

Table 1 Natural frequencies of FOWT

Motion mode	LS-DYNA result (Hz)	FAST result (Hz)
Surge (Sway)	0.0083	0.0076
Heave	0.032	0.032
Roll (Pitch)	0.036	0.032
Yaw	0.12	0.12

the MCOL method. Figure 3 shows the platform surge displacements under different wind velocities, and the results obtained by LS-DYNA are very close to those of FAST. This indicates that the developed model can effectively simulate aerodynamic and hydrodynamic characteristics of FOWT.

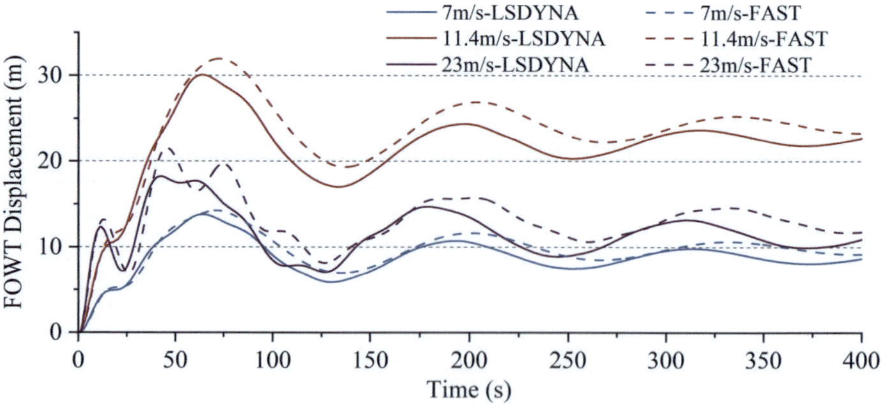

Fig. 3 Comparison of platform surge motion between LS-DYNA and FAST

3 Discussion of Ship-FOWT Collision Results

3.1 Collision Scenarios

Four collision scenarios are considered with respect to different wind velocities: 0 m/s (no wind), 7 m/s (low speed), 11.4 m/s (rated speed), and 23 m/s (high speed). Under the 11.4 m/s condition, the blades experience the maximum wind thrust because of the optimal pitch angle. The moving velocity of the ship is set to 5 m/s, and the wind turbine reaches a stable state before the collision.

3.2 Collision Process

Figure 4 shows the collision forces between the ship and the FOWT under different wind velocities. It can be seen that multiple collisions occur between the ship and the FOWT. The primary collision lasts about 2 s, with the collision force being the greatest under no wind condition and the smallest at the rated speed (11.4 m/s). This is because there is a large pitch angle of the FOWT, resulting in reducing the severity of the collision process. This initial pitch angle is caused by the maximum wind thrust. The secondary collision occurs about 5 s after the first one and lasts longer. Its peak force is 60% of the highest value.

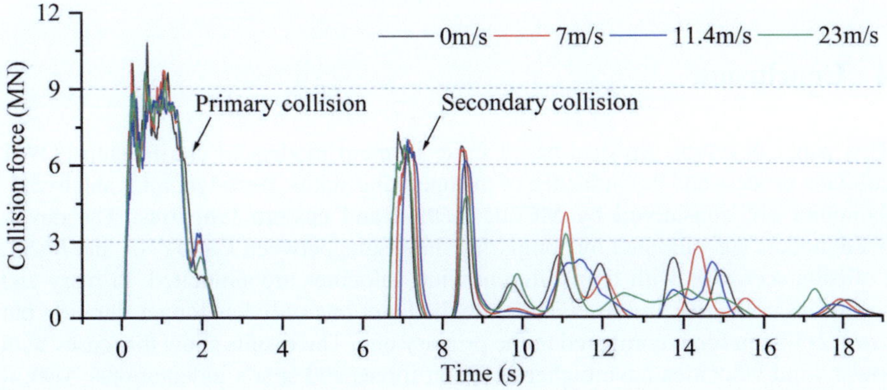

Fig. 4 Comparison of collision force under different scenarios

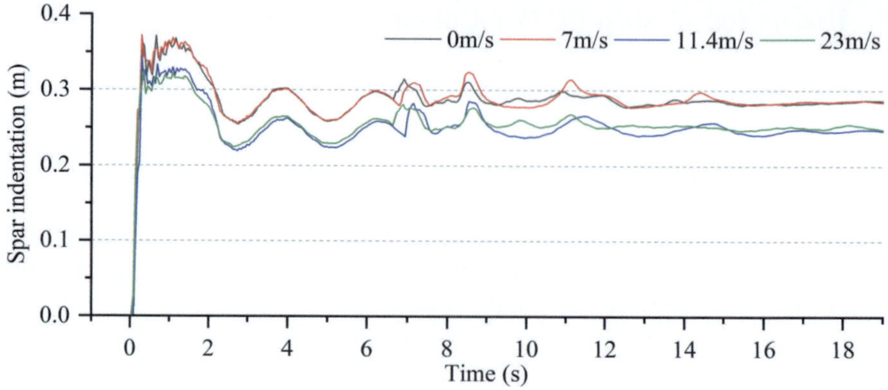

Fig. 5 Comparison of spar's indentation under different scenarios

3.3 Damage Assessment

Figure 5 shows the indentation of the spar during the collision. It can be observed that the spar undergoes significant plastic deformation within the initial 0.2 s, followed by some rebound as the collision force decreases. Due to the collision forces of primary collision under 0 and 7 m/s conditions are higher, the indentations are 0.05 m greater than those of other conditions. Besides, the secondary collision only causes a small fluctuation on the indentation.

4 Conclusion

This paper develops physical-based finite element models to clarify ship-FOWT collision process and the influence of the operating status. Aerodynamics and hydrodynamics are considered by MCOL method and custom functions. The developed models are validated by comparing the results between LS-DYNA and FAST. Collision scenarios with four different wind velocities are simulated. Primary and secondary collisions are observed while the latter one only has longer duration but lower collision force compared to the primary one. The results show that cases with lower wind velocities have higher collision forces and spar's indentations. Also, it should be noticed that the secondary collision has fewer influences on the indentation.

References

1. offshoreWIND.biz. https://www.offshorewind.biz/2023/05/30/cargo-ship-hit-gode-wind-turbine-went-back-into-service-in-24-hours-vessel-said-to-have-been-kilometres-off-course. Last accessed 30 May 2023

2. offshoreWIND.biz. https://www.offshorewind.biz/2024/09/23/service-ship-hits-wind-turbine-at-hornsea-one-offshore-wind-farm. Last accessed 23 Sep 2024
3. Ren YL, Meng QS, Chen C, Hua XG, Zhang ZL, Chen ZQ (2022) Dynamic behavior and damage analysis of a spar-type floating offshore wind turbine under ship collision. Eng Struct 272:114815
4. Zhang YC, Hu ZQ (2022) An aero-hydro coupled method for investigating ship collision against a floating offshore wind turbine. Mar Struct 83:103177
5. Fan W, Zhang ZW, Huang X, Sun WB (2020) A simplified method to efficiently design steel fenders subjected to vessel head-on collisions. Mar Struct 74:102840
6. Jonkman J, Butterfield S, Musial W, Scott G (2009) Definition of a 5-MW reference wind turbine for offshore system development. National Renewable Energy Lab (NREL), Golden, CO (United States)
7. Le Sourne H, Donner R, Besnier F, Ferry M (2001) External dynamics of ship-submarine collision. In: Preprints of 2nd International conference on collision and grounding of ships. Copenhagen, pp 137–144
8. Alsos HS, Amdahl J, Hopperstad OS (2009) On the resistance to penetration of stiffened plates, Part II: numerical analysis. Int J Impact Eng 36:875–887

Stability Analysis of a Concrete Semisubmersible Floating Platform for 10 MW Wind Turbines

Qingshan Zeng, Zuntao Feng, and Jian-Fei Chen

Abstract This paper presents the stability analyses of a tri-columnar concrete semisubmersible floating platform tailored for a DTU 10 MW wind turbine. The stability of the platform is investigated using SESAM software, focusing on the influence of side pontoon diameter, height, spacing, and the width of the bottom horizontal tanks. An analytical model is employed to examine the combined impact of these factors on the stability and their respective sensitivities. Results indicate that the diameter of the side pontoons exerts the most significant influence on the stability of the platform. Furthermore, free decay simulations validate that the natural periods of the platform are adequately separated from typical first-order wave periods, ensuring compliance with design requirements.

Keywords Floating wind turbines · Semisubmersible platforms · Stability · Natural periods

1 Introduction

Fixed foundations are currently predominantly employed for wind turbines in shallow water areas with depths less than 30 m. While more advanced designs may enable their deployment in slightly deeper waters, fixed foundations become economically unviable for water depths exceeding 50 m. Consequently, floating wind turbines emerge as necessary solutions [1]. In recent years, numerous floating wind turbine designs have been proposed, with semisubmersible platforms becoming the predominant

---

Q. Zeng · Z. Feng · J.-F. Chen (✉)
Department of Ocean Science and Engineering, Southern University of Science and Technology, Shenzhen, China
e-mail: chenjf3@sustech.edu.cn

Q. Zeng
Advanced Institute for Ocean Research, Southern University of Science and Technology, Shenzhen, China

X. L. Zhao et al. (eds.), *Proceedings of the Fourth World Conference on Floating Solutions: WCFS 2024; 2–4 December; Hong Kong*, Lecture Notes in Civil Engineering 597, https://doi.org/10.1007/978-981-96-4569-5_61

choice due to their lower construction costs, less technical complexity, and favorable hydrodynamic performance.

Many scholars have carried out in-depth research on the semi-submersible platform floating wind turbines. Shi et al. [2] established a coupled numerical model of hydrodynamic and mooring system is established to study the dynamic response of 10 MW large offshore floating wind turbine under wind and wave action. Tang et al. [3] discussed the viscous damping characteristics of semisubmersible platform and heave plate, and analyzed the motion response of the platform in the frequency domain. Yang et al. [4] studied the hydrodynamic characteristics of semi-submersible wind turbines under the combined action of wind and wave conditions through model tests. Hasan et al. [5] proposed a 10 MW semi-submersible wind turbine and studied the influence of structure size on its stability and hydrodynamic characteristics.

Building upon existing research, and considering the inherent advantages of concrete structures, such as corrosion resistance, high reliability, and extended service life, this study proposed a design concept for a 10 MW concrete semisubmersible floating wind turbine. Numerical simulations, utilising the SESAM software, were conducted to investigate the impact of pontoon distance, height, and diameter on the stability of the platform. Some of the preliminary findings of this study are reported in this paper.

2 A Concrete Semi-submersible Platform for Floating Wind Turbines

This paper presents a three-pontoon concrete semisubmersible floating wind turbine design, comprising a concrete semisubmersible platform, a wind turbine, and a three-chain catenary mooring system, as illustrated in Fig. 1. The key design concepts and advantages of this platform are as follows: (a) The shallow draft of the platform enables operation across a wide range of water depths; (b) The utilisation of a catenary mooring system simplifies the design and reduces construction complexity; (c) The installation and construction processes require minimal technical expertise and incur lower costs; (d) The large waterplane area of the platform contributes to enhanced stability and motion performance; (e) The concrete construction offers superior corrosion resistance and durability.

The semisubmersible platform comprises of three side pontoons, horizontal floating tanks, and heave plates. Buoyancy for the entire floating structure is provided by the three side pontoons and the horizontal floating tanks. The wind turbine is installed on one of the side pontoons, connected to the platform via a flange. The primary platform parameters are detailed in Table 1. The DTU-10 MW [2] horizontal axis wind turbine was adopted for this study. The mooring system consists of three mooring lines, positioned at 120° apart.

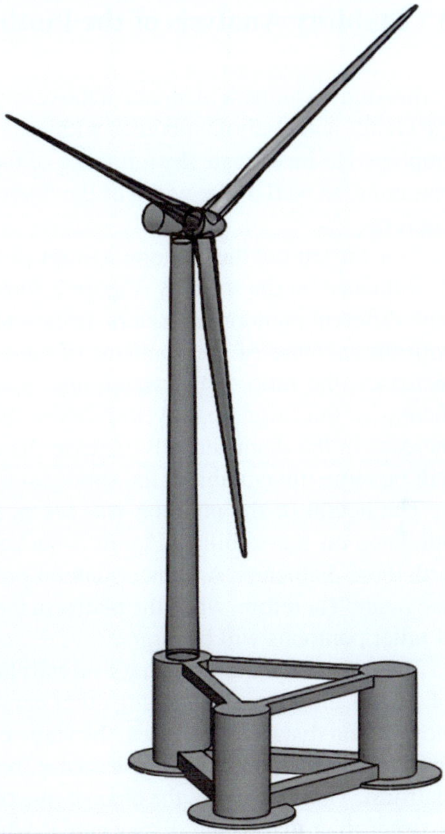

Fig. 1 Conceptual design of a semisubmersible platform for floating wind turbines

Table 1 Parameters of the semisubmersible platform

Parameters	Value
Diameter and height of heave plate (m)	28.0, 1.0
Width and height of the horizontal tanks (m)	6.5, 4.0
Height, diameter and distance of the side pontoons (m)	28.0, 15.0, 55.0
Moment of inertia in roll ($kg \cdot m^2$)	1.2×10^{10}
Moment of inertia in pitch ($kg \cdot m^2$)	7.6×10^9
Moment of inertia in roll ($kg \cdot m^2$)	7.6×10^9
Platform mass (kg)	1.3×10^7
Draft (m)	20.0
Volume of displacement (m^3)	1.6×10^4

3 Stability Analysis of the Platform

In this study, a numerical model of the semisubmersible platform was developed using the GeniE and HydroD modules within the SESAM software suite. This model was employed to investigate the influence of the diameter, height and spacing of the side pontoons, as well as the width of the horizontal tanks, on the overall stability of the platform.

The diameter of the pontoon was set as 13, 15, 16, 18 and 20 m in order to explore its influence on the stability. Figure 2 shows the static stability curves of platforms with different pontoon diameters. It shows that the stability of the platform increases with the increase of the diameter of the pontoon. As the diameter of the pontoon increases, the ratio of the waterplane moment of inertia to displacement volume increases, thus significantly improving the stability of the platform. However, an increase in the diameter will increase the cabin volume, and the damaged stability will be more unfavorable if the subdivision in a pontoon remains unchanged.

The height of the pontoon was set as 31, 33, and 35 m in order to explore its influence on the stability. Figure 3 shows the static stability curves of platforms with these diameters. A higher pontoon provides the platform with more reserves of buoyancy. Therefore, when the platform inclination is the same, the righting moment of taller pontoons will be larger.

The effect of pontoon distance on stability was investigated by setting distances as 54, 56, 58 and 62 m and keeping other parameters unchanged. As shown in Fig. 4, as the pontoon distance decreases, the static equilibrium angle increases, the area ratio and the initial stability height decrease, resulting a reduction in the stability of the platform. This is because the change rate of the moment of inertia of the waterplane is larger than that of the displacement when the distance between the pontoons is reduced.

In addition to the side pontoons, the horizontal floating tanks also provide buoyancy for the platform. The width of the tanks was set to 5.5, 6 and 6.5 m to explore

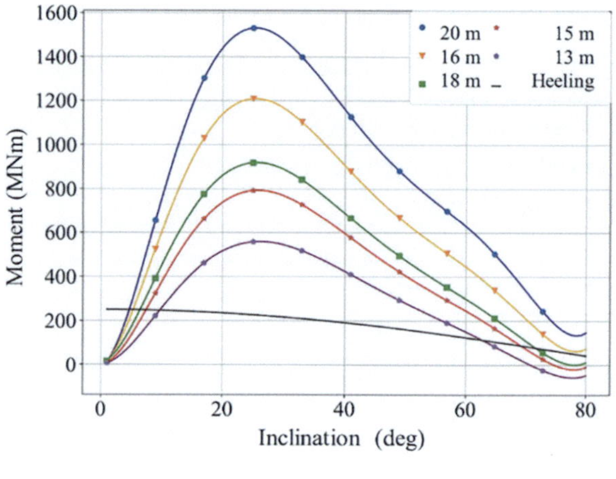

Fig. 2 Static stability curves of platforms with different pontoon diameters

Fig. 3 Static stability curves of platforms with different pontoon heights

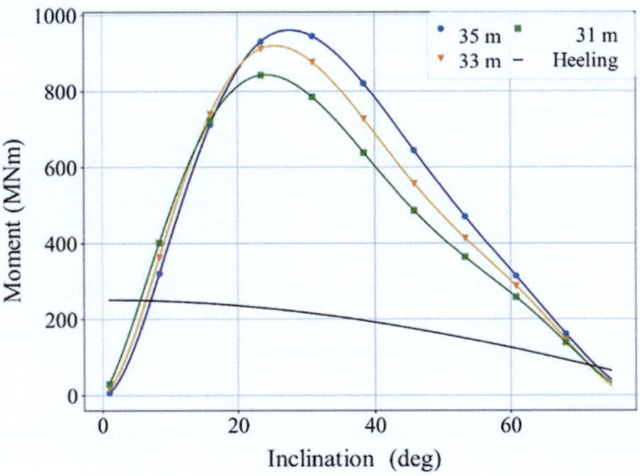

Fig. 4 Static stability curves of platforms with different pontoon distances

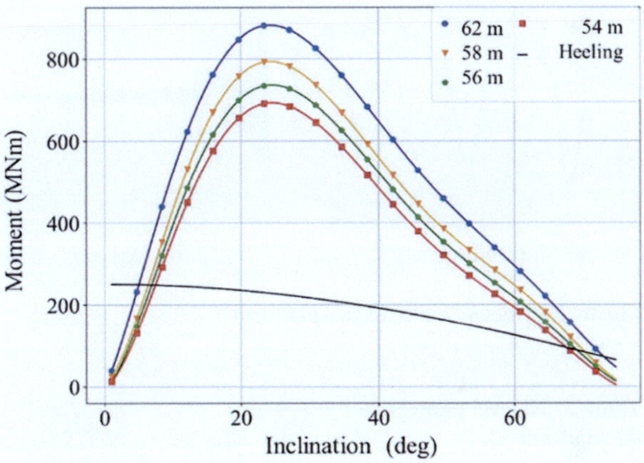

the influence of the width of the bottom horizontal pontoon on the stability. Figure 5 shows the static stability curves of platforms with different width of the horizontal tanks. Since the change of this width does not affect the waterplane moment of inertia, there is little change in the righting moment.

Floating platforms require sufficient stability, but excessive stability is undesirable. Higher platform stability results in shorter natural periods, which may approach wave periods and induce resonance. To investigate this, a free decay simulation of the moored semisubmersible platform was performed under still water conditions. This simulation examines the six degrees-of-freedom motion responses of the platform to an initial excitation in the absence of wind, wave and current loads. The platform was treated as a rigid body in these calculations. The horizontal coordinates of the peak values in the free decay power spectral density curves correspond to the natural frequencies of the wind turbine platform in the respective degrees of freedom.

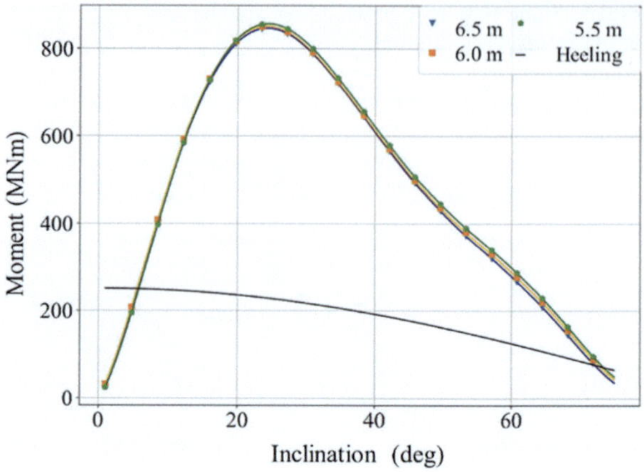

Fig. 5 Static stability curves of platforms with different widths of tank

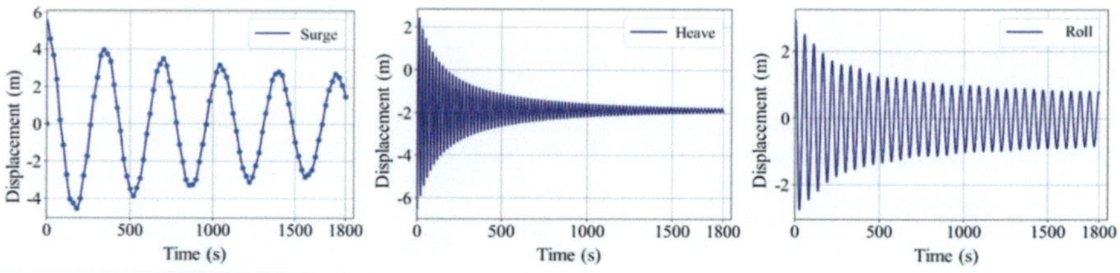

(a) Surge (b) Heave (c) Roll

Fig. 6 Free decay curve of the platform

Table 2 Natural periods of the platform

	Surge	Heave	Roll
Period(s)	350.8	30.0	81.7

The free decay curves for the platform are presented in Fig. 6. Table 2 shows the calculated natural periods. The natural periods of the semisubmersible floating wind turbine platform are all well outside the typical range of 5–20 s for first-order wave loads.

4 Conclusions

This study has proposed a design concept for a 10 MW concrete semisubmersible floating wind turbine. The stability of the platform has been analysed using the SESAM software. The influence of various factors on the stability of the semisubmersible platform has been investigated. An analytical model was employed to examine the combined effect of these factors on the stability and their respective sensitivities. The results demonstrate that the diameter of the side pontoons exerts the most substantial impact on the stability of the platform, primarily by altering the ratio of the waterplane moment of inertia to the displaced volume. Changes in the pontoon diameter significantly affect the waterplane area moment of inertia to displacement volume ratio, and the vertical distance between the center of gravity (COG) and the center of buoyancy (COB), thereby greatly influencing the stability. Furthermore, free decay simulations were performed to confirm that the natural periods of the platform are adequately separated from typical first-order wave periods, ensuring compliance with design requirements.

Acknowledgements The work was financially supported by the National Key Research and Development Program of China (2023YFB3711500), National Natural Science Foundation of China (No. 52178218), and Shenzhen Science and Technology Program (KCXST20221021111409023).

References

1. Shi Z, Wang C, Li Q (2020) Key issues of China's offshore wind power development in the "14th Five-Year Plan." China Power 53(07):8–17
2. Shi W, Xue R, Hou X, Xiu H, Zhang L (2021) Dynamic response on 10 MW semisubmersible floating offshore wind turbine. Ship Eng 43(10):1–9
3. Tang Y, Song K, Wang B (2015) Experiment study of dynamics response for wind turbine system of floating foundation. China Ocean Eng 29(6):835–846
4. Yang C, Xiao F, Cheng Z, Cao Q, Liu Y (2023) Conceptual design and model testing for a semi-submersible floating wind turbine in shallow water. Ocean Eng 41(01):128–140
5. Hasan M, Munbua W, Malta E (2022) Conceptual design of a concrete multi-column floating platform supporting a 10 MW offshore wind turbine. In: International conference on offshore mechanics and arctic engineering. American Society of Mechanical Engineers

Development of a 6-DOF Actuating System for Implementing Real-Time Hybrid Simulation of Floating Bridge Systems Under Wind and Wave Loads

Giheon You, Minyeop Kim, and Yunbyeong Chae

Abstract This study developed both analytical and experimental substructures to perform real-time hybrid simulation (RTHS) for floating bridge systems subjected to wind and wave loads. The upper structure influenced by wind loads is physically tested using a scaled model in a wind tunnel, while the lower structure subjected to wave loads is modeled analytically. The wind loads are measured with a 6-axis load cell, and the corresponding target displacements are calculated assuming rigid body motion. These real-time responses are computed in MATLAB Simulink and applied to the experimental substructure mounted on a 6-degree-of-freedom (DOF) actuating system. The 6-DOF actuating system is controlled using the Speedgoat real-time target machine and the adaptive time series (ATS) compensator. This study proposes a method for experimentally analyzing the dynamic response of the pontoon-bridge pylon system using RTHS.

Keywords Real-time hybrid simulation · Floating bridge system · 6-DOF Stewart platform

1 Introduction

An innovative design for a floating bridge, such as a cable bridge with pylons supported by a TLP (Tension Leg Platform) system, has been proposed in Norway's transportation plans to address the challenge of crossing deep and wide fjords [1]. These floating bridge systems should consider many conditions that are not accounted for in conventional bridges or existing types of floating bridges. Accurate performance evaluation is crucial for the design, construction, and operation of these infrastructures. Consequently, new analytical methods need to be developed to address these challenges. In particular, experimental verification of wind-wave

G. You · M. Kim · Y. Chae (✉)
Seoul National University, Gwanak-gu, Seoul, Republic of Korea
e-mail: ybchae@snu.ac.kr

X. L. Zhao et al. (eds.), *Proceedings of the Fourth World Conference on Floating Solutions: WCFS 2024; 2–4 December; Hong Kong*, Lecture Notes in Civil Engineering 597, https://doi.org/10.1007/978-981-96-4569-5_62

coupled response is necessary for floating bridge systems, which are influenced by both wind and wave loads.

Generally, in model tests of floating bodies, the upper and lower structures follow different kinematic and dynamic similarities, making it difficult to accurately simulate both aerodynamic and hydrodynamic forces. Ransley [2] conducted real-time hybrid model tests in a wave tank to analyze the effects of wind and wave loads on the response of a floating offshore wind turbine. The hydrodynamic loads were applied to a model-scale floating platform, while the aerodynamic effects on the upper structure were analytically modeled and implemented through a fan system.

In this study, real-time hybrid simulation (RTHS) is suggested using wind tunnel, where the structure above the pontoon, influenced by wind loads, is physically tested in the wind tunnel, while the lower structure, subjected to wave loads and other hydrodynamic forces, is analytically modeled.

The motion of a floating body is described with 6-degree-of-freedom (DOF) rigid body motions, assuming that the floating body moves like a rigid structure. For RTHS, a 6-DOF actuating system is required to apply real-time displacements to the scaled model in the wind tunnel. The target displacements of the analytical structure are applied to the scaled model through the real-time displacement control, and the force at the base on the scaled model is measured using a 6-axis load cell and fed into the equations of motion of the entire system, which are solved numerically to obtain the command displacements at the next time step.

2 Development and Verification of Analytical Substructure

2.1 Modeling of the Analytical Substructures

The submerged structure of the floating body was assumed to have six degrees of freedom in motion as a rigid body. Each degree of freedom is referenced to a coordinate system with the origin at the center of gravity of the body, the x-axis in the direction of the incoming waves, and the z-axis directed opposite to the direction of gravity. Translational motions along the x, y, and z axes are referred to as surge, sway, and heave, respectively, while rotational movements about each axis are referred to as roll, pitch, and yaw. In this study, it is assumed that the motion of the floating body is very small, and the slope of incident waves, represented by the product of wave number (k) and amplitude (A), is also very small, allowing the application of linearized theory. Based on this assumption, the motion of a floating body with six degrees of freedom can be described by the frequency-domain equation of motions given in Eq. (1) [3].

$$\sum_{k=1}^{6}\left[-\omega^2\left(M_{jk}+A_{jk}(\omega)\right)+i\omega B_{jk}(\omega)+C_{jk}\right]\overline{\eta}_k(\omega)=F_j^I(\omega)+F_j^D(\omega) \quad (1)$$

where, ω represents the frequency, M_{jk} the inertia, A_{jk} the added mass, B_{jk} the radiation damping, C_{jk} the hydrostatic stiffness, F_j^I the Froude-Krylov force, and F_j^D the diffraction force. Each of these coefficients can be obtained by solving the potential problem with the ocean environment and the floating body as boundary conditions. In this study, the coefficients were obtained through analysis using Ansys AQWA. To apply Eq. (1), which is in the frequency domain, to RTHS, it needs to be transformed into the time domain equation. Cummins proposed the Cummins Equation for the time domain motion of floating bodies, as given in Eq. (2), which includes the convolution integral, by separating added mass and radiation damping using the values at infinite frequency, and then applying inverse Fourier transform as follows [4];

$$[\mathbf{M} + \mathbf{A}^{\infty}]\ddot{\mathbf{x}}(t) + \int_0^{\infty} \mathbf{R}(t-\tau)\dot{\mathbf{x}}(\tau)d\tau + \mathbf{C}\mathbf{x}(t) = \mathbf{F}(t) \tag{2}$$

where, t represents the time, $\mathbf{M}$ is the mass(inertia) matrix, $\mathbf{A}^{\infty}$ is the added mass matrix at infinite frequency, $\mathbf{R}$ is the retardation function, $\mathbf{C}$ is the hydrostatic stiffness matrix, $\mathbf{x}$ is the displacement vector of the body, and $\mathbf{F}$ is the external force. Each of these coefficients can be mathematically derived from the frequency domain coefficients.

The wave excitation force was modeled using the transfer function of amplitude ($P_j(\omega)$) and phase lag ($\varphi_j(\omega)$) of sum of the Froude-Krylov force and diffraction force. For example, the wave force for the j-th degree of freedom due to a regular incident wave $\zeta(t) = A\cos(\omega t + \varepsilon)$ is given by Eq. (3) [5].

$$F_j^I(t) + F_j^D(t) = P_j(\omega)A\cos\big(\omega t + \varepsilon + \varphi_j(\omega)\big) \tag{3}$$

In the case of irregular waves, the JONSWAP spectrum is used to represent the waves as a sum of components with various frequencies. The wave loads generated by irregular waves are calculated as the sum of the wave load component at each frequency.

The wave loads must also consider second-order wave loads, which arise from the superposition of two frequencies. These loads appear at a frequency equal to the sum or difference of the two frequencies. The amplitude of the second-order wave load is given by the product of the amplitudes of each frequency component and the Quadratic Transfer Function (QTF). The second-order wave load generated by two waves, $\zeta_1(t) = A_1\cos(\omega_1 t + \varepsilon_1)$, $\zeta_2(t) = A_2\cos(\omega_2 t + \varepsilon_2)$, is given by Eq. (4);

$$F_j^{2\pm}(t) = P_j^{2\pm}(\omega_1, \omega_2)A_1A_2\cos\Big((\omega_1 \pm \omega_2)t + (\varepsilon_1 \pm \varepsilon_2) + \varphi_j^{2\pm}(\omega_1, \omega_2)\Big) \tag{4}$$

where, $P_j^{2\pm}$ is amplitude QTF matrix and $\varphi_j^{2\pm}$ is phase QTF matrix.

The damping of the floating body is predominantly influenced by the effects of Morison drag. The Morison drag acting on a slender member is exerted in a direction perpendicular to the longitudinal axis of the member. For an object moving in water, the Morison drag is expressed in terms of velocity **u** and the velocity of the object **v** as shown in Eq. (5);

$$d\mathbf{F}_D = \frac{1}{2} C_D \rho dA(\mathbf{u} - \mathbf{v})|\mathbf{u} - \mathbf{v}| \tag{5}$$

where, $d\mathbf{F}_D$ denotes the drag force acting on a small cross-sectional area dA of the slender member, C_D denotes the drag coefficient, and ρ represents the fluid density. The velocities of the fluid and the body, **u** and **v**, include only the components perpendicular to the length of the member. For all Morison members constituting the floating body, the forces were calculated using the Morison equation, and these forces were then converted to forces acting at the center of gravity of the floating body to account for Morison drag.

2.2 *Verification of the Analytical Substructures*

The analytical substructure modeling was verified using the Offshore Code Comparison Collaboration Continuation (OC4) [6]. The geometry and mass properties of OC4 are as shown in Table 1, where SWL stands for Sea Water Level and CM stands for Center of Mass.

Using Ansys AQWA, the diffraction analysis was performed to obtain the frequency dependent coefficients of the structure, including radiation damping, added mass, and wave excitation. The time domain coefficients were obtained by transforming the frequency dependent coefficients, which are shown in Fig. 1. Based on the obtained time domain coefficients, the Cummins equation was formulated and solved using the Newmark method.

The response amplitude operator (RAO) was determined from the structure's steady state response to regular waves and compared with the frequency domain analysis results. Irregular wave responses were verified by comparing the response spectrum, which is the product of the wave spectrum and the square of the RAO. The response spectrum was obtained through Fast Fourier Transform (FFT) and compared

Table 1 Geometry and mass properties of OC4

Geometry	Mass properties	
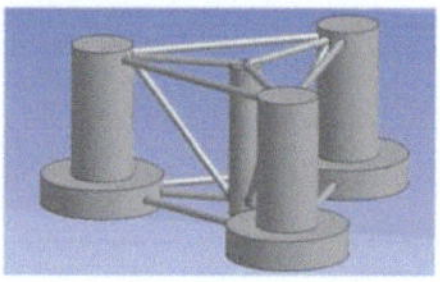	Platform mass	1.3473E + 7 kg
	CM location below SWL	13.46 m
	Roll inertia about CM	6.827E + 9 kg-m^2
	Pitch inertia about CM	6.827E + 9 kg-m^2
	Yaw inertia about CM	1.226E + 10 kg-m^2

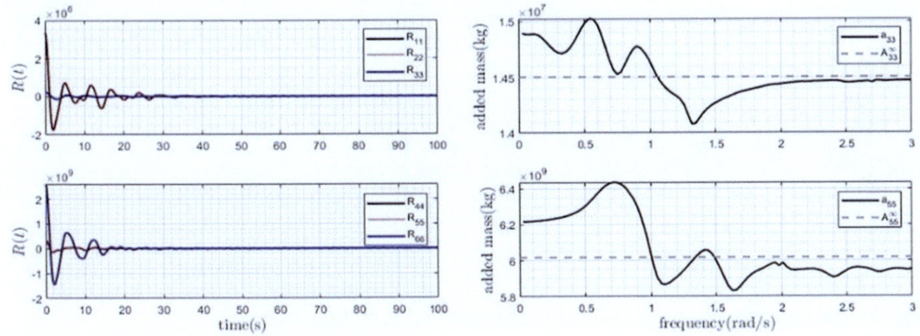

Fig. 1 Time domain coefficients of OC4

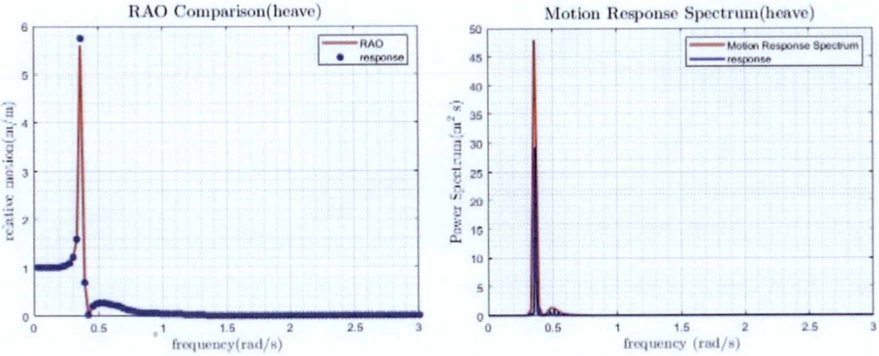

Fig. 2 Verification results for regular wave response (left) and irregular wave response (right)

with Ansys AQWA analysis results, averaged over 30 repetitions. Verification results for the heave responses to regular and irregular waves are shown in Fig. 2.

Since the damping of the floating body's motion is predominantly governed by Morison drag, the modeled Morison element was verified through free decay tests of the floating body. To validate the effectiveness of the modeling, the same parameters were input into the commercial software OrcaFlex, and the results were compared with those from the analytical substructure. The time-displacement results of the heave and pitch free decay are shown in Fig. 3.

As shown in Fig. 3, the result of the free decay test with an initial displacement of 1m in heave and 0.05 rad in pitch closely matched the OrcaFlex results, indicating that the Morison members have been accurately modeled.

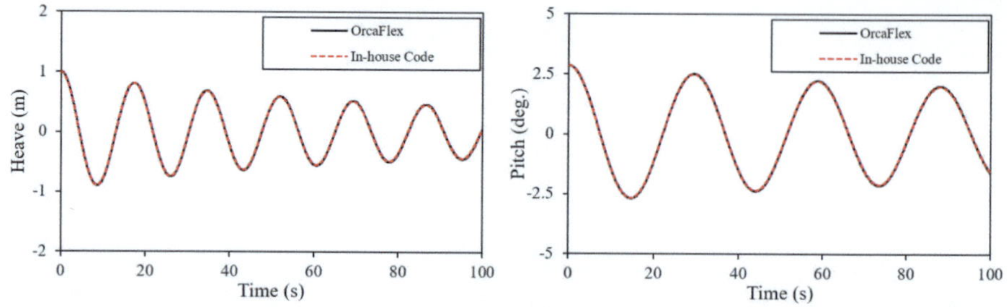

Fig. 3 The time-displacement results of the heave and pitch free decay test

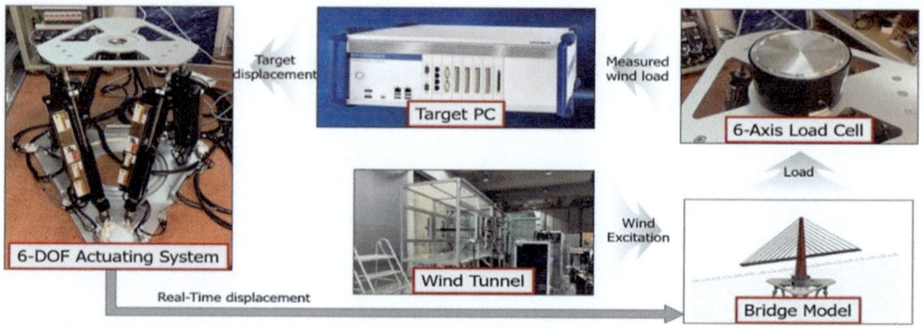

Fig. 4 Configuration of the experimental setup

3 Experimental Substructures

3.1 Experimental Setup

The experimental test setup consists of a 6-DOF actuating system, a 6-axis load cell, a target PC, a test model, and a wind tunnel. Wind loads are generated in the wind tunnel and measured at the base of the model using the 6-axis load cell. Measured loads are input into the target PC and applied as external loads to the analytical structure. All calculations are performed within the target PC, including the numerical integration and the real-time displacement control of the 6-DOF actuating system. This process is illustrated in Fig. 4.

3.2 6-DOF Actuating System

The 6-DOF actuating system, consisting of an upper and base plate connected by six linear motors, adjusts the upper plate's displacement through motor length adjustments, allowing the 6-DOF movement. Inverse kinematics calculates motor lengths from 6-DOF displacements, and forward kinematics computes the 6-DOF displacements from motor lengths. Inverse kinematics uses the reference configuration coordinates to calculate the upper joint coordinates with target displacements, determining the motor lengths from the change of distance between the upper and lower joints. In the reference configuration, the coordinates of the upper and lower joints of the i-th motor are $\mathbf{U}_i^r$ and $\mathbf{B}_i$, respectively, and when the 6-DOF displacement of $\mathbf{u} = u_x, u_y, u_z, \theta_x, \theta_y, \theta_z^T$ is applied, the coordinates of the upper joint $\mathbf{U}_i$ and the length of the i-th motor L_i are determined as shown in Eq. (6);

$$\mathbf{U}_i = \mathbf{R}\big(\mathbf{U}_i^r - \mathbf{H}\big) + \mathbf{T} + \mathbf{H}, \quad L_i = \|\mathbf{U}_i - \mathbf{B}_i\| \tag{6}$$

where, $\mathbf{R}$ is the 3-dimensional rotation matrix due to the rotation of upper plate, $\mathbf{R} = \mathbf{R}_z\mathbf{R}_y\mathbf{R}_x$, and $\mathbf{T}$ is the translational displacement vector $\mathbf{T} = u_x, u_y, u_z^T$.

Determining the 6-DOF displacement from the measured motor lengths requires an iterative calculation process. The gradient descent was utilized to apply forward kinematics. In the n-th iteration, the sum of squared error function $F_n(\mathbf{u})$ is calculated between the estimated length of the j-th motor l_{jn} for the assumed platform displacement u_{in} and the measured length L_j. At this point, to proceed with a new assumption in the direction that minimize $F_n(\mathbf{u})$, the gradient of $F_n(\mathbf{u})$, ∇F_{ni}, was used to calculate the new estimate of $\mathbf{u}_{i(n+1)}$. When $F(\mathbf{u})$ converges to a sufficiently small value, $\mathbf{u}$ at that point was determined as the 6-DOF displacements. This process is described in Eq. (7);

$$F_n(\mathbf{u}) = \sqrt{\sum_{j=1}^{6}\big(L_j - l_{jn}\big)^2}, \quad \nabla F_{ni} = \frac{\partial F_n}{\partial u_i}, \quad u_{i(n+1)} = u_{in} - k_i F_n \nabla F_{ni} \tag{7}$$

where, k_i is the learning rate, which adjusts the magnitude of the gradient to control the speed and stability of the iterative calculation.

Minimizing time delay is crucial for high-accuracy RTHS. The adaptive time series (ATS) compensator was applied as the time delay compensation algorithm [7]. As shown in Fig. 5, the ATS compensator was applied to each motor.

Figure 6 compares the z-direction displacement over time with and without applying the ATS compensator for the target displacement when the platform is subjected to a sinusoidal input with an amplitude of 10mm and a frequency of 1Hz.

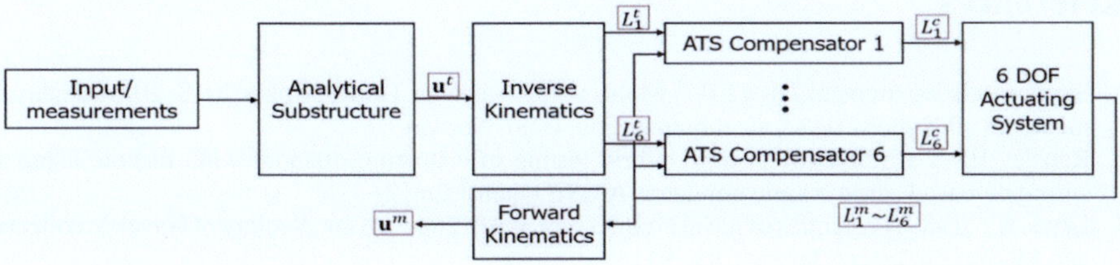

Fig. 5 ATS application process for the 6-DOF actuating system

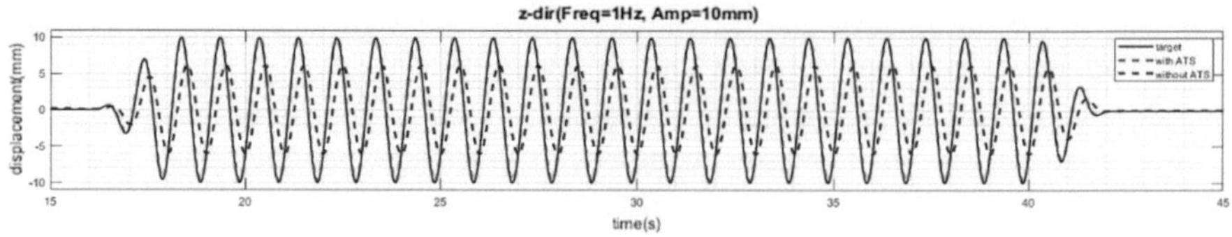

Fig. 6 Comparison of the displacement with the target displacement

As a result of applying the ATS compensator, both the time delay and amplitude errors were significantly reduced compared to the case without the ATS compensator. Before applying the ATS compensator, the time delay was approximately 152 ms with an amplitude error of around 3.8 mm, but after applying the ATS compensator, the time delay reduced to under 2 ms and the amplitude error to approximately 0.1 mm.

4 Conclusions and Further Study

This study developed and verified analytical and experimental substructures for RTHS on a floating structure with upper and lower structures. The analytical modeling of the lower structure, subjected to wave loads, used the time-domain Cummins equation. Coefficients were obtained by transforming frequency-domain coefficients from Ansys AQWA. The Morison drag, calculated using the Morison equation, was considered by defining Morison elements, which consist of the start and end points, cross-sectional area, and drag coefficients. The experimental test setup for simulating the 6-DOF response of the floating system includes the target PC for control, wind tunnel, 6-axis load cell, and 6-DOF actuating system. Inverse kinematics controlled the system, and the ATS compensator minimized time delay and amplitude errors. Future research will consider damping from the Morison effect and secondary wave load effects, as well as the model catenary mooring cables for nonlinear mooring effects. Validation of the experimental substructure will include physical measurements using devices like laser displacement sensors.

References

1. Samferdselsdepartement DK (2017) Meld. St. 33 Nasjonal Transportplan 2018–2029. departementenes sikkerhets- og serviceorganisasjon, Oslo, Norway
2. Ransley EJ et al (2022) Real-time hybrid testing of a floating offshore wind turbine using a surrogate-based aerodynamic emulator. AMSE Open J Eng 2
3. Lewis EV (1989) Principles of naval architecture, vol 3, 2nd revision. Society of Naval Architects and Marine Engineers, Jersey City, NJ
4. Cummins WE (1962) The impulse response function and ship motions. Schiffstechnik 9:101–109
5. Ansys Inc (2022) Aqwa theory manual. Canonsburg, PA

6. Robertson A, Jonkman J, Masciola M, Song H, Coupee A, Coulling A, Luan C (2014) Definition of the semisubmersible floating system for phase II of OC4 (No. NREL/TP-5000-60601). National Renewable Energy Lab. (NREL), Golden, CO (United States)
7. Chae Y, Kazemibidokhti K, Ricles JM (2013) Adaptive time series compensator for delay compensation of servo-hydraulic actuator systems for real-time hybrid simulation. Earthq Eng Struct Dyn 42(11):1697–1715

Creation of Wave-Free Space by Annular Elastic Floats

S. Akasaka, T. Iida, and K. Irifune

Abstract Various facilities such as offshore wind farms have been constructed at sea in recent years. In general, offshore facilities cannot avoid damage caused by waves. This may result in accidents, such as vessels drifting and capsizing as well as collision with marine structures. If wave-free space can be created on the free surface of the sea, we can reduce such wave accidents. Additionally, it leads to further development in offshore utilization such as floating cities. In this study, we aim to create a wave-free space within an annular floating structure consisting of multi-layered elastic plates. We establish the numerical simulation method to solve the wave field around this plate based on the eigenfunction matching method. In addition, we derive the inside wave energy in the annular plate. To achieve wave-free space within the plate, the evolutionary optimization method, Covariance Matrix Adaptation Evolution Strategy (CMA-ES), is employed for seeking the optimum plate design. We designate two parameters associated with each layer of the thin plate as design variables and optimize them to minimize the wave energy inside the circular structure. Optimization is carried out at the specific wave number. We compare the wave displacement around the circular float structure in two cases: (1) when only the wave energy inside the structure is minimized and (2) when both the inside wave energy and the outside scattered wave energy are minimized. Since the plate is the annular ring, the wave-free performance is omnidirectional.

Keywords Wave free space · Annular elastic floats · CMA-ES

1 Introduction

In recent years, various facilities such as offshore wind farms have been constructed at sea and floating cities are conceptualized. There has been increasing interest in marine utilization. However, offshore facilities cannot avoid damage caused by waves. This

S. Akasaka · T. Iida (✉) · K. Irifune
Department of Naval Architecture and Ocean Engineering, Osaka University, Osaka, Japan
e-mail: iida@naoe.eng.osaka-u.ac.jp

X. L. Zhao et al. (eds.), *Proceedings of the Fourth World Conference on Floating Solutions: WCFS 2024; 2–4 December; Hong Kong*, Lecture Notes in Civil Engineering 597, https://doi.org/10.1007/978-981-96-4569-5_63

may result in accidents, such as vessel drifting, capsizing, and collisions with marine structures. Furthermore, one of the reasons why floating cities haven't been realized is the effect of the waves. Then if wave-free space can be created on the free surface of the sea, we can reduce such wave accidents. Additionally, it leads to further development in offshore utilization such as floating cities.

Newman [1] presented the phenomenon called cloaking, wherein the scattering of incident waves by the floats doesn't propagate outside the float array, thereby reducing drift forces using multiple floats. Additionally, Iida et al. [2, 3] achieved cloaking at specific wavenumbers through numerical simulations using a higher-order boundary element method that considers mutual interference and model experiments. However, the cloaking phenomenon depends on the angle of incident waves with multiple floats, Thus, Iida et al. [4] proposed a model in which the cloaking phenomenon is independent of the direction of incident waves by using a multi-layered thin plate structure, thereby reducing the wave drifting force at a specific wave number.

In this study, we aim to create a wave-free space using a multi-layered structure of thin plates arranged annular configuration expanding upon previous research. To achieve this, it is necessary to minimize the wave energy inside the annular elastic floating body. Therefore, in this study, we utilize the CMA-ES optimization algorithm [5]. We consider the two parameters associated with each layer of thin plates as design variables and aim to minimize the wave energy inside the floating body.

2 Formulation of the Problem

2.1 Floats Placement

This research aims to create a wave-free space inside an annular elastic float on the sea. A coordinate system is set up as shown in Fig. 1. The thin plate is made up of N_t concentric layers, and the parameter of the i-th layer counting from the outside is denoted by a superscript (i). The inner radius from the central axis is a, and the outer annular radius of the outermost layer is b. The thickness t_p of the thin plate is assumed to be sufficiently small compared to the wavelength λ. The water depth is assumed to be constant at $z = -h$.

2.2 Governing Equation and Boundary Conditions Equations

This study assumes a linear potential theory. The governing equation and boundary condition equations are presented. These equations are considered on undisturbed free surfaces of the water with the positive z-axis pointing upward and g is the acceleration of gravity.

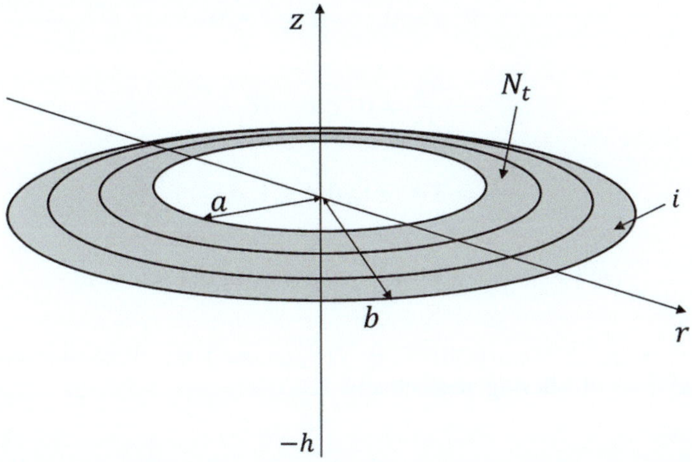

Fig. 1 Coordinate the system of the problem. The plate is assumed elastic and thin. The composite plate consists of N_t horizontally concentric annular rings, and the inner radius from the central axis is a and the outermost radius of the plate is b

$$\nabla^2\phi = 0 \quad -h \le z \le 0 \tag{2.1}$$

$$\frac{\partial\phi}{\partial z} = 0 \quad z = -h \tag{2.2}$$

$$\frac{\partial^2\phi}{\partial t^2} - g\frac{\partial\phi}{\partial z} = 0 \quad z = 0 \tag{2.3}$$

$$\frac{\partial^2\phi}{\partial t^2} - g\frac{\partial\phi}{\partial z} = \frac{1}{\rho_p}\frac{\partial}{\partial z}\left(m\frac{\partial^2\phi}{\partial t^2} + D\nabla^4\phi\right) \quad z = 0 \tag{2.4}$$

In these Eqs. (2.1) is the mass conservation, (2.2) is the kinematic boundary on the bottom, (2.3) is a linearized free surface condition, and (2.4) is the linearized equation for the thin plate. (2.3) is obtained when $p = 0$ for the pressure distribution in the liner-free surface equation, and (2.4) is obtained when the distributed load in elastic waves is applied. Where $D = Et^3/12\left(1 - \nu^2\right)$ is the stiffness modulus, E is Young's modulus, ν is Poisson's ratio, and ρ_p is the density of the thin plate.

Next, Nondimensionalization [6] is performed as follows, using the length a of the inner annular radius of an annular in (2.5).

$$\overline{x} = \frac{x}{a}, \quad \overline{t} = t\sqrt{\frac{g}{a}}, \quad \overline{\phi} = \frac{\phi}{a\sqrt{ag}}, \quad H = \frac{h}{a} \tag{2.5}$$

For simplicity, the bar representing the dimensionless value will be omitted hereafter. Thus, the governing equation and the continuity equation can be rewritten as

$$\nabla^2\phi = 0, \quad -H \le z \le 0 \tag{2.6}$$

$$\frac{\partial\phi}{\partial z} = 0 \quad z = -H \tag{2.7}$$

$$\frac{\partial\phi}{\partial z} = K\phi \quad z = 0(r \le 1, b \le r) \tag{2.8}$$

$$\left(\beta^{(i)}\nabla^4 + 1 - K\gamma^{(i)}\right)\frac{\partial\phi}{\partial z} = K\phi \quad z = 0(1 \le r \le b) \tag{2.9}$$

where $\gamma^{(i)} = \rho_p^{(i)}t_p^{(i)}/(\rho_w a)and\beta^{(i)} = D^{(i)}/\left(\rho_w g a^4\right)$ are dimensionless flexural rigidity and mass of i-th ring, respectively.

2.3 Dispersion Relation and Velocity Potential

Drive solutions that satisfy the governing equation and boundary condition equations. First, the dispersion relation between the water waves on the free surface inside and outside the annular floating and the elastic wave in the floating body is written as:

$$\frac{K}{\beta^{(i)}\kappa_n^4 - K\gamma^{(i)} + 1} = \begin{cases} \kappa_0 & \tanh\kappa_0 H \\ -\kappa_n & \tan\kappa_n H \end{cases} \tag{2.10}$$

K is the dimensionless frequency, and κ is the wave number without distinction between water and elastic waves. In the following, k_n is the wavenumber of the water waves and $\mu_n^{(i)}$ is the wavenumber of the elastic waves. k_n has an infinite number of positive and real solutions, where k_0 indicates progressive waves, and $k_1 k_2 \cdots$ correspond to local waves. Similar, μ_0 has an infinite number of positive and real solutions, where $\mu_0^{(i)}$ indicates progressive waves, $\mu_1^{(i)}, \mu_2^{(i)} \cdots$ indicate local waves, and additionally two complex solutions $\mu_{-2}^{(i)}$ and $\mu_{-1}^{(i)}$ represented damped waves. The dispersion relation in the case of a water wave is when $\beta = 0$ and $\gamma = 0$.

The velocity potential that satisfies these equations is written as

$$\phi(r, \theta, z, t) = \mathrm{Re}\left\{\frac{A}{i\sqrt{K}}\sum_{m=-\infty}^{\infty}\Psi_m(r, z)e^{im\theta}e^{-i\sqrt{K}t}\right\} \tag{2.11}$$

and

$$
\Psi_m = \begin{cases} i^m J_m(k_0 r) f_0(z) + a_{m0} H_m^{(1)}(k_0 r) f_0(z) + \sum_{n=1}^{N} a_{mn} K_m(k_n r) f_n(z) \\ b_{m0}^{(i)} J_m\left(\mu_0^{(i)} r\right) F_0^{(i)}(z) + c_{m0}^{(i)} H_m^{(1)}\left(\mu_0^{(i)} r\right) F_0^{(i)}(z) \\ + \sum_{n=-2, n\neq 0}^{N} b_{mn}^{(i)} I_m\left(\mu_0^{(i)} r\right) F_n^{(i)}(z) + \sum_{n=-2, n\neq 0}^{N} c_{mn}^{(i)} K_m\left(\mu_0^{(i)} r\right) F_n^{(i)}(z) \\ d_{m0} J_m(k_0 r) f_0(z) + \sum_{n=1}^{N} d_{mn} I_m(k_n r) f_n(z) \end{cases} \tag{2.12}
$$

where $J_m(k_0 r)$ is a Bessel function of the first kind, $H_m^{(1)}(k_0 r)$ and $H_m^{(1)}(\mu_0 r)$ are Hankel functions of the first kind, and $I_m\left(\mu_n^{(i)} r\right)$ and $K_m\left(\mu_n^{(i)} r\right)$ are deformed Bessel functions of the first and second kinds. $f_n(z)$, $F_n^{(i)}$ are written as

$$
f_n(z) = \begin{cases} \frac{\cosh k_0(z+h)}{\cosh k_0 h} & n = 0 \\ \frac{\cosh k_n(z+h)}{\cosh k_n h} & n \geq 1 \end{cases}, \quad F_n^{(i)}(z) = \begin{cases} \frac{\cosh \mu_0(z+h)}{\cosh \mu_0 h} & n = 0 \\ \frac{\cosh \mu_n(z+h)}{\cosh \mu_n h} & n = -1, -2, n \geq 1 \end{cases} \tag{2.13}
$$

Velocity potentials are described in three different ways: water waves outside the float, elastic waves on the float, and water waves inside the float. The unknown coefficients a_{mn}, $b_{mn}^{(i)}$, $c_{mn}^{(i)}$ and d_{mn} in the velocity potential are solved by the eigenfunction matching method using the boundary conditions equation around the float [4].

2.4 The Energy Around the Floats

The energy around the float to create a wave-free space is derived. First, the wave energy inside the annular float is derived. The average energy inside the annulus is the sum of the kinetic energy and the potential energy. Considering the time average and applying the velocity potential of the water waves inside the float, the average energy inside the annulus is written as

$$
\overline{E} = \sum_{m=-\infty}^{\infty} |d_{m0}|^2 \times \left[\begin{array}{c} \frac{1}{16} H\left\{ (J_{m-1}{\prime}(k_0))^2 + \left(1 - \frac{(m-1)^2}{k_0^2}\right) J_{m-1}^2(k_0) \right\} \\ + \frac{1}{16} H\left\{ (J_{m+1}{\prime}(k_0))^2 + \left(1 - \frac{(m+1)^2}{k_0^2}\right) J_{m+1}^2(k_0) \right\} \\ + \frac{1}{8}(4 - H)\left\{ (J_m{\prime}(k_0))^2 + \left(1 - \frac{m^2}{k_0^2}\right) J_m^2(k_0) \right\} \end{array} \right] \tag{2.14}
$$

where $H = (1 + 2k_0 h/\sinh 2k_0 h)$. By bringing this close to zero, it is possible to create a wave-free space. Next, consider the energy of the scattered waves [2] generated by the disturbance of the incident waves by the float. The energy of the scattered wave is derived from the principle of conservation of energy. W_D that work done by the fluid on the object from the time-averaged rate of change of the total energy in a given fluid domain is obtained. Applying the velocity potential of the water waves outside the float to it, the energy is written

$$W_D = -\frac{1}{C_0\sqrt{K}} \sum_{n=-\infty}^{\infty} \left\{ \mathrm{Re}\left[i^m a_{m0}^* \right] |a_{m0}|^2 \right\} \tag{2.15}$$

where $C_0 = k_0^2/K + \left(k_0^2 - K^2\right)h$. Note that because a_{m0} is the amplitude of scattered waves, the second term in (2.15) indicates the energy transferred to scattered waves. As a result, the energy of scattered waves W_s is given as

$$W_S = \sum_{n=-\infty}^{\infty} |a_{m0}|^2 \tag{2.16}$$

By bringing this close to zero, cloaking can be achieved.

3 Evolutionary Optimization of the Plate

To create a wave-free space, CMA-ES [5] was used for optimization. The optimization was performed in two cases: the objective function is (1) $F = \overline{E}$, , (2) $F = \overline{E} + 0.05W_s$ and the results are compared. The energy of scatted waves was corrected by a factor of 0.05 in this study due to the larger scale of the scattered wave energy compared to the wave energy inside the annular. In addition, in performing the optimization, the variables are γ and β, which indicated the mass and stiffness of the float. These two variables are subject to the following as $0.01 \leq \gamma \leq 0.5$ and $0.01 \leq \gamma \leq 0.5$. The values of γ and β for each layer are optimized to minimize the wave energy when the frequency of the incident wave $k_{in} = 1.0$, which is dimensionless as the inner annular radius of the float $a = 1$, outer annular radius $b = 5$, number of layers of the thin plate $N_t = 8$, and water depth $h = \lambda$.

4 Results and Discussion

First, the optimized wave energy in case (1) is $\overline{E} = 2.98 \times 10^{-3}$, while the optimized wave energy and scattered wave energy in case (2) are $\overline{E} = 1.31 \times 10^{-2}$ and $W_s = 4.28$. Figure 2 shows the wave displacement of the float and the areas around it. In case (1), a wave-free space is observed inside the float, but the waves around it are disturbed by the scattered waves. On the other hand, in case (2), waves are slightly generated inside the float, but no scattered waves are generated and a phenomenon quite close to cloaking can be observed. Figure 3 shows the variation of variables β and γ in the direction r. As in Iida et al. [4], the β varies from layer to layer, but γ varies only slightly from layer to layer in two cases. This suggests that the stiffness coefficient β has more control over the optimization than the mass coefficient γ. Finally, Fig. 4 shows the variation of the wave energy inside the float and the scattered wave energy versus the wavenumber of the incident wave. The wave energy and scattered wave

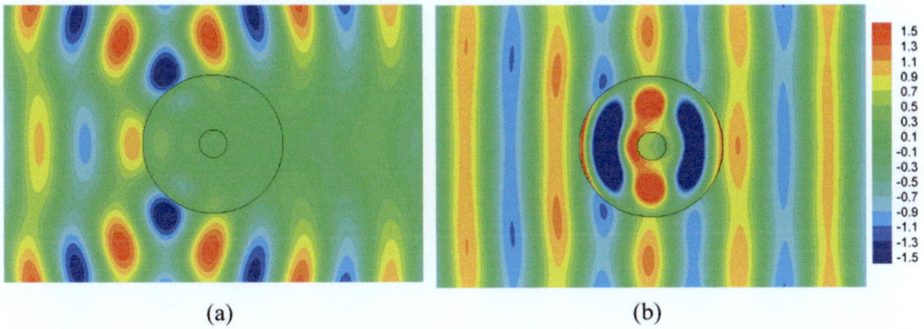

Fig. 2 Wave displacement around the floating body. **a** Show the wave displacement in case (1) when optimized only wave energy inside float. **b** Show in case (2) when optimized wave energy inside float and scattered energy

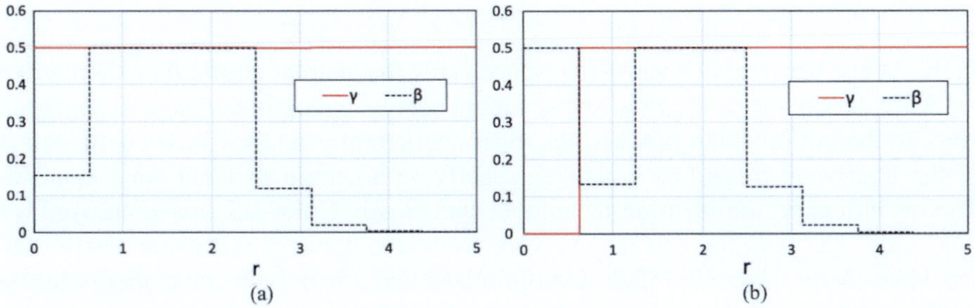

Fig. 3 Spatial distributions of mass γ and flexural rigidities β for the optimized composite plate. **a** Show in case (1) when optimized only wave energy inside float. **b** Show in case (2) when optimized wave energy inside float and scattered energy

energy change as close to zero as possible when the wavenumber of the incident wave $k_{in} = 1.0$. However, the addition of the scattered energy reduces the accuracy of the optimization.

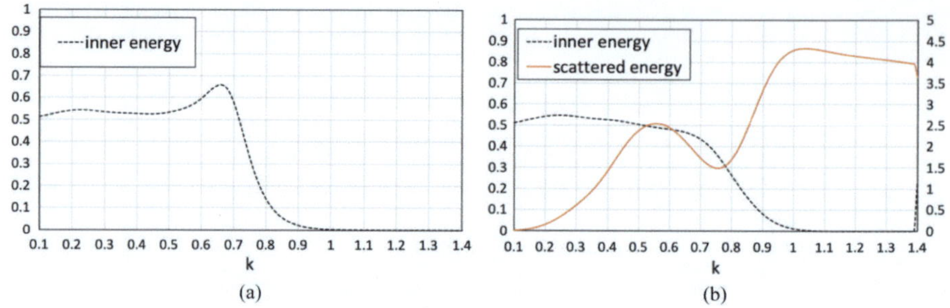

Fig. 4 Variation of energy versus wave number of the incident wave. **a** Show in case (1) when optimized only wave energy inside float. **b** Show in case (2) when optimized wave energy inside float and scattered energy. Left vertical axis: inner energy, right vertical axis: scattered energy

5 Conclusion

In this study, we created a wave-free space using the annular elastic float. We derive the wave energy inside the float and scattered energy and optimize them to approach zero. In the optimization process, the mass coefficient γ and stiffness coefficient β of the float were treated as variables, specifying a certain incident wave number. The evolutionary optimization technique, specifically CMA-ES, was employed for this study. The optimization results were compared between scenarios where only the wave energy inside the float was optimized and where both wave energy inside the float and scattered energy were optimized. In comparison to optimizing only the wave energy inside the float, optimizing both inside wave energy and scattered energy resulted in lower optimization precision. In optimization, there are still problems that need to be improved. Additionally, concerning variables γ and β, while the values of β varied significantly across layers, γ exhibited minimal changes with little discernible variation. It is presumed that β has a significant influence on the optimization process of this study.

References

1. Newman JN (2014) Cloaking a circular cylinder in water waves. Eur J Mech B Fluids 47:145–150
2. Iida T, Kashiwagi M, He G (2014) Numerical confirmation of cloaking phenomenon on an array of floating bodies and reduction of wave drift force. Int J Offshore Polar Eng 24(04):241–246
3. Iida T, Kashiwagi M, Miki M (2016) Wave pattern in cloaking phenomenon around a body surrounded by multiple vertical circular cylinders. Int J Offshore Polar Eng 26(01):13–19
4. Iida T, Zareei A, Alam MR (2023) Water wave cloaking using a floating composite plate. J Fluid Mech 954:A4
5. Hansen N (2016) The CMA evolution strategy: a tutorial. arXiv preprint arXiv:1604.00772
6. Meylan MH (2002) Wave response of an ice floe of arbitrary geometry. J Geophys Res Oceans 107(C1):5–1

Stability and Hydrodynamic Analysis of a Cylindrical Floating Planter

Senqin Zhang, Jinghua Wang, and Xiao Lin Zhao

Abstract Floating coastal structures, considered as an alternative to land reclamation for expanding livable space, are regarded as cost-effective and sustainable solutions for land-scarce cities. Among these structures, floating planters stand out for their ability to enhance the aesthetics of water bodies, provide habitats for aquatic organisms, and reduce land usage. Despite their advantages, there is a dearth of research and engineering guidelines for the design and construction of floating planters. In this research, a preliminary design of a cylindrical floating planter is proposed, and its stability and hydrodynamics in still water is analyzed. The planter consists of an inner steel skeleton and an outer UHPC skin. The cylinder is 5 m in diameter and 5 m in height. The overall weight including ballast, soil and trees is about 699.5kN. For the stability analysis, the wind loadings on the coaxial branched trees are considered. With appropriate weight distribution and ballast, the stability of the structure is analyzed and verified under designed wind condition in still water. The Runge–Kutta Method is used to obtain the motion response of the floating planter by solving the six-degree-of-freedom motion equation. The results showed that the natural periods are 4.2611 s for the heave, 11.0291 s for the sway and 5.5655 s for the roll, which agree well with the theoretical predictions. Further study is still in demand to analyze its motion under design wave conditions.

Keywords Floating structure · Response amplitude operator · Free decay tests

S. Zhang · J. Wang (✉) · X. L. Zhao
Department of Civil and Environmental Engineering, The Hong Kong Polytechnic University, Hong Kong, China
e-mail: jinghua.wang@polyu.edu.hk

J. Wang
Shenzhen Research Institute, The Hong Kong Polytechnic University, Shenzhen, China

X. L. Zhao
Research Institute for Land and Space, The Hong Kong Polytechnic University, Hong Kong, China

X. L. Zhao et al. (eds.), *Proceedings of the Fourth World Conference on Floating Solutions: WCFS 2024; 2–4 December; Hong Kong*, Lecture Notes in Civil Engineering 597, https://doi.org/10.1007/978-981-96-4569-5_64

1 Introduction

Floating planter is a type of auxiliary coastal infrastructure that can be beneficial for the sustainable development of the region [1]. Despite their potential benefits, the design and construction of floating planters, lack adequate research and engineering codes. Particularly, when considering the application of floating planters in locations like Hong Kong, several challenges arise due to the unique environmental conditions. Hong Kong's shallow water depth coupled with significant tidal range (about 1.54 m [4]) presents a hurdle for the deployment of floating structures. Additionally, the region is prone to typhoons, which impose an enormous challenge on the stability and safety of floating planters. To date, guidelines and specific engineering codes for the design and construction of floating planters are still in demand.

This study aims to answer this open question about how to design a floating planter with sufficient stability in still water based on the current knowledge of design practice for floating offshore structures. The paper is structured in the following way: Sect. 2 introduces the environment conditions and describes the design of the planter; Sect. 3 reports the results of stability analysis and free decay tests; Sect. 4 makes concluding remarks with a discussion on the limitations of the present study.

2 Floating Planter Design

2.1 Environment Conditions

The site to be studied is located in the vicinity of Kau Yi Chau, where a land reclamation project is under consideration to meet the increasing need of land space (see Fig. 1 [6]). The mean water depth in this area is about 9 m. The maximum gust wind speed in Hong Kong water is 44 m/s under normal environmental conditions [4]. Under extreme environmental conditions, the maximum gust wind speed in Hong Kong water is expected to be 70 m/s with a return period of about 50 years [4, 5].

2.2 Planter Design

The pontoon of the planter is in cylindrical shape of 2.5 m in radius and 5 m in height. The wall is made of UHPC with a thickness of 100 mm, and the skeleton is made of steel with square hollow section (80 $\times$ 80 $\times$ 8 mm). To enhance stability, the stone ballast is adopted. The bottom of the pontoon is filled with stones, and the top of the ballast is secured with a spacer. The opening at the top of the pontoon provides space for storing soil and plants with a dimension of 1.5 m in radius and 1.2 m in height. The overall design of this floating planter is illustrated in Fig. 2.

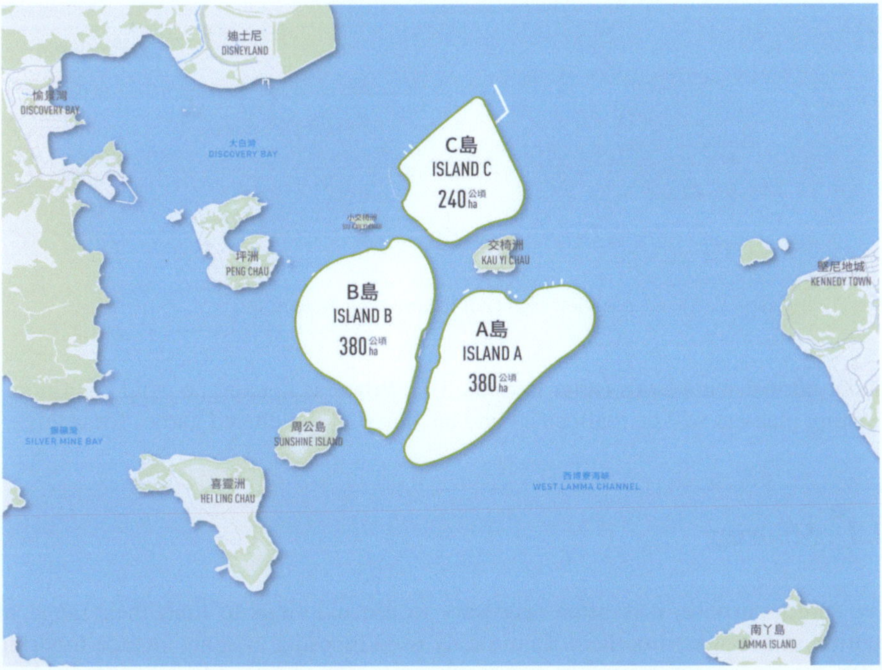

Fig. 1 Location of the land reclamation

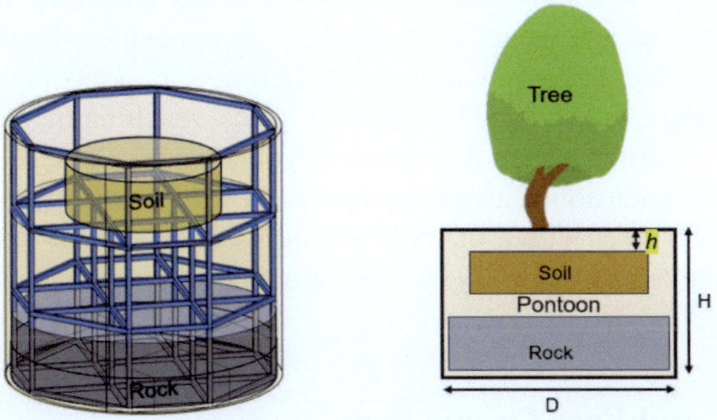

(a) UHPC wall and steel skeleton (b) Side view of the floating planter

Fig. 2 An illustration of the floating planter

We consider a commonly used landscape tree species, the Coaxial Branched Tree, which typically has a height ranging from 2.5 to 5 m and a crown width between 1.5 and 3.0 m [8]. In this study, we take the windward area of the tree to be 4 m^2. In the unmoored state, we need to ensure that the floating platform will remain stable

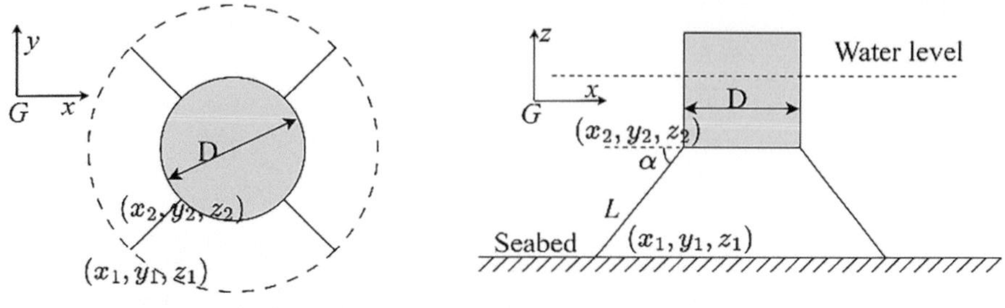

Fig. 3 Top view (left) and side view (right) of the mooring system

under normal wind loads (refer to Sect. 3.1). Whereas in extreme wind conditions, floating planters will be reallocated and anchored in a sheltered harbor for safety.

2.3 *Mooring*

For safety purpose, it is often necessary to use moorings to limit their range of motion in ocean environment. Commonly used mooring systems include mooring piles, catenary mooring system and taut leg mooring system. Mooring piles reduce the lateral movement of the floating body and minimize the risk of capsizing and are often used for near-shore structures. Catenary moorings limit the movement of the float by means of chains under their own weight, which are not in a straight line between the float and the anchor since they do not exhibit much tension or sag, thus allowing the float to move within a certain range. Taut leg mooring system use tensioned mooring ropes to hold the floats in place, resulting in a smaller mooring footprint and a mooring system that uses less cable material [11].

In this paper, a tensioned mooring system is used, which is made of 20 mm diameter polyester fiber with a material stiffness coefficient $\lambda = 653.5$ kN/m [18]. The mooring line length $L = 5.99$ m. The layout of the mooring is shown in Fig. 3.

3 Stability and Free Decay Tests

3.1 *Stability*

Stability is the ability of a floating structure to return to its original equilibrium position after displacement by external forces. This property ensures that the structure will return to equilibrium once these forces cease to act.

When the floating structure tilts at a small angle, the waterline changed from *WL* in still water to W_1L_1 (see Fig. 4). Despite the tilt, the structure's weight remains constant, and its center of gravity does not shift. Consequently, the displacement

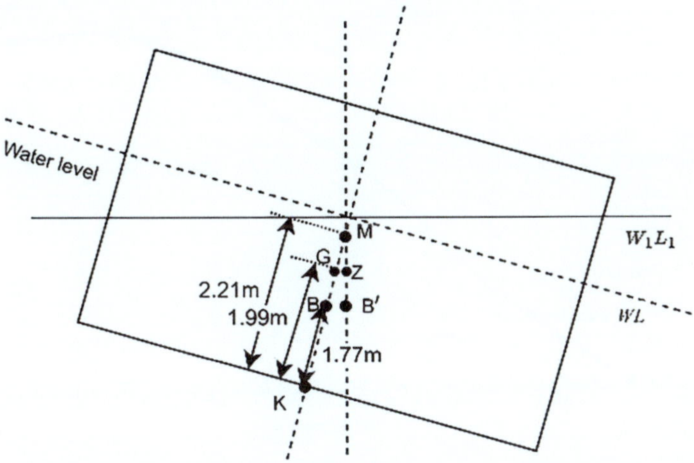

Fig. 4 Cross section and dimension of the planter

volume of the structure stays unchanged. However, the shift in the waterline alters the shape of the submerged volume, causing the center of buoyancy to move from point *B* to *B'*. Buoyancy and gravity form a torque that rotates the structure back to its original equilibrium position.

The center of gravity *G* lies below the metacenter *M*, and the direction of the moment produced by gravity is opposite to the direction of tilt. When the external force is removed, this configuration helps the structure return to its original equilibrium state, signifying a stable equilibrium. Conversely, if the structure cannot return to its original state, it is considered to be in an unstable equilibrium.

The metacentric radius *BM* is a crucial factor in determining stability [14]. It is mathematically expressed as:

$$BM = I/\nabla, \tag{1}$$

where *I* represents the moment of inertia of the waterplane area about the axis of inclination, ∇ denotes the volume of displacement of the floating structure.

According to [4], the design wind speed under normal environmental conditions is specified as 44 m/s. This value is used for stability checks. The wind load can be calculated using equation as below [10]:

$$F_w = \frac{1}{2}\rho_a C_d U^2 A_{tr} \tag{2}$$

which is approximately 13,187 N, where A_{tr} is windward area of trees, ρ_a is the air density, $C_d = 0.769$ is the drag coefficient [12], and U is the incoming wind speed.

The steel skeleton, UHPC shell and soil weighs 28.34 kN, 196.34 kN and 152.68 kN, respectively. A total of 292.12 kN of stones is added at the bottom

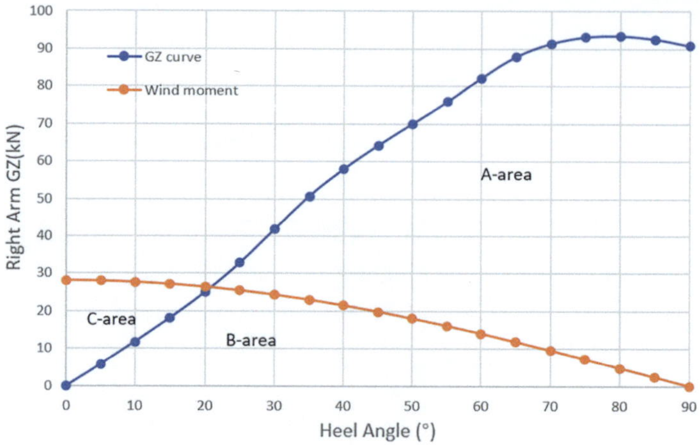

Fig. 5 Righting arm and wind moment in different heel angles

to enhance the stability. The total mass of the floating planter after counterweighting is 699.5 kN, and the height of the center of gravity KM is 2.21 m. The draft is calculated to be 3.55 m, and center of gravity KB = draft/2 = 1.77 m, center of buoyancy KG = 1.99 m and BM = 0.44 m. The recovery moment and the overturning moment curve under wind load are calculated, and result is shown in Fig. 5.

As shown in the Fig. 5, 3 regions are identified, i.e., *A*, *B* and *C*. Note that, the center of gravity is lower than the center of rotation (i.e., $KM > KG$ in Fig. 4). According to [15], dynamic stability can be satisfied when area $A + B > 1.4\,(B + C)$. In the case shown in Fig. 5, area A + B = 94.75 kN·rad is greater than 1.4 (B + C) = 41.016 kN·rad. Hence, the floating planter satisfies the dynamic stability criterion.

3.2 *Free Decay Tests*

Governing Equation and Numerical Scheme

For hydrodynamic analysis, a floating body can be regarded as rigid. Its six motional degrees-of-freedom (DOF) can be described by the equation below [2]:

$$(\boldsymbol{M} + \mathbf{A})\ddot{\eta} + \boldsymbol{B}\dot{\eta} + \mathbf{C}\eta = F_w + F_{rs} \tag{3}$$

where

η, $\dot{\eta}$ and $\ddot{\eta}$ are the displacement, speed and acceleration,
F_{rs} represents reaction forces from the mooring system,
F_w contains the wind-excited forces on the fixed structure,
$\boldsymbol{M}$ are components of the generalized mass matrix for the structure,
$\mathbf{A}$ are hydrodynamic mass or added mass,

B is the hydrodynamic damping or radiation damping,
C is the hydrostatic stiffness,

and

$$C_{22} = 0 \tag{4}$$

$$C_{33} = \rho g A_{wp} \tag{5}$$

$$C_{44} = \rho g \iint_{s} y^2 ds + \rho g V(KB - KG) \tag{6}$$

$$F_{rs} = \mathbf{k}\eta \tag{7}$$

where A_{wp} is the waterplane area, V is the displaced volume of water, ρ is the water density, g is the gravitational acceleration, S is the floating structure submerged surface. The coordinate axes $0xyz$ are shown in the Fig. 3.

The idealized stiffness **k** is derived by assuming that each mooring cable is weightless and perfectly elastic with a known tension and elastic stiffness [13]. Each cable has a constant tension T_{m} an elastic stiffness λ and is strung between coordinates (x_1, y_1, z_1) on the seabed and (x_2, y_2, z_2) at the toe (see Fig. 3). L is the mooring line length. The stiffness factor for each cable can be expressed as:

$$k_{22} = \lambda \cos^2 \beta + \frac{T_m}{L} \sin^2 \beta \tag{8}$$

$$k_{33} = \lambda \cos^2 \gamma + \frac{T_m}{L} \sin^2 \gamma \tag{9}$$

$$k_{44} = k_{22} z_2^2 - 2k_{32} y_2 z_2 + k_{33} y_2^2 \tag{10}$$

where

$$\cos \alpha = \frac{x_2 - x_1}{L}, \quad \cos \beta = \frac{y_2 - y_1}{L}, \quad \cos \gamma = \frac{z_2 - z_1}{L} \tag{11}$$

The cable stiffness matrix is computed for each cable in a mooring system and summed to yield a total mooring stiffness matrix. The fourth-order Runge–Kutta Method is used to solve the Eq. (3) in time domain with a time step size of 0.005 s, which provides high accuracy.

Verifications

In this case, the maximum gust wind speed of 70 m/s under extreme conditions (return period 50 years [4]) is considered for analysis. The results are summarized in

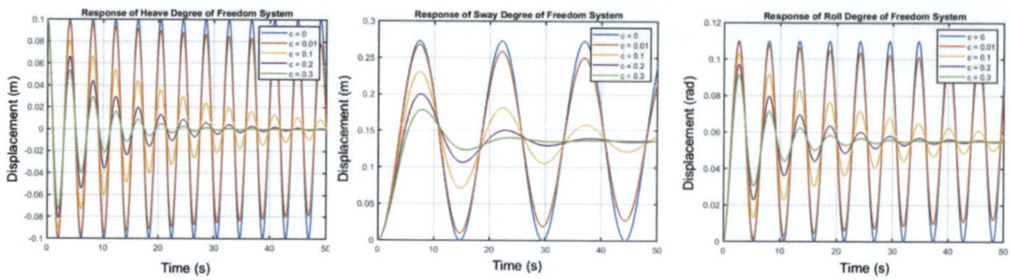

(a) Heave (b) Sway (c) Roll

Fig. 6 Time history of displacement in heave (**a**), sway (**b**) and roll (**c**) under various damping ratios

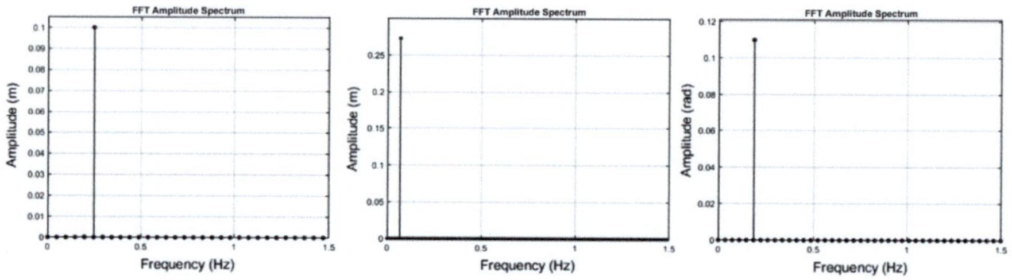

(a) Heave (b) Sway (c) Roll

Fig. 7 FFT amplitude spectrum in heave (**a**), sway (**b**) and roll (**c**)

Fig. 6 including displacements in heave, sway and roll for different damping ratios c. Damping ratios typically ranging from 0.01 to 0.3 [16].

When the damping ratio c is 0, the motions are periodic, and the corresponding natural frequency can be estimated. By performing a Fourier transform on the time signal, the frequency spectra can be obtained as shown in Fig. 7. The natural period can be easily identified and extracted from the figures, i.e., $T_{33} = 4.08$ s for heave, $T_{22} = 14.76$ s for sway, and $T_{44} = 5.56$ s for roll.

To verify the accuracy of the calculations, the intrinsic frequency of the floating body can also be determined analytically through the formula [7]:

$$\omega_n = \sqrt{k_{total}/m_{total}} \tag{12}$$

where k_{total} is the stiffness including mooring line stiffness and hydrostatic stiffness, m_{total} is the generalized total mass including generalized mass and added mass. The natural frequencies obtained are $T_{33} = 4.0788$ s, $T_{22} = 14.7606$ s, $T_{44} = 5.5655$ s. Comparing with the previous calculations obtained from Eq. (3), the results are consistent with a maximum difference less than 1.5%.

3.3 Discussions

In general, floating installations with a length/width ratio of less than 3 should not tilt at an angle of more than 15° [3]. The maximum tilt of the towing structure should not exceed 5° in wind conditions of 16 m/s [7]. In this floating planter design, the maximum inclination angle was 0.075 rad, or 4.30°, which shows that the float is quite stable under mooring.

In designing the floating body, it is crucial to avoid having motion from resonating with the ocean surface wave [9]. While in Hong Kong, the mean peak period ranges from 4.90 to 6.74 s (0.148 to 0.204 Hz) [17]. In further research, to ensure the safety of the floating planter, adjustments to the parameters of the mooring cable (such as diameter, material, etc.) can be made so that the vibration frequency of the floating planter differs from the wave frequency, thereby preventing resonance. This will be carried out in a future study.

4 Conclusions

This research focused on designing a cylindrical floating planter and analyzing its stability and 6 degrees-of-freedom (DOF) motions in still water. Fourier analysis was carried out to verify the calculations of the natural frequencies. From this study, the following primary conclusions can be drawn:

1. It is feasible to design stable cylindrical floating planters through careful mass distribution.
2. Under the mooring condition, the motion of the floating planter in sway and roll is considerably within a small range, indicating its safe deployment at designed location.
3. By performing Fourier analysis on the time domain simulation data, the natural frequency can be accurately calculated, which is useful for the optimization of the design in the future.

The limitation of this study is the neglection of the incident ocean surface waves. A more comprehensive study will be conducted in the follow-up study.

References

1. Barltrop N (1998) Floating structures: a guide for design and analysis, 1st edn. Oilfield Publications Inc., England
2. Bergdahl L (2009) Wave-induced loads and ship motions. Technical report, Chalmers University of Technology, Sweden
3. China Classification Society (2023) Classification specification for offshore mobile platforms. China

4. Civil Engineering Office, Civil Engineering and Development Department (2023) Port works design manual part 1: general design considerations for marine works. Hong Kong, China
5. Civil Engineering Office, Civil Engineering and Development Department (2023) Port works design manual part 2: guide to design of Piers and Dolphins. Hong Kong, China
6. Development Bureau, Civil Engineering and Development Department, Planning Department (2023) Study on the artificial islands in the central waters. Hong Kong, China
7. El-Reedy MA (2019) Offshore structures: design, construction and maintenance, 2nd edn. Gulf Professional Publishing, United States
8. Guangdong Landscape Architecture and Ecological Landscape Association, The Hong Kong Institute of Landscape Architects Association for the Promotion of Landscaping and Greening in Macao (2021) Standard specification of nursery plant materials for Guangdong Hong Kong and Macau. Guangdong Landscape Architecture and Ecological Landscape Association, China
9. Lewis EV (1989) Principles of naval architecture: motions in waves and controllability, 2nd edn. Society of Naval Architects, United States
10. Lin P, Hu G, Tse KT, Song J, Leung AK, Jiang Z, Zou L (2023) Characterizing wind-induced reconfiguration of coaxial branched tree. Agric Forest Meteorol 340
11. Ma KT, Luo Y, Kwan CTT, Wu Y (2019) Mooring system engineering for offshore structures, 1st edn. Gulf Professional Publishing, United States
12. Manickathan L, Defraeye T, Allegrini J, Derome D, Carmeliet J (2018) Comparative study of flow field and drag coefficient of model and small natural trees in a wind tunnel. Urban Forestry Urban Greening 35:230–239
13. Patel MH, Lynch EJ (1983) Coupled dynamics of tensioned buoyant platforms and mooring tethers. Eng Struct 5(4):299–308
14. Sheng ZB, Liu YZ (2003) Ship design, 2nd edn. Shanghai Jiao Tong University Press, China
15. Technical Committee CSB/17 (1989) Code of practice for Maritime structures—Part 6: design of inshore moorings and floating structures. United Kingdom
16. Wang S (2022) Analytical solutions for the dynamic analysis of a modular floating structure for urban expansion. Ocean Eng 266:112878
17. Wave Measurement. https://www.cedd.gov.hk/eng/public-services-forms/port-and-marine/wave-measurement/index.html. Last accessed 03 May 2024
18. Yu X, Lei H, Wang Y (2020) Research on the applicability of multi-points mooring system for floating platform in shallow waters. Naval Sci Technol 42(1):105–110

Materials for Floating Structures

A Green Concrete Material and Its New Biomimetic Sustainable Ocean Floating Structure

Yufei Wu and Xiaoxu Huang

Abstract The authors and their colleagues have recently developed a new concrete technology—compression-cast concrete. Compression casting technology condenses concrete, which makes it stronger and more durable. It is a practical technology that not only greatly improves the performance of concrete materials, but also reduces the cost and greenhouse gas emissions of related concrete structures. The compression-cast concrete technology has been developed for a major research project—a new biomimetic sustainable marine floating structure. The goal of the project is to develop a new type of offshore floating structure characterized by superior structural performance, low construction and maintenance costs, high sustainability, and good survivability in harsh marine environments. To achieve these objectives, the team has developed several new technologies, including compression-cast concrete technology, bionic honeycomb floating unit, flexible floating unit connectors, new mooring systems, new ocean energy harvesting systems, and an effective and practical method for surviving in harsh sea conditions. These new technologies are described in this article.

Keywords Compression cast concrete · Floating structure · Sustainability · Survivability

1 Introduction

Concrete is regarded as the most widely used construction material with significant carbon emissions. It is reported that 400 kg CO_2eq carbon emissions are generated when producing 1 m^3 concrete [1]. It is an urgent task to reduce CO_2 emissions for the construction industry. In the past few decades, many techniques have been proposed to reduce the amount of cement in concrete production, such as partially substituting cement with by-products (e.g., blast silica fume, fly ash, and industrial

Y. Wu (✉) · X. Huang
School of Civil and Transportation Engineering, Shenzhen University, Shenzhen, China
e-mail: yufei.wu@szu.edu.cn

X. L. Zhao et al. (eds.), *Proceedings of the Fourth World Conference on Floating Solutions: WCFS 2024; 2–4 December; Hong Kong*, Lecture Notes in Civil Engineering 597, https://doi.org/10.1007/978-981-96-4569-5_65

Fig. 1 Compression casting process

wastes), and complete replacement of cement with green cement alternatives (e.g., alkali-activated geopolymers and limestone calcined clay cement) [2–4]. However, the high cost hinders their application in practical engineering. Recently, the authors' research group has developed a novel compression casting technology (CCT) (see Fig. 1). With this new technology, the strength and durability of concrete can be improved significantly, thus reducing the cost of concrete [5–8]. Notably, compared to normal casting (NC), lower carbon emission is generated for CCT, providing an effective approach for the sustainable development of the construction industry [1].

To promote the application of CCT in practical engineering, the performances of compression-cast concrete (CCC) are first investigated comprehensively, including mechanical properties, microstructures, durability, and carbon reduction benefits. Then, this green concrete material is applied to floating structures with a biomimetic shape. For this new kind of floating structure, the required information for the design has been investigated including bionic honeycomb floating unit, flexible connectors for units, and new mooring systems. To achieve energy self-sufficiency for floating structures, new ocean energy harvesting systems are studied. In addition, an effective and practical method for surviving in harsh sea conditions is proposed.

2 Compression-Cast Concrete (CCC)

2.1 Mechanical Properties

As shown in Fig. 2, compared with NC, the compressive strength and elastic moduli of CCC increased by 26–83% and 26–69%, respectively [9]. The remarkable improvements in the mechanical properties of CCC are mainly attributed to the reduced air and water contents and the denser microstructure of the concrete [5].

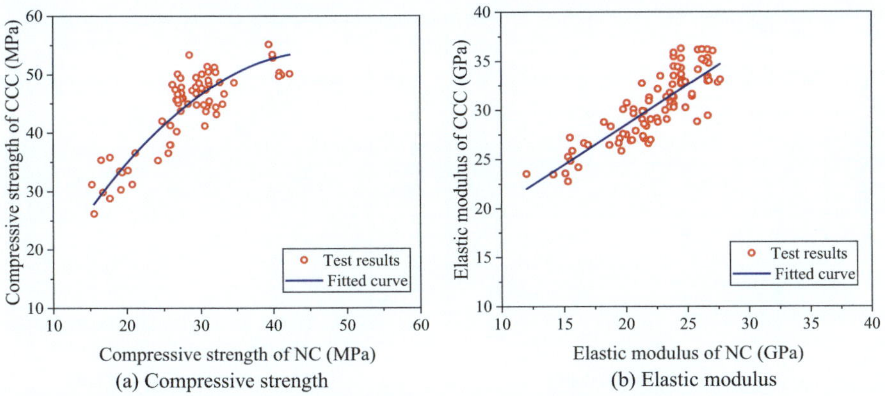

Fig. 2 Improvement of mechanical properties of CCC over NC

2.2 Durability

Carbonation

As shown in Fig. 3a, the carbonation depths are much smaller for CCC specimens compared with NC specimens: after 28 days of CO_2 exposure, the carbonation depth of CCC specimens is 82.6 and 91.7% lower for the casting pressure of 5 and 15 MPa, respectively, compared with NC specimens. The result also shows that the effect of casting pressure on the carbonation depths is insignificant.

Freeze–Thaw (FT) Cycles

Figure 3b presents the retention values of the dynamic elastic modulus of concrete after FT cycles. It can be found that the effect of the casting pressure on the deterioration ratio of the dynamic elastic modulus is obvious. The deterioration ratio decreases as the casting pressure increases. This phenomenon can be explained by the fact that the denser internal structure and higher strength of the cement matrix under higher casting pressure lead to the improved retention of dynamic modulus of elasticity and enhanced FT resistance.

Chloride Penetration

Figure 3c shows that the chloride diffusion coefficients of CCC are smaller than those of NC, indicating that the CCT can postpone the process of chloride penetration. Due to the reduced porosity and densified microstructure of CCC, the inhibition of chloride migration is improved correspondingly.

High Temperature

Figure 3d shows the retention rate of compressive strength (i.e., the ratio of strength at the current temperature to the initial strength at room temperature) for different

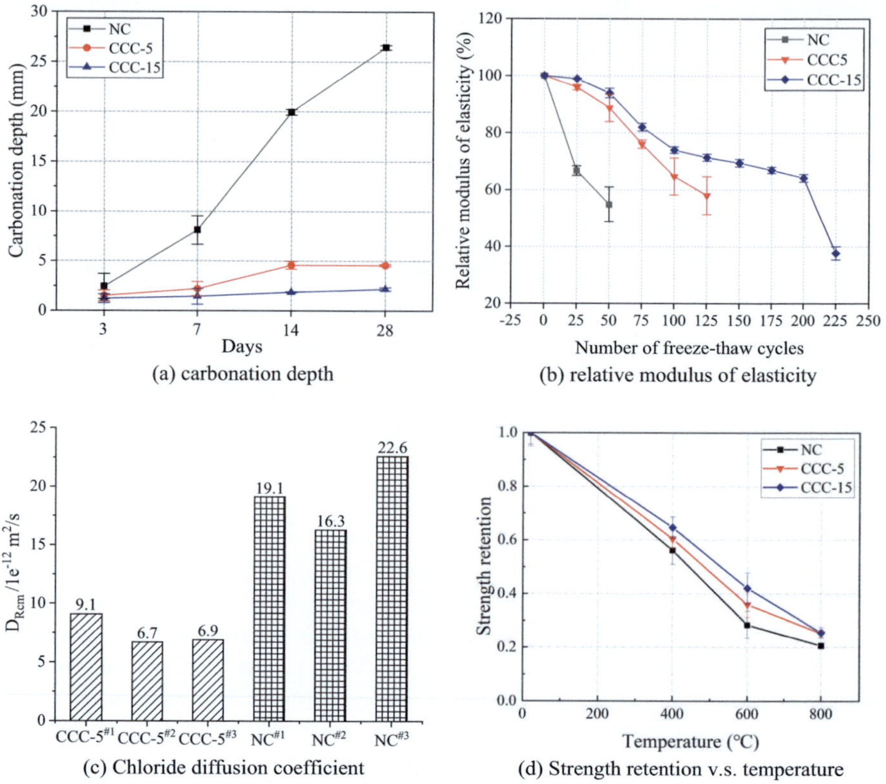

Fig. 3 Durability of CCC

specimens. The strength retention of CCC is larger than NC: compared to NC specimens at 600 °C, the strength retention of CCC increased by 26.4% and 47.9% with casting pressures of 5 MPa and 15 MPa, respectively.

2.3 Cement Saving

Figure 4 presents the experimental results of strength enhancement and cement reduction contributed by CCT. For cement content larger than 253 kg/m^3, the strength enhancement of CCC is significant, as shown in Fig. 4a. The improvement of strength is the basis for the reduction of cement in concrete for the same strength requirement. To achieve the equivalent concrete strength, the cement content in CCC is much lower than that in NC, as shown in Fig. 4b. Thus, CCT can remarkably reduce cement content in concrete production without loss of strength and durability.

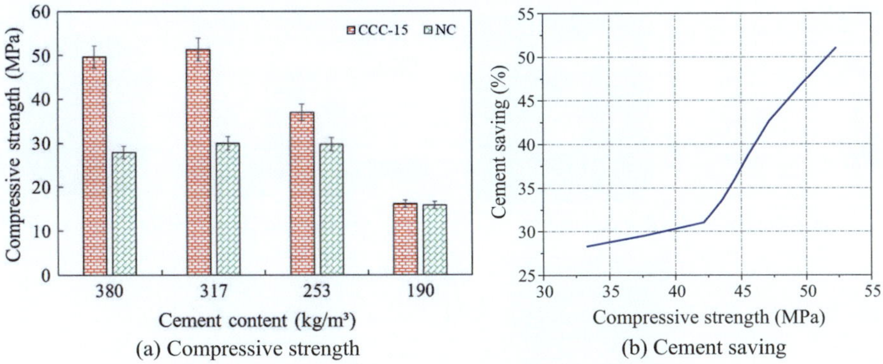

Fig. 4 Strength enhancement and cement reduction contributed by compression cast

3 Marine Floating Structure with CCC

3.1 Bionic Honeycomb Floating Unit

Based on the concept of a bionic honeycomb, hexagonal floating units constructed by CCC are proposed by the author's research group, and the optimal geometric configurations (small-scale unit and medium-large unit) are obtained for different sea states. As illustrated in Fig. 5, the new type of floating unit exhibits several advantages: (a) the floating units are constructed by CCC, which is more economical in manufacturing and maintenance compared to traditional steel structures; (b) different with square modules that are confined to linear expansion, hexagonal floating units allow for dynamic and flexible expansion without gap according to specific requirements (see Fig. 5c); (c) the floating unit is designed as a prefabricated structure, significantly reducing construction and installation time as well as associated costs; (d) the function of diving and floating can be realized for the floating unit as needed.

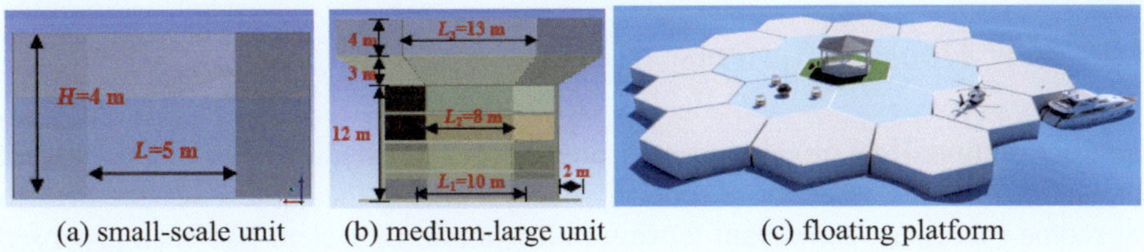

Fig. 5 Honeycomb floating unit and platform

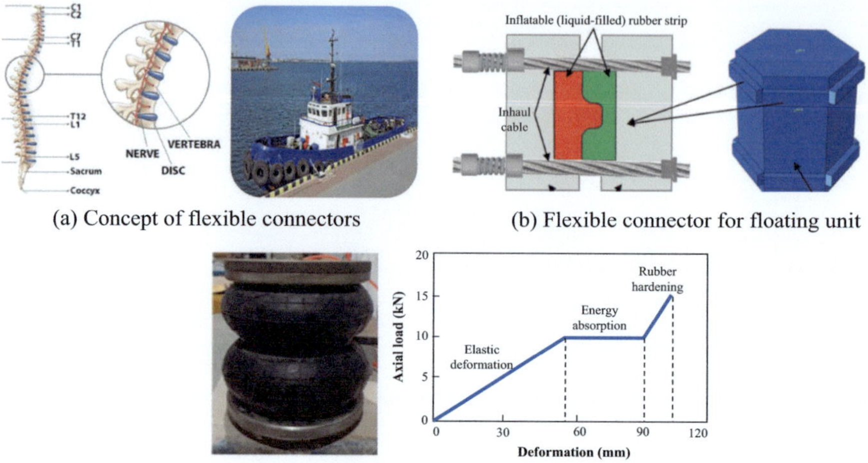

(a) Concept of flexible connectors (b) Flexible connector for floating unit

(c) Experimental results

Fig. 6 Design of flexible connectors for floating units

3.2 *Flexible Floating Unit Connectors*

The existing connections between float units mainly include rigid and flexible connectors. Stress concentration may exist for rigid connectors while most flexible connectors are mechanical mechanisms with inconvenient disassembly and limited buffer capacity. Based on the concept of the intervertebral disc, a new kind of flexible connector is developed with inflatable (or fluid) rubber strips, as shown in Fig. 6. By selecting the fluid type and pressure inside the rubber strip, and setting the fluid in and out of the regulator and the automatic jump valve system, the stiffness, damping and maximum pressure of the connector can be controlled for different working conditions. The proposed ductile connector significantly reduces impact loads.

3.3 *New Mooring Systems*

For a conventional taut mooring system, due to the cyclic loading of waves, significant fluctuations in tension occur in mooring lines, reducing the cable's service life. To overcome this issue, a constant tension pulley mooring system has been proposed for floating structures. As shown in Fig. 7, the proposed mooring system consists of clump weight, pulley, and wires. The advantages of the pulley mooring system are obvious: (1) the force in the mooring line is controlled by weight, avoiding line breaking and reducing anchoring costs; (2) the fluctuation of tension force is small, increasing the fatigue life of the mooring system.

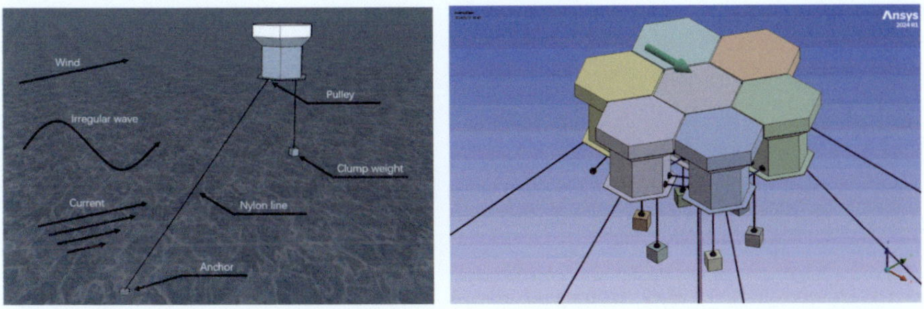

(a) Pulley mooring system for floating unit (b) Pulley mooring system for floating platform

Fig. 7 Design of the pulley mooring system

3.4 New Ocean Energy Harvesting Systems

Based on novel piezoelectric nanomaterials, a new ocean energy harvesting system is proposed [10], as shown in Fig. 8. The system consists of four energy trapping layers: wind and solar energy capture layer on the upper surface; wave energy trapping layer in the middle part; wave energy trapping layers at the bottom and tail. The new floating structure can effectively convert the natural energy of the sea (wave energy, tidal energy, wind energy, solar energy, etc.) into electrical energy storage, and realize the energy self-sufficiency of the floating structure.

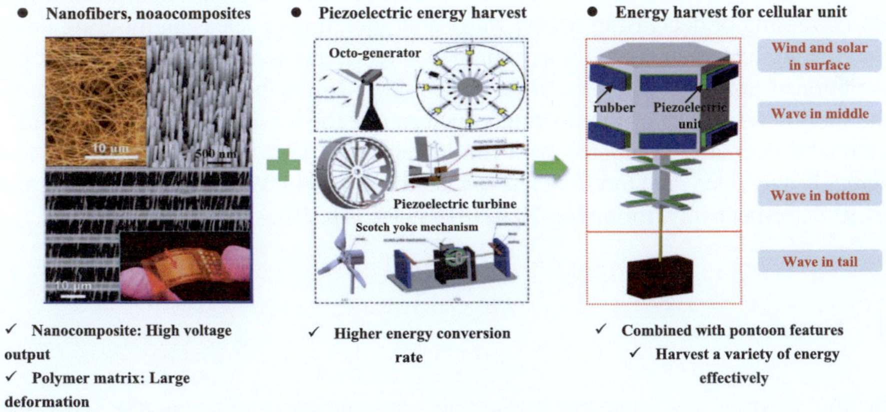

Fig. 8 Design of energy harvesting system for floating structures

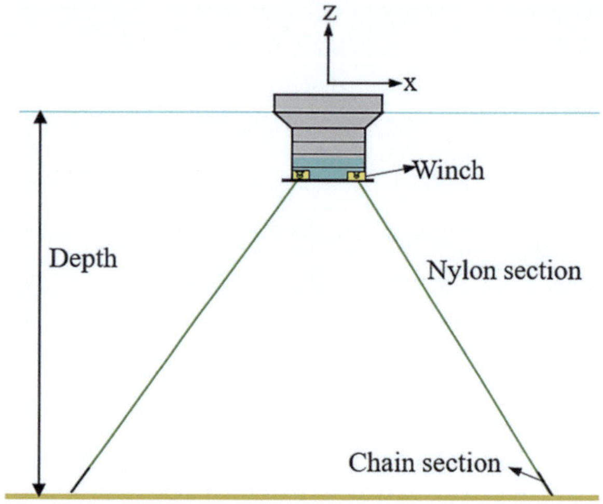

Fig. 9 Submerge for floating units

3.5 *Method for Surviving in Harsh Sea Conditions*

The survivability of floating structures in harsh sea conditions is a critical factor affecting their safety, stability, and large-scale applicability. A method for surviving in harsh sea conditions is developed for floating structures combined with a submersible and floating mooring system. As shown in Fig. 9, the surviving system consists of a taut mooring system and a buoyancy adjustment device (i.e., ballast tank and winch). By increasing or decreasing ballast water and retracting or releasing mooring lines, the floating and submersible states can be switched autonomously to encounter the variation of sea states. The simulation results show that, as the diving depth increases, the motion responses of the floating structure and the tension in the mooring lines gradually decrease: compared with the surface state (S0), the maximum value of surge, heave, pitch, and mooring tension decrease by 94.43%, 66.63%, 91.09%, and 64.91%, respectively, for diving 20 m (SIV) (see Fig. 10).

4 Conclusions

In this study, a new concrete technology—compression-cast concrete (CCC) is introduced, and the comprehensive properties of CCC are investigated, including mechanical properties, microstructures, durability, and carbon reduction benefits. Compared with normal concrete, CCC has much higher strength, better mechanical performance, better environmental impact, and better durability in terms of reduced carbonization depth, less damage under freeze–thaw cycles, smaller chloride penetration depth, as well as greater fire resistance. In addition, this novel technology has been applied to floating structures with a biomimetic shape. The required issues for

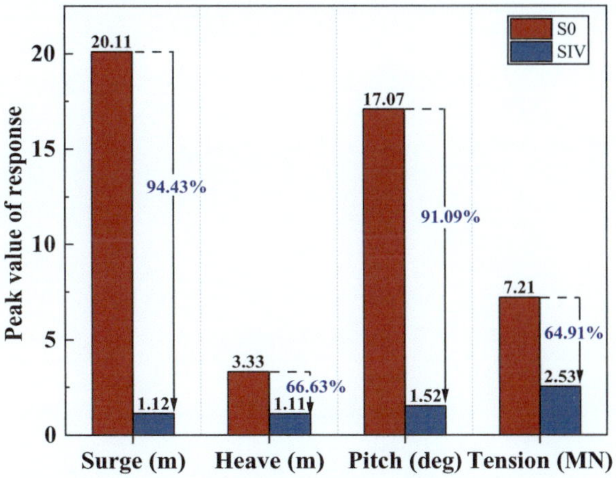

Fig. 10 Comparison of response before and after diving

the design of the floating structures have been investigated including bionic honeycomb floating units, flexible connectors for units, new mooring systems, ocean energy harvesting systems, and methods for surviving in harsh sea conditions. These new technologies provide effective approaches for the sustainable development of floating structures.

References

1. Tang B, Wu H, Wu YF (2024) Evaluation of the carbon reduction benefits of adopting the compression cast technology in concrete components production based on LCA. Resour Conserv Recycl 208:107733
2. Yang KH, Jung YB, Cho MS et al (2015) Effect of supplementary cementitious materials on reduction of CO_2 emissions from concrete. J Clean Prod 103:774–783
3. Scrivener K, Martirena F, Bishnoi S et al (2018) Calcined clay limestone cements (LC3). Cem Concr Res 114:49–56
4. Li C, Sun H, Li L (2010) A review: The comparison between alkali-activated slag (Si + Ca) and metakaolin (Si + Al) cements. Cem Concr Res 40(9):1341–1349
5. Wu YF, Kazmi SMS, Munir MJ et al (2020) Effect of compression casting method on the compressive strength, elastic modulus and microstructure of rubber concrete. J Clean Prod 264:121746
6. Kazmi SMS, Munir MJ, Wu YF (2021) Application of waste tire rubber and recycled aggregates in concrete products: a new compression casting approach. Resour Conserv Recycl 167:105353
7. Wang X, Wang J, Kazmi SMS et al (2022) Development of new layered compression casting approach for concrete. Cement Concr Compos 134:104738
8. Yuan F, Li H, Wu Y (2024) Axial compressive behavior of FRP-confined compression-cast recycled aggregate concrete. J Compos Constr 28(3):04024009
9. Wang X, Munir MJ, Wu YF (2023) Effect of mix design parameters on the behavior of compression cast concrete. J Build Eng 76:107399
10. Liu J, Bao B, Chen J et al (2023) Marine energy harvesting from tidal currents and offshore winds: a 2-DOF system based on flow-induced vibrations. Nano Energy 114:108664

Design Methodology of Bamboo-Raft-Type Floating Structure (BRT-FS)

Chongfeng Xie and Peng Feng

Abstract The authors have developed a novel floating structure—bamboo-raft-type floating structure (BRT-FS). However, the mechanical properties of BRT-FS differ from those of traditional VLFS. The load-bearing capacity and failure mode of BRT-FS are primarily controlled by longitudinal properties, exhibiting strong anisotropy. Facing these unique mechanical properties, it is urgent to establish a new design methodology to guide the engineering application of BRT-FS. This paper surveys the load combinations for concrete floating structures recommended by various standards and establishes load case combinations suitable for the BRT-FS. Based on these three limit states, serviceability limit state (SLS), ultimate limit state (ULS) and accidental limit state (ALS), design methodologies for BRT-FS are proposed.

Keywords BRT-FS · Load combinations · Design methodology

1 Introduction

The demand for floating structures is rapidly increasing, but many experiences from land-based infrastructure construction are not applicable to the more challenging offshore environments. Compared to onshore construction, offshore construction faces unique challenges, such as harsh corrosive environments, extreme load conditions, restricted construction conditions, and high difficulty in maintenance and repair [1–3]. Therefore, it is necessary to explore and study new construction materials and structural configurations applicable to the offshore environments.

The authors have developed a novel floating structure—bamboo-raft-type floating structure (BRT-FS) [4]. This novel floating structure employs lightweight aggregate concrete and fiber reinforced polymer (FRP) as the primary materials for load-bearing components. It includes four main components: concrete tubes, prestressed

C. Xie · P. Feng (✉)
Key Laboratory of Civil Engineering Safety and Durability of Ministry of Education, Department of Civil Engineering, Tsinghua University, Beijing, China
e-mail: fengpeng@tsinghua.edu.cn

X. L. Zhao et al. (eds.), *Proceedings of the Fourth World Conference on Floating Solutions: WCFS 2024; 2–4 December; Hong Kong*, Lecture Notes in Civil Engineering 597, https://doi.org/10.1007/978-981-96-4569-5_66

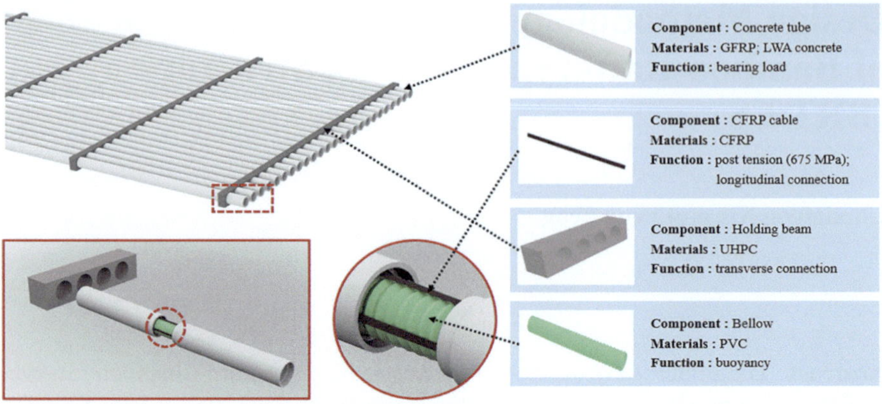

Fig. 1 Main components of BRT-FS

carbon fiber reinforced polymer (CFRP) cables, holding beams, and bellows, as shown in Fig. 1. Compared to traditional very large floating structures (VLFS), this structure offers advantages such as high design flexibility, outstanding durability, reduced cracking risk, high construction and transportation efficiency, and lower overall lifecycle costs.

BRT-FS exhibits completely different mechanical properties in the longitudinal and transverse directions, indicating strong anisotropy. Due to the unique mechanical property, it is urgent to establish a new design methodology to guide the engineering application of BRT-FS. This paper proposes design methodology for BRT-FS under three limit states: serviceability limit State (SLS), ultimate limit state (ULS), and accidental limit state (ALS).

2 Design Methodology

2.1 Load Condition Combination Selection

An investigation was conducted into the standards established by various domestic and international organizations and associations, including China Classification Society (CCS), the International Organization for Standardization (ISO), Det Norske Veritas (DNV), and the European Union (EN). It was found that CCS standards do not explicitly specify the load partial factors for floating platforms. Instead, they provide recommendations primarily on load types and load combinations. In contrast, international standards, based on ISO standards, offer detailed suggestions for load combinations under different limit states. Based on the aforementioned standards, the fundamental load types selected for BRT-FS design include hydrostatic pressure,

Table 1 Load combinations and partial factors

Ultimate state	Permanent load (hydrostatic pressure)	Variable load (topside load)	Environmental load (wave load, current load, and ice load)
SLS	1.0	1.0	1.0
ULS-1	1.3	1.3	0.7
ULS-2	1.0	1.0	1.3
ALS	1.0	1.0	1.0

wave load, current load, topside load, and ice load. The load combinations were considered according to Table 1.

2.2 SLS Design

The design methodology under serviceability limit state (SLS) is illustrated in Fig. 2. The hydroelastic response of the structure is analyzed based on the common wave conditions. The acquired vertical deflection amplitude is used to assess comfort levels. The acquired stress range in the CFRP plates is used to evaluate the long-term performance. The purpose of this step is to ensure the comfort meet personnel requirements and to guarantee that the mechanical performance of CFRP plates does not degrade excessively during service.

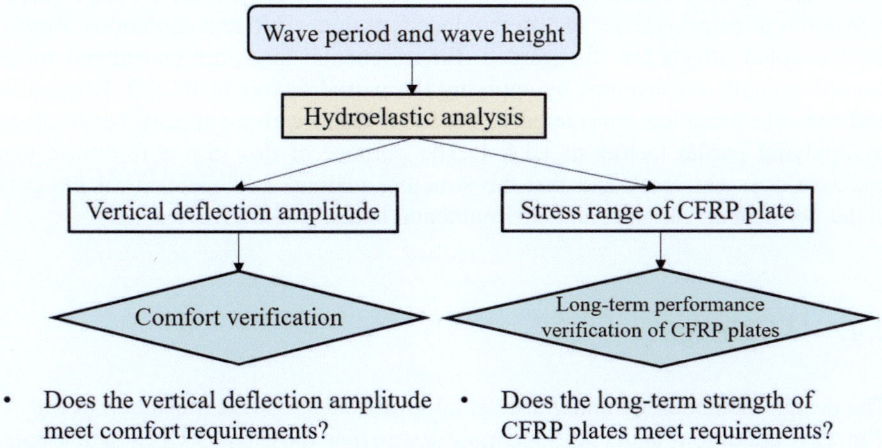

Fig. 2 Design methodology of SLS

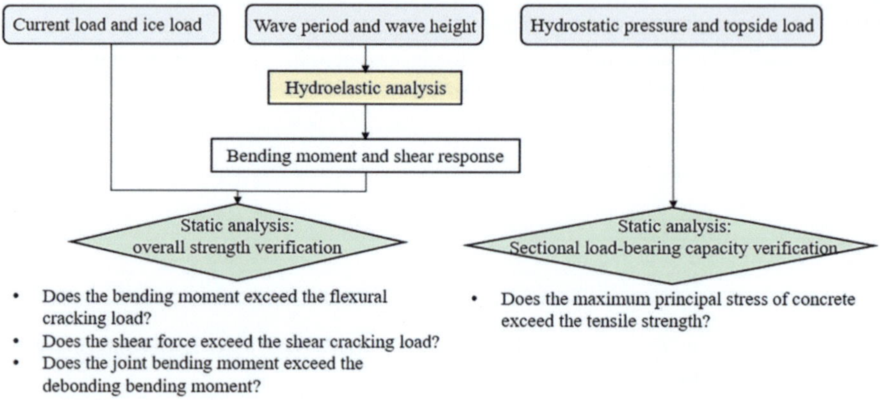

Fig. 3 Design methodology of ULS

2.3 ULS Design

The design methodology under ultimate limit state (ULS) is illustrated in Fig. 3. Hydroelastic analysis is conducted using wave conditions with a recurrence interval of 100 years. The acquired bending moments and shear responses are used to verify the overall strength of the concrete tubes. When lateral loads from current and ice act on the structure, it becomes a critical state requiring verification of the overall strength of the edge concrete tubes. Hydrostatic pressure and topside load have minimal impact on the overall structural performance but affect sectional behavior, thus requiring verification of sectional load-bearing capacity. Concrete experiences maximum principal stresses in different locations under different conditions. Hence, their coupled effects are disregarded. Environmental loads are considered in all overall strength verifications by applying the partial factors of ULS-2. Permanent and variable loads are considered in sectional load-bearing capacity verifications by applying partial factors of ULS-1. The purpose of this step is to ensure that concrete does not crack and that the structure maintains its mechanical integrity under permanent, variable, and environmental loads.

2.4 ALS Design

The design methodology under accidental limit state (ALS) is illustrated in Fig. 4. The wave conditions under ALS are more severe than those under ULS. In this state, the concrete tubes crack. However, the speed of crack propagation must be limited, meaning the cracking process must be controllable. A specific wave period is selected, and the cracking wave height H_{cr} is determined based on the hydroelastic response. At this moment, the bending moment amplitude of each concrete tube is $\mathbf{M}_{cr}$. The historical maximum bending moment amplitude of each concrete tube is denoted as $\mathbf{M}_{\text{max}}$. Initially, the wave height is set as $H = H_{\text{cr}}$ and $\mathbf{M}_{\text{max}} = \mathbf{M}_{\text{cr}}$. H is increased by 0.2 m for each step, and the bending moment response $\mathbf{M}$ is calculated for each wave height. If the bending moment amplitude exceeds the flexural cracking load of concrete tubes, the shear amplitude exceeds the shear cracking load, or the joint bending moment exceeds the debonding bending moment, the structure is considered to have failed and the calculation is stopped. Otherwise, $\mathbf{M}$ is used to update $\mathbf{M}_{\text{max}}$. The bending stiffness of the concrete tubes is corrected based on the stiffness degradation formula for cracked concrete tubes. The corrected stiffness is used for the hydroelastic response calculation of the next wave height. Each concrete tube has two joint bending moments. Since the overall bending stiffness of the concrete tube is controlled by the minimum stiffness, the larger joint bending moment is used for the stiffness correction of the cracked concrete tubes. The purpose of this step is to prevent progressive damage under extreme wave conditions, ensuring the safety of life and property. After enduring harsh conditions, the structure can continue to be used following partial repair and replacement.

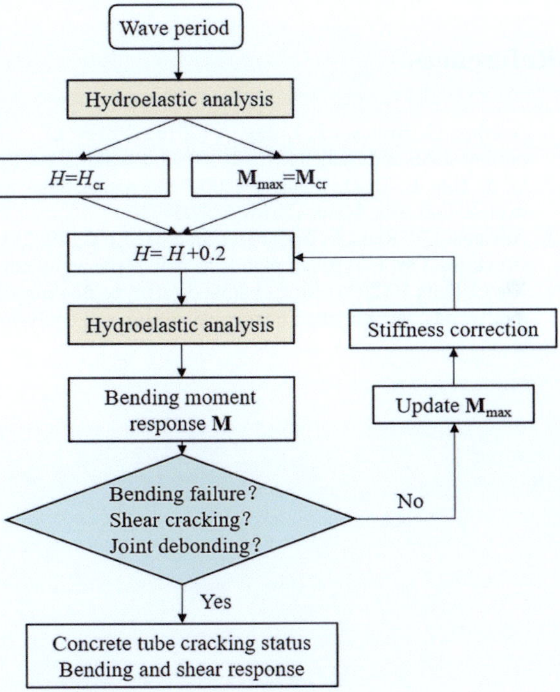

Fig. 4 Design methodology of ALS

3 Conclusion

This paper first conducted a survey of existing standards and found that the standard of CCS does not specify detailed partial factors for floating platforms. Instead, it provides recommendations only for load types and load combination methods. In contrast, the international standards offer detailed recommendations for load combinations under different limit states based on ISO guidelines. Based on the survey results, suitable load case combinations for the BRT-FS are selected. Considering the unique mechanical property of BRT-FS, design methodologies for the three limit states (SLS, ULS, and ALS) are proposed. Under SLS, verification for comfort and CFRP plate durability are performed. Under ULS, overall strength verification is conducted for hydroelastic response, ocean current loads, and ice loads, while sectional load-bearing capacity verification is performed for hydrostatic pressure and topside loads. Under ALS, the overall strength verification is conducted based on the post-cracking bending stiffness of the concrete tubes, aiming to avoid progressive damage under extreme wave conditions.

References

1. Adedipe O, Brennan F, Kolios A (2016) Review of corrosion fatigue in offshore structures: present status and challenges in the offshore wind sector. Renew Sustain Energy Rev 61:141–154
2. Li Y, Hou B, Li H, Zhang J (2004) Corrosion behavior of steel in Chengdao offshore oil exploitation area. Mater Corros 55:305–310
3. Amaechi CV, Reda A, Butler HO, Ja'e IA, An C (2022) Review on fixed and floating offshore structures. Part I: Types of platforms with some applications. J Mar Sci Eng 10:1074
4. Xie C, Feng P (2024) Novel bamboo-raft-type floating structure (BRT-FS) assembled by FRP reinforced concrete tubes: conceptual design and analysis. Eng Struct 318:118705

Durability Evaluation of Maritime Concrete Served in Ocean Environment

Qi Zhao, Keitai Iwama, and Xiao Lin Zhao

Abstract Marine concrete is an economical construction material for durable floating structures. However, marine concrete may experience severe damage during service in the ocean than inland environment. This paper assessed the durability of marine concrete exposed to actual ocean environment after 20 years of service. The chloride ions distribution and CO_2 transport in concrete with varying grades were theoretically simulated using the past 22 years of weather records. The chloride penetration depth and concrete carbonation front were then analyzed and validated by field data. The results indicate that the chloride content and carbonation are highly linked to the microstructures of concrete and the weather conditions at service sites. As W/C ratios decrease, concrete usually becomes denser, and their pore structures are changed. Subsequently, moisture, CO_2 and chlorides from ocean environment become more difficult to penetrate concrete than ever before. The results could be used to determine the critical values of harmful ions and the corrosion initiation time for steel-reinforced concrete structures in service under ocean environment.

Keywords Marine concrete · Durability · Chloride penetration · Carbonation · Ocean environment

1 Introduction

Compared to metallic materials, marine concrete is a preferable choice for massive construction of floating structures due to its low cost and wide availability. Unlike steel, concrete is chemically stable and durable when exposed to marine environments. However, in contrast to inland environments, concrete structures served in

Q. Zhao (✉) · X. L. Zhao
Research Institute for Land and Space, The Hong Kong Polytechnic University, Hong Kong, China
e-mail: qi-cn.zhao@polyu.edu.hk

K. Iwama
Institute for Multidisciplinary Sciences, Yokohama National University, Yokohama, Japan

X. L. Zhao et al. (eds.), *Proceedings of the Fourth World Conference on Floating Solutions: WCFS 2024; 2–4 December; Hong Kong*, Lecture Notes in Civil Engineering 597, https://doi.org/10.1007/978-981-96-4569-5_67

Table 1 Concrete mix with varying grades

Concrete grades	W/C	Cement (kg/m^3)	Sand/Gravel ratio
C35	0.39	370	0.67
C40	0.37	400	0.64
C45	0.35	435	0.61
C50	0.33	470	0.56

Note: W/C denotes the water-cement ratio

ocean environment, usually undergo more coupled interactions from harmful ions, high relative humidity (RH) and frequent wet-dry cycling, of which the chloride salts are the most dangerous for the steel reinforcements in concrete.

Current research has mainly focused on the chloride penetration into concrete through rapid chloride migration (RCM) testing in the laboratory [1]. However, evaluating the chloride content and carbonation process in concrete under actual environments is of significance to assess the long-term durability of maritime concrete structures. Predicting the chloride profile in concrete after decades of exposure remains a challenge. This paper is a part of our recent work to address this knowledge gap.

2 Materials and Dataset

2.1 Concrete

Four concrete grades were used to conduct the simulations. The micro-structures in different concrete vary, and they have a significant influence on the chemical resistance to the environmental attack. Detailed information on concrete mix is listed in Table 1.

2.2 Dataset

The field data from the dry-dock at Pearl Harbor after 18 years of service, including the chloride penetration depth and carbonation front were obtained [2]. The test chloride depth from extracted cores was 39.4 $\pm$ 12.5 mm (about 0.49 in), the carbonation depth was 5.8 $\pm$ 3.6 mm (about 0.14 in). The weather records at Oahu from 2001 to 2023 (i.e., temperature, relative humidity (RH) and gas CO_2), were collected to perform the simulation. We tried to use the true weather data to conduct the analysis and validated by the experimental results.

Table 2 Governing equations regarding mass transport in concrete [3, 4]

Items	Governing equations	Eq. Number
Hydration	$H_i = C_i \cdot H_{i,T_o}(Q_i)\exp\left[-\frac{E_i}{R}\left(\frac{1}{T}-\frac{1}{T_0}\right)\right]$	(1)
Micro-pore structures	$\phi(r) = \phi_{lr} + \phi_{gl}\left[1-\exp\left(-B_{gl}r\right)\right] + \phi_{cp}\left[1-\exp\left(B_{cp}r\right)\right]$	(2)
Moisture transport	$\frac{\partial(\rho_l\phi S)}{\partial t} + divJ(K\nabla P_l) + Q = 0$	(3)
	$K_l = \frac{\rho_l\phi^2}{50\eta}\left(\int_0^{r_c} rdV\right)^2$	(4)
	$K_v = \frac{\phi\rho_v D_0}{2.5}[(1-S)K(h)]\left(\frac{M_w h}{\rho_l RT}\right)$	(5)
CO_2 transport	$\frac{\partial}{\partial t}\left[\phi\left\{(1-S)\rho_{gCO_2}\right\} + S\rho_{dCO_2}\right] + divJ_{CO_2} - Q_{CO_2} = 0$	(6)
	$D_{gCO_2} = \frac{\phi D_0^{gCO_2}}{\Omega}\frac{(1-S)^n}{(1-l_m)/\{2(r_m-t_m)\}}$	(7)
	$D_{dCO_2} = \frac{\phi D_0^{dCO_2}}{\Omega}S^n$	(8)
Chloride transport	$\frac{\partial}{\partial t}(\phi SC_{Cl}) + divJ_{Cl} - Q_{Cl} = 0$	(9)
	$J_{Cl} = -\frac{ØS}{\Omega}D_{Cl}\nabla C_{Cl} + ØS\boldsymbol{u}C_{Cl}$	(10)
	$\boldsymbol{u} = \frac{K\nabla P}{\rho ØS}$	(11)
Concrete strength	$f_c' = f_\infty'\left\{1-\exp\left(-\alpha D_{hyd.out}^{\beta}\right)\right\}$	(12)

Note: Please refer to [3, 4] for more information on the multi-scale transport theory

3 Methodology and Simulation

The moisture, chloride and CO_2 coupled interaction and transport in concrete can be calculated according to the multi-scale mass transport theory, as shown in Table 2. More detailed information can refer to Maekawa et al. [3, 4]**.** The mass transport was calculated using multiscale physico-chemo-mechanical coupling platform.

4 Results and Discussion

4.1 CO_2 Ingress and Carbonation Development

Carbon dioxide (CO_2) from the atmosphere comes into the micro-connected pores of concrete, and the dissolved CO_2 reacts with the hydrated cement products and produces calcium carbonates, decreasing the alkalinity of concrete pore solution. This process is called carbonation, which makes the imbedded steels fragile to corrosion.

In general, when the pH values of concrete pore solutions decrease below 9, the concrete block is usually thought to be carbonated. The pH evolutions in different concrete C35, C40, C45 and C50 are shown in Fig. 1.

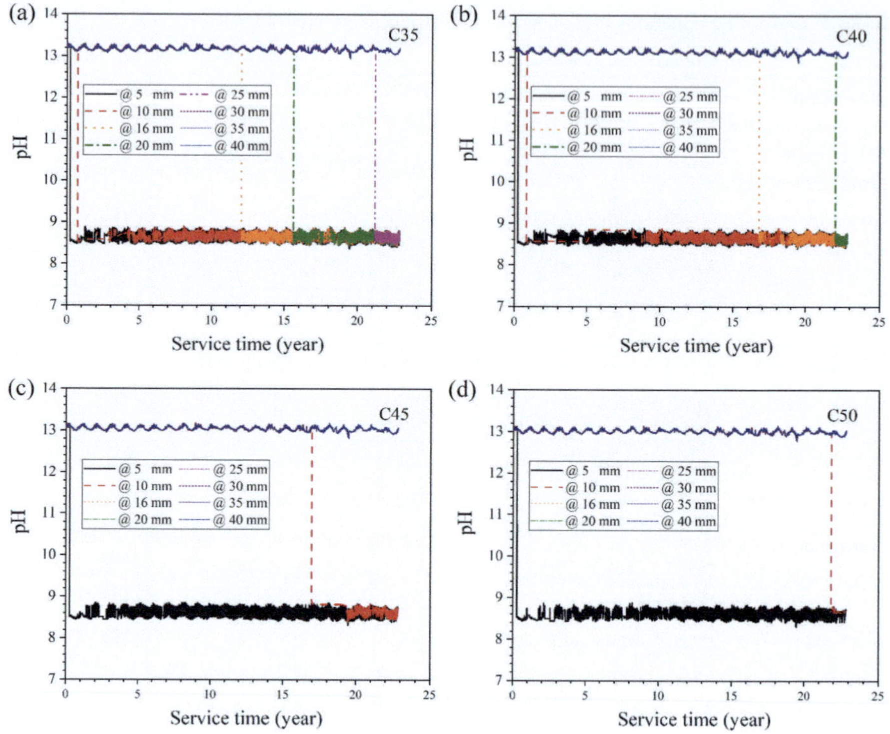

Fig. 1 Concrete carbonation evolution: **a** C35, **b** C40, **c** C45 and **d** C50

From Fig. 1a, the C35 carbonation depth was between 20 mm and 25 mm after 18 years. After only 1 year of service, the carbonated C35 concrete reached 10 mm in depth. As service time increased, for example, the carbonation depth increased to 20.1 mm after 15.6 years. It can be predicted that a 25 mm-in-depth of C35 concrete will be carbonated.

The carbonation depth in C50 is less than 10 mm after 18 years, validated by the field data from extracted samples. Compared to the results above in Fig. 1a–d, it can be concluded that concrete with higher concrete grade has better resistance to carbonation. As concrete grade increases, the carbonation depth of concrete with greater grade becomes smaller than concrete with lower grade after the same exposure period. This is because the higher concrete grades were adopted, the denser concrete and smaller pore structures became.

4.2 Chloride Penetration

The chloride distributions in concrete C35, C40, C45 and C50 during service were simulated, as shown in Fig. 2. From Fig. 2, the chloride contents in concrete at 5 mm depth changed rapidly as the external environment, which can be found in all cases. The chloride value of C35 concrete ranged from 0.02 to 2.92% during the 23 years of service. However, as the depth increased, the fluctuations decreased, which are consistent for the four cases.

It should be noted that C35 concrete had the maximum chloride content near the surface when compared with other three concrete, as illustrated in Fig. 2a–d. In general, the maximum chloride content at 5 mm depth decreased with the increasing of concrete grades. This implies that the chemical resistance is significantly affected by the density and pore structures of concrete.

According to the measured chloride penetration depth using silver nitrate (NT Build 492), the average chloride depth from 16 cores is 39.4 mm with a COV of 31.7%. The free chloride concentration at which the color changes in NT Build 492 is assumed as 0.07 mol/L in pore solution for OPC. However, the single value rather than actual concentrations will lead to significant error [5]. There is no consensus on

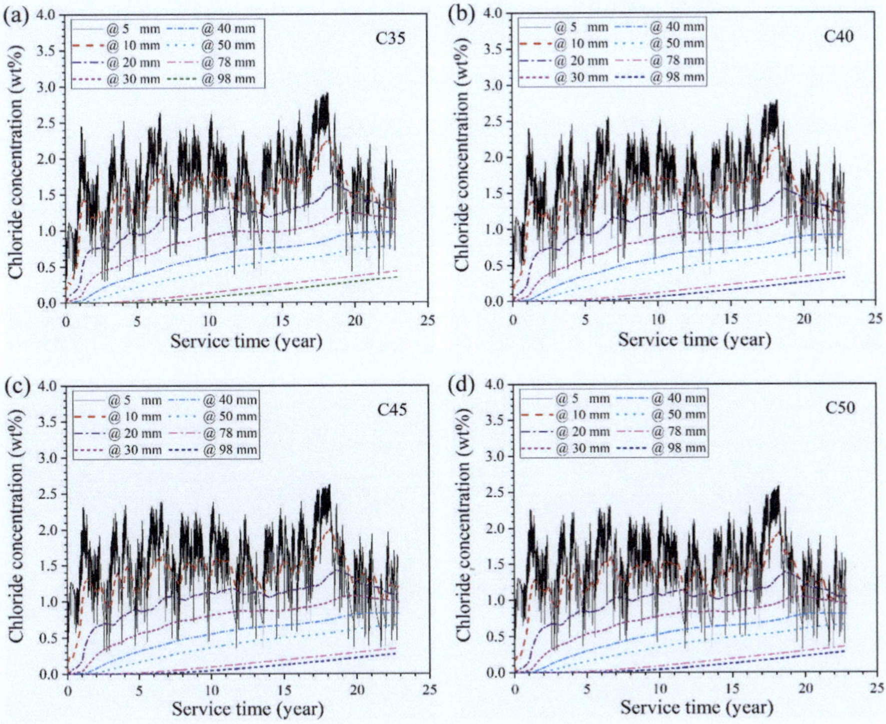

Fig. 2 Chloride content in concrete: **a** C35, **b** C40, **c** C45 and **d** C50

the chloride content by cement weight at color change [6]. In practice, the suggested total chloride range is between 0.5 and 1.0% by the mass of binder [6]. Based on this rule, the simulation results agreed well with test data.

4.3 Moisture Transport

The moisture distributions in difference concrete are summarized in Fig. 3. The RH in pore structures of the near-surface concrete is periodically changing with the atmospheric environment. As concrete depth increases, the RH fluctuations gradually decrease. Furthermore, the RH changes also decreased with the increasing of concrete grades, validated by the data at 98 mm in four cases.

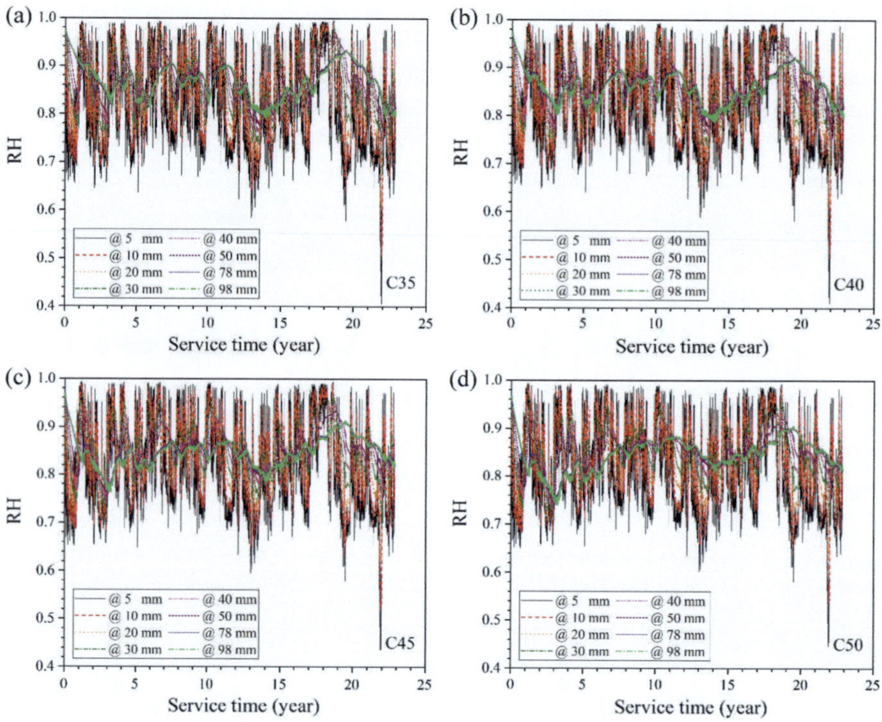

Fig. 3 RH changes in concrete: **a** C35, **b** C40, **c** C45 and **d** C50

5 Conclusion Remarks

This paper assessed the mass transport in several grades of concrete exposed to actual ocean environment for more than 20 years, including chloride ions, CO_2, and RH variations. Some valuable conclusions can be drawn as follows.

- The chloride content and carbonation are highly linked to the microstructures of concrete and the weather conditions at service sites.
- As W/C ratios decrease, concrete usually becomes denser, and their pore structures are changed. Subsequently, moisture, CO_2 and chlorides from ocean environment become more difficult to penetrate concrete than ever before.
- To increase the durability of maritime concrete structures, marine concrete with higher grade should be adopted in construction.
- The chloride profile and carbonation front in concrete can be used to determine the corrosion initiation time for steel-reinforced concrete structures in service under ocean environments. Further analyses are needed to obtain more accurate results.

Acknowledgements The authors thank Dr. Sandra Rayne for her kind guidance during weather data collection. The first author is sponsored by the Innovation and Technology Fund (ITF) Research Talent Hub of Hong Kong.

References

1. Zhao Q, Iwama K, Dai JG, Liu J, Zhang D, Maekawa K, Zhao XL (2024) Deterioration modelling of GFRP-reinforced cement-based concrete infrastructure in service under the natural inland atmospheric environment. Constr Build Mater 447:138005
2. Ramanathan S, Benzecry V, Suraneni P, Nanni A (2021) Condition assessment of concrete and glass fiber reinforced polymer (GFRP) rebar after 18 years of service life. Case Stud Constr Mater 14
3. Maekawa K, Ishida T, Kishi T (2003) Multi-scale modeling of concrete performance. J Adv Concr Technol 91–126
4. Maekawa K, Ishida T, Kishi T (2008) Multi-scale modeling of structural concrete. CRC Press
5. He F, Shi C, Yuan Q, An X, Tong B (2011) Calculation of chloride concentration at color change boundary of $AgNO_3$ colorimetric measurement. Cem Concr Res 41:1095–1103
6. Drenkard H, Fischer C, Sauer V, Gehlen C (2023) On the chloride distribution in concrete and mortar samples after an RCM test. Materials (Basel) 16

Fracture Phase Field Method Based Crack Propagation Prediction for Circumferential Surface Cracks in Subsea Pipelines

Yangyang Qu and Nian-Zhong Chen

Abstract Subsea pipelines are normally subjected to complex environmental loads and structural fracture has been one of predominant failure modes. In recent years, the fracture phase field theory, based on Griffith's variational theory of fracture, has proven to be a valuable tool for simulating material fracture behavior, providing robust capabilities for simulating crack propagation in subsea pipelines. Therefore, within the framework of the fracture phase field method, this study aims to investigate the influence of circumferential surface crack defects on subsea pipelines under axial loads. In this paper, a crack propagation prediction is performed for circumferential surface cracks in subsea pipelines. The corresponding finite element models are then developed using ABAQUS user subroutines UEL and UMAT. A staggered algorithm is employed in the finite element analysis to predict the path of crack propagation in subsea pipelines with circumferential surface cracks. Furthermore, a comprehensive study is conducted to investigate the influence of parameters associated with the initial sizes of circumferential surface cracks on crack propagation and fracture behavior in subsea pipelines. In addition, critical factors affecting the crack propagation performance of subsea pipelines are explored, and insights are gained to improve the structural integrity.

Keywords Subsea pipelines · Fracture phase field · Crack propagation

Y. Qu · N.-Z. Chen (✉)
School of Civil Engineering, Tianjin University, Tianjin, China
e-mail: nzchen2018@hotmail.com

N.-Z. Chen
State Key Laboratory of Hydraulic Engineering Intelligent Construction and Operation, Tianjin University, Tianjin, China

X. L. Zhao et al. (eds.), *Proceedings of the Fourth World Conference on Floating Solutions: WCFS 2024; 2–4 December; Hong Kong*, Lecture Notes in Civil Engineering 597, https://doi.org/10.1007/978-981-96-4569-5_68

1 Introduction

In recent years, as the global demand for energy has continued to escalate, there has been a corresponding increase in the number of pipelines installed [1]. Statistical data reveal that defects in pipe materials and welds are often cited as the primary causes of pipeline incidents [2]. Particularly, the challenges associated with quality control in onsite welding and the complexity of defect detection and rectification frequently lead to the development of cracks during pipeline operation, sometimes resulting in severe incidents such as leaks or ruptures. Such events not only threaten the functionality of the pipelines but may also precipitate significant environmental and economic repercussions. Consequently, a thorough understanding of the crack propagation process is essential for ensuring the safety and reliability of marine engineering structures.

The advancement of computer technology has provided robust technical support for solving numerical simulation issues related to crack propagation prediction, enabling numerous researchers to utilize various numerical analysis or computational mechanics techniques to study the crack propagation process in pipelines. For instance, He et al. [3] employed the Extended Finite Element Method (XFEM) to conduct numerical simulations of pitting crack propagation for surface cracks in monopile foundation; meanwhile, Song and Chen [4] developed an improved cyclic cohesive zone model and predicted the initiation of low-cycle fatigue cracks in submarine pipelines. Currently, the primary methods adopted for numerical calculations of fracture in practical engineering applications include the CZM [5] and the XFEM [6]. The cohesive zone model can simulate the continuous process from crack initiation, evolution to final material fracture [7], however, its application is limited under complex loading conditions and in handling multidirectional crack propagation because it requires the predefined fracture paths using cohesive elements. On the other hand, the XFEM, an improvement on the traditional finite element method, is quite effective in predicting crack propagation issues. Nonetheless, its approach involves treating cracks as mathematical discontinuities, which may encounter difficulties in simulating the intersection of cracks or their propagation in complex materials [8].

Based on the theories of Griffith [9] and Irwin [10], the phase field method conceptualizes crack propagation as an issue of energy balance stability, occurring when the energy release rate reaches a specific threshold. The essence of this method lies in the use of a continuous field function, represented by a phase-field parameter, to depict the material's transition from an undamaged state to complete fracture. This parameter illustrates the degree of internal damage within the material. Compared to traditional approaches, the phase field method can automatically track crack paths through the evolution of the phase-field parameter, thus efficiently simulating complex crack propagation behaviors with reduced computational complexity. This method provides a novel and powerful tool for the fracture analysis of engineering structures.

Consequently, this paper, based on the phase field fracture model undertakes a secondary development within the Abaqus and the Fortran compiler. Utilizing a staggered algorithm for solving [11], it investigates the behavior of crack propagation in pipelines under the influence of axial loads.

2 Fracture Phase Field Model

2.1 *Foundation of Phase Field Model*

The phase field method operates by introducing an additional degree of freedom, denoted as d, which represents a continuum ranging from undamaged to completely fractured material. The non-smooth topology of a crack is typically approximated using an exponential function.

$$d(x) = e^{-|x|/l_c} \tag{1}$$

where, l_c represents the length scale parameter, and $d(x)$ characterizes the regularization of the crack topology. Equation (1) which satisfies $d(x) = 1$, and $d(\pm\infty) = 0$ at the limits, is the solution to the homogeneous differential equation within the domain Ω:

$$d(x) - l_c^2 d''(x) = 0 \text{ in } \Omega \tag{2}$$

The expression for the crack surface within the phase field model can be formalized as follows:

$$\Gamma(d) = 1/l_c \cdot I(d) = \int_\Omega \gamma\left(d, d'\right) dV \tag{3}$$

The one-dimensional crack surface density function, denoted as $\gamma\left(d, d'\right)$, characterizes the localized damage within a linear framework. In multiple dimensions, this function is generalized to accommodate the complexity of the spatial domain Ω, and is expressed as follows:

$$\gamma(d, \nabla d) = 1/(2l_c) \cdot d^2 + l_c/2|\nabla d|^2 \tag{4}$$

To integrate the fracture phase field with the deformation problem, Francfort and Marigo [12] proposed an internal potential energy functional that incorporates both strain energy and crack surface energy.

$$\Pi^{\text{int}} = \int_\Omega \psi(\boldsymbol{\varepsilon}(\boldsymbol{u}), d) dV + \int_\Omega g_c \gamma(d, \nabla d) dV \tag{5}$$

where $\int_{\Omega} \psi(\varepsilon(u), d) dV$ the strain energy, $\psi(\varepsilon(u), d)$ the strain energy density, $\int_{\Omega} g_c \gamma(d, \nabla d) dV$ the fracture energy and g_c the critical energy release rate.

The strain energy density is typically expressed in the following isotropic form:

$$\psi(\varepsilon, d) = g(d) \cdot \psi_0(\varepsilon) \tag{6}$$

where $g(d) = (1 - d)^2 + k$ the parabolic degenerate function, k a small number (10^{-7}) responsible for the stability of the solution.

The elastic energy density can be calculated as:

$$\psi_0(\varepsilon) = \frac{1}{2} \varepsilon^T C_0 \varepsilon \tag{7}$$

where C_0 the linear elastic stiffness matrix of the material, ε the strain.

$$\varepsilon = \frac{1}{2} \left[(\nabla u)^T + \nabla u \right] \tag{8}$$

where u the displacement vector.

Due to damage, elastic energy deteriorates with damage. The true stress tensor in the phase field model considering damage is as follows:

$$\sigma = g(d) \sigma_0 = \left[(1 - d)^2 + k \right] C_0 \varepsilon \tag{9}$$

For stiffness there is

$$C = g(d) \cdot C_0 \tag{10}$$

It is evident that the damage variables, as represented by the phase field, exert a direct influence on the stress and stiffness of the material. When the value of the damage variable reaches 1, it signifies complete material failure at that location, resulting in the stress or stiffness of the element reducing to zero.

The external potential energy is expressed as:

$$\Pi^{ext} = P(u) = \int_{\Omega} \bar{\gamma} \cdot u dV + \int_{\partial\Omega} \bar{t} \cdot u dA \tag{11}$$

where γ and t represent volume force and boundary force respectively.

The Staggered Solution approach [11] was employed to address the phase field fracture problem. This method involves first achieving equilibrium in the displacement field through iterative solutions, followed by iterative solutions of the phase field. Both the displacement field and the phase field utilized the Newton–Raphson algorithm for their iterative solutions.

The function for solving the crack topology is:

$$\Pi^d = \int_\Omega \left[g_c \gamma(d, \nabla d) + (1-d)^2 H\right] dV \tag{12}$$

Introduce history variables as part of the model to capture the evolving damage state:

$$H = \begin{cases} \psi_0(\boldsymbol{\varepsilon}) & \psi_0(\boldsymbol{\varepsilon}) > H_n \\ H_n \end{cases} \tag{13}$$

where H_n represents the energy history at the nth step, calculated as the maximum strain energy density, experienced by the material up to that step. The solution to the phase field fracture problem must satisfy the Karush–Kuhn–Tucker (KKT) conditions:

$$\psi_0 - H \le 0, \quad \dot{H} \ge 0, \quad \dot{H}(\psi_0 - H) = 0 \tag{14}$$

With the damage variable d held constant, the displacement field is computed using the equilibrium equations of the mechanical system:

$$\Pi^u = \int_\Omega \left[\psi(\boldsymbol{u}, d) - \overline{\boldsymbol{\gamma}} \boldsymbol{u}\right] dV - \int_{\partial\Omega} \bar{\boldsymbol{t}} \cdot \boldsymbol{u} dA \tag{15}$$

Once equilibrium in the displacement field is achieved, the phase field is updated iteratively. The iterative solution process for the phase field involves calculating a new phase field, d, at time t_{n+1} based on its state at time t_n.

$$d_{n+1} = \arg\left\{ \inf_d \int_\Omega \left[g_c \gamma(d, \nabla d) + (1-d)^2 H\right] dV \right\} \tag{16}$$

The linear approximate solution for the phase field can be expressed as:

$$\boldsymbol{K}_n^d d_{n+1} = -\boldsymbol{r}_n^d \tag{17}$$

where d_{n+1} represents the unknown variable that contains the new value of the phase field at each integration point, $\boldsymbol{r}$ the residual vector, $\boldsymbol{K}$ the tangential stiffness at time t_n.

In order to calculate the displacement field at time t_{n+1}, the phase field value at time t_n is used:

$$\boldsymbol{u}_{n+1} = \arg\left\{ \inf_u \int_\Omega \left[\psi(\boldsymbol{u}, d_n) - \overline{\boldsymbol{\gamma}} \cdot \boldsymbol{u}\right] dV - \int_{\partial\Omega} \bar{\boldsymbol{t}} \cdot \boldsymbol{u} dA \right\} \tag{18}$$

Similar to the phase field, the approximate linearization solution of the displacement field is:

$$\boldsymbol{K}_n^u \boldsymbol{u}_{n+1} = -\boldsymbol{r}_n^u \tag{19}$$

To implement the scheme in ABAQUS, two element types are used in a hierarchical manner: phase field element and displacement field element, each layer connected at the same node but with different degrees of freedom. The phase field element has only one degree of freedom, the phase field variable d. The degree of freedom of the displacement field element depends on the dimension of the element. Newton–Raphson algorithm was used for iterative solution, and the tangential stiffness matrix and residual vector were iteratively updated:

$$\begin{bmatrix} \boldsymbol{K}_n^d & 0 \\ 0 & \boldsymbol{K}_n^u \end{bmatrix} \begin{bmatrix} d_{n+1} \\ \boldsymbol{u}_{n+1} \end{bmatrix} = -\begin{bmatrix} \boldsymbol{r}_n^d \\ \boldsymbol{r}_n^u \end{bmatrix} \tag{20}$$

3 Numerical Simulation

3.1 Modeling

To explore the impact of initial circumferential surface cracks on pipeline safety under identical loading conditions, a simplified pipeline model featuring initial circumferential surface cracks was developed using ABAQUS. The phase field fracture model discussed in this study does not account for the influence of welding residual stresses and is exclusively solved within the elastic range. Therefore, it is assumed that the weld seam and the pipeline possess equal strength matching, and no specific model incorporating the weld seam is established. Additionally, material plasticity is disregarded, and the pipeline material is treated as a linear elastic material. The specific parameters are detailed in Table 1. Surface cracks are modeled as semi-elliptical and semi-circular surface cracks [13], as illustrated in Fig. 1.

Due to the symmetry of the pipeline, a quarter-section model was utilized for the finite element analysis to streamline computations without sacrificing accuracy. The diameter D of the pipeline was 2 m, the wall thickness t was 0.1 m, the length of the pipeline was $l = 2$ m, and the length of the calculation model was $2D$. The initial crack model was placed on the inner surface at the midpoint of the pipeline. The

Table 1 Model attributes

Elastic modulus E (MPa)	Poisson's ratio	Critical energy release rate g_c (N·mm^{-1})
200,000	0.3	150

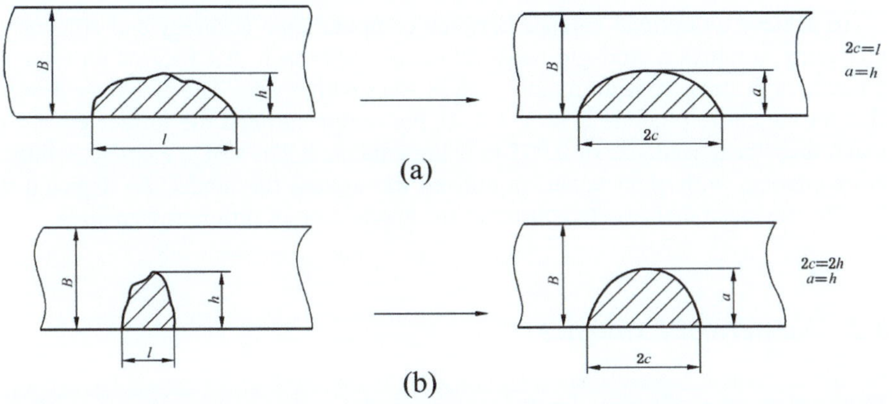

Fig. 1 Regularization of surface defects. **a** For $h < l/2$, the surface defect is represented as a semi-elliptical surface crack. **b** For $h \geq l/2$ (pertinent to fracture considerations), the surface defect is modeled as a semi-circular surface crack

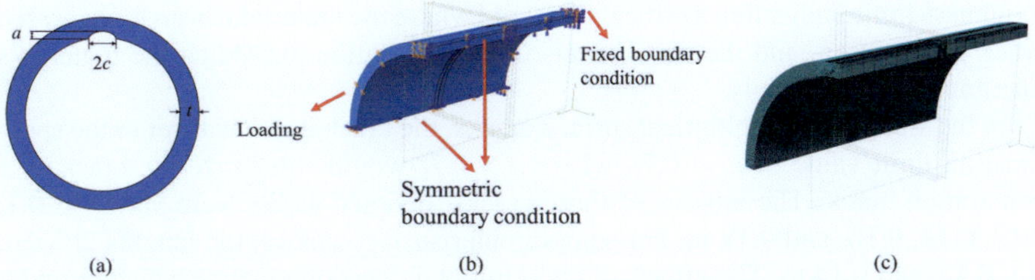

Fig. 2 Pipeline model specification. **a** Schematic diagram of cracks on the inner surface of the pipeline. **b** Boundary conditions. **c** Mesh division

orientation of the initial crack was circumferential, as shown in Fig. 2a, where t was the pipe wall thickness, a was the crack depth, and c was the crack half-length.

One end of the pipeline model is subjected to fixed constraints, encompassing all translational and rotational motions in three directions. At the opposite end from the fixed constraints, displacement loads are applied, which increase uniformly along the pipeline's axial direction. The symmetric plane of the model is equipped with symmetric boundary conditions, as shown in Fig. 2b. The Newton–Raphson algorithm is utilized for the iterative solution of the model. The model is solved in fixed incremental steps within ABAQUS, with a maximum of 200 incremental steps allowed. Each step is sized at 0.01, facilitating detailed tracking of response changes over the loading process. Standard convergence criteria for both force and displacement are employed to ensure that each simulation step accurately reflects the physical behavior of the pipeline under load. The rate of axial displacement applied to the pipeline is set at $\Delta u_x = 0.002$ m per incremental step, mimicking realistic load application rates observed in pipeline operations.

To achieve an optimal balance between computational accuracy and efficiency, this study employs a strategic mesh refinement approach that focuses on regions prone to high stress concentrations, such as areas with cracks. The global mesh seed size for the entire pipeline is set at 0.1 m. For regions around the surface cracks, a much finer local seed size of 0.005 m is implemented. The C3D8 element, a linear brick element with eight nodes, is utilized throughout the model. As depicted in Fig. 2c, the mesh division demonstrates the application of different seed sizes.

3.2 Numerical Examples

To investigate the influence of initial surface cracks on the axial reaction of pipelines under consistent loading conditions, simulations were conducted with pipelines modeled with varying sizes of surface cracks. The models were all subjected to the same loading conditions to isolate the effects of crack size on the pipeline's performance. The criteria for defining failure included the material reaching its fracture toughness limit, indicating a critical threshold where the material can no longer resist crack propagation, and the stress levels in the simulation exceeding the material's maximum stress capacity.

In the study of semi-elliptical surface cracks, the crack depth was set to the specification limit value of $a = 0.7t$, where $a > 0.7t$ would categorize the crack as a penetration crack. The lengths of these cracks, denoted as $2c$, were varied at 0.1, 0.12, 0.14, 0.16, and 0.18 m. For semi-circular surface cracks, the lengths $2c$ were 0.1, 0.12, and 0.14 m. The effects of these initial defects on pipeline behavior under axial loads were documented in Fig. 3a, b, which illustrate the reaction and displacement curves at different pipeline boundaries. The results clearly demonstrate that as the length of the initial surface cracks increases, the pipeline's ability to resist axial loads diminishes, leading to a gradual decrease in the maximum reaction force. This trend highlights the critical impact of crack size on the structural integrity and load-bearing capacity of pipelines.

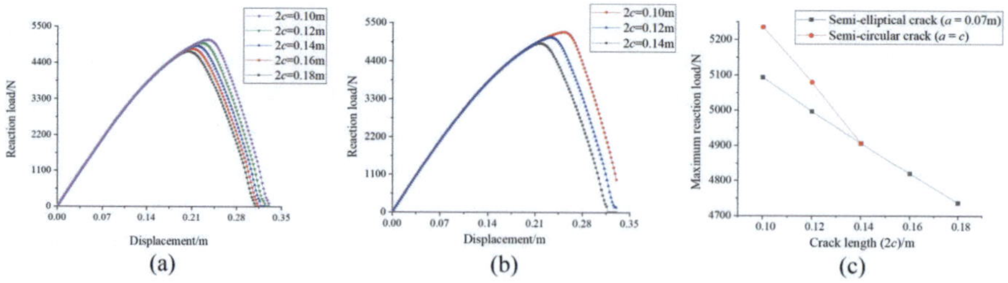

Fig. 3 Numerical simulation result. **a** Reaction-displacement curve of pipe with semi-elliptical crack; **b** reaction-displacement curve of pipe with semi-circular crack; **c** maximum reaction-crack length curve

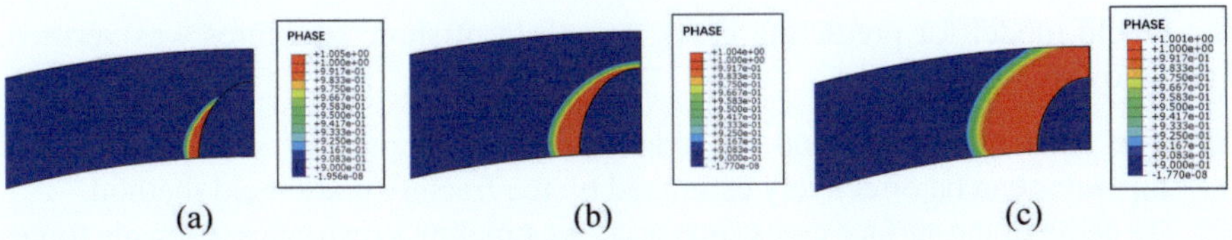

Fig. 4 a = 0.07 m, $2c = 0.1$ m, local evolution of circumferential surface cracks in the pipeline. **a** Initiation of crack. **b** Expansion along the circumferential direction. **c** Expansion through the wall thickness

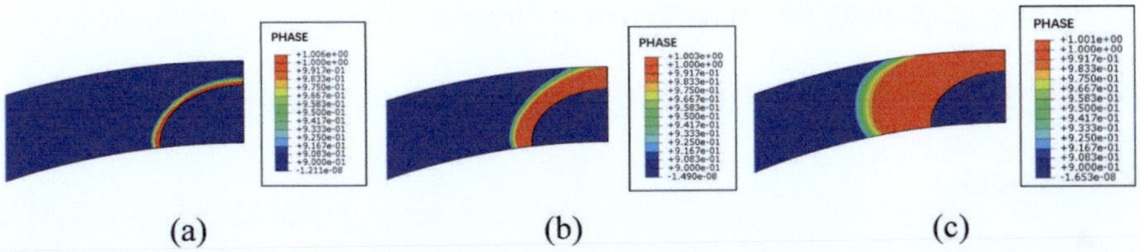

Fig. 5 $a = 0.07$ m, $2c = 0.2$ m, local evolution process of circumferential surface crack of the pipeline. **a** Crack initiation. **b** Expansion through the wall thickness to penetration. **c** Post-penetration expansion along the circumferential direction

As discussed earlier, the phase field variable d in the model quantifies the state of the material: a value of $d = 0$ indicates that the material is intact, whereas $d = 1$ signifies complete material failure. It is generally accepted that when $d \geq 0.9$, cracks are considered to have formed in that region; these areas are highlighted in red in Figs. 4 and 5 to denote the crack zones.

For the local crack growth, the simulation results reveal that with a fixed crack depth at $a = 0.7t$, different behaviors emerge based on the relationship between a and c. When $a \geq c$, the crack initially propagates in the circumferential direction of the pipeline to form a crack with a length equal to half the crack depth. Following this, the crack propagates along the wall thickness direction until it penetrates the wall. Conversely, when $a \leq c$, the crack spreads uniformly in all directions. The penetration occurs first in the direction of wall thickness, followed by rapid expansion along the pipeline's circumferential direction.

4 Conclusions

Based on the fracture phase field method, the behavior of crack propagation in a pipeline with circumferential surface cracks under axial forces was investigated. Numerical simulation was conducted on pipelines with semi-elliptical and semi-circular initial circumferential surface cracks under axial loads. The study focused on the crack propagation and mechanical response of the pipeline with different initial crack sizes. Based on the numerical results, the applicability of the fracture phase field model for predicting crack growth in offshore structures was verified. This study shows:

(1) The fracture process and the crack propagation path of surface cracks of pipelines can be effectively estimated by the fracture phase field method.
(2) As depth of the surface crack increases, the pipeline's resistance to tensile forces is gradually weakened, and the maximum reaction force correspondingly is decreased.
(3) When the depth of the surface crack exceeds its half-length, stress concentration primarily occurs at the circumferential edge of the crack so that the crack tends to propagate along the circumferential direction.

References

1. Zhang Y, Zheng M, An C, Seo JK, Pasqualino IP, Lim F, Duan M (2019) A review of the integrity management of subsea production systems: inspection and monitoring methods. Ships Offshore Struct 14(8):789–803
2. Vishnuvardhan S, Murthy AR, Choudhary A (2023) A review on pipeline failures, defects in pipelines and their assessment and fatigue life prediction methods. Int J Press Vessels Pip 201:104853
3. He M, Pang M, Chen NZ (2023) Engineering critical assessment (ECA) for monopile foundation of an offshore wind turbine subjected to pitting. Ocean Eng 285:115400
4. Song ZH, Chen NZ (2023) A modified cyclic cohesive zone model for low-cycle fatigue crack initiation prediction for subsea pipelines under mode I loading. Ocean Eng 276:114200
5. Wciślik W, Pała T (2021) Selected aspects of cohesive zone modeling in fracture mechanics. Metals 11(2):302
6. Cervera M, Barbat GB, Chiumenti M, Wu JY (2022) A comparative review of XFEM, mixed FEM and phase-field models for quasi-brittle cracking. Arch Comput Methods Eng 29(2):1009–1083
7. Song ZH, Chen NZ (2024) A modified cyclic cohesive zone model for fatigue crack initiation prediction for mooring chain steels. Eng Fract Mech 300:109985
8. Wu JY, Nguyen VP, Nguyen CT, Sutula D, Sinaie S, Bordas SP (2020) Phase-field modeling of fracture. Adv Appl Mech 53:1–183
9. Griffith AA (1921) VI. The phenomena of rupture and flow in solids. Philos Trans Roy Soc Lond Ser A containing papers of a Mathematical or Physical Character 221(582–593):163–198
10. Irwin GR (1958) Fracture. In: Flügge S (ed) Elasticity and plasticity
11. Molnár G, Gravouil A (2017) 2D and 3D Abaqus implementation of a robust staggered phase-field solution for modeling brittle fracture. Finite Elem Anal Des 130:27–38
12. Francfort GA, Marigo JJ (1998) Revisiting brittle fracture as an energy minimization problem. J Mech Phys Solids 46(8):1319–1342
13. The British Standards Institution (2013) Guide to methods for assessing the acceptability of flaws in metallic structures. BS 7910: 2013. London, UK

Hydrogen-Assisted Dislocation Pinning and Depinning on Fatigue Crack Propagation in Subsea Pipeline Steels

Tingsen Zheng and Nian-Zhong Chen

Abstract The resolution of hydrogen embrittlement (HE) and hydrogen-assisted fatigue assessment in subsea pipelines is a critical technology for ensuring the large-scale development of offshore hydrogen industry. Hydrogen-assisted dislocation pinning near the crack tip is normally regarded as one of the key mechanisms of HE. However, there is a lack of a methodology to predict the hydrogen-assisted fatigue for pipeline steel influenced by dislocation pinning. Therefore, a piecewise linear elastic–plastic constitutive equation for pipeline steel in gaseous hydrogen environment is developed, in which the increases of yield strength due to both hydrogen-assisted dislocation pinning and the hardening of the pipeline after depinning are taken into account. To demonstrate the capability of the proposed model, it is introduced into a cyclic cohesive zone model (CCZM) for predicting hydrogen-assisted fatigue crack propagation of pipeline steel. The results shows that the proposed constitutive equation can well reflect the localization of plastic deformation and fatigue crack propagation acceleration caused by hydrogen-assisted HE.

Keywords Hydrogen embrittlement · Fatigue crack propagation · Dislocation pinning · Cyclic cohesive zone model

1 Introduction

The permeation of hydrogen atoms has a significant detrimental effect on the mechanical properties of pipeline steel, which is commonly referred to as hydrogen embrittlement (HE). The initiation and rapid propagation of cracks will occur under loads well below the material's design strength due to HE. Particularly in subsea hydrogen

T. Zheng · N.-Z. Chen (✉)
School of Civil Engineering, Tianjin University, Tianjin, China
e-mail: nzchen2018@hotmail.com

N.-Z. Chen
State Key Laboratory of Hydraulic Engineering Intelligent Construction and Operation, Tianjin University, Tianjin, China

X. L. Zhao et al. (eds.), *Proceedings of the Fourth World Conference on Floating Solutions: WCFS 2024; 2–4 December; Hong Kong*, Lecture Notes in Civil Engineering 597, https://doi.org/10.1007/978-981-96-4569-5_69

pipelines or oil and gas pipelines, hydrogen-assisted fatigue crack stands as one of the primary failure modes [1]. Consequently, the assessment for hydrogen-assisted fatigue crack in subsea pipeline steel has become a critical issue in the subsea oil and gas industry and the development of hydrogen energy.

Nowadays, many theories for HE have been proposed. However, due to the complex mechanisms of HE, significant controversies remain regarding its interpretation [2]. Among the existing HE theories, the widely accepted ones include the hydrogen-enhanced de-cohesion (HEDE) theory and hydrogen-enhanced localized plasticity (HELP) theory [3]. Significant progress has been made in hydrogen-assisted fatigue crack propagation prediction models based on fracture mechanics theory and these HE theories. A series of studies by Cheng and Chen [4–6] proposed corrosion-crack interaction models, which is successfully applied to predict fatigue crack propagation behavior in API pipeline steel under hydrogen and corrosion conditions. In recent years, the cyclic cohesive zone model (CCZM) exhibits good compatibility with the HEDE theory, which provides a new solution for simulating hydrogen-assisted fatigue crack. Song and Chen [7, 8] developed a CCZM for predicting fatigue crack initiation in steel, which is successfully applied to predict fatigue crack initiation life of subsea pipelines and mooring chains. Moriconi et al. [9] introduced an improved hydrogen diffusion model into the cohesive zone model and conducted simulations and experimental comparisons on high-strength steel in hydrogen environments. The results showed good agreement between the model predictions and experimental results in low-pressure hydrogen environments. However, the model struggled to predict changes in fatigue crack propagation rates under high hydrogen concentrations, indicating limitations in the current cohesive zone models for predicting HIFCP.

To effetely reflect the mechanisms of hydrogen-assisted fatigue crack propagation, a key phenomenon is the influence of HE on the plastic zone at the crack tip and the dislocation mobility may be a critical parameter [10]. Nagumo [11] reviewed various HE theories in steel and found that the strain-induced evolution of microdefects such as vacancies and dislocations is affected by hydrogen, which is based on energy stabilization and localized plastic deformation. Xie et al. [12] conducted quantitative mechanical tests under environmental transmission electron microscopy. The results show that strong dislocation pinning occurred in aluminum metal under conditions of excessive vacancy. As seen from the literature review, the pinning-depinning of dislocations has a universal impact on hydrogen-assisted fatigue crack propagation across various metals. However, the pinning-depinning mechanism has not been widely considered in fatigue damage prediction using CCZM.

Therefore, this paper establishes a piecewise elastoplastic constitutive model for pipeline steel in gaseous hydrogen environments, where the hardening of pipeline steel caused by hydrogen-assisted pinning and depinning of dislocation is considered. To validate the effectiveness of proposed model, it is combined with the CCZM developed by Zheng and Chen [13], in which the degradation of cohesive strength and accumulated cohesive length is taken into account. A comparison study between predicated hydrogen-assisted fatigue crack propagation rate of X52 pipeline steel under hydrogen conditions, which are achieve by CCZM with different elastic–plastic constitutive models, is then performed. The results show that the established dislocation pinning-depinning model effectively reflects plastic localization and accelerated fatigue crack propagation under high stress levels induced by hydrogen embrittlement.

2 Methodology

The CCZM used for predicting fatigue crack propagation of pipeline steel, as well as the dislocation pinning-depinning model proposed in this paper, are introduced in this section.

2.1 *Hydrogen Diffusion Model*

In metallic crystals, the total concentration of hydrogen atoms C_{H}, can be expressed as the sum of the lattice hydrogen concentration C_{L}, and the trapped hydrogen concentration C_{T}, within the metallic crystal [14, 15]:

$$C_{\mathrm{H}} = C_{\mathrm{L}} + C_{\mathrm{T}} = \theta_{\mathrm{L}} N_{\mathrm{L}} + \frac{N_{\mathrm{T}}}{1 + \left[C_{\mathrm{L}} \exp\left(\frac{E_B}{RT}\right)\right]^{-1}} \tag{1}$$

where θ_{L} is the lattice hydrogen coverage, N_{L} the lattice hydrogen density, N_{T} the trap hydrogen density, E_{B} is the trap binding energy with a value of 35 kJ/mol in this paper, R the universal gas constant with a value of 8.3145 J/(mol·K), T the absolute temperature. Therefore, the mass transport equation for hydrogen atoms in the metallic crystal can be transformed into a form containing only C_{L} [16]:

$$\frac{D_{\mathrm{L}}}{D_{\mathrm{e}}} \frac{\mathrm{d}C_{\mathrm{L}}}{\mathrm{d}t} = D_{\mathrm{L}} \nabla^2 C_{\mathrm{L}} - \nabla \left(\frac{D_{\mathrm{L}} C_{\mathrm{L}} \overline{V}_m}{RT} \nabla \sigma_{\mathrm{h}} \right) \tag{2}$$

where D_{L} is the lattice diffusion coefficient of hydrogen in the metal, D_{e} the effective diffusion coefficient of hydrogen in the metal, $\overline{V}_m$ is the average molar volume of hydrogen, ∇ is the Hamiltonian, and σ_{h} is the hydrostatic stress. The determination of the diffusion state of hydrogen within the metal under the influence of internal stress fields can be achieved by solved Eq. (2).

2.2 Cyclic Cohesive Zone Model with HE Considering

The cyclic cohesive zone model (CCZM) indicates that the formation of fracture is a progressive process, where the separation of crack surfaces occurs within a cohesive zone at the crack tip. The traction force at the crack tip in this region can be expressed as a function of the crack separation displacement, which is so-call the traction–separation law (TSL). A typical exponential TSL for pure mode I cracks is given as [17]:

$$\frac{T_{\mathrm{n}}}{\sigma_{\max,0}} = \left(\frac{\delta_{\mathrm{n}}}{\delta_0}\right)\exp\left(1 - \frac{\delta_{\mathrm{n}}}{\delta_0}\right) \quad (3)$$

where T_{n} is the normal traction force, $\sigma_{\max,0}$ the initial cohesive strength, which is the cohesive strength of the metal before degradation of material, δ_{n} the normal separation displacement, δ_0 the normal characteristic cohesive. Figure 1 shows the relationship between the cohesive zone and TSL at the crack tip.

The damage evolution equation for CCZM in which impact of HE for fatigue crack is accounted for can be given as [13]:

$$D = \int \frac{\left|\delta_{\mathrm{n},t} - \delta_{\mathrm{n},t-\Delta t}\right|}{\delta_{\Sigma,C}}\left[\frac{T_{\mathrm{n}}}{\sigma_{\max,C}(1-D)} - \frac{\sigma_{\mathrm{f}}}{\sigma_{\max,0}}\right]H\left(\int_0^t\left|\delta_{\mathrm{n},t} - \delta_{\mathrm{n},t-\Delta t}\right|\mathrm{d}t - \delta_0\right)\mathrm{d}t \quad (4)$$

where D is the total damage of the material, $\delta_{n,t}$ the normal separation at time step t, $\delta_{n,t-\Delta t}$ the normal separation at time step t-Δt, Δt the increment of time step, $\delta_{\sum,C}$ is the degraded accumulated cohesive length, $\delta_{\sum,0}$ the initial accumulated cohesive length, $\sigma_{\max,C}$ is the degraded cohesive strength, σ_{f} the endurance limit of the cohesive zone, H the Heaviside function. Failure of the material is deemed to occur when the total damage D is equal to 1.

Degradation of cohesive strength $\sigma_{\max}$ and accumulated cohesive length δ_{Σ} due to hydrogen under gaseous hydrogen conditions is considered in the CCZM to capture the HE on the material properties of pipeline steel. The degradation of $\sigma_{\max}$ is

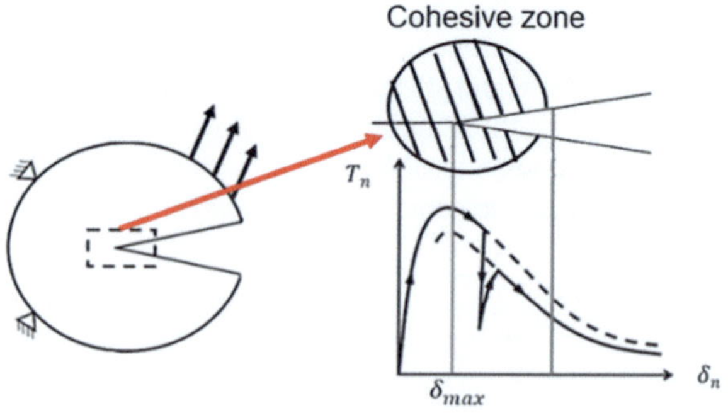

Fig. 1 Diagram of cohesive zone and traction separation law at crack tip

described using the quadratic relationship between $\sigma_{\max}$ and hydrogen coverage θ_{H} in terms of first principles, which is given as [18]:

$$\sigma_{\max,\theta} = \left(0.1687\theta_{\mathrm{H}}^2 - 1.0467\theta_{\mathrm{H}} + 1\right)\sigma_{\max,0} \tag{5}$$

in which

$$\theta_H = \frac{C_{\mathrm{H}}}{C_{\mathrm{H}} + \exp\left(-\frac{\Delta G}{RT}\right)} \tag{6}$$

where ΔG is the Gibbs free energy difference between any microstructural interface and the surrounding material.

Generally, once the hydrogen concentration exceeds a certain critical value, further increases in hydrogen concentration no longer significantly affect the degradation of metal material properties, indicating a saturation effect in HE [19]. The degradation pattern of pipeline steel due to HE is similar to the process described by the Logistic formula, which typically characterizes a process that begins with slow growth, followed by rapid growth, and eventually approaches a plateau. The inverse Logistic-based model for of degradation of δ_{Σ} established by Zheng and Chen [13] is used in this paper, which is given as:

$$\frac{\delta_{\Sigma,C}}{\delta_{\Sigma,0}} = \frac{0.9\left(1 + \frac{\delta_0}{\delta_{\Sigma,0}}\right)}{1 + \frac{\delta_0}{\delta_{\Sigma,0}} e^{sC_L}} + 0.1 \tag{7}$$

where, s the material parameter that is determined based on the hydrogen concentration at which the saturation effect occurs, set at s = 107.62 for pipeline steel.

2.3 Pinning-Depinning Model

Hydrogen atoms will occupy vacancies in the metallic crystal, which impedes dislocation movement and leads to dislocation pinning. Unfortunately, there is no such constitutive model capable of effectively describing the influence of hydrogen-induced dislocation pinning-depinning on material properties. Therefore, this paper proposed a simplified dislocation pinning-depinning model for describing elastic–plastic constitutive behavior of pipeline steel under hydrogen environment. Figure 2 shows the comparison between proposed pinning-depinning model, elastic-perfectly plastic model and Hollomon hardening model [13]. The pinning-depinning model is in terms of the stress–strain relationship of hydrogen-charged metals from a series nanoindentation experiments [20].

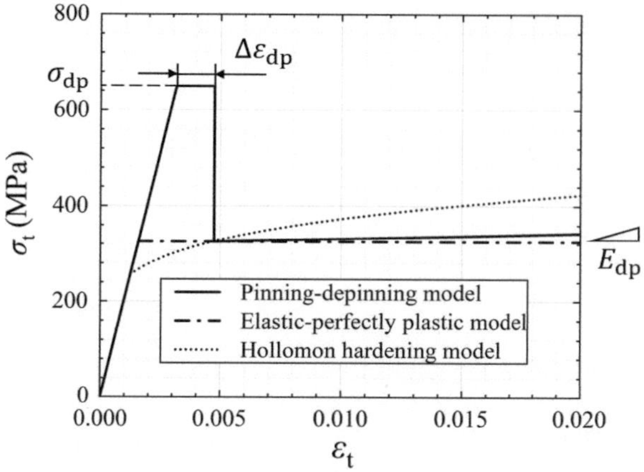

Fig. 2 Dislocation pinning-depinning model for describing elastic–plastic constitutive behavior of X52 pipeline steel

As shown in Fig. 2, the hydrogen-induced dislocation pinning-depinning model indicates a material elastic–plastic behavior consisting of four stages: pinning hardening, plateau, depinning, and depinning hardening. This is consistent with the experimental results of Li et al. [20]. To effectively describe this proposed constitutive model, three key parameters are defined: depinning stress σ_{dp}, increment of depinning strain $\Delta\varepsilon_{dp}$, and depinning hardening modulus E_{dp}, which can be given as:

$$\sigma_{dp} = 2\sigma_y \tag{8}$$

$$\Delta\varepsilon_{dp} = \sigma_y / E \tag{9}$$

$$E_{dp} = E/10 \tag{10}$$

where σ_y the yield strength of pipeline steel, E the Young's modulus of pipeline steel.

3 Finite Element Analysis

A finite element model of a compact tension (CT) specimen made of X52 pipeline steel (as shown in Fig. 3) is established to predict hydrogen-induced fatigue crack propagation. Due to the symmetry of the CT specimen, only half of the model is constructed. The dimensions of the CT specimen are in terms of ASTM E647 standard [21], with a characteristic dimension $W = 25.4$ mm. The CPE4RT element is used to mesh the steel, and a layer of COH2D4 elements is applied along the predetermined fatigue crack growth path. The parameters for the cyclic cohesive zone model (CCZM) are set as follows: $\delta_0 = 0.2$ mm, $\delta_{\Sigma,0} = 3$ mm, $\sigma_{max,0} = 4\sigma_y$, and $\sigma_f = \sigma_y$.

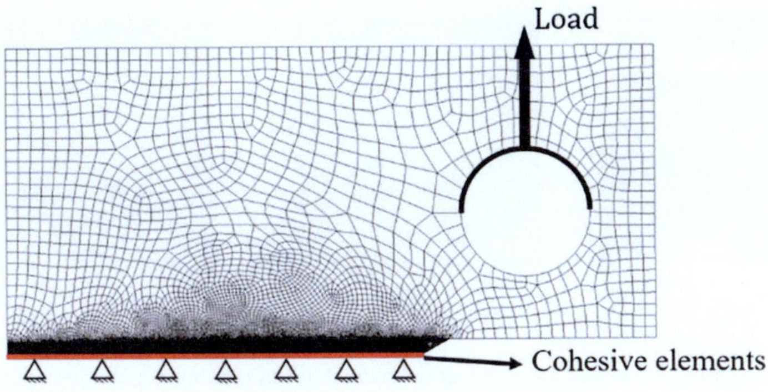

Fig. 3 Finite element model of compact tension specimen that is made of X52 pipeline steel

Table 1 Material properties of X52 pipeline steel

Properties	Value
Ultimate strength (MPa)	526
Density (kg/m^3)	7850
Yield strength (MPa)	325
Young's modulus E (GPa)	206
Poisson's ratio ν	0.3

Different stress intensity factor ranges ΔK at crack tip is selected which is maintained constant during analysis. The load ratio is set to 0.5, and cyclic loading is applied to the upper edge of the pin hole in the CT specimen. The initial hydrogen concentration $C_{L,0}$ is equal to 0.055 mg/kg. The crack growth simulation is performed until the crack propagated beyond the mesh refinement zone. The ΔK values and corresponding hydrogen-induced fatigue crack growth rates da/dN are recorded, where a is the crack length and N the number of load cycles. The material properties for X52 steel are presented in Table 1.

4 Results and Discussion

Figure 4 shows the distribution of equivalent plastic strain at a certain moment during the analysis when different elastic–plastic constitutive models are employed. As shown in Fig. 4, when the proposed dislocation pinning-depinning model is used to simulate the elastic–plastic behavior of X52 pipeline steel, the plastic zone area near the crack tip is smaller than that obtained using the Hollomon hardening model. At the same time, the maximum equivalent plastic strain at the crack tip is increased. This is due to the significant hardening of X52 pipeline steel caused by hydrogen-induced dislocation pinning.

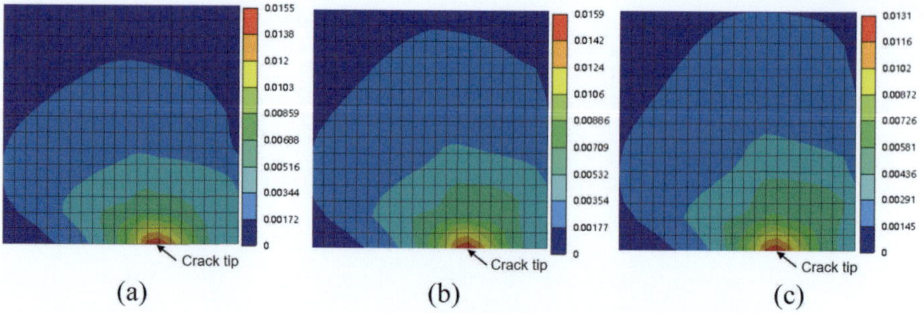

Fig. 4 Distribution of equivalent plastic strain at crack tip: **a** pinning-depinning model; **b** elastic-perfectly plastic model; **c** Hollomon hardening model

However, the hardening effect of the dislocation pinning-depinning model after depinning is lower than that of the Hollomon hardening model. Consequently, the pinning-depinning model predicts a higher maximum plastic strain than the Hollomon model. Due to the lack of consideration for material hardening, the elastic-perfectly plastic model predicts a higher maximum equivalent plastic strain compared to the results predicted by other two models. It can be seen that the proposed hydrogen-induced dislocation pinning-depinning model effectively reflects the hydrogen-induced plastic localization in pipeline steel, which is consistent with the HELP theory.

Figure 5 shows a comparison of predicted hydrogen-induced fatigue crack growth rates using different elastic–plastic constitutive models. As shown in Fig. 5, generally, the elastic-perfectly plastic model provides a relatively conservative prediction due to the lack of consideration for material hardening. The results achieved by using Hollomon hardening model mostly agrees well with the experimental results. However, the predictions of the Hollomon hardening model are slightly lower than the lowest experimental values when there is a higher ΔK, which may be unfavorable for structural safety assessments of subsea pipeline. When the proposed dislocation pinning-depinning model is applied, the predicted hydrogen-induced fatigue crack growth rates are more consistent with the experimental results compared to that predicted by the elastic-perfectly plastic model. Additionally, different from the results of Hollomon hardening model, the predictions of dislocation pinning-depinning model does not yield excessively low predictions at higher ΔK values. In summary, the proposed dislocation pinning-depinning model will better reflect the mechanism of hydrogen-induced fatigue crack growth in X52 pipeline steel.

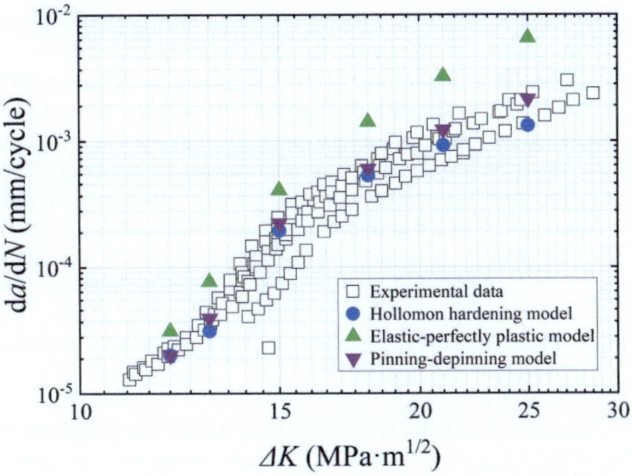

Fig. 5 Comparison between model prediction and experimental data of hydrogen induced fatigue crack propagation

5 Conclusions

A piecewise elastoplastic constitutive model for pipeline steel in gaseous hydrogen environment was established where the hardening of pipeline steel induced by hydrogen-assisted pinning and depinning of dislocation is considered. The proposed pinning-depinning model was combined with the CCZM for predicting hydrogen-induced fatigue crack propagation. A comparison study between rates of hydrogen-assisted fatigue crack propagation of X52 pipeline steel achieved by CCZM with the use of proposed pinning-depinning model, elastic-perfectly plastic model and Hollomon hardening model was then performed. The results show that the proposed dislocation pinning-depinning model can effectively capture the plastic localization and acceleration of fatigue crack propagation under high stress levels due to hydrogen embrittlement.

References

1. Kushida T, Nose K, Asahi H, Kimura M, Yamane Y, Endo S, Kawano H (2001) Effects of metallurgical factors and test conditions on near neutral pH SCC of pipeline steels. In: NACE corrosion (NACE-01213). Houston, Texas
2. Djukic MB, Bakic GM, Zeravcic VS, Sedmak A, Rajicic B (2019) The synergistic action and interplay of hydrogen embrittlement mechanisms in steels and iron: localized plasticity and decohesion. Eng Fract Mech 216:106528
3. Nagao A, Smith CD, Dadfarnia M, Sofronis P, Robertson IM (2012) The role of hydrogen in hydrogen embrittlement fracture of lath martensitic steel. Acta Mater 60(13–14):5182–5189
4. Cheng A, Chen NZ (2017) Fatigue crack propagation modelling for pipeline carbon steels under gaseous hydrogen conditions. Int J Fatigue 96:152–161
5. Cheng A, Chen NZ (2017) Corrosion fatigue crack propagation modelling for subsea pipeline steels. Ocean Eng 142:10–19

6. Cheng A, Chen NZ (2018) An extended engineering critical assessment for corrosion fatigue of subsea pipeline steels. Eng Fail Anal 84:262–275
7. Song ZH, Chen NZ (2023) A modified cyclic cohesive zone model for low-cycle fatigue crack initiation prediction for subsea pipelines under mode I loading. Ocean Eng 276:114200
8. Song ZH, Chen NZ (2024) A modified cyclic cohesive zone model for fatigue crack initiation prediction for mooring chain steels. Eng Fract Mech 300:109985
9. Moriconi C, Hénaff G, Halm D (2014) Cohesive zone modeling of fatigue crack propagation assisted by gaseous hydrogen in metals. Int J Fatigue 68:56–66
10. Barnoush A, Kheradmand N, Hajilou T (2015) Correlation between the hydrogen chemical potential and pop-in load during in situ electrochemical nanoindentation. Scripta Mater 108:76–79
11. Nagumo M (2001) Function of hydrogen in embrittlement of high-strength steels. Iron Steel Inst Jpn Int 41(6):590–598
12. Xie D, Li S, Li M, Wang Z, Gumbsch P, Sun J, Ma E, Li J, Shan Z (2016) Hydrogenated vacancies lock dislocations in aluminium. Nat Commun 7(1):13341
13. Zheng T, Chen NZ (2023) A cyclic cohesive zone model for predicting hydrogen assisted fatigue crack propagation (FCG) of subsea pipeline steels. Int J Fatigue 173:107707
14. Sofronis P, McMeeking RM (1989) Numerical analysis of hydrogen transport near a blunting crack tip. J Mech Phys Solids 37(3):317–350
15. Oriani RA, Josephic PH (1974) Equilibrium aspects of hydrogen-assisted cracking of steels. Acta Metall 22(9):1065–1074
16. Krom AH, Koers RW, Bakker AD (1999) Hydrogen transport near a blunting crack tip. J Mech Phys Solids 47(4):971–992
17. Xu XP, Needleman A (1993) Void nucleation by inclusion debonding in a crystal matrix. Modell Simul Mater Sci Eng 1(2):111
18. Serebrinsky S, Carter EA, Ortiz M (2004) A quantum-mechanically informed continuum model of hydrogen embrittlement. J Mech Phys Solids 52(10):2403–2430
19. Wang M, Akiyama E, Tsuzaki K (2007) Effect of hydrogen on the fracture behavior of high strength steel during slow strain rate test. Corros Sci 49(11):4081–4097
20. Li Y, Morgan TW, Terentyev D, Ryelandt S, Favache A, Wang S, Wirtz M, Hoefnagels JPM, van Dommelen JAW, De Temmerman G, Verbeken K, Geers MGD (2020) Three mechanisms of hydrogen-induced dislocation pinning in tungsten. Nucl Fusion 60(8):086015
21. ASTM E647 (2016) Standard test method for measurement of fatigue crack growth rates. American Society for Testing and Materials Standards

Nanoscale Insight on FRP-Soil Interfacial Mechanical Properties

Pengchang Wei and Zhen-Yu Yin

Abstract Fiber-reinforced polymer (FRP) is extensively utilized across various engineering disciplines for its superior properties, including its roles in protecting, reinforcing, and repairing structural components. In marine geotechnical engineering, the interfacial interactions between FRP and soil are crucial, yet they remain poorly understood at the micro- and nano-scale. Molecular dynamics (MD) simulation method has been employed to explore the interfacial mechanical behavior of the epoxy-quartz interface within the context of FRP-soil systems. The MD models were validated through uniaxial tension tests conducted on bulk epoxy. Subsequent MD nanoindentation simulations examined the epoxy-quartz interface, considering the effects of loading rate and indentation depth. An abrasion model for the epoxy-quartz system was proposed based on the relationship between force and displacement. The simulation results indicated that increased loading rates lead to higher indenter forces and greater indentation hardness of the epoxy substrate, with a power-law relationship observed between indenter force and indentation depth. Additionally, the study determined the overall Young's modulus of the bulk epoxy to be 2.075 GPa, with an average Poisson's ratio of 0.377. For the epoxy substrate, the reduced modulus was 2.419 GPa, and the indentation hardness ranged from 1.492 to 2.174 GPa.

Keywords Sand · FRP-soil interface · Molecular dynamics · Nanoindentation · Viscosity

P. Wei · Z.-Y. Yin (✉)
Department of Civil and Environmental Engineering, The Hong Kong Polytechnic University, Hong Kong, China
e-mail: zhenyu.yin@polyu.edu.hk

Z.-Y. Yin
Research Institute for Land and Space (RILS), The Hong Kong Polytechnic University, Hong Kong, China

X. L. Zhao et al. (eds.), *Proceedings of the Fourth World Conference on Floating Solutions: WCFS 2024; 2–4 December; Hong Kong*, Lecture Notes in Civil Engineering 597, https://doi.org/10.1007/978-981-96-4569-5_70

1 Introduction

Fibre-reinforced polymer (FRP) composites are extensively utilized across diverse fields such as marine engineering, building structures, aerospace, and civil and hydraulic engineering [1]. This widespread adoption is attributed to their exceptional properties, including high strength, low weight, corrosion resistance, electromagnetic transparency, and ease of handling. In civil engineering, FRP composites are effectively employed to protect, reinforce, and repair structures. For example, FRP piles are increasingly preferred over traditional concrete and steel piles in corrosive environments, such as those encountered in marine geotechnical engineering, due to their superior resistance to corrosion and chemicals [2]. As FRP piles become more common in geotechnical engineering, understanding the interfacial properties between FRP composites and soil is essential for maintaining the integrity and stability of the structure-soil system [2, 3]. Shaia et al. [4] indicated that the friction coefficient between FRP composites and sand was higher than that between conventional mild steel and sand. However, the stiffness and surface hardness of FRP composites are notably lower compared to those of concrete and steel structures. Thus, investigating the interfacial mechanical properties of FRP-soil interactions is crucial.

Current research predominantly addresses the macroscopic mechanical properties of FRP and soil/sand mixtures through experimental and numerical analyses [5]. These studies largely emphasized the properties of the mixtures while frequently neglecting the nuances of their interfacial interactions [6]. Consequently, the microscale interfacial behavior between FRP and soil remains insufficiently understood. Additionally, the macroscopic properties of materials are profoundly influenced by their micro- and nano-scale characteristics, where factors such as composition degradation, failure initiation, and microscopic mechanical behavior significantly impact overall performance. Hence, investigating materials at the microscale is crucial for a thorough understanding of their macroscale mechanical properties.

The molecular dynamics (MD) simulation method has proven effective in investigating the interfacial mechanical behavior of materials, offering fundamental insights that are challenging to obtain through experiments. In MD simulations, the atomistic model of the epoxy-quartz system is frequently utilized to represent the FRP-soil interface [2, 3], given that epoxy resin is a critical component of FRP and quartz is a primary constituent of the soil fraction. Research to date has explored various aspects of the epoxy-quartz interface, including adhesion [7], interfacial friction [2], and creep [3] behavior through MD simulations. However, there remains a paucity of studies focused on their interfacial interactions at the microscale.

Nanoindentation is a powerful technique for examining the mechanical properties of materials at the microscale (~10^{-9} m). MD simulations and experimental studies utilizing nanoindentation have predominantly focused on pure epoxy resin, quartz, and mixtures of epoxy composites with various modifications (e.g., nanoclay, fumed silica). These investigations have demonstrated that the interfacial mechanical properties revealed by nanoindentation are significantly influenced by factors such as loading rate, temperature, indenter shape and size, and indentation depth. However, to date, limited research has addressed the interfacial mechanical behavior between epoxy and quartz using MD nanoindentation methods.

In this study, MD nanoindentation simulations were conducted to investigate the interfacial properties of the FRP-soil system at the microscale, specifically focusing on the mechanical behavior between epoxy and quartz. A cross-linked epoxy resin was employed as the substrate, while quartz was the indenter. Initially, uniaxial tension tests were performed on the bulk epoxy to obtain its tensile stress–strain curve, Young's modulus, and Poisson's ratio. The mechanical parameters derived from these tests were compared with existing experimental and simulation data to validate the model's accuracy. Subsequently, MD nanoindentation simulations of the epoxy-quartz system were carried out, considering the effects of loading rate and indentation depth to elucidate the underlying mechanical mechanisms. Moreover, an abrasion model was proposed based on the indenter force F_z and indentation hardness H during the nanoindentation process.

2 Simulation Details

2.1 System Setup

As shown in Fig. 1a, a pyramid-shaped quartz model with dimensions of 35.99 Å × 33.71 Å × 34.05 Å and comprising 1,380 atoms was created using VMD software. Figure 1b shows an epoxy resin model with dimensions of 72.4612 Å × 72.4612 Å × 55.8246 Å, consisting of 500 diglycidyl ether of bisphenol A (DGEBA) and 250 m-phenylenediamine (mPDA). DGEBA serves as the organic component of the epoxy, while mPDA acts as a curing agent. The resulting epoxy resin achieved a conversion percentage of 84% through the cross-linking reaction, indicating the ratio of reacted carbon atoms to the initial number of epoxy groups.

Figure 1c shows the epoxy resin-quartz system comprising 115,380 atoms. In this system, a pyramid-shaped quartz model (Fig. 1a) functions as the indenter, while a 2a × 2b × 1c supercell of epoxy resin (Fig. 1b) acts as the substrate. The quartz indenter was set as a rigid body during whole simulation. The epoxy resin substrate was segmented into three distinct layers: the Newtonian layer, the thermostat layer, and the boundary layer.

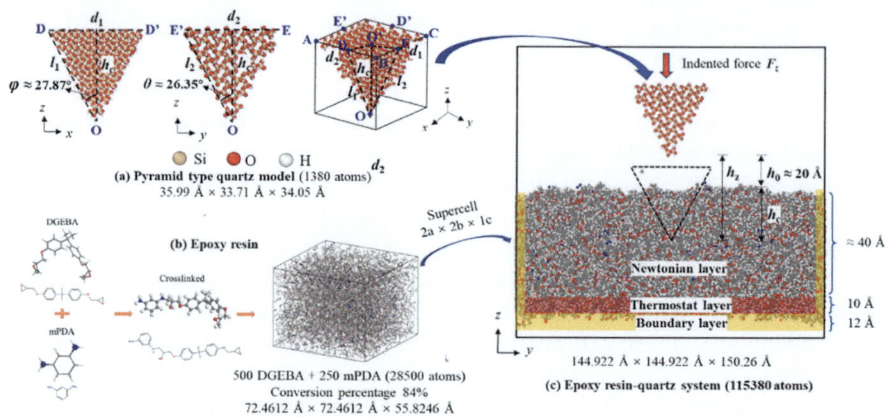

Fig. 1 Epoxy-quartz model: **a** Pyramid type quartz model, **b** Epoxy resin [3], **c** Epoxy resin-quartz system for nanoindentation simulation, where the quartz model is set as an indenter, and the epoxy as a substrate (Note that d_1, d_2, l_1, l_2, φ, θ are corresponding length and angle in the simulation box of quartz (see Fig. 1a), where φ and θ are set as 27.87° and 26.35°, respectively. h_c, h_z, h_o are indentation depth (i.e., contact depth between the epoxy substrate and the quartz indenter), displacement of quartz indenter along the *z*-direction, initial distance between epoxy and quartz, respectively)

2.2 *Force Fields*

The PCFF and CVFF force fields have been effectively used for organic compounds (e.g., epoxy system) and inorganic materials (e.g., quartz) [3], respectively. Thus, the PCFF and CVFF force fields were employed for the epoxy substrate and the quartz indenter, respectively, and their formulation are displayed in Eqs. 1, 2.

Using the 9–6 Lennard–Jones (LJ) parameters for each atom in the epoxy system, the corresponding 12–6 LJ potential parameters were derived through the fitting method we proposed. Subsequently, the non-bonded interactions between the epoxy and quartz were evaluated using the same 12–6 LJ potential, where σ_{ij} and ε_{ij} between atom *i* and atom *j* were calculated by Mixing Lorentz-Berthelot's law [8] (i.e., Eqs. 3, 4).

$$
\begin{aligned}
E_{total} = & \sum_b \left[k_2(b-b_0)^2 + k_3(b-b_0)^2 + k_4(b-b_0)^2\right] \\
& + \sum_\theta \left[H_2(\theta-\theta_0)^2 + H_3(\theta-\theta_0)^2 + H_4(\theta-\theta_0)^2\right] \\
& + \sum_\phi \left\{ \begin{array}{l} V_1\left[1-\cos\left(\phi-\phi_1^0\right)\right] + V_2\left[1-\cos\left(\phi-\phi_2^0\right)\right] \\ +V_3\left[1-\cos\left(\phi-\phi_3^0\right)\right] \end{array} \right\} + \sum_\chi k_\chi \chi^2 \\
& + \sum_b \sum_{b'} (b-b_0)\left(b'-b'_0\right) + \sum_\theta \sum_{\theta'} (\theta-\theta_0)\left(\theta'-\theta'_0\right)
\end{aligned}
$$

$$
\begin{aligned}
&+\sum_{b}\sum_{\theta}(b-b_0)(\theta-\theta_0)\\
&+\sum_{b}\sum_{\phi}(b-b_0)(V_1\cos\phi+V_2\cos 2\phi+V_3\cos 3\phi)\\
&+\sum_{b'}\sum_{\phi}\left(b'-b'_0\right)(V_1\cos\phi+V_2\cos 2\phi+V_3\cos 3\phi)\\
&+\sum_{\theta}\sum_{\phi}(\theta-\theta_0)(V_1\cos\phi+V_2\cos 2\phi+V_3\cos 3\phi)\\
&+\sum_{\phi}\sum_{\theta}\sum_{\theta'}k_{\phi\theta\theta'}\cos\phi(\theta-\theta_0)\left(\theta'-\theta_{0'}\right)+\frac{e^2}{4\pi\varepsilon_0}\sum_{ij}\frac{q_iq_j}{r_{ij}}\\
&+\sum_{ij}\varepsilon_{ij}\left[2\left(\frac{\sigma_{ij}}{r_{ij}}\right)^9-3\left(\frac{\sigma_{ij}}{r_{ij}}\right)^6\right]
\end{aligned}
\tag{1}
$$

$$
\begin{aligned}
E_{total}=&\sum_{b}D_b\left[1-e^{-\alpha(b-b_0)}\right]+\sum_{\theta}H_\theta(\theta-\theta_0)^2+\sum_{\phi}H_\phi[1+s\cos(n\phi)]\\
&+\sum_{\chi}k_\chi\chi^2+\sum_{b}\sum_{b'}F_{bb'}(b-b_0)\left(b'-b'_0\right)\\
&+\sum_{\theta}\sum_{\theta'}F_{\theta\theta'}(\theta-\theta_0)\left(\theta'-\theta'_0\right)+\sum_{b}\sum_{\theta}F_{b\theta}(b-b_0)(\theta-\theta_0)\\
&+\sum_{\phi}F_{\phi\theta\theta'}\cos\phi(\theta-\theta_0)\left(\theta'-\theta'_0\right)\\
&+\sum_{\chi}\sum_{\chi'}F_{\chi\chi'}\chi\chi'+\frac{e^2}{4\pi\varepsilon_0}\sum_{ij}\frac{q_iq_j}{r_{ij}}+\sum_{ij}4\varepsilon_{ij}\left[\left(\frac{\sigma_{ij}}{r_{ij}}\right)^{12}-\left(\frac{\sigma_{ij}}{r_{ij}}\right)^6\right]
\end{aligned}
\tag{2}
$$

$$
\sigma_{ij}=\frac{\sigma_i+\sigma_j}{2} \tag{3}
$$

$$
\varepsilon_{ij}=\sqrt{\varepsilon_i\varepsilon_j} \tag{4}
$$

2.3 Uniaxial Traction and MD Nanoindentation Simulations

All MD simulations were conducted using the LAMMPS code [9]. The simulations employed three-dimensional periodic boundary conditions and utilized the velocity-Verlet algorithm to integrate the motion equations. The cut-off radius for van der Waals forces was set at 10 Å, while the cut-off for long-range electrostatic interactions, calculated using the Particle-Particle Particle-Mesh (PPPM) method, was

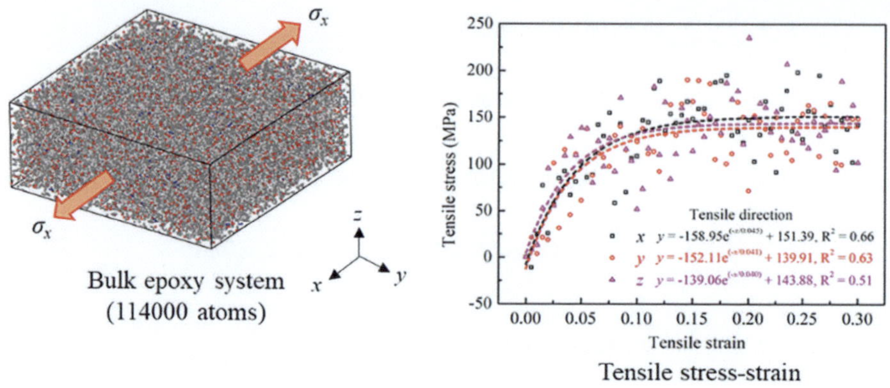

Fig. 2 Evolution of tensile stress of the bulk epoxy system with tensile strain along three directions

8.5 Å. A time step of 1.0 fs was used, and all simulations were conducted at a temperature of 300 K.

The bulk epoxy system, comprising 114,000 atoms (see Fig. 2), was first relaxed using the NPT ensemble for 1.0 ns at the temperature of 300 K and pressure of 1.0 atm, respectively. After relaxation, the tensile mechanical properties of the epoxy system were assessed along three principal directions by applying uniaxial traction at 300 K. This was achieved through a strain-controlled loading approach with a tensile strain rate of 5×10^{-7}/fs.

In the MD nanoindentation simulation of the epoxy-quartz system, as shown in Fig. 1c, the process was divided into two stages: Relaxation Process: The stage was performed with Langevin thermostat at 300 K and NVE ensemble for 2 ns, the quartz indenter kept still during this process. (2) Loading Process: In this stage, the quartz indenter was moved in a displacement-controlled mode with a loading rate ranging from 0.1 to 2.0 Å/ps. Initially, the indenter was positioned 20 Å (h_o) above the epoxy substrate surface (see Fig. 1c). During the loading process, the indenter gradually approached and indented into the epoxy substrate. The final displacement of the indenter along the *z*-direction was set to 45 Å, resulting in a final indentation depth of 25 Å.

3 Results and Discussions

3.1 Uniaxial Traction on Bulk Epoxy

The Young's modulus of the epoxy resin was calculated via linear regression of the tensile stress–strain curve for a strain range of 0–0.04 (as shown in Fig. 2), where the overall Young's modulus ($\overline{E}$) was reported in Table 1 as 2.075 GPa. Table 2 presents the Poisson's ratio of the epoxy resin alongside values from similar materials in previous studies, with an average Poisson's ratio ($\overline{\upsilon}$) of 0.377. These values aligned well with prior experimental and simulation results, indicating that the model developed in this work was effective for simulating the mechanical properties of epoxy resin. Variations in data across Tables 1 and 2 were attributed to differences in epoxy resin compositions, curing agents, test conditions, and simulation parameters such as model size and cross-linking conversion.

Moreover, as shown in Fig. 2, Tables 1 and 2, the tensile stress, Young's modulus, and Poisson's ratio of the bulk epoxy system are relatively close across the three principal directions. This uniformity suggested that the epoxy resin exhibited isotropic behavior, meaning its mechanical properties were similar in all directions. This was supported by previous studies (see Tables 1, and 2), which confirmed that epoxy resins could generally be considered isotropic materials.

Table 1 Young's modulus of epoxy resin in this work and similar materials in previous studies at room temperature

Methods	Young's modulus of epoxy (GPa)	E_x	E_y	E_z	$\overline{E}$	E_r
MD simulations	Epoxy resin (containing 2000 DGEBA + 1000 mPDA) through uniaxial tension using MD simulations in this work	2.203	1.906	2.105	2.075	2.419
	Epoxy resin (containing 200 DGEBA + 100 mPDA) via uniaxial tension using MD simulation [3]	2.61 ± 0.28				
Various experiments in previous works	DGEBA epoxy networks (containing DGEBA/D230, DGEBA/TETA, DGEBA/TEPA) through dynamic mechanical analyses experiments [10]	2.21–2.72				
	Matrix epoxy Epon 862 [11]	2.83				

The overall Young's modulus ($\overline{E}$) is the square average value of Young's modulus along three directions (i.e., E_x, E_y, E_z). E_r is reduced modulus

Notes DGEBA: Diglycidyl Ether of Bisphenol A; mPDA: m-phenylenediamine; DETDA: Diethyltoluenediamine; EPON862: Bisphenol F diglycidyl ether; TETA: triethylenetetramine; TEPA: tetraethylenepentamine; D230: Jeffamine D230

Table 2 Poisson's ratio of epoxy resin in this work and similar materials in previous studies at room temperature

Poisson's ratio of epoxy	ν_{xy}	ν_{xz}	ν_{yx}	ν_{yz}	ν_{zx}	ν_{zy}	$\overline{\upsilon}$
Uniaxial tension using MD simulation in this work	0.351	0.407	0.338	0.420	0.368	0.378	0.377
Epoxy resin composed of EPON862 and curing agent DETDA (conversion degrees of 0–86%) [12]	0.27–0.475						
Epoxy resin (containing EPON862 + DETDA) [13]	0.353 (experimental), 0.448 (MD simulation)						
Matrix epoxy Epon 862 [11]	0.3						

The average Poisson's ratio ($\overline{\upsilon}$) is the geometric average value of the six components of the matrix

3.2 *MD Nanoindentation Behavior*

Figure 3a, b illustrates the evolution of the indentation force (F_z) with the displacement of the quartz indenter (h_z) and the indentation depth (h_c) at various loading rates. The indentation force gradually increased with h_z and the loading rate. Figure 3b shows the relationship between F_z and h_c adheres to a power-law function. This power-law relationship is well-established and commonly used in both molecular dynamics (MD) simulations and experimental nanoindentation studies [14, 15]:

$$F_z = \alpha h_c^m \tag{5}$$

where F_z and h_c are the indentation force of the quartz indenter and the indentation depth, respectively. α and m are constants.

During the MD nanoindentation process, as shown in Fig. 3b, the quartz indenter penetrates the epoxy substrate, causing significant surface deformation and the formation of pits, while no bond or angle breakage occurs throughout the process. The surface wear of the epoxy substrate was characterized by an evident "groove". Figure 4 demonstrates that at a loading rate of 1.0 Å/ps, the displacement magnitude of the atoms within the system increases consistently with the increase in h_z. Concurrently, the surface wear, or deformation area, expanded as h_z increased. This results in a greater number of epoxy atoms being displaced by the quartz indenter, which could lead to increased resistance faced by the indenter, thereby causing the indentation force to rise with h_z.

At a loading rate of 0.1 Å/ps, the indentation force showed considerable dispersion, as illustrated in Fig. 3a, b, with a fitting coefficient R^2 of 0.78. This value was notably lower compared to the coefficients obtained at other loading rates (see Fig. 3b). Consequently, the loading rate of 0.1 Å/ps was excluded from the calculation of the indentation hardness of the epoxy substrate.

As shown in Table 3, the indentation hardness of the epoxy substrate increased with the loading rate from 0.25 to 0.75 Å/ps. Beyond this point, the hardness tended

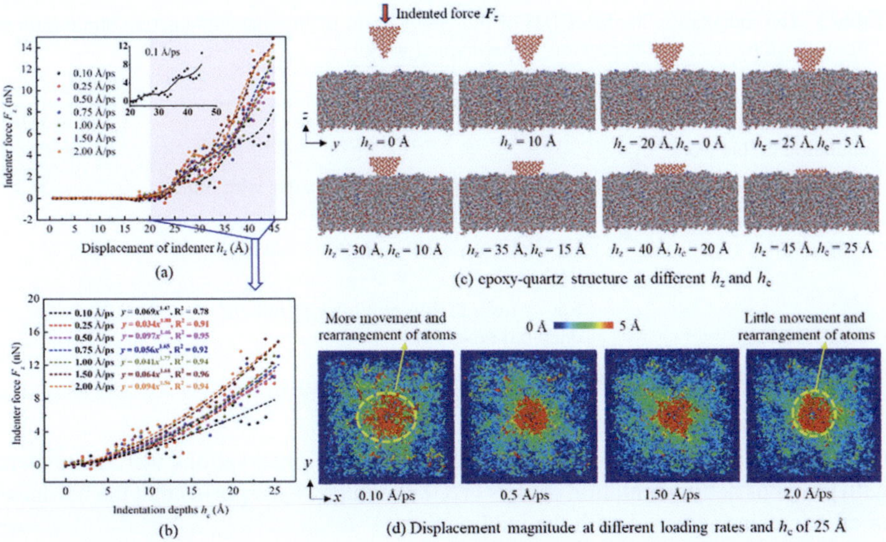

Fig. 3 **a, b** Evolution of the indentation force (F_Z) of the whole system with the displacement of the indenter (h_Z) and the indentation depth (h_C) at different loading rates. **c** Epoxy-quartz structure at different h_Z and h_C. **d** Displacement magnitude of atoms within the whole system at different loading rates and $h_C = 25$ Å

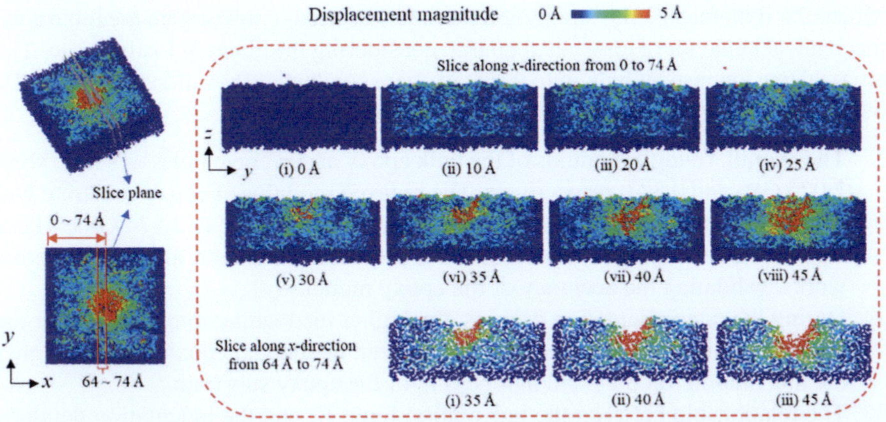

Fig. 4 Displacement magnitude of atoms for the whole system at different displacements of the indenter (h_Z) under two different slices: one slice along x-direction from 0 to 74 Å, and another one from 64 to 74 Å. (The above figures refer to the case of a loading rate of 1.0 Å/ps. The indenter is hidden for better visual clarity)

Table 3 The indentation hardness (H) of the epoxy resin in MD nanoindentation simulation at different loading rates and in previous experimental works

Methods	Conditions and other notes	H (GPa)
MD nanoindentation simulation in this work	Loading rates of 0.25–2.0 Å/ps	1.492–2.266
Nanoindentation experiments in previous works	Epoxy reinforced nanocomposites with fumed silica of 0–3% [18]	0.256–0.318
	RTM6 Epoxy resin with strain rate of 0.01–0.25 and 25 °C [17]	0.23–0.35
	Epoxy resin at different loading rates of 0.25–6.0 mN/s [16]	0.26–0.276

Note RTM: resin transfer moulding

to stabilize within the range of 0.75–2.0 Å/ps. These observations were consistent with previous nanoindentation studies [16, 17], which also reported that the hardness of epoxy resin increased with the loading rate.

4 Conclusions

Molecular dynamics (MD) simulation method was used to investigate the interfacial mechanical behavior of epoxy and quartz, considering the effect of loading rate. The uniaxial tension tests of bulk epoxy system were conducted for validation of the MD models. The main conclusions of this study are as follows:

1. The overall Young's modulus of the bulk epoxy and average Poisson's ratio were 2.075 GPa and 0.377, respectively. The reduced modulus of epoxy substrate was 2.419 GPa, and its indentation hardness ranged from 1.492 to 2.174 GPa. These simulation results were consistent with previous simulation and experimental works, validating the accuracy of the epoxy model.
2. During the nanoindentation process, the higher the loading rate, the smaller the distance of surface connecting atoms between quartz and epoxy, and the higher the indentation force F_z and hardness H of the epoxy substrate.
3. The relationship between the indentation force F_z and the indentation depth h_c followed a power-law for the nanoindentation process, and the surface wear of the epoxy substrate was an obvious "groove".

This work could provide molecular insight into the interfacial mechanical behavior of epoxy and quartz, reflecting their local deformation behavior. Furthermore, the nanoscratching behavior of epoxy-quartz, the effect of indenter shape and size, and their adhesion effects are worth developing in the future.

References

1. Aliasghar-Mamaghani M, Khaloo A (2021) Effective flexural stiffness of beams reinforced with FRP bars in reinforced concrete moment frames. J Compos Constr 25(1)
2. Wan J et al (2024) Molecular dynamics simulations of the interface friction behavior between fiber-reinforced polymer pile and sand. Tribol Int 192:109288
3. Xu WQ, Yin ZY, Zheng YY (2023) FRP-soil interfacial mechanical properties with molecular dynamics simulations: Insights into friction and creep behavior. International J Numer Anal Methods Geomech
4. Shaia HA, Al-Asadi AK, Ramadan SH (2018) Evaluation of the interface friction between fiber-reinforced polymers and granular materials using modified shear apparatus. Int J Civ Eng Technol 9(13):1010–1016
5. Abyaneh MJ, El Naggar H, Sadeghian P (2020) Numerical modeling of the lateral behavior of concrete-filled FRP tube piles in sand. Int J Geomech 20(8)
6. Zappalorto M et al (2015) Mechanical behaviour of epoxy/silica nanocomposites: experiments and modelling. Compos A Appl Sci Manuf 72:58–64
7. Yaphary YL et al (2017) Molecular dynamics simulations on adhesion of epoxy-silica interface in salt environment. Compos B Eng 131:165–172
8. Frenkel D, Smit B (2001) Understanding molecular simulation: from algorithms to applications. Academic Press
9. Plimpton S (1995) Fast paraller algorithms for short-range molecular dynamics. J Comput Phys 117:1–19
10. Amaral CR et al (2014) Impact of aliphatic amine comonomers on DGEBA epoxy network properties. Polym Eng Sci 54(9):2132–2138
11. Muller A et al (2010) Influence of particle diameter on coefficient of thermal expansion of SiO_2/epoxy particulate composites. In: Process SAMPE fall technology conference. Salt Lake City
12. Li C, Strachan A (2011) Molecular dynamics predictions of thermal and mechanical properties of thermoset polymer EPON862/DETDA. Polymer 52(13):2920–2928
13. Tack J (2006) Thermodynamic and mechanical properties of Epon 862 with curing agent detda by molecular simulation. Texas A&M University
14. Ma Z, Gamage RP, Zhang C (2021) Mechanical properties of α-quartz using nanoindentation tests and molecular dynamics simulations. Int J Rock Mech Min Sci 147:104878
15. Oliver WC, Pharr GM (1992) An improved technique for determining hardness and elastic modulus using load and displacement sensing indentation experiments. J Mater Res 7(6):1564–1583
16. Fu K et al (2014) Effect of loading rate on the creep behaviour of epoxy resin insulators by nanoindentation. J Mater Sci Mater Electron 25(8):3552–3558
17. Frontini P et al (2015) High temperature nanoindentation response of RTM6 epoxy resin at different strain rates. Exp Mech 55(5):851–862
18. Tzetzis D et al (2013) Nanoindentation measurements of fumed silica epoxy reinforced nanocomposites. J Reinf Plast Compos 32(3):160–173

Evaluation of Bond Strength Between FRP Bars and Concrete Under Seawater Immersion Using Explainable Machine Learning Models

Heng Cao, Xiao Lin Zhao, Mudassir Iqbal, Daxu Zhang, and Pei-Fu Zhang

Abstract Floating structures are required to mitigate the issues of overcoming the land shortage in Hong Kong and other densely populated coastal cities. For this purpose, conventional steel pontoons need to be replaced by fiber-reinforced polymer (FRP) from the standpoint of durability and economy. Among various issues of FRP and concrete in coastal areas, the bond performance between FRP bars and concrete is critical in structural design and application. This study compiled 219 real pull-out test records with seven input parameters to investigate the bond performance between FRP bars and concrete under seawater immersion. Based on the dataset, three machine learning algorithms, namely Decision Tree, Random Forest, and XGBoost, were employed to construct predictive models for the bond strength between FRP bars and concrete under seawater immersion. Evaluation metrics, including R^2, MAE, and RMSE, were exploited to calculate the performance of the developed models. Due to the superior predictive performance of the Random Forest model, this study further employed SHAP analysis to investigate the effects of influencing parameters on the bond strength. The developed models and the SHAP analysis provided a reference for assessing and designing bond performance between FRP bars and concrete under seawater immersion conditions.

Keywords FRP bars · Bond strength · Seawater immersion · Machine learning · SHAP analysis

H. Cao · M. Iqbal (✉) · D. Zhang · P.-F. Zhang
School of Ocean and Civil Engineering, Shanghai Jiao Tong University, Shanghai, China
e-mail: mudassiriqbal29@sjtu.edu.cn

X. L. Zhao
Research Institute for Land and Space, The Hong Kong Polytechnic University, Hong Kong, China

X. L. Zhao et al. (eds.), *Proceedings of the Fourth World Conference on Floating Solutions: WCFS 2024; 2–4 December; Hong Kong*, Lecture Notes in Civil Engineering 597, https://doi.org/10.1007/978-981-96-4569-5_71

1 Introduction

Introducing floating structures as a complementary approach to land reclamation is necessary to overcome the land shortage challenge and sustainable growth of densely populated areas [1]. The use of conventional steel is prone to durability issues on account of the corrosion of steel bars, resulting in substantial economic losses. Instead, fiber-reinforced polymer (FRP) is a high-performance material characterized by its lightweight, high strength, and excellent corrosion resistance. Substituting FRP bars for steel bars in concrete is considered a proficient solution to mitigate the corrosion issues of traditional steel-reinforced concrete, particularly in harsh environments.

The bond performance between FRP bars and concrete is crucial in their synergistic operation, and the bond strength between them is a significant attribute to consider in structural design and application. While researchers have generally agreed on bond strength under ordinary environmental conditions [2], the bond performance in aggressive environments, such as seawater immersion conditions, remains inadequately understood. Despite several experimental studies [3–6], the bond performance between FRP bars and concrete in such environments remains to be comprehensively explicated. Notably, machine learning, an artificial intelligence method with a distinct advantage in solving complex and non-linear problems, has been increasingly employed in civil engineering research, demonstrating high feasibility and effectiveness [7, 8]. In light of the limitations of traditional research methods and the excellence of machine learning approaches, this study leveraged three explainable machine learning models in the domain of FRP bars and concrete to introduce a novel method for evaluating bond strength between fiber-reinforced polymer (FRP)bars and concrete under seawater immersion.

2 Methodology

2.1 Machine Learning Algorithms

Machine learning is a subfield of artificial intelligence wherein a computational system processes a given dataset to generate a model capable of addressing specific problems [9]. In this study, three tree-based machine learning algorithms, namely Decision Tree (DT), Random Forest (RF) and XGBoost, were employed to construct regression models to predict the bond strength between FRP bars and concrete subjected to seawater immersion solutions. Among them, a Decision Tree is a typical supervised learning algorithm that utilizes a hierarchical structure to classify data or predict outcomes. It operates through a recursive division of the dataset into more diminutive subsets based on the input features' values and then uses the majority vote of the instances in each subset to make a prediction. Random Forest is an ensemble learning technique that augments the precision and robustness of predictions by amalgamating an array of decision trees. Each tree is generated from a stochastic selection of the training dataset and the feature set. XGBoost is an extreme gradient boosting algorithm that iteratively trains decision trees on the residuals of the preceding trees to correct their errors, thereby improving model performance.

2.2 Evaluation Criterion

For a thorough assessment of the machine learning regression models' efficacy, three metrics are utilized for evaluation: Mean Absolute Error (MAE), Root Mean Squared Error (RMSE) and the coefficient of determination (R^2). Equations (1)-(3) detail the calculation methods for these criteria, where $\hat{y}_i$ represents the predicted value, y_i denotes the true value and n signifies the total number of the samples. In model evaluation, lower values of MAE and RMSE indicate better predictive performance, while an R^2 value closer to 1 signifies a higher degree of prediction accuracy. Moreover, it is also crucial to consider the discrepancy in evaluation criteria between the training and testing sets, as substantial discrepancies typically suggest model overfitting, which needs to be addressed through hyperparameter optimization to enhance generalizability.

$$MAE = \frac{1}{n}\sum_{i=1}^{n}|\hat{y}_i - y_i| \tag{1}$$

$$RMSE = \sqrt{\frac{1}{n}\sum_{i=1}^{n}(\hat{y}_i - y_i)^2} \tag{2}$$

$$R^2 = \frac{\sum_{i=1}^{n}(y_i - \hat{y}_i)^2}{\sum_{i=1}^{n}(y - \overline{y})^2} \tag{3}$$

2.3 SHAP Analysis

SHAP (SHapley Additive exPlanations) analysis is a robust method for interpreting machine learning models [10]. In SHAP analysis, the machine learning model generates the corresponding predicted value for each data containing all input parameters. The SHAP value is calculated according to Eq. (4), where y_i denotes the predicted value of the ith sample, y_{base} is the base value (typically the average value of the predicted variable), and $f\left(x_{i,j}\right)$ represents the SHAP value, which measures the impact that the jth parameter exerts on the forecasted value for the ith sample. A positive $f\left(x_{i,j}\right)$ implies that the jth parameter increases the predicted value of the ith sample, while a negative $f\left(x_{i,j}\right)$ indicates a decrease. The absolute value of $f\left(x_{i,j}\right)$ signifies the magnitude of the impact. SHAP analysis offers a unique advantage over other interpretability methods like Permutation Importance and tree-based Feature Importance by not only ranking input parameters' impact but also visualizing their effects on predictions through SHAP value calculations.

$$y_i = y_{base} + f\left(x_{i,1}\right) + f\left(x_{i,2}\right) + \cdots + f\left(x_{i,j}\right) + \cdots + f\left(x_{i,k}\right) \tag{4}$$

3 Datasets

A comprehensive dataset is paramount in machine learning, as the accuracy and versatility of the developed models are contingent upon the quality of the dataset [9]. To ensure a robust and accurate dataset of bond performance between FRP bars and concrete subjected to seawater immersion, this study extensively reviewed relevant literature and established four rules for data collection:

(1) Only tests conducted under seawater immersion, including natural seawater immersion and artificial seawater immersion according to standard configuration, were selected.
(2) Only tests conducted through cube direct pull-out tests were selected, while tests carried out through beam or cylinder pull-out tests were filtered out.
(3) Only tests whose failure mode was pull-out failure were selected, while tests with failure modes of splitting failure were filtered out.
(4) Only tests with sufficient parameters, including test materials, procedures, and results, were selected, while tests with missing parameters were filtered out.

According to the mentioned conditions, a total of 219 experimental datasets were collected from relevant research papers [3–6, 11–14]. The statistical overview of the parameters is encapsulated in Table 1. Given that the parameters "FRP bar type" and "Surface treatment" are categorical variables that cannot be directly received by the machine learning models, these two parameters were processed by sequential coding detailed in Table 2. The Pearson coefficient was calculated to gain a preliminary understanding of the internal relationships between the parameters, as shown in Fig. 1. The Pearson coefficient measures the correlation between two parameters, ranging from −1 to 1. The closer the absolute value is to 1, the stronger the linear correlation between the parameters. As observed in Fig. 2, all the parameters in this study exhibit weak linear correlations, indicating a complex internal nonlinearity and underscoring the necessity to develop advanced machine learning models.

Table 1 Statistical overview of the parameters

Parameters	Type	Max	Min	Average	Std
FRP bar type	Input	3	1	1.6	0.6
Surface treatment	Input	6	1	2.9	2.0
FRP bar tensile modulus (GPa)	Input	152.1	42.8	61.0	27.1
FRP bar diameter (mm)	Input	12.0	8.0	10.0	1.6
Concrete compressive strength (MPa)	Input	66.9	26.6	50.4	8.7
Corrosion age (d)	Input	360	0	85.3	95.9
Corrosion temperature (°C)	Input	60	25	35.7	21.2
Bond strength (MPa)	Output	30.51	10.21	18.98	4.21

Table 2 Sequential coding of categorical parameters

FRP bar type	BFRP	GFRP	CFRP
Coding	1	2	3

Surface treatment	Helically wrapped (HW)	Ribbed (RB)	Sand coated (SC)	Helically grooved (HG)	Ribbed and sand coated (RBSC)	Helically wrapped and sand coated (HWSC)
Coding	1	2	3	4	5	6

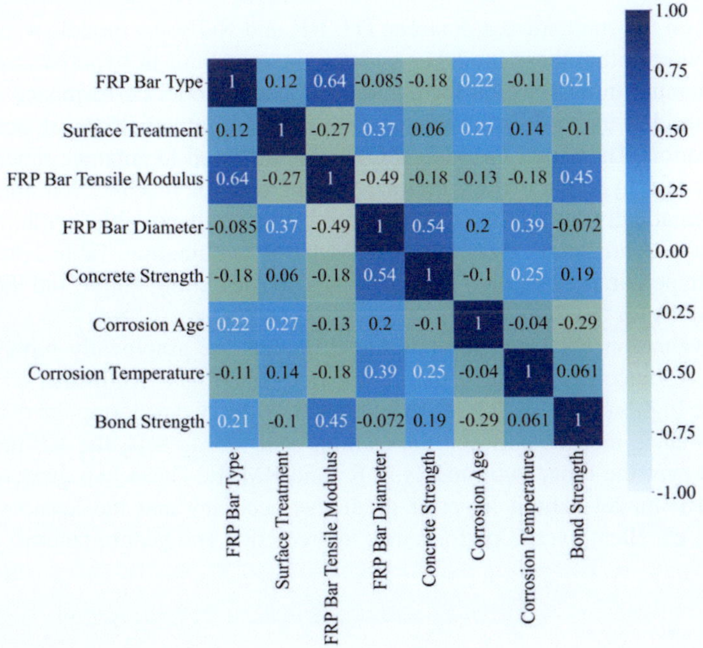

Fig. 1 Pearson correlation coefficient heatmap of the parameters

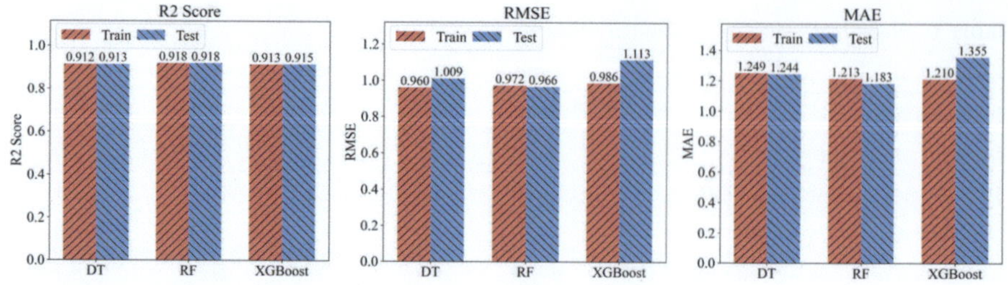

Fig. 2 Performance of the models

4 Model Development and Parameter Analysis

4.1 Model Development

Following the determination of datasets, the data was randomly partitioned to two subsets: 80% for training the model and 20% for testing. The input parameters were standardized to mitigate the influence of the magnitude and dimension of the input parameters on model training. Equation (5) illustrates the method for standardization.

$$X_{\text{stand}} = \frac{X-\mu}{\sqrt{\sigma^2}} \tag{5}$$

Based on the standardized datasets, DT, RF, and XGBoost models were trained. It is noteworthy that these models possess numerous built-in hyperparameters that require manual tuning, which significantly impacts model performance. Machine learning models using default hyperparameters often underperform in accuracy or generalization. Therefore, GridSearchCV was employed to enhance model performance. It is a hyperparameter optimization method for machine learning models that systematically uses cross-validation to evaluate all possible combinations of selected hyperparameters to determine the optimal combination. Table 3 presents the selected hyperparameters, their default values, the searching range, and the optimal combinations for the three models.

The evaluation results of the optimized models are graphically represented in Fig. 2. The outcomes revealed that, compared to DT and XGBoost, RF demonstrated the highest R^2, lowest RMSE, and lowest MAE. Considering the discrepancy in performance metrics between the training and testing sets, the RF model also prevailed over the other two models in R^2 and RMSE. These two aspects suggest that the RF model exhibit superior predictive accuracy and the least overfitting, indicating excellent overall performance in prediction and generalization.

Table 3 Hyperparameters for model tunning in GridSearchCV

Models	Hyperparameters	Default	[Searching range] (increment)	Optimal
DT	max_depth	None	[1,100] (1)	8
	max_features	None	[1, 7] (1)	6
RF	n_estimators	100	[10, 20, 50, 100, 200]	200
	max_features	auto	[1, 7] (1)	6
	max-depth	None	[5, 7, 10, 15, 20]	7
XGBoost	n_estimators	100	[20,200] (1)	21
	min_child_weight	1	[2, 10] (1)	7
	gamma	0.0	[0.0, 0.1, 0.2, 0.3, 0.4, 0.5]	0.2

4.2 Parameter Analysis

To comprehensively evaluate the bond performance between FRP bars and concrete subjected to seawater immersion conditions, parameter analysis on the strength of the SHAP analytical approach was conducted to assess the impact of each input parameter on the output variable, namely, bond strength. Given the superior performance of the RF model, SHAP analysis was carried out on the predictive outcomes yielded by the RF model.

The summary plot of SHAP values is depicted in Fig. 3a, where each point represents a sample, the color of the point indicates the value of the parameters, and the abscissa represents the magnitude of the SHAP value. Figure 3b shows the ranking of each parameter based on feature importance calculated by average SHAP values. It can be seen that, among all the seven input parameters, FRP bar tensile modulus, concrete strength, and surface treatment exert a pronounced effect on the bond strength between FRP bars and concrete under seawater immersion, while the impact from FRP bar type, corrosion temperature, and FRP bar diameter is relatively insignificant.

The effect of each input parameter on bond strength is detailed in Fig. 4. Concerning FRP bar tensile modulus and concrete strength, SHAP values gradually increase as their values increase, which can be seen in Fig. 4a, b, showing higher FRP bar tensile modulus and concrete strength lead to higher bond strength between FRP bars and concrete subjected to seawater immersion. An explanation for this phenomenon could be that concrete with high compressive strength often has denser internal microstructure and better compactness, which can inhibit the generation of cracks, thereby reducing the corrosion effect of seawater on FRP bars [8].

For corrosion age and temperature, Fig. 4d, f show that as their values increase, SHAP values decrease, indicating a negative effect on the bond strength. On one hand, longer immersion time leads to more micro-damage caused by seawater corrosion. On the other hand, the coefficients of thermal expansion for FRP bars and concrete are discrepant. Upon temperature fluctuations, the rates of thermal expansion and contraction of FRP bars exceed those of concrete [15]. This phenomenon generates

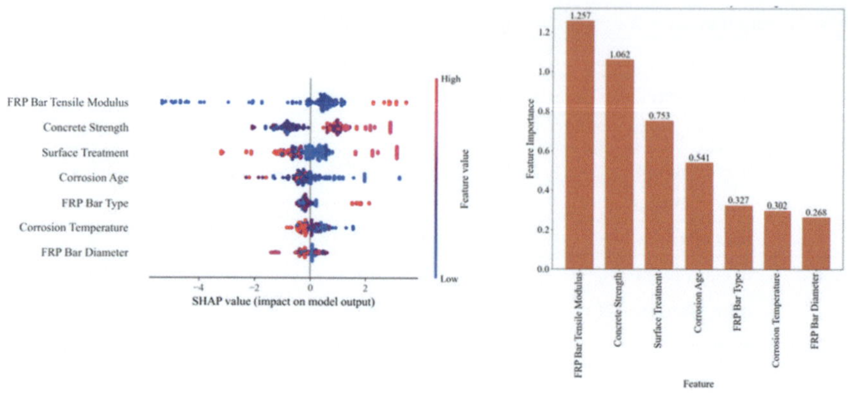

(a) Summary plot of SHAP values (b) Feature importance of input parameters

Fig. 3 Results of SHAP analysis

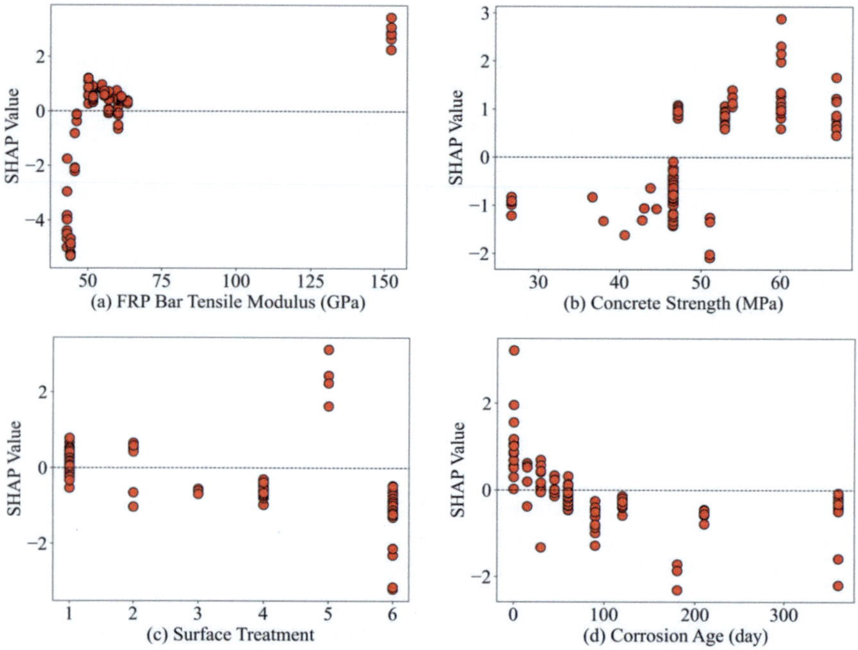

Fig. 4 SHAP value distribution of each input parameter

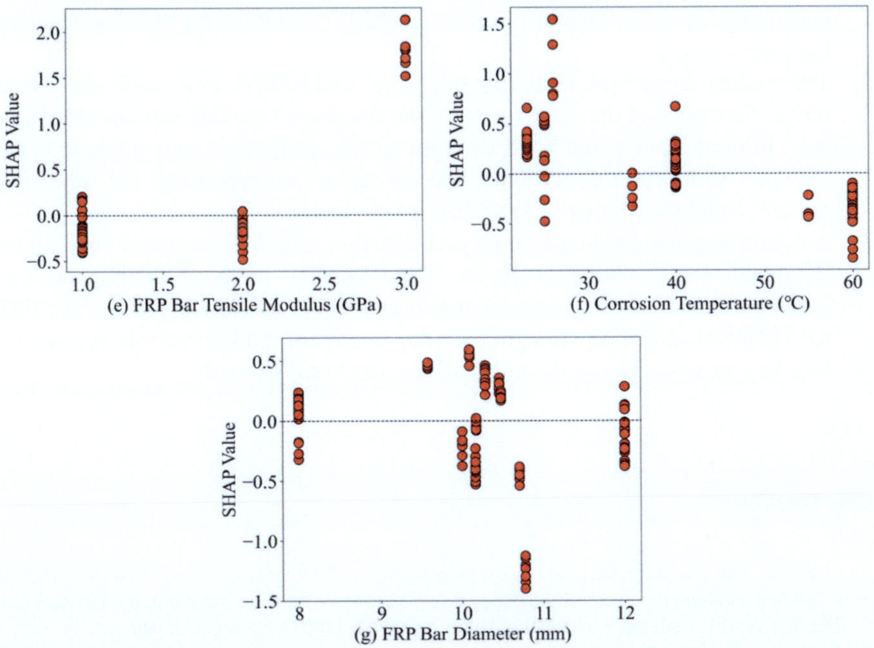

Fig. 4 (continued)

radial stress and produces micro-cracks and voids at the interface between FRP bars and concrete, decreasing the bond strength [16]. Additionally, Fig. 4c, e show that the surface treatment of 5, i.e., ribbed and sand-coated, along with the FRP type of 3, i.e., CFRP bars help improve the adhesive strength between FRP bars and concrete subjected to seawater immersion, while Fig. 4g shows the impact pattern of FRP bar diameter is not significant, corresponding to the minimum influence in Fig. 3.

5 Conclusions

In this study, a comprehensive dataset of bond performance between FRP bars and concrete in seawater immersion solutions was established, and based on the dataset, three explainable machine learning models, namely DT, RF, and XGBoost, were developed for predicting the bond strength between FRP bars and concrete. Additionally, the SHAP analysis method was employed to assess the impact of different input parameters on bond strength. The subsequent specific conclusions can be delineated:

(1) A dataset comprising seven input parameters and one output parameter was established. The dataset comprises 219 real experimental data points from direct pull-out tests collected from previously published research papers. This accurate

and comprehensive dataset provides a solid foundation for training machine learning models.

(2) The evaluation metrics, including R^2, MAE, and RMSE, were utilized to assess the performance of the three models. The developed models demonstrated efficacy in prediction of the bond strength between FRP bars and concrete under seawater immersion conditions. The RF model outperformed the other two models, exhibiting superior predictive performance.

(3) A comprehensive parameter analysis based on predictive outcomes yield by the RF model was conducted using the SHAP analysis method. The impact of all seven parameters was visualized and discussed. It was concluded that the FRP bar modulus, concrete strength, surface treatment, and corrosion age are the four key parameters significantly influencing bond strength.

References

1. Zhao XL, Dai JG, Ding XL et al (2023) Developing a sustainable and smart floating structure solution for enhancing liveability in Hong Kong's crowded built environment. In: Tabeta S (ed) The 3rd World conference on floating solutions (WCFS2023). Springer, Japan
2. Yin S, Hua Y, Xu S (2021) A review on research progress and application of concrete structures internally reinforced with FRP bars. Jianzhu Jiegou Xuebao/J Build Struct 42:134–150
3. El Refai A, Abed F, Altalmas A (2015) Bond durability of basalt fiber-reinforced polymer bars embedded in concrete under direct pullout conditions. J Compos Constr 19:04014078
4. Ruiz Emparanza A, Kampmann R, De Caso F et al (2022) Durability assessment of GFRP rebars in marine environments. Constr Build Mater 329:127028
5. Ruiz Emparanza A, De Caso Y Basalo F, Kampmann R et al (2018) Evaluation of the bond-to-concrete properties of GFRP rebars in marine environments. Infrastructures **3**:44
6. Lu Z, Su L, Lai J et al (2021) Bond durability of BFRP bars embedded in concrete with fly ash in aggressive environments. Compos Struct 271:114121
7. Ganaie MA, Hu M, Malik AK et al (2022) Ensemble deep learning: a review. Eng Appl Artif Intell 115:105151
8. Zhang P-F, Iqbal M, Zhang D et al (2024) Bond strength prediction of FRP bars to seawater sea sand concrete based on ensemble learning models. Eng Struct 302:117382
9. Baduge SK, Thilakarathna S, Perera JS et al (2022) Artificial intelligence and smart vision for building and construction 4.0: machine and deep learning methods and applications. Autom Constr **141**:104440
10. Lundberg S, Lee S-I (2017) A unified approach to interpreting model predictions, pp 4–7. Preprint at https://doi.org/10.48550/arXiv.1705.07874
11. Taha A, Alnahhal W (2021) Bond durability and service life prediction of BFRP bars to steel FRC under aggressive environmental conditions. Compos Struct 269:114034
12. Wu G, Dong Z, Xu B et al (2016) Bond performance and calculation method for the basic anchorage length of BFRP bar with concrete in marine environment. Tumu Gongcheng Xuebao/China Civ Eng J 49:89–99
13. Xu B, Study on bond behavior between FRP bar and concrete in marine environment. Ph.D. Thesis, Southeast University, China
14. Altalmas A, El Refai A, Abed F (2015) Bond degradation of basalt fiber-reinforced polymer (BFRP) bars exposed to accelerated aging conditions. Constr Build Mater 81:162–171

15. Belarbi A, Wang H (2012) Bond durability of FRP bars embedded in fiber-reinforced concrete. J Compos Constr 16:371–380
16. Masmoudi R, Masmoudi A, Ouezdou MB et al (2011) Long-term bond performance of GFRP bars in concrete under temperature ranging from 20 °C to 80 °C. Constr Build Mater 25:486–493

Tensile Behaviour of Inter-modular Self-locking Connections of Trussed Tubular Hybrid Structures

Xialu Liu and Chao Hou

Abstract Advanced marine structures are in demand for the efficient acquisition of marine resources. The trussed tubular hybrid structure incorporating composite members such as concrete-filled steel tubes (CFST) exhibits excellent structural performance and easy-to-take anti-corrosion measures, indicating its great potential to be utilized in floating structures such as offshore platforms for wind turbines. Applying modular connection technology to trussed tubular hybrid structures effectively improves the on-site construction speed, and reduces the impact of harsh marine environments on the construction quality. In this study, three types of novel inter-modular self-locking connection configurations with better sealing performances are proposed, which would be suitable for marine modular trussed tubular hybrid structures. The connection design is optimized through pre-splicing tests, and the mechanical performance of the connection under axial tension is then revealed through experimental tests. The test results demonstrate that the self-locking connection exhibits excellent tensile behaviour, where the pressing blocks and the shear keys in the connection can jointly resist the tensile force as a whole. Fracture at the upper plate of the sliding groove is identified as the typical failure mode. It is found that increasing the thickness of the upper plate at the sliding groove can effectively improve the tensile strength of the proposed connection.

Keywords Self-locking connection · Trussed tubular hybrid structure · Pre-splicing test · Tensile test

---

X. Liu · C. Hou (✉)
Department of Ocean Science and Engineering, Southern University of Science and Technology, Shenzhen, People's Republic of China
e-mail: houc@sustech.edu.cn

X. L. Zhao et al. (eds.), *Proceedings of the Fourth World Conference on Floating Solutions: WCFS 2024; 2–4 December; Hong Kong*, Lecture Notes in Civil Engineering 597, https://doi.org/10.1007/978-981-96-4569-5_72

1 Introduction

The sustainable development and utilization of marine resources are essential for maintaining ecological balance and promoting sustainable development. Advanced marine structures are increasingly sought after for efficient resource acquisition. Floating structures, in particular, offer an economical solution for deep-sea energy collection and can be designed for diverse applications, including wind energy development, natural gas extraction, oil exploration, aquaculture, and land supplementation for airport facilities, as illustrated in Fig. 1 [1–3].

As the demand for marine resources and energy development continues to grow, large-scale offshore infrastructure has become a critical focus in structural engineering. Advanced marine structures are essential for the efficient extraction of these resources. However, the harsh and corrosive marine environments necessitate infrastructure that is high-performance and easy to construct. Traditional structural designs often fail to meet the requirements for rapid construction and long-term reliable service in such conditions. As illustrated in Fig. 1, certain offshore platform structures employ truss systems to provide effective load-bearing capacity and mitigate wave impacts. Trussed tubular hybrid structures incorporating composite members such as concrete-filled steel tubes (CFST) have successfully been used in large-scale bridges, towers, and airport terminals, which are notable for high material efficiency, excellent bending and torsional resistance, and easy-to-take anti-corrosion measures [4]. Compared to hollow tubular trusses, the performance of trussed tubular hybrid structures is significantly improved due to the interaction of materials and the constraint effect of the steel tube on the chord members [5–7].

Trussed tubular hybrid structures must also facilitate on-site installation when applied to large-scale floating structures. Inter-module connections (IMCs), crucial for the rapid construction of modular buildings, are responsible for load transfer and deformation coordination, significantly impacting the overall structural performance. IMCs can be mainly categorized into six types based on the adopted connecting methods: welded, bolted, prestressed, rotary, self-locking, and grouted. As an efficient connection method, self-locking connection avoids complex construction processes and demands on additional construction space [8], which is more suitable for marine structures with harsh construction conditions.

(a) Oil exploitation platform [1] (b) Wind turbines [2] (C) Offshore fish farms [3]

Fig. 1 Typical applications of floating structures

Implementing IMCs in trussed tubular hybrid structures enhances connection quality assurance during processing and construction. It effectively addresses challenges posed by harsh construction conditions, high on-site costs, and quality control issues in marine engineering applications of trussed tubular hybrid structures. However, current IMCs are predominantly tailored for modular steel construction with square column sections, with fewer options proposed for other column configurations. When such a connection is applied to composite columns with higher strength, higher requirements are placed on its mechanical properties, which increases the design difficulty.

Therefore, this study proposes three types of novel inter-modular self-locking connection (SLC) configurations, which would be suitable for marine modular trussed tubular hybrid structures. Given the corrosive marine environment, traditional IMC elements like bolts and perforations are avoided. Instead, self-locking connections with strong sealing properties are proposed to minimize assembly complexities. The feasibility of these designs is ensured through pre-splicing tests, and their mechanical performance under axial tensile load is studied through experimental testing.

2 Design and Performance of Novel SLC

2.1 Design of Novel SLCs

Three types of inter-modular SLCs have been developed, emphasizing excellent sealing properties and minimized on-site assembly. These SLCs were designed for symmetric cross-sections and adaptable trussed tubular hybrid structures, eliminating the need for surface holes and on-site welding, thereby reducing construction errors. The types included circumferential SLC, flange-reinforced SLC, and rod-stiffened SLC, as illustrated in Fig. 2. All three types of SLC forms are suitable for circular CFST module columns.

The circumferential SLC, depicted in Fig. 2a, comprises key components including a perforated sleeve, shear keys, pressing blocks, plug-in short tube, and positioning tube. The shear keys, crucial for self-locking functionality, featured a trapezoidal cross-section with outer sloped and inner arc surfaces connected to springs. The perforated sleeve was a steel sleeve with sliding grooves aligned to the shear keys arrangement, housing springs, and triangular sections of the shear keys outside these sliding grooves. At the upper steel tube end, pressing blocks, triangular in shape, were positioned according to the shear keys layout. Plug-in short tubes were centrally placed within the upper column to enhance assembly efficiency and prevent relative rotation and horizontal displacement between modular columns. The tube was internally connected to the upper steel tube and filled with concrete, featuring external positioning bars. Positioned at the bottom modular column, the positioning tubes were connected to the perforated sleeve, incorporating grooves

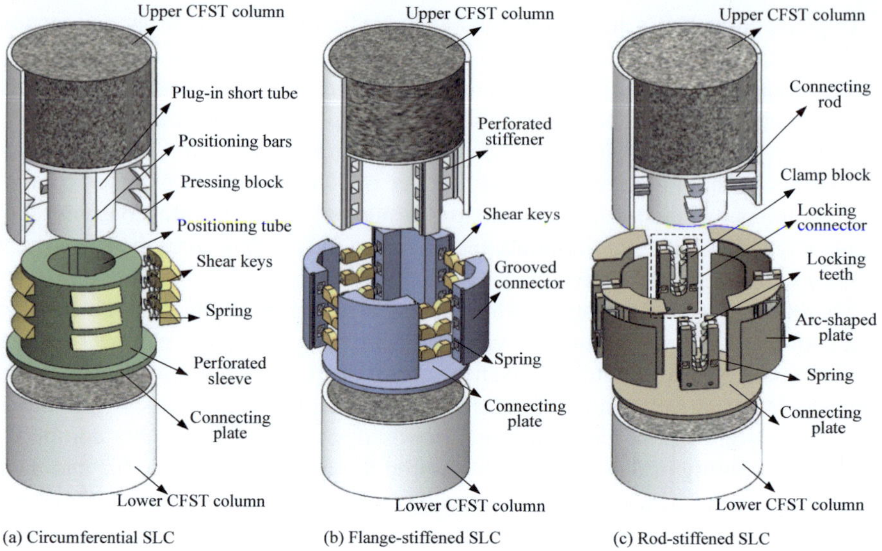

Fig. 2 Three novel inter-modular SLCs suitable for trussed tubular hybrid structure

inside for the positioning bars. During assembly, the upper column moves downwards, and the plug-in short tube inserts into the positioning tube, restricting the horizontal displacement of modular columns. As the upper column continues downward, the sloped surfaces of the pressing block and shear key make contact, exerting pressure on the shear key. This pressure causes the shear key to move inward along the sliding grooves until fully seated. Subsequently, the pressing block advances over the shear keys, which spring back to their original positions due to the spring force. Once the pressing blocks pass all shear keys, the upper and lower modular columns achieve a complete self-locking connection.

The flange-reinforced SLC, shown in Fig. 2b, comprises essential components such as perforated stiffeners, shear keys, and grooved connectors. Perforated stiffeners were positioned on both sides at the upper column end, connecting to the inner and outer steel tubes. Grooved connectors were fixed to the lower column's connection plate, featuring grooves on each side to accommodate shear keys fixed with springs. The shear keys outside the grooves have curved surfaces to facilitate compression and rebound. During assembly, as the upper column moves downward, the perforated stiffeners insert between the grooved connectors, compressing the shear keys into the grooves on both sides. As the holes in the perforated stiffeners align with the sliding grooves, the shear keys spring back and engage with the perforated stiffeners, completing the self-locking connection.

Figure 2c illustrates a developed rod-stiffened SLC featuring a connecting rod between the inner and outer steel tubes of the upper modular column. The connecting rod included teeth on both sides and a curved surface on the lower side. In the lower column, locking connectors were installed, comprising locking teeth, springs, and clamp blocks. These locking connectors were linked by arc-shaped plates and secured to the lower column's connection plate. During assembly, as the connecting rods move downward, pressure causes the locking teeth to open sideways. When the connecting rod reaches the bottom of the connector, the locking teeth rebound under the action of the spring, engaging with the connecting rod to complete the connection.

2.2 Pre-splicing Tests of Novel SLC

To validate the feasibility of the developed SLCs, initial research focused on the circumferential SLC. Due to its effective sealing properties, a pre-splicing test was conducted to assess internal engagement before mechanical testing. Figure 3 illustrates the engagement status of shear keys within the connection area post-assembly. In column C1, the middle shear key was fully ejected, while the upper and lower keys rebounded by 1/2. Column C2 showed complete ejection of both upper and lower shear keys, with the middle key rebounding 1/3. Column C3 exhibited full ejection of the upper shear key, minimal ejection in the middle, and approximately 1/3 ejection in the lower. In column C4, the shear key was fully ejected, while the middle and lower rebounded 1/2. Overall, the shear keys indicate that 5/12 fully ejected, 6/12 rebounded to 1/3–1/2 positions, and 1/12 did not eject. This result may be caused by underutilized spring force or manufacturing errors, where the height of the pressing block's top surface exceeds that of the shear key's bottom surface. Hence, these findings highlight the need for further optimization in SLC design to enhance performance and reliability.

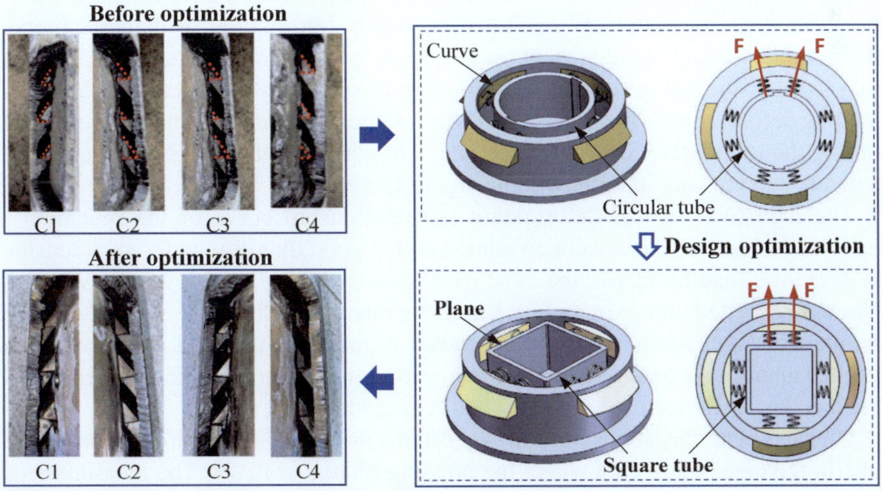

Fig. 3 Pre-splicing tests and design optimization of SLC

Upon analyzing the connection design, observed that the spring force direction does not align consistently with the shear key's rebound direction, hindering complete rebounding. Therefore, the cross-section of the positioning tube was changed from circular to square, and the vertically curved surface on the shear key, connected to the spring, was replaced with a flat surface. This modification ensured that both sides connected to the spring were vertical planes, optimizing the utilization of spring stiffness during the self-locking process. Following these adjustments, a second pre-splicing test was conducted on the improved SLC. As depicted in Fig. 3, the shear keys exhibited nearly complete rebounding, successfully achieving full engagement with the pressing blocks.

2.3 Axial Tensile Test of SLC

To reveal the tensile performance of the optimized circumferential SLC, an axial tension test was conducted on the single-layer shear key SLC. The SLC was positioned in the middle of the specimen, with upper and lower circular CFST columns. The test setup is shown in Fig. 4a. After loading, a noticeable gap was observed at the connection of the upper and lower module columns of the specimen, with no significant deformation on the steel tube surface. To examine the internal connection damage and the core concrete of module columns, the outer steel tube of the specimen was cut open. The failure mode is shown in Fig. 4b, with cracking damage occurred at the upper plate of the sliding groove. Under tensile force, the shear keys resisted the force transmitted by the pressing block and conveyed it to the top plate of the sliding groove, causing upward bending. When the top plate reached a certain displacement, the corners cracked and could no longer sustain the tensile force from the shear key. The shear key exhibited no major deformation or damage, though distinct indentations were present on the upper surface ends and small notches on both sides of the lower surface. The pressing blocks showed noticeable downward bending deformation at both ends, while the middle part remained undeformed. Upon cutting the outer steel tube to inspect the core concrete of the CFST columns, the concrete surface was smooth, with no significant cracks.

The load (N)–vertical deformation (Δ) relationship curve of the specimen is depicted in Fig. 5a. The specimen demonstrated good ductility under axial tension. In the initial loading phase, the axial load increased rapidly with the growth of Δ. When Δ exceeded approximately 1.5 mm, the rate of load increase slowed, entering the plastic stage. As deformation continued to increase, the load rose slowly and reached an ultimate load of 415.5 kN at $\Delta = 10.1$ mm. Thereafter, the load remained and then began to decrease rapidly when Δ exceeded 12.2 mm.

Four sets of strain gauges were arranged along the height direction of the outer steel tube in the connection area where the pressing blocks were set. The measurements from these gauges were essentially the same, indicating that under tensile force, the deformation of the circumferential pressing blocks was generally uniform, allowing them to jointly resist the tensile force. The typical longitudinal N–ε relationship

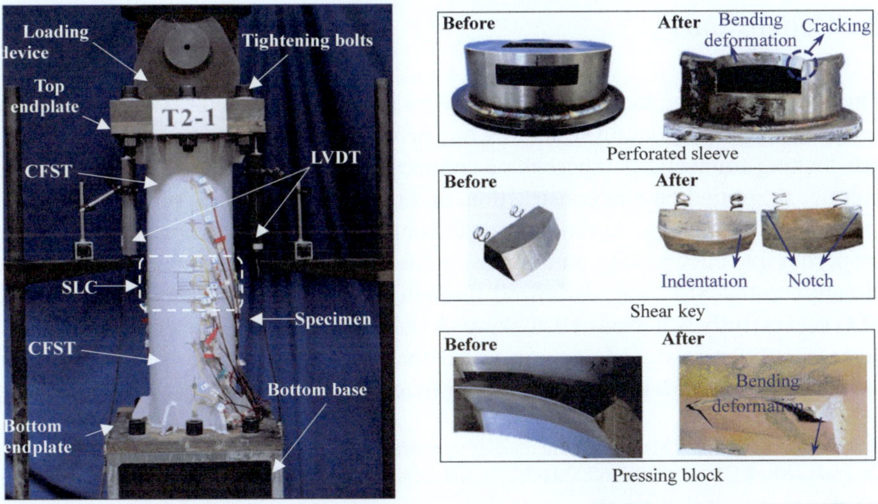

(a) Test setup (b) Failure modes

Fig. 4 Tensile test setup and failure modes

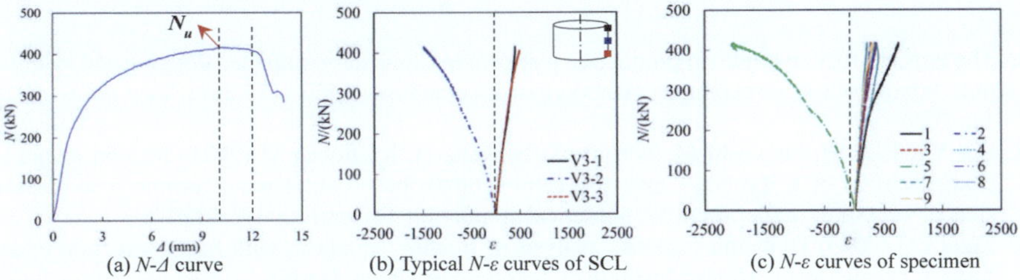

(a) *N*-*Δ* curve (b) Typical *N*-*ε* curves of SCL (c) *N*-*ε* curves of specimen

Fig. 5 Typical relationship curves

curve for the connection is presented in Fig. 5b. The longitudinal strain development exhibited a consistent trend. During loading, the shear keys resisted the upward movement of the pressing block. The pressure applied to the bottom of the pressing block caused it to expand outward, leading to local compression of the outer steel tube. Consequently, the middle strain recorded by the blue strain gauge was negative. The *N*-*ε* relationship curve for the specimen is shown in Fig. 5c. Strain gauge No. 1 was located at the loading end, with the gauges numbered sequentially towards the fixed end, and strain gauge No. 9 was positioned at the fixed end. When the tensile force was less than 250 kN, the strain values generally increased linearly with the tensile force. Besides the compressive deformation around the pressing block, other measurement points showed tensile deformation.

3 Conclusions

This paper addressed the challenges faced by trussed tubular hybrid structures when applied to offshore floating structures, and proposed three new types of inter-module self-locking connections. These three connections avoided the need for welding and bolting, simplified on-site construction, and reduced the risk of construction errors in harsh environments. Additionally, the connection designs featured good sealing properties and easy-to-take anti-corrosion measures. The construction design of the circumferential SLC was optimized through pre-splicing tests, and the optimized SLC successfully achieved self-locking. Axial tension tests on the optimized single-layer circumferential SLC demonstrated that the connection possessed excellent tensile behavior, with the shear keys jointly resisting tensile forces.

Acknowledgements This work is financially supported by the Shenzhen Science and Technology Program (No. RCYX20210706092044076). The financial support is highly appreciated.

References

1. The largest offshore crude oil production platform in China has completed docking and installation. http://www.sz.gov.cn/cn/xxgk/zfxxgj/zwdt/content/post_8777159.html. Last accessed 18 Jun 2024
2. Liu YY, Hu CH, Sueyoshi M, Yoshida S, Iwashita H, Kashiwag M (2021) Motion response characteristics of a Kyushu-University semi-submersible floating wind turbine with trussed slender structures: experiment vs. numerical simulation. Ocean Eng 232:109078
3. Zhou LQ (2023) Dynamic response analysis of floating aquafarm with frame structure under wave & current loads. Harbin Institute of Technology, China, Harbin
4. Han LH (2016) Concrete filled steel tubular structures—theory and practice, 3rd edn. Science Press, Beijing
5. Chen SL, Hou C, Zhang H, Han LH (2019) Structural behavior and reliability of CFST trusses with random initial imperfections. Thin-Walled Struct 143:106192
6. Fong M, Chan SL, Brian U (2011) Advanced design for trusses of steel and concrete-filled tubular sections. Eng Struct 33(12):3162–3171
7. Xu W, Han LH, Tao Z (2014) Flexural behaviour of curved concrete filled steel tubular trusses. J Constr Steel Res 93:119–134
8. Chen ZH, Wang J, Liu JD, Khan K (2021) Seismic behavior and moment transfer capacity of an innovative self-locking inter-module connection for modular steel building. Eng Struct 245:112978

New Materials for Floating Concrete Structure: Development of Ultra-High Performance Concrete Based on New Challenges

Rui Yu, Yuan Feng, Kangning Liu, and Zhonghe Shui

Abstract As commonly known, concrete, in comparison to materials such as steel and polymers, offers superior corrosion resistance and lower cost in developing floating structure, making it a prime choice in marine construction industry. However, due to the critical environments of marine, especially the area close to the seas of southern China, the floating structure should resist the impact of marine wave and high concentration erosion ions, which may exceed the load-bearing capacity of ordinary concrete. Therefore, to overcome the obstacles mentioned above, this manuscript provides a comprehensive overview of a newly developed Ultra-High Performance Concrete (UHPC) with advanced mechanical properties, volume stability and corrosion resistance. To be specific, a new model considering the effect of liquid phases and chemical reaction on the particle packing is proposed to obtain a UHPC matrix with high density. Additionally, a new strategy is utilized to optimize the orientation and distribution of steel fibers, which is beneficial for improving the toughness of concrete. Furthermore, a multiply defense method is proposed to resist the penetration and corrosion of harmful ions. In general, based on the approaches developed in this study, an advanced composite material can be produced, which can effectively support the construction of floating concrete structure served in marine environment.

Keywords Floating concrete structure · Ultra-high performance concrete · Dense particle packing theory · Corrosion resistance

R. Yu · Y. Feng · K. Liu · Z. Shui (✉)
State Key Laboratory of Silicate Materials for Architecture, Wuhan University of Technology, Wuhan, China
e-mail: zhshui@whut.edu.cn

R. Yu · Y. Feng
International School of Materials Science and Engineering, Wuhan University of Technology, Wuhan, China

K. Liu
Department of Civil and Environmental Engineering and Research Centre for Resources Engineering Towards Carbon Neutrality (RCRE), The Hong Kong Polytechnic University, Hung Hom, Kowloon, Hong Kong, China

X. L. Zhao et al. (eds.), *Proceedings of the Fourth World Conference on Floating Solutions: WCFS 2024; 2–4 December; Hong Kong*, Lecture Notes in Civil Engineering 597, https://doi.org/10.1007/978-981-96-4569-5_73

1 Introduction

The development of infrastructure in maritime areas is currently a key driver in the exploration of marine resources and the construction of a maritime power. The construction of floating platforms is an important technical means to achieve this blueprint. Among the options, concrete materials, which are cost-effective, widely sourced, and durable, have gradually become one of the main choices for developing floating structures, as opposed to steel structures that are prone to rust and high-cost organic structures. In the face of the complex marine environment, concrete floating platforms need to withstand long-term erosion by common ions in seawater such as chloride ions and sulfate ions, and damage caused by the unstable scouring of seawater leading to surface structure peeling [1–3]. The damage to structures requires significant maintenance costs in the later stages, potentially harming socioeconomic benefits. Therefore, enhancing the durability of concrete is crucial for the stability and sustainable development of floating platforms.

To improve the durability of concrete in the marine service environment, the primary task is to reduce the transport channels and transmission rate of corrosive ions like chloride ions within the concrete, lower the concentration of these ions as much as possible, and enhance mechanical properties such as compressive and impact resistance to mitigate damage caused by seawater scouring [4, 5]. This requires concrete materials to possess excellent mechanical properties, toughness, and dense structure. Ordinary concrete, due to its high porosity, is difficult to meet these requirements and cannot achieve long-term stability against ion erosion. Consequently, some researchers have begun to incorporate traditional Ultra-high-performance concrete (UHPC) with dense skeletal structure and fiber reinforcement characteristics into floating platform structures to improve overall durability [6, 7]. However, most current studies only focus on the adaptability of UHPC compared to ordinary concrete in the marine environment, neglecting issues such as the impact of liquid phase components on the stability of UHPC's packing structure, the potential introduction of pore structures by disordered fiber arrangement that may reduce durability, and the corrosion protection of reinforcing fibers. These factors can all contribute to the unstable development of UHPC's durability.

Based on this, to achieve stable development of UHPC durability, this study incorporates liquid phase components into the densest packing model to achieve stable volume development and densification of the skeletal structure. Additionally, optimizing the arrangement of steel fibers aims to achieve synergistic development of toughness and dense structure through orderly fiber distribution. Furthermore, the introduction of layered double hydroxides (LDHs) is intended to further prevent the diffusion of corrosive ions, maintaining the high corrosion resistance of UHPC. Ultimately, a highly stable and long-lasting UHPC will be developed.

2 Method

2.1 Design of the Densest Packing Considering the Liquid Phase

Firstly, based on the commonly used MAA model (Andreasen and Andersen model) in UHPC design [8, 9], multiple mix proportions are designed through the particle size distribution of raw materials (Eqs. 1, 2), and the relationship between water film thickness (WFT, Eq. 3) and material composition is constructed. Based on the constructed component model, a gradient water film thickness is selected and applied to the particle size distribution to obtain a new particle size distribution. This is then incorporated into the MAA model to design a component volume ratio considering liquid phase thickness. Subsequently, the optimal liquid phase component ratio is determined by comparing the mechanical properties and pore structures of each group.

$$P(D) = \frac{D^q - D^q_{\min}}{D^q_{\max} - D^q_{\min}} \tag{1}$$

$$RSS = \sum_{i=1}^{n} \left(P_{mix}\left(D_i^{i+1}\right) - P_{tar}\left(D_i^{i+1}\right)\right)^2 \tag{2}$$

$$WFT = \frac{\mu_{ew}}{A} \tag{3}$$

where D is the particle diameter; $P(D)$ is the proportion of particles smaller than D; $D_{\min}$ and $D_{\max}$ are the minimum and maximum particle diameters, respectively; q is the distribution modulus; P_{mix} and P_{tar} are the grading curves of the mixture and the target, respectively; WFT is the water film thickness; μ_{ew} is the ratio of excess water volume to solid particles; A is the specific surface area of solid particles.

2.2 Regulation of Steel Fiber Orientation and Distribution

Based on the optimal mix ratio mentioned above, by varying the volume fraction of steel fibers (0, 0.5, 1, 1.5, 2, and 2.5%) and the pouring method, the average area and dispersion of fiber distribution in the sample cross-section are detected using binarization processing and other techniques [10, 11]. This allowed for a quantitative analysis of fiber orientation and distribution characteristics and their impact on mechanical properties (Fig. 1).

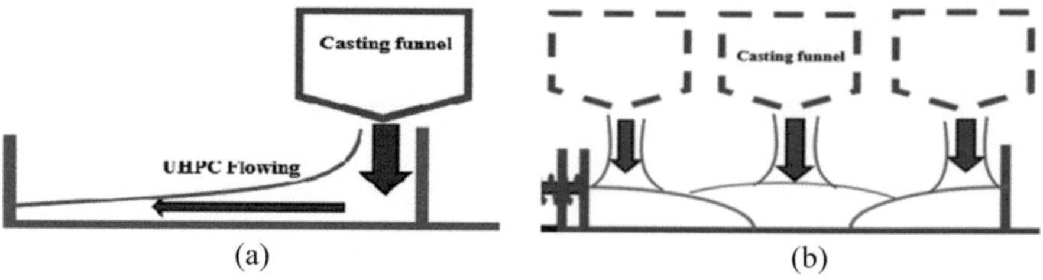

Fig. 1 The pouring methods: **a** cast at one side of the mold with flowing process; **b** cast randomly without flowing process [10]

2.3 Investigation of Ion Capture Technology

In terms of enhancing the corrosion resistance of the paste, cement paste is mixed with NaCl solutions of different concentrations (0, 0.3, 0.6, 0.8, 1, and 1.5 mol/L), and inorganic intercalation materials such as LDHs (8.5 wt%) with properties of adsorbing chloride and sulfate ions are added [12]. By testing the content of water-soluble Cl^- in the NaCl mixed paste, the adsorption isotherm of LDHs is obtained to quantify the ion adsorption capacity of LDHs and to clarify the ability of the LDHs structure to capture corrosive ions in cement-based materials (Eq. 4).

$$Q_E = \frac{m_{cl} - m_{c0}}{m} \tag{4}$$

where Q_E is the adsorption capacity of LDHs for Cl^- ions at equilibrium (mg/g); m is the mass of LDHs; m_{c0} and m_{cl} are the masses of Cl^- in the control group and the LDHs-added group, respectively, after solidification.

3 Results and Discussion

3.1 New Densest Packing Model Considering the Liquid Phase

Experimental results of the model optimization process are shown in Fig. 2. Figure 2a, b depict the relationship model between WFT and raw materials based on MAA ratio design. The results indicate that, compared to cementitious materials, superplasticizer and water content have a more significant impact on WFT. Higher water content or increased use of superplasticizer results in thicker WFT, highlighting that water content is a critical factor determining WFT thickness. Based on this, by adjusting the dosage of superplasticizer and water, a ratio shown in Fig. 2c is designed on the basis of the constructed WFT and component model, and the mechanical properties of different groups are compared (Fig. 2d), along with pore structures (Fig. 2e).

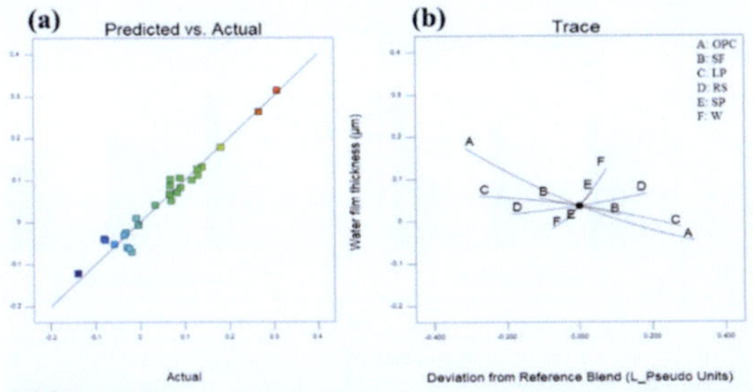

(c)

Groups	OPC	SF	LP	RS	SP	W
SP25	1	0.246	0.257	1.555	0.036	0.271
SP30					0.043	0.265
SP35					0.051	0.259
SP40					0.058	0.254
SP45					0.065	0.248
SP50					0.072	0.242
SP55					0.080	0.236
SP60					0.087	0.230

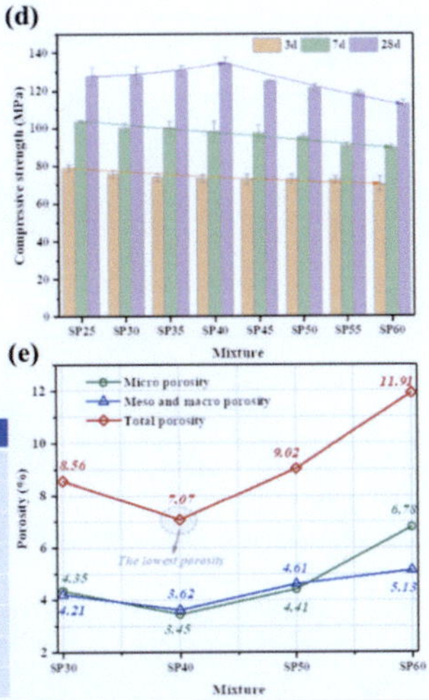

Fig. 2 Model optimization results: **a** model prediction accuracy; **b** effect of components on WFT; **c** optimized ratio design; **d** compressive strength and **e** porosity. OPC: Ordinary Portland cement; LP: Limestone powder; SF: Silica fume; W: Water; SP: superplasticizer

Results show that the SP40 group exhibits superior mechanical properties and a denser pore structure with a total porosity of 7.07%, compared to SP30, SP50, and SP60 groups with total porosities of 8.56%, 9.02%, and 11.91%, respectively, representing an increase of approximately 70% at its highest. Furthermore, at later stages (28 days), the compressive strength of UHPC initially increases and then gradually decreases with an increase in superplasticizer dosage. The compressive strength peaks at 134.2 MPa with a superplasticizer dosage of 40 kg/m^3 (optimized design ratio).

In summary, after appropriately adjusting the dosage of superplasticizer, it is possible to achieve a dense packing between the liquid phase and solid particles, further enhancing skeletal density and mechanical properties, thereby ensuring the stable development of the accumulation structure.

3.2 *Impact of Steel Fiber Orientation and Distribution on Performance*

Experimental results of the mechanical properties of UHPC under different volumes of steel fiber addition and pouring methods, as well as fiber orientation distribution, are shown in Fig. 3. The results indicate that, with consistent volume fractions of steel fibers, UHPC cast with single-sided pouring consistently exhibits higher flexural strength (Fig. 3a) compared to random pouring methods, with an approximately 20% increase. Evaluation of fiber efficiency K_f (Fig. 3b) shows higher efficiency in

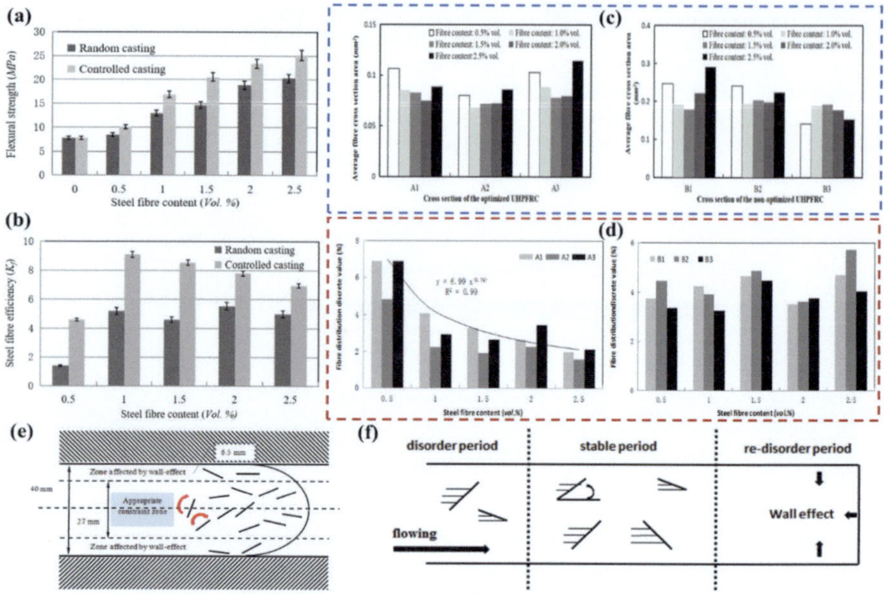

Fig. 3 Steel fiber orientation and distribution: **a** flexural strength; **b** steel fiber efficiency; **c** orientation evaluation; **d** distribution evaluation; **e** flow process under optimal fiber dosage and **f** fiber orientation movement process

unilateral pouring, with an optimal fiber dosage yielding maximum efficiency.

$$K_f = \frac{(f_i - f_2)}{n_0} \tag{5}$$

where f_i represents the flexural strength of UHPC prepared with different volumes of steel fiber, f_0 denotes the flexural strength of UHPC without steel fibers, and n_0 indicates the volume fraction of steel fibers.

To further clarify the reasons for the impact of pouring method and fiber dosage on efficiency, quantitative evaluations of fiber curve and distribution are conducted using image processing techniques, as shown in Figs. 3c and 4d. The results indicate that the average fiber area at the cross-section of UHPC formed by single-sided pouring is smaller than that achieved by random pouring, with a ratio of approximately 0.3, indicating superior fiber orientation in single-sided pouring. Moreover, the dispersion of fibers decreases regularly with increasing dosage under single-sided pouring (Fig. 3d), while random pouring shows greater fluctuation in dispersion, suggesting that single-sided pouring and increasing fiber dosage contribute to improving the fiber distribution within UHPC.

In summary, the flow of the paste is hindered by the mold walls, which disrupts fiber orientation. Additionally, a large quantity of fibers can cause disorderly distribution due to their own rotation or arrangement. Therefore, single-sided pouring and a

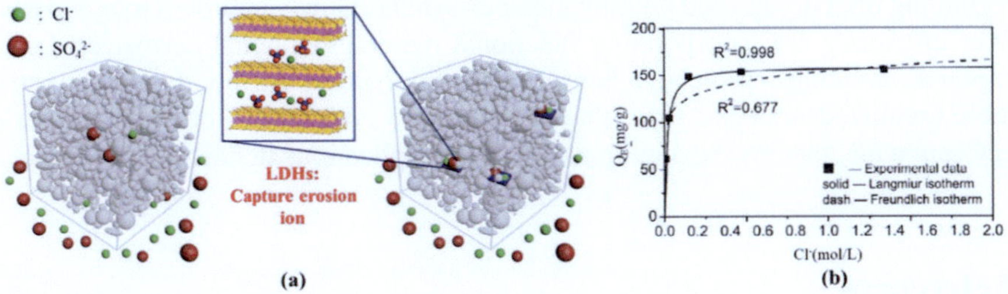

Fig. 4 **a** LDHs adsorption mechanism; **b** ion adsorption isotherms of LDHs in paste and theoretical models

steel fiber dosage of 1 vol% optimize orientation and distribution, thereby enhancing fiber efficiency and overall toughness.

3.3 LDHs Adsorption Characteristics of Corrosive Ions

LDHs, as layered double hydroxides, possess a main layer structure with positive charges that require intercalation of balancing anions to maintain charge neutrality. Therefore, LDHs exhibit a certain degree of adsorption for anions such as SO_4^{2-} and Cl^-, making them theoretically suitable as functional materials in concrete to protect steel reinforcement against chloride ion corrosion. Figure 4 shows the quantified results of LDHs adsorbing chloride ions in cement-based materials. The results indicate that as the concentration of Cl^- ions increases within the matrix, the adsorption capacity of LDHs is gradually enhanced, with a maximum adsorption capacity reaching 159.7 mg/g. Comparing with common charge adsorption processes described by Langmiur and Freundlich theoretical models, it is found that in cement-based materials, some LDHs ions' adsorption capacity even exceeds theoretical values. This suggests that within the matrix, LDHs not only capture corrosive ions through charge adsorption but also physically adsorb them via their structural properties, thereby reducing the concentration of corrosive ions and protecting the internal stability of steel fibers against permeation and corrosion.

4 Conclusion

This study explores methods such as optimizing the dense packing structure of solid–liquid phases, controlling fiber orientation and distribution, and introducing functional materials to enhance the durability of UHPC. Specifically, through optimizing the packing structure, it achieves further enhancement of matrix density and reduction of porosity, thereby blocking migration pathways for corrosive ions. Simultaneously, regulating fiber dosage and pouring methods synchronously improves toughness and fiber efficiency. By incorporating functional materials, internal corrosive ions are captured, creating a protective barrier against corrosion for the fibers. Ultimately, these optimized strategies strengthen the durability and stability of UHPC in marine environments, thereby expanding its potential applications in floating platforms.

References

1. De Weerdt K, Orsáková D, Geiker MR (2014) The impact of sulphate and magnesium on chloride binding in Portland cement paste. Cem Concr Res 65:30–40
2. Oh BH, Jang SY (2007) Effects of material and environmental parameters on chloride penetration profiles in concrete structures. Cem Concr Res 37(1):47–53
3. Yin S, Jing L, Yin M, Wang B (2019) Mechanical properties of textile reinforced concrete under chloride wet-dry and freeze-thaw cycle environments. Cem Concr Compos 96:118–127
4. Yi Y, Zhu D, Guo S, Zhang Z, Shi C (2020) A review on the deterioration and approaches to enhance the durability of concrete in the marine environment. Cem Concr Compos 113:103695
5. Li K, Zhang D, Li Q, Fan Z (2019) Durability for concrete structures in marine environments of HZM project: Design, as-sessment and beyond. Cem Concr Res 115:545–558
6. Sun M, Yu R, Jiang C, Fan D, Shui Z (2022) Quantitative effect of seawater on the hydration kinetics and microstructure development of ultra high performance concrete (UHPC). Constr Build Mater 340:127733
7. Li J, Yu Z, Xu F, Guo Z, Ding Q, Hou D (2024) The evolution mechanism of anti-abrasive UHPC microstructure under impact and erosion. Constr Build Mater 411:134282
8. Yu R, Spiesz P, Brouwers HJH (2014) Mix design and properties assessment of ultra-high performance fibre reinforced concrete (UHPFRC). Cem Concr Res 56:29–39
9. Fan D, Tian W, Yu R (2022) Incorporation of liquid phase into solid particle packing model for precise design of low water/binder cement-based composites (LW/B-CC): modelling and experiments. Compos B Eng 242:110070
10. Song Q, Yu R, Shui Z, Wang X, Rao S, Lin Z (2018) Optimization of fibre orientation and distribution for a sustainable ultra-high performance fibre reinforced concrete (UHPFRC): experiments and mechanism analysis. Constr Build Mater 169:8–19
11. Lee Y, Lee S, Youn J, Chung K, Kang T (2002) Characterization of fiber orientation in short fiber reinforced composites with an image processing technique. Mater Res Innov 6(2):65–72
12. Yang H, Xiong C, Liu X, Liu A, Li T, Ding R et al (2021) Application of layered double hydroxides (LDHs) in corrosion resistance of reinforced concrete-state of the art. Constr Build Mater 307:124991

Exploration and Application of Cement Based Floating Structures

Xu Gao, Zhonghe Shui, and Xibo Qi

Abstract Faced with the challenge of developing cement based floating structure for marine applications, 12 years of exploration has been carried out. First of all, it is necessary to fully understand the marine environment, in order to determine what performance of cement based materials can be qualified for the "floating" work. The impact of wind and waves and severe corrosive environment put forward high requirements for cement based materials. Through the attempts of various types of concrete from C50 to C200, some relationships between the floating structure and the performance of cement based materials were found. The characteristics of advanced cementitious material make it possible for complex floating structural forms, such as fabricated floating wharf, tubular floating body, tubular floating foundation, pipe rod plate fabricated structure. The square floating box is the first form of floating structure that developed, but its environmental adaptability is limited. Therefore, the floating structure of double-layer shell with ultra high performance concrete wave plate is developed to improve the stability. The application of cement based floating structures has also expanded rapidly, from small floating bridges and wharfs to large offshore wind turbines and photovoltaic bases, to complex marine ranching facilities and offshore buildings, and has penetrated into all aspects of human life. Cement based floating structure has a bright and long-term future.

Keywords Floating structure · Ultra-high performance concrete · Cement based floating structure

X. Gao
School of Civil Engineering and Architecture, Wuhan University of Technology, Wuhan, China

Z. Shui (✉) · X. Qi
State Key Laboratory of Silicate Materials for Architecture, Wuhan University of Technology, Wuhan, China
e-mail: zhshui@whut.edu.cn

Z. Shui
Wuhan University of Technology Advanced Engineering Technology Research Institute of Zhongshan City, Zhongshan, China

X. L. Zhao et al. (eds.), *Proceedings of the Fourth World Conference on Floating Solutions: WCFS 2024; 2–4 December; Hong Kong*, Lecture Notes in Civil Engineering 597, https://doi.org/10.1007/978-981-96-4569-5_74

1 Introduction

Marine related industries such as aquaculture, tourism, and resource exploitation all rely on the structure of marine floating platforms, thus the development of floating marine facilities has important strategic significance [1]. Since the 1990s, the United States, Japan and other countries have carried out research on large floating structures including offshore airports, civil buildings, military bases and offshore oil storage silos. For example, the U.S. military public the plan to build an afloat forward staging base (AFSB) in the Gulf region to provide logistical support. In response to the risks caused by sea level rise, the Netherlands has developed a large number of floating residential buildings, and formed a new construction industry [2].

Researches on floating facilities has also been carried out in China as early as the 1980s. For example, the three-dimensional hydroelastic theory has been put forward in the early 1980s, and studied the hydroelastic response of super large floating structures since the early 1990s. The stress analysis of multi column super large floating structure in wind and waves [3, 4]. Since 2012, the research and development of large and super large floating platforms have been carried out, of which the most representative is the research and construction of super large offshore platforms [5, 6].

Steel floating structure is the most commonly applied traditional floating structure in the ocean. However, the relevant industries must meet the three requirements of "safety, practicability and economy" at the same time, of which "economy" is the key issue at present [7]. For example, in the aquaculture industry, the cost of an aquaculture boat is hundreds of millions of yuan, equivalent to 5000 yuan/m^3. Truss and gravity steel cages can reduce the investment to 500–600 yuan/m^3, but due to the high maintenance cost of anti-corrosion, the final comprehensive cost will also rise significantly.

Using high strength and high toughness cement based materials as the main structure to manufacture marine floating structure facilities can effectively reduce the one-time investment of marine infrastructure, and also reduce the maintenance cost. Taking aquaculture as an example, the cost be reduced to less than 100–300 yuan/m^3, showing obvious advantages.

The research team of the authors is one of the first teams in China to propose the application of cement based materials in marine floating facilities. The research and development of cement based materials floating platform have been carried out, accumulated technology has been obtained in a series of aspects such as hydrodynamic research, material design, structural design, construction technology and berthing safety. The constructed cement based floating structures have carried out offshore floating and mooring tests in coastal areas, and achieved the expected goals. This paper briefly introduces the exploration and practice experiences of the research team in developing cement based floating structures.

2 Development of Marine UHPC

2.1 Design Principles

Compared to normal concrete, the harsh marine environment puts forward higher performance requirements for concrete materials, such as mechanical properties and durability. Therefore, ultra high performance concrete (UHPC) especially designed for marine environment is necessary. In order to produce UHPC that qualifies floating structure applications, the following design principles are highly suggested: (1) Using particle pack models to physically design the composition of all solid particles, which would help to achieve an ideal dense matrix. (2) Adding suitable type and dosage of fibers to improve flexural strength, tensile strength, fracture energy, toughness, and avoid crack. (3) Chemically design the binder composition and additives to improve the resistance of chloride attack. 4) When steel fiber is used, special attention should be paid to the electrochemical reactions and corrosion due to fiber exposure [8, 9].

2.2 Properties of Marine UHPC

Following the above mentioned principles, UHPC that suitable for marine environment and floating structures can be produced. Figure 1 briefly exhibits an example of using particle packing model and numeral analysis for the concrete mix design. In this manner, UHPC with compressive strength between 150 and 200 MPa, flexural strength between 20 and 50 MPa, elastic modulus between 40 and 60 MPa and fracture energy between 20 and 40 kJ/m^2 can be achieved. Besides, excellent long term properties can also be obtained, some properties are exampled in Fig. 2.

3 Application of Marine UHPC Based Floating Structures

Based on the development of marine UHPC, different types/generations of cement based floating structures are designed, produced and applied, for different application purposes, including the original concrete floating box, wave dissipation floating box, mortise-bolt connection box, special shaped box-pipe structure, tube-rod-plate structure, and prefabricated pipe frame structure.

3.1 Floating Concrete Boxes

In the early stage, floating concrete boxes are firstly developed, Fig. 3 shows the full size trials of multi-functional concrete floating cover, the function include modular

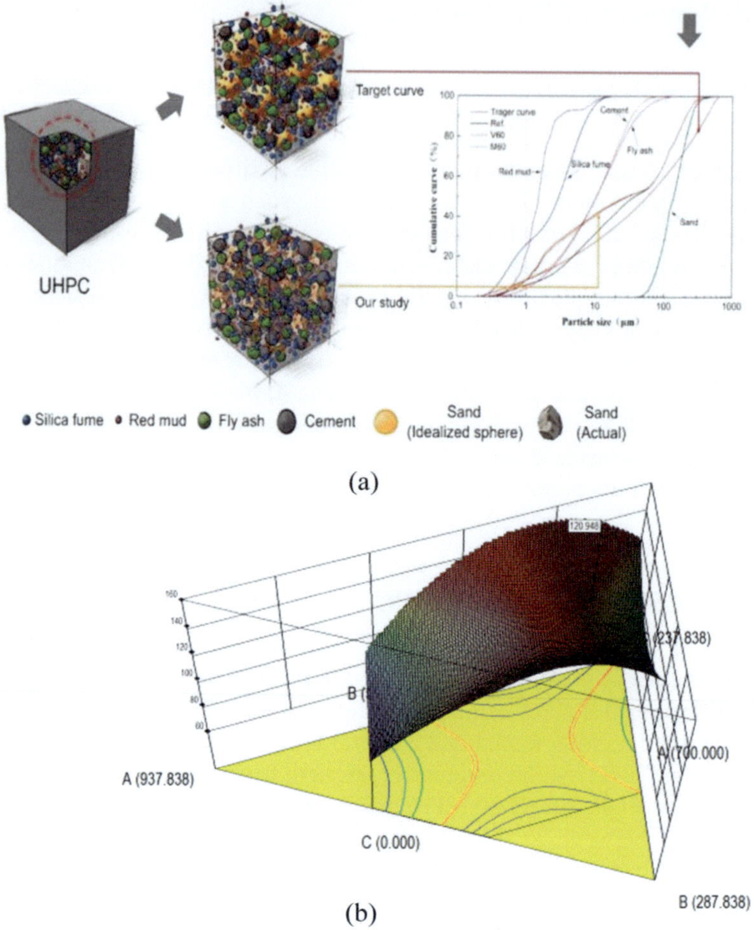

Fig. 1 UHPC mix design: **a** particle pack model; **b** numeral analysis

assembly, heat storage and mulching planting. The unit size of this floating box is 5 m × 2.5 m × 1 m, with buoyancy of 6.25 t. This floating structure was to be used for the thermal insulation top cover of a solar hot water pool in a winter stadium.

In order to reduce the influence of waves on human comfort and the potential safety issues caused by collision, a double shell floating concrete boxes is then developed, UHPC perforated plates are used for impact resistance and wave dissipation. This floating box is 14 m × 8 m × 2 m and has a buoyancy of more than 60 t. This structure was used as a functional platform during the construction of the Shenzhen–Zhongshan Sea crossing bridge, see Fig. 4.

For the purpose of improving the construction efficiency and the convenience of transportation and assembly, a mortise bolt connection box structure is proposed, which is characterized by low requirements for construction site and can be assembled

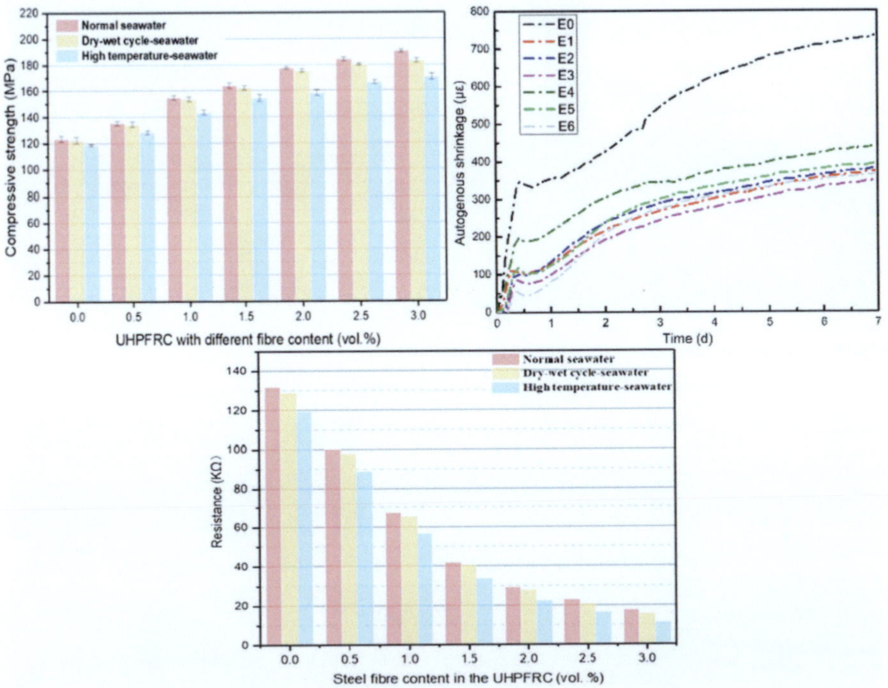

Fig. 2 Example of UHPC properties: **a** strength loss; **b** shrinkage; **c** electrochemical properties

Fig. 3 The original version of the floating concrete box

Fig. 4 The double shell wave dissipation floating box

Fig. 5 The mortise bolt connection floating box

freely according to the actual demand. Figure 5 shows the application of this structure as a manual mobile wharf in a marine construction project.

3.2 Special Shaped Box-Pipe Floating Structure

The floating structures shown in Sect. 3.1 are all platforms, their main function is load (human, cargo and etc.) bearing. In order to achieve different application scenarios, the research team also proposed the frame type floating structure, this type of structure is usually semi-submersible, as shown in Fig. 6. The advantage of this type of structure [10, 11] is its large loading space. The key points of manufacture this type of floating structure include the preparation technology of high-performance cement based prefabricated components, and the corresponding assembly technology. The research team is now conducting hydrodynamic analysis, construction and assembly related work of a special shaped box-pipe floating structure. It is expected that it will be launched at the end of this year and serve as a cement based aquaculture cage for aquaculture.

3.3 Tube-Rod-Plate Floating Structure

Another structure type proposed by the research team is the tube-rod-plate floating structure [12, 13]. This type of floating structure has the characteristics of fast construction and assembly, standardization of the prefabricated components, long maintenance cycle, fast construction speed and high structural stability. Figure 7 shows some examples of this type of structure, some of the following structures are now under construction guided by the research team, and will be used as small offshore management and protection platform.

4 Future Prospective of Cement Based Floating Structures

4.1 Challenge and Replace the Basic Materials for Marine Infrastructure

Due to the advantages of cement based materials in material performance, construction and maintenance costs, it is believed that cement based material, especially marine UHPC, will be widely used in marine infrastructure in the future. For example in large-scale offshore platforms, the basic material for main structure will gradually change from steel to cement-steel composite, while the small-scale operation platforms will change from polymers to cement based materials.

4.2 Bring New Application Directions

Cement based floating communities have become normal in some areas, cement based wharfs and floating tunnels are gradually becoming a reality. Maritime tourism and leisure resorts are increasingly concerned by commercial capitals. Cement-based offshore new energy platforms, such as wind power, photovoltaic, nuclear energy, thermoelectric energy, etc., are also gradually promoted.

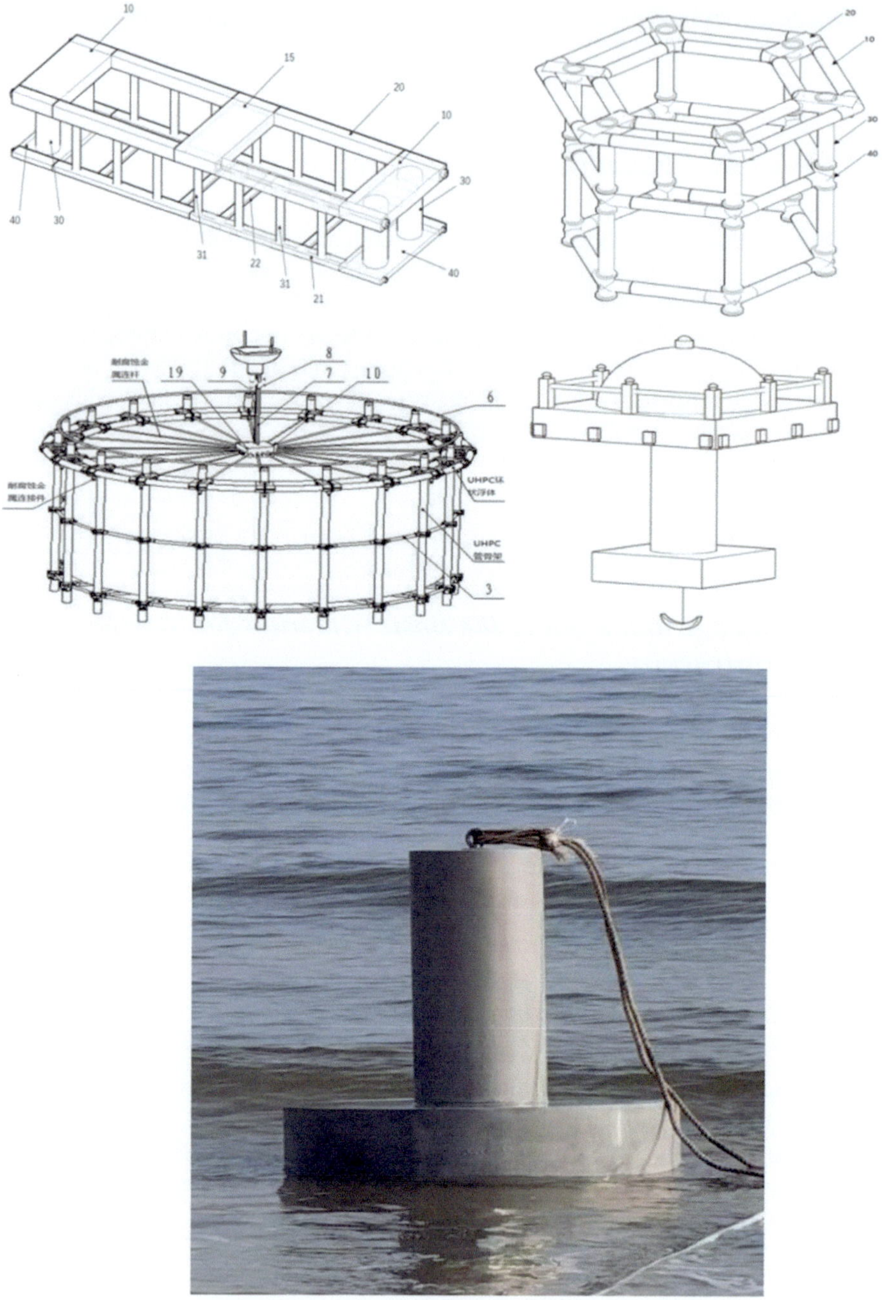

Fig. 6 Examples of special shaped box-pipe floating structures, and a practical application as a cement buoy

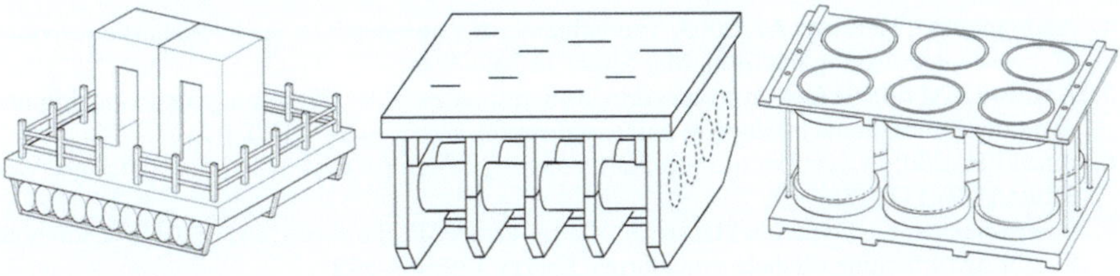

Fig. 7 Examples of the tube-rod-plate floating structure

4.3 Bring New Industrial Chain

The promotion and application of cement based floating structure will bring new industrial chain. At the material level, it includes cement based raw materials, anti-corrosion additives, multi-functional modifiers, etc.; As for prefabricated components, including a series of concrete prefabricated components, connectors, supporting materials and components; At the structural level, it includes fabricated aquaculture cage, floating platforms, fixed platforms, etc.; In terms of the software systems, it includes facility intelligent control system, safety monitoring system, floating community management system, etc. The relevant industrial chain is expected to generate a market scale of 100 billion yuan within ten years.

5 Conclusion

This paper briefly introduces the research and practice history of the research team on cement based floating structures in the past 12 years, including the research and development of ultra-high performance marine concrete, the design of different structural forms of cement based floating structures, and the application of cement based floating structures. It is believed that the cement based floating structure has broad application prospects and is expected to bring new development direction and industrial chain in the future, forming a "cement revolution" in the construction of marine infrastructure.

References

1. Tian J (2010) A brief analysis of the industrialization value of floating marine structures. Resour Habitant Environ 10:72–74
2. Wei C, Tang Y, Ni Z (2007) Floating houses—hotel design in Rotterdam, Netherlands. Huazhong Archit 09:22–24
3. Korobkin AA, Khabakhpasheva TI (2007) The construction of exact solutions in the floating-plate problem. J Appl Math Mech 71:287–294

4. Andrianov AI, Hermans AJ (2003) The influence of water depth on the hydroelastic response of a very large floating platform. Mar Struct 16:355–371
5. Kashiwagi M (2000) Research on hydroelastic responses of VLFS: recent progress and future work. In: International offshore & polar engineering conference, vol 10, p 2
6. Suzuki H (2005) Overview of megafloat: concept, design criteria, analysis, and design. Mar Struct 18(2):111–132
7. Castro-Santos L, Filgueira-Vizoso A, Carral-Couce L, Formoso JÁF (2016) Economic feasibility of floating offshore wind farms. Energy 112:868–882
8. Song Q, Yu R, Shui Z, Rao S, Fan D, Gao X (2020) Macro/micro characteristics variation of ultra-high performance fibre reinforced concrete (UHPFRC) subjected to critical marine environments. Constr Build Mater 256:119458
9. Song Q, Yu R, Shui Z, Rao S, Wang X, Sun M, Jiang C (2018) Steel fibre content and interconnection induced electrochemical corrosion of ultra-high performance fibre reinforced concrete (UHPFRC). Cem Concr Compos 94:191–200
10. Shui Z, Meng X, Wang L, Zheng W (2023) Chinese Patent ZL20210846768.6
11. Shui Z (2024) Chinese Patent ZL 2020 110846768.6
12. Shui Z, Wang X (2024) Chinese Patent ZL 202211556528.8
13. Shui Z, Wang L, Liu X et al (2024) Chinese Patent ZL 202110016322.5

Experimental Study on the Mechanical Properties of Limestone Calcined Clay Cement (LC^3) Seawater Sea-Sand Concrete

Junpeng Zhang and Xiaodan Teng

Abstract This study explores the development of a chloride-resistant, eco-friendly concrete by partially replacing Portland cement with limestone and calcined clay at varying proportions of 0, 30, 45, 52.5, and 60%. The concrete was produced using seawater, sea sand, and gravel. Macro and micro experiments were conducted to evaluate the effects of different replacement rates on the chloride ion content and compressive strength of LC^3 seawater sea sand concrete. The main findings are as follows: the concrete with 30% and 45% replacement showed lower strength than the 0% replacement concrete at 1–7 days. However, by 28 days, the compressive strengths of these mixes had improved to 50 MPa and 45.6 MPa, 13.6% and 3.6% higher than 0% replacement rate, respectively. Beyond 45% replacement, a significant reduction in strength was observed. Additionally, concrete with 30 and 45% replacement showed a notable decrease in free chloride ion content from 1 to 3 days compared to the 0% mix. The higher the replacement rate, the more the free chloride ion content was reduced. These results provide valuable theoretical guidance for developing building materials and structural systems for marine structures.

Keywords Limestone calcined clay cement · Seawater · Sea sand · Concrete · Mechanical properties

---

J. Zhang · X. Teng (✉)
School of Civil Engineering and Architecture, Guangxi University, Nanning, People's Republic of China
e-mail: xdteng@gxu.edu.cn

X. Teng
State Key Laboratory of Featured Metal Materials and Life-Cycle Safety for Composite Structures, Nanning, People's Republic of China

X. L. Zhao et al. (eds.), *Proceedings of the Fourth World Conference on Floating Solutions: WCFS 2024; 2–4 December; Hong Kong*, Lecture Notes in Civil Engineering 597, https://doi.org/10.1007/978-981-96-4569-5_75

1 Introduction

In marine engineering, the use of traditional concrete faces challenges related to the scarcity of freshwater and river sand resources. As a result, researchers have begun to utilize seawater and sea sand as filling materials in aluminum alloy pipes [1, 2]. However, the high chloride ion environment within seawater and sea sand concrete can lead to corrosion of the aluminum alloy pipes, thereby reducing the durability of the composite structure. To reduce the corrosion of the aluminum alloy pipes caused by the internal concrete, it is necessary to lower the chloride ion content in seawater and sea sand concrete from the perspective of cement materials.

Limestone Calcined Clay Cement (LC^3) is a new type of low-carbon blended cement that benefits cement manufacturers by reducing carbon dioxide emissions during production. The main innovation of LC^3 lies in replacing traditional cement with calcined clay, limestone, and gypsum in a ratio of 6:3:1. It has been estimated that using LC^3 to replace conventional types of cement can achieve a reduction in carbon emissions of up to 40%, indicating significant future potential. Moreover, the materials in LC^3 exhibit a good synergistic effect, enabling it to achieve strength comparable to ordinary cement at specific proportions. Additionally, LC^3 cement possesses excellent properties, such as low alkalinity and high chloride ion binding capacity. Existing research has shown that even at a cement replacement rate of 50% with Limestone Calcined Clay (LC^3-50), similar compressive strength to ordinary Portland cement (OPC) can be achieved at a water-to-binder ratio of 0.5 after 7 days [3]. However, Ruan et al. [4] have increased the replacement rate of calcined clay and limestone to 50–80%. Due to the rapid consumption of calcium hydroxide by calcined clay, a significant amount of unreacted calcined clay and limestone remains in the system, leading to a noticeable loss in compressive strength compared to OPC and LC^3-50. On the other hand, Wu's [5] research indicates that increasing the replacement rate of calcined clay and limestone effectively reduces the calcium hydroxide content in the paste. Building on this foundation, researchers managed to enhance the compressive strength by lowering the water-to-binder ratio to 0.4; however, it still remained lower than that of OPC at the same water-to-binder ratio.

When preparing LC^3 cement-based materials using seawater, LC^3 effectively consumes calcium hydroxide and binds chloride ions during the hydration process, thereby reducing the alkalinity and chloride ion content in seawater sea-sand concrete. Research by Shi et al. [6] found that the calcium hydroxide content in seawater-based LC^3 paste was reduced by 41.7%, 74.5%, 84%, and 89.4% at 1, 3, 7, and 28 days, respectively, compared to OPC paste. Additionally, the chloride ion content in seawater-based LC^3 paste was reduced by −5.8, 68.3, 62.9, and 72.5% at the same intervals compared to OPC paste. This indicates that LC^3 has superior capability to reduce the alkalinity of concrete and immobilize chloride ions compared to ordinary OPC, making it more suitable for use in seawater and sea sand concrete. Liu et al. [7] have already studied the workability, mechanical properties, drying shrinkage behavior, and microstructure of seawater and sea sand concrete containing LC^3. The findings indicate that the seawater and sea sand concrete with added LC^3

exhibits higher early strength and better long-term strength; LC^3 can suppress the drying shrinkage behavior of seawater and sea sand concrete; and replacing OPC with LC^3 can improve the pore structure of self-consolidating concrete (SCC).

Currently, there is limited research on the application of LC^3 cement in seawater and sea sand concrete, with existing studies primarily focusing on the strength of LC^3 seawater and sea sand concrete in comparison to ordinary concrete. There is a lack of studies addressing factors affecting the durability of aluminum alloys, such as the chloride ion content in LC^3 seawater and sea sand concrete. This study will focus on investigating the effects of partially replacing Portland Cement with calcined clay and limestone at replacement ratios of 0, 30, 45, 52.5, and 60% on the chloride ion content and compressive strength of LC^3 seawater and sea sand concrete from 1 to 28 days. The aim is to develop a concrete with low chloride ion content and a 28-day strength reaching 40 MPa using LC^3 cement, seawater, and sea sand. This concrete will be applied in aluminum alloy pipes to ensure strength while reducing the corrosion of the aluminum alloy pipes by the internal concrete environment, thereby enhancing the overall lifespan of the structure.

2 Materials and the Experimental Methods

2.1 Material

The cement used in this study was Portland Cement (P.O II) from Ningguo Cement Plant, Anhui Conch Cement Co. The compressive strength after 3 days is 43.8 MPa, and the compressive strength after 28 days is 54.8 MPa, which meets the requirements of the GB 175-2007 [8] standard. The seawater is sourced from the Ocean Chemical Industrial City in Xingan County, Jiangsu Province, China, and is mixed with freshwater to create seawater with a salinity of 3.5%. The sea sand is secondary sand sourced from Beihai City, Guangxi Zhuang Autonomous Region, with an apparent density of 2540 kg/m^3. The coarse aggregate has a particle size of 5–20 mm and is graded to meet the requirements of GB/T 14685-2001 [9]. The calcined clay (MK) and limestone (LS) are sourced from Henan Hengyuan New Materials Co., Ltd. and Henan Yixiang New Materials Co., Ltd., respectively; their chemical compositions are shown in Tables 1 and 2. The raw material samples are depicted in Fig. 1.

Table 1 Composition of the calcined clay

Material	Al_2O_3 (%)	SiO_2 (%)	Fe_2O_3 (%)	TiO_2 (%)
MK	42.5	55.7	0.80	1.00

Table 2 Composition of the limestone

Material	$CaCO_3$ (%)	SiO_2 (%)	Fe (%)	MgO (%)	Al_2O_3 (%)
LS	98.5	0.02	0.003	0.11	0.01

Fig. 1 Raw material samples

2.2 *Design of Mix Proportion*

In this study, a water-to-binder ratio of 0.4 was designed, using seawater (S) as the mixing water and sea sand (S) as the fine aggregate. Calcined clay and limestone were used to partially replace Portland Cement at proportions of 0, 30, 45, 52.5, and 60%. The test groups were labeled as SS-LC^3-XX, where SS represents seawater and sea sand, and LC3-XX indicates the proportion of Calcined clay and limestone used to replace Portland Cement by XX%. The detailed mix proportions are shown in Table 3. The concrete was prepared according to the GB/T 50080-2016 [10] standard, and the fresh concrete was cast into cube specimens measuring 100 mm × 100 mm × 100 mm. All specimens were placed in a standard curing room maintained at 25 °C and a relative humidity of 90% or above for curing.

Table 3 Mix proportion of the concrete specimens (kg/m^3)

Grade	Sample	Water	P.O II	Calcined clay	Limestone	Sand	Coarse aggregates
M1	SS-LC^3-0	163	408	0	0	673	1203
M2	SS-LC^3-30	163	285.6	81.6	40.8	673	1203
M3	SS-LC^3-45	163	224.4	122.4	61.2	673	1203
M4	SS-LC^3-52.5	163	193.8	142.8	71.4	673	1203
M5	SS-LC^3-60	163	163.2	163.2	81.6	673	1203

2.3 Experimental Method

Compression Test

The compressive strength tests of the concrete were conducted according to GB/T 50081-2002 [11], with curing ages of 1 day, 3 days, 7 days, and 28 days. The samples were tested after curing for the specified periods. Each test group consisted of three concrete specimens, and the compressive strength was calculated based on the average values. The mechanical performance tests were performed using a WAW-1000 microcomputer-controlled electro-hydraulic servo universal testing machine.

Chloride Ion Content Test of Concrete

According to the method for measuring the chloride ion content in concrete by Yang et al. [12], the free chloride ion content was measured using an improved Volhard method. The testing process is illustrated in the Fig. 2. Samples were taken from the crushed concrete specimens at 1 day and 3 days of curing, which were then ground and passed through a 0.08 mm sieve. After drying at 105 °C for 2 h, the powdered samples were obtained. Eight grams of the processed sample powder were combined with 100 mL of deionized water. After stirring for 24 h, the free chloride ion solution was filtered. To the filtered free chloride ion solution, 5 mL of nitric acid was added, followed by 10 mL of a silver nitrate standard solution until the addition of the first drop of potassium thiocyanate resulted in no color change. The solution was shaken until precipitation, then 5 mL of dibutyl phthalate was added and mixed thoroughly to consolidate the precipitate. Finally, 2 mL of ammonium iron sulfate was added as an indicator, and the potassium thiocyanate standard solution was titrated until a stable brick-red color appeared, recording the amount of potassium thiocyanate solution used.

3 Experimental Results and Analysis

3.1 Results of the Compressive Strength Test

The failure mode of LC^3 seawater sea-sand concrete is illustrated in Fig. 3. During the failure process of the LC^3 seawater sea-sand concrete cube specimens, there was no sudden collapse or spalling. After compressive failure, the shape remained almost intact, exhibiting good deformation capacity, achieving the expected load-bearing capacity, and demonstrating higher ductility compared to ordinary concrete.

The results of the compressive strength for seawater sea-sand concrete after replacing Portland Cement with calcined clay and limestone at different replacement rates are shown in Fig. 4. For concrete with replacement rates of 30 and 45%,

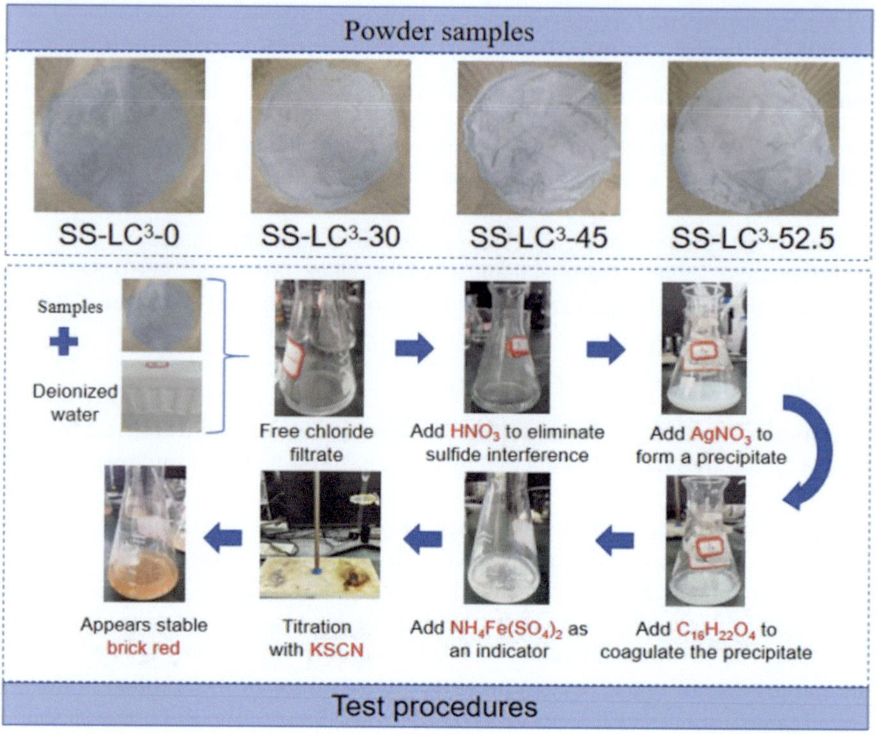

Fig. 2 Flow chart of chloride ion content test in concrete

Fig. 3 Comparison of damage patterns between silicate cement seawater sea-sand concrete (left) and LC^3 seawater sea-sand concrete (right)

although the early strength from 1 to 7 days was lower than that of the 0% replacement rate, by 28 days, the strength of both exceeded that of the concrete with a 0% replacement rate. Specifically, the strength of the concrete with a 30% replacement rate reached 50 MPa, while the strength of the concrete with a 45% replacement rate reached 45.6 MPa. This is similar to the conclusions of the study by Wu et al. [5]: the initial hydration reaction rate of calcined clay is slower than that of cement, but under the influence of volcanic ash, the later strength development is greater than that of ordinary cement. This is especially true when the limestone content is between 10 and 15% and the calcined clay content ranges from 15 to 50%. The LC^3 binding system shows higher strength development compared to cement. However, when the substitution rate continues to increase, there is a rapid decline in strength. At substitution rates of 52.5 and 60%, the 28-day strength of LC^3 seawater and seawater-sand concrete reaches only 37.1 MPa and 26.1 MPa, respectively. This is attributed to the rapid consumption of calcium hydroxide by calcined clay, leading to a significant amount of unreacted calcined clay and limestone in the system, resulting in a clear decline in strength of the LC^3 system compared to Portland cement [4]. Moreover, the higher the replacement rate of calcined clay and limestone, the lower the strength.

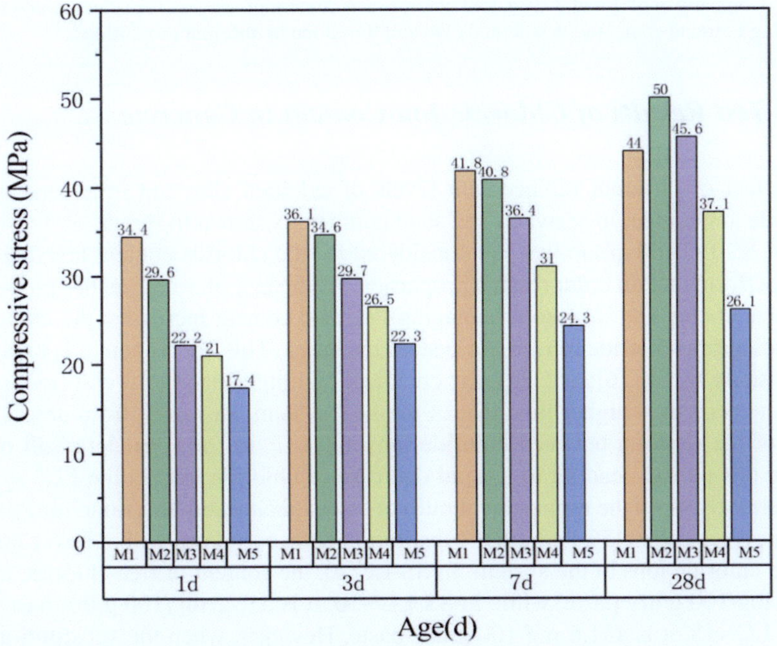

Fig. 4 Comparison of 1, 3, 7 and 28 days strength of seawater sea sand concrete after replacing Portland cement with calcined clay and limestone in different proportions

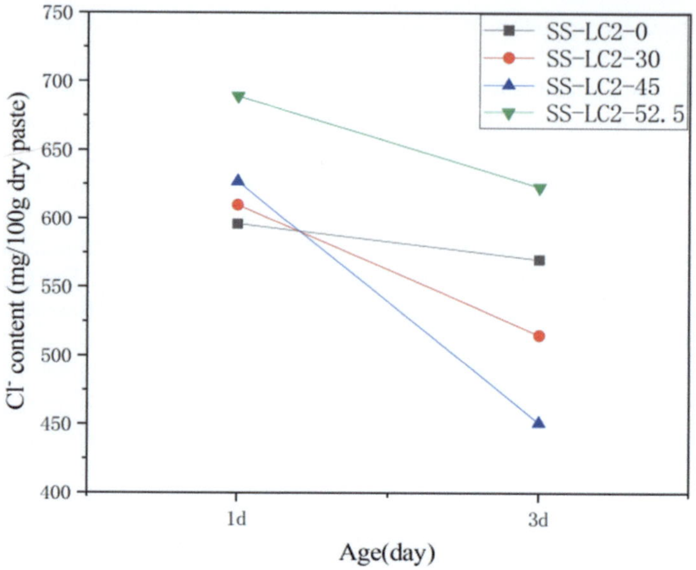

Fig. 5 Comparison of 1 and 3 days free chloride ion content of seawater sea-sand concrete after replacing Portland cement with calcined clay and limestone in different proportions

3.2 *Test Results of Chloride Ion Content in Concrete*

The effect of different replacement levels of calcined clay and limestone on the chloride ion content in seawater sea-sand concrete is shown in Fig. 5.

The SS-LC^3-60 group was not considered for the chloride content test due to its poor performance in compressive properties. At the age of 1 day, as the proportion of calcined clay and limestone replacing Portland cement increases, the content of remaining free chloride ions in the system increases. This phenomenon is explained in the study by Shi [6]: at 1 day, the chemical binding capacity for chloride ions of Portland cement is higher than that of the LC^3 system. However, from days 1 to 3, most of the reaction between chloride ions and LC^3 to form Friedel's salt occurs during this period, leading to a rapid decrease of chloride ions in the LC^3 system. Furthermore, when the replacement ratio of calcined clay and limestone for Portland cement is less than 45%, a higher substitution ratio correlates with a lower content of free chloride ions in the system. In SS-LC^3-0, the content of free chloride ions is 607.5 mg/100 g dry paste, while in SS-LC^3-30, it is 515.2 mg/100 g dry paste, and in SS-LC^3-45, it is 451.6 mg/100 g dry paste. However, when the substitution rate continued to increase, the level of free chlorine ions in group SS-LC^3-52.5 was even higher than that in group SS-LC^3-0. This is similar to the pattern found in the study of Yang et al. [13]. When the substitution of metakaolin is too high, the amount of cement used decreases, and the CH produced by cement hydration decreases insufficiently to stimulate the large dosage of metakaolin and thus lead to a decrease in the curing rate.

4 Conclusion

This study investigates the performance of LC^3 seawater and seawater-sand concrete from 1 to 28 days through tests of compressive strength and chloride ion content. By optimizing the design of the new LC3 seawater sand concrete, the main influencing factors on its mechanical properties are analyzed. The following conclusions are drawn.

1. After being subjected to compressive failure, LC^3 seawater sea-sand concrete retains its shape almost intact, without exhibiting the cracking and collapse typically seen in ordinary concrete. The overall failure mode is relatively uniform, demonstrating higher ductility than ordinary concrete.
2. During the initial period from 1 to 7 days, the strength of the SS-LC^3-30 and SS-LC^3-45 groups is lower than that of the SS-LC^3-0 group due to the slower reaction rate of calcined clay compared to cement. However, by the age of 28 days, under the influence of volcanic ash reactions, the strengths of both the SS-LC^3-30 and SS-LC^3-45 groups exceed that of the SS-LC^3-0 group, reaching 50 MPa and 45.6 MPa, respectively. When the replacement rate continues to increase, the strengths of the SS-LC^3-52.5 and SS-LC^3-60 groups remain significantly lower than that of the SS-LC^3-0 group from 1 to 28 days, and the higher the calcined clay and limestone replacement rate, the lower the strength.
3. At the age of 1 day, Portland cement generates more Friedel's salt and binds more chloride ions than LC^3. However, from 1 to 3 days, most of the reaction between chloride ions and LC^3 to form Friedel's salt occurs during this period. As a result, the free chloride ion content in the SS-LC^3-0 group is 607.5 mg/100 g dry paste, while the free chloride ion contents in the SS-LC^3-30 and SS-LC^3-45 groups are 15.2% and 25.7% lower than that of the SS-LC^3-0 group, respectively. However, as the LC^3 replacement rate continues to increase, it appears to inhibit the system's ability to bind chloride ions. The method used in this study to measure free chloride ions, which involved drying samples at 100 °C and stirring for 24 h, may have led to an underestimation of the LC^3 system's capacity to bind chloride ions.

References

1. Chen ZP, Song CM, Mo LL et al (2024) Calculation of axial compressive properties and bearing capacity of short columns of seawater sea sand concrete with aluminum alloy pipe spiral reinforcement. J Build Struct 45(02):136–147 (in Chinese)
2. Yang TQ et al (2019) Research on mechanical properties of CFRP-constrained recycled coarse aggregate-sea sand concrete combination short column. Shenzhen University (in Chinese)
3. Avet F et al (2016) Development of a new rapid, relevant and reliable (R3) test method to evaluate the pozzolanic reactivity of calcined kaolinitic clays. Cem Concr Res 85:1–11

4. Ruan YX et al (2022) Microstructure and mechanical properties of sustainable cementitious materials with ultra-high substitution level of calcined clay and limestone powder. Constr Build Mater 314:125416
5. WU L, Bao R, Dai J et al (2024) Hydration kinetic modeling of limestone-calcined clay-cement (LC3) system [J/OL]. Materials Herald 1–12, 13 Aug 2024 (in Chinese)
6. Shi C, Yin J, Hu C (2023) Microstructure, hydration, and chloride binding behavior of limestone calcined clay cement prepared using seawater. J Mater Civ Eng 35(12):04023458
7. Liu J et al (2022) Effects of w/b ratio, fly ash, limestone calcined clay, seawater and sea-sand on workability, mechanical properties, drying shrinkage behavior and micro-structural characteristics of concrete. Constr Build Mater 321:126333
8. The Ministry of Construction of the People's Republic of China. GB 175-2007, Common Portland Cement
9. Bai C (2001) GB/T14685-2001 Pebble and gravel for building. Fujian Build Mater 04:19 (in Chinese)
10. Standard on test methods for properties of ordinary concrete mixes GB-T 50080-2016 (in Chinese)
11. Standard test method for mechanical properties of ordinary concrete GB/T50081-2002
12. Yang LF, Chen Z, Feng QG et al (2010) A method for determining chloride ion content in high performance concrete. Guangxi Zhuang Autonomous Region: CN200910114256.4, 01 Dec 2010
13. Yang CH, Yan Y, Ou ZW (2010) Study on the chloride ion binding properties of metakaolin cement net paste. Concrete 10:1–3+7 (in Chinese)

Structural Design of an Innovative Floating Pontoon Using Ultra-High-Performance Concrete and Steel Skeleton

Haien Xue, Yinglei Li, Xiao Lin Zhao, and Siwei Liu

Abstract Floating coastal structures offer viable alternatives to land reclamation as global populations increase, particularly in land-scarce cities such as Hong Kong. Steel, known for its high strength and low weight, has been extensively utilized in long-span structures on land. However, its use in marine environments is restricted due to poor corrosion resistance. Concrete, specifically grades of C60 or higher, has provided a more durable solution for pontoons but still poses challenges in self-weight reduction. This paper introduces a novel floating pontoon combining ultra-high-performance concrete (UHPC) with a steel skeleton. The corrosion resistance of UHPC effectively protects the internal steel structure with a lower self-weight of pontoon unit can be achieved. Shell and beam-column elements are utilized to numerically simulate the proposed pontoon unit. An advanced second-order analysis by employing the constant load Newton–Raphson method is used. Finally, the section capacity analysis of the steel skeleton and stress analysis of the UHPC panels are subsequently presented.

Keywords Floating structure · Second-order analysis · Steel skeleton

1 Introduction

Floating structures provide a sustainable alternative to traditional land reclamation for expanding coastal megacities, particularly in Hong Kong, which faces a severe land shortage [1–3]. Moreover, anthropogenic climate change has accelerated polar ice melt to unprecedented levels, with glacial discharge rates increasing sixfold since

H. Xue · Y. Li · X. L. Zhao · S. Liu (✉)
Department of Civil and Environmental Engineering, The Hong Kong Polytechnic University, Hong Kong, China
e-mail: si-wei.liu@polyu.edu.hk

X. L. Zhao
Research Institute for Land and Space, The Hong Kong Polytechnic University, Hong Kong, China

X. L. Zhao et al. (eds.), *Proceedings of the Fourth World Conference on Floating Solutions: WCFS 2024; 2–4 December; Hong Kong*, Lecture Notes in Civil Engineering 597, https://doi.org/10.1007/978-981-96-4569-5_76

the 1980s [4, 5]. In response to land scarcity and climate risks, the United Nations convened its first High-Level Round Table on Sustainable Floating Cities in 2019 [6]. Currently, floating structures are gradually gaining popularity in engineering and are frequently used in forms such as floating houses and bridges. Pilot applications, including floating buildings in Singapore and Amsterdam, have demonstrated satisfactory and promising results [7, 8]. Although concrete is widely used for more permanent pontoon installations in commercial settings, a grade of C60 or above with increased thickness is required due to corrosive marine environment [9]. Consequently, structural design of lightweight and durable floating pontoons presents significant challenges, with limited codes and specifications available to guide their construction.

This study aims to propose a new floating system of pontoon which is composed of ultra-high-performance concrete (UHPC) skin and a steel skeleton. UHPC is a lightweight and durable material that provides excellent protection for the inner steel skeleton. Conversely, the steel skeleton reduces the span length of the UHPC skin, resulting in a more lightweight and high-strength floating pontoon. Additionally, an advanced second-order analysis using the constant load Newton–Raphson method is employed to analysis the proposed pontoon subjected to design loads, which also brought the P-Δ and P-δ effects of the steel skeleton into consideration. The proposed structural design has also been submitted for a Hong Kong patent application [15].

2 Structural Design of Floating Pontoon

2.1 Structural Layout

The floating pontoon consists of an ultra-high-performance concrete (UHPC) skin and a steel skeleton (as shown in Fig. 1). The UHPC skin provides protection against environmental corrosion. The steel skeleton acts as structural support for the UHPC skin, the UHPC skin and steel skeleton is connected by bolts.

The design environment is located near Kau Yi Chau, where a land reclamation project is under consideration to meet Hong Kong's increasing demand for land space [10]. The structural design of pontoon is characterized by varying dimensions

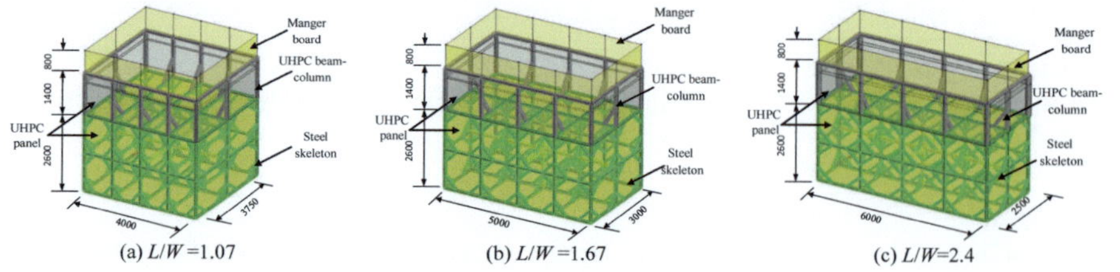

Fig. 1 Illustration of the proposed floating pontoon unit with different aspect ratio (mm)

Table 1 Comparison of different self-weight in different structural type

Structural type	Geometry size of pontoon	Self-weight (t)
Concrete skin–Concrete interior	6 m × 2.5 m × 4 m × 0.2 m	40.22
Concrete skin–Steel skeleton interior	6 m × 2.5 m × 4 m × 0.2 m	31.51
UHPC skin–Concrete interior	6 m × 2.5 m × 4 m × 0.08 m	26.15
UHPC skin–Steel skeleton interior	6 m × 2.5 m × 4 m × 0.08 m	17.26

by changing aspect ratios of length (L) to width (W), while maintaining a constant occupied area ($A = LW$), providing a usable area of 15 m^2 for the upper structure. The pontoon has a height (H) of 4 m and a draft of 3.8 m. The pontoon includes a steel skeleton at the base and a UHPC beam-column structure on top. The outer wall, made of UHPC, has a thickness of 80 mm, and a 0.8 m high wave wall (manger board) is incorporated to prevent water ingress. For comparison, a pontoon with an aspect ratio of $L/W = 2.4$ is selected, and the unit mass of the traditional concrete pontoon is also brought into consideration. According to Hong Kong's latest design standards [9, 11, 12], a minimum concrete grade of C60 is required for marine works with a design life of 100 years, necessitating a 200 mm thick concrete wall. Table 1 compares the self-weight of three manufacturing types based on these pontoon specifications, illustrating the advantages of the proposed floating structural system in reducing self-weight of the floating pontoon.

2.2 *Plate Element for Ultra-High-Performance Concrete (UHPC)*

The design strengths of UHPC skin shall be determined according to NF P 18-710 [13] and from the experimental strengths on UHPC cylinders (for compression) and coupons (for direct tension) according to relevant standards. The characteristic strength is defined as the strength of UHPC below which not more than 5% of the test results are expected to fall. The design compressive strength shall be defined as:

$$f_{cd} = \frac{0.85 f_{ck}}{\gamma_C} \tag{1}$$

where f_{cd} is the compressive design strength, f_{ck} is the characteristic compressive strength, and γ_C is the material partial factor.

The UHPC skin was simulated by plate element, as shown in Fig. 2(a), the strain energy U^e of an element with area A_e and thickness t is the sum of the bending strain energy, ${U_b}^e$, and the shear strain energy ${U_s}^e$, and given by:

$$U^e = U_b^e + U_s^e = \int\limits_{Ae} \left(\frac{1}{2}[\chi]^T[D_b][\chi] + \frac{1}{2}[\gamma]^T[D_s][\gamma] \right) dxdy \tag{2}$$

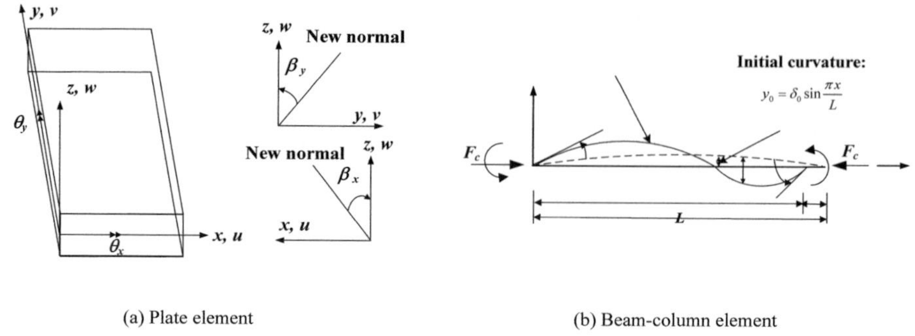

Fig. 2 Plate element and beam-column element used in the analysis of UHPC and steel skeleton

in which the $[\chi]$ and $[\gamma]$ are the bending curvatures and transverse shear strains respectively given by:

$$[\chi] = [\,\partial\theta_y/\partial x \ -\partial\theta_x/\partial y \ \partial\theta_x/\partial x + \partial\theta_x/\partial y\,]\,[\gamma] = [\,\theta_y - \partial w/\partial x \ -\theta_x - \partial w/\partial y\,] \tag{3}$$

$[D_b]$ and $[D_s]$ are the elasticity matrices respectively given by:

$$[D_b] = \frac{Et^3}{12\,(1-\nu^2)}\begin{bmatrix} 1 & \nu & 0 \\ \nu & 1 & 0 \\ 0 & 0 & (1-\nu)/2 \end{bmatrix} [D_s] = \frac{5Et}{12\,(1+\nu)}\begin{bmatrix} 1 & 0 \\ 0 & 1 \end{bmatrix} \tag{4}$$

2.3 *Beam-Column Element for Second-Order Analysis of Steel Skeleton*

Steel Skeleton was simulated by beam-column Element, as shown in Fig. 2b, where the initial curvature of the element was also considered. The equilibrium equation of the element can be established by the principle of stationary potential energy and the forces and nodal displacements relationship can be obtained as:

$$M_{z1} = \frac{EI}{L}\left[\left(4 + \frac{2F_cL^2}{15EI}\right)\theta_1 + \left(2 - \frac{F_cL^2}{30EI}\right)\theta_2\right] \tag{5}$$

$$M_{z2} = \frac{EI}{L}\left[\left(2 - \frac{F_cL^2}{30EI}\right)\theta_1 + \left(4 + \frac{2F_cL^2}{15EI}\right)\theta_2\right] \tag{6}$$

$$F_c = EA\left[\frac{e}{L} + \frac{1}{40}(\theta_1 + \theta_2)^2 + \frac{1}{24}(\theta_1 - \theta_2)^2\right] \tag{7}$$

in which F_c (taken as positive for tension) is axial force, M_{zi} ($i = 1, 2$) is the end moment, e is axial shortening.

Second-order analysis of constant load Newton–Raphson method is employed to account for both P-Δ and P-δ effects for thestructural design of the steel skeleton interior. In the second-order elastic (P-Δ-δ) design, the section capacity check is used to determine the resistance of the structure. A plastic hinge method in a section is assumed when the following interactive equations are satisfied, and the following equations are adopted for structural uses of steel [11]:

$$\frac{F_c}{A_g p_y} + \frac{M_x + F_c(\Delta_x + \delta_x)}{M_{cx}} + \frac{M_y + F_c\left(\Delta_y + \delta_y\right)}{M_{cy}} = \chi \leq 1 \quad (8)$$

$$\frac{F_c}{A_g P_y} + \frac{m_{LT}[M_x + F_c(\Delta_x + \delta_x)]}{M_b} + \frac{m_y\left[M_y + F_c\left(\Delta_y + \delta_y\right)\right]}{M_{cy}} = \chi \leq 1 \quad (9)$$

in which $F_c(\Delta_x + \delta_x)$ and $F_c(\Delta_y + \delta_y)$ are the additional moment due to frame and member deflections and they include the effects of initial imperfections; member lateral-torsional buckling check is carried out by replacing M_{cx} in Eq. (1) by the buckling resistance moment M_b. The section capacity check is carried out using Eqs. (8) and (9) with a plastic hinge is assumed when χ is greater than 1.

3 Structural Analysis of the Pontoon Unit

3.1 *Boundary Conditions*

An aspect ratio (*L/W*) of 2.4 is selected for the pontoon, with boundary conditions are illustrated in Fig. 3. It should be noted that a water spring is applied at the bottom of the pontoon to simulate the interaction between the pontoon and water. The stiffness of the spring is defined as:

$$k = \rho g A \quad (10)$$

where ρ is the density of water, g is the gravitational acceleration, and A is the area, calculated as $A = L \times W$.

3.2 *Loading Conditions*

A series of loading conditions were considered in the second-order direct analysis of the proposed pontoon unit, including dead load (DL, self-weight and superimposed dead load on the pontoon), live load (LL, upper structure and maintenance loads),

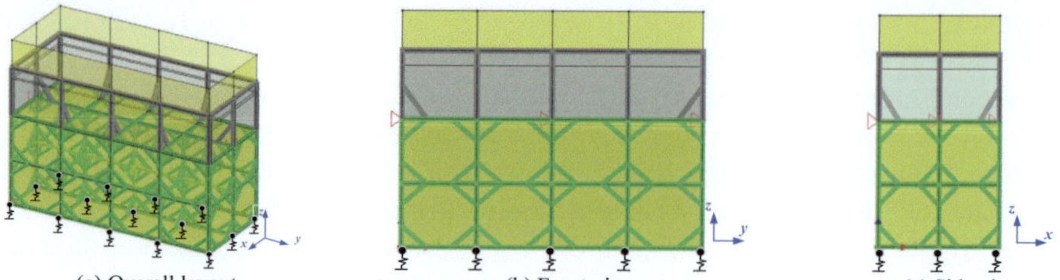

(a) Overall layout (b) Front view (c) Side view

Fig. 3 Boundary conditions of the pontoon unit

Table 2 Summary of load types applied on the pontoon unit

Load types	Load type	Values	Load case	Load type	Values
Self-weight:	DL	16.94 t	Wind load	WL	3.0 kPa
Superimposed dead load	DL	1.2 kPa	Wave load	WL	0.6–2.1 kPa
Sand Load inside the pontoon	LL	11.33 kPa	Buoyancy and water pressure	WL	0–37 kPa
Upper structure load	LL	23.46 kPa	Thermal loads	TL	T = −20/ + 35 °C
Maintenance load	LL	0.75 kPa			

thermal load (TL), wind load, and wave load (WL). The total applied loads on the pontoon are listed in Table 2, with values calculated based on the Hong Kong code [11, 12].

3.3 Section Capacity Analysis of the Steel Skeleton

To simulate variations in wind direction, three wind and wave orientations are considered in this paper, at angles of 0°, 45°, and 90° relative to the pontoon. Consistency is maintained between the directions of the wave and wind loads to achieve the most extreme combined effects.

The Non-linear Integrated Design and Analysis (NAF-NIDA) software, version 10, is employed for second-order elastic analysis in this study due to its significant advantages in the non-linear analysis of steel structures [14]. Additionally, load combinations are conducted according to the relevant specifications [9], based on the load types considered in Sect. 3.1. For accurate simulation of the steel skeleton, a total of 456 steel members were modeled in the software. The section capacity factor (SCF) for each member was investigated, and the SCFs for different members under varying wind and wave conditions are presented in Fig. 4.

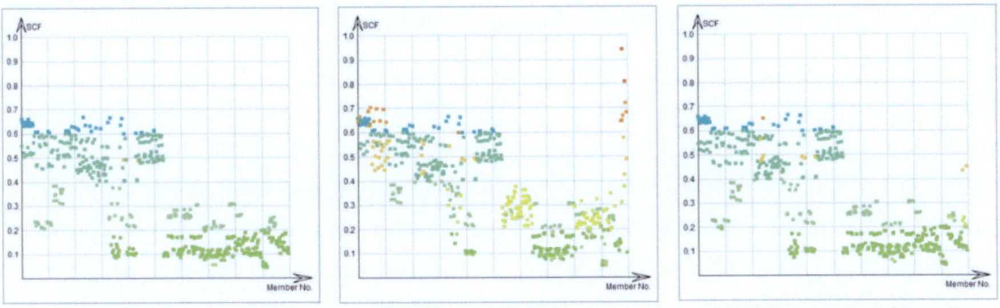

(a) 0 degrees of wind (wave) (b) 45 degrees of wind (wave) (c) 90 degrees of wind (wave)

Fig. 4 Scatter chart of section utilization

The scatter chart distributions of SCF are very similar across all loading conditions, with maximum SCF values of 0.67, 0.94, and 0.66 for wind (wave) directions of 0, 45, and 90°, respectively. A relatively higher SCF value is observed at 45°, which could be attributed to the increased contact area between the UHPC skin and wind (wave) when the orientation is 45°.

3.4 Stress Analysis of the UHPC Panel

The UHPC panels of the pontoon are divided into different pieces according to the boundary condition. There are five different dimensions of UHPC panel in this model, including one type in the x–z plane (1.5 m × 1.3 m), two types in x–y plane (1.5 m × 1.25 m in the corner and middle), two types in y–z plane (1.3 m × 1.25 m in the corner and middle). The locations of these panels are shown in Fig. 5. The detailed boundary conditions of these UHPC panels are also summarized in Table 3.

An extreme water pressure of 40 kPa was applied on these panels. The stress contours of UHPC panels are shown in Fig. 6. The maximum tensile stresses are

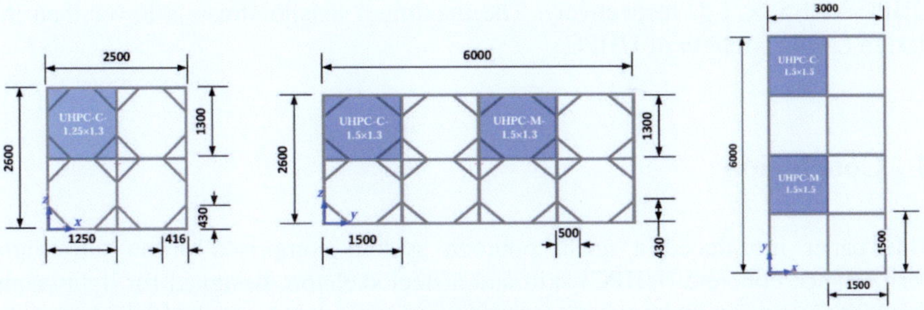

(a) x-z plane (b) y-z plane (c) x-y plane

Fig. 5 Different panels in the analysis model (mm)

Table 3 List of UHPC panels in different boundary conditions

Nomenclature	Dimension (mm)	Boundary conditions
UHPC-C-1.25 × 1.3	1250 × 1300	Two adjacent sides fixed; the other two pinned
UHPC-C-1.5 × 1.3	1500 × 1300	Two adjacent sides fixed; the other two pinned
UHPC-M-1.5 × 1.3	1500 × 1300	Three sides fixed; one side pinned
UHPC-C-1.5 × 1.5	1500 × 1500	Two adjacent sides fixed; the other two pinned
UHPC-M-1.5 × 1.5	1500 × 1500	Two adjacent sides fixed; the other two pinned

Note "C" denotes the panel at the corner of the plane, "M" denotes the panel at the middle of the plane

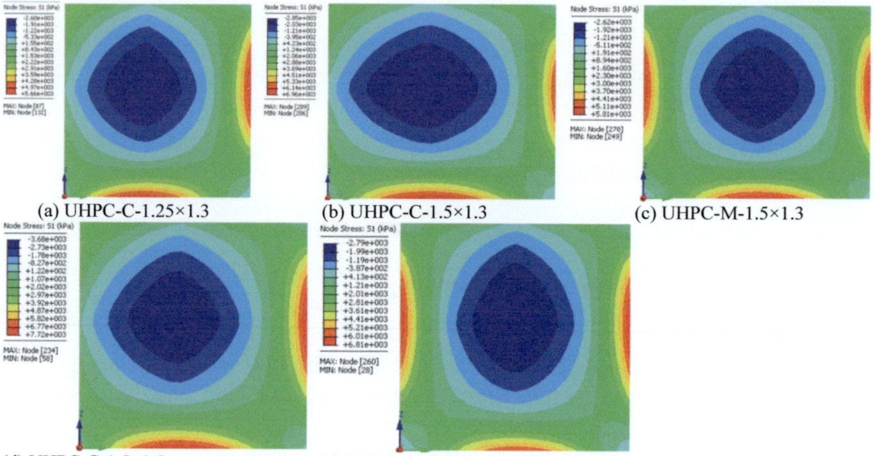

(a) UHPC-C-1.25×1.3 (b) UHPC-C-1.5×1.3 (c) UHPC-M-1.5×1.3 (d) UHPC-C-1.5×1.5 (e) UHPC-M-1.5×1.5

Fig. 6 Stress contour of UHPC panels

5.66 MPa, 6.96 MPa, 5.81 MPa, 7.72 MPa and 6.81 MPa, for the panel UHPC-C-1.25 × 1.3, UHPC-C-1.5 × 1.3, UHPC-M-1.5 × 1.3, UHPC-C-1.5 × 1.5 and UHPC-M-1.5 × 1.5, respectively. The maximum tensile stress is lower than the design tensile strength of UHPC.

4 Conclusion

This paper introduced a novel pontoon system comprised of an ultra-high-performance concrete (UHPC) skin and a steel skeleton, designed for lightweight durability in marine environments. An advanced second-order analysis employing the constant load Newton–Raphson method is used to analyze the proposed pontoon unit.

Section capacity analysis and stress analysis are subsequently conducted, demonstrating the reliability and adaptability of the proposed pontoon under extreme design loads in Hong Kong.

Acknowledgements The work described in this paper was partially supported by grants from the Research Grant Council of the Hong Kong Special Administrative Region through the project "Second-order direct analysis for the structural design of steel members with irregular cross-sections (PolyU/21E/15203121)". Additional support was provided by the Department of Civil and Environmental Engineering at The Hong Kong Polytechnic University through the project "Development of Next-generation Sustainable Steel Construction using Built-up CFS Members (G-UAP3)".

References

1. Zhao XL, Dai JG, Ding XL, Graaf-van Dinther R, Wang CM, Wang B (2023) Developing a sustainable and smart floating structure solution for enhancing liveability in Hong Kong's crowded built environment. In: Proceedings of the World conference on floating solutions (WCFS2023). Springer Nature Singapore, Singapore, pp 3–18
2. Isobe E (1999) Research and development of mega-float. In: Proceedings of the 3rd international workshop on very large floating structures, pp 7–13
3. Watanabe E, Utsunomiya T, Wang CM (2004) Hydroelastic analysis of pontoon-type VLFS: a literature survey. Eng Struct 26(2):245–256
4. Mouginot J, Rignot E, Bjørk AA (2019) Forty-six years of Greenland Ice Sheet mass balance from 1972 to 2018. Proc Natl Acad Sci 116(19):9239–9244
5. Rignot E, Mouginot J, Scheuchl B, Van Den Broeke M, Van Wessem MJ, Morlighem M (2019) Four decades of Antarctic Ice Sheet mass balance from 1979–2017. Proc Natl Acad Sci 116(4):1095–1103
6. Mohammed A (2019) Sustainable floating cities can offer solutions to climate change threats facing urban areas
7. Lin FY, Spijkers O, Van der Plank P (2022) Legal framework for sustainable floating city development: a case study of the Netherlands. In: Proceedings of the second world conference on floating solutions (WCFS2020). Springer Singapore, Rotterdam, pp 433–460
8. Nguyen HP, Dai J, Wang CM (2018) Reducing hydroelastic responses of pontoon-type VLFS using vertical elastic mooring lines. Mar Struct 59:251–270
9. Port Works Design Manual Part 1 (2002) General design considerations for marine works. CEDD, Hong Kong
10. Development Bureau, Civil Engineering and Development Department, Planning Department (2023) Study on the artificial islands in the central waters. Hong Kong, China
11. Code of Practice for Structural Use of Concrete, Buildings Department, Hong Kong (2013)
12. Code of Practice for the Structural Use of Steel, Buildings Department, Hong Kong (2011)
13. NFP 18-710 (2016) National addition to Eurocode 2—design of concrete structures: specific rules for ultra-high performance fibre-reinforced concrete (UHPFRC). France
14. Liu YP, Du ZL, He JW (2018) Recent development and implementation of direct analysis in NIDA. In: 9th International conference on advances in steel structures (ICASS 2018). Hong Kong Institution of Steel Construction
15. Liu SW, Li YL, Zhao XL, Wang JH, Zhang SQ, Xue HE. A hybrid floating body structure using ultra-high-performance concrete and steel skeleton, Hong Kong Patent (No.: HK30111777)

Modular Floating Structures

Towards a Comprehensive Evaluation of Sediment Transport Caused by Off-the-Coast Modular Floating Structures' (MFS) Breakwaters

Shai Asher and Gil Wang

Abstract Modular floating structures (MFS) emerge as a sustainable solution for creating additional useable spaces in the marine environment. However, positioning MFS near the coast can harm the natural sedimental balance and coastal morphodynamics. The environmental impact of floating structures on sediment transport can be evaluated with numerical models that rely heavily on structure-specific transmission coefficients as one of their primary inputs. Using an XBeach model, the current research examines the effect of a multimodule floating seawall configuration on the natural sedimental balance off the eastern Mediterranean coast. The current study uses surveyed bathymetry and over 30 years of measured wave data to detail the environmental condition of the model. In addition, the specific transmission coefficients of the floating seawall used in this research are obtained through physical testing. Although the current research is site-specific, the results from this study can help optimize the distance of the MFS from the adjacent coastline to minimize effects on the sediment balance or to be used as part of a coastal protection scheme.

Keywords Sediment transport · Floating breakwater · Wave energy

S. Asher (✉) · G. Wang
CAMERI—Coastal and Marine Engineering Research Institute, Technion Israel Institute of Technology, Haifa, Israel
e-mail: shaiashe@cv.technion.ac.il

G. Wang
e-mail: wang@cv.technion.ac.il

X. L. Zhao et al. (eds.), *Proceedings of the Fourth World Conference on Floating Solutions: WCFS 2024; 2–4 December; Hong Kong*, Lecture Notes in Civil Engineering 597, https://doi.org/10.1007/978-981-96-4569-5_77

1 Introduction

1.1 Floating Versus Fill-Material Based Offshore Islands

This research study continues the line of research on sustainable urban development offshore [1], motivated by climate change and demographic events that cause a scarcity of land in coastal areas. This is a fundamental issue for the quality of life of almost half of the global population [2]. The proposed study focuses on the floating breakwaters shielding potential urban development from propagating ocean waves, which is the primary obstacle to the feasibility of almost any offshore construction. The currently available breakwater technologies (e.g., rubble-mound, caissons, or composite) can, in theory, provide the protection needed for offshore development. However, in the current context, they become obsolete due to adverse environmental impacts and technological limitations [3]. This study focuses on the coastal morphodynamics of sandy shorelines influenced by a new floating breakwater system (seawall), which is constructed from decommissioned large-scale ocean-going vessels. The 'circular economy' suggested design approach reduces building cost, fabrication time, material waste, and the environmental impact associated with shipbreaking [4]. However, little is currently known about the long-term influence of offshore structures on the local beach nourishment climate.

More than 25 years have passed since the joint "Declaration of Intent on Artificial Islands" was signed between the Prime Minister of Israel, Mr. S. Peres, and the Prime Minister of the Kingdom of the Netherlands, Mr. W. Kok. This initiative was born from an earlier Israeli-Dutch research collaboration (1993–1995) that investigated an offshore urban development of the metropolitan region of Tel Aviv, which proposed a fill-material-based offshore island model (a one-square-kilometer drop-shaped island) [5]. The "Declaration of Intent on Artificial Islands" was followed by a Dutch/Israeli Steering Committee formed to investigate the feasibility of artificial islands off the Israeli coastline. The duration of the research was almost two years (seven quarters of the year), with a budget of two million dollars. The committee concluded that fill-material-based artificial islands are not feasible off the Israeli coast due to the lack of fill material (only 4% of the required volume was available) and the environmental impact associated with sediment transport interference.

There are several studies on the effect of Wave Energy Converter farms on sandy coasts through case studies (e.g., [6, 7] among many others). However, the beach response is site-and-device-specific.

The objective of the current research is to help optimize the distance of the MFS from the adjacent coastline to minimize effects on the sediment balance or to used it as part of a coastal protection scheme such as natural beach nourishment.

2 Methods

2.1 Limiting Bathymetry ≧ 28 m and Distance from the Shoreline

The bathymetry is a critical factor (physical constraint) for the deployment of floating structures, as the water must be deep enough to accommodate their draft. Typically, the draft ranges from a few meters [1, 8] to around 25 m [9]. In this research, a water depth of at least 28 m is required, which aligns with the dimensions of the floating seawall [4]. The main bathymetric feature of the Israeli coastline close to shore follows the land's contour (offset) with a gradual slope (ca. 1:100), descending until it reaches a depth of about 100 m at 10 km offshore, which is the edge of the continental shelf. part of the continental shelf. From there a steep slope (ca. 1:2) starts, ascending to 1000 m at 12 km offshore.

2.2 Model Principal Spatial Layout and the Floating Seawall

This study concentrates on the floating breakwater setup located approximately four kilometers offshore, at a water depth of 28 m, west of Tel Aviv-Yafo. The floating breakwater system, known as the Floating Seawall, is constructed entirely from decommissioned large ocean-going vessels, aligned with the circular economy principle for sustainable development. The Floating Seawall is essential for facilitating any floating developments in open sea areas, as it mitigates incoming waves and lowers the maximum wave height on its sheltered side for various applications. In the current study the Floating Seawall is constructed from all 400 DWT ore carriers. This layout has been previously introduced as part of a physical testing study conducted in The Ship Hydrodynamic Laboratory at TU Delft [4]. The main particulars of the ORE carrier are given in Fig. 1, showing both full scale layout and the box-shape model used in the physical experiment to determine the wave transmission coefficient. In addition to the module size of the Floating Seawall, Fig. 1 shows the spatial arrangement of the four km Floating Seawall and its closest to the shoreline at 28 m water depth, allowing the under-keel clearance of ≧ 5 m.

2.3 Offshore Wave Data

Wave properties were recorded by Datawell's directional Waverider buoys at Ashdod (166,997 E, 642,614 N, New Israeli Grid coordinates) and Haifa (194,504 E, 749,688 N) at a water depth of ca. 24 m, beginning from 1994 (CAMERI, Israel). The wave properties are converted to deep water values using an approximation method proposed by [10].

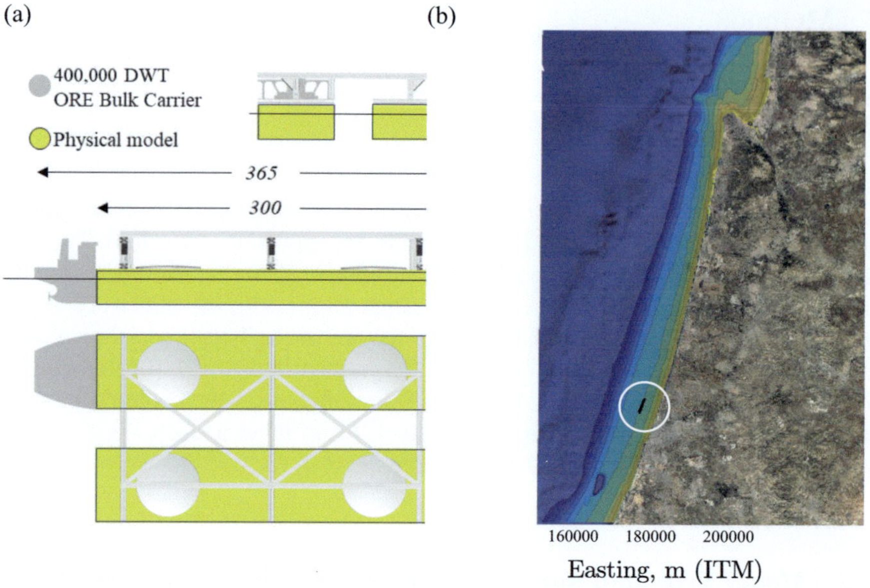

Fig. 1 **a** Model layout, main particulars of a typical Floating Seawall module [ref]. **b** Plan view of the spatial Floating Seawall arrangement, showing 11 modules (inside the white circle) with a total length of four km

2.4 Wave Climate Schematization

The expected impact zone of a ca. 4 km wide breakwater system located ca. 4 km seaward of the shore is O (10 km). Therefore, the time frame for the impacted coastline to reach a new equilibrium is measured in decades [11]. Wave-climate schematization was applied to reduce the computational resources required for a decades-long shoreline morphology simulation. Due to its appealing relative simplicity, the "Energy Flux Method (EFM)" was selected as the schematization technique [12]. A novel adjustment to the method was applied to overcome EFM's purported sequencing issue [11]. The offshore records of daily maximal significant wave height (and corresponding mean direction) were first categorized according to the region's four most prevalent weather systems, the occurrence of which roughly corresponds with the seasons [13]. Next, in each weather system type, the wave data were separated into bins of equal energy flux such as shown in Eq. (1) below,

$$E = \rho g H_{m0}^2 / 8 C_g. \tag{1}$$

where ρ is the water density (1025 kg/m^3), g is the gravitational acceleration (9.81 m s^{-2}), H_{m0} is the significant wave height and C_g is the wave group celerity, in deep water.

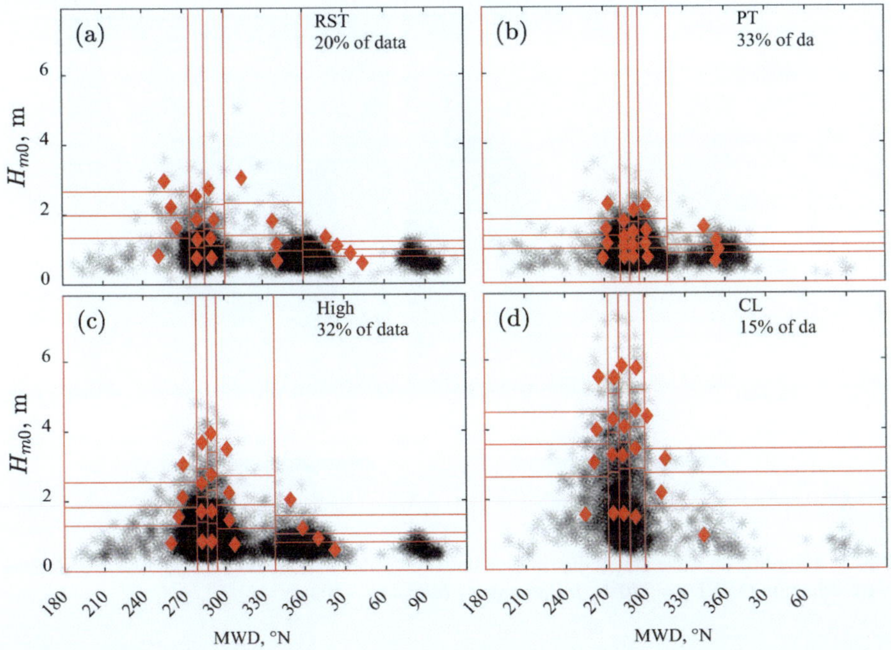

Fig. 2 Directional distribution of significant wave height, sorted to equi-energy bins (who's edges are marked by the red lines) and four weather system types: **a** 'Red Sea Trough', **b** 'Perssian Trough', **c** 'High', and **d** 'Cypress Low'

Figure 2 shows the wave conditions distribution for each weather system type and notes the percentage of time (Days Per Year, DPY) for each. In total, eighty wave conditions are used to construct the "representative" year, with each condition applied for DPY/20 days.

2.5 Nearshore Wave Boundary Conditions

DHI's MIKE21 FM SpectralWave module was utilized to generate the wave boundary conditions for the nearshore morphological simulations. For each of the eighty wave conditions, a domain was constructed, spanning 123 km from Haifa in the north to Ashdod in the south and 51 km seaward to a water depth of ca. 1000 m. The bathymetry for the domain is based on the GEBCO 2024 global terrain model, as shown in Fig. 3. The seaward boundary conditions were set at the northern (Haifa) and southern (Ashdod) edges using the transformed off-shore wave data, with linear interpolation applied in between. Non-reflective boundary conditions were applied on the domain's lateral edges. Lateral shadowing effects near the study site were minimized by keeping the nested nearshore domain at least 20 km away from the edges of the outer domain. Each wave condition was applied with and without the

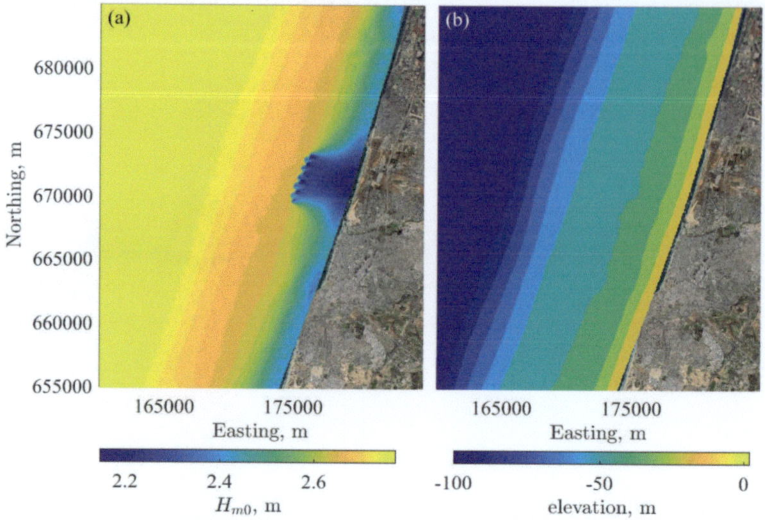

Fig. 3 **a** Significant wave height distribution under an offshore condition of $H_{m0} = 3$ m and mean wave direction (MWD) = 270°N. **b** Bathymetry (down to −100m)

array of MFS breakwaters. The effect of the breakwaters is modeled using a sink term in the wave action conservation equation, where a transmission coefficient controls its magnitude. The transmission coefficients, given in Table. 1, were specified according to laboratory measurements, as detailed in a previous study [4]. An iterative procedure was used to determine the significant wave height at the MFS breakwater's location and adjust the transmission coefficient accordingly. An example of the significant wave height field is shown in Fig. 3.

2.6 Nearshore Morphological Simulations

The open-source 'ShorelineS' (© 2020 IHE Delft & Deltares) was used to evaluate the effects of the floating breakwater array on sediment transport and future shoreline position. In 'ShorelineS', as in other similar 1D 'one-line' models, the shoreline evolution is based on the solution of a one-dimensional sediment mass balance equation. The domain extends ca. 34 km, centered around the planned position of the 4 km wide MFS breakwater array. The mass balance equation is discretized in space with a 1st-order accurate upwind scheme with 400 m-wide segments (in the longshore direction) and an explicit time-integration scheme with a time step of 12 h. The longshore sediment transport rate, Q_s m^3 $s^{(-1)}$ is calculated using the [14], shown in Eq. (2).

$$Q_s = 2.33 H_{m0}^2 T_p^{1.5} m_b^{0.75} D_{50}^{-0.25} \sin^{0.6}(2\phi). \tag{2}$$

Table 1 Transmission coefficient for each floating breakwaters (Floating Seawall [4])

T_p	3	6	7	8	9	10	11	11.5	11.75	12	12.25	12.5	12.75	13	14
H_{m0}	0.5	0.69	1.12	1.7	2.46	3.42	4.07	4.45	4.65	4.85	5.05	5.26	5.47	5.69	6.6
C_t	0	0.38	0.24	0.15	0.23	0.14	0.64	0.64	0.48	0.32	0.19	0.18	0.38	0.63	0.68

where, H_{m0} (m) is the significant wave (taken at the point of breaking), T_p is the peak wave period (s), D_{50} is the median grain diameter (0.18 mm), m_b is the mean bed slope in the breaking zone, ϕ is the relative angle of waves at the breaking point. Wave conditions at the breaking line were extracted from the outer, regional spectral wave simulations. Existing small-scale marine structures (e.g., marinas, groyns, etc.) were not included here. The annual longshore net sediment transport rate was calibrated to a value range of 100 to 80 $\times$ 10^3 m^3 $year^{-1}$ (northward) in the modeling region [10].

3 Results

Since measured historical coastline position data are not available for this study, the coastline response is evaluated as the difference between the response induced by the MFS breakwaters array and the present beach response. Figure 4 presents the residual shoreline evolution trends (Δ) of the Tel-Aviv region. Induced by the impact of the MFS array, accretion tendencies are observed to its south, and erosion to its north. The shoreline nearly reaches a new equilibrium position by the end of the simulation (10 years). However, a wider domain is required to reduce edge effects in a longer simulation. Therefore, any further testing remains outside the scope of this study.

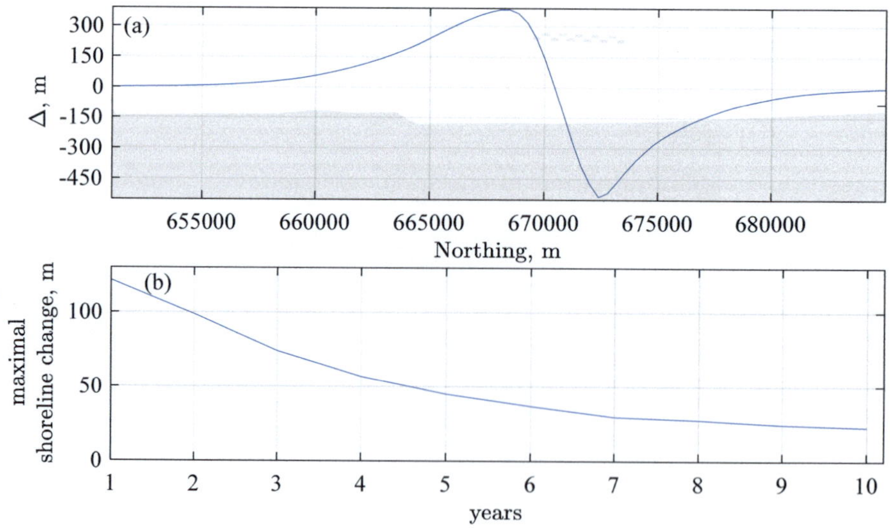

Fig. 4 Shoreline (relative) response to the MFS array for ten repetitions of the representative year. **a** The grey patches are the coastal outline and MFS array. **b** Maximal shoreline position changes over the duration of the simulation

4 Discussion

Given the highly pronounced erosion tendency north of the MFS breakwaters array (Fig. 4), and despite the existing coastal protection structures, it is likely that the MFS should be located at some optimal position further seaward than proposed here, using the closest location from the shore at water depth of 28 m.

References

1. Wang G, Goldfeld Y, Drimer N (2019) Expanding coastal cities—proof of feasibility for modular floating structures (MFS). J Clean Prod. https://doi.org/10.1016/j.jclepro.2019.03.007
2. United Nations (2014) World urbanization prospects: the 2014 revision: highlights
3. Wang CM, Watanabe E, Utsunomiya T (eds) (2007) Very large floating structures. CRC Press, London, New York
4. Wang G, Bar D, Schreier S (2024) The potential of end-of-life ships as a floating seawall and the methodical use of gap resonance for wave attenuation. Ocean Eng 298:117246. https://doi.org/10.1016/j.oceaneng.2024.117246
5. Stive MFJ, Dronkers J, Buurman R, de Haan HJ, Waterman RE, Zwemmer D, Brachya V, Derin D, Golan G, Schwartz-Millner B, Seginer A, Stav D (1998) Artificial islands off the Mediterranean Coast of Israel—feasibility R&D study (Phase I), Interim report by the Dutch/Israeli Steering Committee—Submitted to The Ministry of National Infrastructure, Jerusalem, The Ministry of Transportation. Public Works & Water Management, Den Haag
6. Palha A, Mendes L, Fortes CJ, Brito-Melo A, Sarmento A (2010) The impact of wave energy farms in the shoreline wave climate: Portuguese pilot zone case study using Pelamis energy wave devices. Renew Energy 35:62–77. https://doi.org/10.1016/j.renene.2009.05.025
7. Ozkan C, Perez K, Mayo T (2020) The impacts of wave energy conversion on coastal morphodynamics. Sci Total Environ 712:136424. https://doi.org/10.1016/j.scitotenv.2019.136424
8. Fujikubo M, Suzuki H (2015) Mega-float. In: Wang CM, Wang BT (eds) Large floating structures. Springer Singapore, pp 197–219. https://doi.org/10.1007/978-981-287-137-4_8
9. Drimer N, Gafter R (2017) Delta-type VLFS—hydrodynamic aspects. Ships Offshore Struct 1–14. https://doi.org/10.1080/17445302.2017.1384440
10. Perlin A, Kit E (1999) Longshore sediment transport on Mediterranean Coast of Israel. J Waterw Port Coast Ocean Eng 125:80–87. https://doi.org/10.1061/(ASCE)0733-950X(1999)125:2(80)
11. Luijendijk AP, de Schipper MA, Ranasinghe R (2019) Morphodynamic acceleration techniques for multi-timescale predictions of complex sandy interventions. J Mar Sci Eng 7:78. https://doi.org/10.3390/jmse7030078
12. Benedet L, Dobrochinski JPF, Walstra DJR, Klein AHF, Ranasinghe R (2016) A morphological modeling study to compare different methods of wave climate schematization and evaluate strategies to reduce erosion losses from a beach nourishment project. Coast Eng 112:69–86. https://doi.org/10.1016/j.coastaleng.2016.02.005
13. Hochman A, Harpaz T, Saaroni H, Alpert P (2018) The seasons' length in 21st century CMIP5 projections over the eastern Mediterranean. Int J Climatol 38:2627–2637. https://doi.org/10.1002/joc.5448
14. Kamphuis JW (1991) Computation of coastal morphology. Department of Civil Engineering Queen's University at Kingston Kingston, Ontario, Canada

Characteristic Lengths of Modular Floating Structures

Lingzhen Li, Kaiming Bi, and Xiao Lin Zhao

Abstract The characteristic length of a very large floating structure (VLFS) refers to the area influenced by a local perturbation, such as a concentrated load. Previous studies have revealed that the relative sizes of VLFSs over their characteristic lengths significantly influence their structural behavior. The current study proposes two analytical models to assess the influence of local perturbations on modular floating structures and quantify the characteristic lengths. The bending model considers finite connector bending stiffness, while disallowing the connector shear deformation. On the contrary, the shear model considers finite connector shear stiffness while disallowing the connector bending deformation. Both models assume the floaters to be infinitely stiff. The proposed analytical models are validated by numerical simulations based on hydroelastic analysis, accounting for finite connector bending and shear stiffness.

Keywords Modular floating structures (MFS) · Characteristic length · Connector stiffness

1 Introduction

Climate change is increasing the sea level height, reducing the living area of coastal countries and regions, endangering human safety and living conditions. On the other hand, the population is continuously increasing, making it very difficult to further expand land for living. Floating structure is regarded as an ideal solution to overcome these challenges, offering further benefits such as earthquake and tidal insensitivity

L. Li · K. Bi · X. L. Zhao (✉)
Department of Civil and Environmental Engineering, The Hong Kong Polytechnic University, Hong Kong, China
e-mail: xiao-lin.zhao@polyu.edu.hk

X. L. Zhao
Research Institute for Land and Space, The Hong Kong Polytechnic University, Hong Kong, China

X. L. Zhao et al. (eds.), *Proceedings of the Fourth World Conference on Floating Solutions: WCFS 2024; 2–4 December; Hong Kong*, Lecture Notes in Civil Engineering 597, https://doi.org/10.1007/978-981-96-4569-5_78

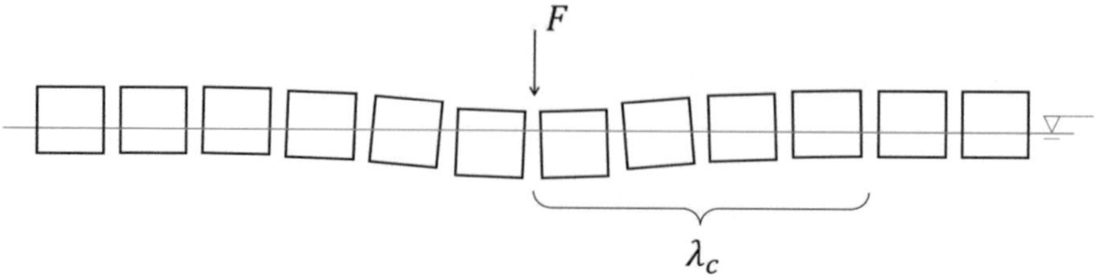

Fig. 1 Schematic of the influence of a local perturbation on a modular floating structure. Rectangles represent floaters, while the blue horizontal line represents the stationary water surface. Adjacent two floaters are connected by connectors (not shown)

[1]. Modular floating structures offer several advantages, including (1) efficient and high-quality fabrication due to modularity, (2) ease of fabrication and deployment thanks to the small size of floaters, and (3) the ability to be re-designed and re-deployed for changing functions. Previous studies [2–4] have highlighted that the structural behavior of mat-like (continuous) very large floating structures (VLFSs) is significantly impacted by their sizes relative to their characteristic lengths. The characteristic length refers to the area influenced by a local perturbation, such as a concentrated load. This effect may exist in modular floating structures as well, whose characteristic length analysis has, however, not been reported yet. To address this research gap, the current study proposes two analytical models, based on the schematic shown in Fig. 1, to assess the influence of a local perturbation on modular floating structures, and quantify the characteristic length.

2 Analytical Models

Before applying the external load, the gravity of floaters is balanced by buoyancy at the original equilibrium. Therefore, we exclusively focus on the additional buoyancy resulting from the external load, as depicted in Fig. 2a. Due to symmetry, only half of the modular floating structure is analyzed. Two analytical models, one accounting for connector bending deformation and the other for shear deformation, are respectively proposed.

2.1 Bending Model

Model Setup. In the bending model, the connector bending stiffness is finite, while the connector shear stiffness and floater stiffness are assumed to be infinitely large. Thus, the equilibrium of a single floater, which is illustrated in Fig. 2b, reads Eqs. (1) and (2).

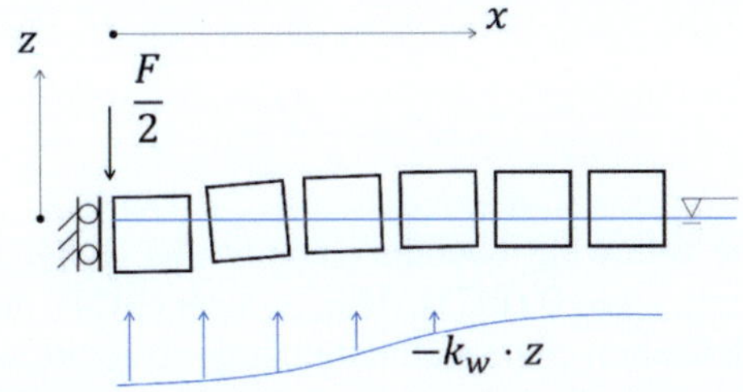

(a) Half of the modular floating structure, k_w represents the hydroelastic stiffness per unit length

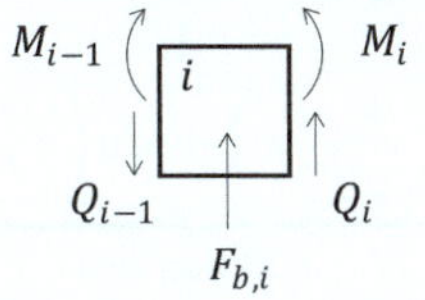

(b) Equilibrium of a floater in the bending model

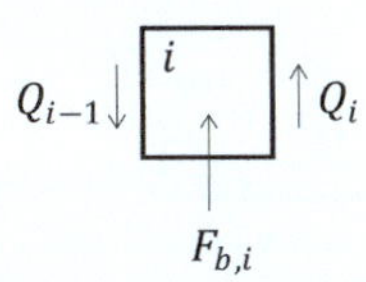

(c) Equilibrium of a floater in the shear model

Fig. 2 Schematics of a modular floating structure subjected to a concentrated load

$$Q_{i-1} - Q_i = F_{b,i} \tag{1}$$

$$M_{i-1} - M_i = Q_i \cdot L + F_{b,i} \cdot \frac{L}{2} \tag{2}$$

where Q_i and M_i are the shear force and bending moment acting on the right-hand side of the i th floater; Q_{i-1} and M_{i-1} are the corresponding values on the left-hand side of the floater; L represents the length of a single floater; $F_{b,i}$ denotes the additional buoyancy resulting from the concentrated load.

Connector bending moment (M) and shear force (Q) have the relationships shown in Eqs. (3) and (4), and the additional buoyancy reads Eq. (5).

$$M = k_m \cdot L \cdot z'' \tag{3}$$

$$Q \approx -M' \tag{4}$$

$$F_b = -k_f \cdot z \tag{5}$$

where z represents deflection of the modular floating structure; k_m denotes the connector bending stiffness; $k_f = \rho \cdot g \cdot B \cdot L$ represents the hydroelastic stiffness per floater; B is the floater width.

Substituting Eqs. (2)-(5) into Eq. (1) leads to the following governing equation for the bending model:

$$z^{(4)} + \omega^4 \cdot z = 0 \tag{6}$$

where

$$\omega^4 = \frac{k_f}{k_m \cdot L^2} \tag{7}$$

By imposing the following boundary conditions (BCs): $Q_{x=0} = F/2$ (BC1), $z'|_{x=0} = 0$ (BC2), $z|_{x=\infty} = 0$ (BC3), $M|_{x=\infty} = 0$ (BC4), and $Q|_{x=\infty} = 0$ (BC5), one can derive the deflection of the modular floating structure, as expressed by Eq. (8).

$$z = \underbrace{-\frac{F}{2k_w} \cdot \frac{\omega}{\sqrt{2}}}_{\text{deflection at concentrated load}} \cdot \underbrace{e^{-\frac{\omega}{\sqrt{2}} \cdot x}}_{\text{exponential decay term}} \cdot \left[\underbrace{\cos\left(\frac{\omega}{\sqrt{2}} \cdot x\right) + \sin\left(\frac{\omega}{\sqrt{2}} \cdot x\right)}_{\text{oscillatory term}} \right] \tag{8}$$

Characteristic Length. Equation (8) indicates that the deflection of a modular floating structure due to a local perturbation decays exponentially and oscillates around zero deflection. An inspection suggests that beyond the second intersection of the deflection profile (i.e., the location of zero deflection, $x|_{z=0}$), the oscillating deflection becomes negligible. Therefore, the distance between this second intersection and the local perturbation, expressed in Eq. (9), is considered as the characteristic length.

$$\lambda_c = \left(2 - \frac{1}{4}\right)\pi \cdot \sqrt{2} \cdot \sqrt[4]{\frac{k_m \cdot L^2}{k_f}} \tag{9}$$

2.2 *Shear Model*

Model Setup. In the shear model, the connector shear stiffness is finite, while the connector bending stiffness and floater stiffness are assumed to be infinitely large. Thus, the equilibrium of a single floater, which is illustrated in Fig. 2c, reads Eq. (10).

$$Q_{i-1} - Q_i = F_{b,i} \tag{10}$$

$$Q_i = k_q \cdot (z_{i+1} - z_i) \tag{11}$$

where $F_{b,i}$ is expressed as Eq. (5); Q_i denotes the shear force acting on the right-hand side of the i-th floater; k_q represents the connector shear stiffness.

Substituting Eqs. (5) and (11) into Eq. (10) leads to the governing equation of the shear model, which reads Eq. (12).

$$z'' - \mu^2 \cdot z = 0 \tag{12}$$

where

$$\mu^2 = \frac{k_f}{k_q \cdot L^2} \tag{13}$$

By imposing the following boundary conditions: $Q_{x=0} = F/2$ (BC1) and $z\,|_{x=\infty} = 0$ (BC2), one can derive the deflection of the modular floating structure as follows:

$$z = \underbrace{-\frac{F \cdot \mu}{2 \cdot k_w}}_{\text{deflection at concentrate load}} \cdot \underbrace{e^{-\mu \cdot x}}_{\text{exponential decay term}} \tag{14}$$

Characteristic Length. Equation (14) suggests that the deflection of a modular floating structure due to a local perturbation decays exponentially, and it becomes zero only at an infinite distance. An inspection suggests that beyond the characteristic length expressed as Eq. (15), the deflection becomes negligible (no greater than 0.2% of the maximum deflection).

$$\lambda_c = 2\pi \cdot L \cdot \sqrt{\frac{k_q}{k_f}} \tag{15}$$

3 Validation

A series of numerical simulations were performed to analyze the influence of a concentrated load on modular floating structures and validate the developed analytical models of bending and shear types. The rigid floaters were interconnected by connectors with finite bending and shear stiffness. Buoyancy acting on floaters was modelled by hydroelastic springs. The equilibrium state, involving the applied concentrated load and associated buoyancy, is expressed as Eq. (16). It should be noted that inertia and damping terms, which are essential for dynamic analysis, are unnecessary in the current static analysis.

$$\left\{K_{str} + K_{hyd}\right\} \cdot [z; \theta] = f \tag{16}$$

where K_{str} represents the connector stiffness matrix, involving both connector bending and shear stiffness, while K_{hyd} denotes the hydroelastic stiffness matrix; z and θ are arrays of floater heave and pitch motions at equilibrium; f represents an array of external load, where only the term of concentrated load is none zero.

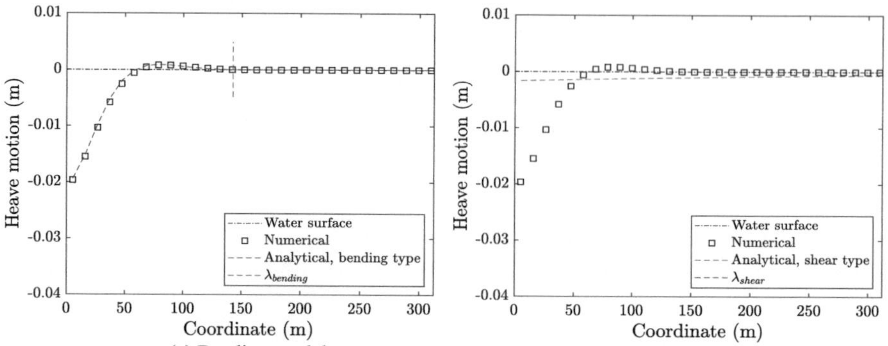

(a) Bending model (b) Shear model

Fig. 3 Analytical solution versus numerical solution, with $k_m = 10^9$ N · m/rad and $k_q = 10^9$ N/m as an example. The black dashed line represents still water surface; the hollow rectangles denote floater positions assessed through the numerical model; the blue and red dashed curves are results assessed through the analytical models of bending and shear types, respectively. The estimated λ_{shear} is located outside of the displayed coordinate

Two categories were analyzed: (1) finite connector bending stiffness and negligible connector shear deformation; (2) finite connector shear stiffness and negligible connector bending deformation. In the numerical modelling, negligible connector bending or shear deformation was realized through defining a large stiffness value correspondingly. Each floater was assumed to be 10 m long, 10 m wide, and 4 m high. The concentrated load was assumed to be 100 kN. Since the entire system is linear, the exact value of the concentrated load does not alter the shape of deflection profile nor the value of characteristic length. It only affects the deflection values (y-axis) in Figs. 3 and 4.

3.1 Intermediate k_m and Large k_q

Figure 3 illustrates an example in the first category, where the deflection is primarily due to connector bending deformation. It showcases that, when connector shear deformation is negligible, the deflection of the modular floating structure assessed by the analytical model of the bending type closely approximates the numerical result. In addition, Eq. (9), pertaining to the bending model, accurately captures the characteristic length (Fig. 3a). However, the performance of the analytical model of the shear type is unsatisfactory in this category, as shown in Fig. 3b.

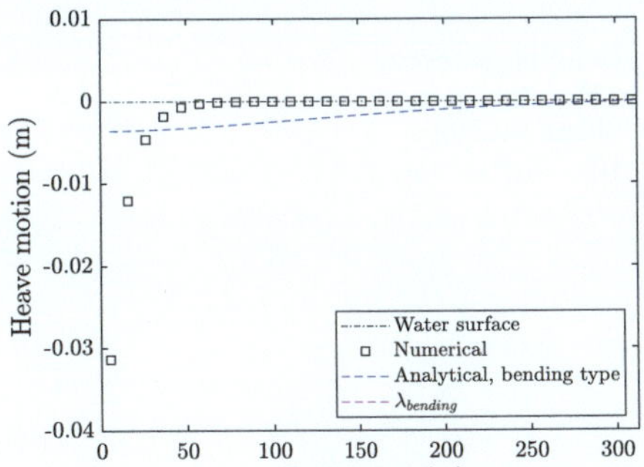

(a) Bending model

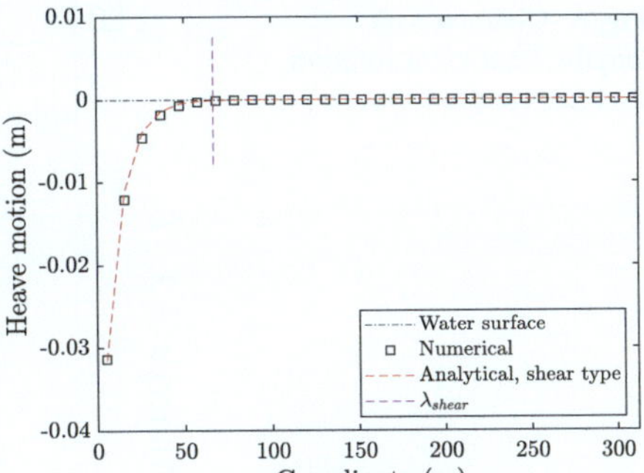

(b) Shear model

Fig. 4 Analytical solution versus numerical solution, with $k_q = 10^6$ N/m and $k_m = 10^{12}$ N · m/rad as an example. The black dashed line represents still water surface; the hollow rectangles denote floater positions assessed through the numerical model; the blue and red dashed curves are results assessed through the analytical models of bending and shear types, respectively. The estimated $\lambda_{bending}$ is located outside of the displayed coordinate

3.2 *Intermediate k_q and Large k_m*

Figure 4 illustrates an example in the second category, where the deflection is primarily due to connector shear deformation. It showcases that, when the connector bending deformation is negligible, the deflection of the modular floating structure assessed by the analytical model of the shear type closely approximates the numerical result. In addition, Eq. (15), pertaining to the shear model, accurately captures the characteristic length (Fig. 4b). However, the performance of the analytical model of the bending type is unsatisfactory in this category, as shown in (Fig. 4a).

4 Characteristic Lengths Vary with Connector Stiffness

Figure 5 corresponding to the special cases detailed in Figs. 3 and 4, demonstrates that the proposed analytical models of bending and shear types accurately capture the characteristic lengths of modular floating structures, in which the connector bending and shear stiffness vary across a wide range of magnitudes.

5 Concluding Remarks

The current study proposes two analytical models for assessing the deflection of modular floating structures due to local perturbation, i.e., a concentrated load. The bending model incorporates a finite connector bending stiffness, while excluding

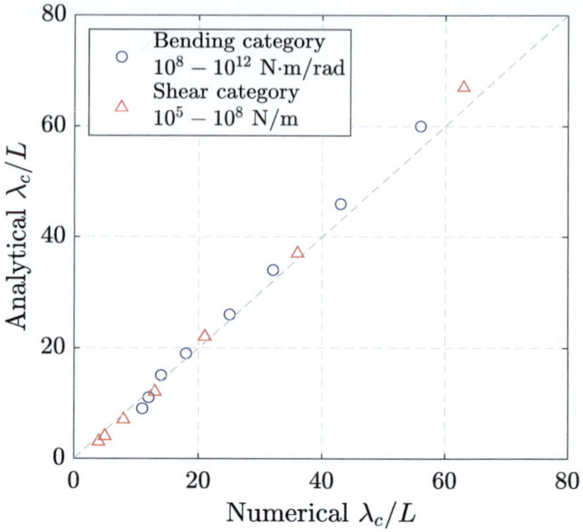

Fig. 5 Characteristic lengths, analytical solution vs. numerical solution

connector shear deformation. Conversely, the shear model considers a finite connector shear stiffness, while disallowing connector bending deformation. The performance of the two analytical models is validated by numerical simulations based on the hydroelastic analysis.

Acknowledgements The first author is sponsored by the Innovation and Technology Fund (ITF) Research Talent Hub of Hong Kong.

References

1. Wang CM, Wang BT (2015) Large floating structures. Springer, Heidelberg
2. Suzuki H, Yoshida K, Iijima K (1996) A consideration of the structural design of a large-scale floating structure. J Mar Sci Technol 1:255–267
3. Suzuki H, Riggs HR, Fujikubo M, Shugar TA, Seto H, Yasuzawa Y, Bhattacharya B, Hudson DA, Shin H (2007) Very large floating structures. In: International conference on offshore mechanics and arctic engineering, pp. 597–608
4. Zhang M, Schreier S (2022) Review of wave interaction with continuous flexible floating structures. Ocean Eng 264:112404

Experimental Investigation of a Novel Concept of Modular Multi-purpose Floating Structures

Filippo Giorcelli, Francesco Niosi, Bruno Paduano, Mattia Glorioso, Oronzo Dell'Edera, Marco Fontana, Gabriella Gaeta, Renata Archetti, and Sergej Antonello Sirigu

Abstract Modular Multi-purpose Floating Structures (MMFS) provide a possible solution to the growing need for space resulting not only from the rapidly growing global population but also from the expanding blue economy sector. The desired space is generated in a more sustainable way than land reclamations methods, by interconnecting together standardized floating modules, making this technology adaptable and suited to a broad range of possible open sea application. Herein, this study aims to investigate the dynamical response of a system composed by three floating hexagonal modular platforms connected via semi-rigid connectors and moored at the seabed with a taut mooring configuration. An experimental investigation has been carried out on a 1:50 prototype at the wave basin of the Laboratory of Hydraulic Engineering (LIDR) of Università degli Studi di Bologna. The objective of the paper is the experimental validation of the proposed technological concept, through the assessment of the loads acting on the system and the evaluation of comfort indices for various representative sea states. The preliminary results show the trends of the loads acting on the connection system and the mooring system, which are essential for the correct design of the full-scale system. Additionally, the comfort and safety indices are met for all the analyzed sea states, demonstrating the validity of the proposed design.

Keywords Modular multi-purpose floating structures · Experimental campaign · Mooring system · Multi-body hydrodynamics · Very large floating structures

---

F. Giorcelli · F. Niosi · B. Paduano · M. Glorioso · O. Dell'Edera · M. Fontana · S. A. Sirigu (✉)
Marine Offshore Renewable Energy Lab, Department of Mechanical and Aerospace Engineering, Politecnico di Torino, Torino, Italy
e-mail: sergej.sirigu@polito.it

G. Gaeta · R. Archetti
Department Civil, Chemical, Environmental and Materials Engineering, Università Degli Studi di Bologna, Bologna, Italy

X. L. Zhao et al. (eds.), *Proceedings of the Fourth World Conference on Floating Solutions: WCFS 2024; 2–4 December; Hong Kong*, Lecture Notes in Civil Engineering 597, https://doi.org/10.1007/978-981-96-4569-5_79

1 Introduction

In the blue economy sector, Very Large Floating Structures (VLFS) are an attractive way to potentially create space on water and sustain the blue economy activities. VLFS consist of floating modules that can be either rigidly interconnected or linked via semi-rigid connections and anchored to the seabed using a mooring system. For example, they can serve as logistics support bases for O&M operations and the installation of renewable offshore plants, facilitate aquaculture, seaweed farming, and vertical farming for countries facing land shortages, and help mitigate shipping traffic through the creation of floating port terminals [1]. Additionally, VLFS hold potential in the residential sector, offering a sustainable alternative to traditional land reclamation practices, which rely on environmentally harmful dredging techniques [2]. VLFS technology development has been primarily hindered by the high costs and technological limitations. The principal technical challenges involve the high loads acting on the structures and the harsh, corrosive marine environment [3], which significantly impact the comfort and usability of VLFS for the applications discussed above. In particular, two major VLFS's problems to be addressed are the reduction of hydro-elastic response of the system [4, 5] and the high loads on the connection and mooring systems mainly due to waves action, both related to comfort and safety during operative and extreme events [6].

Within this context, this paper aims to describe a novel modular multi-purpose floating structure concept and the preliminary experimental results obtained on a scaled prototype. The new system concept is composed of interconnected floating modules linked to each other via semi-rigid connectors (which provide an optimum balance between the rigidity of the connections and the relative motion between platforms) and anchored to the seabed using a taut mooring, which provides better station keeping characteristics than catenaries. Additionally, the taut mooring system has a reduced impact on the marine ecosystem compared to catenary mooring, due to the absence of chains on the seabed and the consequent erosion and destruction of the seafloor. The main objective of this paper is to present the experimental campaign performed and the obtained outcomes, with a focus on analyzing comfort parameters for two configurations of the system (single platform and three "in line" floaters layout) in terms of maximum inclination, accelerations and loads acting on the mooring and connections.

2 Experimental Campaign Setup

The experimental campaign has been carried out in the wave basin of the Laboratory of Hydraulic Engineering (LIDR) of Università degli Studi di Bologna, illustrated in Fig. 1, whose dimensions are 12.00 (m) $\times$ 8.00 (m) $\times$ 0.60 (m) respectively in length, width, and water depth. Figure 1 also describes the layout of the novel MMFS

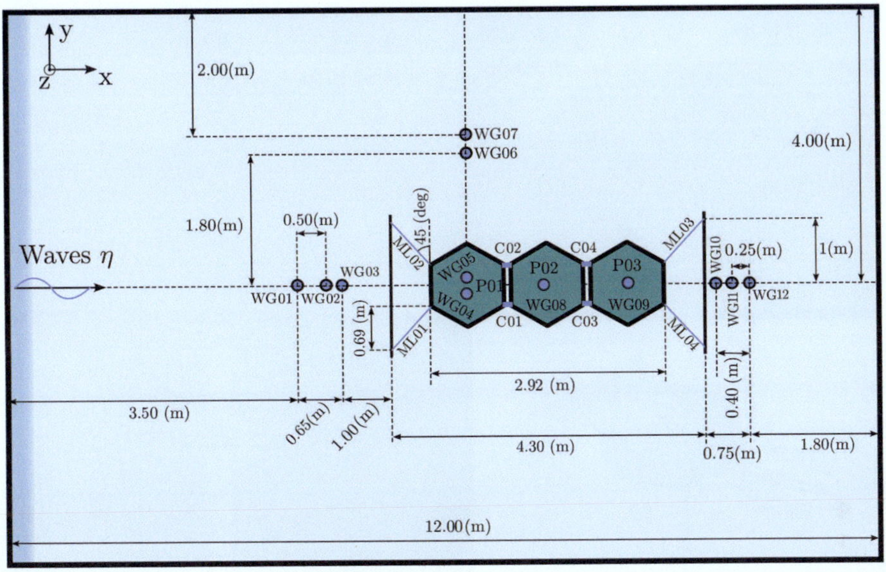

Fig. 1 Experimental setup of the LIDR wave basin facility. In the figure WG describes the position of the 12 wave gauges (shaded circles refers to WGs removed during the tests), ML refers to the four mooring lines, while C to the mechanical connectors between platforms. Figure adapted from [7]

system under test, which consists of three hexagonal floating modules arranged in a line.

Each module is made of stainless steel and appropriately ballasted internally, and it is shown in Fig. 2a. The semi-rigid mechanical connectors consists in mechanical piston. The connection system consists of 5 mechanical pistons, which contain compression springs inside (see Fig. 2b, d). The mechanical pistons are connected to the ends of the structures through spherical joints to prevent shear forces and moments at the joints. The mooring line geometry is shown in Fig. 2c and has been realized using very stiff polyester ropes and an extension spring. Each mooring line and mechanical connector are equipped with a load cell.

For the present experimental tests, the connection system has been specifically designed to emulate the stiffness properties of the full-scale connector system, representing the axial, bending and shear stiffness of the system. A previous work concerning the analysis and optimization of the semi-rigid connector stiffness has been carried out [8]. During these experiments, both the isolated floater the three platforms in line layout has been tested. The Froude's law is used to scale the experimental design parameters from the full-scale to the prototype one (1:50), which are reported in Table 1. The system response is investigated under various regular and irregular sea states representative of moderately sheltered areas, and the complete test matrix is given in Table 2. Each sea state is generated with regard to the JONSWAP spectrum (peak enhancement factor $\gamma = 3.3$). The data acquisition system campaign

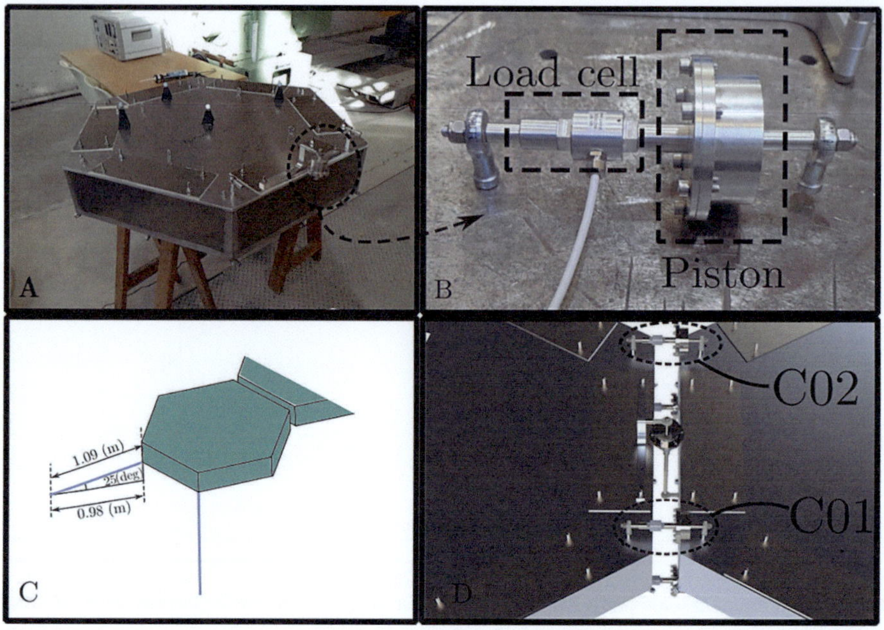

Fig. 2 MMFS tested floater prototype (**a**), semi-rigid mechanical connection system with sensor for load measurement (**b**), mooring configuration and geometry (**c**), connection system layout (**d**)

consists in: a National Instruments C-RIO9042 module used to acquire signals from the load cells of the mooring and connector systems, a motion capture system (Qualisys) consisting in four cameras managed by a standalone PC, which starts recording after receiving a trigger signal from the C-RIO9042 module, and a separate data acquisition system handles the acquisition of the 12 wave gauges signals (WG01 to WG12).

3 Preliminary Experimental Results

In this section the outcomes regarding the comfort indexes in terms of maximum displacement and acceleration are discussed. Regarding the nomenclature adopted in this work to designate the degrees of freedoms (DoFs) of the system, the first subscript stands for the DoF (1–6), while the second (n) refers to the platform number of the MMFS: 1 represents the bow platform, 2 the center, and 3 the stern platform.

$$\mathrm{x} = \left[\mathrm{surge},\ \mathrm{sway},\ \mathrm{heave},\ \mathrm{roll},\ \mathrm{pitch},\ \mathrm{yaw}\right] = \left[x_{1,n}, x_{2,n}, x_{3,n}, x_{4,n}, x_{5,n}, x_{6,n}\right] \tag{1}$$

Table 1 Full-scale and 1:50 scaled design parameters

Parameter	Units	Prototype scale (1:50)	Full-scale
Bathymetry	(m)	0.60	30.0
Floater diameter	(m)	1.08	54.0
Floater height	(m)	0.18	9.0
Floater edge length	(m)	0.54	27.0
Draft	(m)	0.12	6.0
Three platform layout length	(m)	2.92	146
CoG distance from the deck	(m)	0.04	2.12
Distance between platform	(m)	0.06	3.0
Floater pitch inertia	(kgm^2)	8.26	2.58×10^9
Floater roll inertia	(kgm^2)	8.26	2.58×10^9
Floater yaw inertia	(kgm^2)	13.34	4.16×10^9
Floater mass	(kg)	93.20	1.16×10^9
Mooring line length	(m)	1.09	54.5
Mooring line stiffness	(N/m)	130	325×10^3
Connector stiffness	(N/m)	88,500	221×10^3

Table 2 Sea states tested during the experimental campaign (prototype scale) and their reference full-scale value

Sea State	Steepness	Prototype scale (1:50)		Full- scale	
		H_s (m)	T_p (s)	H_s (m)	T_p (s)
IRR01	1:40 Low	0.020	0.71	1.00	5.00
IRR02	1:50 Low	0.022	0.85	1.10	6.00
IRR03	1:50 Low	0.031	1.00	1.50	7.00
IRR04	1:50 Low	0.040	1.13	2.00	8.00
IRR05	1:30 High	0.026	0.71	1.30	5.00
IRR06	1:30 High	0.037	0.85	1.90	6.00
IRR07	1:30 High	0.051	1.00	2.60	7.00
IRR08	1:30 High	0.067	1.13	3.30	8.00

In order to investigate the comfort parameters of the novel MMFS proposed, the outcomes presented are those regarding the maximum pitch motion and the root mean square (RMS) of the measured acceleration for surge and heave DoFs (Fig. 4). Since the experimental tests have been carried out with long crested frontal waves (as sketched in Fig. 1), only the excited degrees of freedom are considered in the analysis: surge, heave, and pitch. In order to assess the system's comfort performance, two references were evaluated for determining comfort criteria: the NORDFOLK criteria for comfort on cruise ships [9], and the criteria presented by the Horizon2020 Space@Sea project study [10].

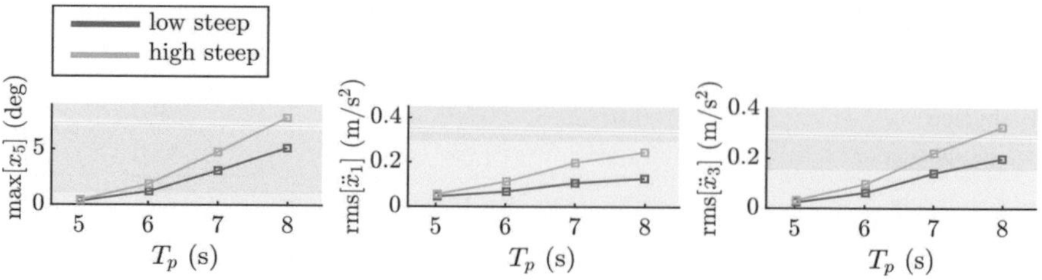

Fig. 3 Comfort indexes for the isolated module layout. The red background color highlights the comfort limit criteria with respect to Space@Sea indications [10]

The experimental results in terms of comfort indexes are shown in Fig. 4, for the three-module configuration and Fig. 3 for the isolated module configuration. The results demonstrate excellent comfort performance for the analyzed sea states for the three-module layout. As expected, the surge and pitch motions of the three platforms are very similar because the overall system behaves similarly to a rigid body due to the stiffness of the connections. In the multi-body configuration, the most critical conditions are represented by the IRR07 and IRR08 sea states for the heave acceleration of stern and bow platforms. On the other hand, referring to the naval cruise boundaries of [9] (second row of Table 3), the only acceleration non eligible becomes the one of the extreme sea state IRR8. Therefore, the system provides excellent comfort and safety performance even for harsh environments and waves up to 2.60 (m) significant height. Instead, for the single floater layout, the comfort thresholds are severely exceeded both with respect to pitch motion and heave acceleration. However, these results highlight how to increase the number of modules for a chosen configuration positively impacts the comfort levels achievable by the system.

In conclusion, the measured forces acting on the mooring lines and connectors are analyzed for the extreme IRR8 sea state, which is representative of a high stress condition, are shown in Fig. 5. The mooring system, appropriately pretensioned, responds satisfactorily and the condition of slack mooring, which is to be avoided for taut elastic moorings, does not occur. As expected, the highest loads on the connections occur between the first and second platforms (C01 and C02), as these platforms are more significantly stressed by the wave action. Therefore, the system shows a good behavior for the analyzed extreme conditions. Moreover, it is worth emphasizing how these promising results in terms of loads acting on critical systems (such as mooring and connections between different modules) are essential for future system design and optimization.

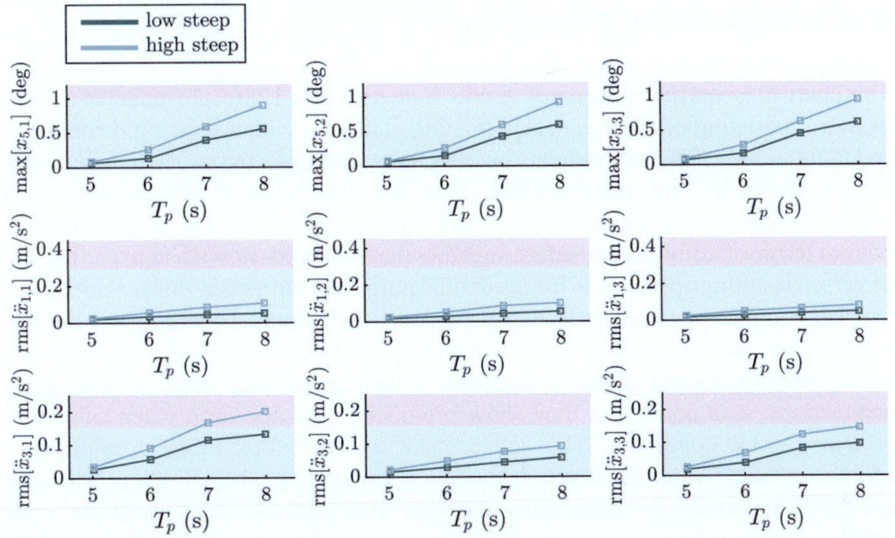

Fig. 4 Comfort indexes for the three-modules in line layout. The red background color highlights the comfort limit criteria with respect to Space@Sea indications [10]

Table 3 Comfort criteria for floating platforms

Thresholds reference	Maximum inclination (°)	RMS [$\ddot{x}_3$] (m/s^2)	RMS [$\ddot{x}_1$] (m/s^2)
Space@Sea [10]	1.00	0.15	0.30
NORDFOLK [9]	2.00	0.20	0.30

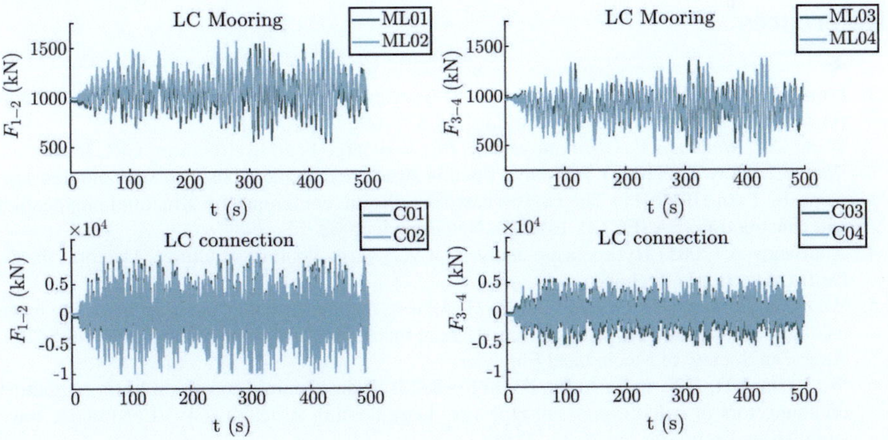

Fig. 5 Measured loads acting on mooring lines and connectors during the extreme IRR8 sea state test

4 Conclusions and Future Works

This study presents the preliminary results from tests on a 1:50 scale model of a novel MMFS, consisting of interconnected floating platforms with a semi-rigid connector and taut mooring. The experiments investigate two layouts: the isolated module and three modules "in line" configurations. The outcomes reveal that as the number of connected platforms increases, the system provides progressively better performance both in terms of comfort and safety, meeting the thresholds described in [9, 10] and therefore resulting appropriate for residential purposes. Moreover, this is also true for harsh environments, characterized by waves outside the typical operating conditions, reaching full-scale H_s values equal to 2.60 (m). Then, observing the values of loads acting on critical components of the MMFS such as mooring lines and semi-rigid connections, it is noted that they show promising behavior even when subject to critical working conditions. This achievement is essential for future developments of this research, which will include system optimization. These findings will also inform future studies for calibrating and validating numerical models of the novel MMFS system design.

Acknowledgements Acknowledgements and Funding
The authors are grateful for the funds received concerning the presented study which received funding from the Ministry of University and Research and PNRR's PhD scholarship program defined in DM-352 in collaboration with Eni S.p.A. This publication is part of the PNRR-NGEU project, which has received funding from the MUR—DM 352/2022.

References

1. Lamas-Pardo M, Iglesias G, Carral L (2015) A review of very large floating structures (VLFS) for coastal and offshore uses. Ocean Eng 109:677–690
2. Wang CM, Watanabe E, Utsunomiya T (2006) Very large floating structures. CRC Press
3. Wang CM, Tay ZY (2011) Very large floating structures: applications, research and development. In: Lam HF (2011) The twelfth east Asia-Pacific conference on structural engineering and construction (EASEC12). Elsevier, Netherland, pp 62–72
4. Andrianov A (2005) Hydroelastic analysis of very large floating structures. Doctoral thesis, Delft University of Technology
5. Waals OJ, Bunnik THJ, Otto WJ (2018) Model tests and numerical analysis for a floating mega island. In: International conference on offshore mechanics and arctic engineering, vol 51203. American Society of Mechanical Engineers
6. Wu L, Jiang H, Ji X, Ju X, Xiang Z, Gu M (2023) Hydrodynamic constraint loads estimation on connectors of column-pontoon type very large floating structure (CP-VLFS) under wave stimulation. Front Mar Sci 10:1113555
7. Giorcelli F, Niosi F, Glorioso M, Paduano B, Sirigu SA (2024) Experimental investigation of a novel modular multi-purpose floating structure concept. In: Proceedings of 6th renewable energies offshore RENEW 2024 (paper under review). Lisbon, Portugal
8. Sirigu SA, Boccanfuso V, Carapellese F (2023) Multi-body analysis of modular floating islands: optimisation of connector stiffness. In: Proceedings of the third world conference on floating solutions. Tokyo, Japan

9. NORDFORSK (1987) Assessment of ship performance in a seaway. Results of a Nordic co-operative project on seakeeping performance of ships
10. Schay J, Hüsken C. Space@Sea—D10.5, updated requirements formulation. https://spaceatsea-project.eu/public-information/public-deliverables. Last accessed 18 Jun 2024

Dynamic Behavior of Multi-module Floating Structures with Different Connector Types

Yoonho Choi, Kilje Jung, Wonjoon Lee, and Jangseok Yi

Abstract In today's world, the rise in sea levels due to global warming presents a significant threat to environments worldwide. One solution to mitigate this issue is the implementation of Modular Floating Structures (MFS), which can provide sustainable and expandable urbanized components for cities. In this study focuses on the development of a time-domain multi module floating model aimed at addressing the complexities of the floater-connector-mooring coupled system, considering various connector types such as rigid, single hinged, and double hinged. Additionally, the serviceability of multi-module floating structures based on connector type and the arrangement of the mooring system, considering aspects of scalability and potential applications for floating structures.

Keywords Floating structure · Multi-module · Concrete floater · Module connector

1 Introduction

In recent decades, the escalating challenges posed by global warming have intensified the urgency for sustainable solutions in coastal infrastructure and maritime industries. As the climate of Earth continues to warm, the consequences of rising sea levels have become increasingly apparent, necessitating innovative approaches to address the vulnerabilities of coastal regions. Among these solutions, the concept of floating structures holds promise due to its adaptability and potential to mitigate the impacts of sea level rise. Floating structures, characterized by their ability to float on water surfaces while accommodating various functions such as residential, commercial, and industrial purposes, present a compelling alternative in the face of rising sea levels. These structures offer flexibility in location and resilience against inundation,

Y. Choi (✉) · K. Jung · W. Lee · J. Yi
Hyundai Engineering and Construction, Seoul, South Korea
e-mail: yoonho@hdec.co.kr

X. L. Zhao et al. (eds.), *Proceedings of the Fourth World Conference on Floating Solutions: WCFS 2024; 2–4 December; Hong Kong*, Lecture Notes in Civil Engineering 597, https://doi.org/10.1007/978-981-96-4569-5_80

thereby safeguarding coastal communities and infrastructure from the adverse effects of cli-mate change.

For the above reasons, there are many researchers in the world try to develop more efficient and adoptable multi body floating structures. MARIN in Netherlands developed floating island, SPACE@SEA which is to provide for smart and sustainable growth based on future needs, by the development of standardized, affordable modular islands with a low ecological impact [1]. Especially for develop multi module floating structures, connection condition is very important and critical issues for hydrodynamic behavior for fully motion of floating system. Ren et al. [2] investigated the effect of different outermost connector types on the hydrodynamic responses of a modular multi-purpose floating structure (MMFS) system. Wang et al. [3] investigated influence of connector stiffness and damping on dynamic behavior for very large floating structure (VLFS). Shi et al. [4] developed stiffness combination of a flexible-base hinged connector (FBHC) for very large floating structures compare with a rigid hinged connector (RHC) and find out the FBHC can significantly reduce the connector loads with a confined flexibility. Vincentius Rumawas et al. [5] researched human factors evaluation in ship design which is well listed regarding criteria of global standards for human performance in the ships.

In this paper, the dynamic behavior of multi-body floating structures with different types of connectors are analyzed, and serviceability criteria was examined according to the viscous damping of the structures, the type of connection, and application of mooring facilities. Since there are no serviceability standards for large floating structures, the criteria for large cruise ships are set considering manned conditions and are based on NORDFORSK (1987) and BV NR 636. The human performance, especially comfort and sea-sickness, will be necessary considered in operational part of floating structures.

Some specific applications have been assumed for initial design of floating module. Because of the size for unit module should be fabricated easily on a large scale using dry or floating dock or fabrication yard, the shape of unit module is adopting a square and concrete material is used for considering efficiency of structural system and cost effectiveness for installation.

The average payload on deck space is more than 3.0 t/m^2 which is considered living condition with 4–5 floor building and freeboard of the floating structure has to be secured at least 2.5 m considering design wave height 2.0 m.

2 Hydrodynamic Analysis for Unit-Module Floating Structure

2.1 *Evaluation Unit Module with Different Damping Conditions*

For evaluation of unit module per height of structure, there are two model, M1(H = 11.0 m) and M2(H = 9.5 m), is considered M1 has 3.0 m freeboard and can be resisted payload on deck space within 4.0 t/m^2 and M2 has 2.5 m freeboard and maximum pay load is 3.5 t/m^2 (Fig. 1).

The important influence of serviceability evaluation for floating module is pitch RAO (Response Amplitude Operator) and pitch RMS (Root Mean Square), heave and surge acceleration RMS. Both modules are not satisfied requirement because of numerical analysis software ANSYS AQWA could not reflect viscous damping automatically. In this research, the total damping (i.e. potential damping and viscous damping) is assumed 10% to 15% and re-analyzed of hydrodynamics. Figure 2 is shown wave direction and inspection point on floater, Fig. 3 is shown pitch RAO of both unit modules M1 and M2 following different damping conditions.

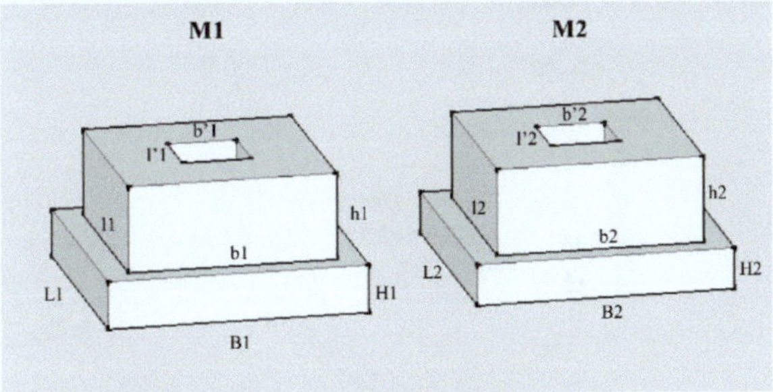

Fig. 1 Shape of unit module M1 (left), M2 (right) and size of both module

Simbol	Size (m)
B1, B2, L1, L2	60.0
H1	11.0
H2	9.5
b1, b2, l1, l2	48.0
h1, h2	20.0
b'1, b'2, l'1, l'2	16.0

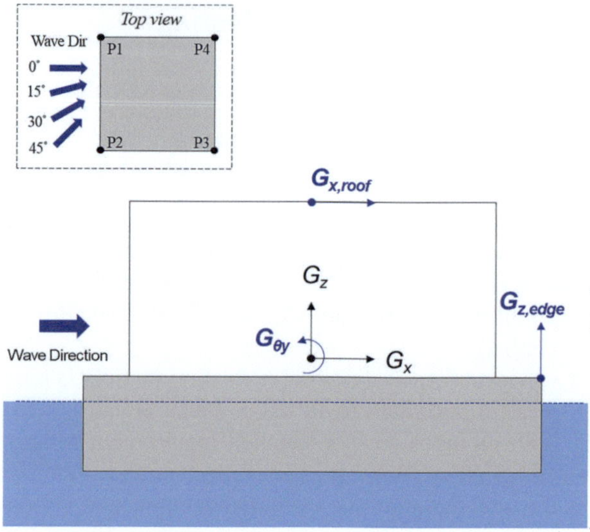

Serviceability Requirements		
RAO pitch	$\leq 5^\circ$	
RMS pitch	$\leq 2^\circ$	$G_{\theta y}$
RMS heave	$\leq 0.3\text{m/s}^2$	$G_{z,edge}$
RMS Surege	$\leq 0.25\text{m/s}^2$	$G_{x,roof}$

Fig. 2 Serviceability requirements at inspection points

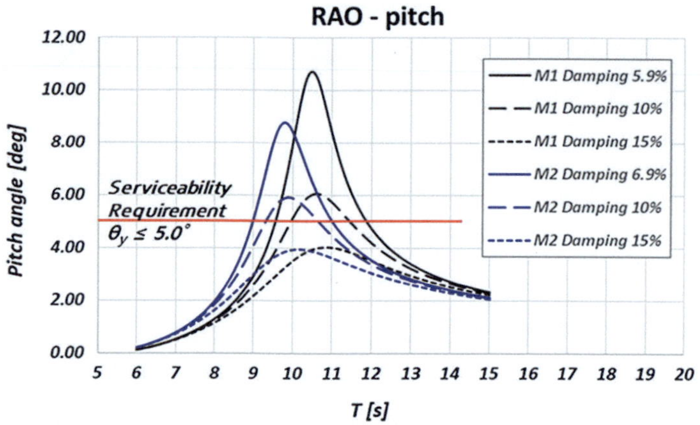

Fig. 3 Pitch RAO result with different damping conditions

2.2 *Evaluation Unit Module with Different Mooring System*

One of the securing methods for serviceability requirement is that using fixed mooring solution like dolphin mooring system. It was conducted at Incheon new harbor of floating platform constructed by Hyundai Engineering & Construction. Fixed mooring system should be constructed at low water depth condition, in deep water condition, the cable mooring is more cost effective than this method. But in this research, water depth assumed about 20 m, dolphin mooring system can be conducted for motion reduction of floating module (Fig. 4).

Figure 5 shown RMS results following different contact stiffness with floater and dolphin fender. A strong mooring constraint is highly effective in the motion reduction of floating structure, but it is expected that significant structural reinforcement will be required due to the large load and moment acting on the fender, mooring guide, and dolphin piles. Therefore, to conduct the flexible connection stiffness of the fender can be one effective way to not only achieve motion reduction for serviceability requirement but also to complement for the excessive structural design.

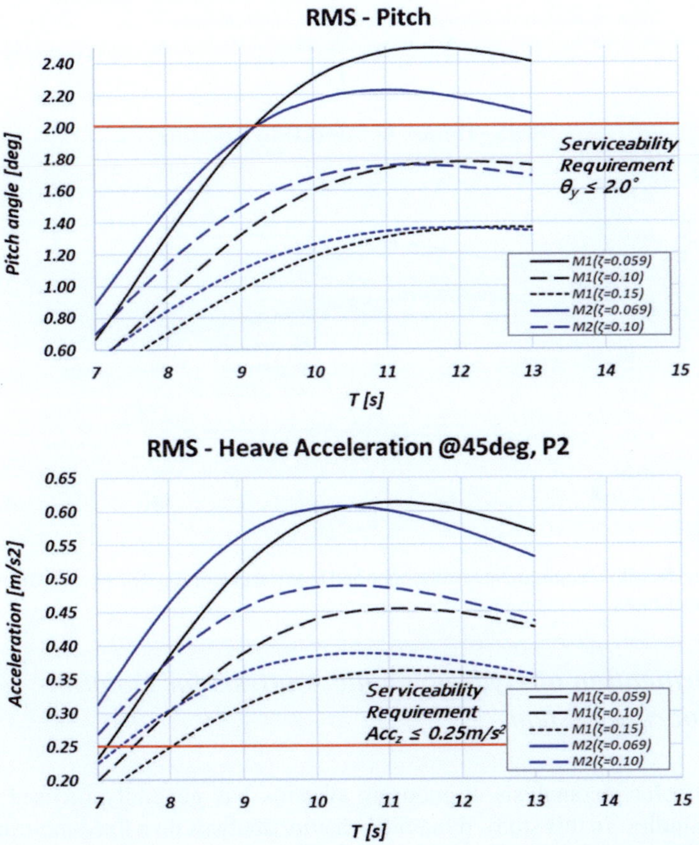

Fig. 4 Serviceability requirement results, pitch and heave acceleration RMS with different damping conditions

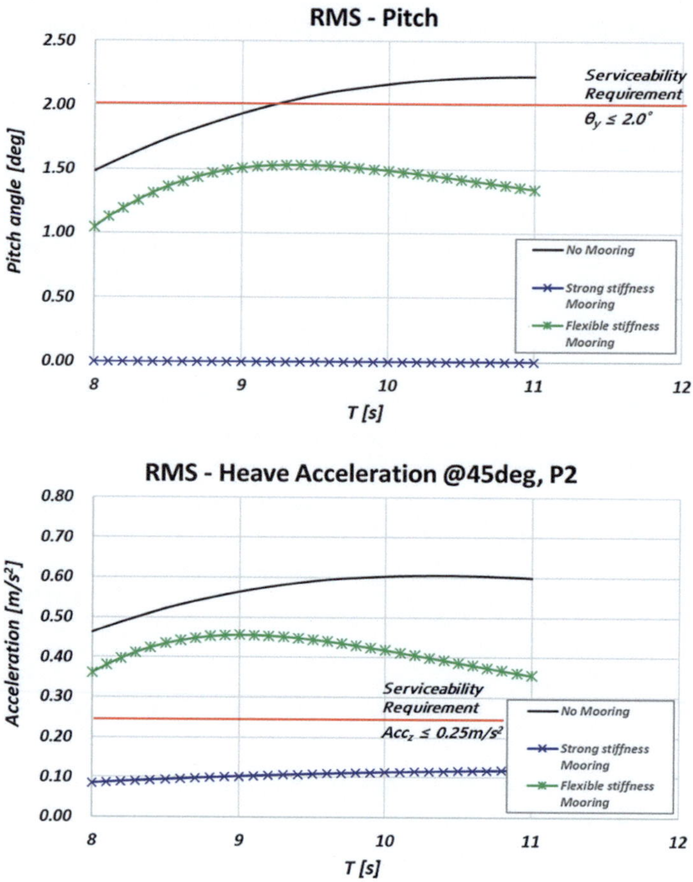

Fig. 5 Serviceability requirement results, pitch and heave acceleration RMS with different mooring connection stiffness

2.3 *Verification of Hydrodynamic Analysis for Dolphin Mooring System*

Dynamic behavior analysis of mooring systems has generally focused on cable mooring studies. In this study, dynamic behavior analysis on a fixed mooring system [6] is performed. To verify the results, AQWA and Orcaflex were utilized to validate the reliability of the analysis outcomes. The structural modeling for the interaction analysis of the mooring system, fender, and floating structure is shown in the Fig. 6, and detailed modeling methods were applied by referring to the manuals of each program.

To verify reliability, the time history analysis is conducted on regular waves under the same connection stiffness conditions and the response results for the 6 degrees of freedom (pitch, surge, heave, etc.) are similar, without significant discrepancies.

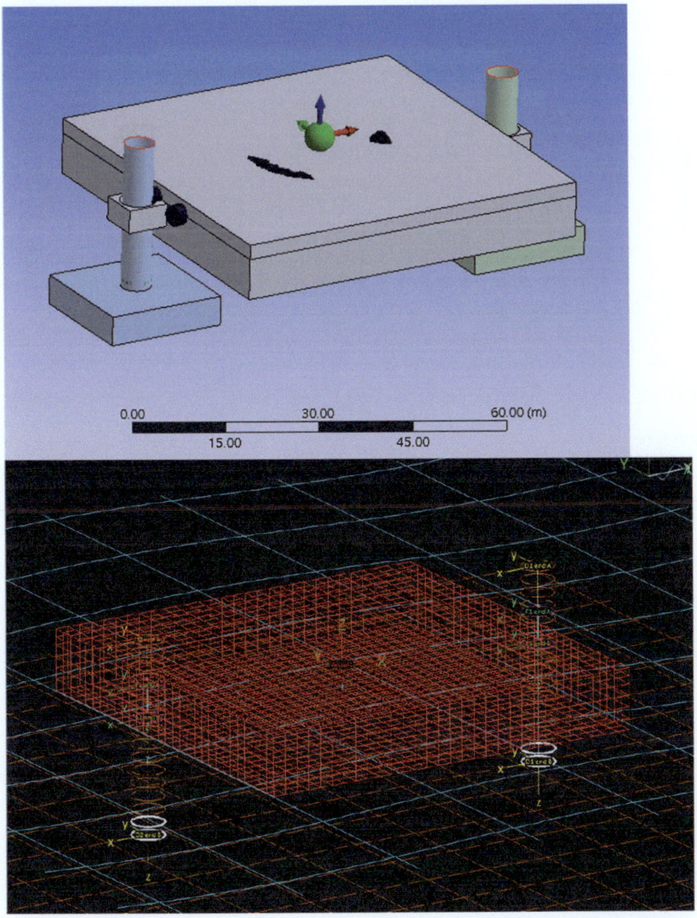

Fig. 6 Numerical modeling for verify interaction behavior with mooring, fender and floater (L: ANSYS AQWA, R: OrcaFlex)

In addition, it was confirmed that the response results of RAO were significantly reduced due to the influence of the mooring system, and it is expected that the hydrodynamic results that can fully satisfy the serviceability criteria can be derived if the effects of damping, mooring, and multi-module are all considered. For more realistic behavioral results, further verification of various cases, such as considering fender friction and applying multiple fixed mooring systems to multi-module floating structures, is needed (Fig. 7).

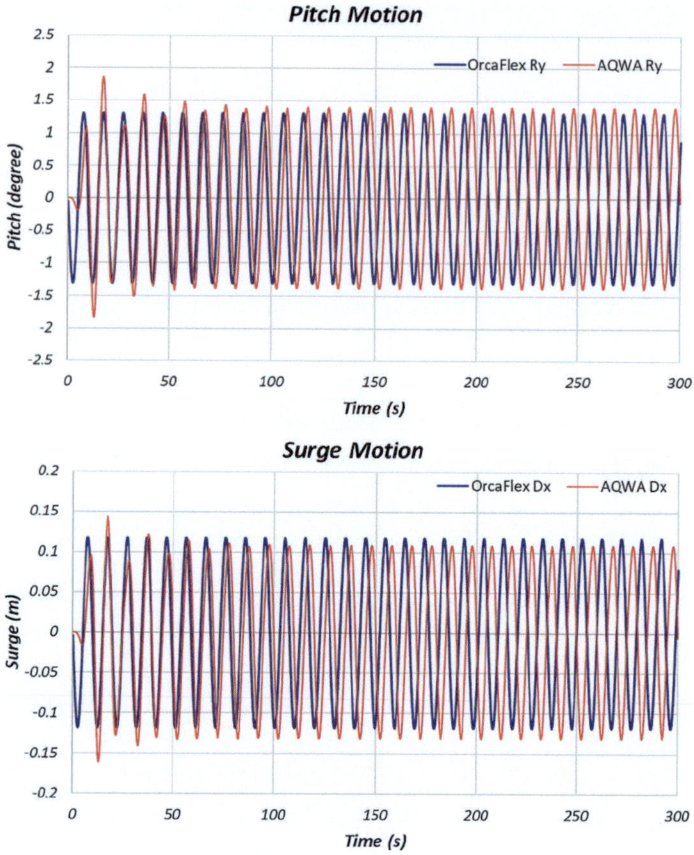

Fig. 7 Verified result of hydrodynamic behavior compare ANSYS AQWA with OrcaFlex

3 Design of Multi-module Floating Structures

3.1 Concept of Various Type of Connectors

Arrangement of multi-module floating structures has to be considered in order to be expanded for a mega cluster. In this study, the minimum component size multi-module structures are analyzed for the evaluation of hydrodynamic motion. To connect each module, various type of connecters such as fixed connection, universal joint, hinge and rope & fender are modeled. The type of connectors has to be chosen to expand the mega floater, being flexible yet resistance to tension in both horizontal and vertical direction, as well as reducing rotational movement caused by wave load (and hogging, sagging) (Figs. 8 and 9).

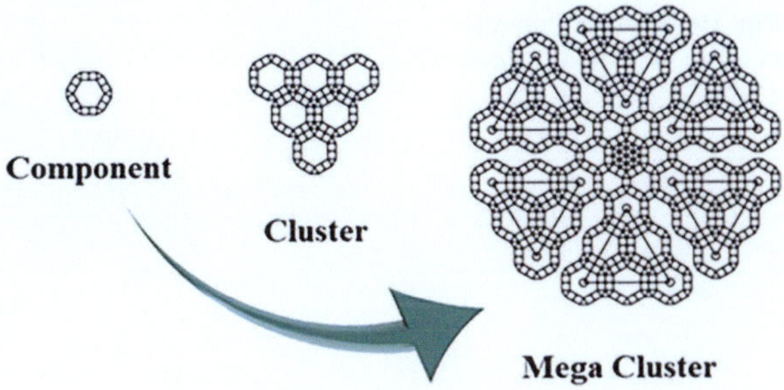

Fig. 8 Expand plan of multi-module floating structures

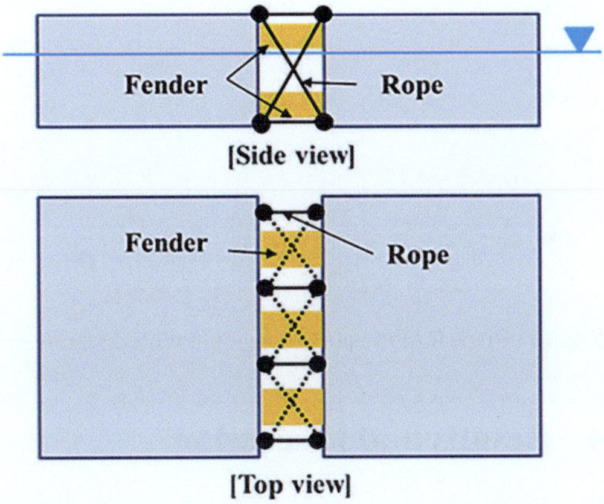

Fig. 9 Model connector concept design: rope and fender

3.2 Hydrodynamics Results with Different Type of Connectors

The hydrodynamic analysis is performed using ANSYS AQWA for multi-module floating structures in 4 × 2 arrangement with various types of connectors such as fixed, hinge, and rope & fender. Each type of connector is modeled using the connection option in the program and evaluation of the feasibility of the connection modeling, is compared to the results when the module connection conditions are interpreted as a connection stiffness matrix and found to yield very similar results.

Fixed connections are very effective at reducing motion, but it can be difficult to ensure structural safety due to the increased concentrated load on the connection structure. Hinge connection has a large difference in motion reduction effectiveness by module, and some modules have larger motion than unit module. Rope and Fender showed an overall motion reduction effect, and the motion reduction effect is expected to be larger depending on the stiffness of the rope (Figs. 10 and 11).

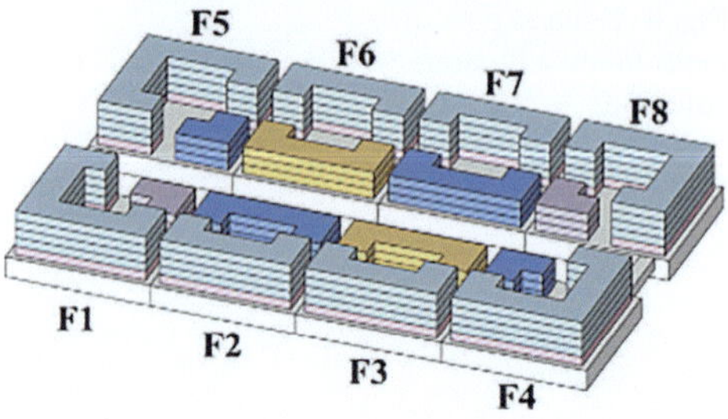

Fig. 10 Arrangement of multi-module floating system

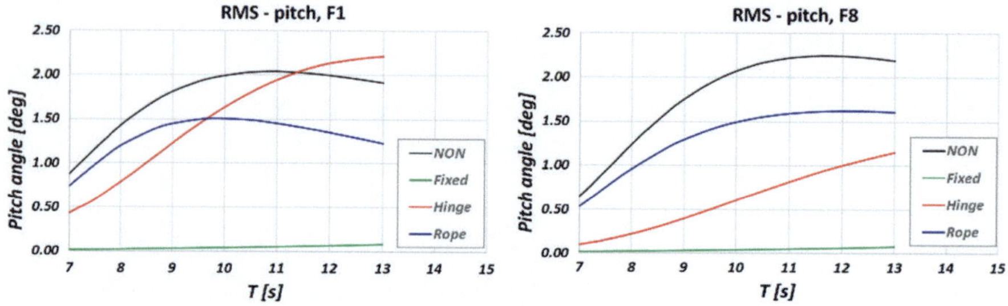

Fig. 11 Pitch RAO results of module with different connector type

4 Results and Discussion

This study examined whether it is possible to reduce motion by applying practical methods to secure the serviceability requirements of floating structures, each method showed some effectiveness in reducing motion. And it is expected that the usability requirements can be satisfied sufficiently if hydrodynamic analysis of the entire system of multi-module floating structures is performed.

Acknowledgements This work is supported by the Korea Agency for Infrastructure Technology Advancement (KAIA) grant funded by the Ministry of Land, Infrastructure and Transport (Grant RS-2023-00250727) through the Korea Floating Infrastructure Research Center at Seoul National University.

References

1. Flikkema M, Waals O (2019) SPACE@SEA The floating solution. Front Mar Sci 6. Article 553
2. Ren N, Zhang C, Magee AR, Hellan Ø, Dai J, Ang KK (2019) Hydrodynamic analysis of a modular multi-purpose floating structure system with different outermost connector types. Ocean Eng 176:158–168
3. Wang Y, Wang X, Xu S, Wang L, Ding A, Deng Y (2019) Experimental and numerical investigation of influences of connector stiffness and damping on dynamics of a multi-module VLFS. Int J Offshore Polar Eng 427

4. Shi QJ, Xu DL, Zhang HC, Zhao H, Wu YS (2018) Optimized stiffness combination of a flexible-base hinged connector for very large floating structures. Mar Struct 60:151–164
5. Vincentius R, Bjørn Egil A, Christian K (2016) Human factors evaluation in ship design: a case study on offshore supply vessels in the Norwegian Sea, Part 1: theoretical background and technical constructs. Naval Eng J 128
6. Kwanghoe J, Sanghyu L, Heesung K, Yoonho C, Sara K (2019) Design and construction of the concrete floating pier in Golden Harbor Incheon. In: World conference on floating solutions

Study on the Hydrodynamic Response of a New Floating Multi-module Platform with Bionic Conceptual Connection

Yanjun Mao, Xiaozhou Ma, Guohai Dong, and Yufei Wu

Abstract With the development and utilization of ocean renewable energy, the role of floating large-scale multi-module composite foundations becomes more prominent. This study explores the design of multi-module floating platforms at sea, focusing on a composite platform composed of seven hexagonally interconnected modules with flexible connectors using steel cable-fender connectors. The research provides the wave frequency response characteristics of each module in the composite platform. The flexible connectors reduce the pitch motion of the modules and primarily bear loads influenced by the pitch motion of the modules. Additionally, the study finds that connector loads are greater in the parallel wave directions compared to oblique directions, providing insights for connector design.

Keywords Offshore floating solution · Multi-module · Flexible connectors · Bionic concept

1 Introduction

Looking towards future strategic development, concepts such as floating airports, floating port terminals, offshore floating cities, and floating islands are gaining attention among researchers. Offshore photovoltaic and wind energy are currently the most feasible and promising types of renewable energy, both requiring stable and reliable floating foundations. Compared to rigidly connected large floating platforms, flexible connected multi-body composite platforms are more suitable as offshore floating photovoltaics platform. Compared to the rigid connectors, the flexible connectors can

Y. Mao · X. Ma (✉) · G. Dong
The State Key Laboratory of Coastal and Offshore Engineering, Dalian University of Technology, Dalian, Liaoning, China
e-mail: maxzh@dlut.edu.cn

Y. Wu
Guangdong Provincial Key Laboratory of Durability for Marine Civil Engineering, Shenzhen University, Shenzhen, China

X. L. Zhao et al. (eds.), *Proceedings of the Fourth World Conference on Floating Solutions: WCFS 2024; 2–4 December; Hong Kong*, Lecture Notes in Civil Engineering 597, https://doi.org/10.1007/978-981-96-4569-5_81

release the relative motion between modules that can decrease the internal bending stress significantly. Some researchers have begun studying the structural design and hydrodynamic performance of flexible connected multi-body platforms. Research based on the Mobile offshore base (MOB) concept has examined the hydrodynamic performance of serially connected multi-body composite structures [1, 2]. However, research on large-scale three-dimensional composite platforms remains relatively limited. For instance, the Space@Sea project [3] studied physical and numerical models of platforms composed of triangular and quadrilateral floating units, exploring the hydrodynamic performance and stability of flexible multi-body composite platforms, and suggesting applications such as offshore photovoltaics and offshore energy islands that are relatively insensitive to structural motion responses [4]. The structure of flexible connected multi-body composite platforms belongs to multi-body dynamic systems, characterized by complex motion properties. Hydrodynamic interactions between floating units can alter the dynamic characteristics of each floating unit [5]. The nonlinearity of flexible connectors can induce nonlinear response phenomena in multi-body dynamic systems, such as the added mass (AD) effect [6]. Therefore, detailed research into the hydrodynamic characteristics of flexible connected multi-body composite platforms is necessary. To explore more design options for offshore multi-body composite platforms, this study adopts a biomimetic approach, using hexagonal floating units combined in a honeycomb structure. The connectors are steel cable-fender type flexible connectors.

2 Model Setup and Verification

2.1 Model Setup

This study utilizes a time-domain numerical method based on potential flow theory to compute the hydrodynamic parameters of the composite platform and its time-domain dynamic response characteristics under wave action. The floating units are hexagonal with a side length of 20 m, constructed using prestressed concrete materials with internal bulkheads to strengthen unit integrity as shown in Fig. 1. The geometric parameters of the unit are shown in Table 1.

The composite model adopts a 3D arrangement of 7 floating units, with 6 peripheral units connected around a central unit. This configuration represents a fundamental form of hexagonal floating unit combination. The platform model specifics are depicted in Fig. 2, with coordinates centered on M1, where the X-axis is positive to the right, Y-axis positive upwards, and the XY plane at the waterline. The modules are spaced 2.5 m apart and interconnected with steel cable-fender connectors as shown in Fig. 3. Each adjacent floating unit is connected by 4 elastic cables and buffered by 2 rubber fenders, with a cable stiffness of 5×10^6 N/m and fender stiffness of 6.03×10^7 N/m.

Fig. 1 The structure of the floating module

Table 1 The geometric parameters of the unit

Parameters	Values	Units
Side length	20	m
Module height	10	m
Mass	4.99×10^6	kg
Draft	4.685	m
Center of mass	4.925	m
Moment of inertia I_{xx}	6.90×10^8	$kg \cdot m^2$
Moment of inertia I_{yy}	6.90×10^8	$kg \cdot m^2$
Moment of inertia I_{zz}	1.01×10^9	$kg \cdot m^2$

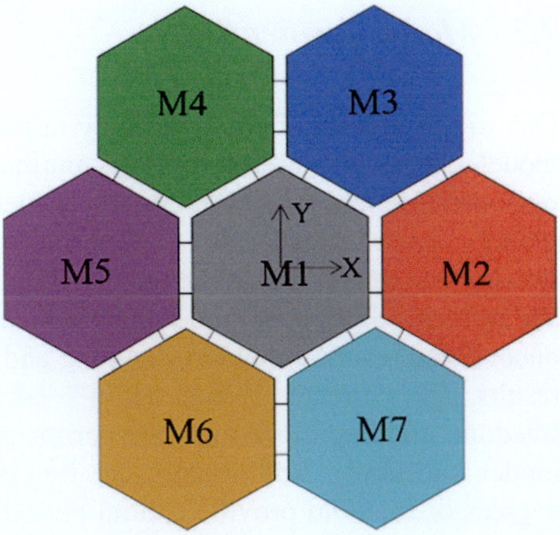

Fig. 2 The sketch of the composite platform

The multi-module composite platform employs a catenary mooring system with 18 anchor chains arranged as shown in Fig. 4. The system includes angles between C1 and C2 and C3 and C4 set at 30°, with C2 aligned parallel to C3, a mooring radius of 165 m, cable lengths of 180 m, and a dry weight per meter of 90 kg.

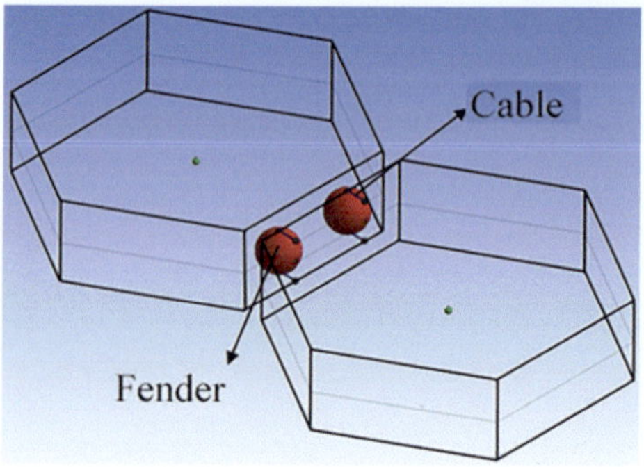

Fig. 3 The schematic of the cable-fender connector

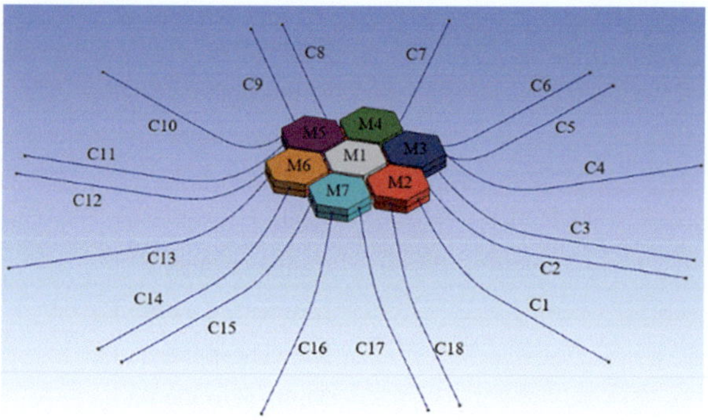

Fig. 4 The arrangement of the mooring system

2.2 *Model Verification*

Due to the neglect of fluid viscosity in the potential flow theory-based numerical model, structural motion damping is minimal, often overestimating structural motion responses. Hence, physical model experimental data are used to correct the numerical model. Figure 5 shows the free decay curves of surge, pitch, and heave for the floating unit, including uncorrected and viscosity-corrected numerical model results and referenced physical model experimental data. The comparison indicates that the uncorrected results show low damping and significant deviation from experimental results. The corrected results add 4% critical damping to each motion degree of freedom, aligning closely with experimental values and confirming the numerical model's accuracy. Additionally, the free decay curves of the floating unit's three degrees of freedom provide natural period characteristics, heave and pitch natural periods are around 7.5 s, while surge periods are longer, approximately 97.23 s.

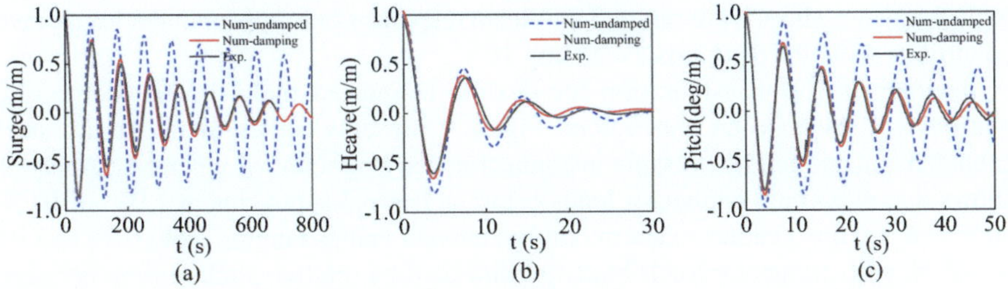

Fig. 5 Free decay curves of 3DOF: **a** Surge, **b** heave, **c** pitch

3 Results

To assess the hydrodynamic performance of the multi-module composite platform, wave frequency response characteristics of the composite floating units are computed under regular wave conditions. The calculations simulate regular waves with a constant wave height of 1.5 m and wave periods T ranging from 5 to 17 s (wave frequency ω = 1.256 rad/s to 0.369 rad/s), propagating along the positive X-axis at an operational water depth of 40.0 m.

The wave frequency response curves for each floating unit in the composite platform are shown in Fig. 6. Due to structural symmetry, results are shown only for M1 to M7. The shaded area on the graph represents the frequency domain response results for the unconnected composite platform. The presence of connectors reduces the peak heave motion of the overall platform from 11.4° to 5°, with similar responses observed for each floating unit, indicating a single-peak response characteristic.

Differences in heave responses are apparent among modules, particularly with the first module M5 in the wave-facing side showing resonance response near its natural period. Compared to other modules, M5 exhibits significantly larger heave responses due to less constraint in the heave direction from the connectors and its location in the wave-facing side. Other modules do not show significant resonance effects near their natural periods. Given the connectors linking seven modules together, the wave frequency response characteristics of surge motion show reduced values in the

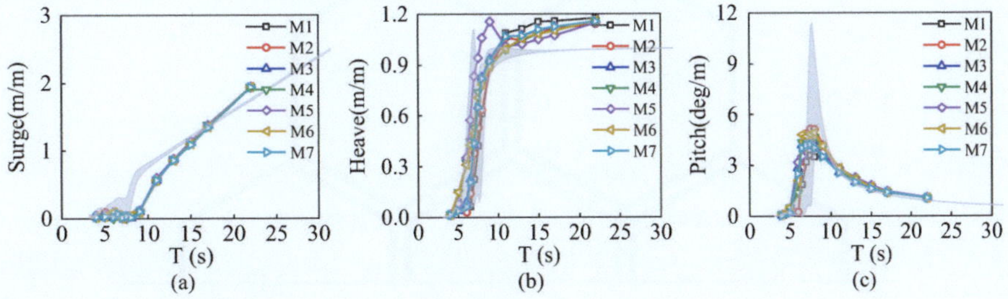

Fig. 6 The wave frequency response cures of 3DOF: **a** Surge, **b** heave, **c** pitch

high-frequency short-wave range. Significant responses only appear under long-wave conditions with wave periods exceeding 10 s.

Another critical consideration for flexible connected multi-module composite platforms is the load on connectors. Figure 7 presents curves of maximum steel cable tension and fender pressure in connectors varying with wave periods. Analysis shows that maximum connector loads occur at T = 6.5 s, peaking at 1.93 × 10^6 N. At T = 6.5 s, most cables reach maximum tension values ranging from 0.89 to 1.93 × 10^6 N, with connector loads heavily influenced by relative pitch motion between modules. Therefore, the connector load response curve exhibits similar characteristics to the overall module pitch motion response curve, both peaking at T = 6.5 s. Increased longitudinal stiffness due to connectors shifts peak load periods towards higher frequencies compared to the floating unit's natural period at T = 7.5 s. Furthermore, Fig. 8 shows the distribution of cable forces at T = 6.5 s, with parallel-to-wave direction connector loads significantly larger than those in non-parallel directions.

Finally, attention to forces of the mooring system, as shown in Fig. 9, reveals peak cable forces near resonance periods. The wave-facing chain C10 experiences the highest forces. Cable forces primarily correlate with platform average drift. Figure 10 displays average drift displacement, indicating significant average drift displacement near resonance periods for large composite platform design references.

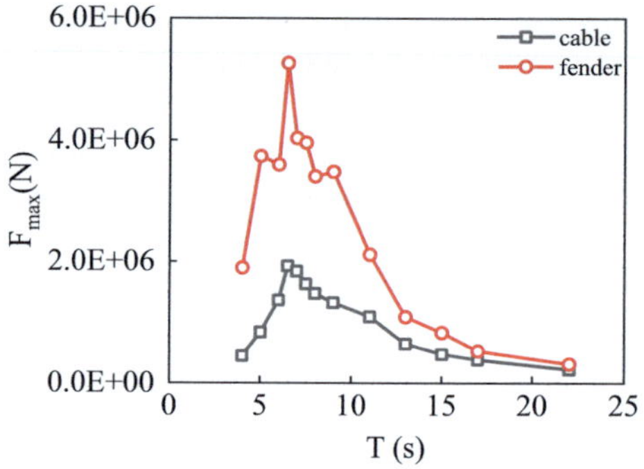

Fig. 7 The variation of cable and fender load with the wave periods

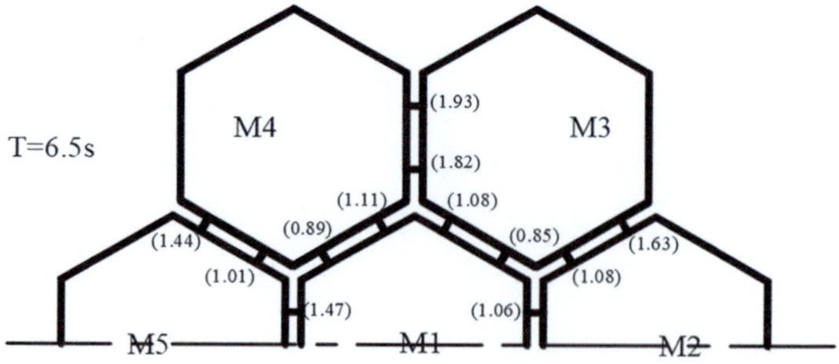

Fig. 8 The distribution of the maximum cable forces at T = 6.5 s (×10^6 N)

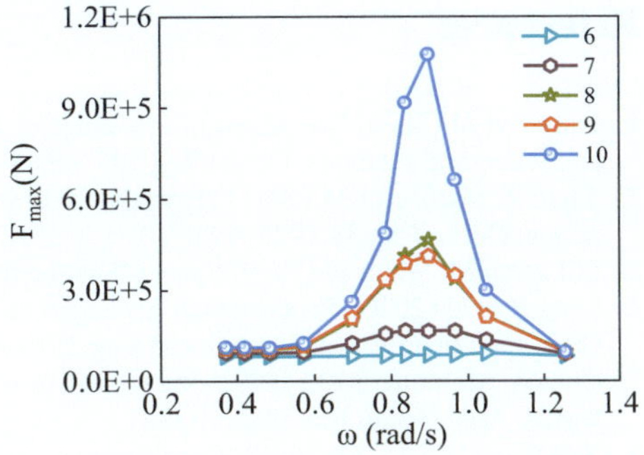

Fig. 9 The variation of mooring force with the wave frequency

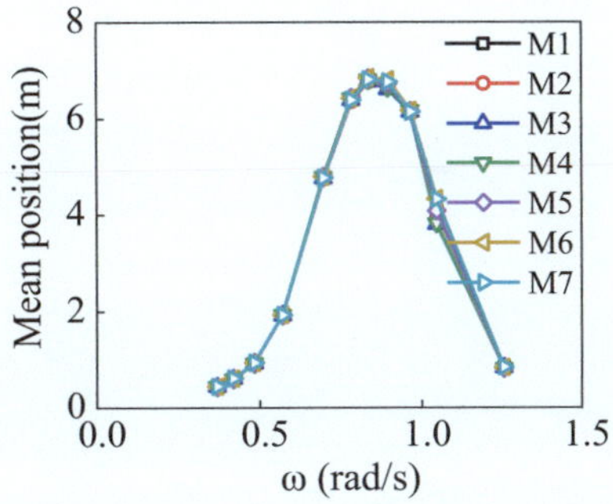

Fig. 10 The mean drift of the platform near the natural wave frequency

4 Summary

This project studies the hydrodynamic characteristics of a seven-module composite platform with flexible connectors. It provides wave frequency response characteristics for the three degrees of freedom of each module within the composite platform. The pitch motion of modules is reduced under the influence of connectors. While connector loads are significantly affected by relative pitch between modules, resulting in similar wave frequency response curves to module pitch motion response curves. Connector loads are notably larger in the parallel wave direction compared to no-parallel ones. In heave motion, significant resonance response is observed in the first module M5 in the wave-facing side, while other modules show less pronounced resonance responses near their natural periods. The conclusions drawn from this study on flexible connected multi-module composite platform provide valuable references for the design of such platforms. The future works need to be focus on the improvement of the flexible connectors to enhance their stability.

References

1. Ding J et al (2020) Investigation of connector loads of a 3-module VLFS using experimental and numerical methods. Ocean Eng 195:106684
2. Tajali Z, Shafieefar M (2011) Hydrodynamic analysis of multi-body floating piers under wave action. Ocean Eng 38:1925–1933
3. Flikkema M, Waals O (2019) Space@Sea the floating solution. Front Mar Sci 6
4. Ding R et al (2020) Experimental investigation on characteristic change of a scale-extendable chain-type floating structure. Ocean Eng 213:107778
5. Chen S, Mahrenholtz O (1992) Interaction of water waves and floating twin cylinders in beam waves. Appl Ocean Res 14:371–379
6. Xu DL et al (2014) On study of nonlinear network dynamics of flexibly connected multi-module very large floating structures. In: Vulnerability, uncertainty and risk, pp 1805–1814

SEAform's Modular Floating Platforms: A Comprehensive Environmental Impact Evaluation

Letizia Pincetti, Sergej A. Sirigu, Alice Rosiello, Carola Chicco, Fabio Viotti, and Paolo C. Priarone

Abstract The rising sea level and scarcity of available land in rapidly growing coastal cities have led to the necessity of expanding towards water, in a process known as land reclamation. This market has been primarily dominated by dredging, which leads to marine and coastal ecosystems degradation. In order to meet the demand of land reclamation for different purposes, while also aiming to reduce the resultant environmental effects, the SEAform project introduces an alternative approach to dredging, using multi-use interconnected modular floating platforms. This study aims to evaluate the potential environmental impacts associated with the floating platforms developed by SEAform through a comparative analysis with the effects of dredging alongside an evaluation of the platforms' carbon footprint. The analysis reveals the environmental advantages of floating platforms, illustrating their significantly lower impact compared to traditional dredging techniques. By comprehensively understanding its environmental impact, the SEAform project can effectively and progressively mitigate its effects on the marine ecosystem, positioning itself as an environmentally conscious choice for coastal development and offshore strategies.

Keywords Floating platform · Environmental impact · Carbon footprint

L. Pincetti (✉) · S. A. Sirigu
Department of Mechanical and Aerospace Engineering, Politecnico di Torino, Turin, Italy
e-mail: letizia.pincetti@polito.it

A. Rosiello
SEAform s.r.l., Turin, Italy

C. Chicco
Department of Environment, Land and Infrastructure Engineering, Politecnico di Torino, Turin, Italy

F. Viotti
Research and Technology Centre West Coast, Kiel University, Büsum, Germany

P. C. Priarone
Department of Management and Production Engineering, Politecnico di Torino, Turin, Italy

X. L. Zhao et al. (eds.), *Proceedings of the Fourth World Conference on Floating Solutions: WCFS 2024; 2–4 December; Hong Kong*, Lecture Notes in Civil Engineering 597, https://doi.org/10.1007/978-981-96-4569-5_82

1 Introduction

Major coastal cities are globally expected to face, or are already experiencing, two main challenges: population growth and climate change. By the year 2050 over 80% of the anticipated global population of 9 billion individuals is projected to reside in urban areas [1], while, by the year 2100, the global mean sea level is expected to increase between 43 cm (RCP2.6) and 84 cm (RCP 8.5) [2]. These issues will likely result in insufficient land to accommodate the increasing number of people moving to urban areas and heightened vulnerability to the risk of flooding due to more extreme weather conditions and rising sea levels [3].

A solution to these problems is the expansion of urban areas over open water surfaces through the creation of new land, a process commonly referred to as land reclamation [4]. Currently, one of the most widely used techniques for land reclamation is dredging, while more innovative and less impactful alternatives, such as floating solutions, have been emerging in recent years. In this context SEAform, a Politecnico di Torino spin-off, is developing an innovative and sustainable solution involving modular floating platforms.

The aim of this paper is to assess the potential environmental impacts of SEAform's floating platforms by comparing them to the effects of dredging and providing an initial estimation of their carbon footprint.

2 Comparative Analysis of Environmental Impacts

2.1 Dredging

Dredging is currently the most commonly used technique for creating artificial land, with the market valued at over 12 billion US dollars as of 2022, representing nearly all land reclamation activities. Dredging involves excavating and removing underwater sediments, such as sand, silt or gravel, to create navigable channels, maintain certain depths, or build new land [5], as shown in Fig. 1. A. When the dredged sediments are placed in a controlled manner in the designated area to gradually build up the land, the technique is called "hydraulic fill". If the dredged sediments are used to fill portions of the sea enclosed by barriers, it is known as "infill" [5].

Dredging has significant environmental impacts during the construction phase, including:

- Partial or complete destruction of benthic habitats, such as coral reefs and seagrass beds, both in the excavation and deposition areas. This habitat loss leads to a reduction in biodiversity and suitable areas for shelter and breeding for many species, with consecutive negative effects on the trophic web [6];
- Increased water turbidity and sediment resuspension, which causes a reduction in water transparency, light availability, and a potential release of contaminants.

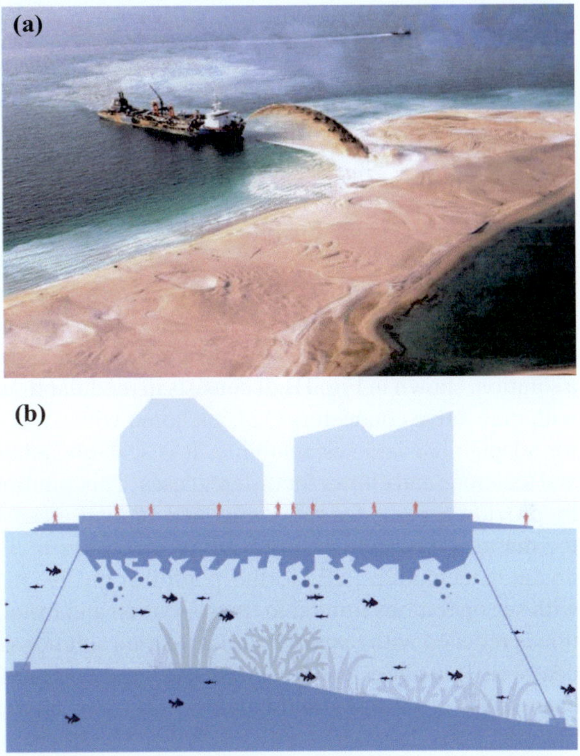

Fig. 1 **a** Dredger depositing material to build new land [21]. **b** Scheme of a SEAform floating platform

These effects significantly impact benthic organisms, reducing their feeding and reproduction rates and potentially burying or smothering them [7];

- Noise produced by dredgers employed to excavate, transport and deposit sediments, affecting in different ways a wide spectrum of marine organisms. Depending on the level and time of exposure noise can cause both behavioural and physiological responses, such as temporary hearing loss or chronic stress [8].
- Risk of collisions between marine organisms and equipment [9].

During the operational phase, dredged land interferes with the entire water column, majorly altering the hydrodynamic regime, diverting currents and waves, modifying water circulation, wave motion and bathymetry [7], causing coastal erosion and changing the water's chemical-physical parameters such as temperature, dissolved oxygen and pH [10].

2.2 Floating Solutions and SEAform

The concept of living on water is not new, with floating houseboats common worldwide, especially in the Netherlands [4]. However, there is renewed interest in expanding this idea for a sustainable future city development.

Floating structures have many benefits compared to traditional inland buildings, such as additional construction ground, possibility to be relocated, efficient use of renewable energy resources and resilience and adaptation to climate change [11].

In this scenario SEAform, a spin-off from Politecnico di Torino, born from the project of Marine Offshore Renewable Energy Laboratory (MOREnergy Lab), a research centre at Politecnico di Torino specialized in offshore renewables, proposes a solution, shown in Fig. 1B. It consists in modular floating platforms, interconnected with innovative connectors [12], anchored with a taut mooring system and designed for adaptability and sustainability. It is scalable, adaptable to various bathymetries and locations and allows for different uses. This solution aims to support coastal cities by providing space for residential and support areas while integrating itself within the marine ecosystem, minimizing its environmental impacts [13]. With respect to dredging techniques, SEAform's platforms involve mainly land-based construction, with sea operations limited to transportation and installation. Moreover, noise pollution is reduced with respect to other floating solutions by avoiding drilling and using deadweight anchors, and by employing synthetic mooring ropes instead of catenary chains. Water turbidity and benthic ecosystem alteration is minimized as the taut mooring lines do not lie on the seabed, avoiding sediment resuspension.

However, introducing floating platforms into the water inevitably impacts the marine environment. These impacts can be mitigated and minimized with proper understanding and management:

- The platform prevents solar radiation from reaching the water underneath, reducing photosynthesis [14];
- Alteration of superficial currents, which can be slowed down or sped up by the platforms [14];
- Slight changes in water's physical–chemical parameters due to shading and altered water movement [15];
- Risk of entanglement of marine organisms on mooring lines [16].

New habitat creation. It has been documented that artificial hard surfaces, like floating platforms, provide substrates for sessile organisms, with filter-feeding macrobenthic species such as mussels and sponges often attaching underneath them [17]. These organisms gather food by filtering suspended particles in the water column. This occurs because artificial platforms, such as concrete platforms, offer structurally complex hard substrates unlike the typically featureless seabed. This phenomenon is known as the "artificial reef effect" and concrete is recognized as one of the best materials for attracting marine life due to its structural complexity, which resembles natural corals reefs, and to its strength, that provides a stable support even in challenging conditions [18]. The colonization of sessile organisms provides shelter from currents and predators, an ideal environment for reproduction and spawning, and an increased food availability for mobile invertebrates, fish and even marine mammals, which have even been observed resting on, aggregating around, or foraging near subsea structures [19]. By establishing new habitats and attracting a wider variety of fish, SEAform's platforms contribute to a more biodiverse marine environment and promote ecosystem regeneration. This effect can also extend to the concrete deadweight anchors resting on the seabed, and it can be further enhanced through the application of Nature Inclusive Designs, which is an approach to design that emphasizes the integration of nature and natural processes into human-created products [20].

2.3 Comparison

To quantify and compare the environmental impact of these two land reclamation techniques, an extensive literature review has been conducted in a previous work, focusing on the measured or expected effects of numerous dredging and floating platform projects [22]. Based on this analysis, the spatial extent and duration indexes for the main identified impacts were estimated, assigning weighted values according to Table 1:

Furthermore, since the dredging process involves extracting material from one location, transporting it, and depositing it at the designated site, it can be estimated that, for the same surface of reclaimed land, the dredging process duration is twice as long as that of installing floating platforms, which mainly involves transport and anchor system installation. In terms of spatial extent, dredging has an impact on two locations, the excavation site and the deposition site, resulting in a disturbance that, for the same surface of reclaimed land, can be assumed to be twice that of floating platforms. It is important to note that while floating platforms mainly impact the water surface, land created through dredging affects the entire water volume. Consequently, the actual difference in spatial extent between these two technologies is likely greater, but it is site-specific, since it depends on the depth at which the land is created, which influences the affected water volume. Therefore, doubling the spatial extent can be regarded as a conservative estimate.

As a result, the weighted values for dredging are doubled to take these differences into account. Lastly, considering the potential for reduced environmental impact

Table 1 Spatial extent and duration indexes as a percentage relative to the dredged land/platform

Value	1	2	5	10
Spatial extent	0%–100% dredged land/ platform	100%–200% dredged land/ platform	200%–1000% dredged land/ platform	> 1000% dredged land/platform
Duration	0%–25% phase duration	25%–75% phase duration	75%–100% phase duration	> 100% phase duration

provided by SEAform technology compared to other floating solutions, such as taut mooring with synthetic ropes, the index values for the main impacts of this solution on the marine ecosystem were estimated [22]. The results of this comparison are presented in Table 2:

It is important to note that, while some impacts of dredging and floating platforms may be related to the same outcome, the underlying causes, phases in which they occur, and relative implications can vary considerably between the two technologies.

The resultant estimated reduction in environmental impact on the marine environment using the floating platforms developed by SEAform instead of the dredging technique can be calculated according to Eq. 1:

$$\frac{\Sigma dredging\ indexes - \Sigma SEAform\ solution\ indexes}{\Sigma dredging\ indexes}\% = impact\ reduction \quad (1)$$

which is equal to a 78% impact reduction.

Table 2 Spatial and temporal extent index values for dredging and floating platform impacts, estimated from literature review of similar projects, and SEAform solution, assumed considering the design measures aimed at minimizing its impacts

Impact	Dredging		Other floating platforms		SEAform solution	
	Spatial extent	Duration	Spatial extent	Duration	Spatial extent	Duration
Turbidity and sediment resuspension	20	20	2	5	1	5
Noise	20	10	5	5	2	5
Benthic habitat destruction/shading	2	20	2	2	1	2
Hydromorphological changes	20	20	2	5	2	5
Collisions/ entanglement	2	10	1	5	1	5
Changes in physical–chemical water parameters	4	10	2	5	1	5

3 Carbon Footprint Estimation

3.1 *Main General Assumptions*

Due to the novelty of SEAform technology and the challenges of obtaining comprehensive data, the initial carbon footprint assessment focuses on a cradle-to-gate approach. This approach is commonly used for business-to-business (B2B) products [23] and considers the embodied carbon emissions associated with the most impactful lifecycle stages, raw material extraction and manufacturing processes, up to the point where the product leaves the factory. These stages are frequently the dominant contributor to a product's overall carbon footprint, especially for construction materials like concrete and steel [24]. However, this approach excludes the subsequent lifecycle stages of the product, including transportation, installation, use phase, maintenance and end-of-life scenarios. This choice has been made for mainly two reasons: limited data availability and high uncertainty regarding these phases, and site and project specificity. Factors like transport distances, platform usage and end-of-life scenarios widely vary depending on the specific location and project goals. A comprehensive carbon footprint assessment would require more project-specific data. This initial cradle-to-gate assessment provides a starting point for understanding the environmental impact of SEAform platform production and its embodied carbon.

3.2 *Carbon Footprint Assessment*

For the carbon footprint calculation presented in this paragraph, the system has been modeled as 2 platforms, anchored with 4 deadweight anchors connected to the platforms with one taut nylon mooring line each.

Each platform has a regular hexagonal shape, with an edge L of 9 m, for a total surface S of 210.44 m^2, and it has an height H of 3 m. The structure of the platform's hull is composed by a lightweight reinforced concrete, with a reinforcement ratio, calculated as the mass of the reinforcement steel over the total mass of the reinforced concrete, of 7.02%, and the total mass of the hull is 278.30 t.

The material used for the 4 deadweight anchors is a lightly reinforced concrete with a reinforcement ratio of 3%. The dry mass of a single deadweight anchor is 68.40 t.

The mooring lines connecting the platforms to the deadweight anchors are 8-strands nylon lines. Each line has a length of 14.74 m and a diameter of 70 mm.

The following assumptions regarding the embodied carbon of each material have been made:

- Concrete: a conservative assumption has been made considering the value indicated in [25] for a non-reinforced concrete with a comparable strength class (C40/50);

Table 3 Main components of the described system and their embodied carbon

Component	Material	Mass [t]	Emission factor [t CO_2e/t]	References	Resultant emission of single component [t CO_2e]
Platform hull	Concrete	258.77	0.159	[25]	78.25
	Steel	19.53	1.9	[25]	
Deadweight anchor	Concrete	66.35	0.159	[25]	14.45
	Steel	2.05	1.9	[25]	
Mooring line	Nylon	0.26	10.66	[26]	2.76

- Steel: the embodied carbon value for steel reinforcement bars indicated in [25] has been considered;
- Nylon: for the mooring lines, nylon 6,6 and its embodied carbon value [26] has been chosen instead of nylon 6 due to its higher durability [27].

The main components and their embodied carbon are listed in Table 3:

Therefore the estimated cradle-to-gate carbon footprint $(t\ CO_2e)_{total}$ for the described configuration, which includes 2 floating platforms, 4 deadweight anchors and 4 mooring lines, can be calculated by summing the contributions of the various components (listed in Table 3), according to Eq. 2:

$$2 \cdot (tCO_2e)_{hull} + 4 \cdot (t\ CO_2e)_{deadweightanchor} + 4 \cdot (tCO_2e)_{mooringline} = (tCO_2e)_{total} \tag{2}$$

which is equal to 225.34 t CO_2e. The resulting carbon footprint per square meter of new land created, calculated as the total carbon footprint of the system divided by the area of the two platforms, is equal to 53.54 kg CO_2e/m^2. It has to be noted that these results originate from a small-scale pilot project, and more accurate carbon footprint calculations will be obtained from larger case studies.

4 Conclusions and Further Developments

This paper provides an evaluation of the potential environmental impacts associated with the floating platforms developed by SEAform, focusing on their effects on the marine ecosystem and on their cradle-to-gate carbon footprint. The findings highlight that these floating platforms have a significantly lower environmental impact on marine ecosystems with respect to the traditional dredging technique, making them a more sustainable alternative in the land reclamation context.

Future research should focus on specific locations and project uses to assess the environmental impacts more accurately. This includes conducting a comprehensive carbon footprint analysis that follows a cradle-to-grave approach, considering the stages excluded in this initial assessment. Efforts should focus on reducing the total.

t CO_2e resulting from this cradle-to-gate assessment. This could be achieved by selecting a low-carbon concrete, which is the primary contributor to this carbon footprint evaluation. Additionally, using recycled steel can further decrease the embodied carbon of SEAform technology. Moreover, it would be useful to conduct a sensitivity analysis to account for the uncertainty of input data in carbon footprint calculations, particularly regarding emission factors.

In conclusion, by comprehensively understanding its environmental impact, the technology presented can progressively mitigate its effects on the marine ecosystem, positioning itself as an environmentally conscious choice for coastal development and offshore strategies.

References

1. Hoballah A, Peter C (2012) Sustainable, resource efficient cities: Making it happen! UNEP, Nairobi
2. Intergovernmental Panel on Climate Change (IPCC) (2019) IPCC Special report on the ocean and cryosphere in a changing climate. Cambridge University Press
3. Härtwich H (2016) The impact of floating platforms on the benthic community structure in Dutch freshwater ecosystems. Universiteit van Amsterdam
4. Foka E (2014) Water quality impact of floating houses. A study of the effect on dissolved oxygen levels. Master Thesis, Technische Universiteit Delft, Delft
5. van't Hoff J, van der Kolff AN (2012) Hydraulic fill manual: for dredging and reclamation works. CRC press, Vol 244
6. Levine-Fricke LFR (2004) Framework for assessment of potential effects of dredging on sensitive fish species in San Francisco Bay—Final report. US Army Corps of Engineers. Report, pp 001–09170
7. Nightingale B, Simenstad CA (2001) Dredging activities: marine issues. Washington State Transportation Center, University of Washington, Seattle, WA, 98105
8. Todd VL, Todd IB, Gardiner JC, Morrin EC, MacPherson NA, DiMarzio NA, Thomsen F (2015) A review of impacts of marine dredging activities on marine mammals. ICES J Mar Sci 72(2):328–340
9. Laist DW, Knowlton AR, Mead JG, Collet AS, Podesta M (2001) Collisions between ships and whales. Mar Mamm Sci 17(1):35–75
10. Demir H, Otay EN, Work PA, Börekçi OS (2004) Impacts of dredging on shoreline change. J Waterw Port Coast Ocean Eng 130(4):170–178
11. Porporato G (2023) FloatScapes: a floating adaptive pavilion. A proposal to lessen the environmental impact of floating architecture on aquatic ecosystems. A study in the Venice Lagoon. PhD dissertation, Politecnico di Torino, Turin
12. Sirigu SA, Boccanfuso V, Carapellese F (2023) Multi-body analysis of modular floating islands: optimisation of connector stiffness. World conference on floating solutions (pp 313–328). Singapore: Springer Nature Singapore
13. SEAform, https://seaform.it. Last accessed 2024/06/26
14. Czapiewska K, Roeffen B, Dal Bo Zanon B, Lin FY, Härtwich H (2017) Environmental assessment framework for floating development—French Polynesia. Blue front
15. de Lima RLP, de Graaf-Van Dinther RE, Boogaard FC (2022) Impacts of floating urbanization on water quality and aquatic ecosystems: a study based on in situ data and observations. J Water Climate Change 13(3):1185–1203

16. Maxwell SM, Kershaw F, Locke CC, Conners MG, Dawson S, Johnson AF (2022) Potential impacts of floating wind turbine technology for marine species and habitats. J Environ Manag 307:114577
17. Cole VJ, Glasby TM, Holloway MG (2005) Extending the generality of ecological models to artificial floating habitats. Mar Environ Res 60(2):195–210
18. Baine M (2001) Artificial reefs: a review of their design, application, management and performance. Ocean Coast Manag 44(3–4):241–259
19. Russell DJ, Brasseur SM, Thompson D, Hastie GD, Janik VM, Aarts G, McConnell B (2014) Marine mammals trace anthropogenic structures at sea. Curr Biol 24(14):R638–R639
20. Hermans A, Bos OG, Prusina I (2020) Nature-Inclusive Design: a catalogue for offshore wind infrastructure: Technical report (No. 114266/20–004.274). Witteveen + Bos
21. Rethinking The Future, https://www.re-thinkingthefuture.com/city-and-architecture/a3146-the-manmade-islands-of-dubai/. Last accessed 2024/06/29
22. Pincetti L (2023) Sustainable coastal development: a comparative analysis of environmental impacts between floating platforms and dredging. Master thesis, Politecnico di Torino, Turin
23. Carbon Trust, https://www.carbontrust.com/what-we-do/product-carbon-footprint-labelling/product-carbon-footprint-label. Last accessed 2024/06/25
24. Institution of Structural Engineers (2011) A short guide to embodied carbon in building structures. Inst Struct Eng, Great Britain
25. Circular Ecology, https://circularecology.com/embodied-carbon-footprint-database.html. Last accessed 2024/06/25
26. Oekobaudat, https://oekobaudat.de/OEKOBAU.DAT/datasetdetail/process.xhtml?uuid=d62953d2-61e8-427b-ada2-8a3c4e9756c1&version=20.23.050&stock=OBD_2023_I&lang=en, 2024/06/25
27. Essentra Components, https://www.essentracomponents.com/en-gb/news/solutions/wire-cable/the-differences-between-nylon-6-and-nylon-6-6. Last accessed 2024/06/25

An Innovative Design Concept of Modular Pneumatic Floating Platform Part II

Toshio Nakajima, Yoshihiko Yamashita, and Shingo Saito

Abstract This paper presents the progress of the study on a pneumatic floating body following the previous paper presented at the WCFS 2023 Japan (Nakajima et al. 2023). It describes the theoretical background of an innovative design concept of the pneumatic floating structure that can be used for a Very Large Floating Structures (VLFS) like a floating city and/or floating airport located in the open sea. The pneumatic floating body has a unique feature of reducing rocking and tilting in rough waves caused by strong winds as it has an open bottom. The rigid connection unit such as 3D truss beams connect floating modules in which the open space can be used for housing lifeline. The pneumatic floating modules control their buoyant force by adding or reducing inner air pressure so that they can cope with variable loadings on the modules. Accordingly, any buildings having different weights can be built on a floating platform without the stress differences between the modules that are connected by 3D truss beams. This paper introduces so-called "Air Pocket Factor" that plays a key role in designing the pneumatic platform and proposes an image of architectural complex to visualize this structural system as an example.

Keywords Very large floating structure (VLFS) · Pneumatic floating module · Air pocket factor · Sea city

1 Introduction

Experiment of the Mega-Float Project was conducted in Tokyo Bay, Japan from 1995 to 2000, having a 1000 m-long floating runway [2] which was pursued by major shipbuilding companies, steel companies, and many other companies. As the results of experiment, useful knowledge on very large floating structures (VLFS) was

---

T. Nakajima (✉) · Y. Yamashita
Waterfront Real Estate Co., Ltd, Kobe, Japan
e-mail: tmunakajima@gmail.com

S. Saito
Tokyo University of Science, Tokyo, Japan

X. L. Zhao et al. (eds.), *Proceedings of the Fourth World Conference on Floating Solutions: WCFS 2024; 2–4 December; Hong Kong*, Lecture Notes in Civil Engineering 597, https://doi.org/10.1007/978-981-96-4569-5_83

obtained. The Mega-Float was bult into a large scale by connecting many box-type modular floating structures. There was a concern that the structure may be damaged by slamming in rough waves, hence, breakwaters or other structures need to be placed around the Mega-Float to prevent it from the damage by waves. On the other hand, a VLFS which connects semi-submersible-type modular floating structures such as the US Navy's MOB scheme [3] can reduce the impact of strong waves as it has an excellent seakeeping ability. Unfortunately, as the semi-submersible-type modular floating structures are expensive as compared with those of barge-type or pneumatic-type, they are not suitable for a VLFS such as a floating sea city. In contrast, the pneumatic floating structure presented in this paper prevents slamming as the wave pressure acting on the structure is uniformly distributed such that the floater reduces vertical wave forces. One of the best features that makes pneumatic floating structure suitable for the foundation of a sea city is that it is economical as compared with conventional barge-type platform since it has no bottom plate.

As described in Ref. 1, many patents are certified on air floats, however, it was only Dr. Seidl who disclosed so-called the "Air Pocket Factor" [4]. He demonstrated that by introducing the Air Pocket Factor, the vertical motions such as heave, roll and pitch of the air float can be significantly damped as compared with those of barge-type. This development opened the way to design air float structures.

2 Principles of Pneumatic Floating Structure

As Dr. Seidl has indicated, the buoyancy of an air-cushioned float is the sum of atmospheric air pressure and difference in water pressure between the external and internal water surfaces (water pressure). It is important to note that hydrodynamic characteristics between a scale model and a prototype pneumatic float cannot be linearly converted simply by scale. Since the magnitude of atmospheric air pressure inside of the scale model of pneumatic float cannot be scaled down, atmospheric air pressure inside of the scale model significantly contributes on the total buoyancy force as compared with that of the prototype.

As stated above, the similarity rule between a scale model and actual float does not hold on to atmospheric pressure. It is unreasonable to change the atmospheric air pressure to the level of a scale model in the wave tank experiment. In other words, effect of the wave force on the scale model in a water tank experiment is not similar to that of a prototype float. Buoyancy of a pneumatic floating body corresponding to the water pressure (ρ g h), where h is difference in the mean water levels between inside and outside of the body is similar between a scale model and a prototype, namely similarity rule is applied to the water pressure.

Dr. Seidl considered that the air-cushioned floating body can utilize the compressibility of air and in theory, the vertical wave force can be reduced to zero by making the volume of the air floating body infinite. He was the first engineer to derive the formula of the Air Pocket Factor which is an essential factor for designing a pneumatic floating structure.

The Air Pocket Factor is expressed by two independent variables (α and β) that are derived from the Boyle's law (Gas Law). Dr. Seidl found that Air Pocket Factor directly affects the natural period of the heave motion of a floating air-cushioned body. The Gas Law of similarity is applicable because the variable α increases as the relative volume of the air pocket (the height of the float if the area is the same) increases. On the other hand, the variable β is the ratio between the atmospheric pressure and the water pressure supporting the float. Thus, the variable β has a direct relationship with the atmospheric pressure. There is no linearity between the scale model in the water tank experiment and full-scale floating body on the sea. The Air Pocket Factor directly affects the natural period of the heave motion of the air-cushioned float, T_{heave} as expressed by the equation below. In fact, the effect of air volume in the floating body against natural period of heave motion, T_{heave} is significantly large in the full-scale floating body. In contrary, it is extremely difficult to reproduce T_{heave} that is compatible with the full-scale body by using scale-model in a water tank experiment.

The approximate natural period of heave motion can be calculated as shown below,

$$T_{heave} = 2\pi\sqrt{\frac{M + M_h}{C33}} \tag{1}$$

where

T_{heave}: Natural period of heave motion,

M: Mass of the float,

M_h: Added mass of the float,

$C33$: restoring coefficient of the float.

According to Reference [4], restoring coefficient can be obtained by $C33 = \rho g A_W \times$ F air.

$$(A_W : waterplanearea)$$

Dr. Seidl [4] derived the equation of Air Pocket Factor (Fair) as follows.

$$\mathrm{F\ air} = \frac{d_h}{df} = 1\boldsymbol{\alpha}/(1 + \boldsymbol{\alpha} + \boldsymbol{\beta}) \tag{2}$$

where $\boldsymbol{\alpha} = c/h$, $\boldsymbol{\beta} = \mathrm{P}*/\mathrm{h}$; P* is the head of atmospheric pressure (10.33 m).

3 Comparison of Heave Natural Periods Between the 1/47 Scale Model and the Full- Scale Model

A series of motion tests by pneumatic scale model were conducted at the Institute of Industrial Science Ocean Engineering Basin of the University of Tokyo. The 1/47 scale model was used for the free decay tests to obtain heave and pitch natural periods. The test results are reported in paper [5]. The total length of the 1/47 scale model is 2.0 m and on a side of each hexagonal module unit is 0.4 m that correspond to the full-scale body of 94 m and 18.8 m, respectively. The difference of the mean water lines between inside and outside of the 1/47 scale model is 0.05 m (h).

The natural period of the heave motion of the full-scale air-cushioned floating body was compared with that of the scale model. Figure 1 shows a picture of the 1/47 scale model in which six hexagonal air-cushioned floating modules are shown. In this water tank experiment, the natural period of the heave motion was measured as 1.07 s. Figure 2 shows a plan view and a cross-sectional view of the same air-cushioned floating model.

Tabulated curves of Air Pocket Factor with respect to the variable α are shown in Fig. 3. The variable β is 206.2 for 1/47 scale model whereas that of full-scale body is 4.4 in this figure. It is apparent that the variable β changes significantly for different variable α in the full-scale body. In contrary, the variable β of the 1/47 scale model does not change much against the variable α.

Figure 4 summarizes the results of calculations on the Air Pocket Factor with respect to the independent variables (α and β) which shows the effect of the volume

Fig. 1 Photos of the 1/47 scale model of pneumatic floating structure

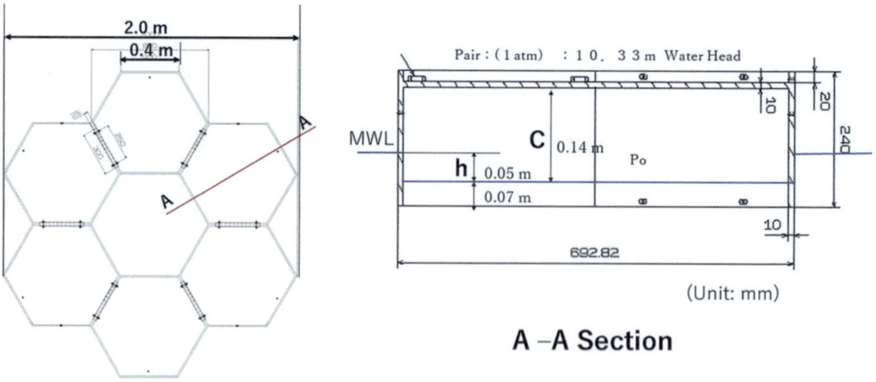

Fig. 2 Plan and sectional views of the 1/47 scale model of pneumatic floating body

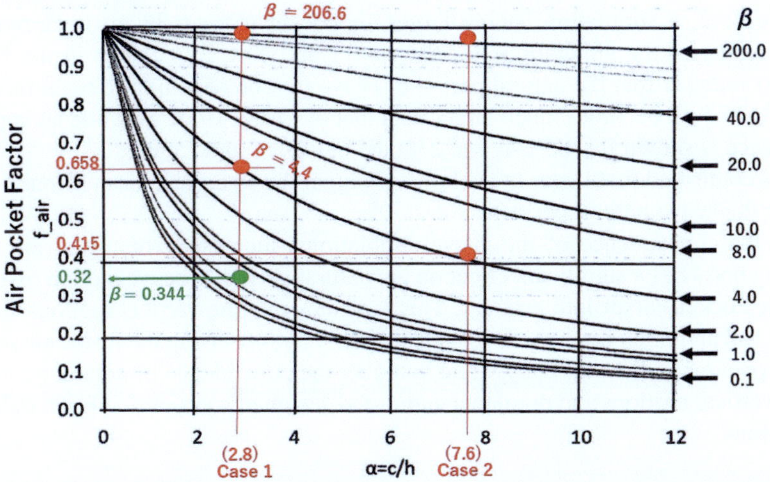

Fig. 3 Tabulated curves of Air Pocket Factor respect to the independent variables α and β

of the air in a floating body on the motion of the air-cushioned floating body. In the case 1, a wave tank experiment was conducted by using the 1/47 scale model after which natural period of the heave motion was measured as 1.07 s. As this value has an influence of the Air Pocket Factor (APF) of 0.987, calculation was carried out to estimate the natural period of the heave motion that is not affected by APF. Then, similarity rule (square root of 47) was applied to convert the natural period of heave motion in the full-scale floating body to be 7.33 s. Finaly, APF of the full-scale model (0.658) was applied to find out the correct natural period of heave motion, i.e., 9.02 s. This is the reason why similarity rule does work between the scale model and full-scale model for the air-cushioned floating body.

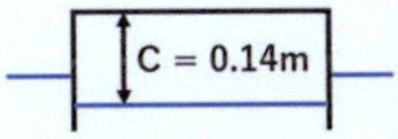

[Case 1] (original) h = 0.05m, c = 0.14m (1/47 Model)

	1/47 Model	Full Scale
α (c/h)	2.8	2.8
β (P air / P o)	206.6	4.4
f air	0.987	0.658
T heave	1.07 sec.	7.33 sec. ⇒ 9.02 sec.

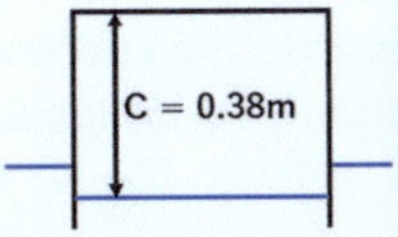

[Case 2] h = 0.05m, c = 0.38m (1/47 Model)

	1/47 Model	Full Scale
α (c/h)	7.6	7.6
β (P air / P o)	206.6	4.4
f air	0.964	0.415
T heave	1.08 sec.	7.43 sec. ⇒ 11.5 sec.

Fig. 4 A comparison of air pocket factors and heave natural periods between 1/47 scale model and prototype for Case 1 and Case 2

In case 2, a simulation having larger air pocket (C = 0.38 m) is carried out. Comparison of the simulations for the case 1 and case 2 is shown in the Fig. 4. Results indicate that the natural period of heave motion of a full the scale-model in Case 2 (11.5 s.) becomes much larger than that of Case 1 (9.02 s.) while no apparent difference is shown for the 1/47 scale model in Case 1 and Case 2.

It is confirmed that larger the volume or bigger the size of an air-cushioned body, longer the heave natural periods.

As the consequence of the above simulations, increased volume of air in the floating body has a significant effect on the natural period of heave motion by taking the Air Pocket Factor into account. This indicates that the vertical motions such as heave, roll and pitch can be reduced considerably by increasing the inside air volume of the pneumatic floating body while avoiding the coincidence of natural periods of these vertical motions and dominant and/or the designed wave period in irregular sea conditions.

4 Innovative Sea City Concept by Introducing the Pneumatic Floating Foundation

In general, cities on land face water-related disasters due to climate change, such as water shortages arising from the reduced rainfall and flooding from concentrated heavy rain. Lower-lying regions are especially vulnerable to the sea-level rise or flooding. On the contrary, the sea city in open sea that is constructed on innovative pneumatic floating foundation of several kilometers scale is safe and unaffected by these disasters. A super-large pneumatic floating structure is constructed by connecting many floating modular units by truss beams.

When the size of the pneumatic floating structure is large such as a super-large pneumatic floating structure, the variable β becomes negligibly small, then the equation of Air Pocket Factor becomes as follows.

$$\text{F air} = 1/(1+\boldsymbol{\alpha}) \tag{3}$$

Figure 5 shows the plan view of the proposed floating city in which the twelve hexagonal floating modules are connected by rigid truss beams. Each floating module can be removed by disconnecting the truss beams, if necessary. The total area of the proposed floating city is 194.85 ha in which the area of each floating module unit is about 16.24 ha. The total length in the longitudinal and transverse directions is about 2.0 km and 1.2 km, respectively. In the transverse direction (a total length of 1.2 km), even if the condition of the case 1 (short freeboard; refer to Fig. 4) is considered where the variables $\alpha = 2.8$ and $\beta = 0.344$ (the green arrow in Fig. 3), the Air Pocket Factor and the natural heave period are obtained as 0.32 and 46.02 s. respectively. This confirms that the effect of air cushion on the seakeeping capability of a super-large pneumatic floating structure is significant.

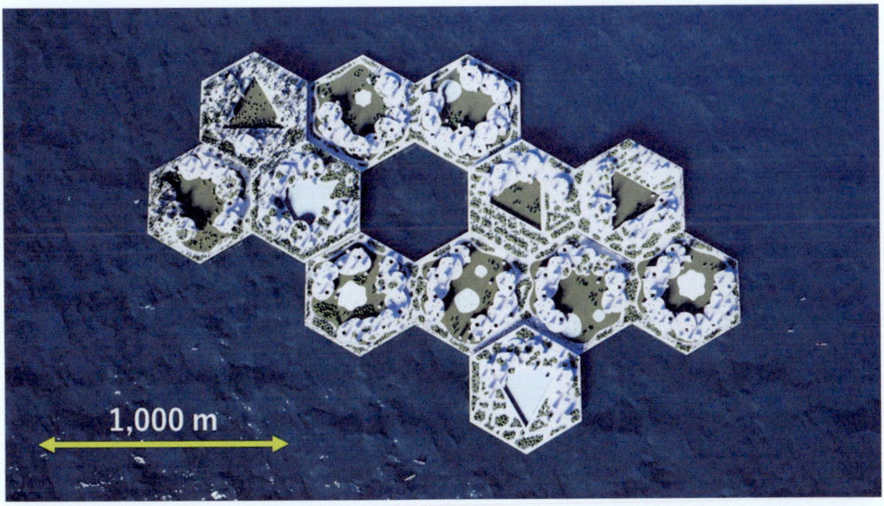

Fig. 5 Plan view of the sea city using the pneumatic floating foundation

An estimated population in the proposed sea city is 24 thousand that includes 11 thousand households with respective population density of 123 persons/ha. Each residential unit of 8,000 people forms a total of 24,000 residential community that is divided into three residential zones. One junior high school and three elementary schools are allocated in the city. In addition, shopping complex, private offices, government offices, and amusement facilities are constructed in the same zone. The internal transportation system is mainly composed of walkways; however, some designated areas furnish moving walkways, small electric cars, and flying cars. The bird eye's view of the sea city is shown in Fig. 6.

5 Concluding Remarks

This paper tried to provide technical evidence of so-called Air Pocket Factor (APF) that is supported by experiment. Governing equations of the APF is presented on the ground of the Boyle's Law to assess the effect of the confined air that can reduce vertical motions in a pneumatic floating body. A practical image of a pneumatic floating structure is presented as an example.

To realize an offshore sea city, a floating foundation having high stability and comfort that stays still even in rough weather conditions is essential. At the same time, floating foundation is expected to be structurally strong such as having high resistance against slamming in strong waves as well as economically feasible. The pneumatic floating structure shown in this paper can cope with the slamming force exerted by strong waves and lessen vertical motions (including pitching and rolling motions) caused by vertical wave forces. Further, breakwater is not necessary which

Fig. 6 Bird- eye's view of the sea city using the pneumatic floating foundation

is required for a large barge-type floating structures. However, mooring system is required to control the horizontal forces such as drift force. Another major feature of the pneumatic floating structure is that it can be built with reasonable cost as compared with conventional barge-type or semi-submersible type structures. The biggest feature of a pneumatic floating structure is that it utilizes the atmospheric pressure and its compressibility confined in the floating body as such it can cope with wave forces significantly. At the same time, it can considerably reduce not only the heave motion but also pitching and rolling motions of the floating body. It should be mentioned that rigid body assumption is not adequate on a very large pneumatic floating structure as the hydro-elastic behaviors should be taken into account [6]. Further extensive studies are necessary to realize the construction of various types of pneumatic floating platforms in future.

References

1. Nakajima T, Yamashita Y, Imai M, Yamashita Y (2023) An innovative design concept of modular pneumatic floating platforms, lecture notes in civil engineering, Proceedings of the third world conference on floating solutions, Japan, WCFS2023, Springer
2. Sato C (2003) Result of 6 years' research project of Mega-float, Ocean Space Utilization Technology, Ministry of Land, Infrastructure and Transport of Japan, pp 436–442, January 28–31
3. Taylor R (2003) MOB project summary and technology SPIN-OFFS, Ocean Space Utilization Technology, Ministry of Land, Infrastructure and Transport of Japan, pp 29–36, January 28–31
4. Seidl LH (1980) Development of an ASP (air stabilized platform), DO-A01-78-00-3104, U.S. Dept. of Commerce/Maritime Administration/Office of Commercial Development

5. He Y, Hirabayashi S, Tabeta S, Nakajima T, Yamashita Y, Imai M (2022) Hydrodynamic Moion behavior of air-cushion-supported hexagonal floating platform for offshore wind turbine, Proceedings of the ASME 2020 41st interntional conference on ocean, offshore and arctic engineering, OMAE2022,Hamburg, Germany
6. Ikoma T, Masuda K, Maeda H, Rheem CK (2002) Hydroelastic behavior of air-supported flexible floating sturcures, Proceedings of the 21st International Conference on OMAE'02, ASME

Experimental Study on Hydrodynamic Response of a Modular Floating Structure System with Chain-Type Expansion

Yanwei Li, Nianxin Ren, Yaqiong Liu, and Jinping Ou

Abstract Modular floating structures provide a feasible way to solve the land scarcity in and around coastal cities. The present work investigates the hydrodynamic responses of a proposed modular floating structure system through the scale model test. The proposed system consists of several inner TLP hexagonal floating structure modules and outermost floating artificial reef modules, forming a chain-type configuration. The outermost floating artificial reef modules can play the role as a floating breakwater, meanwhile providing habitat for marine organism. The scale model test was carried out in the wave flume of Hainan University. Effects of the chain-type modular expansion on the system's dynamic response have been extensively studied through both regular and irregular wave tests. Experimental results indicate that chain-type modular expansion notably reduce surge response and enhances the system's survival capability under extreme sea condition. The primary experimental results in this article can offer valuable guidance for the practical application of modular floating structures in engineering.

Keywords Modular floating structure · Model test · Hydrodynamic response · Chain-type expansion · Floating artificial reef

Y. Li · N. Ren (✉) · Y. Liu
School of Marine Science and Engineering, Hainan University, Haikou, China
e-mail: rennianxin@hainanu.edu.cn

N. Ren
State Key Laboratory of Marine Resource Utilization in South China Sea, Hainan University, Haikou, China

J. Ou
School of Civil and Environmental Engineering, Harbin Institute of Technology, Shenzhen, China

X. L. Zhao et al. (eds.), *Proceedings of the Fourth World Conference on Floating Solutions: WCFS 2024; 2–4 December; Hong Kong*, Lecture Notes in Civil Engineering 597, https://doi.org/10.1007/978-981-96-4569-5_84

1 Introduction

Global warming is leading to continued shoreline erosion in and around coastal cities. However, the majority of the world's population is distributed in coastal areas, which urgently requires expanding available space into the ocean [1]. Modular floating structure has been proved to be a kind of marine structure for effective exploitation of ocean space, and its characteristic of modular construction provides convenience for building long-life marine floating platforms [2].

Modular floating structure is a complex multi-body system, and the hydrodynamic coupling effect among modules significantly influences its behavior. Several numerical and physical model experiments have been developed to study the hydrodynamics of modular floating structures. Bispo et al. [3] formulated an analytical model for modular floating structures based on linearized wave theory and small structure response. They obtained the analytical solution through the velocity decomposition method, validating its high consistency with the results generated by the AQWA code. Ding et al. [4] conducted an experiment analyzing the responses of an 8-module modular floating structure under complex wave conditions. The results indicated that an uneven seabed exerts a significant influence on motion responses and connector loads.

The conventional anchor chain mooring method used for modular floating structures occupies a considerable sea area, hindering the expansion of the modules [5]. The TLP-type floating wind turbine offers valuable guidance for positioning modular floating structures. The large vertical stiffness of the tension leg markedly decreases the structure's responses in heave, pitch, and roll [6]. So far, several numerical and experimental studies have explored the conceptual design and dynamic characteristics of modular floating structures. Nonetheless, there remains a scarcity of research reports focusing on TLP-type modular floating structures and its modular expansion, particularly concerning model experiments.

In present work, effects of the chain-type modular expansion on the system's hydrodynamic responses have been extensively studied by a series of model tests under regular and irregular waves. The floating artificial reefs, located as the outermost modules in the system, aim to serving as a floating breakwater and enhancing the surrounding marine environment [7].

2 Experimental Setup

The proposed modular floating structure system comprises a chain-type arrangement of hexagonal floating structures interconnected by hinged connectors, with the floating artificial reef modules integrated on the outermost side. Each hexagonal module utilizes four symmetrically moored tension legs, with a displacement designed to be greater than its mass to ensure adequate pretension. Hexagonal modules are color-coded in red and yellow to aid in locating the waterline. The scale model test was conducted in the wave flume (50 m * 2.5 m * 2 m) of Hainan University, and the scale ratio is set to be 1:80 based on Froude scaling principle. The general layout of the scale model test is shown in Fig. 1, where each module is labeled as Mi ($i = 1 \sim 7$). Detailed parameters of the proposed system are listed in Table 1. The representative photos of the proposed system with different hexagonal modules are shown in Fig. 2.

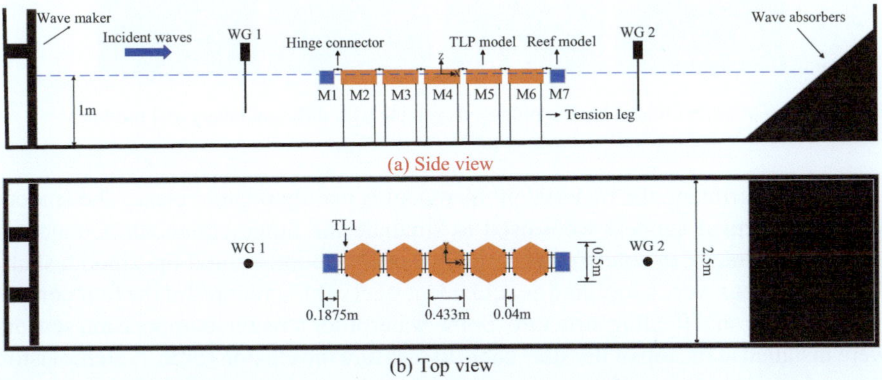

Fig. 1 Schematic diagram of the general layout of the scale model test

Table 1 Detailed design parameters of the proposed modular floating structure system

Parameters	Full scale	Scaled ($\lambda = 1/80$)	Units
Hexagonal TLP module			
Side length	20	0.25	m
Mass; displacement	6,600,000; 10,650,000	12.89; 20.80	Kg
Water depth; draft	80; 10	1; 0.125	m
Z_CoG	– 5	– 0.0625	m
Ixx = *Iyy*, *Izz*	9.95×10^8; 1.25×10^9	0.304; 0.381	Kg·m^2
Adjacent distance	3.2	0.04	m
Floating artificial reef module			
Length; breadth; draft	20; 15; 10	0.25; 0.188; 0.125	m
Mass; displacement	800,500; 800,500	1.563; 1.563	Kg
Z_CoG	– 2	– 0.025	m
Ixx, *Iyy*, *Izz*	1.2×10^8; 9.8×10^7; 1.0×10^8	0.037; 0.030; 0.031	Kg·m^2
Cut-opening ratio	20%	20%	
Tension leg dimension			
Length; outer diameter; thickness	70; 1.2; 0.04	Equivalent steel cable	m
Axial stiffness	437,309,697.4	68,329.64	N/m

Fig. 2 Photo of scaled model for the proposed systems with different hexagonal modules

In the experiment, the material of the model is mainly organic glass. The smooth shaft and vertical support were used to simulate the hinge connector. Connected adjacent modules can solely exhibit relative pitch motion around the smooth shaft. The tension legs were simplified as equivalent steel cables, moored at the four corners of each hexagonal floating structure. Some waterproof tension–compression sensors were installed at the top of the steel cable to measure the tension force. A stereovision measurement system with two cameras and some targets was used to measure the 6-DOF motions of multiple modules. In addition, wave gauges (WG) were placed on both sides of the model system to measure the wave information, as shown in Fig. 1.

3 Experimental Results and Discussion

Effects of chain-type modular expansion on the dynamic response of systems has been extensively studied through both regular and irregular wave tests. Due to the significant vertical stiffness of the tension leg platform, its most prominent response occurs in surge under unidirectional incident waves [8, 9]. Meanwhile, there is minimal relative surge motion between adjacent modules because of the hinge connection. Consequently, solely the surge results of the up-wave hexagonal module M2 and its maximum tension leg load of the TL1 are presented. All results are presented in full scale to provide a more comprehensive understanding.

The dynamic response results of systems with different numbers of modules under some representative regular wave sea conditions (wave height of 3 m, wave period of 7 s, 10 s, 17 s) are shown in Fig. 3. Figure 3a illustrates a decreasing trend in surge amplitude as the number of modules increases, suggesting that augmenting the modules can enhance the system's stability. At a wave period of 10 s, the surge amplitude of the 3-module system exhibits relatively significance. Comparatively, the surge amplitudes of the 5-module, 7-module, and 9-module systems have decreased by 43.7%, 48.6%, and 62.6%, respectively, in contrast to the 3-module system. Hence, to enhance the comfort and stability of the system under operational sea conditions, it is advisable to avoid having an excessively small number of modules within the system. Figure 3b shows that at a wave period of 17 s, each system's tension leg experience maximum force with no significant variation among them. This suggests that the increase in the number of modules has a relatively minor impact on the tension leg force. Hence, it can be concluded that the chain-type modular expansion enhances the system's motion response characteristics without notably amplifying the tension leg load.

To assess the system's safety, the extreme responses of the system with varying numbers of modules under extreme sea condition (JONSWAP, $\lambda = 3.3$, Hs = 4 m, Tp = 10 s) were further investigated. Table 2 lists the extreme responses characteristic information.

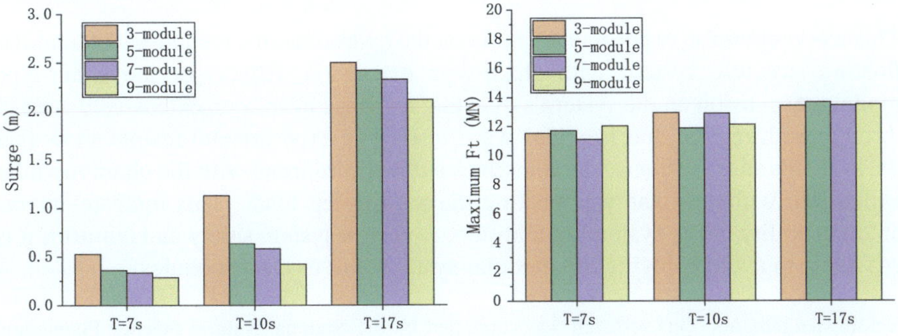

(a) Surge of the M2 (b) Maximum tension leg force of the TL1

Fig. 3 Comparison of the main dynamic responses of the systems with different numbers of modules

Table 2 Main extreme response characteristic information of the systems with different numbers of modules under extreme sea condition

		Surge of the M2 (m)	Ft of the TL1 (MN)
9-module	Max	0.631	13.098
	Mean	–	9.754
	STD	0.135	1.347
7-module	Max	0.854	14.113
	Mean	–	9.730
	STD	0.155	1.642
5-module	Max	1.276	13.578
	Mean	–	9.745
	STD	0.257	1.357
3-module	Max	3.140	14.822
	Mean	–	9.797
	STD	0.659	1.769

According to Table 2, compared with the 3-module system, the maximum surge amplitude of the 5-module, 7-module and 9-module systems decreased by about 59.4%, 72.8% and 79.9% respectively, indicating that the increase in the number of modules can improve the system's survivability under extreme sea condition. Furthermore, in comparison to other systems, the 3-module system displays the highest maximum value and standard deviation of tension leg load, signifying an increased risk of tension leg failure. To ensure the system's safety, it's advisable not to have too few modules within the system.

4 Conclusion

This work reports an experimental study on the hydrodynamic response of a modular floating structure system with chain-type expansion. Effects of the chain-type modular expansion on the system's dynamic response have been extensively studied through both regular and irregular wave tests. The experimental results show that the system's surge response demonstrates a downward trend with the chain modular expansion, without notably impacting the tension leg loads. This indicates favorable scalability of the system. In addition, to ensure system safety and comfort, it is advisable to avoid applying the modular system with too few modules.

Acknowledgements This research was supported by the National Natural Science Foundation of China (Grant No. 52161041), the Natural Science Foundation of Hainan Province (Grant NO. 520RC552). The financial supports are greatly acknowledged.

References

1. Wang C, Tay Z (2011) Very large floating structures: applications, research and development. Proced Eng 14:62–72
2. Suzuki H (2005) Overview of Megafloat: concept, design criteria, analysis, and design. Mar Struct 18(2):111–132
3. Bispo I, Mohapatra S, Soares C (2022) Numerical analysis of a moored very large floating structure composed by a set of hinged plates. Ocean Eng 253:110785
4. Ding J, Geng Y, Xu S, Yang W, Xie Z (2021) Experimental study on responses of an 8-module VLFS considering different encounter wave conditions. Mar Struct 78:102959
5. Jae I, Osman A, Yenduri A, Nizamani Z, Nakayama A (2022) Optimisation of mooring line parameters for offshore floating structures: a review paper. Ocean Eng 247:110644
6. Tian C, Liu M, Xiao L, Wei H (2023) Investigation for the effect of the column step on the flow-induced motion of a TLP. Mar Struct 90:103400
7. Wang G, Wan R, Wang X, Zhao F, Lan X, Cheng H, Tang W, Guan Q (2018) Study on the influence of cut-opening ratio, cut-opening shape, and cut-opening number on the flow field of a cubic artificial reef. Ocean Eng 162:341–352
8. Li Y, Ren N, Li X, Ou J (2022) Hydrodynamic Analysis of a novel modular floating structure system integrated with floating artificial reefs and wave energy converters. J Marine Sci Eng 10(8):1091

9. Ren N, Ma Z, Shan B, Ning N, Ou J (2020) Experimental and numerical study of dynamic responses of a new combined TLP type floating wind turbine and a wave energy converter under operational conditions. Renew Energy 151:966–974

Hydroelastic Analysis of a Modular Very Large Floating Structure in a Focused Wave Group

Zhiqiang Li, Xingya Feng, Dengshuo Chen, and Jian-Fei Chen

Abstract This paper investigates the hydroelastic behavior of modular floating structures (MFS) made from ultra-high performance concrete (UHPC) reinforced with prestressed fiber-reinforced polymer (FRP) for floating photovoltaic (FPV) systems. The study analyzes an equivalent plate model to assess the impact of peak frequency and focused location on structural response. Results indicate a limited variation in structural response with peak frequency, except for notable peaks at 0.78 rad/s and 1.38 rad/s, corresponding to the structure's first two natural frequencies, signifying resonance. Higher natural frequencies have a negligible effect on the overall response variation. Furthermore, the variation of the focused location does not significantly alter the internal force response throughout the focusing wave generation process. Energy distribution across the structure and transfer between modules mitigate disparities in force loading, making the MFS design adaptable and resilient to varying wave conditions.

Keywords Hydroelastic analysis · Modular floating structure (MFS) · Ultra-high performance concrete (UHPC) · Focused wave

1 Introduction

With the increasing global focus on renewable energy, solar power has emerged as a popular option due to its abundance and sustainability. Land-based photovoltaic systems, especially in China, have made significant advancements [1]. However, limited land resources restrict further expansion, thereby highlighting the potential of Floating Photovoltaic (FPV) systems, which install solar panels on water bodies like lakes, reservoirs, and offshore environments. The development of FPV technology has driven the rise of Very Large Floating Structures (VLFS) [2, 3]. Selecting suitable

Z. Li · X. Feng (✉) · D. Chen · J.-F. Chen
Department of Ocean Science and Engineering, Southern University of Science and Technology, Shenzhen, China
e-mail: fengxy@sustech.edu.cn

X. L. Zhao et al. (eds.), *Proceedings of the Fourth World Conference on Floating Solutions: WCFS 2024; 2–4 December; Hong Kong*, Lecture Notes in Civil Engineering 597, https://doi.org/10.1007/978-981-96-4569-5_85

materials for these structures is crucial for ensuring safety and cost-effectiveness. High-density polyethylene (HDPE) is low-cost but lacks strength and durability [4], while steel, though robust, is expensive [5]. Concrete floating structures offer a balanced solution, combining cost-effectiveness and safety.

Concrete has been widely used in marine structures, such as floating bridges, platforms, and breakwaters [6]. Prestressed concrete addresses cracking issues in traditional reinforced concrete, allowing for the design of longer, thinner, and more cost-effective structures [7]. In recent years, Ultra-High Performance Concrete (UHPC) has increased the attractiveness of floating structures due to its superior mechanical properties, durability, and corrosion resistance [8]. UHPC has been effectively utilized in bridge structures, reducing size and weight while maintaining strength and lowering life-cycle costs [9]. These benefits make UHPC an appealing option for FPV systems.

While practical engineering applications of UHPC in floating structures are not found, some studies explored design concepts utilizing UHPC as a fundamental material in FPV systems. One design concept of double-segment floating sandwich structures composed of two steel panels and a honeycomb UHPC core layer was proposed, and the mechanical performance optimization of the floating structure has been conducted by structure-fluid coupling method [2]. Another lightweight and high-stiffness floating structure composed of UHPC surface panel was designed as the upper support for solar panels [10]. A hydrodynamic-structural-material approach has been proposed for designing these large floating structures.

Our previous research proposed a modular box-pontoon structure using UHPC and prestressed FRP bars, featuring thin plates that enhance flexural performance and durability [11, 12], balancing cost and crack resistance. This modular design, combined with the use of UHPC and FRP, leads to a structure that is not only cost-effective but also highly resilient, stable, and environmentally friendly. These studies examined the hydroelastic response of the proposed MFS in regular waves. Building upon this work, this paper further investigates the hydroelastic analysis of the MFS in focused waves. An equivalent plate model is first employed to conduct the hydroelastic analysis. Then the effects of the focused position and the peak frequency of the focused wave are then discussed. discussed.

2 Equivalent Plate Model of MFS

In this study, the proposed MFS designed for FPV systems is composed of multiple box-pontoon modules connected by hinge connectors. Each module is constructed using UHPC-CFRP composites, as shown in Fig. 1. The deck, bottom plate, and side walls are made of UHPC plates reinforced with prestressed CFRP bars. To complete the hydroelastic analysis, an equivalent plate model of the box-pontoon structure must first be developed.

Based on the principle of modal equivalence, a comparative analysis of the vibration frequencies and mode shapes of the equivalent plate model and the box-pontoon

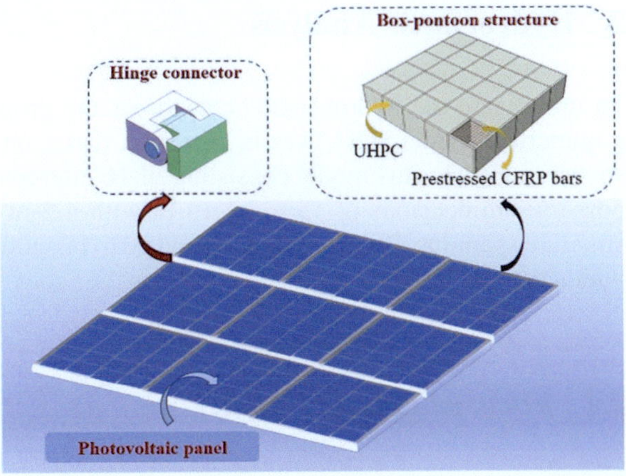

Fig. 1 Diagram and illustration of a modular floating structure using UHPC-CFRP

model was conducted to ensure consistency in the vibration responses across all modes. Figure 2 shows the comparison of mode shapes between the box-pontoon model and the equivalent plate model from the 7th to the 10th mode shapes. It can be observed that the mode shapes of the box-pontoon structure model closely match those of the equivalent plate model, and the frequency values of each mode are also similar, indicating that the equivalence principle is satisfied (see Fig. 2).

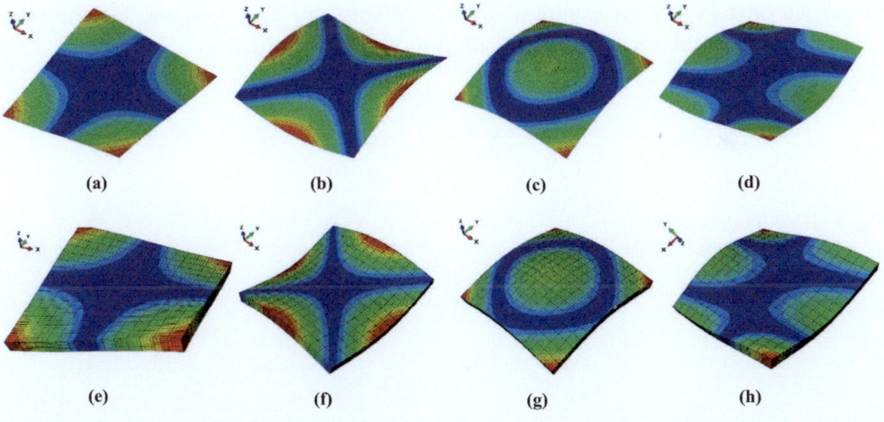

Fig. 2 The 7th, 8th, 9th, and 10th modes of equivalent plate and actual structural models. **a**, **b**, **c**, and **d** represent the 7th, 8th, 9th, and 10th modes of the equivalent plate model, respectively; **e**, **f**, **g**, and **h** correspond to the 7th, 8th, 9th, and 10th modes of the actual structure model

3 Hydroelastic Analysis

To investigate the hydroelastic behavior of the proposed MFS in focused waves, a numerical model HN4, was investigated based on the equivalence principle, as shown in Fig. 3. This model consisting of 16 interconnected modules, each module possesses dimensions of 15 m by 15 m with a 2 m depth, resulting in an overall structure spanning 60 m in length (L) and 60 m in width (B). The structure is designed with a draft of 0.8 m.

3.1 Focused Pressure Model

Based on a random wave spectrum, a linear New wave model was proposed to generate a focused wave group [13]. This model has been widely used in the ocean engineering field to represent extreme waves. For a given wave energy spectrum, the focused wave elevation is given by:

$$\eta(x, y, z) = \sum_{n=1}^{N} A_n \cos[k_n(x - x_c)\cos\theta + k_n(y - y_c)\sin\theta - \omega_n(t - t_c)] \tag{1}$$

where A_n, k_n and ω_n are the wave amplitude, wave number and wave circular frequency of the nth wave component, respectively. are the coordinates of the focal position and t_c is the focal time. The amplitude is represented as

$$A_n = \frac{S(\omega_n)\Delta\omega}{\sum_{n=1}^{N} S(\omega_n) \bigtriangleup \omega} \tag{2}$$

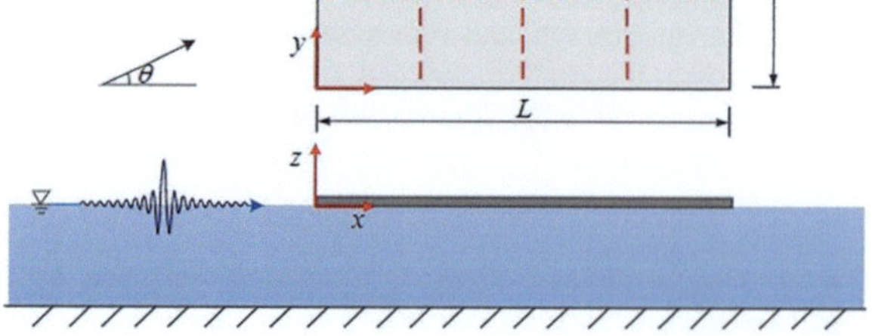

Fig. 3 HN4 model under focused wave groups with an incident angle θ

where the total number of wave components N is supposed to be large enough to regenerate the desired wave spectrum. The random waves are generated by 3-parameter JONSWAP wave spectrum which can be expressed as

$$S(\omega) = \frac{5}{16}(1 - 0.287 \ln \gamma)H_s^2 \frac{\omega_p^4}{\omega^5}(-\frac{5}{4}(\frac{\omega}{\omega_p})^{-4})\gamma^{\exp(-0.5(\frac{\omega-\omega_p}{\sigma\omega_p})^2)} \quad (3)$$

where γ is the peak enhancement factor, Hs the significant wave height, and ω_p the peak frequency. The peak shape parameter is defined as

$$\sigma = \begin{cases} 0.07, \omega \leq \omega_p \\ 0.09, \omega > \omega_p \end{cases} \quad (4)$$

In this paper, a modified pressure focused wave model is proposed to represent the action of focused wave group.

$$P(x, y, t) = \sum_{n=1}^{N} A_n P_n(x, y) \cos(\omega_n t + \varphi_n) \quad (5)$$

where $P_n(x, y, t)$ and φ_n are the amplitude and phase of the hydrodynamic pressure induced by the nth wave component. The former can be obtained by the procedure for regular waves proposed in our previous studies [10, 11] is expressed as $\varphi_n = -\omega_n t_c - k_n(x - x_c) - k_n(y - y_c)\varphi_n = -\omega_n t_c - k_n(x - x_c) - k_n(y - y_c)$.

3.2 *Effect of Peak Frequency*

In this study, the focused waves are generated using the focused pressure model, which modulates the phases of the hydrodynamic pressures caused by individual wave components under random sea conditions so that they converge at a designated location and time. The random sea conditions are described using the three-parameter JONSWAP spectrum, where different sea states have varying significant wave heights and peak frequencies, which are crucial factors in analyzing structural response. Since the model assumes a linear relationship, the wave height is proportional to the amplitude of the structural response. Therefore, this study primarily examines the impact of peak frequency on structural response under different sea conditions. As shown in Table 1, shows the range of peak frequency (ω_p) from 0.55 to 1.49 rad/s, corresponding to peak period ranging from 4.20s to 11.39s. The significant wave height (H_s) is set as 2.5 m. The ω_{low} and ω_{high} represent the lower and higher frequency limits of the selected wave components.

As shown in Fig. 4, the maximum values of the internal force responses of the structure vary with the peak frequency. The most significant responses are the

Table 1 The frequency range of numerical cases

Case	$H_s(m)$	$T_p(s)$(s)	ω_p(rad/s)	ω_{low}(rad/s)	ω_{high}(rad/s)
H2.5 ω 1.49	2.5	4.20	1.49	0.75	3.72
H2.5 ω 1.38	2.5	4.56	1.38	0.69	3.45
H2.5 ω 1.25	2.5	5.04	1.25	0.63	3.13
H2.5 ω 1.14	2.5	5.52	1.14	0.57	2.85
H2.5 ω 1.02	2.5	6.12	1.02	0.51	2.55
H2.5 ω 0.92	2.5	6.83	0.92	0.46	2.30
H2.5 ω 0.78	2.5	8.03	0.78	0.39	1.95
H2.5 ω 0.66	2.5	9.47	0.66	0.33	1.65
H2.5 ω 0.59	2.5	10.55	0.59	0.30	1.48
H2.5 ω 0.55	2.5	11.39	0.55	0.27	1.38

maximum longitudinal bending moment M_y and the maximum shear force F_z, which are analyzed as the variables of internal force response with respect to peak frequency changes. It can be observed that the maximum values of the internal force responses of the structure do not significantly vary with the peak frequency. However, in the frequency ranges of 0.78 rad/s and 1.38 rad/s, the internal force responses reach relatively higher peaks.

By analyzing the natural frequencies of the structure, it is found that the first six elastic modes of the structure correspond to natural frequencies of 0.80 rad/s, 1.48 rad/s, 2.13 rad/s, 3.52 rad/s, 3.93 rad/s, and 4.00 rad/s, respectively. It can be seen that the frequencies of the first two elastic modes (0.80 rad/s, 1.48 rad/s) are very close to the peak frequencies (0.78 rad/s, 1.38 rad/s) at which the structure's responses show significant peaks. Thus, it can be inferred that the significant peaks in the internal force responses are closely related to the natural frequencies of the structure. At these frequencies, the structure is prone to resonance, leading to a significant increase in internal force responses. Additionally, apart from the first two modes, the other natural frequencies of the structure (2.13 rad/s, 3.52 rad/s, 3.93 rad/s, and 4.00 rad/s) are significantly higher than the range of wave peak frequencies (0.55 rad/s–1.49 rad/

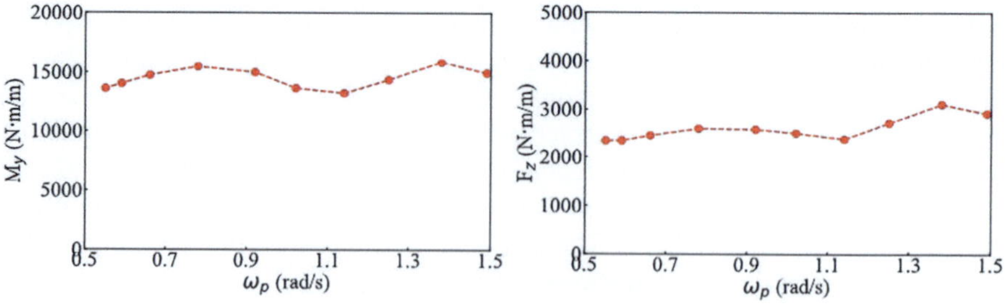

Fig. 4 The structural responses with the peak frequency. **a**M_y; **b** F_z

s). Hence, considering the contributions of other modes to the structural response, the overall trend of structural response variation with peak frequency is not obvious.

3.3 *Effect of Focused Location*

This section explores the influence of different focusing locations on the structure's internal forces. To achieve this, we systematically positioned four focusing points (P1, P2, P3, P4) along the x-axis at equal intervals, effectively dividing the axis into three sections. Figures 5 and 6 illustrate the contour plots of internal force response amplitudes (M_y and F_z) for each of the 16 modules under these varied focusing conditions. Figure 5 reveals that the largest bending moments are concentrated at the modules positioned at the bow and stern. In contrast, the bending moment amplitudes in the middle modules are significantly lower. This observation suggests that the bow and stern modules act as "breakwaters," shielding the interior modules from the brunt of the bending forces.

Further analysis of the shear forces (Fig. 6) indicates that the areas with the highest shear force intensities primarily reside near the hinge positions of the modules at the bow and stern. This phenomenon aligns with the observed concentration of bending

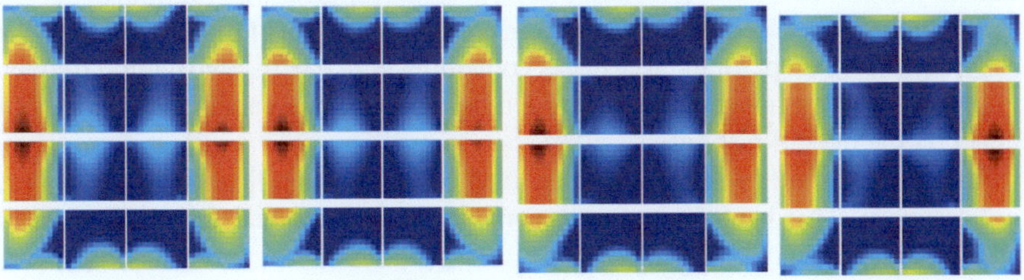

(a) P1 (b) P2 (c) P3 (d) P4

Fig. 5 Contour plots of bending moment amplitudes for each of the 16 modules under focusing conditions

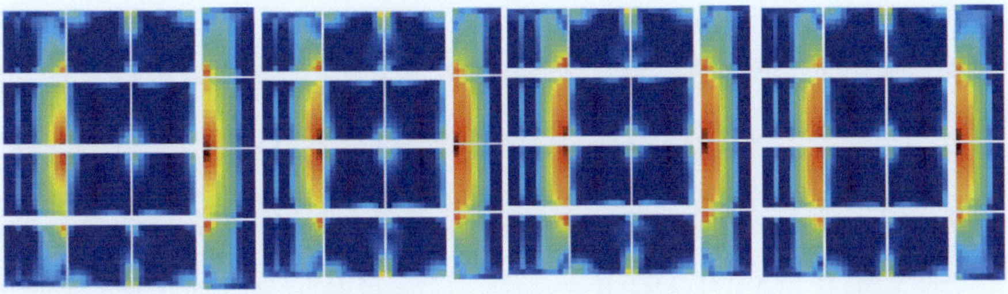

(a) P1 (b) P2 (c) P3 (d) P4

Fig. 6 Contour plots of shear force amplitudes for each of the 16 modules under focusing conditions moments at these locations. The combined effects of wave diffraction and radiation, intensified at the bow and stern, generate greater wave forces in these regions. This leads to higher internal forces within the bow and stern modules, while the internal forces in the middle modules remain comparatively lower.

Furthermore, comparing the structural response diagrams at different focus locations, it can be observed that the variation in focus location does not significantly impact the internal force response throughout the entire focusing wave generation process. During the focusing wave's interaction with the structure, wave energy might be transferred and dispersed across different components, leading to minimal overall differences in the internal force response. This effect is particularly pronounced in larger or multi-modular structures, where energy distribution is more widespread. The energy transfer among modules can further mitigate disparities in force loading at specific locations.

4 Conclusion

This study explored the hydroelastic analysis of modular floating structures (MFS) constructed from UHPC reinforced with prestressed FRP, specifically designed for FPV systems. By analyzing an equivalent plate model, the influence of peak frequency and focused wave location on structural response was examined.

The analysis revealed that the structural response exhibits limited variation with peak frequency, except for notable peaks coinciding with the first two natural frequencies, indicating resonance phenomena. Higher natural frequencies had a negligible impact on the overall response.

Furthermore, the variation in focused location did not significantly alter the structural response. This is attributed to the energy distribution and transfer within the multi-modular structure, effectively mitigating variations in force loading across different sections. These findings highlight the inherent adaptability and resilience of the proposed MFS design, offering a promising solution for FPV systems exposed to various wave conditions. Further research could explore the optimization of structural parameters and material properties to enhance the performance and stability of these systems in engineering applications.

References

1. Gorjian S, Sharon H, Ebadi H et al (2021) Recent technical advancements, economics and environmental impacts of floating photovoltaic solar energy conversion systems. J Clean Prod 278:124285
2. Chen Y, Ma H, Li A et al (2022) Hydroelastic analysis of double-segment floating sandwich structures under wave action. Ocean Eng

3. Karpouzoglou T, Vlaswinkel B, Molen JVD (2020) Effects of large-scale floating (solar photovoltaic) platforms on hydrodynamics and primary production in a coastal sea from a water column model. Ocean Sci 16(1):195–208
4. Dai J, Zhang C, Lim HV et al (2020) Design and construction of floating modular photovoltaic system for water reservoirs. Energy 191(15):116549
5. Claus R, López M (2022) Key issues in the design of floating photovoltaic structures for the marine environment. Renew Sustain Energy Rev, 164
6. Wang CM, Tay ZY (2011) Very large floating structures: applications, research and development. Proced Eng 14(3):62–72
7. Meng G, Jia J, Wang J (2013) Study on flexural behavior of prestressed ultra-high strength concrete beams. J Harbin Eng Univ 34(5):575–580
8. Bajaber MA, Hakeem IY (2020) UHPC evolution, development, and utilization in construction: a review. J Mater Res Technol, 10(3)
9. Li C, Feng Z, Pan R et al (2020) Experimental and numerical investigation on the anchorage zone of prestressed UHPC box-girder bridge. J Bridge Eng
10. Jin Z, Liu Y, Zhu M, Fang H, Li A (2023) A novel analytical model coupling hydrodynamic-structural-material scales for very large floating photovoltaic support structures. Ocean Eng 275:114113
11. Li Z, Chen D, Feng X, Chen JF (2023) Hydroelastic analysis and structural design of a modular floating structure applying ultra-high performance fiber-reinforced concrete. Ocean Eng 277:114266
12. Li Z, Chen D, Feng X (2023) Hydroelastic and expansibility analysis of a modular floating photovoltaic system with multi-directional hinge connections. Ocean Eng 340289:116218
13. Chen D, Feng X, Hou C et al (2022) A coupled frequency and time domain approach for hydroelastic analysis of very large floating structures under focused wave groups. Ocean Eng, 111393

Numerical Investigation on Motion Response of Modular Semi-Submersible Floating Platform in Waves

Lidong Zhang, Jia Wang, Xuanlie Zhao, Jing Geng, Zhiqiang Hu, and C. Guedes Soares

Abstract Modular semi-submersible floating platforms have recently gained popularity in offshore renewable energy applications. In this paper, the hydrodynamic responses of such platforms are investigated numerically. A constraint matrix is derived that characterizes the motion relation of the interconnected multi-module system. The frequency-domain motion equation is formulated using the Lagrange multiplier method. Parametrical investigations are conducted to explore the hydrodynamic performance of the modular semi-submersible floating platform in waves. The effects of incident wave angle and the spacing between adjacent modules are further investigated. The numerical results of the hydrodynamic responses of the multi-module system are presented, and the forces acting on the connectors are analyzed. It is found that the maximum responses in heave and pitch degree-of-freedom are observed for the module at the lee side. The motion responses of the modules are relatively larger at the lower frequency range. In general, the vertical forces acting on connectors at the middle and high-frequency ranges are larger than at the low-frequency range. The findings can provide valuable references for designing and constructing of floating modular structures.

Keywords Modular floating platform · Lagrange multiplier method · Motion response · Potential flow theory

L. Zhang · J. Wang · X. Zhao (✉) · J. Geng
College of Shipbuilding Engineering, Harbin Engineering University, Harbin, China
e-mail: xlzhao@hrbeu.edu.cn

X. Zhao · J. Geng · C. Guedes Soares
International Joint Laboratory of Naval Architecture and Offshore Technology between Harbin Engineering University and University of Lisbon, Harbin Engineering University, Harbin, China

Z. Hu
School of Engineering, Newcastle University, Newcastle, UK

C. Guedes Soares
Centre for Marine Technology and Ocean Engineering (CENTEC), Instituto Superior Técnico, Universidade de Lisboa, Lisbon, Portugal

X. L. Zhao et al. (eds.), *Proceedings of the Fourth World Conference on Floating Solutions: WCFS 2024; 2–4 December; Hong Kong*, Lecture Notes in Civil Engineering 597, https://doi.org/10.1007/978-981-96-4569-5_86

1 Introduction

Developing of offshore airports, terminals and multi-purpose platforms are crucial to explore marine resources. However, as the structural size increases, the structural deformation exerts a pivotal influence on the motion response of the floating body. Many investigations have been conducted to explore the hydrodynamics of very large floating structures (VLFS) or multi-body systems. Wu [1] developed a three-dimensional linear hydroelastic theory to analyze the motion and deformation of an arbitrary-shaped floating body in waves. Kiyokazu and Shoichi [2] developed a numerical analysis code to estimate the elastic motions and the connecting loads of box-shaped VLFS. Watanabe et al. [3] performed a hydroelastic analysis of a pontoon-type circular VLFS using the modal expansion method. Bispo et al. [4] presented a numerical model associated with wave interaction with a moored articulated very large floating structure and analyzed the hydrodynamic performance on the floating structure for different design parameters.

Du and Ertekin [5] numerically investigated four-module floating structure's motion response and loading in multi-directional random waves. Lee and Newman [6] proposed a frequency-domain motion equation for a floating system consisting of multiple modules, and the response and loading of the floating system are analyzed. Sun et al. [7] employed the Lagrange multiplier method to formulate the motion equation of the interconnected floating bodies. Jin et al. [8] and Han et al. [9] used the open-source tool WEC-Sim to model a two-body hinged raft wave energy converter and investigated the wave power extraction performance of the device.

This paper focuses on a floating system consisting of four interconnected modules (see Fig. 1). The frequency-domain motion equation is formulated based on the Lagrange multiplier method. Using a numerical method, parametrical investigations focus on the analysis of the hydrodynamic responses and forces on connectors.

2 Numerical Method

Considering the hydrodynamic coupling of a multi-module system, the 6-degree-of-freedom (6DoF) motion equation of a modular semi-submersible platform without the connection can be formulated as:

$$\left(-\omega^2(\mathbf{M}+\mathbf{a})-\mathrm{i}\omega\mathbf{b}+\mathbf{C}\right)\boldsymbol{\varepsilon}=\mathbf{f} \tag{1}$$

Here i is the imaginary unit; ω is the wave angular frequency; $\mathbf{M}$, $\mathbf{a}$, $\mathbf{b}$, $\mathbf{C}$, $\boldsymbol{\varepsilon}$ and $\mathbf{f}$ denote mass matrix, added mass matrix, radiation damping matrix, restoring force matrix, 6DoF motion response vector and wave excitation force vector, respectively. $\mathbf{a}$, $\mathbf{b}$ and $\mathbf{f}$ are obtained by using the potential-flow solver NEMOH [10].

Motion Eq. (1) can be simplified as $\mathbf{K}\varepsilon=\mathbf{f}$, where $\mathbf{K}=-\omega^2(\mathbf{M}+\mathbf{a})-\mathrm{i}\omega\mathbf{b}+\mathbf{C}$, $\boldsymbol{\varepsilon}=[\boldsymbol{\varepsilon}_1,\ \boldsymbol{\varepsilon}_2,\ \ldots,\ \boldsymbol{\varepsilon}_n]^{\mathrm{T}}$, $\boldsymbol{\varepsilon}_i=\left\{\varepsilon_i^{(1)},\ \varepsilon_i^{(2)},\ \ldots,\ \varepsilon_i^{(6)}\right\}$, $i=1,2,\ldots,n$, n represents the

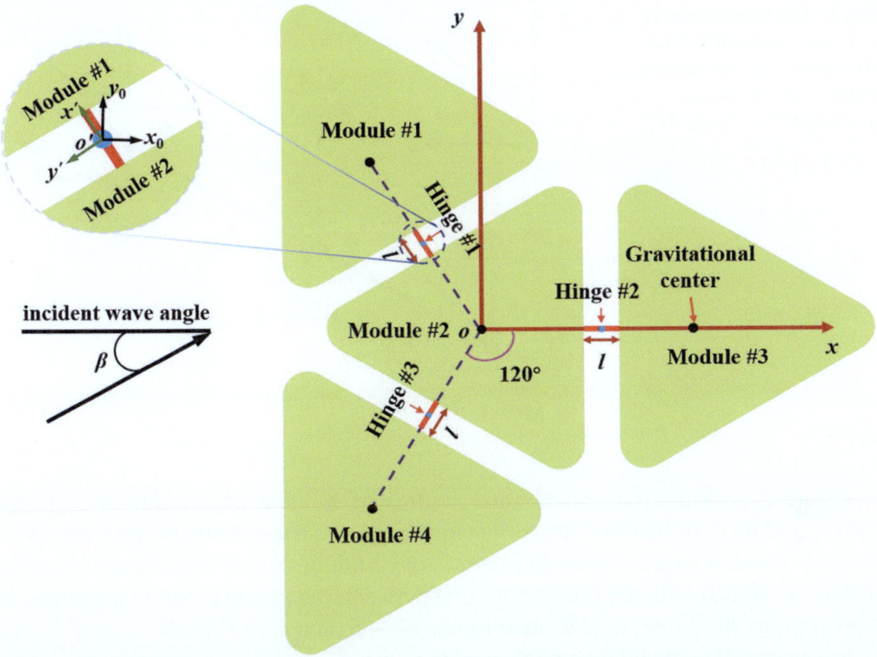

Fig. 1 Top-view of the modular semi-submersible floating platform. β is incident wave angle, l denotes the spacing between adjacent modules. o-z, $o' - z_0$ and $o' - z'$ axes point upwardly

number of the modules, $\varepsilon_i^{(j)}$ corresponds to the complex response of the i-th module in the j-th DoF. $j = 1, 2, \ldots, 6$ corresponds sequentially to the surge, sway, heave, roll, pitch and yaw DoF, respectively.

The sketch of the multi-module system is shown in Fig. 1. The central module (i.e., Module #2) is surrounded by three modules (i.e., Module #1, #3, #4). Hinge #1 denotes the connector between Module #1 and Module #2. The role of Hinge #1, #2 and #3 can be found in Fig. 1. Note that the connectors considered in this paper are column hinges. Based on the Lagrange multiplier method [7], the corresponding motion equation can be derived as:

$$\begin{bmatrix} \mathbf{K} & \mathbf{D}^{\mathrm{T}} \\ \mathbf{D} & \mathbf{O} \end{bmatrix} \begin{bmatrix} \boldsymbol{\varepsilon} \\ \boldsymbol{\lambda} \end{bmatrix} = \begin{bmatrix} \mathbf{f} \\ \mathbf{0} \end{bmatrix} \tag{2}$$

Here $\mathbf{D}$ denotes the constraint matrix; the force vector of the hinges is expressed as $\boldsymbol{\lambda} = [\boldsymbol{\lambda}_1, \boldsymbol{\lambda}_2, \ldots, \boldsymbol{\lambda}_m]^{\mathrm{T}}$, where $\boldsymbol{\lambda}_i = \left\{\lambda_i^{(1)}, \lambda_i^{(2)}, \ldots, \lambda_i^{(5)}\right\}$, $i = 1, 2, \ldots, m$, m stands for the hinge number; $\lambda_i^{(j)}$ corresponds to the complex load of the i-th hinge in the j-th DoF. For instance, $\lambda_1^{(j)}, j = 1, 2, \ldots, 5$ corresponds sequentially to $x'-$, y'-, and z'-direction complex forces, as well as the x'- and z'-direction complex torques of the local coordinate system, $o' - x'y'z'$, acting on the Hinge #1, respectively.

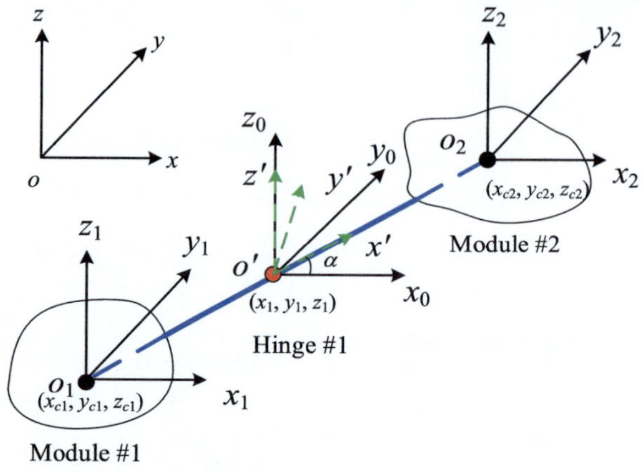

Fig. 2 The coordinates of two adjacent modules (i.e., Module #1 and Module #2) and the corresponding connector (i.e., Hinge #1)

Figure 2 presents the gravitational centres of Module #1, Module #2 and the corresponding coordinate of the connector Hinge #1. The coordinate system of $o' - x_0y_0z_0$ is parallel with the global system *o-xyz*. And the $o' - x'$ axis in the $o' - x'y'z'$ system is aligned with the connecting line between the two gravitational centres. In this instance, the angle, α, is formed by the $o' - x_0$ and $o' - x'$ axes.

It is known that the relative pitch motion in the $o' - x'y'z'$ system is released under action of the connectors. The constraint matrix **D**, which characterizes such kind of motion, can be derived as:

$$\mathbf{D} = \begin{bmatrix} \cos\alpha & \sin\alpha & & & & \\ -\sin\alpha & \cos\alpha & & & & \\ & & 1 & & & \\ & & & \cos\alpha & \sin\alpha & \\ & & & & & 1 \end{bmatrix} \cdot \begin{bmatrix} 1 & 0 & 0 & 0 & -(z_{c1}-z_1) & (y_{c1}-y_1) & -1 & 0 & 0 & 0 & (z_{c2}-z_1) & -(y_{c2}-y_1) \\ 0 & 1 & 0 & (z_{c1}-z_1) & 0 & -(x_{c1}-x_1) & 0 & -1 & 0 & -(z_{c2}-z_1) & 0 & (x_{c2}-x_1) \\ 0 & 0 & 1 & -(y_{c1}-y_1) & (x_{c1}-x_1) & 0 & 0 & 0 & -1 & (y_{c2}-y_1) & -(x_{c2}-x_1) & 0 \\ 0 & 0 & 0 & 1 & 0 & 0 & 0 & 0 & 0 & -1 & 0 & 0 \\ 0 & 0 & 0 & 0 & 1 & 0 & 0 & 0 & 0 & 0 & -1 & 0 \\ 0 & 0 & 0 & 0 & 0 & 1 & 0 & 0 & 0 & 0 & 0 & -1 \end{bmatrix} \tag{3}$$

Here (x_{c1}, y_{c1}, z_{c1}) and (x_{c2}, y_{c2}, z_{c2}) denote the gravitational centre coordinates of Module #1 and Module #2, respectively, and the coordinate (x_1, y_1, z_1) represents the hinge point of the connector.

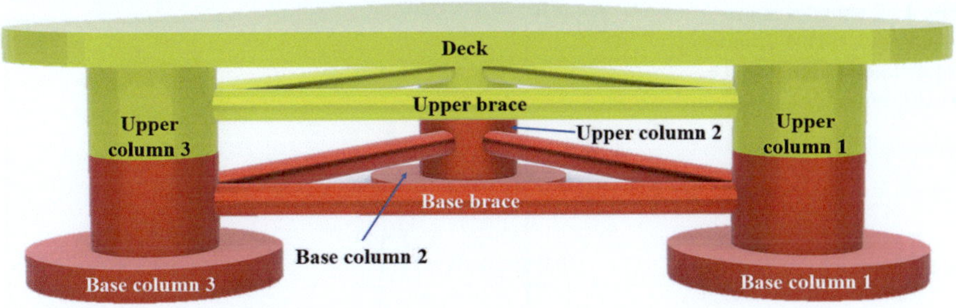

Fig. 3 The identical module unit of a modular semi-submersible floating platform

Table 1 The geometrical parameters of the floating platform

Term	Dimensions (m)
Diameter of upper column 1, upper column 2 and upper column 3	4.0
Diameter of base column 1, base column 2 and base column 3	8.0
Diameter of upper brace	1.0
Diameter of base brace	1.0
Height of deck	1.0

3 Description of Modular Semi-Submersible Floating Platform

As shown in Fig. 1, the plane of the modular semi-submersible floating platform contains four identical floating modules. The sketch of the module is depicted in Fig. 3, and the corresponding parameters are shown in Table 1.

4 Results and Discussion

4.1 Effect of Incident Wave Angle

In this subsection, the effect of incident wave angle on the hydrodynamic responses of the multi-module system is investigated by considering $\beta = 0°$, $30°$ and $60°$. Figure 4 presents the results of heave and pitch DoF responses for an isolated platform. It can be demonstrated that the incident wave angle slightly influences the trend of heave and pitch motion. As for the heave motion, it is observed that the peak value is located at $\omega = 0.65$ rad/s. A spike is observed at $\omega = 0.50$ rad/s in the trend of pitch motion against frequency.

Fig. 4 Results of heave **a** and pitch **b** DoF responses for isolated module in $\beta = 0°$, 30° and 60°

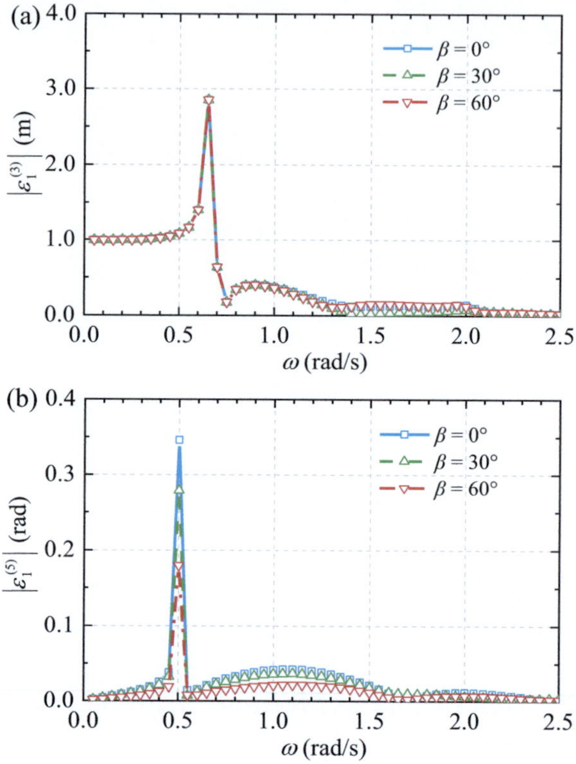

Figure 5 demonstrates the results of the heave and pitch DoF responses of each module for the multi-module system when $\beta = 0°$ and $l = 3$ m. Compared to the isolated case, the hydrodynamic interference among different modules modifies the trend of heave and pitch responses, although to a limited extent. Multiple peaks are found at $\omega = 0.55$ rad/s and 0.65 rad/s in Fig. 5b. Furthermore, it is found that the maximum motion responses in heave and pitch DoF are observed at Module #3, which is located on the lee side.

Figure 6 shows the impact of incident wave angle on the hydrodynamic responses of Module #3. It is observed that the peak values of the heave and pitch DoF responses decrease with the increase of incident wave angle. It is noteworthy that the amplitude of heave motion corresponding to $\beta = 60°$ is slightly larger than that corresponding to $\beta = 0°$ and 30° within the frequency range of 0.80 rad/s < ω < 1.50 rad/s. However, the pitch DoF response of Module #3 decreases as the incident wave angle increases.

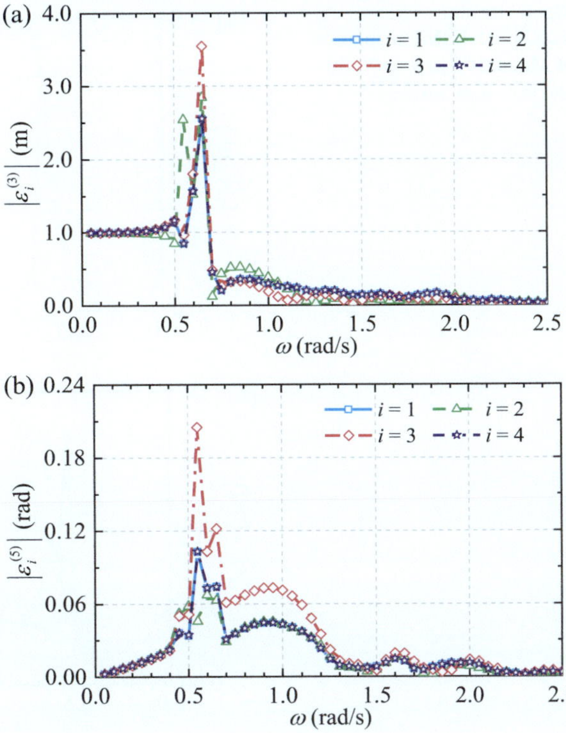

Fig. 5 Results of heave **a** and pitch **b** DoF responses for modular semi-submersible floating platform. ($\beta = 0°$, $l = 3$ m)

4.2 Effect of Module Spacing

The effect of module spacing ($l = 1$ m, 3 m and 5 m) is investigated in this subsection, where $\beta = 0°$ is considered here. Figure 7 shows us the variation of heave and pitch motion for different module spacings. The modifications of the heave and pitch DoF responses are primarily reflected within the frequency range of 0.50 rad/s < ω < 0.80 rad/s. It is noted that the peak values of heave DoF response are 3.47 m, 2.55 m and 4.55 m for $l = 1$ m, 3 m and 5 m, respectively. Correspondingly, the peak values of pitch DoF response are 0.08 rad, 0.10 rad and 0.54 rad.

Figure 8 illustrates the results of vertical forces acting on connectors by considering $\beta = 0°$ and $l = 3$ m. Due to the symmetrical arrangement, the vertical forces acting on Hinge #1 and Hinge #3 are identical. It's worth noting that the force acting on Hinge #2, which is located at the lee side, is larger than that on Hinge #1/Hinge #3 at the tested frequency range.

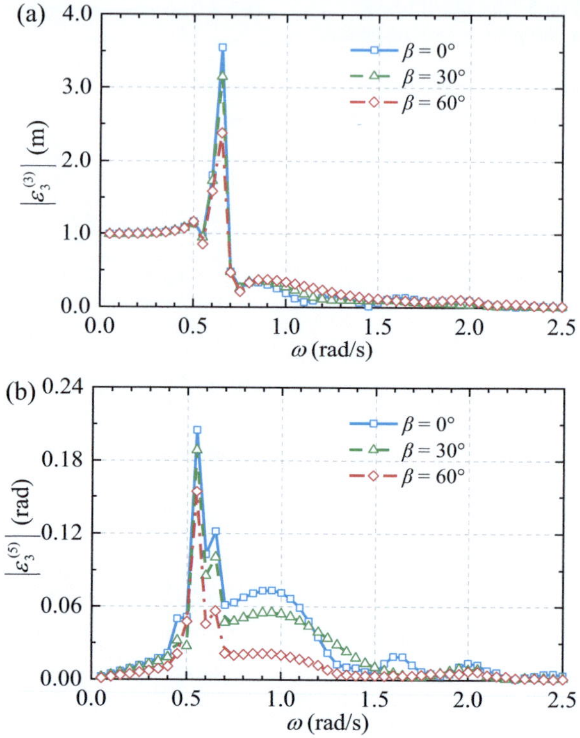

Fig. 6 Results of heave **a** and pitch **b** DoF responses of Module #3 in $\beta = 0°$, 30° and 60°. ($l = 3$ m)

Figure 9 describes the vertical forces acting on Hinge #1 (i.e., $\left|\lambda_1^{(3)}\right|$) and Hinge #2 (i.e., $\left|\lambda_2^{(3)}\right|$) for $l = 1$ m, 3 m and 5 m. A spike is found for both $\left|\lambda_1^{(3)}\right|$ and $\left|\lambda_2^{(3)}\right|$ at $\omega = 0.55$ rad/s for $l = 5$ m. However, such kind of spikes cannot be found for case of $l = 1$ m and 3 m. With the exception of the spike observed at the lower frequency range, it can be found that $\left|\lambda_1^{(3)}\right|$ and $\left|\lambda_2^{(3)}\right|$ at the middle and high-frequency range (i.e., 1.00 rad/s < ω < 2.50 rad/s) are larger than those observed at the lower frequency range (i.e., 0.05 rad/s < ω < 1.00 rad/s). It can be concluded from the calculations that medium and short waves may trigger a greater force acting on the connector.

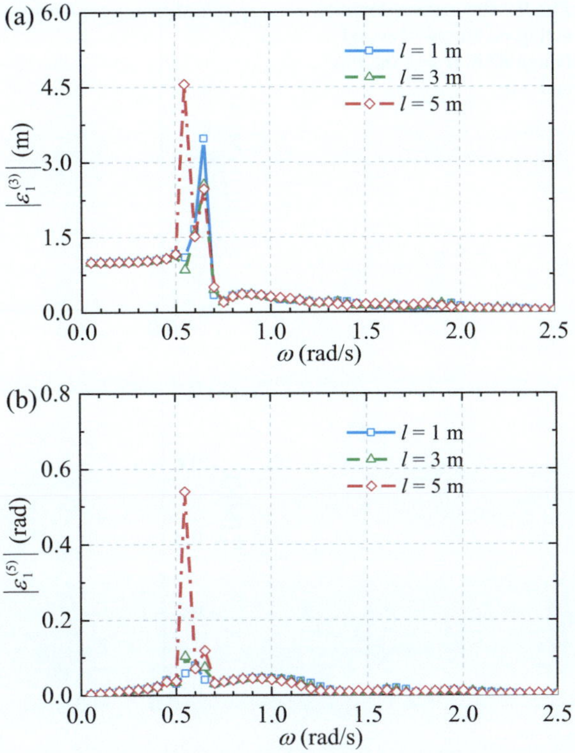

Fig. 7 Results of heave **a** and pitch **b** DoF responses for Module #1 in $l = 1$ m, 3 m and 5 m. ($\beta = 0°$)

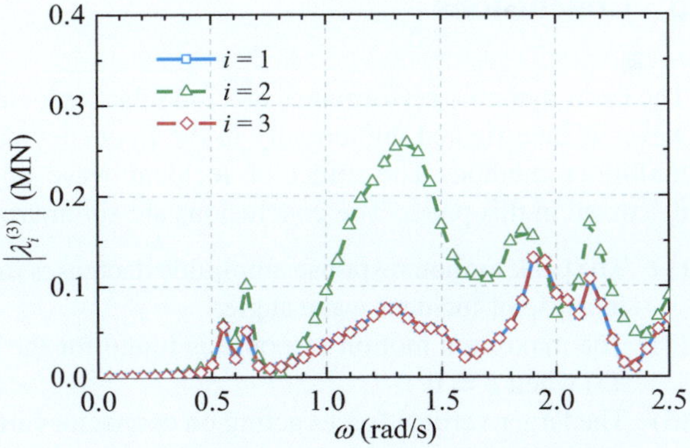

Fig. 8 Results of vertical forces acting on different connectors. ($\beta = 0°$, $l = 3$ m)

Fig. 9 The vertical forces acting on Hinge #1 **a** and Hinge #2 **b**. ($l = 1$ m, 3 m and 5 m)

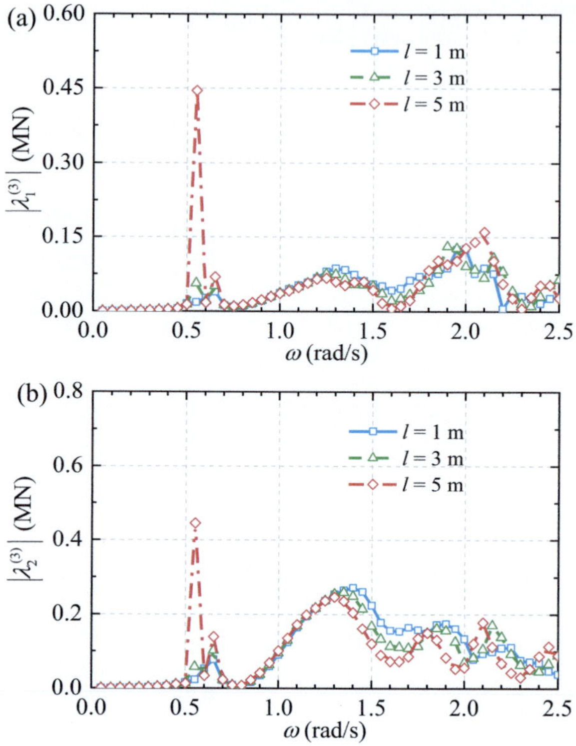

5 Conclusions

The hydrodynamic performance of a modular semi-submersible floating platform in waves is investigated numerically in the frequency domain based on the Lagrange multiplier method. The effect of incident wave angle and module spacing are discussed in this paper. The conclusions are summarized as follows:

(1) The peak motion response amplitude decreases in heave and pitch DoF with the increase of incident wave angle;
(2) The maximum motion response is found for the lee side module (i.e., Module #3) when $\beta = 0°$;
(3) The larger vertical forces acting on connectors are found at the middle and high frequency range.

Acknowledgements This work is supported by the National Key Research and Development Project of China (Grant No. 2023YFC3106904) and the Fundamental Research Funds for the Central Universities, China (Grant No. 3072023JC0101).

References

1. Wu Y (1984) Hydroelasticity of floating bodies. Brunel University, Thesis
2. Kiyokazu Y, Shoichi H (1998) On the hydroelastic response of box-shaped floating structure with shallow draft. J Soc Naval Archit Jpn 183:227–237
3. Watanabe E, Utsunomiya T, Wang CM, Hang LTT (2005) Benchmark hydroelastic responses of a circular VLFS under wave action. Eng Struct 28:423–430
4. Bispo IBS, Mohapatra SC, Guedes SC (2022) Numerical analysis of a moored very large floating structure composed by a set of hinged plates. Ocean Eng 253:110785
5. Du S, Ertekin RC (1991) Dynamic response analysis of a flexibly joined, multi-module very large floating structure. In: Proceedings of the institute of electrical and electronics engineers, pp 1286–1293. Honolulu, Hawaii, USA
6. Lee CH, Newman JN (2000) An assessment of hydroelasticity for very large hinged vessels. J Fluids Struct 14(7):957–970
7. Sun L, Taylor RE, Choo YS (2011) Responses of interconnected floating bodies. IES J Part A Civil Struct Eng 4(3):143–156
8. Jin S, Wang D, Hann M, Collins K, Conley D, Greaves D (2023) A designed two-body hinged raft wave energy converter: from experimental study to annual power prediction for the EMEC site using WEC-Sim. Ocean Eng 267:113286
9. Han Z, Jin S, Greaves D, Hann M, Shi H (2024) Study on the energy capture spectrum of a two-body hinged-raft wave energy converter. Energy, 132057
10. Babarit A, Delhommeau G (2015) Theoretical and numerical aspects of the open source BEM solver NEMOH. 11th European wave and tidal energy conference, Nantes, France

Policy, Social Acceptance and Legal Matters

Environmentalisms Clashing: Social Acceptance Problems of Large Floating Structures in Urban Development from Buckminster Fuller Until Today

Stefan Huebner

Abstract The Intergovernmental Panel on Climate Change (IPCC) recently evaluated large floating structures as an experimental yet noteworthy strategy for urban climate adaptation, advancing urban elements to the surfaces of aquatic bodies. In 2021, the Global Center on Adaptation, located in Rotterdam, Netherlands, moved to its floating office, exemplifying floating adaptation. Currently, UN-Habitat, the metropolitan government of Busan, South Korea, and the ocean technology company Oceanix are collaborating on a large floating settlement designed to adapt to sea level rise and enhance the city's flood resilience. Floating structures for urban development have emerged as a tool in the repertoire of technology-focused environmentalist thinking. Focusing on the question of social acceptance, this paper explores how different forms of environmentalist thought have either legitimized or delegitimized the use of floating structures as a form of urban development from Buckminster Fuller's Triton City in the 1960s to the present day. Based on examples from the 1960s–1970s and from the 2010s–2020s, this paper argues that floating structures, including residential ones, can serve as tools in an advance strategy in places where local conditions enable their installation as part of urban flood risk management and climate adaptation. However, conflicts between technology-focused and techno-critical environmentalist "schools" of thought, clashes over incompatible aesthetic perceptions of floating structure designs and urban waterfronts, and particularly right-libertarian views of floating structures as mobile means to establish new legal frameworks beyond governmental environmental legislation's reach continue to impact their political and consequently market and community acceptance. The success of floating structures in urban development therefore hinges not only on engineering and architectural considerations but also on situating them in environmentalist and socio-economic development strategies.

Keywords Urban development · Social acceptance · Environmentalism

---

S. Huebner (✉)
National University of Singapore, Singapore, Singapore
e-mail: arihust@nus.edu.sg

Society of Floating Solutions (Singapore), Singapore, Singapore

X. L. Zhao et al. (eds.), *Proceedings of the Fourth World Conference on Floating Solutions: WCFS 2024; 2–4 December; Hong Kong*, Lecture Notes in Civil Engineering 597, https://doi.org/10.1007/978-981-96-4569-5_87

1 Introduction

Buckminster Fuller (1895–1983) was a controversial US designer with multiple personal flaws [1], including the repeated theft of academic ideas and providing many wrong descriptions of events in his publications. These issues aside, he is also well-known for his *Triton City* design (1967–1969), which was the first high-tech floating settlement proposal that received at least initial governmental support. Based on an unrealistic, attention-seeking proposal for Tokyo Bay called *Tetrahedronal City* (1966), the strongly downscaled *Triton City* design, looking like a truncated tetrahedron and supposed to accommodate about 5,000 people per modular unit, received a research grant from the US Department of Housing and Urban Development (HUD) to create a design study and two models [2]. In November 1968, the US Navy confirmed the technical feasibility of *Triton City* [3]. However, no *Triton City* prototype was constructed, in contrast, for example, to Japanese architect Kikutake Kiyonori's *Aquapolis*, a floating settlement used to host the Japanese exhibition at the 1975 Ocean Exposition in Okinawa, very likely visited by several million visitors [4]. This paper asks why and how Fuller's floating settlement design failed to have an immediate impact. It argues that a serious lack of social acceptance, illustrated in multiple occasions, prevented any prototype construction. The main reason was the clash between Fuller's libertarian and techno-optimist form of environmentalism with a different form of environmentalism, focused on governmental reform and techno-criticism. Fuller and his countercultural adherents made some of the original contributions to what today is called "bright green environmentalism," whereas his opponents represented a nascent "dark green environmentalism."

2 "Bright Green" and "Dark Green" Environmentalisms

Currently, "bright green" and "dark green" environmentalisms, though no monolithic entities, represent two major facets of a broad spectrum of environment-related thought. Contrasting with dark green environmentalism, which distrusts technological responses to problems and views them as temporary fixes eventually leading to larger problems, bright green environmentalism embraces technology and innovation as key solutions to environmental crises. Dark green environmentalists, often Malthusian in their approach, advocate for austerity, world population reduction, and enforcing pollution limits nationally and globally through government-driven reform. In contrast, bright green environmentalists favor consumer choice to enable individuals to purchase tools to reduce their ecological footprint. Many adherents of dark green environmentalism see themselves on the left side of the political spectrum, critiquing capitalism's obsession with economic growth and the commodification of nature. Bright green environmentalists, on the other hand, often view themselves as pragmatic and libertarian-leaning, with less trust in centralized government. They endorse economic growth, as low-income countries will not allow people in the West to deny them their socio-economic development. Bright green environmentalists see economic security as foundational for adopting postmaterialist values like environmental concern. They also pragmatically support largely free markets but, unlike neoliberals, argue for the inclusion of ecosystem service values and pollution costs in pricing through eco-capitalist valuations, such as CO_2 pricing and carbon emission trading [5].

3 Libertarianism and Government-Driven Reform

US senator Gaylord Nelson (1916–2005), the co-founder of "Earth Day," and Fuller clashed over the role of government-driven reform versus Fuller's libertarian idea of an "outlaw area," meaning an unregulated space like outer space—or the ocean before the Law of the Sea conferences—in which designers could work without any restrictions but those provided by nature. For Fuller, the free competition of designs in this "outlaw area" served as a tool to create market choice-driven demand for the best designs being transferred to government-regulated "law areas," resulting in efficiency increases and ecological footprint reductions [6]. In contrast, Nelson's approach of government-driven reform aimed for national and global harmonization of law, meaning the closing of any "outlaw area". After the Santa Barbara oil spill of 1969, Nelson considered "floating cities" and other floating structures, like proposed floating airports, as serious sources of water pollution, if built in international waters beyond the reach of governmental legislation [7]. By the late 1960s, the United States, with a few federal state exceptions, claimed only a territorial sea of three nautical miles (about 5.56 km) and had *no* contiguous zone of limited control before changes in the 1970s and 1980s [8]. Moreover, multiple right-libertarian projects during the 1960s and early 1970s of building independent entities in international waters, all but one (the Principality of Sealand in the North Sea) unsuccessful, had illustrated to adherents of government-driven reform that territorial waters had to be extended, contributing to the creation of Exclusive Economic Zones (EEZs), usually extending 200 nautical miles (370 km) from coastlines [9]. The "dark green" lack of social acceptance for Fuller's design ideas and assigning environmental responsibility to individuals instead of the state, not the *Triton City* design itself, therefore was one the reasons for it not seeing any prototype implementation.

4 Aesthetics and Affordability

Liz Carpenter (1920–2010) and Fuller clashed over the aesthetics of urbanization and population-dense architectural forms whose scale was beyond that of the individual human. Carpenter was the press secretary to First Lady Lady Bird Johnson (1912–2007). Her heavy criticism of the *Triton City* design in May 1968 at a meeting with HUD and Fuller resulted in the withdrawal of HUD support in all but name and a delayed HUD press release regarding *Triton City* only in October 1968. Fuller's ecological design science subordinated aesthetics to affordability and the efficiency of geometric shapes like tetrahedrons. In contrast, Carpenter represented a terracentric focus on single-family homes in suburbs for the white US middle class, ignoring Fuller's focus on affordability for (often black) low-income families. Fuller, however, used the truncated tetrahedron shape to combine a layered apartment block design with the availability of garden space and sky view that otherwise was limited to single-family homes, as apartment blocks usually stacked balconies on top of each other and left no space for private gardens. Carpenter's dark green environmentalist ideas also rejected the construction of floating settlements in urban waterfronts, considering them to be a kind of pristine aquatic environment that should not undergo urban development. Fuller's focus was not on urban wetlands but on port spaces that due to containerization became available for new purposes, hardly representing any form of pristine environment [10]. In any case, the combination of dark green environmentalist and white US middle class aesthetics, which partially overlapped, were another illustration of the lack of social acceptance for a floating settlement.

5 Technology and Technocracy

A major clash between Fuller and public intellectual Lewis Mumford (1895–1990) illustrates the conflict over technology between bright green and dark green environmentalisms. Mumford, a historian of technology and urbanization, is not typically associated with dark green environmentalism, as his primary concerns were technological and political threats rather than environmental degradation. However, his views had noteworthy public outreach, which affected dark green environmentalist thought regarding technocracy and technology. In his 1970 book, *The Pentagon of Power*, Mumford linked Fuller with the most contentious notions of technocracy, indirectly stigmatizing the *Triton City* design. The book presented one of Fuller's photomontages, showcasing a land-based *Tetrahedronal City* in Shōriki Matsutarō's Yomiuriland amusement park in Tokyo, without specifying the (Disneyland-like) location and juxtaposed it with an image of the Giza pyramid complex. A subchapter titled "Autocratic Technocracy" was found beneath these images [11]. There, Mumford cautioned readers that technocracy and autocracy, just as in antiquity, were "twins". He elaborated on how architectural skills used for socially beneficial projects like irrigation canals and flood protection, through centralized-authoritarian decision-making could result in the destruction of cities, extermination of populations, or "mercilessly exploiting the mass of workers whose forced labor, disciplined to machinelike precision, made these feats possible" [12]. Mumford directly linked Fuller to such anti-humanistic forms of technocracy, certainly evoking images of the twentieth century's most devastating scientific and technological advancements, including atomic, biological, and chemical weapons. Fuller, calling himself a technocrat eager to supplant elected politicians with technological experts, but promoting libertarian ideas that strongly differed from the social visions of the technocracy movements in North America and Western Europe, and even more from those in communist-ruled China and the Soviet Union, therefore meant that the *Tetrahedronal City* idea became associated with an idea of technological advancements serving to oppress and control people, which reduced social acceptance. In addition to this problem, the limited flexibility of the much smaller *Triton City* design, incorporating a form of functional separation of space that limited the freedom of inhabitants by prescribing them where to live, where commercial and public buildings had to be located, etc., and reducing their ability to adjust their apartments to their own needs, further undermined social acceptance.

6 Racial Segregation

Other distinct factors, such as continued racism and interest in residential segregation in parts of US society, which Fuller's embryonic bright green architectural ideas intentionally challenged, also adversely affected plans to trial *Triton Cities* in Boston Bay as part of an unsuccessful proposal to host the 1976 World Exposition there. In the context of the 1960s Civil Rights movement, this complex conflict arose from affordable housing construction as an early bright green idea of social justice colliding with the lack of social acceptance among political leaders in parts of white Boston, who were inclined toward maintaining residential segregation. The World Exposition proposal that they successfully undermined foresaw the use of *Triton City* modules to host parts of the exposition in the waters off South Boston, which afterwards were to be turned into public housing for low-income, often black families, a plan that therefore would have challenged segregation in that area [13]. Fuller therefore faced a series of challenges that in some cases were rather unrelated to the floating *Triton City* design itself but challenged the environmentalist and social ideas that he hoped to implement with it. More recent floating settlement projects from the 2010s and 2020s illustrate significant changes, as inter-governmental institutions like the United Nations and the Intergovernmental Panel on Climate Change, but also the governments of the Maldives and Saudi Arabia, have begun to reconsider such ideas and given them some social acceptance in the context of transformative urban adaptation to climate change and sea level rise through an "advance" of urbanization to aquatic surfaces.

References

1. For a more detailed version of this paper, see my book manuscript titled "Earth's Amphibious Transformation"
2. Triton foundation, Inc. (ed) (1968) A study of a prototype floating community. Prepared for the Department of Housing and Urban Development. Triton Foundation, Inc., Cambridge, MA
3. Robert A, Frosch DH (1968) in M1090: R. Buckminster fuller papers, Series 2, Box 178, Folder 7, Stanford University Library

4. Huebner S (2020) Tackling climate change, air pollution, and ecosystem destruction: how US-Japanese Ocean industrialization and the metabolist movement's global legacy shaped environmental thought (c. 1950s–present). Environ Hist 25(1):35–61
5. Matthew C, Nisbet (2014) Disruptive ideas: public intellectuals and their arguments for action on climate change, WIREs climate change 5, no. 6: 809–823; Alex Steffen (2006) Worldchanging: A User's Guide for the 21st Century. Abrams, New York
6. Tomkins C (1966) Profiles in the Outlaw Area. The New Yorker 35–97. Zelov C, Cousineau P (1997) Design Outlaws on the Ecological Frontier. Knossus, Philadelphia
7. 91st Cong., 1st sess., Congressional Record 115 (29 October 1969): 32156–32159. The text was a reprint of Gaylord A. Nelson, Our Polluted Planet, The Progressive 33, 11 (1969): 13–17
8. Maritime Zones and Boundaries, https://www.noaa.gov/maritime-zones-and-boundaries. Accessed 23 Sept 2023
9. In the United States, such libertarian exit projects ended with United States v. Ray, 294 F. Supp. 532 (S.D. Fla. 1969); United States v. Ray, 423 F.2d 16 (5th Cir. 1970)
10. Oral History Transcript, Sharon Francis, Interview 3 (III), 6/27/1969, page 10, by Dorothy Pierce (McSweeny), LBJ Library Oral Histories, Lyndon B. Johnson Presidential Library, Austin, Texas, USA
11. Mumford L (1970) The myth of the machine, vol 2: The pentagon of power. Harcourt Brace Jovanovich, New York. Picture page 10 (between 179 and 180)
12. Mumford, The myth of the machine, vol 2, picture page 10
13. Šimková P (2021) Urban Archipelago. An environmental history of the Boston Harbor Islands. Massachusetts University Press, Amherst, pp 131–133; United States Bicentennial World Exposition Co. et al., ed., United States Bicentennial Exposition, Boston 1976. Technical Report, iv-12 and iv-15, in: Collection "Fairs Never Held," ref. no. AMD 2003, Special Collections Research Center, Henry Madden Library, California State University, Fresno, CA, USA

Is Land Reclamation, Ethical, Equitable or Defensible? A New Touchstone for Space at Sea

Soon Heng Lim

Abstract In recent decades there is a growing awareness that it is not only climate change, important as it is, that needs attention for the long-term viability of planet Earth. The damage to biodiversity is just an important an issue to address. The literature on ethical rights now covers the rights of all species including those in the air, on land and in the oceans. Those rights need to be protected. Creation of space to accommodate an expanding economy result in the degradation of the biosphere. This is especially true of land reclamation. Politicians, city planners and developers need a new benchmark to determine the viability of land reclamation. They need to walk the talk about the preservation of the marine eco-system. The government of Penang, Malaysia, turns a deaf year to the protests of 6000 fishermen when it decided to reclaim 1800 ha which would turn the rich fishing ground, and its vibrant marine ecosystem of corals, mollusks, seaweeds. The collateral damage to reclamation goes unreported. The livelihood of farmers and fishermen are compromised. The "pit stops" of migratory birds, and the places where sea turtles return to breed are destroyed heedlessly. The unwitting introduction of invasive species has the potential to cause damage far beyond the reclaimed area such as the aggressive beach vitex, the green crab, and the killer algae. Many unidentified species are lost through reclamation escape the attention of humanity and science. As the world population rockets to more than 8 billion, it is inevitable that more space is needed for humanity. Land reclamation cannot be the solution for a planet in crisis. It is unethical, inequitable and harmful to all living species including our own. The new touchstone for space creation is the use of floating structures.

Keywords Marine biodiversity and ecology · Ethical · Equitable · Sustainability · Floating structures

---

S. H. Lim (✉)
Society of Floating Solutions, Singapore (SFSS), Singapore, Singapore
e-mail: wcfs2019@gmail.com

X. L. Zhao et al. (eds.), *Proceedings of the Fourth World Conference on Floating Solutions: WCFS 2024; 2–4 December; Hong Kong*, Lecture Notes in Civil Engineering 597, https://doi.org/10.1007/978-981-96-4569-5_88

1 Introduction

According to a recent report 106 cities have reclaimed 2530 km^2 of land in the last two decades. This is an area equivalent to Luxembourg. As the threat of rising sea levels becomes imminent more resources are poured in to raise the elevation of the land and to protect the land-water interface from erosion by waves, currents and floods. Everybody knows this but decides to turn a blind eye just as we do about burning fossil fuels.

For more than 30 years the sinking of the runaways and terminal building at Kansai Airport has proved vividly the enormous challenges of using reclaimed land to create space. In Dubai the famed Palm Jumeirah is losing the battle against the forces of waves and current aggravated by its shape, which looks pretty but is a hydrodynamic disaster.

This paper will highlight other issues that should put the ethical dimension under the spotlight. The ethics of land reclamation has never featured in serious studies. The ethical consideration should address the impact on the rights and livelihood of coastal communities and the protection of biodiversity upon which all life depends.

It brings into focus why creating space on water using the principles of buoyancy is better for all creatures on earth.

2 Biodiversity, an Ethical Imperative

Biodiversity and climate change are inextricably linked. Plants and grass convert CO_2 into oxygen. Kelps and seaweeds stabilize the acidic levels in the ocean, protecting marine life by absorbing carbon dioxide using sunlight for photosynthesis.

Biodiversity is the building block of proteins and carbohydrates. Birds and insects pollinate our crops. Snails, slugs and worms decompose dead matter into nutrients for grains, vegetables, fruits and everything else the world puts on its dinner table. "Coral reefs only occupy 0.1% of the total ocean surface area–but a quarter of known marine life is dependent on them for survival."

"In a marine ecosystem, or any type of ecosystem, a keystone species is an organism that helps hold the system together. Some ecosystems might not be able to adapt to environmental changes if their keystone species disappeared. That could spell the end of the ecosystem, or it could allow an invasive species to take over and dramatically shift the ecosystem in a new direction" [1].

Food for humanity is possible only because we have birds and bees pollinating plants and earth worms and slugs enriching the soil. Bacteria consume deadly microplastics and increase the nitrogen content of soil. Corals are crucial to biodiversity and as such a vital resource for many where it is found.

As the most intelligent creature on the planet man is duty bound to protect biodiversity. Sadly, this is not the case. "Since 1970, there has been a staggering 68% decrease in population sizes of mammals, birds, amphibians, reptiles, and fish globally. … we could be headed into the sixth mass extinction" [2].

Wetlands, marshlands, mangroves are habitats of a rich diversity of creatures that is found nowhere on land. Unfortunately, they are often described as "wasteland" of low commercial value and are easy targets when developers seek to reclaim land for sale. In pursuit of GDP growth, governments have ignored ethics and the need to treat man and nature with respect.

In the name of economic progress, trees that take decades to mature are bulldozed, wetlands are dewatered. Millions of tons of fill material are dumped into to reclaim land heedless of the symbiotic relationship between man and marine ecology.

"The collapse in the population of many migratory birds is an indicator of the rapid loss of biodiversity and particularly wetlands, which not only sustain these remarkable species but are also our life support systems," according to the Asian Flyways Initiative (AFI).

The UN has promulgated 17 Sustainable Development Goals. Goal 15 urges us to manage forests, combat desertification, and halt and reverse land degradation to stop biodiversity loss. Goal 14 obliges us to conserve the sustainable use of the oceans, seas and marine resources.

3 Land Reclamation, a Morally and Unstainable Practice

The extent of the problem can be gleaned from a 2015 study, which finds more than 7500 km^2 of coastal land was reclaimed in China between 1985 and 2010. Another study finds that the world has reclaimed more than 2500 km^2 of coastal land since 2000. Such aggressive reclamation is unlikely to decelerate even as the UN has declared a red line for the export of sand [3].

Land reclamation is the maritime version of deforestation of the rain forest. In a way it is worse, as the probability of the recovery of deforested vegetation is higher than a seabed destroyed by land fill and sediments.

"Marine sand is the first line of defense against sea level rises, which, if the predictions of scientists are correct, will soon be inundating Malaysia's and Indonesia's low-lying shores, putting its people and property at high risk and incurring significant economic losses," Yet the two countries are planning to lift their self-imposed ban on sand export [4].

We highlight the following challenges of land reclamation. Investors are often inadequately briefed by their engineers who prefer not to venture to something they have not done before. The following are some of the possible pitfalls which can be avoided:

a. Land subsidence is unpredictable as we see in Kansai Airport which continues to sink after 30 years.
b. With more frequent storms, and floods, it would be more costly to protect reclaimed land.

c. Sediment transport also known as sediment flux kill marine life well beyond the reclaimed area.
d. Sink holes (blue holes) and soil liquefaction can and do occur in the seabed below the reclaimed land.
e. Reclaimed lands are vulnerable to earthquakes and tsunamis.
f. Reclaimed land needs years to consolidate and that increases the financing cost.
g. Reclaimed land is a poor heat sink. It aggravates rather than mitigates the Urban Heat Island effect.
h. Sand export are from time to time banned by successive governments, causing the security of its supply chain to be vulnerable.
i. Importation of invasive marine species that could result in killing native ecosystems.

The amount of material that needs to be imported is staggering. The Kansai Airport consumes "189 million cubic meters of rock and sand," and its runways and terminal buildings are still sinking after 30 years. The damage to the marine ecology boggles the mind. Kansai would have bankrupted a country like North Korea or even the Philippines.

The principles of buoyancy offer a more biodiversity friendly and climate resilient solution. It offers new pathways to space creation which avoids all the above shortcomings and offer employment opportunities to coastal communities.

3.1 Roads, Drains Sewage and Utilities for a Floating City

The city of Venice (est. in 451 AD, current pop. 261,000) is not a floating city but provides many interesting features which reflects what a floating city could look like. For starters, all roads and bridges are for pedestrians. People are healthier because of the need to walk. They have gondolas and vaporetto if you need to get around in a hurry. The air is less polluted as waterborne transport is more fuel efficient.

There are no drains. All rain water falls on the surface and find its way into the sea. The sewage system is centuries old but it works. In modern times there are better solutions using on board treatment systems. Venice has piped water supply and electrical network. Today floating solar and wind turbine farms may be used.

Electricity can be supplied by undersea cables from existing land-based networks. However, micro SMRs (small modular reactors) are a possible alternative. China and Russia already have them. They may be mounted on a pontoon and can be deployed anywhere. The risk of a meltdown of a floating nuclear reactor is near zero since there is always water around the reactor to remove any decay heat in an unplanned event.

4 Floating Energy, Food Farms and More

At OMAE 2024 held in June in Singapore, it was inspiring to see academics, industry leaders and young researches presenting concepts based on the principles of floatation. These include floating fish farms, floating seaweed farms, and of course wind and solar farms out at sea. This awakening to the fact that the sea is the planets biggest resource for food and energy. It is accessible not by reclamation, which is fortunate, but by using the principle of buoyancy. 70% of the planet is covered by water and it is rightly so that the event generated so much interest.

Floating structures do not degrade coastlines caused by dredging and mining filled material. As they float, they do not damage the marine ecology and biodiversity. As no foreign material is imported the risk of depositing invasive alien species from another part of the world is practically eliminated.

The construction of floating structures required by wealthy countries can be performed in poorer countries thereby providing employment where it is needed. An example is a floating concrete dock designed in the US for a client in Hawaii which was constructed in Batam, Indonesia. Upon completion it was transported across the Pacific Ocean.

At the OMAE 2023 Conference in Melbourne I presented a paper titled "Floating Nuclear Power Plants, a safe way out of the Energy and Climate Conundrum." It is a concept that would enable small countries such as Singapore to use nuclear energy without the fear of radioactive pollution.

Food security is a major concern in the event of a supply chain disruption such as during the Covid-19 pandemic. Singapore put in place a policy to be 30% self-sufficient in food by 2030. It is now clear that the policy faces serious challenges. Several farm startups, in spite of governmental support, could not achieve commercial viability due to the high cost of land in the city state. A viable solution would be to build floating farms that may be deployed to areas where production cost is significantly lower e.g. Indonesia or Malaysia. Being deployable such farms can be towed back to and moored in Singapore waters when sea and air freights grind to a halt as it did in 2020. Such farms may be built on concrete platforms for durability and stability with water desalination and storage capacity below deck. Solar energy is used to desalinate water.

With floating power plants, refueling is possible out at sea alleviating the need for berths and storage tanks. The sea provides the sink needed to dissipate waste heat alleviating the need for cooling towers and long discharge culverts.

Other massive structures that can benefit from the principles of floatation are cruise centers, ports, airport runways, water parks and swimming pools. Organisations which are considering/promoting such ideas include MARIN/Blue21 (Netherlands), MEYER (Finland), NEOM (Saudi Arabia), Oceanix Busan (South Korea) and Shimizu (Japan). Images of the concepts are on the Internet. It is worth noting that all these countries were major land reclaimers. Their effort to transition to floatable structures is good for the environment and commendable.

4.1 Resources to Build Floating Structures

The infrastructure to build floating structures are simple and readily available. A drydock would be nice but prohibitively expensive. The 165,000 ton floating berth in Monaco was built literally in a hole in the ground which was flooded when the structure was ready to launch. Hundreds of caissons for Singapore's port was built under a simple shed. The caisson is launched by skidding into the water with the aid of a floating dock, a procedure that takes minutes.

The skills required to build a concrete floating structure are easily acquired. Primarily it involves the cutting and fixing of steel rebars. The concrete batching and pouring requires higher level of supervision. Highly experience quality control staff are deployed to the job site to ensure specifications are met. Owners may also engage third party quality inspectors.

There are several reasons why reinforced concrete is a better option (than steel) for the part of the floating structure than is immersed in water. Ease of construction by semi-skilled work force is a major consideration. The superstructure may be manufactured with a lighter material such as steel, aluminium or timber.

5 Ethics and Equity

When a first world nation expands their territorial land by simply acquiring the necessary fill material from a third world nation, the act is hardly any different from an unlawful acquisition of sovereign land from the latter. One may argue that it is commercial transaction, but it is not. Loss of biodiversity is not quantifiable and its impact stretches across generations.

"Biodiversity is our life-support system. Access to biodiversity's benefits is a human rights issue. And communities of color and low-income communities suffer disproportionately due to biodiversity loss, sometimes in a targeted manner. Yet the mainstream environmental movement often does not articulate how environmental racism drives unequal access to biodiversity's benefits" [5].

Even if the fill material is mined from within the state instead of imported, there are ethical issues to consider. A Malaysian state not long ago bulldozed its way against protest from its citizens to reclaim a heritage fishing ground. The proposed reclamation was originally to spread over 4500 acres, but subsequently reduced to 2300. Nevertheless, it is a still a severe blow to the marginalized coastal communities and a biodiversity disaster.

The possibility of building floating large concrete structures in countries with limited engineering skill has been proven in the delivery of the floating concrete dock by workers in Batam to an owner in Hawaii. First world countries may source floating structures in third world countries. This would be more ethical than mining sand as it would generate income for coastal communities and preserve biodiversity. It is an equitable solution as both parties benefit from such a relationship. It is more defensible against criticism as it measures favorably when benchmarked against the Sustainable Development Goals of the United Nations which was adopted by all United Nations Member States in 2015, to provide "a shared blueprint for peace and prosperity for people and the planet, now and into the future".

6 Conclusion

The transition from land reclamation to the use of mega floats to create space for development has many advantages and is compatible with the UN Sustainable Development Goals. Mega floats are suitable to support any kind of development are founded on reclaimed land. They are safe sustainably designed and does minimal damage to the adjacent ecosystem.

Biodiversity is as important as climate and the damage to one impacts the other. The world has seen five mass extinction before and environmentalists are warning that anthropogenic degradation of biodiversity is driving the sixth. There are multiple causes for the degradation massive land reclamation is one which is particularly severe for coastal wetlands and coral reefs.

If we are serious about maintaining a vibrant marine bio-diversity and protecting the Blue Planet for humanity we need to protect our seas and oceans, our wetlands and corals from the onslaught of developers. We need to acquaint them with the fact there are other more sustainable and more ethical and equitable ways to balance the demand for space and the protection of biodiversity for the good of humanity. This as important an issue as the use of clean energy to stop global warming.

References

1. National geographic education keystone species. https://education.nationalgeographic.org/resource/keystone-species. Last accessed 20 Aug 2024
2. Joe's bloom biodiversity collapse and the sixth mass extinction phase. https://www.joesblooms.com/biodiversity-collapse-and-the-sixth-mass-extinction-phase. Last accessed 20 Aug 2024
3. Reclamation a flawed solution. https://www.beneaththesands.earth/reclamation. Last accessed 20 Aug 2024
4. Why the UN should regulate Singapore's dangerous sand imports. https://thediplomat.com/2023/06/why-the-un-should-regulate-singapores-dangerous-sand-imports. Last accessed 20 Aug 2024
5. Justly biodiverse: nature's lifeline for the health of all communities. https://www.endangered.org/justly-biodiverse/. Last accessed 20 Aug 2024

Floating Cities and Buoyant Urbanism in International Law: Inundation, Territory and *Erga Omnes* Obligations

Brydon T. Wang

Abstract The increasing vulnerability of Small Island Developing States (SIDS), and in particular low-lying South Pacific nations, due to climate change necessitates innovative and sustainable adaptation strategies. One such strategy is the development of floating cities, which offers the promise of mitigating the effects of rising sea levels and the incursion of water in urban centres, while ensuring the continued socio-economic stability in the region. This paper argues that the creation and enforcement of legal frameworks supporting floating cities needs to be examined at international law. While steps have been taken to fix maritime baselines to set out the territorial waters of island states subjected to inundation, the legal concept of Statehood in international law requires us to address the fundamental question of a 'defined territory' and how the ambit of the definition of 'territory' could be addressed by floating structures. This paper argues that the recognition of floating structures as 'territory' is crucial as it allow states to avoid costly and significantly delayed responses through land reclamation. To aid this recognition of floating structures as 'territory', this paper argues that such recognition should be treated as an *erga omnes* obligation—an obligation owed by states to the international community as a whole. This paper examines the international legal precedents, proposed legal frameworks, and the technical solution of floating cities to argue for the legal necessity of such urban interventions. It concludes by asserting that the recognition of legal support for floating cities as an *erga omnes* obligation is not only a legal necessity but also a moral imperative to ensure the survival and prosperity of vulnerable populations in the face of global climate change.

Keywords Floating cites · International law · Statehood · Erga Omnes

B. T. Wang (✉)
Queensland University of Technology, Brisbane, QLD, Australia
e-mail: Brydon.Wang@qut.edu.au

X. L. Zhao et al. (eds.), *Proceedings of the Fourth World Conference on Floating Solutions: WCFS 2024; 2–4 December; Hong Kong*, Lecture Notes in Civil Engineering 597, https://doi.org/10.1007/978-981-96-4569-5_89

1 The Pressing Need for Legal Recognition of Floating Settlements

Scientific projections suggest that low-lying island states could become uninhabitable within the next few decades due to rising sea levels, which threaten to submerge significant portions of their landmass [1, 2]. The South Pacific region is acutely vulnerable to the adverse effects of climate change, including sea-level rise, extreme weather events, and shifts in ecological systems [3]. These changes in rainfall patterns and ocean temperatures are disrupting local ecosystems and food security, significantly threatening the socio-economic stability and cultural heritage of the nations in this region. The problem is compounded by the 'saltwater intrusion into rivers and underground aquifers' that are tied to the negative effects of sea level rise [4]. As such, these landmasses will either be eventually submerged or be rendered uninhabitable [5].

According to the most recent 2023 IPCC *Synthesis Report*, climate change has already led to increased frequency and intensity of severe weather events, resulting in devastating impacts on vulnerable communities globally [6]. The report highlights that some regions are nearing the limits of their adaptive capacity, making urgent and innovative 'climate resilient development action' crucial. Accordingly, these environmental challenges compound existing socio-economic issues, exacerbating poverty and threatening the livelihoods of island populations. As viable options to continue remaining on increasingly threatened landmasses are exhausted, communities are being displaced.

Significant amounts of tax dollars have been spent to address in-situ risk reduction and hazard prevention strategies. For example, the low-lying atoll nation of the Maldives sought to construct an artificial island *Hulhumanle* and a six-foot tall seawall, the *Great Wall of Male*, to protect the Maldivian population from submergence [7, 8]. However these prevention strategies along with subsequent emergency disaster responses are no longer sufficient to hold at bay climate-driven erosion, coastal flooding and the ongoing incursion of water in communities. While some proponents have suggested that the 'only feasible solution to permanently protect the inhabitants of these communities' is to accept climate-induced community relocations [5], the concept of floating cities, or buoyant urbanism, emerges as a promising adaptation strategy.

These innovative urban structures offer a proactive alternative to the displacement of these vulnerable coastal populations by allowing communities to preserve their socio-economic and cultural links to the existing footprint of their urban environments while adapting to new environmental realities through resilient and sustainable floating environments [9]. These floating solutions align with the emphasis within the *United Nations Environment Programme* (UNEP) on prioritising ecosystem-based adaptation measures, which can provide co-benefits such as improved biodiversity, health outcomes, and economic resilience.

However, under customary international law, the concept of Statehood is traditionally linked to the possession of a 'defined territory'. The 1933 *Montevideo Convention on the Rights and Duties of States* outlines four criteria for statehood: a permanent population, a 'defined territory', government, and the capacity to enter into relations with other states. The first three criteria (including the criteria of a defined territory) predate the Montevideo Convention and were articulated by the tribunal in the *Deutsche Continental Gas Gesellschaft* case [10].

Accordingly, the impending loss of landmass (or 'territorial extinction' [8]) due to sea level rises challenges the applicability of these four criteria to island states facing inundation and has a wider impact on the maritime zones for SIDS. Under the *United Nations Convention on the Law of the Sea* (UNCLOS), territorial seas (up to 12 nautical miles from the maritime baseline), contiguous zones (up to 24 nautical miles), exclusive economic zones (EEZs) (up to 200 nautical miles), and the continental shelves are all calculated from maritime baselines. These baselines are ambulatory boundaries; that is, where sea level rises, the low-water line would shift inland (that is, a 'landward shift' of the coastline) and cause the surrounding maritime zones to shrink [11]. For SIDS, these zones are critical. The United Nations observes that for SIDS, the EEZ is 'on average, 28 times the country's land mass. Thus, for many SIDS the majority of the natural resources they have access to comes from the ocean' [12]. Accordingly, the ability for these island states to access and thrive from ocean development and the blue economy is deeply entwined with their maritime boundaries.

Armstrong and Corbett observe that four options arise at international law in relation to these maritime baselines: first, to allow these baselines to remain ambulatory; that is, to allow baselines to shift with the low water mark, meaning that SIDS will cease to have any baselines where their landmass is fully submerged. Second, to use 'straight baselines' that will enable states with low-lying delta regions to maintain current maritime boundaries for where their coastlines are partially submerged. However, under this second option, maritime boundaries are still ambulatory and thus this solution does not aid low-lying SIDS. Third, to affix current maritime baselines, allowing the maritime zones of SIDS to expand to fill in partially submerged areas of current landmass. A number of South Pacific Island States have adopted domestic legislation 'purporting to 'fix' their maritime limits' while engaging in international advocacy though the *Pacific Maritime Boundaries Project.* Collectively, [SIDS in the South Pacific] have advocated for or asserted a right to fix baselines in public statements such as the 2010 *Framework for a Pacific Oceanscape*, the 2015 *Taputapuatea Declaration on Climate Change* and the 2018 *Delap Commitment on Security on Common Wealth of Oceans*' [13]. In August 2021, the Pacific Islands Forum Leaders issued the *Declaration on Preserving Maritime Zones in the Face of Climate Change-related Sea-Level Rise* [11]. Fourth, to affix current maritime baselines to the degree where the land mass is fully submerged. In doing so, the deterritorialised SIDS retain their EEZs and other maritime zones. However, Armstrong and Corbett note that there is still no agreement on which of these options 'is the most legitimate and justifiable choice [8].

This paper argues that provided floating structures are located within the affixed maritime baselines—such that they float over the inundated original landmass of a nation state—there should be development and implementation of legal frameworks at international law to support the recognition of floating cities as 'defined territory' under the Montevideo definition of 'Statehood'. Such recognition of territory is positioned as an *erga omnes* obligation—an obligation owed by states to the international community as a whole—that then allows SIDS to preserve their EEZs even when their landmasses are submerged.

This paper begins with an introduction to how international law has been engaging with the inundation of island states and the threat to their statehood and sovereignty. It then proceeds to examine the status of floating structures in international law before discussing the legal concept of *erga omnes* duties and its relevance to recognising territory in floating structures. The severe and immediate threats faced by low-lying SIDS, and South Pacific nations in particular, necessitate a collective international response, underscoring the moral and legal imperative for global solidarity and support.

2 International Law Approaches to Aid Island States Facing Inundation

While a nation state still has land territory—regardless of degree of loss through coastal erosion or inundation from sea level rises—it continues to possess territory and 'there can be no challenge to its statehood' [4]. However, where a state loses its entire land territory through inundation, this has profound implications at international law as its statehood can be extinguished. International law gives primacy to states as regulated subjects. Willcox observes that '[t]he destruction or disappearance of a state without an immediate successor—and its implications for statehood, sovereignty, self-determination, and the protection of basic human rights—represents an unprecedented challenge to the international community and contemporary human rights framework' [14]. Likewise, the primary sources of international law—treaties and customary international law—can only be created by states, with United Nations membership [15] and the right to maritime zones under UNCLOS inherently tied to a coastal nation's statehood.

The *International Law Commission* (ILC) initiated work on the topic of sea-level rise in relation to international law in 2018, including its long-term programme of work. In 2019, the ILC incorporated the topic into its formal programme and established an open-ended Study Group that has since produced several key issues papers [11]. The first issues paper, discussed in 2021, addressed matters related to the law of the sea. This was followed with a supplementary paper in 2023. In 2022, the ILC reviewed the second issues paper, focusing on statehood and the protection of individuals affected by sea-level rise. Following further work on statehood and the protection of person, a memorandum highlighting relevant texts and final reports from the Study Group was produced that illustrated the ILC's approach to addressing the challenges posed by sea-level rise to SIDS [11].

There are exceptions to the need for 'defined territory' in the form of legal personality given to international organisations as subjects of international law. This mechanism recognises that groups of people have rights to self-determination and attract human rights obligations. Legal scholars and international bodies have argued that the principle of state continuity, which allows a state to retain its legal personality despite changes in government or territory, could be extended to states that lose their territory due to environmental changes. Thus, the emergent concept of 'de-territorialised statehood' has been proposed as a potential solution, where a state could maintain its sovereignty and international rights despite the physical disappearance of its land territory [16]. This could involve the relocation of government functions to an offshore territory or perhaps even the use of floating structures to maintain a physical presence within the boundaries of the submerged land mass.

However, the practical and legal implications of a de-territorialised approach to Statehood remains unclear, and there is a need for further legal innovation and consensus within the international community. For example, the obligations of states under the *International Covenant on Civil and Political Rights* (ICCPR) are tied to those 'within its ***territory*** and subject to its jurisdiction' [emphasis added]. Similarly, despite the potential solution of floating structures, there remains the question of how maritime zones are to be calculated from maritime baselines set around an inundated land mass. The solution of the affixation of maritime baselines raises complex legal questions, particularly regarding its consistency with the existing provisions of UNCLOS [11]. While some argue that such a measure is necessary to protect the rights of vulnerable states, others caution that it could set a precedent that undermines the stability and predictability of the international legal order [11]. Although affixation of maritime boundaries has gained some support, it remains controversial and has not yet been codified in international law, leaving the future of maritime rights in a changing climate uncertain. Despite these challenges, the issue is gaining increasing attention in international forums, and there is a growing recognition of the need to adapt international law to address the unique challenges posed by climate change.

Given the unprecedented challenges posed by climate change, there is a growing recognition of the need for legal innovation to address the specific needs of island states and the question of 'territory'. Proposals include the recognition of new forms of territory, such as territory demarcated by buoyant urbanism (the agglomeration of floating structures) or artificial islands, as a means of preserving statehood and sovereignty for nations facing the loss of their land. The next section considers the legal status of floating cities in international law.

3 Legal Status of Floating Structures in International Law

Floating cities are essentially large, floating pontoon or semi-submersible platforms designed to support permanent human settlements, complete with residential, commercial, and recreational spaces [17]. They are envisaged to be self-sufficient, incorporating renewable energy sources, waste management systems, and sustainable food production to reduce their environmental impact. Floating cities are designed to be environmentally sustainable, often incorporating renewable energy systems, waste recycling, and sustainable food production to minimise their ecological footprint [9]. These cities can serve as models for sustainable urban development, demonstrating how human habitation can be adapted to environmental changes without exacerbating the underlying problems. By integrating these principles, floating cities can contribute to the achievement of several *Sustainable Development Goals* (SDGs), including SDG 11 (Sustainable Cities and Communities), SDG 13 (Climate Action), and SDG 14 (Life below Water).

In contrast, artificial islands are man-made landforms constructed in bodies of water, often by piling sand or other materials to create new land above the waterline. Unlike floating cities, which are buoyant and can move with the tides, artificial islands are fixed structures attached to the seafloor (that is, 'bottom-founded') [18]. As artificial islands are static in nature, they do not offer the same benefits as floating structures that can keep a pace with rising sea levels and float above any incremental change in sea levels. They also take a long time to form and settle before they can be used to support urban development.

Floating cities represent a significant shift in how we conceptualise habitation in response to environmental changes [19]. The notion of buoyant urbanism challenges the traditional view of land as the primary and permanent basis for human settlement, pushing the boundaries of urban planning and international law [20]. Under UNCLOS, the legal status of artificial islands and floating structures is addressed in a limited capacity. Article 11 of UNCLOS excludes 'offshore artificial islands, installations and structures (including floating structures) from the calculation of the low-water line' that is used to calculate maritime zones [17]. Accordingly, artificial islands and floating cities do not generate territorial seas or EEZs of their own, and their existence cannot alter the baselines from which a state's maritime zones are measured. Moreover, UNCLOS does not provide a comprehensive framework for the governance or recognition of floating cities, which are inherently mobile and may exist beyond the territorial jurisdiction of any single state [17]. This omission leaves questions about their status in international law, particularly in areas beyond national jurisdiction, where no state has sovereignty. As such, for low-lying islands under threat of inundation, the adoption of floating structure technology as a way of retaining habitation over traditional waters or over their submerging land mass will not allow them to retain their claim over their territorial waters without some adjustments at international law, such as the affixation of current maritime baselines. This need to interpret UNCLOS 'in light of changing circumstances and/or taking into account the interests of States affected by sea-level rise' has been discussed at some length in the Study Group and the ILC during its seventy-second session in 2021 [11].

4 *Erga Omnes* Duties

International environmental law has increasingly recognised the need for a collective response to global environmental challenges, with certain obligations now being classified as *erga omnes*, or obligations owed by all states to the international community as a whole [21]. The protection of the environment, particularly in the context of climate change, is emerging as an *erga omnes* obligation due to the universal impact of environmental degradation and the shared responsibility of states to mitigate and adapt to these changes.

The concept of *erga omnes* obligations in international law refers to duties that states owe to the international community as a whole. *Erga omnes* obligations encompass norms such as the prohibition of genocide, slavery, and racial discrimination. These norms have a higher status in international law, often overlapping with *jus cogens* norms (fundamental principles accepted by the international community that no state is permitted to derogate from). The enforcement of *erga omnes* obligations allows any state to invoke the responsibility of another state for breaches, even if it is not directly affected by the violation, and has been considered for populations that may lose their statehood due to inundation [14, 22]. Accordingly, the principle of *erga omnes* obligations as adopted into international environmental law reinforces the moral and legal duty of all states to support measures that mitigate global threats and protect vulnerable populations. This collective responsibility ensures that no single nation bears the burden of global climate impacts alone, fostering international solidarity and cooperation.

The *International Court of Justice* (ICJ) has recognised certain environmental obligations as being of an *erga omnes* nature, particularly those related to the preservation of the global environment. For example, in the Advisory Opinion on the *Legality of the Threat or Use of Nuclear Weapons* (1996), the ICJ underscored that the environment is a concern of the entire international community, highlighting the duty of states to ensure that activities within their jurisdiction or control do not cause environmental harm beyond their borders (or the 'entire ecosystem of the planet') [23, 24].

This paper argues that the recognition of floating structures as 'territory' in support of statehood should be constructed as an *erga omnes* obligation. Such structures can be 'defined territory' as they would be a 'physical area where sovereign powers can be exercised over a permanent population' [25]. This recognition would be limited to states (not just SIDS) that, where fully inundated, would have their land territory extinguished and potentially lose their maritime zones. This recognition of territory in support of statehood recognition is a fundamental human right intimately tied with the right to nationality. It is supplemented by the right to remain connected to traditional land territory and waters, and is so fundamental to SIDS communities that there are clear examples where vulnerable populations, such as the Tuvaluans, have repeatedly rejected the idea of climate-induced relocation and resettlement (in 2008 and 2010) [22].

The right to nationality, as a fundamental human right, could be argued to fall within the scope of obligations *erga omnes*, particularly in the context of preventing statelessness. The ICJ's dictum in the *Barcelona Traction* case emphasised that the protection of basic human rights, which includes the right to nationality, engages the legal interest of all states [26]. Three arguments are advanced to support this. First, given the potential for entire populations to be rendered stateless due to the effects of climate change, it is arguable that the international community has a collective responsibility to ensure the preservation of nationality and prevent the human rights violations associated with statelessness. This impacts on vulnerable populations and their ability to fully enjoy their right to self-determination—a fundamental right that is captured in the International Bill of Rights as well as the UN Charter, requiring the 'respect for the principle of equal rights and self-determination of peoples' [15]. This is further supported by the principle of non-discrimination, which underpins many human rights obligations recognised as *erga omnes*. Statelessness often leads to discriminatory treatment, loss of access to health care, adequate standard of living, freedom of movement, as stateless individuals are denied equal access to rights and services available to citizens of states. The prevention of statelessness, therefore, is not only a matter of protecting the right to nationality and self-determination but also of upholding the principle of non-discrimination, which is essential to the broader protection of human rights under international law.

Second, by recognising floating cities as defined territory, the international community would be taking a concrete step towards ensuring that the rights of those displaced by climate change are upheld, thereby fulfilling its common *erga omnes* obligations. The principle of collective responsibility is central to addressing the severe and immediate threats posed by climate change, particularly to South Pacific nations. These nations are disproportionately affected by climate impacts despite contributing minimally to global greenhouse gas emissions. The urgency of their situation necessitates a concerted international response, ensuring that no single nation bears the burden of global climate impacts alone. Recognising the development and support of floating cities as an *erga omnes* obligation aligns with the moral and legal duty of all states to mitigate global threats and protect vulnerable populations. The collaborative efforts required to implement and sustain floating cities highlight the necessity of international solidarity and shared responsibility.

Third, floating cities are a proactive and innovative adaptation strategy that may address the multifaceted challenges posed by climate change, particularly in vulnerable regions like the South Pacific. Floating cities hold global significance as a model for climate resilience [9]. They represent a forward-thinking approach to urban adaptation, designed to withstand rising sea levels and extreme weather events, and may even offer ways of increasing marine habitats and aid with carbon capture [27]. By showcasing how human habitats can be reimagined to coexist with changing environmental conditions, floating cities offer a template that can be adapted and replicated worldwide. This universality transcends national borders, making their success a matter of global concern and interest. As Tams elucidates, *erga omnes* obligations are those that protect interests fundamental to the international community [21], underscoring the collective benefit derived from pioneering climate adaptation measures like floating cities. As the IPCC emphasises the critical role of ecosystem-based adaptation strategies in building resilience and providing co-benefits such as improved health outcomes and economic resilience, there is a pressing need to integrate sustainable construction technologies and buoyant urbanism to allow floating cities to alleviate pressures on coastal lands, promote sustainable living practices, and enhance biodiversity conservation efforts [9].

5 Conclusion

Recognising floating structures and floating cities as part of a state's territory requires a reinterpretation of existing international law, particularly UNCLOS. While UNCLOS currently does not grant territorial status to floating structures, there is scope for legal adaptation to accommodate the unique challenges posed by climate change and the threat of inundation that SIDS face with repeated sea level rises. As mentioned above, such recognition would be framed as an obligation *erga omnes*, limited to floating cities deployed within the existing maritime boundaries of SIDS where they are installed above existing submerged land territory. This approach reflects the need for collective responsibility of the international community to prevent statelessness and protect the rights of affected populations. Moreover, this approach could serve as a precedent for addressing other emerging challenges related to climate change and human rights. As environmental changes continue to reshape the global landscape, the international community will increasingly face the need to adapt its legal frameworks to protect vulnerable populations. The recognition of floating structures as sovereign territories represents a proactive and forward-thinking solution that not only addresses the impending risk of statelessness but also sets the stage for broader legal innovations in response to climate change.

References

1. Intergovernmental Panel on Climate Change (IPCC) (1996) Climate change 1995. In: Watson RT, Zinyowera MC, Moss RH (eds) Impacts, adaptations and mitigation of climate change: scientific-technical analysis. Cambridge University Press, Cambridge
2. IPCC, IPCC Fourth Assessment Report (AR4) (2007) Climate change 2007—synthesis report
3. Mimura N (1999) The vulnerability of island countries in the South Pacific to sea level rise and climate change. Clim Res 12:137–138
4. Jain AG (2014) The 21st century Atlantis: the international law of statehood and climate change-induced loss of territory. Stanford J Int Law 50:1–27, 6
5. Bronen R (2011) Climate-induced community relocations: creating an adaptive governance framework based in human rights doctrine. New York Univ Rev Law Soc Change 35:357
6. IPCC (2023) Summary for policy makers. In Lee H, Romero J (eds) Climate change 2023: synthesis report. IPCC, Geneva, Switzerland, pp 1–34
7. Gagain M (2012) Climate change, sea level rise, and artificial islands: saving the Maldives' statehood and maritime claims through the constitution of the oceans. Colorado J Int Environ Law Policy 23:77

8. Armstrong C, Corbett J (2020) Climate change, sea level rise, and maritime baselines: responding to the plight of low-lying atoll states. Glob Environ Polit 21(1):89
9. Wang CM, Wang BT (2020) Floating solutions for challenges facing humanity. In: ICSCEA 2019: Proceedings of the international conference on sustainable civil engineering and architecture, pp 3–29
10. Germano-Polish Mixed Arb. Trib., Deutsche continental gas Gesellschaft v Polish State, Ann. Dig. (5) 1929/30, 11
11. United Nations International Law Commission (2023) Sea-level rise in relation to international law
12. United Nations Office of the High Representative for the Least Developed Countries, Landlocked developing countries and Small Island developing states. About Small Island developing states. https://www.un.org/ohrlls/content/about-small-island-developing-states
13. Scott K (2018) Rising seas and Pacific maritime boundaries. Australian Institute of International Affairs. https://www.internationalaffairs.org.au/australianoutlook/rising-seas-and-pacific-maritime-boundaries/
14. Willcox S (2012) A rising tide: the implications of climate change inundation for human rights and state sovereignty. Essex Hum Rights Rev 9(1):1–19
15. Charter of the United Nations Art 4(1)
16. Rayfuse R (2011) International law and disappearing states—maritime zones and the criteria for statehood. Environ Policy Law 41:281
17. Wang B (2024) Development of the legal definition of the floating city: judicial interpretation of structural characteristics of floating homes and developments. In: Ikoma T, Tabeta S, Lim SH, Wang CM (eds) Proceedings of the third world conference on floating solutions: WCFS2023, 28–29 August, Tokyo, Japan. Springer, Singapore, pp 139–157
18. Soons A (1974) Artificial islands and installations in international law. Law of the Sea Institute, University of Rhode Island (Occasional Paper 22)
19. Wang CM, Wang BT (2014) Great ideas float to the top. Large floating structures: technological advances. Springer, Singapore, pp 1–36
20. Cradduck L, Wang B, Weinert KD (2024) Underpinning the value of buoyant urbanism: returning Torrens to the seas. In: Proceedings of the ASME 2024 43rd International conference on ocean, offshore and arctic engineering (OMAE): volume 9: Philip Liu honoring symposium on water wave mechanics and hydrodynamics; Blue economy symposium. American Society of Mechanical Engineers (ASME), United States of America
21. Tams CJ (2005) Enforcing obligations Erga Omnes in international law. Cambridge Uviersity Press
22. Wong D (2013) Sovereignty sunk? The position of 'Sinking States' at international law. Melbourne J Int Law 14:1–46
23. Black-Branch JL (2021) Obligations Erga Omnes: the missing link for nuclear nonproliferation and disarmament compliance. The treaty prohibiting nuclear weapons. Cambridge University Press
24. Tanaka Y (2021) The legal consequences of obligations Erga Omnes in international law. Netherlands Int Law Rev 68:1–33
25. Simmons BA (2006) Trade and territorial conflict in Latin America: international borders as institutions. In: Kahler M, Walter BF (eds) Territoriality and conflict in an area of globalisation. Cambridge University Press, Cambridge, p 255
26. Barcelona Traction, Light and Power Company, Limited (Belgium v Spain) (Judgment) [1970] ICJ Rep 3, 32, [33]
27. Wang B (2023) Regulatory reform and the potential use of floating artificial reefs under blue carbon and ocean carbon methodologies. In: 42nd International conference on ocean, offshore and arctic engineering (OMAE2023), June 11–16. Melbourne, Australia

Addressing San Francisco's Housing Crisis Through Floating Solutions

Carolina Jiménez Amador, Greg Henderson, Kirstin Weeks, Runo Okiomah, and Gordon Atkinson

Abstract The San Francisco Bay Area, despite its rich culture and technological innovation, is experiencing a significant housing crisis with a lack of space for traditional housing and the undeniable social costs of increased homelessness and displacement of minorities. Additionally, it is highly susceptible to flooding due to low elevations at the Bay Shore and anticipated sea level rise. Given these factors, floating housing communities rise as an opportunity to mitigate these issues. This paper presents a comprehensive approach to integrating a floating homes community within the San Francisco Bay Area's urban development strategy. Hence, the aim is to create an innovative housing solution targeting local development needs, environmental considerations, and social equity. To achieve this, we proposed the BAFH method which involves six pillars: Expertise; Affordable Housing; Geographical Suitability; Community Integration and Social Impact; Sustainability and Ecological Impact; and Technology and Design. We discuss the application of this method, highlighting the selection of top sites like Treasure Island, which offer unique advantages for sustainable urban expansion. We expect that this can enhance the project's feasibility and community acceptance. Furthermore, the paper explores how these six pillars can be generalized to other developments, enhancing urban resilience and adaptability.

Keywords Floating houses · BAFH method · Resilient

1 Introduction: Assessing the Current Urban Housing Landscape

San Francisco celebrated for its rich culture, famous landmarks, and technological innovation, is grappling with a significant housing crisis exacerbated by geographic and environmental constraints. These challenges include limited developable land,

C. J. Amador (✉) · G. Henderson · K. Weeks · R. Okiomah · G. Atkinson
Bay Area Floating Homes Inc., Sausalito, CA, USA
e-mail: slhcarolina@gmail.com

X. L. Zhao et al. (eds.), *Proceedings of the Fourth World Conference on Floating Solutions: WCFS 2024; 2–4 December; Hong Kong*, Lecture Notes in Civil Engineering 597, https://doi.org/10.1007/978-981-96-4569-5_90

high susceptibility to flooding, and seismic risks. With areas experiencing notable local land subsidence, a key factor that increases the area's vulnerability to flooding, it is projected to impact 98–218 km^2 due to relative sea level rise, with potential inundation of 413–429 km^2 considering potentially higher rates of ice loss in Antarctica and Greenland [1]. This situation is further aggravated by soaring housing costs, leading to increased homelessness and displacement [2].

As in other major cities, the lack of space for traditional housing exacerbates the crisis and leads to an increase in the cost of housing. San Francisco comprises about 122 km^2 of land and is the second most densely populated city in the U.S. According to the San Francisco Chronicle, the median price of a home in San Francisco as of June 2024 is 1.57 million dollars [3], a cost unaffordable to the majority of its residents. The social costs are undeniable, with growing numbers forced to live in unconventional dwellings and facing displacement due to development pressures. The most recent data indicates that 8323 homeless people were living in San Francisco, the majority unsheltered [4].

Given the traditional housing limitations of the San Francisco Bay Area and its flooding potential, it is reasonable to suggest floating architecture as a feasible solution to mitigate these constraints. Floating home communities could reduce construction and maintenance costs and increase social adhesion. Nevertheless, despite the growing interest in building on water globally, several challenges such as acquiring entitlements, adapting model health and safety codes, creating new templates for property ownership, financing, and insurance, and others, need to be addressed so the projects can be feasible and effective.

In direct response to the region's severe housing crisis, this paper outlines a comprehensive approach to integrating floating homes within the urban development strategy of the San Francisco Bay Area. We aim to create innovative housing solutions that align with local needs for development, environmental sustainability, and social equity. By considering these factors, we can ensure the project's feasibility and create a pathway toward a more resilient and inclusive future for San Francisco.

2 BAFH Method

2.1 Expertise

Bay Area Floating Homes Inc. (BAFH) was founded as a non-profit, public benefit corporation in direct response to the pressing housing and environmental challenges identified in the San Francisco Bay Area. With a vision to innovate urban living through sustainable housing solutions, our company harnesses the collective expertise of a multidisciplinary team, including environmental scientists, urban planners, architects, and engineers. This team's formation was strategically aligned to create not just homes, but sustainable, integrated community environments on water.

2.2 *Affordable Housing*

The "California Statewide Housing Plan 2022" outlines the state's strategic commitments to alleviate the housing shortfall [5]. This ambitious plan includes initiatives to construct 2.5 million new homes by 2031, focusing on accelerating housing production and enhancing affordability through substantial funding support and streamlined regulations. Specifically, the plan identifies innovative construction methods and sustainable development practices as crucial to overcoming the physical and regulatory constraints currently impeding housing development.

By aligning the development of floating homes with these state efforts, the potential for cost-effective, scalable housing solutions becomes tangible. These homes, utilizing innovative construction methods such as modular building techniques, promise significant advantages in terms of both time and resource efficiency. Additionally, by incorporating these homes into mixed-use communities, the plan facilitates a comprehensive urban solution that supports not only residential but also commercial growth, aligning with broader socioeconomic goals.

2.3 *Geographical Suitability*

In our journey to select optimal sites for floating homes, we conducted comprehensive evaluations of the physical characteristics of potential locations, such as flood-prone areas, protected land, protected water, water depth, wave protection, and the proximity to critical habitats like wetlands and eelgrass. Our selection process incorporated hydrographic surveys and environmental impact assessments to ensure that the proposed floating structures could be safely and sustainably integrated into the aquatic ecosystem.

We also aligned our site selection with the City and County planning goals, including the San Francisco Sea Level Rise Action Plan and the Bay Conservation and Development Commission's preferred development zones (BCDC) for land adjacent to potential sites for our project. Recognizing the importance of these factors, we intend to continue coordinating our future development efforts with these established plans. This strategy ensures that our floating homes are more than just housing solutions; they are a sustainable and integral part of the city's growth strategy. By promoting community engagement and enhancing public perception, our projects align with the broader urban development goals. This alignment not only helps streamline regulatory approvals and infrastructure integration, thereby boosting urban development and ecological health but also strengthens community ties. Such synergy enhances our ability to provide essential services seamlessly, stimulates economic growth, and improves community and environmental well-being, delivering benefits for both developers and the broader community.

2.4 Community Integration and Social Impact

Our floating home projects aim not only to provide housing but to forge inclusive, sustainable communities for residents currently facing displacement due to planned urban developments. By offering affordable housing units and co-housing opportunities within our floating communities, we directly address the social equity challenges posed by the city's growth, providing stable housing alternatives for lower-income families at risk of displacement. Each community is designed with ample communal spaces and amenities to enhance social interaction and build a strong sense of community. Moreover, our development plans will be designed to integrate existing and proposed public open spaces such as parks, walkways, gardens, and playgrounds, to foster community interaction and ensure our homes are accessible and actively enhance the environments where they are built. The strategic design also emphasizes seamless connectivity between the floating homes and shore-based facilities, promoting a unified urban fabric that enhances the overall quality of life.

2.5 Sustainability and Ecological Impact

Our floating homes project prioritizes sustainability throughout the site selection process, rigorously assessing each location for its ecological impact and potential for environmental improvement. This meticulous evaluation covers the effects on local marine life and the feasibility of habitat restoration, ensuring that each site not only minimizes ecological disturbances but also actively enhances local biodiversity and environmental health. Leveraging our team's expertise in urban ecology, we plan to implement biophilic design innovations such as artificial scaffolding for oyster and mussel beds and floating gardens to restore the health of marine ecosystems. Additionally, hydroponic gardens on communal floating platforms, which could serve as a sustainable food source for the community, and green roofs on buildings will aim to reduce carbon footprints and support local wildlife.

We specifically target underdeveloped, post-industrial, and ecologically degraded areas, as well as sites in floodplains or at risk from sea-level rise. These locations present prime opportunities for regeneration and sustainable growth, allowing our project to facilitate ecological recovery while minimizing the impact on untouched natural environments.

2.6 Technology and Design

We aim to leverage a variety of technologies to develop cost-effective, durable, and sustainable floating homes that are in harmony with natural forces. Our commitment to integrating advanced technologies and eco-friendly practices ensures that our designs promote sustainability and affordability. Our projects will strive for net-zero energy efficiency with the use of solar panels, and water-source heat pumps, alongside energy-efficient solutions that enhance affordability and environmental responsibility.

We are engaging local partners who have already demonstrated proficiency in off-site construction and marine flotation systems. This collaboration enables us to leverage their specialized skills to achieve significant cost savings in construction, enhancing the overall efficiency and affordability of our projects.

The SAFE Foundation System (Self Adjusting Floating Environment), crafted by Greg Henderson of Bay Area Floating Homes Inc. (a WCFS 2020 contributor), ingeniously merges nautical architecture with conventional land-based construction techniques to redefine building foundations. This innovative system transforms flood-prone shorelines into high-value real estate by allowing structures to float and adapt to fluctuating water levels and ground conditions. The system consists of three key components: a containment vessel, a buffer medium, and a prefabricated construction platform. These elements work together to create a broad, stable base that naturally mitigates flood risks by rising and falling with the water levels, making it ideal for development on previously unsuitable land due to flooding and poor soil quality [6].

Thanks to the SAFE Foundation, our plans extend to developing high-density, mixed-use projects along these challenging shorelines, thereby broadening urban housing options in areas once deemed unusable. The adaptability of the SAFE Foundation not only meets the challenges of urban expansion but also aligns with sustainable development goals by diminishing environmental impacts and enhancing disaster resilience. For instance, the buffer medium effectively decouples the construction platform from the earth, offering seismic protection similar to traditional base isolation but more cost-effectively [6]. Also, the precast concrete containment vessels, equipped with drain rock and filtration systems, are designed for permeability, ensuring environmental compatibility [6]. This approach is pivotal for transforming neglected waterfronts into vibrant residential and commercial centers, thereby enriching community life and ecosystem health.

3 Analysis and Implications

Thanks to the BAFH method, we were able not only to locate several suitable sites for approximately a 200 floating homes community but also to objectively determine the most suitable among them. Although currently this project is at a design stage, we propose that our method can also be applied to other developments. This approach offers a more holistic and comprehensive strategy that enhances the overall value of both the new floating community and its existing hosting location.

Treasure Island has been selected as the preferred site due to its advantageous location and natural characteristics that favor the development of floating homes (see Fig. 1). This site offers protection from wave and storm action and provides direct access and views to downtown San Francisco, making it a highly desirable location. Additionally, it is adjacent to a planned mixed-use residential and commercial zone currently being developed by the Treasure Island Development Authority. Due to its artificial origins from the late 1930s, the existing ecosystem of the island and its surroundings certainly offers room for improvement, which simplifies development considerations related to environmental impact. The island also has several pre-existing post-industrial structures that add to its developmental potential. These features, coupled with its sufficient depth and natural shelter from harsh conditions, underscore Treasure Island's strategic advantages and suitability for sustainable floating home development.

By integrating our innovative methods into various development projects, we can create enriched living environments that benefit all stakeholders. Similarly, by synergizing with strategic plans and regulatory frameworks, the floating homes project not only addresses the pressing housing shortage but also promotes sustainable practices that benefit the local environment. This comprehensive approach positions the project as a leader in innovative urban housing solutions, setting a benchmark for future developments and ensuring its role as a sustainable and integral component of the city's growth strategy.

This initiative underscores the need for policymakers to adjust policies, as projects like this demonstrate that they are viable and highly beneficial options. They enhance societal adaptability while ensuring safe urban growth that aligns with the needs of all stakeholders involved. Such projects prompt a reevaluation of urban development strategies, advocating for more inclusive and sustainable approaches to city planning.

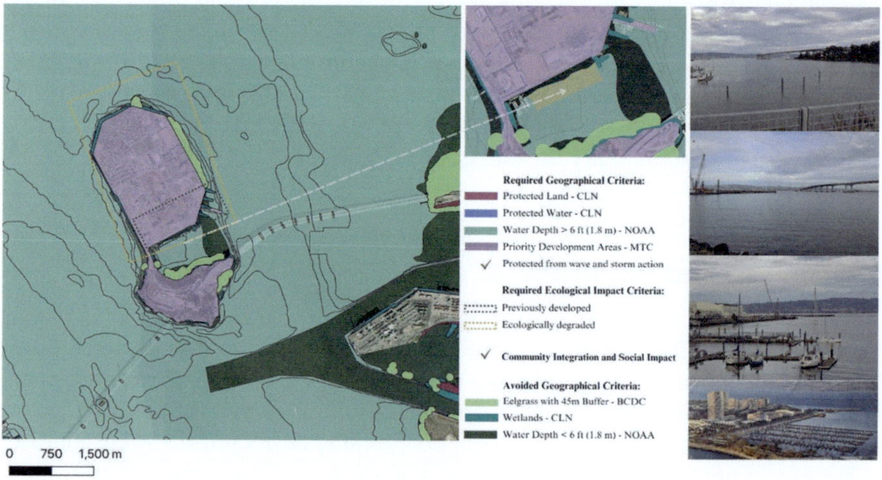

Fig. 1 Treasure Island site analysis based on the BAFH method

4 Future Directions

Recognizing the urgency of California's housing crisis, California has allocated billions of US dollars through the California Department of Housing and Community Development (HCD) 2022 Statewide Housing Plan. In addition to direct funding and tax incentives, California and local governments have recently passed legislation to streamline the development process and to provide technical assistance to help plan for more housing [5].

This project not only addresses local affordability concerns but also serves as an opportunity for the immediate adaptation of the surrounding communities, potentially acting as a template for other areas in the bay as well as coastal cities globally with similar challenges. Our vision encompasses modern communities that unite the goals of affordability, diversity, equal opportunity, ecological responsibility, net-zero carbon footprint, and emergent technologies.

5 Conclusions

The BAFH method ensures that floating communities are more than just housing alternatives; they are sustainable, integral components of the city's growth strategy, creating a cohesive and resilient urban fabric. Through this proactive framework, Bay Area Floating Homes Inc. addresses social and environmental challenges in urban landscapes, demonstrating a commitment to sustainable development that benefits both the community and developers and promises lasting advantages for future generations.

As we look to the future, the scalability of this model presents opportunities as well as challenges. Regulatory hurdles, technological advancements, and financial modeling are areas that require ongoing attention to ensure these innovative solutions can be replicated both locally and globally. To overcome these challenges, continued collaboration with policymakers, community stakeholders, and industry experts will be essential. By proactively addressing these challenges, floating homes can serve as a blueprint for sustainable urban development, potentially transforming housing policy and planning in coastal cities around the world. Shirzaei, M., Bürgmann, R.: Global climate change and local land subsidence exacerbate inundation risk to the San Francisco Bay Area. Science Advances, 4, eaap9234 (2018).

References

1. U.S. Department of Housing and Urban Development. Annual homeless assessment report to Congress. https://www.huduser.gov/portal/datasets/ahar.html. Last accessed 15 May 2024
2. Leonard C, Arevalo A (2024) California's housing market: data for homebuyers and sellers in every ZIP code. San Francisco Chronicle 1–2

3. City and County of San Francisco. https://www.sfgov.org/scorecards/safety-net/homeless-population. Last accessed 10 Jul 2024
4. California Department of Housing and Community Development: Statewide Housing Plan. https://statewide-housing-plan-cahcd.hub.arcgis.com/. Last accessed 25 May 2024
5. Hill K, Henderson G (2020) Pond urbanism: floating urban districts on shallow coastal groundwater. In: Piątek Ł, Lim SH, Wang CM, de Graaf-van Dinther R (2020) Second world conference on floating solutions (WCFS2020), pp 62–72. Springer, Singapore
6. The San Francisco Bay Conservation and Development Commission. https://bcdc.ca.gov/. Last accessed 12 Jun 2024

Research on Onna Village Regional Management of Marine Tourism

Isei Ran and Xiaobo Lou

Abstract Onna Village in Okinawa is a popular tourist destination surrounded by rich natural environments, attracting visitors both from Japan and abroad. The village is blessed with mountains, rivers, and the sea, with marine tourism playing a central role in supporting the regional economy. However, the rapid growth of tourism has brought to the forefront issues such as environmental degradation and the sustainable management of local resources. This paper examines the development of marine tourism in Onna Village, the challenges that have emerged from this growth, and the regional management strategies designed to address these challenges.

Keywords Regional management · Marine tourism · Tourism and sustainability

1 The Development of Marine Tourism in Onna Village and Policy

Onna Village was historically a fishing community. However, in 1964, landmarks such as Manzamo and Cape Maeda were designated as tourist attractions, marking the beginning of its transformation into a tourism hotspot [1]. The Okinawa International Ocean Expo in 1975 further accelerated the construction of resort hotels and tourism facilities, shifting the village's economy towards a tourism-centric model. Marine tourism became the driving force behind the village's economic growth, offering a range of activities such as diving, snorkeling, kayaking, and fishing experiences [1].

From an industrial perspective, the village's agriculture-based structure has remained largely unchanged. In 1970, the Onna Village Fisheries Cooperative was established, and fishing ports gradually opened, leading to an increase in fishing

I. Ran (✉)
Applied Environment System, Tokyo University of Science and Technology, Tokyo, Japan
e-mail: ranisei0310@gmail.com

X. Lou
Marine Policy and Culture, Tokyo University of Science and Technology, Tokyo, Japan

X. L. Zhao et al. (eds.), *Proceedings of the Fourth World Conference on Floating Solutions: WCFS 2024; 2–4 December; Hong Kong*, Lecture Notes in Civil Engineering 597, https://doi.org/10.1007/978-981-96-4569-5_91

activities. Mozuku farming began in 1978, and sea grape cultivation followed in 1989. These farming activities have since become integral to the development of Onna Village's fishing system. Additionally, the Tourism Nation Promotion Basic Act was enacted on January 1, 2007, and the Tourism Nation Promotion Basic Plan was approved by the Cabinet on March 30, 2012, establishing a five-year plan.

As tourism flourished, both the population and employment in Onna Village increased. Although the village aimed to revitalize the region through tourism, the rapid expansion of the industry brought about serious challenges, such as environmental degradation and the strain on local resources. As a result, regional management aimed at balancing tourism with environmental protection has become critical.

2 Challenges Facing Onna Village

2.1 *Red Soil Runoff*

One of the most severe environmental issues in Onna Village is red soil runoff. As tourist facilities, golf courses, and vacation homes are developed, exposed soil is easily washed into the sea by rain, causing widespread coral bleaching and severe damage to seaweed and Mozuku farming. In 1990, 95.5 tons of Mozuku were destroyed due to red soil runoff, resulting in a financial loss of approximately 14.23 million yen [1].

Coral reefs, which form the backbone of marine ecosystems, are also crucial resources for the tourism industry. Tourists flock to Onna Village to enjoy its beautiful sea and coral reefs, but the damage caused by red soil runoff threatens these resources and poses a risk to the tourism industry itself. Additionally, the runoff reduces the clarity of the water, negatively impacting coastal fishing.

2.2 *Tourism Pollution*

Tourism pollution refers to the environmental and social problems caused by an influx of tourists. As the number of visitors to Onna Village has increased, issues such as illegal dumping, unauthorized parking, and traffic congestion have worsened the living conditions of local residents. Furthermore, along the coastal areas, leisure activities such as diving and jet skiing often interfere with fishing, leading to growing conflicts between tourism operators and fishermen.

For example, tourism operators sometimes encroach on fishing zones without permission, affecting the livelihood of the fishermen. Environmental destruction and the misuse of resources by tourists have also become critical issues, necessitating prompt countermeasures to preserve the appeal of the village as a tourist destination.

2.3 Environmental Conservation

Bruno and Selig [2] and Burkr et al. [3] pointed out that human activity and global warming have degraded coral reefs worldwide, with 75% of them at risk of extinction. In the summer of 2016, sea temperatures in Onna Village exceeded 30 °C, leading to coral bleaching that persisted even after temperatures dropped below 30 °C in September. Significant coral bleaching occurred, but some recovery was seen by October, although areas with higher bleaching levels experienced high mortality. Higa et al. [4] proposed three approaches for coral reef recovery: reducing local and global stress, actively transplanting corals, and designating well-preserved reefs as marine protected areas. Omori [5] noted that despite 40 years of coral transplantation efforts, the success rate remains uncertain due to high mortality rates within 3–4 years of transplantation. While large-scale transplantation can potentially aid coral reef regeneration, challenges remain in restoring ecosystem services on a broader scale.

2.4 Conflicts Between Fisheries and Tourism

Traditional fishing has long been the backbone of Onna Village's economy. However, the rapid growth of tourism has intensified conflicts between fishermen and tourism operators over the use of marine resources [6]. Fishermen rely on marine resources for their livelihood, while tourism operators use the same sea areas for leisure and tourism activities, leading to frequent clashes.

Activities such as diving, snorkeling, and kayaking often disrupt fishing grounds and aquaculture zones, reducing the income of fishermen and threatening the harmony of the community. To achieve coexistence between the tourism and fishing industries, clear rules and coordination mechanisms are necessary.

3 Regional Management

3.1 Red Soil Runoff Prevention Measures

To address the issue of red soil runoff, Okinawa Prefecture enacted the "Red Soil Runoff Prevention Ordinance" in 1996, requiring large-scale development projects to implement measures to prevent runoff. In Onna Village, the local government and the fisheries cooperative have been actively exploring methods to reduce red soil runoff, such as planting vegetation in areas prone to runoff and installing barriers to prevent soil erosion.

Fishermen themselves have also become involved in conservation efforts, contributing to the resolution of the red soil issue through environmental protection activities. Cooperation between the government, local residents, and fishermen is essential in protecting the marine environment of Onna Village.

3.2 Countermeasures Against Tourism Pollution

Addressing tourism pollution requires the active participation of local residents. In Onna Village, tourism operators and residents collaborate on beautification projects and environmental conservation activities. The local government has also issued guidelines for tourist behavior to raise environmental awareness among visitors.

Specific measures include the installation of trash bins at beaches and public facilities used by tourists, along with regular cleanup activities to prevent illegal dumping. Parking and restroom facilities have also been improved to encourage orderly behavior among tourists. Additionally, a local "Tourism Pollution Countermeasures Committee" has been established to strengthen cooperation with tourism operators and promote sustainable tourism practices.

3.3 Collective Participation for Environmental Conservation

Onna Village is recognized for its beautiful natural environment, which includes coastlines, forests, and sunsets. This natural beauty is a shared asset for villagers, tourism businesses, and visitors, and its preservation is vital for the village's tourism industry. In response to environmental concerns, Onna Village enacted the "Onna Village Environmental Conservation Ordinance" in 1991 to regulate land use and development.

Efforts such as coral breeding and transplantation by the local fisheries cooperative, which began in 1998, aim to protect coral reefs. Despite a mass coral bleaching event in 2016 due to rising water temperatures, the survival rate of farmed corals was higher than that of natural corals.

Environmental conservation cannot rely solely on regulations; it requires the active participation of residents and tourists. For example, restaurant owners must maintain cleanliness, and local authorities should provide guidelines. The village also needs to address issues like waste disposal and environmental protection through various laws, including the Environmental Conservation Ordinance and the Onna Village Landscape Ordinance.

Discussions are ongoing about establishing a fee system for visitors to contribute to coral conservation and other environmental efforts.

3.4 *Coordination for Coexistence Between Fisheries and Tourism*

To promote coexistence between tourism and fishing, Onna Village established the "Marine Use Coordination Council," where fishermen, tourism operators, and local government representatives work together to create rules for marine use [7]. For example, specific areas are designated for tourism activities, while other areas are reserved for fishing to prevent interference between the two industries.

Strict regulations have been implemented to prevent tourism operators from entering fishing zones without permission, and these rules are also communicated to tourists. This ensures the safety of fishing activities while allowing tourism to operate sustainably under local guidelines. A system has gradually been developed to balance the benefits of both industries.

4 Applicability to Other Regions

In the previous chapters, we explored the key challenges faced by Onna Village and the regional management strategies implemented to address these issues. To provide a clear understanding of these strategies and their potential applicability to other regions, Table 1 summarizes the main problems identified in Onna Village and the corresponding solutions that have been implemented. This table will serve as a reference point for discussing the adaptability and effectiveness of these management measures in different regional contexts.

The initiatives undertaken by Onna Village can serve as a model for other tourist areas in Okinawa and coastal regions across Japan. In particular, the "collaborative management" model aimed at balancing tourism with environmental conservation offers effective solutions to many of the challenges faced by tourist destinations. This model can be applied to other regions in the following ways.

Cooperation Among Various Stakeholders: In Onna Village, the local government, fishermen, tourism operators, and residents work together to tackle issues such as red soil runoff and tourism pollution. This collaborative approach is highly effective in areas where multiple industries and stakeholders coexist. For instance, other coastal tourist destinations could develop rules and frameworks for joint management to ensure the coexistence of fisheries and tourism.

Sustainable Management of Tourism Resources: Onna Village's efforts to prevent red soil runoff and address tourism pollution serve as valuable examples of how to balance environmental conservation with economic growth. These initiatives provide a model case for other regions aiming to develop sustainable tourism that is mindful of the environment. Particularly, promoting "ecotourism," which leverages local resources without placing a burden on the environment, is crucial for regions where natural environments are the foundation of tourism.

Table 1 Summary of key challenges and management solutions in Onna village

Problem	Corresponding solution/management measures
Soil runoff	– Implement the "Red Soil Runoff Prevention Ordinance" to regulate development projects – Introduce self-regulatory measures led by fishermen to reduce red soil runoff
Tourism pollution	– Establish and strictly enforce the "Onna Village Coastal Management Ordinance" – Strengthen visitor guidance and education on the proper use of coastal areas and fishing port facilities
Environmental conservation	– Introduce and enhance coral transplantation, promoting the "Coral Sea" initiative – Implement the "Onna Village Environmental Conservation Ordinance" to limit development and encourage sustainable growth
Conflicts between traditional fisheries and marine tourism	– Establish a Marine Use Coordination Association to mediate among various stakeholders – Develop and implement the "Onna Village Fisheries Revitalization Plan" to enhance management and cooperation

Flexible Policies and Regulations: The "Red Soil Runoff Prevention Ordinance" enacted by Okinawa Prefecture is a flexible policy model that promotes environmental conservation without hindering the development of the tourism industry. Such ordinances and regulations provide a crucial framework for introducing region-specific solutions to environmental challenges faced by tourist destinations. Tailoring policies to the unique characteristics of each region is key to balancing tourism development with environmental protection.

5 Conclusion and Discussion

This paper has analyzed the challenges that accompany the development of marine tourism in Onna Village and the regional management efforts to address them. Marine tourism has made significant contributions to the village's economy, attracting many tourists each year. However, the rapid growth has also led to environmental problems and conflicts between industries, notably the degradation of coral reefs due to red soil runoff, tourism pollution, and clashes with traditional fishing.

In response, Onna Village has established management models such as the "Red Soil Runoff Prevention Ordinance" and the "Marine Use Coordination Council," aimed at balancing tourism growth with environmental conservation and promoting the sustainable use of local resources. Moreover, specific measures have been implemented to foster cooperation between fishermen and tourism operators, aiming for coexistence between both sectors.

The Onna Village model has the potential to evolve and be applied to more regions in the future. As tourism continues to drive economic growth, managing local resources sustainably will become increasingly important. In today's world, where climate change and natural disasters are on the rise, flexible management capable of addressing environmental risks is essential. Onna Village must continue to strengthen the cooperation between local residents, fishermen, and tourism operators to enhance the sustainability of the region as a tourist destination while addressing new environmental challenges.

References

1. Lou X (2013) The era of marine industries: challenges toward revitalizing fishing villages, 1st edn. Rural Culture Association, Japan
2. Bruno JF, Selig ER (2007) Regional decline of coral cover in the Indo-Pacific: timing, extent, and subregional comparisons. PLoS ONE 2(8):711
3. Burkr L, Reytar K, Spalding M, Petty A (2011) Reefs at risk revisited. World Resources Institute, Washington DC, USA, p 116
4. Higa Y, Niizato H, Zaan Y, Nagata S et al (2018) Progress in coral seedling production and transplanting techniques using asexual reproduction in the coral conservation and restoration project of Okinawa prefecture. J Jpn Coral Reef Soc 20(1):21
5. Makoto O (2016) The goal is a survival rate of over 40% three years after transplanting: current status, challenges, and directions for coral transplantation. Midoriishi 27:2–3
6. Hiyane Y (2017) A study on the utilization order of coastal areas: a case study of Onna village in Okinawa prefecture. J Reg Ind Stud 3:39–53
7. Uehara M (2012) Success factors of Onna village fisheries cooperative's 'regional fisheries revitalization plan': Mozuku, Sea grapes cultivation, resort adaptation, and human resources. Fish Prom 46(12):1–52

Urban Planning

Optimal Locations for Seasteading: Geostrategic Opportunities on the High Seas

Joerg Baumeister

Abstract Seasteading initiatives often express a hopeful vision of achieving autonomous living in new, uninhabited sea locations. This aspiration for self-sufficiency and freedom fuels exploration into untapped waters, where possibilities appear boundless. However, while political autonomy exists in international waters, the question of where to establish these aquatic settlements remains unanswered. To address this challenge effectively, a systematic approach is necessary. Instead of relying on random selection or personal preferences, scientific research is conducted to map out various critical factors. These include wave height, sea surface temperature, and photosynthetically active radiation for efficient aquaculture, as well as accessibility to international airports in order to link seasteaders with global transportation networks. By analyzing these layers comprehensively, optimal locations for seasteading can be identified. These areas offer the best prospects for autonomous living on the high seas. Through this methodical step-by-step process and recommendations about the next steps, the path to realizing the dream of seasteading becomes clearer, offering a promising vision for the future of human habitation beyond the limitations of traditional land-based societies.

Keywords Seasteading · Strategic development · Floating cities · Floating structures · High seas · International waters

1 Utopias

Traditional thinking and existing jurisdiction are primarily based on land rather than the sea. The reason for this is the absence of established methods for placemaking, fostering economic and social progress, and asserting territorial control on the sea,

---

https://www.joergbaumeister.com, https://www.seacities.org.

J. Baumeister (✉)
SeaCities Lab, Griffith University, Gold Coast, Australia
e-mail: j.baumeister@griffith.edu.au

X. L. Zhao et al. (eds.), *Proceedings of the Fourth World Conference on Floating Solutions: WCFS 2024; 2–4 December; Hong Kong*, Lecture Notes in Civil Engineering 597, https://doi.org/10.1007/978-981-96-4569-5_92

all of which are typical features of land governance. Consequently, the sea is often viewed as an empty and uncontrollable expanse, with only limited connections to human civilization [1].

It is no wonder that this vast ocean serves as the setting for Thomas More's satirical description of a happy society in a 'no-place', translated from Greek into Latin as 'utopia'. This introduction of the first so-called utopia in the sixteenth century is a work of speculative fiction that discusses an ideal state of a commonwealth including the towns and lives of the Utopians [2].

In the eighteenth century, another fictional novel titled 'A general history of the robberies and murders of the most notorious pirates' [3] describes the opposite concept of living in a non-governed free colony with its capital named Libertatia, derived from the Latin word 'liberi,' meaning 'free'. The inhabitants, known as Libertanians, aspire to create a society that serves as a barrier against the rich and powerful. All individuals of all races, religions, and beliefs can live in Libertatia without judgment or the control of authoritarian state institutions.

Whereas both the ideal state and non-state concepts remained at a utopian level, another utopia named Próspera became a reality on the Caribbean island of Roatán in 2021. Following the neo-liberal Nobel laureate in economics Paul Romer, who had proposed designating special economic zones in poorer countries to be managed by companies, this private island state was founded by investors and venture capital entrepreneurs [4]. The owner of this special economic zone in Honduras is a US company registered in the tax haven of Delaware taking on state functions and collaborating with the landowners, for instance, to determine the rate of a 5% income tax.

The use of cryptocurrency does not diminish the appeal of this private state to those who might be considered 'outlaws', as well as to honorable wealthy individuals seeking tax advantages or a combination of both. Unfortunately, there might not be enough time to further develop Próspera's experimental framework because Honduras' new president is currently implementing her promise to end this private neo-liberal experiment [5].

Other idealists, such as members of the Seasteading Initiative, aim to establish a different utopia driven by humanitarian and ecological motives. They advocate for the creation of communities as groundbreaking solutions to global challenges such as overcrowding, ineffective governance, and related issues [6].

Having learned from Próspera's failure as a special economic zone due to its connection to a specific state, seasteaders aim to create floating communities in international waters. These so-called High Seas cover almost half of the Earth's surface and offer unrestricted space to select the ideal location for the realization of a utopia that combines the best aspects of both the original utopian and libertarian worlds [7].

2 Factors to Locations

To find the optimal spot for a seasteading community in the high seas, it is essential to understand the necessary factors for its implementation and operation. Therefore, potential factors will be analyzed and combined with geographical locations in this Part 2, using a matrix method described in the publication 'Predicting Geostrategic Developments and Conflicts in the High Seas' [8]. The following application of the matrix method uses thereby an approach that progressively eliminates less suitable locations, leaving only the best sites for seasteading developments in the end.

Seasteading is the creation of permanent, self-sufficient communities on the ocean, similar to homesteading but located on the high seas [9]. From a systems-based perspective, the entire environment should be encompassed, including the seastead's (A) anchoring, (B) production, (C) construction, and (D) community [10]. Therefore, the following elements are considered essential and are consequently linked to specific locations.

A. Anchoring: The structure needs to be either free-floating or anchored in the high seas. A free-floating structure requires some form of propulsion to control its position, while an anchored structure needs a suitable seabed with a limited depth for anchoring. Examining the bathymetry of the high seas reveals an appealing fact: there are many areas in the middle of the ocean with limited depth. The vastness of the oceans is only perceived by terrestrial beings. In reality, so-called seamounts often create areas of very shallow water, making them ideal locations to anchor future seasteading structures. This is why the locations of seamounts in the high seas are indicated in Fig. 1 as a first step.

B. Production: To achieve self-sufficiency, the basic needs of seasteaders must be met. Energy can be generated through renewable sources like wind, waves,

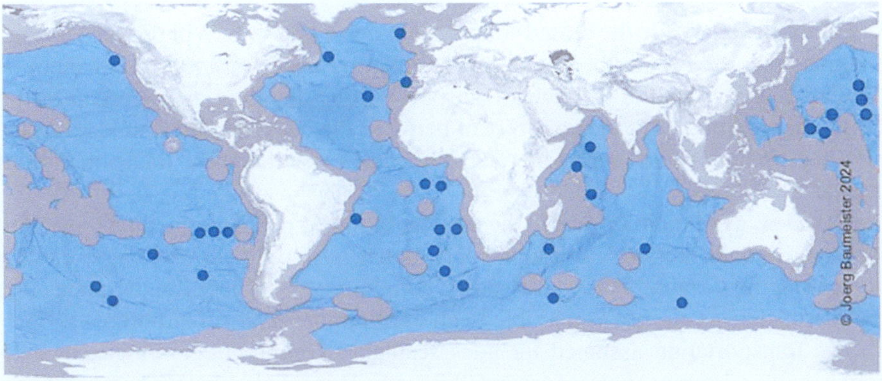

Fig. 1 The high seas, or international waters, are depicted in light blue. Shallow offshore areas, known as seamounts, which have depths of less than 100 m, are marked with dark blue dots. The map was created by the author based on GIS data

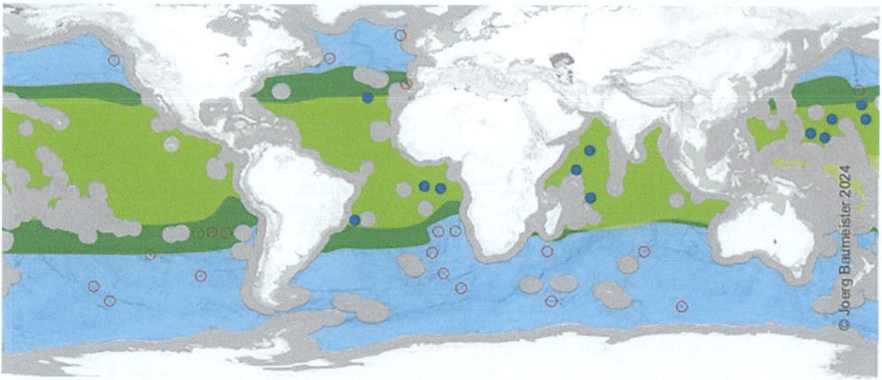

Fig. 2 Seamounts are marked with dark blue dots in zones with preferred factors for aquaculture, such as an average sea surface temperature of 24–32 °C shown in dark green including a mean photosynthetically active radiation of 45–55 E/m^2/day shown in bright green [11]. Red circles indicate seamounts that did not meet the criteria and were excluded. The map was created by the author based on GIS data

and water. This enables the possibility of desalinating seawater if rainfall fails to provide sufficient drinking water. Food, another basic need of seasteaders, can be produced far from the coast through aquaculture, which achieves its highest efficiency with elevated sea-surface temperatures and increased photosynthetically active radiation [11]. As a result, seamount areas with high sea surface temperatures and photosynthetically active radiation are chosen in step 2, as shown in Fig. 2.

C. Construction: The floating structure's design requires a limit on wave height to prevent disruptive motions caused by high waves [12]. Consequently, the absence of hurricanes [13] and low wave activity become selection criteria. This reduces the number of locations from 12 in step 2 to 6 in this step 3, as illustrated in Fig. 3.

D. Community: While the closed-loop system described in B. will produce enough energy, water, and food, there will still be a strong need for goods that cannot be produced on the floating structure. Transport will be therefore an important factor. Given the minimum distance of 360 km between the coast and the boundary of the high seas, ports should ideally be relatively close to facilitate the transportation of goods.

However, even more critical is the transportation of people to facilitate economical, educational, cultural, spiritual, recreational, and political exchange, thereby preventing the isolation of the seasteading community. This necessitates connecting the seasteaders as closely as possible to an existing international airport.

The transportation assumed includes seaplane, ground-effect vehicle, or helicopter, with an average speed of around 250 km/h. This ensures that the best locations, all within 500 km of major international airports, can be reached within a 2-hour travel time resulting in the selection of three optimal locations in this final step presented in Fig. 4.

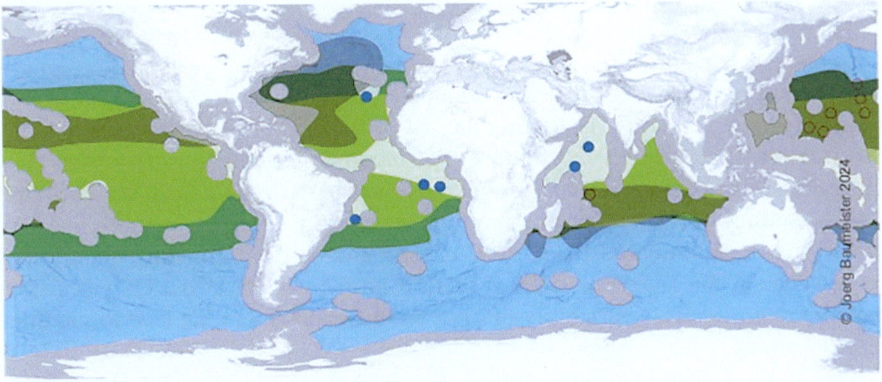

Fig. 3 The remaining seamounts are marked with dark blue dots in zones with preferred factors for aquaculture and construction, such as a maximum mean significant wave height of 2 m [12] and locations outside of tropical cyclone risk zones 2–5 [13]. Red circles indicate seamounts that did not meet the criteria and were excluded. The map was created by the author based on GIS data

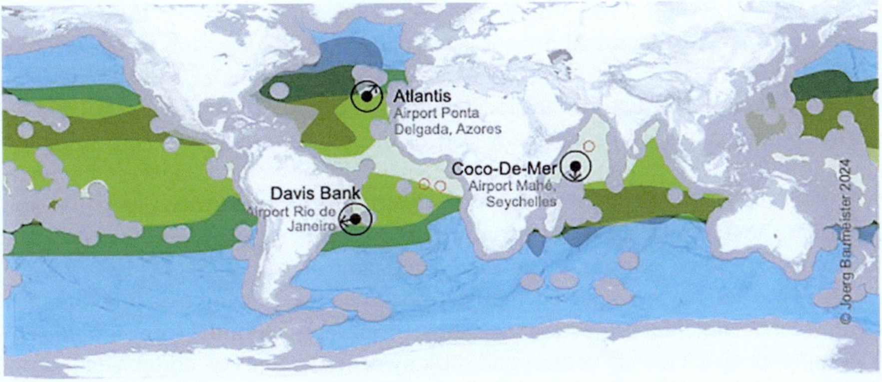

Fig. 4 The three final locations are named after the associated seamounts Davis Bank, Coco-De-Mer, Atlantis and are situated in international waters. They were selected based on optimal aquacultural and construction factors, as well as their excellent connectivity, being within 500 km of the international airports in Rio de Janeiro, Ponta Delgada, and Mahé. Red circles indicate seamounts that did not meet the criteria and were excluded in this final round. The map was created by the author based on GIS data

These three prime locations for seasteading—Davis Bank, Coco-De-Mer, and Atlantis—are named for their associated seamounts, with the Atlantis Seamount deriving its name from the mythical Atlantis.

3 Opportunities

Seasteading communities will need to compete with both themselves and land-based states to attract residents. Therefore, the list of factors can vary depending on the requirements. Each seasteading initiative can incorporate additional elements crucial for successful implementation, such as direct access to sea trade, construction requirements, conflict avoidance, or proximity to other seasteading communities. At the same time, location-based risks, such as piracy or territorial claims, should also be considered, the latter especially if there is an increase in UNCLOS's decision-making authority. This variability of opportunities and risks may lead to modified ideal locations.

Due to the selection criteria applied, the three proposed locations Davis Bank, Coco-De-Mer, and Atlantis are all situated in climatically attractive areas surrounded by the open sea, characterized by warm water and low waves. In addition to conventional concepts such as residential buildings and tourism offerings like casinos, hotels, and cruise liners, it will be crucial to explore more innovative ideas. These include developing new aquaculture products, applying marine biotechnology, establishing green technology hubs, and initiating floating real estate ventures. If desired, this approach enhances functional diversity and adds valuable assets to the seasteading concept.

The seasteading structures are located above seamounts, which are hotspots of marine biodiversity. By being positioned in these areas, they will help safeguard the marine environment and implement the new UN High Seas Treaty, which aims to protect 30% of international waters as Marine Protected Areas [14]. To avoid generating adverse effects, the environmental protection measures of the seasteading structures should be constantly considered and controlled.

Finally, after having selected the best position, the best mix of functions and the best environmental protection measures, discussion about the form should start. A comprehensive view of natural and built system elements as components of the environmental system enables thereby the development of eco-friendly seasteading structures. The development of the form is directly influenced by specific conditions unique to each location. Therefore, finalizing the choice of location is necessary before beginning the form development process.

Specialized aquatectural expertise is essential for this project, given the unique challenges posed by the marine environment, such as waves and buoyancy effects. These challenges are not typically within the scope of expertise for conventional architects trained in terrestrial design, a fact that becomes particularly evident when reviewing some of the initial seasteading simulations [6].

How the decisions are made during the planning and implementation process will also depend on whether they adhere to an original utopian bottom-up approach, a prosperitarian top-down approach, or if the seasteaders truly embrace a libertarian approach, shaping their own ideal societal decision structure together.

References

1. Steinberg P, Peters K (2015) Wet ontologies, fluid spaces: giving depth to volume through oceanic thinking. Environ Plan D: Soc Space 33(2):247–264
2. More ST (1869) Utopia: originally printed in Latin, 1516. Palala Press
3. Johnson C, Cordingly D (2017) Pirates: a general history of the robberies and murders of the most notorious pirates. Osprey Publishing
4. Dymond LH (2015) A recent history of recognized economic thought: contributions of the Nobel laureates to economic science. Lulu.com.LNCS Homepage. http://www.springer.com/lncs. Last accessed 2016/11/21
5. Bellinghausen, Yves: Der Staat sind wir. Die Zeit (2024) under https://www.zeit.de/2024/21/privatstadt-honduras-prospera-steuern-libertaeres-experiment. Last accessed 2024/06/23
6. Seasteading Institute 2024 Homepage http://www.seasteading.org. Last accessed 2024/06/23
7. Baumeister J (2023) Developing aquatic urbanism: a taxonomy describing 7 strategies and 35 tactics. In: Baumeister J, Giurgiu IC, Linaraki D, Ottmann DA (eds) SeaCities. Cities Research Series, Springer
8. Baumeister J (2024) Predicting geostrategic developments and conflicts in the high seas. In: Baumeister J, Giurgiu IC, Ottmann DA (eds) At the edge of the sea: the evolution of amphibious concepts. Cities Research Series, Springer
9. Friedman P, Taylor B (2012) Seasteading: competitive governments on the ocean, KYKLOS, vol. 65—May 2012—No. 2, pp 218–235
10. Baumeister J, Ottmann DA (2015) Urban ecolution—a pocket generator to explore future solutions for healthy and ecologically integrated cities. UWAP, Crawley, Australia
11. Baumeister J, Giurgiu IC (2022) SeaOasis—floating aquaculture for smallholders' global food security. Springer Nature
12. Map based on Meteorological Service of New Zealand (2022) Hindcast maps https://app.metoceanview.com/hindcast/. Last accessed 2024/01/05
13. MunichRE Nathan world map of natural hazards (2011) https://catalogue.unccd.int/Map_NATHAN%20-%20World%20map%20of%20natural%20hazards.pdf. Last accessed 2024/08/08
14. Baumeister J, Morrison A (2023) SeaSurveyor: an innovative floating solution for establishing marine protection areas in shallow international waters. In: World conference on floating solutions WCFS2023 Japan

Urbanizing the Smaller Islands Sustainable Floating Towns for the Singapore of Tomorrow

Giuseppe Torrisi, Daniela Procopio, and Lim Soon Heng

Abstract Issues such as climate change, energy crises, population growth and the increasing concentration of residents in urban areas are the challenges of the future, and some solutions to these challenges are already on the way. New plans for floating cities are one of these answers. Starting from a pilot project in the city of Singapore the study proposes viable urban planning solutions applicable to many coastal cities of the world. Singapore serves as an exemplary case study, having expanded its territory through land reclamation in recent decades, a practice that, though effective, is environmentally detrimental and unsustainable in the long run. As an island surrounded by smaller islands, Singapore is a city-state grappling with limited land availability, a burgeoning population, and a flourishing economy that continues to attract many. These factors underscore the urgency of our study, aiming to address future challenges and alleviate concerns about what lies ahead.

Keywords City and town planning · Floating architecture · Environmentally-friendly architecture and design · Urban planning · Floating cities

1 Introduction

An uncertain future awaits us and the challenges that humanity must prepare to face are difficult and at the same time stimulating. Issues such as climate changes, renewable energy, population growth and constant concentration of residents in urban areas are the challenges of the future and some answers to these challenges are already on the way.

New plans for floating cities could be one of these answers and it seems they are a viable solution to address climate changes as well as population growth in land-scarce cities. Currently there are some ongoing developments for floating cities in Korea and in the Maldives. Other floating developments are currently explored near

G. Torrisi (✉) · D. Procopio · L. S. Heng
Torrisi & Procopio Architetti, Rome, Italy
e-mail: studio@torrisi-procopio.com

X. L. Zhao et al. (eds.), *Proceedings of the Fourth World Conference on Floating Solutions: WCFS 2024; 2–4 December; Hong Kong*, Lecture Notes in Civil Engineering 597, https://doi.org/10.1007/978-981-96-4569-5_93

Panama for seasteading communities and a huge industrial floating development is planned in Saudi Arabia.

Far from discussing utopian visions or far-off ideas, this exploration focuses on innovative solutions poised to materialize sooner than anticipated, leveraging existing technology. Starting from a pilot project for the city of Singapore, we present practical urban planning solutions that hold relevance for numerous coastal cities worldwide.

Singapore serves as an exemplary case study, having expanded its territory through land reclamation in recent decades, a practice that, though effective, is environmentally detrimental and unsustainable in the long run. As an island surrounded by smaller islands, Singapore is a city-state grappling with limited land availability, a burgeoning population, and a flourishing economy that continues to attract many. In such a context, the first questions we wanted to address with our study were:

- How to face population growth in a typical land-scarce city like Singapore?
- What solutions can be explored for the planning of our future?

Being aware of the complexity of the issues and that it is not possible to give definitive solutions, our proposal should be therefore considered as an input for further studies, as a first step towards open debates involving multiple parties, not just for operators in the planning and architecture industry, but for all those who wish to feel involved in rethinking our cities and who want to contact us to propose solutions, ideas and criticisms.

Therefore, to answer these crucial questions we have identified two ambitious ways to support Singapore's long-term development needs:

1. A better use of the areas on the mainland through the relocation offshore of some industrial and military activities (floating industrial developments) with the aim of freeing up areas for some new high-density mixed-use developments.
2. The possibility of urbanizing the smaller islands around Singapore by means of mixed-use developments, both on land and floating through the use of very large floating structures (platforms).

Our proposal aims to promote the possibilities of urbanizing nearby islands into sustainable floating towns for the Singapore of tomorrow. The use of clustering small islands with floating structures gives rise to the ability of achieving varied urban design of high and low-rise building forms giving variety of spaces for housing and commercial uses. The advantage of the use of islands is the avoidance of conflict with Singapore Port navigation fairways.

The new expansion of the city should be planned according to the same principles and rules that have characterized the urban planning of modern Singapore over the past 40 years. Continuity with the plans of the past is essential to ensure that the new developments are harmonious and that the new parts of the city are vibrant and liveable.

Not everyone is aware of the criteria and principles that have been used to plan modern Singapore over the past 40 years. So, although there is no shortage of publications that illustrate every detail of the "Prototype of a city", the conceptual model that incorporates all the above criteria and principles, we will try to describe it briefly so that our proposed pilot project inspired by it is understandable to all readers.

2 Prototype of a City

The prototype of a city that inspired the planning of modern Singapore is an organic system consisting of multiple hierarchical cells.

According to this model Singapore was divided in 5 regions from 600,000 to 1 million inhabitants (Fig. 1), in turn divided in 26 new towns from 150,000 to 300,000 inhabitants (Fig. 2), then in neighborhoods from 4000 to 8000 inhabitants and precincts from 700 to 1000 inhabitants.

In this system each region, new town, neighborhood and precinct has its own centre. Daily necessities are available in neighborhood centres which are accessible by foot, higher-level needs are available in new town centres or regional centres that

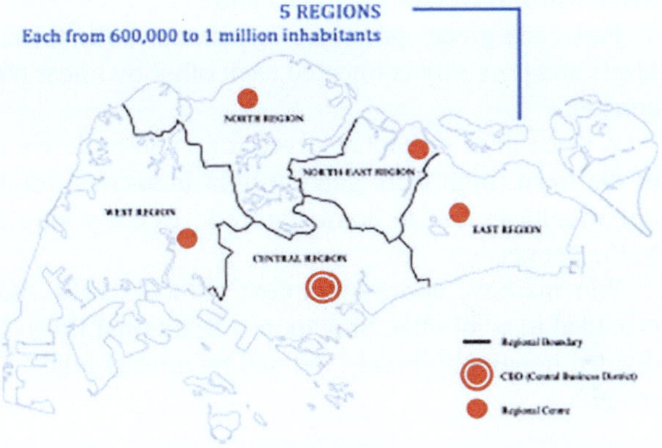

Fig. 1 The 5 regions of Singapore

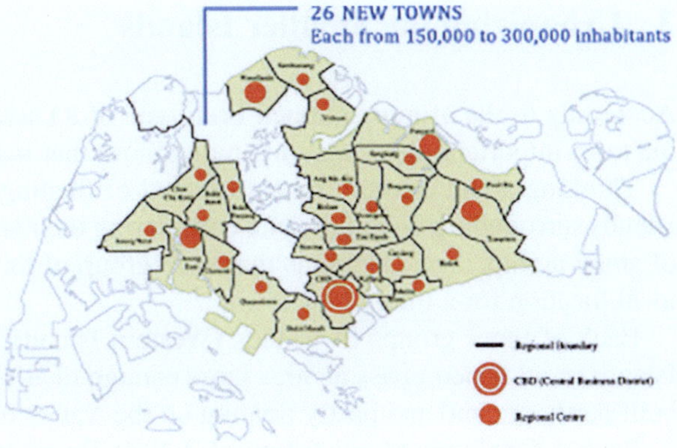

Fig. 2 The 26 new towns of Singapore

are accessible by bicycle or bus, leaving the activities of the highest level for the CBD (Central Business District) which is accessible by MRT, bus.

These decentralized structures achieve the twin goals of a good quality of life and reduced traffic congestion. Thanks to this prototype, Singapore was built according to the vision of a "unique tropical city of excellence. A city that is healthy and beautiful, unique and gracious, with a distinctive character" (1).

We believe that this model should be followed for future plans of the city, and in particular, the new town planning unit should be considered as a reference model for the future expansion of the city.

The self-sufficient New Towns are the core concept of the prototype. Unlike other part of the world, where the satellite towns generally refer to small isolated towns on the edge of a metropolitan area, in Singapore's plan a New Town is a major planning unit, the key to the success of an urban plan that works very well.

According to the Prototype, a new town consists of:

A minimum land area of 15 sq.km and a population from 150,000 to 300,000 persons.

A road system with major and minor arterials, local roads, expressways and the MRT which traverses the town centre.

Parks and green spaces are distributed at the town, neighborhoods and precincts levels and possibily connected each other by linear public green belts to form green corridors.

A percentage of land reserved for a non-pollutive industries located at the fringe of the town (high tech, general light industries, food factories). This provides job opportunities close to home and reduces daily long-distance commute, preventing traffic congestion.

Wet markets, sports complex, outdoor cafes and food courts are necessarily provided to symbolize Singapore's traditional lifestyle. Each new town was in turn divided in neighborhoods, defined by arterial roads, and precincts, defined by local roads.

3 Urbanizing the Smaller Islands

According to the aforementioned Prototype of a City, a self-sufficient new town is the most important planning unit among those that make up the city.

Thinking about a future mixed on land or floating development for the smaller islands surrounding Singapore and considering their size, shape and their peculiarity of small archipelagos, it seems that each group of these smaller islands can be the ideal location for a future new town.

Each of the 3 groups of islands (Western Islands, Southern Island and Eastern Islands) could encompass an area large enough to allow the planning of a new town built partly on land and partly floating on the waters of the archipelago.

Such a development consisting of 3 New Towns would ensure accommodation for at least 720,000 new inhabitants. Minimum, as it is worth remembering that a floating development is always easily scalable and flexible according to the future needs.

3.1 Southern Islands Development

New Development Total Area (On Land and Floating): 8.3 sq.km

On Land Area (Islands): 1.55 sq.km

On Water Area: 6.75 sq.km

Estimated Population: 70,000 inhabitants distributed in 7 neighborhoods of 10,000

Each neighborhood is in turn divided in 10 precincts of 1000 inhabitants.

The Southern Islands (Fig. 3) are part of the central region of Singapore and include Sisters Islands, Tekukor, St. John, Lazarus and Kusu Island. The new development concerns the creation of a new urban area built both on land and on water (Fig. 4).

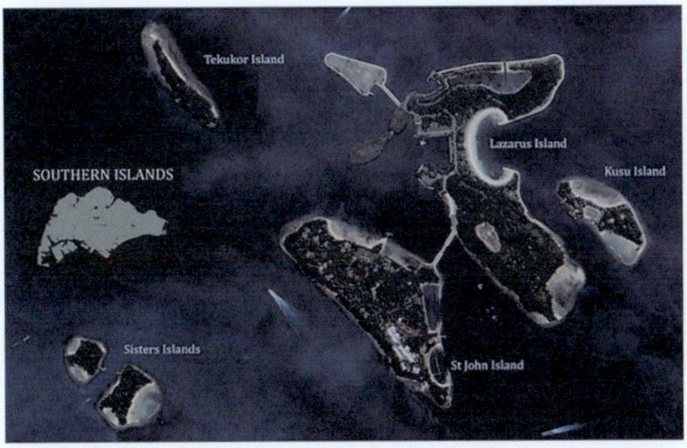

Fig. 3 The Southern islands

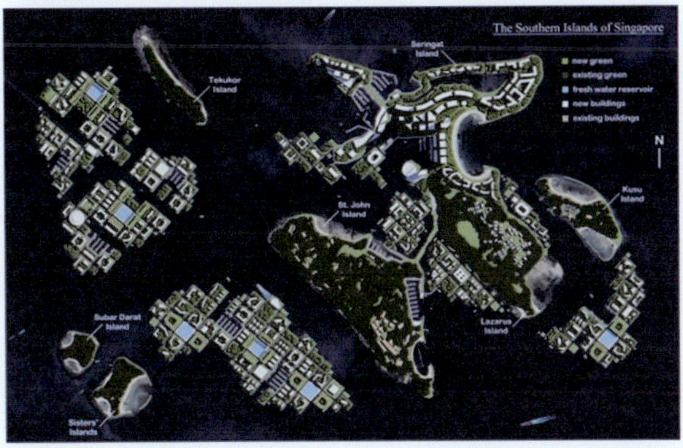

Fig. 4 The new urbanization

The future New Town will be connected to the mainland by the extension of the Sentosa Express and by the already existing ferry service from Marina Bay, which should be appropriately enhanced.

As part of the central region (Fig. 5) and due to the proximity to the Central Business District and to Sentosa, this development should be strongly characterized by touristic, economic activities and by medium–high level residential areas.

Of the new 7 neighborhoods envisaged by the urban plan (Fig. 6), three are built entirely on floating platforms like independent islands (N3, N4 and N5), while the other four (N1, N2, N6 and N7) are mainly developed on land. The New Town centre (N1) is located on the northern part of Lazarus Island, where the high-density commercial, tourist and residential buildings are most concentrated.

The main marinas and the ferry station, as well as museums, theaters and leisure facilities are also located in this area.

The floating neighborhoods are instead characterized by a lower population density and low/medium-rise buildings, however they are equally provided with

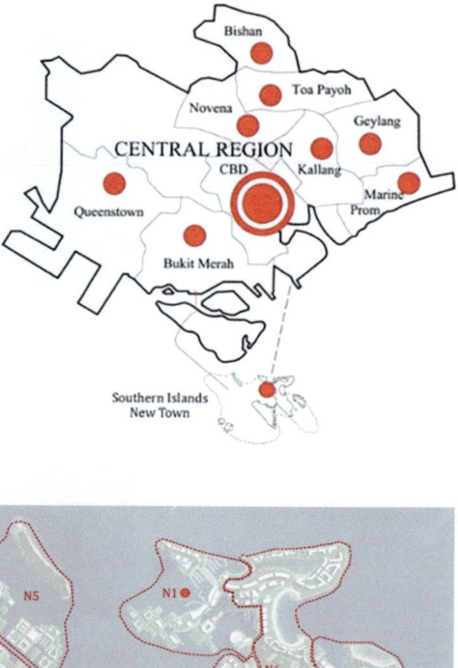

Fig. 5 The Central region

Fig. 6 The 7 new neighborhoods

all the necessary infrastructures and services such as sports complexes, parks, shops and markets.

Tekukor Island, Sisters Islands, Kusu Island, and St. John Island as well as the southern part of Lazarus Island, are not affected by the new developments and their lush vegetation as well as their existing buildings are entirely preserved.

3.2 *Western Islands Development*

New Development Total Area (On Land and Floating): 15.7 sq.km

On Land Area (Islands): 5.7 sq.km

On Water Area: 10 sq.km

Estimated Population: 150,000 inhabitants distributed in 15 neighborhoods of 10,000

Each neighborhood is in turn divided in 10 precincts of 1000 inhabitants.

The Western Islands development (Figs. 7, 8) is part of the west region of Singapore and includes Bukom Island, Palau Ular, Semakau Island, and Sebarok Island, while the Western Islands Industrial development of about 10 sq.km includes Sudong Island, Pawai Island, and Senang Island, and will house all the industrial structures removed from Bukom and Sebarok Islands.

Both the industrial and the mixed-use developments will be connected to the mainland by the existing ferry service from Pasir Panjang and a ferry service from the southern new town will be also provided.

The Western Islands mixed-use development consists of 15 new neighborhoods both on land and on water (Fig. 9). The New Town centre (N1) is located in the central part of Palau Ular and features the largest marina of this group of islands, a ferry terminal, the greatest concentration of leisure facilities as well as a large number of high-rise buildings.

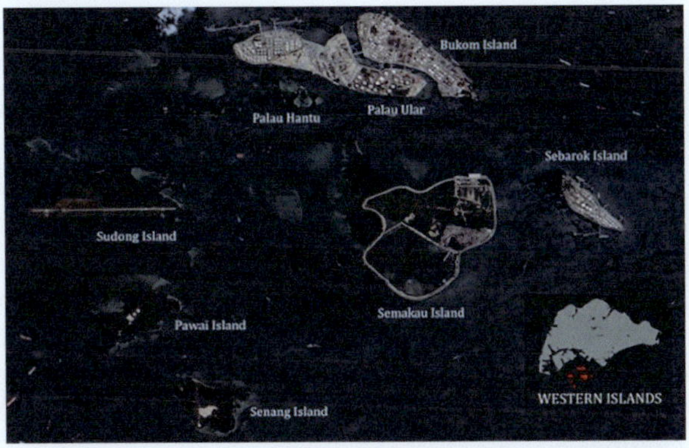

Fig. 7 The Western islands

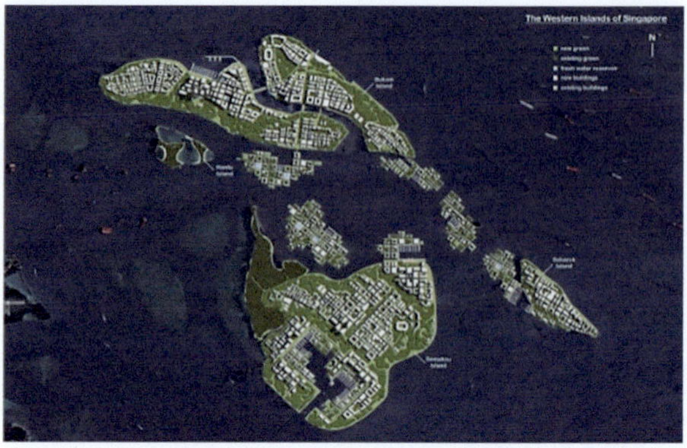

Fig. 8 The new urbanization

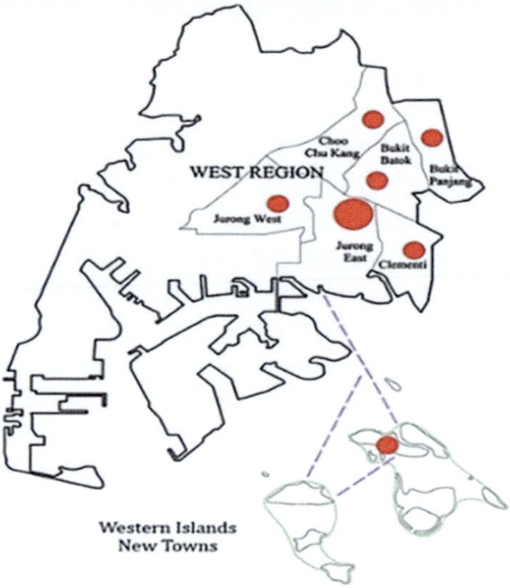

Fig. 9 The West region

At Palau Ular there are other two neighborhoods, N2 and N3, planned entirely on land. These three neighborhoods share the extensive park that crosses the whole island along the west–east axis.

Bukom Island, connected to Palau Ular by three bridges, consists of only two neighborhoods: N4 and N5. Built entirely on land, these neighborhoods share a large park developed along the NW–SE axis of the island.

Semaku Island consists of 6 neighborhoods. Of these, N8, N9 and N12 are entirely on land, while N10, N11 and N13 are built partly on land and partly on water. Also in Semaku Island an interconnected system of parks and green belts allows people to reach all neighborhoods through pats surrounded by lush vegetation and leisure facilities.

At Sebarok Island N14 is built partly on land and partly on water.

Lastly, the only neighborhoods built entirely on floating platforms are N6, N7 and N15, each of them with its own parks, floating reservoir and marina.

3.3 *Eastern Islands Development*

Palau Ubin and Ketam Island

New Development Total Area (On Land and Floating): 13.7 sq.km

On Land Area (Islands): 10.4 sq.km

On Water Area: 3.3 sq.km

Estimated Population: 100,000 inhabitants distributed in 10 neighborhoods of 10,000

Each neighborhood is in turn divided in 10 precincts of 1000 inhabitants.

Tekong Island

New Development Total Area (On Land and Floating): 58 sq.km

On Land Area (Islands): 37.8 sq.km

On Water Area: 20.2 sq.km

Estimated Population: 400,000 inhabitants distributed in 40 neighborhoods of 10,000

Each neighborhood is in turn divided in 10 precincts of 1000 inhabitants.

The Eastern Islands development is part of the north-east region of Singapore and includes Palau Ubin and Tekong Island. The future new towns will be connected to the mainland by the KPE extension and by the existing ferry services from Changi. The construction of a bridge linking Palau Ubin and Tekong each other would also be desirable. (Figs. 10, 11, 12).

Palau Ubin. The new development at Palau Ubin (Fig. 13) consists of 10 new neighborhoods built partly on land and partly on water. It is a low-impact urban plan with

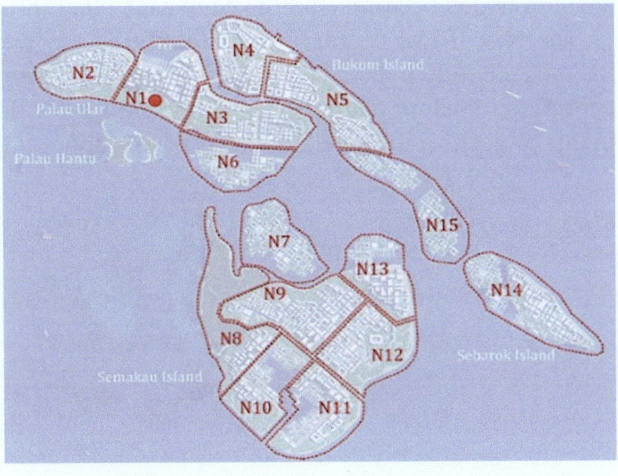

Fig. 10 The 15 new neighborhoods

Fig. 11 The Eastern islands

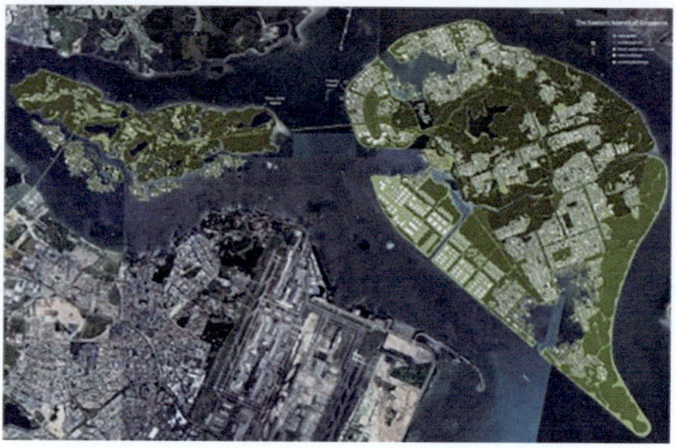

Fig. 12 The new urbanization

low population density, conceived with the aim of preserving the lush nature of the island and with the ambition of creating a modern garden city where all the districts are surrounded by parks and linked to each other through green paths.

The New Town centre (N1) is located in the western area of the island, near the most densely populated neighborhood of the new urban development: N3, built entirely on floating platforms, which is the largest floating development envisaged in this project for the smaller islands of Singapore (Fig. 14).

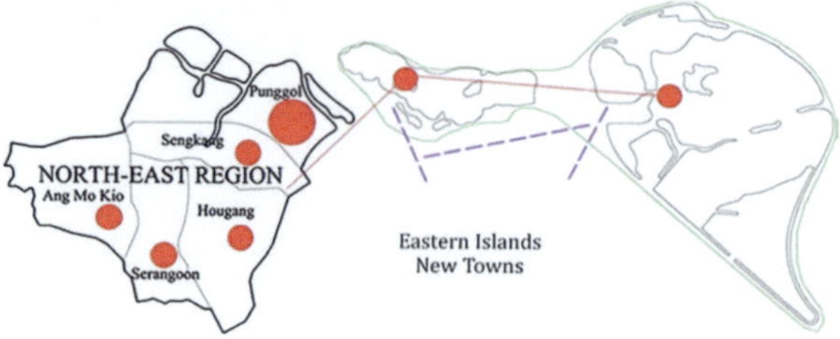

Fig. 13 The North-East region

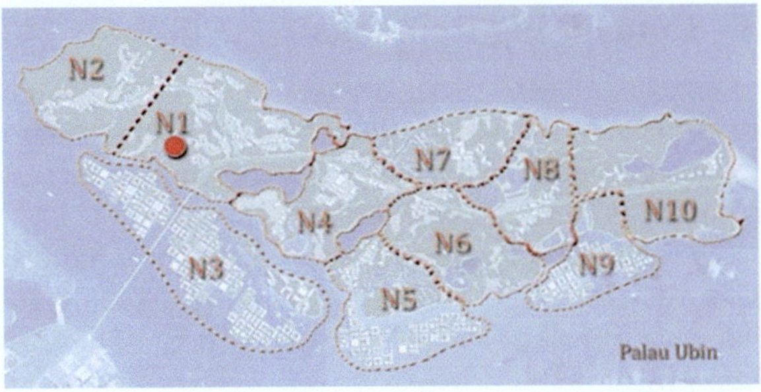

Fig. 14 The 10 new neighborhoods

Tekong Island. The huge development at Tekong Island (Fig. 15) consists of 40 new neighborhoods mainly developed on land. The floating developments are located in the two large basins to the north–west and south–east of the island. The coastal neighborhoods in the north–west area (from N34 to N40 and also N8), the closest as the crow flies to the Changi airport area, are mainly characterized by industrial, commercial and storage typologies and only minimally by residential typologies.

The New Town centre (N1) is located in the northern area of the island and it overlooks the large basin which is the main access to the island. Just like the other islands, Tekong has a system of parks that links the neighborhoods to each other. Given the size of the island, this means kilometers of connections and attractions in the lush greenery, again in accordance with a model of a modern garden city.

Fig. 15 The 40 new neighborhoods

4 Concluding Remarks

In essence, our vision for the urbanization of smaller islands surrounding Singapore revolves around the concept of establishing three additional new towns, seamlessly blending both land and water elements.

We deliberately steered away from the notion of creating a floating city far removed from the coast, recognizing that such isolated entities might struggle to integrate effectively with the existing urban fabric and, consequently, might prove impractical.

Our alternative approach involves envisioning each smaller island's development as a well-connected new town, complete with distinct neighborhoods and precincts. Some of these neighborhoods or precincts are built on water as floating developments. By adopting this strategy, each floating development becomes an integral component of the larger new town, intricately linked to the mainland. Simply put, our conviction lies in the belief that the success of a pilot floating development hinges on its integration within a broader land-based urban framework.

What characterizes the city we envisage through this proposed development? Undoubtedly, it mirrors the vibrant, livable essence of present-day Singapore, a tropical garden city.

As the Singapore Master Plan undergoes continual review and updates to adapt to evolving circumstances and future trends, its foundational planning principles and structures remain steadfast. It is our sincere hope that our proposal aligns with the plan's fundamental principles, prompting necessary updates to regulations that will accommodate the widespread incorporation of floating structures, addressing the future requirements of the city. It's also our hope that our proposed planning approach can be applicable to other coastal cities in contexts similar to that of Singapore.

Reference

1. Ker LT (2015) Challenges and reforms in urban governance—insights from the developments experience of China and Singapore, pp 82–148. Centre for Liveable Cities Singapore and Development Research Center of the State Council of the People's Republic of China

Future Marine Urban Habitation—Concepts and Technologies

Z. Zhou, N. Zhang, W. Gao, and Q. Su

Abstract The impact of climate change and sea level rise on coastal regions, such as Hong Kong and Singapore, has become increasingly severe. Over the next 100 years, the land area of these coastal regions will drastically shrink due to sea level rise. Additionally, rapid urban development and population growth in these regions further increase the demand for land. One potential and effective solution to address these challenges is the construction of a new system for human habitation in the remote and open seas. In this study, the future marine urban system is categorized into interrelated subsystems, including safety and protection, green ecological environment, self-sustaining energy, self-sustaining nutrition, transportation, and marine industry chains. The safety and protection subsystem, located on the periphery of the marine urban system, includes new protective structures such as eco-friendly wave-absorbing breakwaters to provide a stable and safe marine environment for habitation. The green ecological environment system requires zero carbon and zero waste emissions from marine cities. The self-sustaining energy subsystem consists of offshore clean energy capture facilities (wind, wave, solar, etc.). The self-sustaining nutrition subsystem is primarily achieved through hydroponic food and large-scale offshore aquaculture. The transportation system includes the internal transportation and logistics network of the marine cities, as well as air and bridge connections to the mainland cities. This article elaborates on the engineering concepts for future marine urban habitation, elucidates potential construction approaches for the related technological systems and subsystems, and explores the future planning of marine urban living and industries.

Keywords Sea-level rise · Marine habitation · Sea space utilization · Very-large floating structures

---

Z. Zhou (✉) · W. Gao · Q. Su
Marine Hydrodynamic Research Facility, The Hong Kong University of Science and Technology (Guangzhou), Guang-Zhou, China
e-mail: zhuoweizhou@hkust-gz.edu.cn

N. Zhang
State Key Laboratory of Coastal and Offshore Engineering, Dalian University of Technology, Dalian, China

X. L. Zhao et al. (eds.), *Proceedings of the Fourth World Conference on Floating Solutions: WCFS 2024; 2–4 December; Hong Kong*, Lecture Notes in Civil Engineering 597, https://doi.org/10.1007/978-981-96-4569-5_94

1 Introduction

Coastal cities are increasingly vulnerable to the threats of rising sea levels and more intense storms [1]. Throughout the twentieth century, a significant increase in global sea levels was observed, a trend that is expected to not only continue but also to accelerate in the twenty-first century and beyond, driven by the impacts of global warming [2–4]. According to projections under the lower Representative Concentration Pathway (RCP) 2.6 scenario, a global mean sea-level rise of 0.30–0.65 m is likely (with a central 66% probability) by 2100, relative to the baseline period of 1986–2005. Under the higher RCP 8.5 scenario, this rise could be more severe, ranging from 0.63 to 1.32 m by 2100 [5]. This increase in sea levels will reduce the amount of habitable land in coastal areas, which are among the most densely populated regions globally. Approximately 10% of the world's population currently resides in low-elevation coastal zones, defined as areas below 10 m in elevation [2, 6]. Moreover, coastal cities worldwide are expanding rapidly, particularly in developing countries, with a growing population [6]. It is projected that by 2030, half of the world's population will live along coastlines [1]. The rising sea level, combined with the fast-paced development of coastal cities, places an increasing number of people at risk of coastal flooding, heightening their vulnerability to this threat [7].

Several studies have highlighted the predominantly negative socioeconomic impacts associated with sea-level rise [1, 2, 8, 9]. The acceleration of sea-level rise poses significant threats to coastal zones, leading to immediate issues such as the submergence of coastal lands and increased flooding, as well as saltwater intrusion into surface water bodies. Over the longer term, these areas will face further challenges, including heightened coastal erosion and additional saltwater intrusion into groundwater supplies. Coastal ecosystems, such as salt marshes and mangroves, are also at risk. These ecosystems may decline unless they can accumulate sediment at a rate that keeps pace with sea-level rise. Coastal regions, including notable examples like Hong Kong and Singapore, are identified as particularly vulnerable to the exacerbating effects of sea-level rise on coastal flooding [8–11].

In response to the challenges posed by sea-level rise and urbanization, Hong Kong and Singapore have initiated ambitious coastal land reclamation projects. The Kau Yi Chau Artificial Islands project in Hong Kong [12] aims to create approximately 1000 hectares of new land. This project is part of the broader Harbour Metropolis initiative, designed to meet the territory's medium to long-term land needs. Similarly, Singapore's Long Island project [13], which involves land reclamation off the East Coast, seeks to fulfill multiple national objectives. These include enhancing coastal protection, improving flood resilience, bolstering water resilience, and generating additional land for future development. The project also aims to offer new recreational opportunities (Fig. 1).

However, it's important to note that coastal land reclamation technology is generally suitable only for shallow waters and is associated with significant environmental and sustainability concerns [7, 14]. Therefore, there's a pressing need for a cautious approach to the environmental implications of such projects.

As a viable alternative approach, Very Large Floating Structures (VLFSs) offer several advantages. These include lower construction costs, flexibility in site selection, high resistance to waves, rapid construction, straightforward installation, ease of expansion, and minimal environmental impact [15, 16]. VLFSs hold the potential to fulfill a variety of roles, such as maritime airports, logistic hubs, floating bridges, artificial islands, and even floating cities [15–18]. This versatility has generated significant interest among architects, city planners, and engineers, making VLFSs a compelling option for future coastal and marine infrastructure development (Fig. 2).

In this study, we introduce a novel mega-city concept for Marine Urban Habitation (MUH) situated in remote ocean areas, comprising several VLFSs and supporting structures. This concept integrates critical aspects such as safety, self-sustaining energy systems, nutritional provisions, ecological balance, transportation, and marine industry systems, aiming to offer a new paradigm for sustainable living in harmony with the ocean. This approach underscores the potential of future advancements in coastal and offshore engineering to address the pressing challenges posed by sea-level rise and urbanization. Previously, we have presented our concepts during in-house symposia [19, 20], where they were shared through oral reports. In this paper, we provide a more detailed exposition of our innovative MUH concept.

2 Functional Objectives of MUH

Adapting to the challenges of sea-level rise and increased urbanization requires strategic use of marine environments for human sustainability. This effort demands collaboration across various disciplines within marine science and technology, aiming to address several key goals. These include adjusting to changing human living conditions, extending living spaces into marine areas, supporting maritime strategies like the Maritime Silk Road, enhancing national capabilities in marine science and technology, creating new marine industries, transforming excess production capacities into new productive uses, and promoting cross-disciplinary work in marine science and engineering.

To achieve these goals, developing marine spaces for human habitation must meet several functional objectives:

(1) **Creation of Habitable Marine Spaces**: Developing areas that can support both permanent and temporary living at sea, ensuring safety, security, efficient logistics, and advanced communication and digital infrastructure.
(2) **Development of Self-sustaining Environments**: Designing spaces that harmoniously blend ecological, residential, energy, and artificial elements.
(3) **Achievement of Energy Self-sufficiency**: Utilizing a combination of constant and intermittent energy sources, like wave, current, wind, and solar energy, supported by robust energy infrastructures.
(4) **Attainment of Nutritional Self-sufficiency**: Implementing seawater desalination, hydroponics, and aquaculture to ensure food independence.

(5) Promotion of Marine Industry Growth: Laying the groundwork for marine industry sectors, including marine chemicals/biopharmaceuticals and marine equipment and materials.

Addressing these challenges requires envisioning a new type of marine urban system that integrates interconnected subsystems focused on safety and protection, self-sustaining energy/nutrition/ecological balance, efficient transportation, and marine industry growth (Fig. 3).

3 Technologies and Designs

MUH can be utilized as satellite cities to existing coastal megacities. But more ambitiously, it is conceptualized as standalone marine cities located in the remote and open seas. This study envisions a marine city designed to support a population of approximately 0.5 million, spanning an area greater than 100 km^2. The living spaces could be created by several VLFSs each exceeding 3 km in length, which is within the realm of feasibility [16]. Setting aside the financial aspects for the moment, this discussion focuses on the design perspectives of MUH, employing innovative and advanced technologies.

To realize the functional objectives of MUH, the proposal is segmented into six subsystems: safety and protection, green ecological environment, self-sustaining energy, self-sustaining nutrition, transportation, and marine industry chains.

Safety and Protection: This subsystem includes the construction of protective barriers like floating breakwaters and seawalls to reduce wave impact and stabilize core areas of MUH. Incorporating wave energy conversion systems [21, 22] within these structures can also aid in electricity generation, while floating structures can adapt to changing water levels.

Green Ecological Environment: Aiming for zero wastewater discharge and carbon neutrality, this subsystem employs seawater and waste recycling, carbon capture, and the creation of floating green spaces [23, 24] and artificial reefs [25] to maintain ecological balance.

Self-sustaining Energy: This subsystem focuses on strategically positioning energy generation facilities within protected zones, using technologies such as floating offshore wind turbines [26, 27], floating photovoltaic [28, 29], and floating nuclear reactors [30, 31] to ensure a reliable and sustainable energy supply.

Self-sustaining Nutrition: This subsystem involves the establishment of seawater desalination plants [32], large-scale offshore aquaculture [33] zones, and hydroponic food production [34] within protected areas to ensure a steady supply of fresh water, protein, and vegetables.

Transportation: This subsystem covers the development of offshore transportation infrastructure, including airports, urban air mobility solutions [35], immersed tunnels

[36], floating bridges [37], and submerged floating tunnels [38], facilitating efficient movement within and outside the marine habitat.

Marine Industry Chains: This subsystem focuses on developing various marine-related sectors, such as tourism, wellness, survival experiences, and vertical space utilization, supported by engineering solutions tailored for sea-based human survival.

The MUH design process is envisaged to unfold in three steps:

(1) **Initial Construction**: Begin with an outer safe wall approximately 11 km in diameter, enclosing an area around 100 km^2. An inner safe wall will be constructed following the completion of the outer wall.
(2) **Functional Zone Development**: Between the two safe walls, zones dedicated to energy and nutrition production will be planned and developed.
(3) **Habitat Creation**: Within the inner safe wall, calm and protected water areas will host living spaces, workspaces, and transportation infrastructure.

This comprehensive strategy outlines a vision for creating sustainable human habitats within marine environments, leveraging state-of-the-art engineering and technology to establish self-sufficient, ecologically balanced living spaces at sea (Fig. 4).

The construction of MUH leverages a broad spectrum of technologies, from tried-and-true methods to the forefront of innovation. Emerging technologies are steering towards more sustainable and innovative approaches, especially in energy acquisition, where marine renewable energy sources like wind, wave, tidal, and ocean thermal energy conversion, lead research and development efforts. The exploration of soilless plant cultivation in marine settings introduces new possibilities for food production, while extending environmental protection and digital technologies into marine spaces significantly broadens the engineering horizon. Technologies aimed at disaster prevention and mitigation are also evolving, emphasizing the importance of protecting coastal and offshore infrastructures from natural disasters.

However, certain areas remain underexplored, offering room for groundbreaking progress. These include the development of efficient offshore waste management and recycling techniques, the construction of VLFSs, and improving permanent safety measures for buildings in marine environments. Additionally, establishing comprehensive offshore healthcare services and developing international legal frameworks to support human habitation at sea pose significant challenges that must be addressed to facilitate the sustainable expansion of human activities into maritime settings.

The transition from well-established practices to cutting-edge innovations is crucial for promoting sustainable development, environmental conservation, and the expansion of human habitation and activities in coastal and marine regions. As the field continues to evolve, interdisciplinary collaboration and innovation will be essential in navigating the challenges and seizing the opportunities presented by the vast and dynamic marine environment (Fig. 5).

4 Concluding Remarks

MUH presents a forward-looking solution to the increasing demand for land in coastal regions, exacerbated by challenges such as rising sea levels and urban expansion. This study highlights the critical need for exploring marine spaces as viable options for human settlement, advocating for a multidisciplinary approach that merges advancements in coastal and offshore engineering, sustainable living practices, and environmental conservation. The envisioned marine urban system proposes a holistic strategy for establishing self-sufficient, ecologically harmonious living areas at sea, with a focus on ensuring safety, achieving energy and nutritional independence, maintaining ecological balance, and fostering the growth of marine industries. The MUH domain is marked by a wide range of technologies, from established methods of energy and nutrient procurement to burgeoning practices in environmental safeguarding and digital innovation, signifying a promising future for sustainable marine development. The continuous evolution of this field, powered by interdisciplinary cooperation and inventive efforts, is essential for surmounting current obstacles and leveraging the immense potential of the marine environment. This progress lays the groundwork for a sustainable proliferation of human activities into maritime territories, demonstrating the pivotal role of MUH in shaping the future of coastal and offshore living.

(a)

(b)

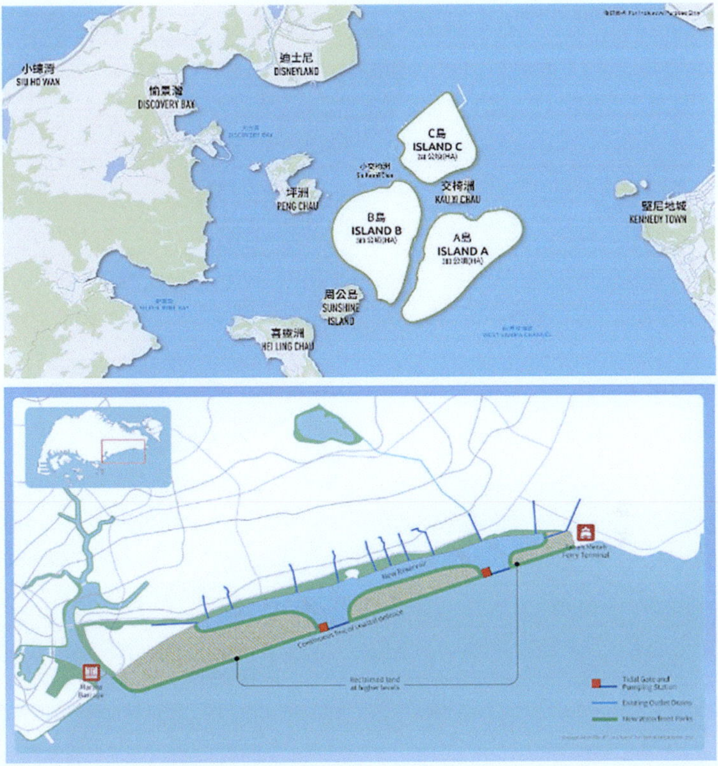

Fig. 1 Newly land reclamation proposals for adaptation urban development and sea level rise: **a** Kau Yi Chau Artificial Islands in Hong Kong (*source* www.centralwaters.hk) and **b** Long Island in Singapore (*Source* www.ura.gov.sg/)

Fig. 2 Original components of a Mega-Float System (*Source* [17])

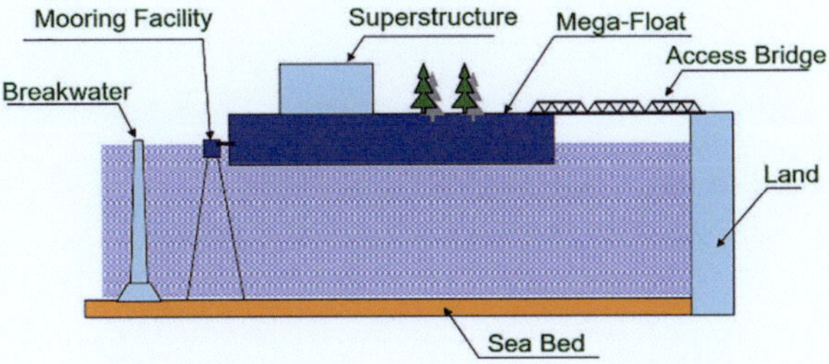

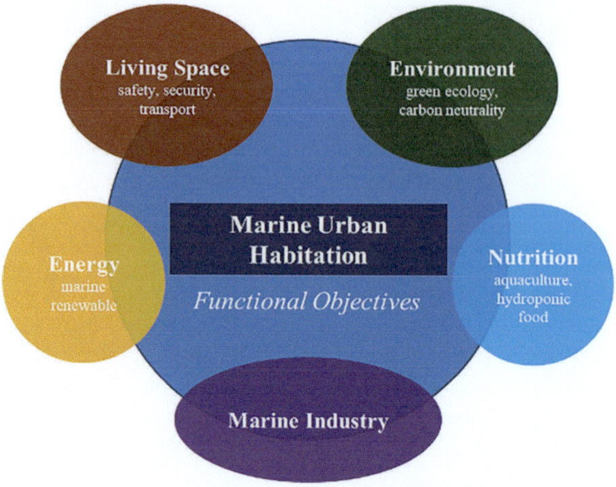

Fig. 3 Functional objectives of MUH

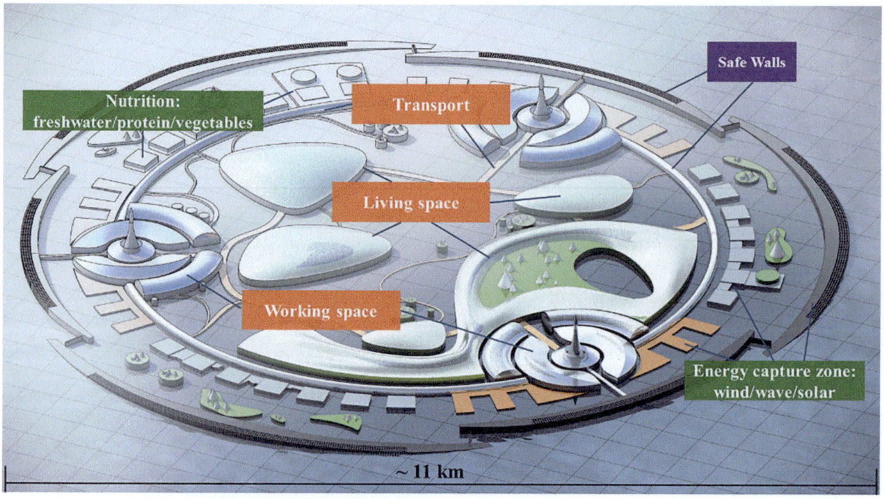

Fig. 4 A sketch design of MUH: aerial view

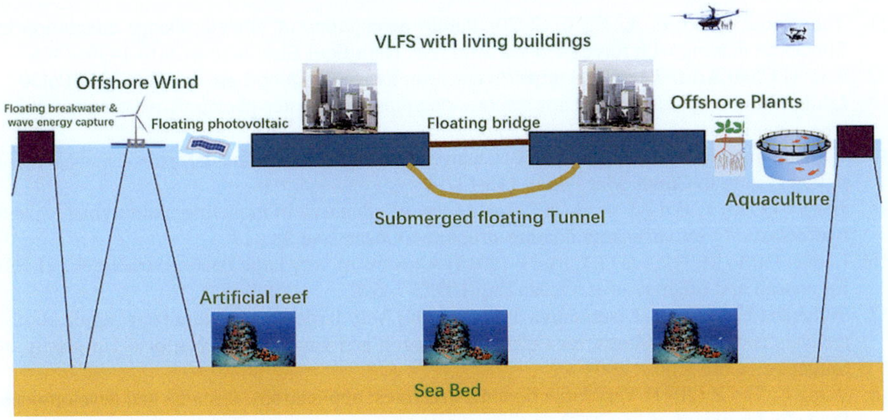

Fig. 5 A sketch design of MUH: landside view with technologies

Acknowledgements We extend our sincere thanks to **Weico Architect** for their valuable contributions to Fig. 4 in this research. Their support has significantly enriched the quality of our study, and we are deeply grateful for their support.

References

1. Shen P, Wei S, Shi H et al (2023) Coastal flood risk and smart resilience evaluation under a changing climate. Ocean-Land-Atmos Res 2:0029
2. Nicholls RJ, Cazenave A (2010) Sea-level rise and its impact on coastal zones. Science 328(5985):1517–1520
3. Frederikse T, Landerer F, Caron L et al (2020) The causes of sea-level rise since 1900. Nature 584(7821):393–397
4. Wang J, Church JA, Zhang X et al (2021) Reconciling global mean and regional sea level change in projections and observations. Nat Commun 12(1):990
5. Horton BP, Khan NS, Cahill N et al (2020) Estimating global mean sea-level rise and its uncertainties by 2100 and 2300 from an expert survey. npj Clim Atmos Sci 3(1):18
6. Neumann B, Vafeidis AT, Zimmermann J et al (2015) Future coastal population growth and exposure to sea-level rise and coastal flooding—a global assessment. PLoS ONE 10(3):e0118571
7. AlQahtany AM, Dano UL, Abdalla EME et al (2022) Land reclamation in a coastal metropolis of Saudi Arabia: environmental sustainability implications. Water 14(16):17
8. Cannaby H, Palmer MD, Howard T et al (2016) Projected sea level rise and changes in extreme storm surge and wave events during the 21st century in the region of Singapore. Ocean Sci 12(3):613–632
9. Yu Q, Lau AKH, Tsang KT et al (2018) Human damage assessments of coastal flooding for Hong Kong and the pearl river delta due to climate change-related sea level rise in the twenty-first century. Nat Hazards 92(2):1011–1038
10. Parker A (2018) Sea level oscillations in Japan and China since the start of the 20th century and consequences for coastal management—part 2: China pearl river delta region. Ocean Coast Manag 163:456–465

11. Yousefpour R, Prinz A, Ng C (2020) Public perceptions of climate change adaptation in Singapore dealing with forecasted sea level rise. Hum Ecol Risk Assess 26(6):1449–1475
12. Kau Yi Chau Artificial islands https://www.centralwaters.hk/en/. Last accessed 2024/06/30
13. Long island https://www.ura.gov.sg/corporate/planning/master-plan/draft-master-plan-2025/long-island. Last accessed 2024/06/30
14. Li FX, Ding DD, Chen ZJ et al (2020) Change of sea reclamation and the sea-use management policy system in China. Mar Policy 115:12
15. Jiang CQ, Xu P, Bai XL et al (2023) A review of advances in modeling hydrodynamics and hydroelasticity for very large floating structures. Ocean Eng 285:17
16. Lamas-Pardo M, Iglesias G, Carral L (2015) A review of very large floating structures (VLFS) for coastal and offshore uses. Ocean Eng 109:677–690
17. Watanabe E, Wang C, Utsunomiya T et al (2004) Very large floating structures: applications, analysis and design. Centre for Offshore Research and Engineering. National University of Singapore. Core report, 2004–02
18. Wang C, Tay Z (2011) Very large floating structures: applications, research and development. Procedia Engineering 14:62–72
19. The first symposium on latest advances in marine land development technology, https://ce.hkust.edu.hk/news/ocean-we-want-liangxiaoxieshoujubanyantaohui-gongyi-haishangrenjugongcheng. Last accessed 2024/06/30
20. The second symposium on latest advances in marine land development technology https://www.hkust-gz.edu.cn/2024/06/09/港科大与港科大(广州)联合举办第二届未来海上/. Last accessed 2024/06/30
21. Mustapa MA, Yaakob OB, Ahmed YM et al (2017) Wave energy device and breakwater integration: a review. Renew Sustain Energy Rev 77:43–58
22. Li YW, Ren NX, Li X et al (2022) Hydrodynamic analysis of a novel modular floating structure system integrated with floating artificial reefs and wave energy converters. J Mar Sci Eng 10(8):18
23. Vo TKQ, Vo TDH, Ntagia E et al (2023) Pilot and full scale applications of floating treatment wetlands for treating diffuse pollution. Sci Total Environ 899:15
24. Yeh N, Yeh P, Chang YH (2015) Artificial floating islands for environmental improvement. Renew Sustain Energy Rev 47:616–622
25. Paxton AB, Shertzer KW, Bacheler NM et al (2020) Meta-analysis reveals artificial reefs can be effective tools for fish community enhancement but are not one-size-fits-all. Front Mar Sci 7:11
26. Ha K, Truong HVA, Dang TD et al (2021) Recent control technologies for floating offshore wind energy system: a review. Int J Precis Eng Manuf-Green Technol 8(1):281–301
27. McMorland J, Collu M, McMillan D et al (2022) Operation and maintenance for floating wind turbines: a review. Renew Sustain Energy Rev 163:13
28. Zhang M, Schreier S (2022) Review of wave interaction with continuous flexible floating structures. Ocean Eng 264:13
29. Dai J, Zhang C, Lim HV et al (2020) Design and construction of floating modular photovoltaic system for water reservoirs. Energy 191:15
30. Wang QW, Zhang Y, Zhang H (2023) The development of floating nuclear power platforms: special marine environmental risks, existing regulatory dilemmas, and potential solutions. Sustainability 15(4):20
31. Buongiorno J, Jurewicz J, Golay M et al (2016) The offshore floating nuclear plant concept. Nucl Technol 194(1):1–14
32. Curto D, Franzitta V, Guercio A (2021) A review of the water desalination technologies. Appl Sci-Basel 11(2):36
33. Naylor RL, Hardy RW, Buschmann AH et al (2021) A 20-year retrospective review of global aquaculture. Nature 591:551–563
34. Lee S, Lee J (2015) Beneficial bacteria and fungi in hydroponic systems: types and characteristics of hydroponic food production methods. Sci Hortic 195:206–215

35. Gao ZY, Yu Y, Wei QS et al (2024) Noise-aware and equitable urban air traffic management: an optimization approach. Transp Res Part C-Emerg Technol 165:25
36. Su QK, Zhu YL, Chen Y et al (2022) Hong Kong Zhuhai Macao bridge-tunnel project immersed tunnel and artificial islands—from an owners' perspective. Tunn Undergr Space Technol 121:18
37. Fenerci A, Kvåle KA, Xiang X et al (2022) Hydrodynamic interaction of floating bridge pontoons and its effect on the bridge dynamic responses. Mar Struct 83:103174
38. Zhang HQ, Yang ZW, Li JZ et al (2021) A global review for the hydrodynamic response investigation method of submerged floating tunnels. Ocean Eng 225:13

A Planning of Floating Habitat Complex in Consideration of Feng-Shui (风水) Criteria

Sang W. Pak and Han-seok Lee

Abstract In light of designing a harmonious and permanent oceanic space for living, here we superpose some of key Feng-shui criteria to a floating oceanic residence. Although oceanic application of Feng-shui, in the absence of its key subject of mountain(s), allows only a weak connection, this approach is self-evidently effective to plan more habitable atmospheric and hydrographic environment of an oceanic residence. In this designing of floating habitat complex, in conjunction with the previous article ' WCFS2023 Japan-12', we take some key Feng-shui design criteria into account for planning an ideal residential space, '明堂', of the Ulsan floating complex within nearshore coastal water of the southeastern sea of Korean peninsula. Five(5) selected Feng-shui criteria that we extend their land-origin domains of application into the oceanic environment are (1) northernly-backing mountain ranges, (2) southernly meandering water mass, (3) south façade of a structure and the directional orientation of a house; 坐向论, (4) water gain; 得水法, and, newly argued, (5) Feng-shui for oceanic residence; 海上住居风水.

Keywords Coastal planning · Floating oceanic architecture · Oceanic Feng-shui

1 Introduction; Feng-Shui for a Floating Oceanic Architecture on the Open Sea

Ancient oriental wisdom of Feng-shui, especially following instructions of Form school, suggests key conditions in selecting a building-site and, further, provides general architectural design guidelines of a building on the site. As the traditional geological theory for surveying residential sites and planning cities, Feng-shui has

S. W. Pak (✉)
Korea Ocean Space & Settlement Co., Ltd., Busan, Republic of Korea
e-mail: dive8752@gmail.com

H. Lee
Sangji Environment & Architecture Inc., Busan, Republic of Korea

X. L. Zhao et al. (eds.), *Proceedings of the Fourth World Conference on Floating Solutions: WCFS 2024; 2–4 December; Hong Kong*, Lecture Notes in Civil Engineering 597, https://doi.org/10.1007/978-981-96-4569-5_95

its goal to deliver an optimal and secured living environment that brings ultimate harmony between residents and the natural environment [1].

From a utilitarian perspective, the life system of Feng-shui is defined by boundaries within specific mountain ranges with sufficient elevation and a defined water basin to guarantee meteorological and hydrological advantages for the operations of life therein. In practice, Feng-shui is keen to exploit the thermal advantages of geological formations. The bitter cold intensity of the winter westerly winds is tempered by the high northwestern mountains, and the local atmospheric boundary layer is kept warm by being nestled by the main backing mountain to the north and two hills to the east and west, called 'Dragon' and 'Tiger', both running parallel to the south. Further to the south, a meandering body of water brings more utility and, at the same time, some disutility. The ease of irrigation and the likelihood of flooding increase by getting close to the water boundary. The south-facing orientation of a house, as a microcosm of Feng-shui's site selection, adds thermal advantages to the daily indoor life. And by adding an in-house water source, these Feng-shui survey objectives are closing in with the achievement of securing one of those life's security resources for permanent residency. The abundance of water resources within a home will bring unprecedented convenience and privacy to everyday life. The desire for a high-quality life and the striving for advanced living system are never-ending [2] as we are seeking, in the following, for a permanent oceanic residence within coastal nearshore with design objectives of passive indoor thermal control, based on relatively stable thermal property of water, and blue economic sustainability in light of oceanic Feng-shui criteria. Utilitarian advantages of ocean in resources are exploited while limiting its systematic boundaries of sustainable life-making mostly onboard a floating architecture and within a proximate ocean of artificial atoll [3]. A semi-autonomous coastal community built across nearshore Ulsan may signal the beginning of the necessary transition of anthropogenic life systems to fully sustainable ones.

2 A Floating Architecture Moored Within Storm-Induced Shoaling Zone Inside Submerged Breakwaters as '明堂' Have Its Northernly-Backing Mountain Ranges

The typhoon-induced wave climate in the nearshore continental shelf of Ulsan poses a major threat of environmental forcing to all oceanic structures within, whether fixed or floating. As ancient people of Feng-shui countries have searched for reliable protection from cold northwest winds, which was critical for their safe winter sheltering, and settled on the foot slopes of the northernly-backing mountain, an oceanic residential site is surveyed against the local wave climate in the nearshore Ulsan around water depths between 30 and 50 m, defined as the storm wave shoaling zone. Designing an artificial underwater atoll around the site will alter the bathymetry and, therefore, the local wave climate, allowing a floating oceanic architecture more favorable conditions for weathering storms.

As sedentism is well-preserved by favoring thermodynamics within the surrounding terrain of a Feng-shui selected site, a well-planned oceanic site shall provide secured advantages in hydrodynamic stability. With its designed maximum wave breaking height of 15 m, the sites are set between a 30 m water depth contour at the shallowest and outer boundary of 50 m at the deepest water depth, while the general marine operability diminishes as the working water depth increases (See Fig. 1).

A moored floating oceanic architecture of 100 m diameter is protected from storm-induced high energy wave fields within an artificial atoll of 1000 m diameter. Both intact and failure stability analysis are required for its mooring plan, because a failure of mooring system could cause the architecture to drift and be stranded on shore.

The high heat content of seawater makes it to be utilized as heat source of 16 °C during winter and as heat sink of 26 °C during summer in nearshore Ulsan. To plan energy passivity on this architecture cool air above the water–air boundary is collected and distributed to indoor spaces for ventilation. (See Fig. 2. Left) For the public indoor spaces, atmospheric circulation is designed to be initiated by warm air rising, assisted by preheated thermal fins on the rooftop by insolation. Stronger wind speed above the ocean surface compared to on-land requires an effective methodology that combines the weathertightness of the superstructure, indoor air circulation and air quality control. The vacuum applied to private indoor spaces is vented out to rooftop funnels by a primarily passive ventilation management system, while in public spaces a general openness is designed without compromising weathertightness to withstand storms (See Fig. 2 Right).

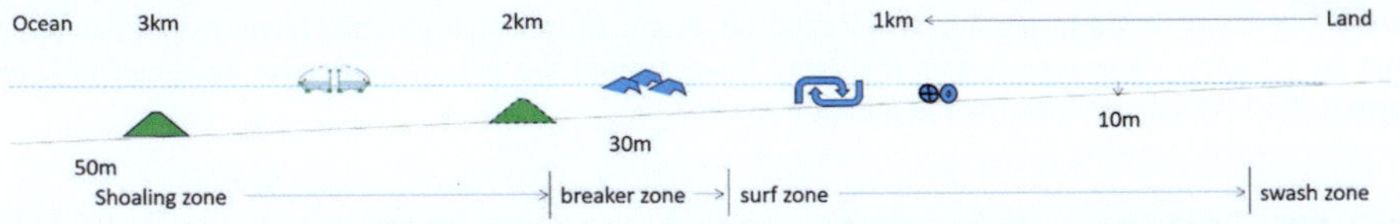

Fig. 1 Shows the simplified bathymetry (brown line) of coastal nearshore Ulsan with a floating architecture moored within an atoll (green trapezoids) set in shoaling area less than 50 m deep. With the designed wave breaking depth of 30 m during a typhoon, a rip current zone (thick arrows whirling) and a longshore current zone (⊕⊙) are symbolized across the surf zone

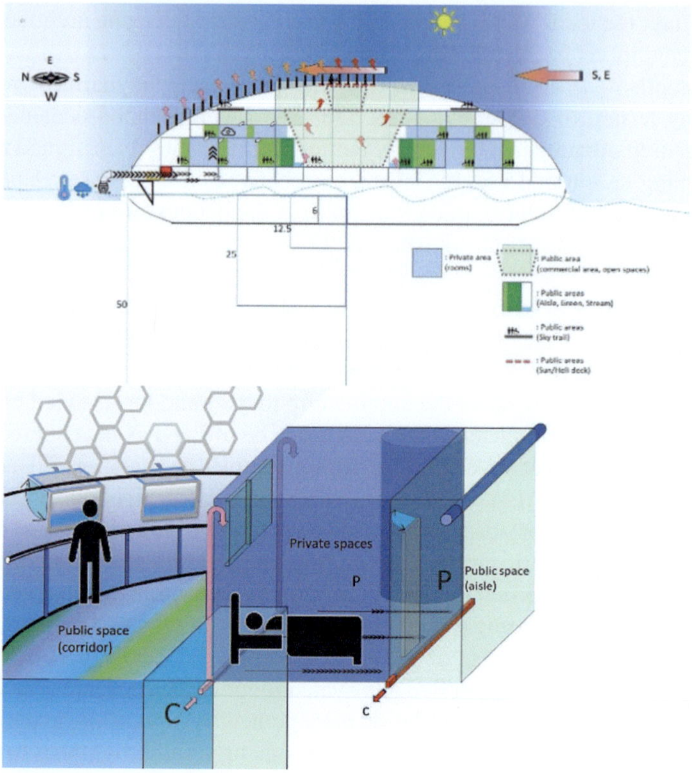

Fig. 2 Left shows that, during summer when typical south-to-east winds blow, the preheated fins between funnels cause warm indoor air to rise, while cooler air is collected from the water–air boundary layer to aid in passive air circulation. Right shows that air sinks have installed to keep the air pressure of private spaces relatively lower, while the public spaces are designed to keep atmospheric openness towards the outdoor space

3 Flood Safety and Outdoor Water Accessibility of a Floating Oceanic Residence as '明堂' Have Planned with Its Southernly Meandering Water Mass

For communities on a mountain slope, having meandering water mass nearby implies a lot from a variety of perspectives, including from safety risks to utility. The design criteria for a floating architecture and its oceanic communities aboard share the same aspects of both protection from flooding and efficient outdoor water accessibility including above and on the water as well as underwater spaces.

The artificial wave barrier, designed as a submerged atoll with its assumed freeboard of 10 m and crest width of 15 m, is protecting a floating architecture within the atoll from storms with designed maximum breaker height of 15 m in sea state 5 (wave height less than 4 m) or lower states without the risk of flooding. The 11 m draft and 4 m freeboard of the architecture provide the first line of flood defense while general watertightness designed for the hull and 1st floor of the superstructure with the drainage features of camber, sheer, bulwark and freeing ports enhance seaworthiness of open deck spaces. The superstructure is, in general, required to be weathertight, while the convertible rooftop at Sun & Wind deck on 6th floor allows this space to become semi-outdoor with jogging tracks and gardens. More openings and general openness design in public spaces are desired for permanent residence, allowing sufficient ambient light and air to come in, while active canopy opening and closing control is mandated. The exterior of superstructure provides safe aerodynamic response to storm wind and reliable weathertightness, keeping indoor spaces dry and free from the risk of consequential flash flooding.

Water accessibility inside the atoll is guaranteed under normal sea states within calm to smooth water surfaces, posing no wave or current-induced difficulty for surface and underwater activities. To enhance safer water accessibility, water activity ports and an airlock are planned. (See Fig. 3) An airlock at the central keel area and supplementary spaces located at its surrounding will guarantee entries and exits into the water column without wave slaps or wave pushes and pulls. Planning of pre- and post-activity service areas with thermal support of shower, sauna, drying, change clothes, refreshments and toilets will expand the residents' horizons of daily activities from indoor spaces toward the water column.

Outdoor space shall have designed features for safe navigation and homing guidance, although neighborhood design features are not available yet. Wearable gears, including informative electronics and personal propulsion devices, add more safety aids to work and leisurely activities on and under the sea surface. Effective waterborne transportation aids, such as guiding ropes and diver propulsion vehicles, shall be utilized in conjunction with real-time positioning and other geomatic information. For activity planning and management reasons, sea states at different access points and activity destinations shall be monitored automatically and distributed to users in real time. Well-preserved or restored underwater ecosystem within and outside the atoll will add to the variety of underwater flora and fauna, which would accelerate this transition of anthropogenic living system from land-based large market-dominated ones toward sustainability-dominated local blue economy of those oceanic communities.

4 West Façade of a Floating Oceanic Residence as Its South Façade of '明堂' (坐向论)

The typical Feng-shui south-facing façade of a house built on a northernly-backing mountain slope shall have good matching properties with an iconic west-facing façade of a floating oceanic architecture in nearshore Ulsan. This is partly because oceanic building sites do not exhibit strong specificity in the thermal directionality of seawater. Instead, the strongest environmental forcing of storm-induced wave field

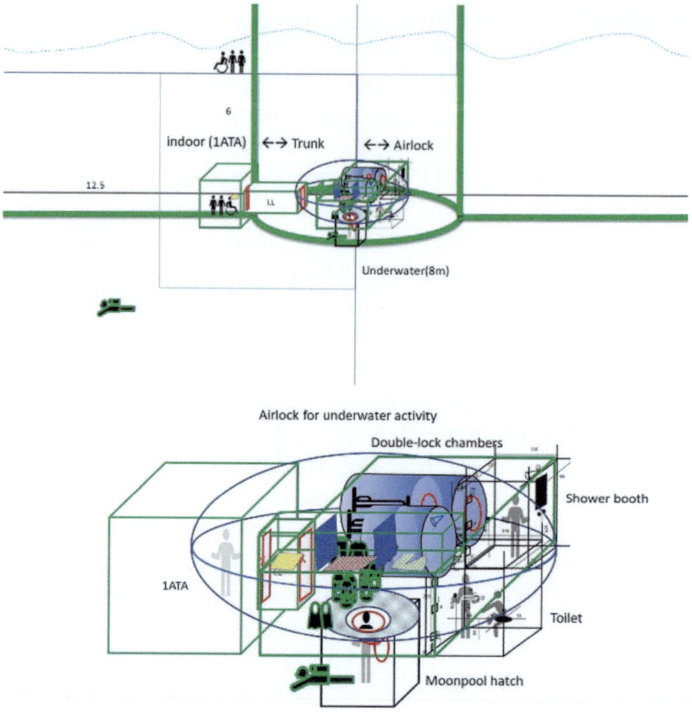

Fig. 3 Left shows an airlock installed at the center of hull bottom for accessing underwater space. Right shows the conceptual formation of the airlock unit with two double-lock chambers, a shower, a toilet and a moonpool hatch

has very specific directional orientation and depth characteristics in the region. The directional orientation of west-facing façade provides passive protection from most destructive storm winds and waves across the nearshore. This directionality of main façade also comes with a wide opening section of the submerged atoll, through which main navigational route is planned to various destinations including Ulsan, Busan, Fukuoka, Shanghai to the south and Pohang, Donghae and Vladivostok to the north (See Fig. 4 Left). The façade is equipped with a helideck and a boat landing below it (See Fig. 4 Right).

To maintain indoor thermal habitability a superstructure and its exterior are to have sufficient thermal insulation in addition to mandatory weathertightness. The angle of insolation in Ulsan reaches up to 78° in the summer and 31° during the winter. The section of rooftop which makes close to a right angle with incoming sunlight shall be protected with metallic shield film or photovoltaic film to effectively avoid or drain insolation. On the Sun & Wind deck of 6th floor the convertible section of rooftop will contribute to indoor thermal control with designed sliding and folding opening and closing mechanism. Considering its 25 m height above the sea level and 100 m diameter, the daytime visual impression of this architecture is assumed to be

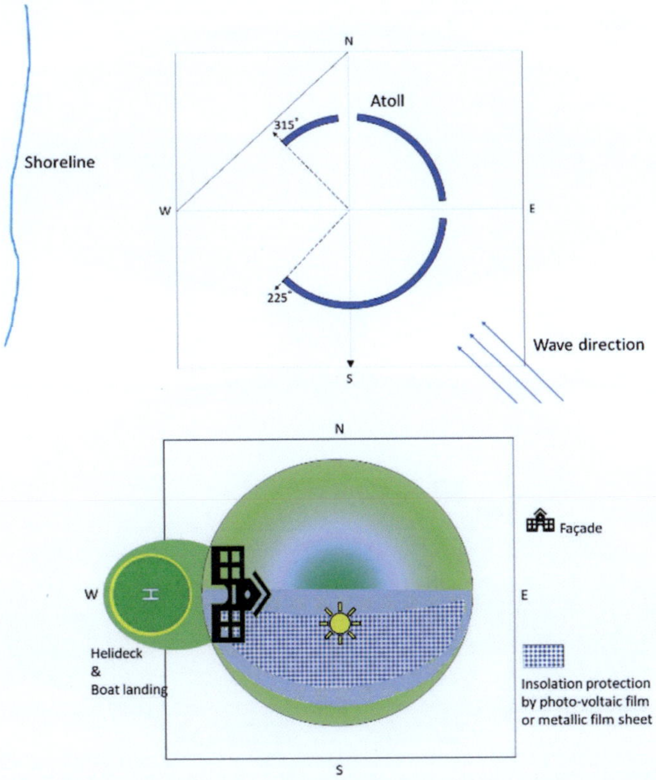

Fig. 4 Left shows sections of the submerged atoll upon completion with a large opening to the west. Right shows a west-facing façade (Top view) animated with the arrival areas of a helideck and a boat landing (hidden below the helideck)

minimal, except for its symmetric appearance and color observed from the coastline 2 km away.

In line with the utilitarian design principles, a façade shall be designed to provide socio-psychological impression as an iconic spatial reference for both residents and visitors. Some of the design features of the façade, including the multiple pillars and reception entrances, are illustrated in the following images, in which the symmetry of exterior and emptiness of the freeboard deck spaces (4 m above waterline) are emphasized for an observer who just came on board. (See Fig. 5) Hull colors and exterior symmetry, or asymmetry for some opened section, get apparent for a visitor arriving to a helideck or a boat landing. For new arrivals, the façade serves as the starting point of one's mental map of her or his onboard orientation, which provides them with vital information about other exits and ports.

The windows in the corridors of public spaces are designed for both weathertightness and ease of opening and closing by either hinged or folding mechanism. All external openings in the façade shall be planned for weathertightness except for the

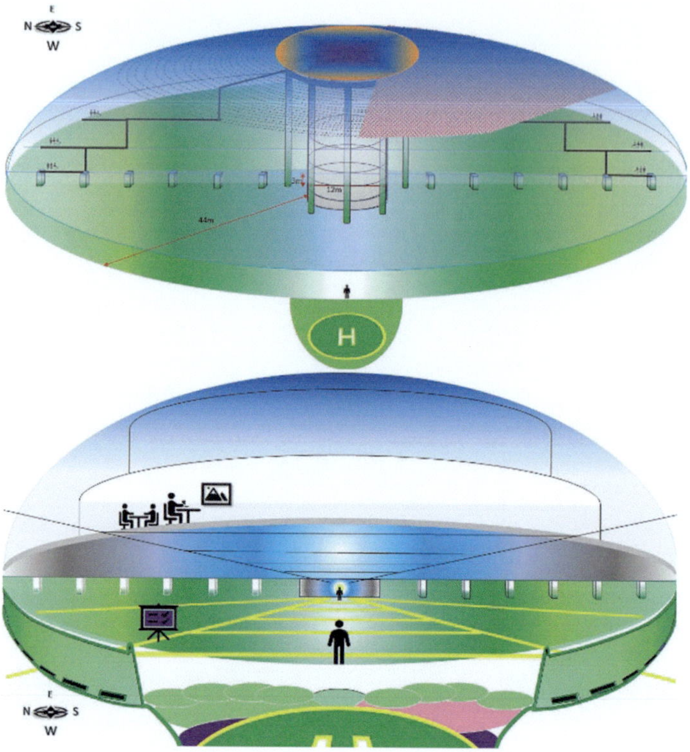

Fig. 5 Left shows an arrival impression of a person approaching the residence by helicopter at an altitude of approximately 200 m. Right shows another impression of a person standing on the helideck looking out at the façade

1st floor and below, where watertight design is mandatory. Weathertight design of the superstructure is based on two-layered interior pressure management of private and public spaces, creating pressure boundary envelopes between the exterior and public spaces, for a storm-induced flood management, and between public and private spaces, for indoor air quality management.

Each residential unit is provided with atmospheric ventilation sinks to eliminate air circulation into public spaces. A vacuum of 2 psi or less is applied for floor dust filtration and water vapor ventilation in sanitary and cooking areas. Private spaces are designed to self-regulate their micro-climates. Ventilation in public spaces is primarily designed and operated mostly passively by ambient pressure changes. With the designed basic airtightness, air circulates within open spaces through doors, windows, and hatches, allowing each space randomly acting as air sources, sinks or passages. In principle however, outdoor air enters through windows in the corridor or through inlets near water–air boundary, heats up on its way to the central area of the architecture, then rises and be sucked outside through the rooftop openings with the help of thermal fins and high-velocity winds overhead.

To increase local moisture level, indoor waterfalls and fountains are installed. Although the superstructure shall maintain weathertight envelope anytime it is required, especially over storms, however, long-term confinement is not the primary design goal of public spaces. Adding windows based on passive air circulation guidelines, intaking fresh breezes, and increasing the level of outdoor exposure shall become the main design philosophy of the ocean district lifestyle in Ulsan.

The extended version of façade may include other floating structures around the residence, including lighting buoys and signature statues, either permanently or temporarily installed, that represent the residential seascape. An array of wind and wave energy converters exposed above sea level will form the backdrop for the façade. Planting vegetation in the indoor and outdoor spaces on the ground floor, freeboard deck, will give the façade a variety of looks and shapes. Indoor plants in public spaces are subject to stricter control, including species and planting location with monitoring of pollen distribution, while plants in private spaces are grown solely based on residents' personal preference.

4.1 Semi-Autonomous Life Resources Gain at a Floating Oceanic Residence as '明堂' Completes Its 'Water Gain' (得水法)

For modern oceanic floating residences, the need for various resources remains the same, just as the spontaneous flow of water is desired from wells found by a Feng-shui surveyor. A floating oceanic residence shall first be passive to energy losses before planning active energy use to make up for those losses. The high thermal conductivity of seawater acts as a permanent heat sink for the hull, so the insulation is planned by applying the required design thickness of geopolymer concrete. Before planning the seasonal heating and cooling capability, any indoor mechanical heat source shall be analyzed and incorporated into heat budget calculations.

It is reasonable to extrapolate the Feng-shui 'water gain' to cover other life-sustaining resources such as food and energy that can be effectively supplied, stored, and eventually grown or generated near the surrounding artificial reefs. Outside the atoll, various types of oceanic energy converters will be installed to power winds, waves and currents. Wind and wave energy generation shall be from purpose-built floating devices installed outside the atoll without compromising the mooring stability of floating residential structures nearby (See Fig. 6).

For the energy security of those oceanic residences, redundancy is planned with a quadruple energy supply mixture combining photo voltaic, current, wave and wind energy in proportions of 20, 30, 30 and 20 percent respectively. The energy farms will be located outside atolls in optimal locations for each power convert group, according to the generation requirements of high current, wave, and wind quality. Power storage facilities with battery chambers are modularized and installed offboard to avoid the risk of potential fire fatality but within atolls for the ease of maintenance.

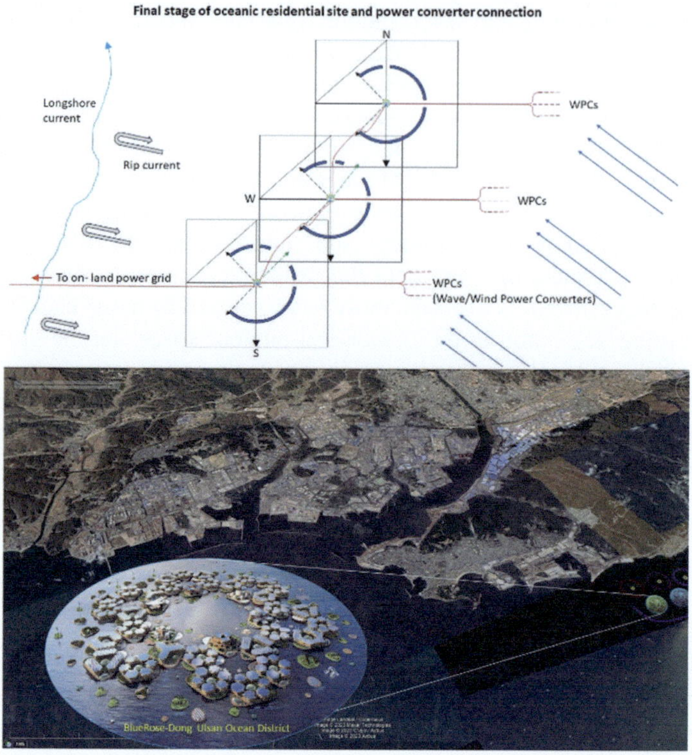

Fig. 6 Left shows a group of floating oceanic residences with submerged atolls, which partially surround the residences, and the local power connection routes to wind and wave power converters installed on the outside of the atolls. Right shows Ocean District of Ulsan city with an enlarged sub-district of BlueRose-dong, an adopted floating community illustrated by Oceanix.com [5]

Freshwater is distilled from seawater by an active energy process of reverse osmosis. The distillation also forms the initial stage of brim production, followed by an electrolysis process to maximize the concentration levels of bicarbonates in the brim until they precipitate spontaneously. Condensation and filtration of atmospheric moisture at indoor drinking fountains produces additional water for consumption, which forms a practical part of the local atmospheric humidity control system, making location-specific indoor habitability more comfortable.

The quality of seawater within the atoll shall be kept clean enough to be pumped into the in-house calcium carbonate plant within the hull. Modules for gas (hydrogen and carbon dioxide) storages are installed offboard and within atolls to increase safety factor of the residences. Data centers could be designed in other offboard modules but would be installed in isolated locations inside atolls, taking advantage of the sufficient thermal cooling of seawater convection and avoiding sediment buildup or scouring. Other large gas storage modules could be installed in deeper waters outside the atolls.

5 Feng-Shui for Oceanic Residence (海上住居风水)

Any fair and precise Earth system can be an offshoot model of the Feng-shui philosophy [5] that we have applied or are trying to apply to understand these new oceanic habitats more holistically. Since Feng-shui (风水) literarily stands for two Chinese words, 'wind' and 'water', most Feng-shui practices have been to better understand their behavior within specific geological boundaries, namely Feng-shui sites.

It is clear that wind and water are still the obvious main players in global general circulation, and their abundant presence in the oceanic environment has a great impact on the well-beings of oceanic residents, as the most destructive environmental forcing in coastal nearshore is a storm with strong wind and high waves. A fusion of science and engineering that provides guidelines for oceanic residents, villagers, and Ulsanians floating on the edge of continental shelf to protect their livelihoods, practical habitats, and sustainable oceanic lifestyles from sea level rise, flash floods, droughts, heat waves, and especially typhoons or hurricanes without having to flee their new living habitats, we simply argue it is Oceanic Feng-shui.

References

1. Xu P (1990) Feng-shui: a model for landscape analysis
2. ChoiC-J (1997) Korean native Feng-shui. Minumsa, Seoul
3. Pak S, Lee H-S (2023) A study on planning of sustainable floating architecture outside a bay utilizing submerged breakwater reef that sequesters CO_2. In: 3rd World conference on floating solutions on proceedings. Springer, London, pp 1–2
4. OCEANIX Homepage. https://oceanix.com/media__uploads/images/01_BIG_SFC_OceanixCity_Aerial_Image-by-BIG-Bjarke-Ingels-Group-1.jpg. Accessed 22 June 2024
5. Choi C-J (2007)City geomancy. Minumsa, Seoul

Location Selection Method for Floating MiC Factories

Yushu Yang, Wenkang Guo, Mingyu Zhang, and Heng Li

Abstract Given Hong Kong's coastal location and ample floating area, the establishment of a floating MiC (Modular Integrated Construction) factory is a viable solution to reduce the dependence on imported MiC modules from mainland China, which is necessitated by the high land costs in Hong Kong. The floating MiC factory's location can have varying degrees of societal and environmental implications. The study utilized the PESTEL-AHP method to ascertain the weights of the elements that influence the outcome and construct a mathematical model. The process involves utilizing GIS spatial analysis and AHP weighting criteria to produce thematic maps. These maps are then combined using weighted overlay to create a comprehensive map. The appropriateness index is then employed to identify the optimal location decision. According to the location selection research, the sea region to the west of Lamma Island is the top priority for location selection.

Keywords Location selection · Floating platform · Weighted overlay by GIS · PESTEL-AHP

1 Introduction

1.1 Research on MiC Technology and Floating Factories

Faced with the challenges of high construction costs and future labor shortage and aging in Hong Kong, MiC (modular integrated construction) technology is gradually being applied in the construction industry in Hong Kong. MiC technology can effectively improve engineering efficiency, shorten construction period, and reduce construction costs, thereby alleviating many problems faced by the construction

---

Y. Yang · W. Guo (✉) · M. Zhang · H. Li
Department of Building and Real Estate, The Hong Kong Polytechnic University, Hong Kong, China
e-mail: xewsylvan@163.com

X. L. Zhao et al. (eds.), *Proceedings of the Fourth World Conference on Floating Solutions: WCFS 2024; 2–4 December; Hong Kong*, Lecture Notes in Civil Engineering 597, https://doi.org/10.1007/978-981-96-4569-5_96

industry in Hong Kong [1]. At present, most of the engineering projects using MiC technology in Hong Kong rely on the supply of MiC modules from other regions. Due to the lack of local MiC module production capacity and the immaturity of the industrial chain, the transportation cost remains high, which partially offsets the cost advantage brought by MiC technology. There are currently more than 80 projects using MiC in the construction industry in Hong Kong, and Hong Kong will build at least 200,000 MiC units in the future. With the increasing demand for MiC projects and modules, it is imperative to establish a MiC module production factory in Hong Kong. This will not only help reduce transportation costs and improve the economic benefits of MiC technology, but also cultivate the local MiC industry and contribute to the transformation and upgrading of Hong Kong's construction industry [2].

There are two options for building a MiC factory: to build an onshore Mic factory, or to develop a floating factory to transfer the whole process to the sea. However, the problem of establishing a local MiC module factory in Hong Kong is the high land cost and excellent geographical location of Hong Kong. Hong Kong has complex terrain, scarce land resources, and many areas have large terrain fluctuations [3]. The traditional land-based construction model of MiC factories is difficult to promote on a large scale in Hong Kong, which poses challenges to both construction costs and construction efficiency. The development of floating MiC factories can make full use of Hong Kong's abundant marine resources and get rid of the constraints of land resource shortages and high costs. Building factories at sea can significantly reduce land costs and improve the economy of factory construction and operation. At the same time, floating MiC factories have the characteristics of strong flexibility and good mobility [4]. In 2024, China's shipbuilding orders have far exceeded that of Japan, which is the second largest in the world, which means that shipyards in China's Greater Bay Area can provide all-round construction to maintenance services for floating MiC factories, and can ensure that the performance of its ships is stable enough. Therefore, factories can choose suitable water locations for adjustment and redeployment according to actual conditions to avoid terrain restrictions. Flexible location adjustment can shorten the transportation distance and time from MiC factories to construction sites, thereby reducing transportation costs and risks faced by long-distance transportation [5].

1.2 Factors Affecting Site Selection of Floating MiC Plants

According to the above content, the research goal of this paper is to develop a method applicable to the geographic information system (GIS) and PESTEL-AHP method to plan the location of the floating MiC factory, and determine the appropriate location of the floating MiC factory by determining the weight of a set of parameters as the address selection. The PESTEL analysis method can deeply explore the impact of six major factors on site selection, including politics, economy, society, technology, environment and law. The analytic hierarchy process (AHP) can quantitatively evaluate these factors and find out their relative importance [6]. Through the comprehensive application of PESTEL-AHP, key influencing factors can be objectively identified, various site selection options can be scientifically weighed, and the optimal water location can be finally selected. This systematic site selection analysis method can not only improve the scientificity and rationality of site selection, but also help to enhance the participation and recognition of all stakeholders. The choice of floating MiC plant needs to take into account the marine space planning and marine policy of Hong Kong and the Greater Bay Area to determine the optimal location of the plant. In addition, the construction of floating MiC factories in areas close to fisheries and aquaculture may cause dissatisfaction among relevant practitioners, so it is also necessary to consider the impact of floating MiC factories on fisheries and aquaculture in Hong Kong and the waters around the Greater Bay Area [7].

2 Experimental Content

The experiment proposes to set up an offshore jack-up MiC factory in Hong Kong waters with an area of about 5000 m^2. All the comprehensive factors considered for site selection can be assembled in a single map (Fig. 1).

In this experiment, ArcMap software was used to visually calibrate the various impact factors. The geographical data are derived from those released by the Hong Kong Government. To avoid obstructing marine traffic, offshore assembly of synthetic construction method factories need to avoid ship channels and passages. Meanwhile, much of the seabed in Hong Kong waters is now occupied by other facilities. For example, there are buried communication cables, silt dumping areas, sand mining areas, explosives dumping areas, mooring areas, submarine gas pipelines and sewage outlets. Although we are a floating platform, we should still avoid these submarine installations when our support columns are lowered to the seabed.

On the other hand, Hong Kong's waters have extensive marine protected areas including marine parks, coastal protection zones, fish farms and artificial fishing reefs. The location of the planta should avoid all the above coastal protection zones and maintain a safe distance. Meanwhile, the floating plant formulated in this experiment requires the use of pipelines to transport industrial water and wastewater, so only areas where the length of the laid pipeline is less than 2 km are selected for this experiment.

3 Experimental Methods

3.1 PESTEL Analysis

Political

If the traditional method of purchasing these products from mainland China or abroad is chosen, the choice of module sizes is limited due to the constraints of customized entrances. If the offshore plant is allowed to relocate, the plant operator can dock the plant to work near those sites that require special sizes and continue to export MiC modules to the site. The choice of area should be based on geographical policy, strict adherence to routing arrangements and port usage norms.

Economical

Some construction sites are in the deep Hong Kong and far from the ports in Mainland China. At the same time, the maritime freight in HK has more strengths. Disposable income level: Types of workers here and their demand for wages. If conditions permit, investigate the existing costs of purchasing MIC products from the mainland and the transportation, avoid the areas with low costed construction sites that have reached cooperation with the mainland.

Social

For this experiment, one of the reference elements was the choice of an address close to a densely populated area with many buildings. These places could not use more space for new facilities, and they wanted shorter time constraints for the project, so they chose MIC. Another reason is that the traffic situation is not good here because these places are very congested. Therefore, it was important to set a cheaper, more stable and faster mode of transport. It is worth noting whether there are some special interest groups.

Technological

There are two aspects to this topic. The first is if there are some property developers who need to complete their projects in this new way. The second one is what kind of technology we have. Where do we source our raw materials, our maintenance patterns and how much noise and pollution range can we control during the operation of our plant.

Legal

When siting an offshore plant, international and national laws and regulations need to be strictly adhered to ensure the legality, safety and sustainability of the project. Internationally, the United Nations Convention on the Law of the Sea (UNCLOS), the Convention of the International Maritime Organization (IMO) and the Convention of the International Labor Organization (ILO) should be complied with to regulate maritime rights and inter-sets, maritime safety and labor conditions. Domestically (in China, for example), compliance with the Marine Envi-ornamental Protection Law, Sea Area Use Management Law, Maritime Traffic Safety Law, Environmental Impact As-assessment Law and Water Pollution Prevention and Control Law is required to ensure marine environmental protection, the rational use of sea resources and compliance with environmental impact assessments of projects.

Environmental

This experimental site should avoid swamps, wetlands, ecological reserves and military management areas. There is no doubt that working on offshore platforms is climate-dependent, and to ensure the safety of employees, floating platforms need to be sited in waters where the climate and ocean currents are relatively stable. Then there is energy and resources. Floating platforms must be sited near energy centers, waste treatment plants, ports and construction sites, especially the first two. In addition, due consideration should be given to media attention to have better prospects for development.

3.2 AHP Analysis

The experiment allows for the removal of some unnecessary factors and their materialization based on the results of the PESTEL analysis. Therefore, the above analyses need to be condensed into several factors that can be made publicly available in government databases. Table 1 shows the results of the initial generalization.

But obviously this kind of result is still somewhat general, and it is difficult to be used as a factor directly displayed on the map. After consideration, the Table 2 is the elements to be used in AHP and Euclidean distance analysis.

In addition to this, factors such as ecological reserves, artificial reef areas, sewage intake points, distribution reservoirs, roads, etc. are also marked on the map. Normally, AHP analysis methods require the use of matrix, vector and function operations to check consistency. The main question is to what extent these factors will affect the results. Figure 2 shows the weights occupied by the different types of factors in Table 2.

Table 1 Preliminary results based on PESTEL analytical model

Types	Factors
Policy	Preferential policies, route arrangement, government regulation, industry regulations, labor market, future planning of the government
Economic	Fiscal policy, disposable income level, import and export factors, labor productive level, labor and capital output, existing economic environment and development trend
Social	Social demand for MIC, population density, special interest groups, attitude towards environmental pollution, population structure, education level, distance from the construction site group, project location of the commercial cooperation company
Technology	Extent of demand for MIC technology by external real estate developers and other companies, source of raw materials, strength level of relevant technology companies
Environment	Avoid the special area, nature environment, climate, pipeline transportation area (like power/water plant, waste disposal plant), port, construction site, media
Legal	Relevant laws (like Environmental Legislation, maritime law, etc.)

Table 2 Factors used in AHP and Euclidean distance analysis

Types	Factors
Port	Mid-stream site, Kwai Tsing container terminals, public cargo working areas, etc
Public space	Public-relax-spaces, public facilities
Ecological	Marine parks
Residence	Residential areas
Private land	Purchased lands
Scenic spot	Country parks, campsites, beaches

3.3 Data Collection and Processing

Data is the basis for analytical decision-making. After determining the factors affecting the site selection of offshore factories, the relevant data were collected. And the basic map data includes Hong Kong land use status map, administrative planning map and road traffic map. The data collected above are processed accordingly. Spatial data must be under the same spatial reference, and attribute data must correspond one-to-one with related spatial data. Along these lines, this experiment first requires the construction of a spatial database for site selection. Since the site selection for the experimentally formulated project was defined in the sea, some of the data that are not related to land, such as woodlands, valleys, streams, and other areas, I set their colors to grey so that we can highlight the factors we need. Figure 3 shows the map processed by the GIS software.

3.4 Regions Selection

Due to the need to use pipelines to transport industrial water and wastewater, areas where the length of the underwater pipeline is less than 2 km were chosen for this experiment to control costs. A buffer area of 2 km was created for the influence factor of the pipe connection, and then the overlap of the two surfaces was intercepted using the "crop" tool. The buffer tool was then used to delineate the area to exclude areas where the offshore plant site could not be located, such as within 1 km of a residential area.

Following completion of the master preparation, the overlap between the two surfaces was removed using the "erase" tool and the process was repeated to complete the modification of all elements. For instance, sites that are 1 km from a residential area, 500 m from a harbor, 1 km from a beach attraction, 2 km from a coastal protection zone, 1 km from a fish farming area and a reef ecosystem, and so on. The buffer range can be changed, and this can be repeated if the plant noise data and image data of the surrounding water ecosystem from future field tests differ from the buffer data created this time. Limitations based on this idea using a multitude of factors. Create multiple data range maps, overlay them, and then remove the areas to get the range processing results.

3.5 Buffer Zone Analysis and Euclidean Distance

Figure 4 shows the buffer analysis diagram for several important factors. Buffer analysis of important factors is an important analysis method for address selection. The production of buffers requires the use of the Multi-Ring Buffer feature in the analysis tool. After making a map box, select the desired area to create coordinates and export the map. Having obtained the actual data and specific address of the plant, this method must be used to study and analyze the extent of the buffer zone for some important factors before the plant is built in order to finalize the justification of the address.

It is not enough to just mark the location and delineate the area, the exact location needs to be visualized on a map. The establishment of Euclidean distance map can intuitively express the influence of various factors on the straight-line distance of the selected site. Different from the general spatial analysis results, this experimental proposed project has some special characteristics. Therefore, a separate Euclidean distance map is first created for each factor for the whole territory of Hong Kong using the factors in Table 2. Color scales for this map are not necessarily detailed as these maps are only used for spatial analysis. Figure 4 shows two Euclidean distance maps. Upon completion of the above experimental process, the complete Euclidean distance map needs to be superimposed and a large-scale superimposed map is drawn within Hong Kong.

4 Experimental Results

In order to gain more acceptance and relatively low cost, offshore plants need to be built in selected areas close to shipping lanes and avoid subsea facilities. Figure 5 shows four suitable addresses that fulfil the above analysis. The calculated areas A-D are 6512.6 m^2, 7095.3 m^2, 5083.1 m^2 and 10,218.3 m^2, respectively, which all meet the requirements for the establishment of an offshore plant. Finally, Fig. 5a gives the results of the software evaluation for these four sites. The assessment results obtained in Fig. 5 are the result of a weighted overlay. The weighted overlay map was converted to a raster map and then reclassified and assigned rating values.

Nevertheless, the priority rating does not directly lock the sea area where the factory will eventually be built. Site selection is a comprehensive issue with many factors to consider. Therefore, the content analyzed in this section is established in Table 3 for further analysis.

Based on the above analysis, this paper draws a comprehensive evaluation table for the alternative points of the jack-up offshore MiC plant, and the alternative point

Table 3 Comprehensive evaluation form for offshore plant alternative sites

Result	Spatial extension relationship		Main priorities derived from software analysis
	Water transportation	Attraction	
A	√	√	5
B	√	×	5–6
C	×	√	5
D	√	√	3

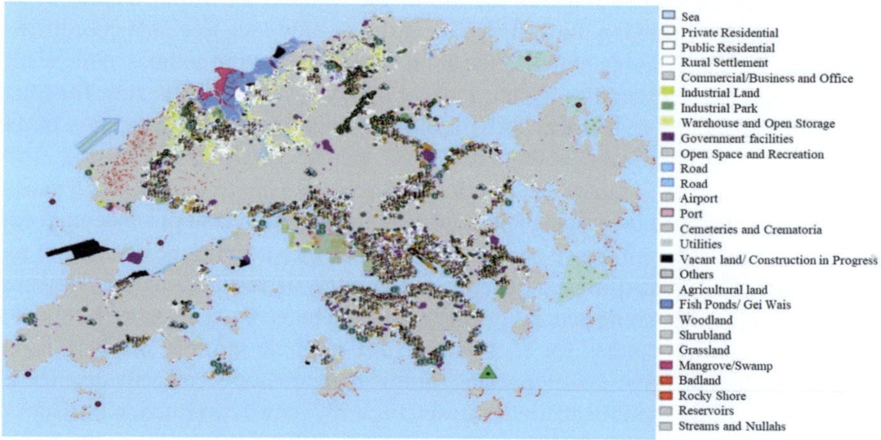

Fig. 1 Comprehensive factor map

A becomes the final recommended scheme. Import selected a results into ArcGIS and add coordinate scaling analysis. This type of map provides a clearer and more direct view of the area's landmarks, surroundings, and latitude and longitude. The standard used for the latitude and longitude grid is the Asian Longitude and Latitude System.

5 Conclusion

This paper provides a unique discussion on the siting of an offshore MiC plant and develops a project element for testing. The experiment used the AHP-PESTEL methodology to weight the factors and then framed the range of influence of each factor through a geospatial analysis system. The site selection results were finally obtained through weighted stacking and scoring.

Address	Port	Public Space	Ecological	Residence	Private Land	Scenic Spot	Wi
Port	1	1/3	1/7	1/6	1	1/4	0.0419
Public Space	3	1	1/6	1/6	2	1/6	0.0669
Ecological	7	6	1	2	7	1	0.3349
Residence	6	6	1/2	1	6	1	0.2540
Private Land	1	1/2	1/7	1/6	1	1/5	0.0416
Scenic Spot	4	6	1	1	5	1	0.2608

Fig. 2 Weights of the factories

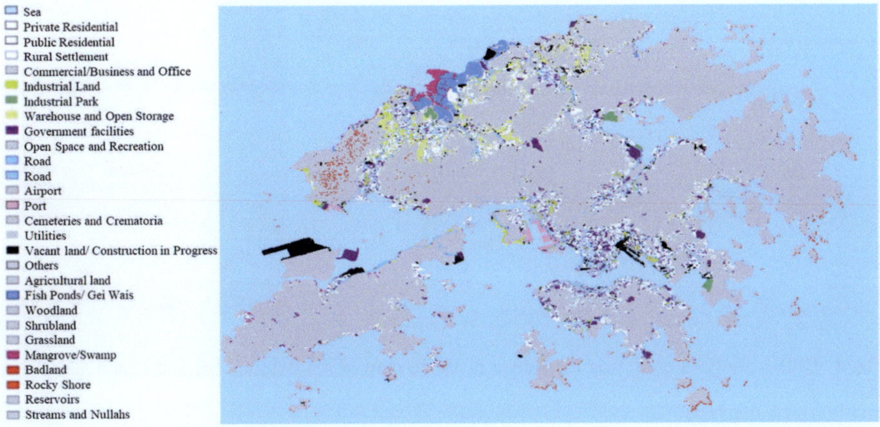

Fig. 3 The raw data map by ArcMap

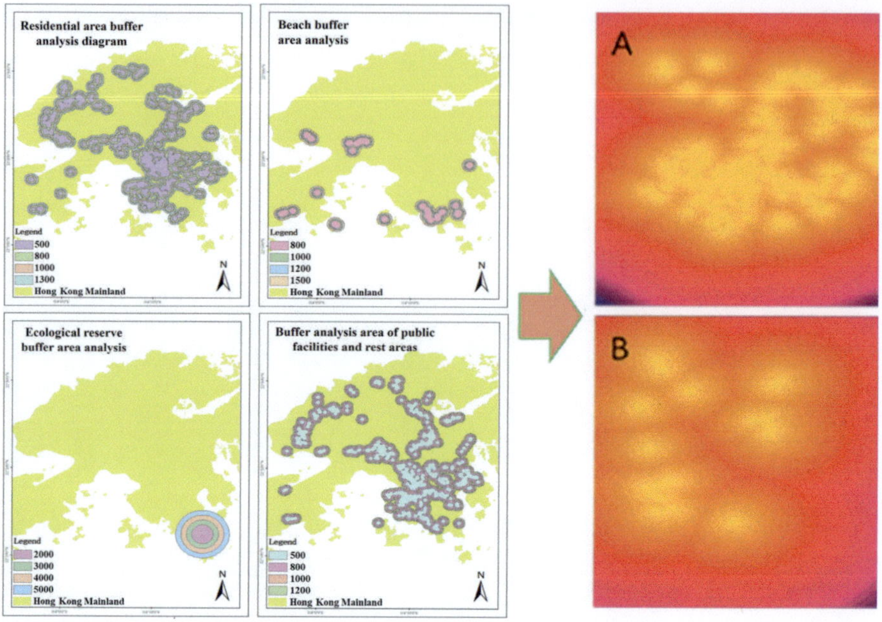

Fig. 4 Buffer analysis chart and Euclidean distance map of residentials (**a**) and beach attractions (**b**)

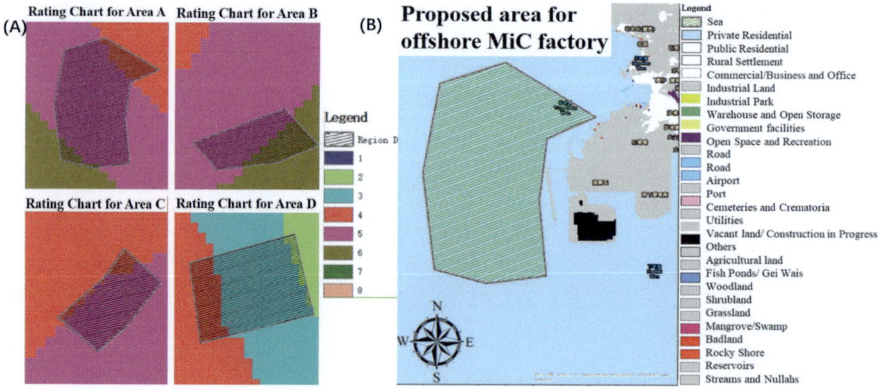

Fig. 5 Regional rating map (**a**) and proposed area for the jack-up offshore floating MiC factory (**b**)

Acknowledgements This work was supported by the RISUD Interdisciplinary Research Scheme [grant number 1-BBWV]

References

1. Hyun M-K, Jin S-J, Yoo S-H (2024) Evaluating a floating offshore wind farm: households' willingness to pay in South Korea. Ocean Coast Manag 255
2. Boehlert GW, Gill AB (2015) Environmental and ecological effects of ocean renewable energy development: a current synthesis. Oceanography 23:68–81
3. Brussa G, Grosso M, Rigamonti L (2023) Life cycle assessment of a floating offshore wind farm in Italy. Sustain Prod Consum 39:134–144
4. Maslov N, Claramunt C, Wang T, Tang T (2017) Method to estimate the visual impact of an offshore wind farm. Appl Energy 204:1422–1430
5. Gilpin A (1995) Environment impact assessment: cutting edge for 21st century, 1st edn. Cambridge University Press, Cambridge
6. Díaz H, Guedes Soares C (2020) An integrated GIS approach for site selection of floating offshore wind farms in the Atlantic continental European coastline. Renew Sustain Energy Rev 134
7. Gill AB (2005) Offshore renewable energy: ecological implications of generating electricity in the coastal zone. J Appl Ecol 42:605–615

Floating Urbanism, the Bold Vision of Streetless Cities

Tania Feldzer, John-Salvator Liotta, and Emmanuel Dufrasnes

Abstract To build upon the discussions initiated by Rutger de Graaf on floating urbanism at the recent WCFS conferences, this document aims to not only delve into the potential of reimagining street spaces in floating cities but also to shed light on the innovative concept of "floating urbanism," which envisions cities free from traditional streets. It critically examines and reimagines the foundational principles of urban planning, positing the provocative idea that cities might not require streets to serve as their backbone. Thought an experiment made during the SIP (Pedagogical Week Innovation) in ULB (University Libre de Bruxelles) in March 2024, the students were invited to work on conceptual exercises to rethink street functions in urban design, based on the idea that the absence can create new perspectives of mobility. From Tim Ingold's work, students were tasked to reimagining urban mobility and space in existing cities built without streets, Catalhoyuk in Turkey, Ganvié in Benim and Uros in Peru. The methodology consists of three steps: first, developing the DNA of each given city model. Then, by using creative thinking techniques, such as mind mapping and SCAMPER, generate innovative ideas on 5 key mobility concepts: mobility of memory, mobility in time, mobility for networks and generative infrastructure, and mobility for division. Finally, by overlayering the 3d models obtained, we have produced a series of prototypes of innovative mobility in cities. The text shows how floating urbanism offers new potentialities, emphasising ecological management, community spaces, adaptability to climate change, and integrated mobility networks in time motion.

Keywords Floating urbanism · Mobility · Network

T. Feldzer (✉) · J.-S. Liotta · E. Dufrasnes
Research Laboratory 'Habiter', School of Architecture La Cambre Horta, ULB (Université Libre de Bruxelles), Bruxelles, Belgium
e-mail: tania.feldzer@ulb.be

E. Dufrasnes
School of architecture, ENSAS, Strasbourg, France

T. Feldzer
Architecture Department, KadirHas University, Istanbul, Turkey

X. L. Zhao et al. (eds.), *Proceedings of the Fourth World Conference on Floating Solutions: WCFS 2024; 2–4 December; Hong Kong*, Lecture Notes in Civil Engineering 597, https://doi.org/10.1007/978-981-96-4569-5_97

1 Introduction

For the architectural conception of a holiday village in Tunisia in the beginning of the 1960s, Cantal Dupart (Cantal Dupart is a French urban architect, professor emeritus at the CNAM of Paris, former member of ANRU, Nice Modernist Prize 2002, Director of the Urbanisme magazine in 1990–1991, Co-initiator of Grand Paris in 2010), a young architecture student at the time in *Les Beaux-Arts de Paris*, had proposed to not build roads between the sleeping rooms and the shared spaces. His strategy was to let the visitor themselves trace the path that would suit them best. Thus, an area of earth and sand was deliberately left on the ground so that over weeks and years a vernacular path emerged, traced by spontaneous acts or so-called "desire line." What is remarkable in this initiative is not so much that the visitor became an actor in the production of public space but that circulation was not the founding act of the project's layout. It outcome a street, traced without determinism and initially conceived by voluntary omission. Streets are not just means of communication but also integral elements of the urban fabric, shaping the interactions and experiences of city dwellers. Cantal Dupart's bold experiment in Tunisia highlights an organic and participatory approach to urban design, questioning the role of the architect and planner in the creation of public space. Reconsider the street or its absence is the theme we would like to address in this article by asking the provocative question: can we conceive urbanism without streets? If the question arises, it is because such configurations already exist or have existed in the past in cities like Çatalhöyük in Turkey or Ganvie in Benin. According to our hypothesis, the street is thought of as a circulation before being a space, consumed and consuming, often dedicated to the technology of the car. The exercise of an urbanism without streets allows us to rethink spatial organization logic, movements, mobilities, and social interactions. It structures the city, determines a spatial organization, a social policy, organizes voids and solids, addresses the question of density, and the ecological footprint of public space. To build upon the discussions initiated by Rutger de Graaf on floating urbanism at recent WCFS conferences, this document aims not only to explore the potential of reimagining street spaces in floating cities but also to shed light on the innovative concept of "floating urbanism," which envisions cities free from traditional streets. De Graaf, in his work on urban water management, emphasizes the importance of adapting existing infrastructures to climate challenges. The proposed research aims to further explore the reflections on new cities by Blue21 by reexamining key notions of "urban structure" and "property rights."

This text proposes to explore how we can think differently about our movements in models of existing cities built without streets and to test our hypotheses through an experiment conducted during the pedagogical innovation week organized by the author at ULB (Université Libre de Bruxelles) in March 2024.

2 Methodology

To address the questions posed in this article, we first propose to explore, through a historical and analytical study, how urban issues have been treated in terms of street design and its limits of understanding. Then, we will detail the experimentation developed during the workshop and the proposed method to rethink the characteristics of the street differently. Finally, we will analyse the results of the experimentation and the insights they may provide on urban design and public spaces.

3 Literature Review

3.1 The Street as a Structuring Space

Eric Alonzo, in his bibliographic work on the history and theory of the way [1], sheds light on their evolution and how streets have shaped our cities. As early as the fifteenth century, in his treatise *De Re Aedificatoria*, Leon Battista Alberti assigned a particular status to the street, considering it a work of art to be treated as an object of convenience and beauty [2]. However, Alonzo criticizes the so-called "modern" conception of streets, which he sees as a technological alienation in service of the car. The invention of the car led to the functional separation of flows and the rectification of routes, creating a spatial and conceptual isolation of main roads, kept at a distance from built-up areas. This isolation, as David Mangin points out, is rendered impassable by the proliferation of support infrastructures for trains, cars, buses, and bicycles [3]. Mangin highlights how this configuration leads to sectoral urbanism, contributing to social segregation and urban sprawl [4].

3.2 The Absence of Streets: An Experiment from the 60 s

The work of Aldo Van Eyck and Piet Blom in the 1960s questions the solid/void duality by freeing itself from the space of the street [5]. Piet Blom viewed each house as a cell within a larger urban organism, promoting social interactions and community while maintaining the intimacy of personal space. Blom's concept of circulation manifests through the creation of public spaces as places for meetings and exchanges [6]. These spaces are actively created by the community, opening a reflection on the possibility of rethinking land use and the place of public space [7] (Fig. 1).

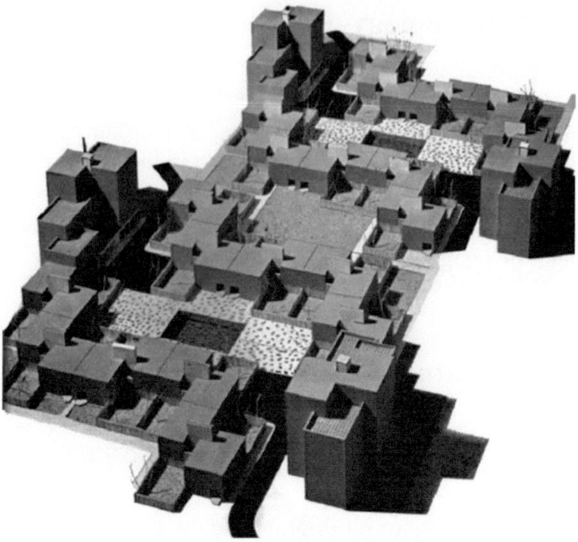

Fig. 1 Pratical planning exercice, study project 1959, Piet Blom, published in forum, vol.15 no.5

3.3 Conclusion of the Analytical Research

Imagining the street as an impermanent and de-hierarchized space allows us to question the notions of mobility and circulation. Rethinking the land and real estate status of private and public spaces can open up new perspectives in urban design.

4 Experimentation During the SIP

The methodology is based on the idea of considering the street by its absence and viewing it not as a structuring element but as an intrinsic and singular entity. Inspired by key concepts from the anthropologist Tim Ingold, we explore the street as memory, as social interaction, as the development of a new temporality, as a network, and as a division of property [8]. Each work group was assigned a city and where engage to create new hypothesis of mobility. The steps included creating the city's DNA, formulating planning hypotheses, and analyzing the combined results.

4.1 Hypotheses of Mobility by Function

Mobility as a Space of Memory

Lines of movement weave knowledge and memories, constituting living paths. Urban spaces serve as receptacles for collective and individual memory [9]. considers displacement as public spaces impregnated with memorial elements that transcend their immediate use [10].

Mobility as a Space of Social Interaction

Social spaces emerge from streets and communication networks [11]. These streets actively constitute social space by connecting people and ideas, fostering encounters and neighborhood life [12].

Temporary Mobility

Streets represent pathways through time, where each movement and trace is a manifestation of temporality. Lines of movement inscribe time into space, creating continuity between past, present, and future [13]. Mobility is both a physical and temporal experience, influencing how spaces are lived and perceived [14].

Mobility as a Network

Streets facilitate understanding of connectivity and networks, forming a complex web of relationships and exchanges. Connectivity is essential in contemporary urban experiences. Students hypothesized that streets can evolve based on network necessities [15].

Mobility as a Guide and Spatial Division

Lines delimit and structure space, creating distinct zones [16]. They guide movement and interaction, influencing power dynamics and identity [17].

4.2 Research Supports and Proposals

Catalhöyük

The Neolithic city of Catalhöyük is characterized by the absence of streets and movement over rooftops. The study questions the hierarchy of spaces and the relationship between private and public spaces. Mobility could be perceived as an embodiment of its "life lines," moving vertically rather than horizontally. This perspective suggests a form of connectivity and social network based not on physical configurations but on interpersonal and spatial relationships.

Towards a De-hierarchized Architecture

Historically, around 7500 B.C., spaces were not hierarchized. Projecting this observation to 2100, we could envisage a standardization influenced by our contemporary housing models. These models, powered by generative artificial intelligence like "Architectures AI," could favor the return to non-nominative and non-hierarchized habitats accessible to all. Mobility in these future habitats would be vertical over

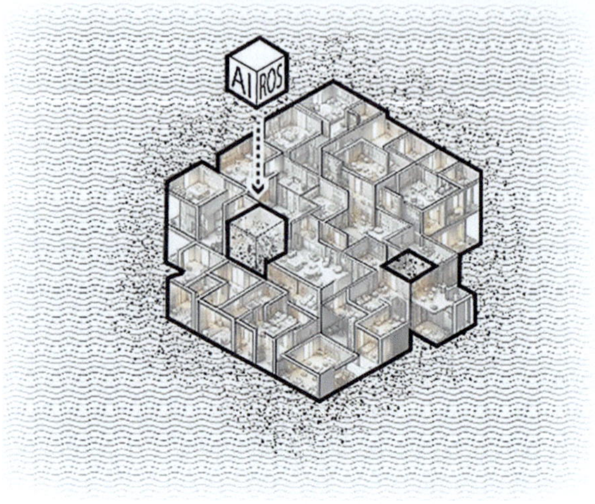

Fig. 2 Drawings of long-term mobility recompositions in Catalhöyük, *Source* author based on the work of students Christopher Luamba, Jeremy Mukala, Alina Maria Petreścu, Alix Revaz

prolonged durations, adapted to changes in family units, while horizontal mobility would be facilitated by universal accessibility to public and outdoor spaces (Fig. 2).

Ganvié

The lake city of Ganvié is built on stilts, with pirogue-based movement. Adaptation to seasonal variations and gender-based community distribution are key aspects. The transformation of current practices offers innovative opportunities, notably through the use of organic waste as resources for creating new circulation spaces. This approach rethinks mobility by integrating territorial changes, where stilt constructions allow an expansion of contact surface between dwellings, optimizing space usage. A reversal of the traditional relationship between built structures and land is envisaged through topographical manipulations that adapt the territory to the built environment. These topographical interventions aim to harmonize architecture with its natural environment, promoting a more organic and sustainable integration.

Rethinking Functions

It is crucial to design functions related to mobility. The establishment of floating platforms, whose functions are not predefined, offers a framework for participatory mobility management. These modular and flexible structures allow space to adapt to the changing needs of users, facilitating a more collective and dynamic approach to urban planning (Fig. 3).

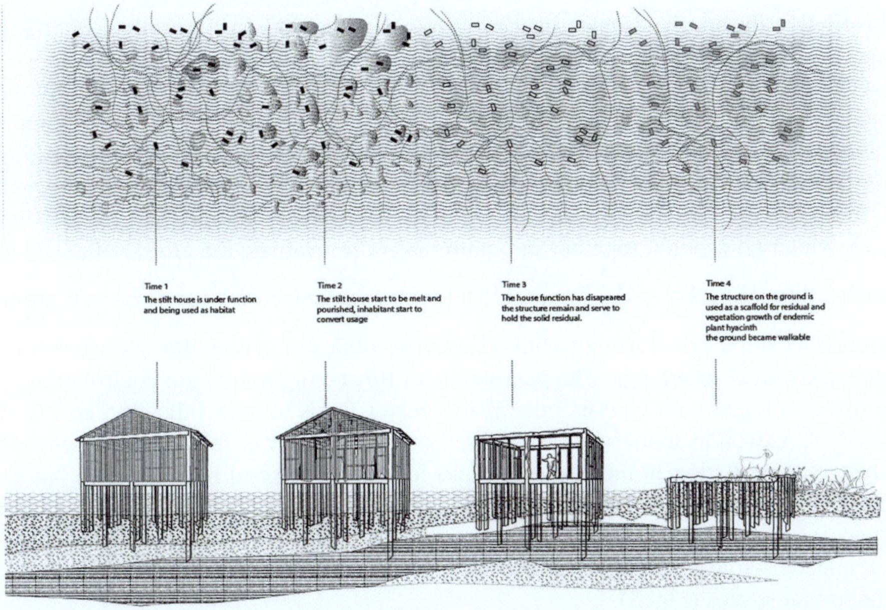

Fig. 3 Territorial plan of Ganvié, source: author based on the work of ULB students Inès Francois, Alexis Gazel, Valentina-Claudia Popa, Kristine Sampang during the SIP2024, ULB, Bruxelles, Belgium

4.3 *Uros*

The floating city of Uros is composed of micro-islands made of reeds, with variable movement. Mobility is rethought without a sense of immobility, promoting social interactions and increased connectivity (Fig. 4).

The Free Plan of Uros

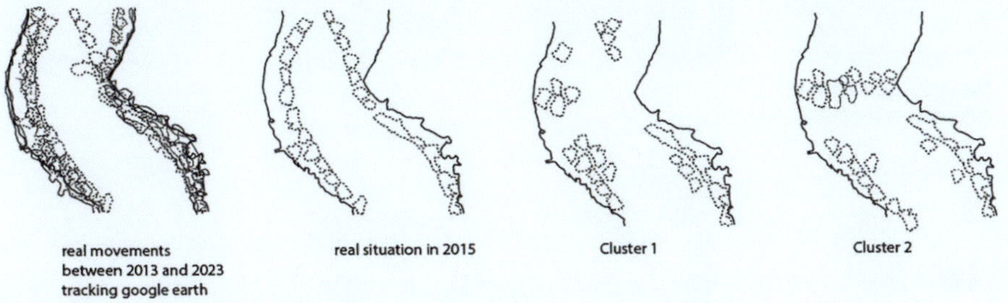

Fig. 4 Territorial plan of Uros, Scan of the movement of islands over the years with sattelite google earth tracking between 2013 and 2023. *Source* author, 2024

Inspired by the burolandschaft theory developed in the 1960s by the brothers Eberhard and Wolfgang Schnelle, "Spaces were defined by the functions of the people who worked in them and their relations to colleagues." [18]. The project drawn by the students is inspired by the free plan office, proposing a game where the participant is invited to create their own scenario. The project extends the idea further by integrating the formation of clusters based on urban functions. In 1974, Yona Friedman proposed a formula to rethink the uses according to their distances (d) and their frequencies (w), which he brought together in a path matrix to evaluate the effort value (E) of each entity. Thus, $Ex = \sum_{y=1}^{y=n} dxy.wy$ [19]. In the proposal, each clusters Ex = 0 group around a central axis, forming a hub of activity, while circulation flows are redefined to bypass these groupings. The parameters of this configuration are controlled by a script developed with Grasshopper, using vector fields to adjust the size, radius of clusters, as well as numerical proximities and the number of relationships between clusters. Circulation in these spaces is no longer determined by fixed axes but by clusters of constantly evolving functions, reflecting a more flexible and responsive approach to spatial organization. This model offers an adaptive response to changing user needs, while promoting increased interaction and connectivity between different functional spaces (Fig. 5).

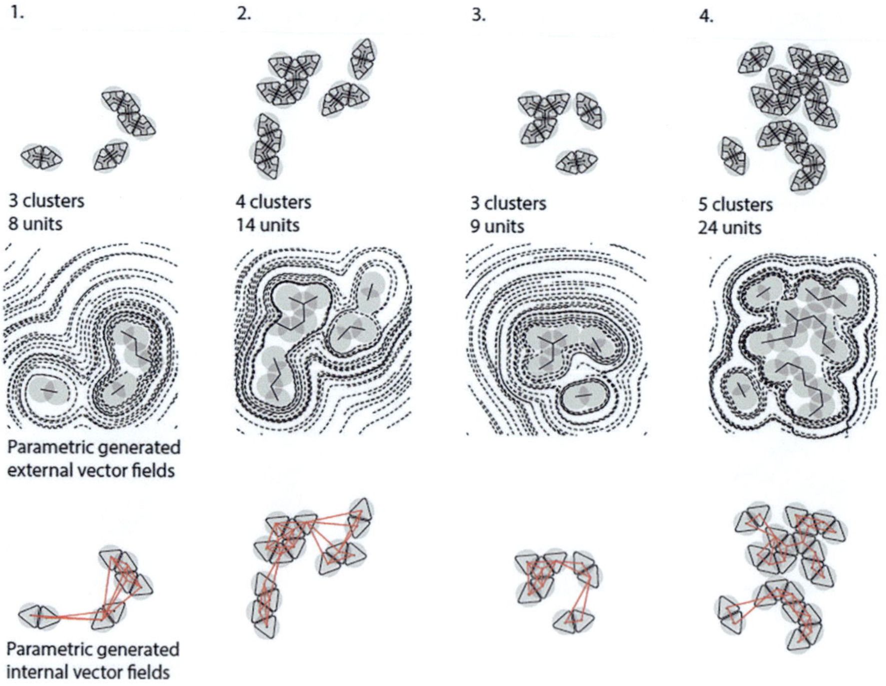

Fig. 5 Cluster of islands generated by parametric drawing in Grasshopper, field attractor. *Source* author, 2024

5 Conclusion

The results raise obvious technical and political feasibility questions while opening up the field of possibilities. They also question the fundamental idea: should we reproduce the same city model on water as on land? These approaches encourage reinventing the notions of "plots," "islets," real estate, and privacy in urban composition. They also promote more active user participation in configuring their environments, fostering spatial democracy where each individual can influence the form and function of their living space. These models prefigure a future where architecture and urbanism not only meet functional requirements but also aspire to create spaces that promote collective and individuality while respecting ecological imperatives. Can mobilities be design by a new evaluation of function not only dictated by use but by frequency, usage and distance based on an improvement of Yona Friedman theory of space organisation?

References

1. Alonzo É (2018) L'architecture de la voie: histoire et théories. École d'architecture de la ville & des territoires ; Éditions Parenthèses, Champs-sur-Marne , Marseille
2. Alberti LB, Rivera J (1991) De re aedificatoria. Akal, Madrid
3. Les flux (2006) l'architecture et la ville: Entretien avec David Mangin, architecte-urbaniste, par Éric Charmes et Benjamin Pradel. Flux. no 63–64, pp 89–99
4. Mangin D (2006) La rue en toutes franchises: Pouvoirs. no 116, pp 131–147
5. Di Palma V, Periton D, Lathouri M eds (2009) Intimate metropolis: urban subjects in the modern city. Routledge, London , New York
6. Gaffikin F, Mceldowney M, Sterrett K (2010) Creating shared public space in the contested city: the role of urban design. J Urban Des 15:493–513
7. Merx S (2011) Public pie performing public space. Perform Res 16:132–137
8. Ingold T (2016) Lines: a brief history. Routledge, London, New York
9. Lynch K (2008) The image of the city. Press, Cambridge, Mass, M.I.T
10. Aceska, A (2012) Loose space: possibility and diversity in urban life—edited by Karen A. Franck and Quentin Stevens. Int J Urban Regional Res 36, 411–412
11. Gehl J (2010) Cities for people. Island Press, Washington, DC
12. Jacobs J (1992) The death and life of great American cities. Vintage Books, New York
13. Cresswell T (2006) On the move: mobility in the modern Western world. Routledge, New York, NY
14. Amin A, Thrift NJ (2012) Cities: reimagining the urban. Polity Press, Cambridge
15. Hillier B, Hanson J (2005) The social logic of space. Cambridge University Press, Cambridge
16. Massey DB (2012) For space. SAGE Publications Ltd
17. Soja EW (1996) Thirdspace: journeys to Los Angeles and other real-and-imagined places. Blackwell, Cambridge, Mass
18. How the 21st century office was born in post war Europe. https://workplaceinsight.net/how-the-21st-century-office-was-born-in-post-war-europe/. Accessed 25 May 2024
19. Friedman Y (1975) Toward a scientific architecture. MIT Press, Cambridge, Mass

A Data-Driven Approach for Urban Transitions in Delta Regions Towards Living with Water Through Floating and Mobile Infrastructure

Sridhar Subramani, Mathilde Marengo, Iacopo Neri, and Koen Olthuis

Abstract Delta cities are complex systems characterized by continuous change and unpredictable behavior due to global (e.g. sea level rise) and local (e.g. flooding and urban densification) agents. The "living with water" paradigm offers a framework to address these challenges by deploying modular floating infrastructure within urban planning strategies. These solutions boast inherent adaptability and flexibility, materialized by dynamic platforms, allowing for spatial reconfiguration and functional transformation in response to evolving user needs and environmental pressures. The paper discusses the research developments of a planning methodology to facilitate the transition towards "living with water" in delta cities. The methodology leverages scenario creation to enable cities to respond and coexist with uncertainty. The three-stage process begins with designing a planning system based on user activities within functional spaces. Rotterdam serves as a testbed due to its strategic advantages and knowledge of floating solutions for adapting to rising sea level threats. Time-use surveys inform the allocation of functional types on floating platforms, with their distribution determined by frequency of use. Overlaying user flows with accessibility identifies the demand for a specific function over time. In the second stage, the planning system uses reinforcement learning to simulate responses of floating infrastructure to varying demand locations. Then, the system measures de-densification, proximity, and functional object availability over time. This evaluation tool redefines thresholds for functional distribution and demand locations by incorporating scenarios such as seasonal flooding. Such a data-driven performance evaluation facilitates the identification of implementation strategies for cities to develop on water, optimizing urban planning in delta regions to foster resilience, adaptation, and long-term sustainability.

M. Marengo · I. Neri
Institute for Advanced Architecture of Catalonia, Sant Marti, Barcelona, Spain

S. Subramani (✉) · K. Olthuis
Delft University of Technology (TU Delft), Delft, Netherlands
e-mail: s.subramani@tudelft.nl

Waterstudio.nl, Rijswijk, Netherlands

X. L. Zhao et al. (eds.), *Proceedings of the Fourth World Conference on Floating Solutions: WCFS 2024; 2–4 December; Hong Kong*, Lecture Notes in Civil Engineering 597, https://doi.org/10.1007/978-981-96-4569-5_98

Keywords Living with water · Data-driven design · Mobile floating solutions · User activity

1 Introduction

1.1 Delta Urbanism

Balancing urbanization and natural environments, Delta Landscapes house about 10% of the world's population, including major cities like Amsterdam, London, Miami, New York, Mumbai, Bangkok, and Shanghai [1]. Climate emergencies, particularly sea-level rise, pose significant threats. From 2006 to 2022, sea levels have risen at an average rate of 3.7 mm/year, marking a concerning trend. Greenhouse gas emissions and uncertainties in ice-sheet processes may cause sea levels to rise by up to 5 m by 2150 [2]. These vast uncertainties in the projection significantly increase the challenges for investment planning in coastal management strategies in low-lying countries like the Netherlands. Since these coastal defense projects lack flexibility for adaptation during their development, they will need to be redone periodically once they reach their limits. These strategies demand an adaptive approach with short-term and long-term implementation options depending on how the future unfolds [3]. Urban planning in delta regions must address these complexities flexibly, requiring new relationships between design, engineering, science, and governance.

The Netherlands exemplifies these challenges, being a low-lying country in the Rhine, Meuse, Scheldt, and Ems river deltas. About 26% of its territory is below sea level, and 60% is vulnerable to flooding. Major urban areas, including Amsterdam, The Hague, and Rotterdam, rely on an extensive flood protection system. This system includes dams, seawalls, storm surge barriers, dikes, and dunes, costing about 1.2 billion euros annually for maintenance [4]. The Delta Programme, initiated in 2010, aims at protecting the Netherlands from flooding, secure fresh water supplies, and make the country climate-proof [5]. With a growing population expected to reach 18 million by 2025, the Netherlands faces increased pressure on urban areas and infrastructure. Sea levels are projected to rise by 1.5 m by mid-century, with annual rainfall between 900 and 1000 mm affecting major urban regions. These threats impact civic infrastructure and water infiltration in urban areas. The Dutch Delta Programme has developed strategies to address sea-level rise, focusing on protection and adaptation, categorized as 'Protect-Closed,' 'Protect-Open,' 'Advance,' and 'Accommodate' [6]. 'Protect-Closed' involves shielding inland areas with dams, navigation locks, and sluices to prevent sea-level rise and salinization. 'Protect-Open' maintains a sea connection, requiring extensive dike improvements to manage flooding. 'Advance' extends the coastline seaward to build a stronger coastal flood defense. 'Accommodate' adapts to rising water levels by raising buildings, creating mounds, constructing ring dikes, or developing floating structures. The 'living with water' concept, derived from 'Accommodate,' promotes floating infrastructures as an adaptive response.

1.2 Urban System

Rotterdam, one of the Netherlands' largest ports and fast-growing urban regions, has integrated climate change adaptation approaches, such as Rotterdam Climate Proof (2008) and Rotterdam Climate Change Adaptation Strategy (2013). These strategies strengthen flood defenses, adapt urban spaces, and increase city resilience through measures like water squares, infiltration zones, dikes, coastal protection, evacuation routes, water-resistant buildings, and floating structures. Projects like the Stadshavens (City-Ports) district 2004 are exemplary of the city's intention to showcase 'urban delta technology' by focusing on adaptive building and floating construction [7, 8].

This research by design explores how a dynamic planning system can address urban uncertainties with floating solutions, how mobile elements adapt to human behavior and scenarios, and whether data-driven methods can enhance our understanding of urban dynamics. The goal is to develop a tool for strategizing resilient urban developments in climate sensitive delta regions, using Rotterdam as a testbed.

To begin, a systems study of Rotterdam illustrates relationships between key elements, with nodes representing system elements and networks depicting influences. This research evaluates resilience based on adaptability and growth under uncertain conditions. In this framework, key nodes include not only urban growth predictions, but also lead time to address changes in demand, water authority, accessibility to facilities, and user distribution. Urban growth is constrained not only by space availability within the city but also by the potential for expansion beyond city limits, negatively impacted by sea-level rise, and heavily utilized port activities. The water authority faces challenges in expanding and maintaining defense systems. Rapid population growth strains space and response times, while increased densification affects accessibility. Translating research objectives into tangible nodes, the proposed systems map evaluates the impact of the floating infrastructure on the availability and adaptability of the surrounding space as they enhance response time and functional proximity.

2 Methodology

The methodology is classified into three essential topics that together define a data-driven evaluation tool: planning system, scenario testing, and evaluation criteria for impact assessment (see Fig. 1).

Initially, a planning system is established based on the current frequency of use (see Fig. 2). This system translates user demand for particular functions into objects that can move based on time. It shows how many people will engage in a particular activity and the locations of people shifting from one activity to another at a given time, helping identify the movement of floating structures. Overlaying the Time Use Survey data of the Netherlands with land use and population density distribution provides a guideline for the flow and distribution of mobile infrastructures

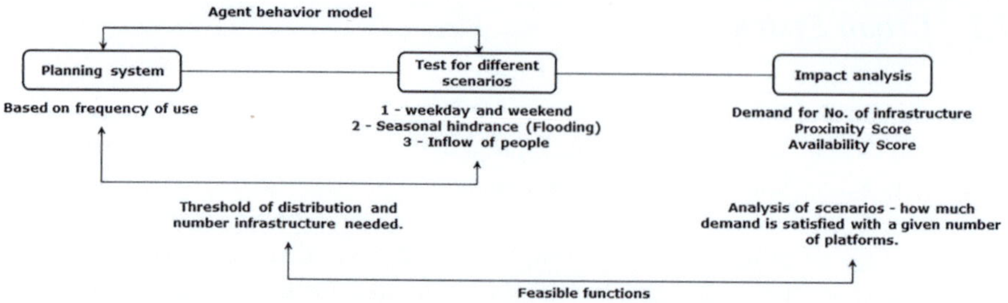

Fig. 1 The methodology for determining transition phases through the evaluation process

over time. Furthermore, in order to validate the system's ability to accommodate different scenarios, a computational model based on Reinforcement Learning is built to predict the most efficient movement pattern of these structures. Based on the inferred behavior, the number of floating elements can be tuned and evaluated to improve performance. The final step is the impact analysis, which assesses the availability of spaces for different functions and how much demand is satisfied based on some floating elements. Thus, the performance evaluation defines the transition phases for moving from land to water.

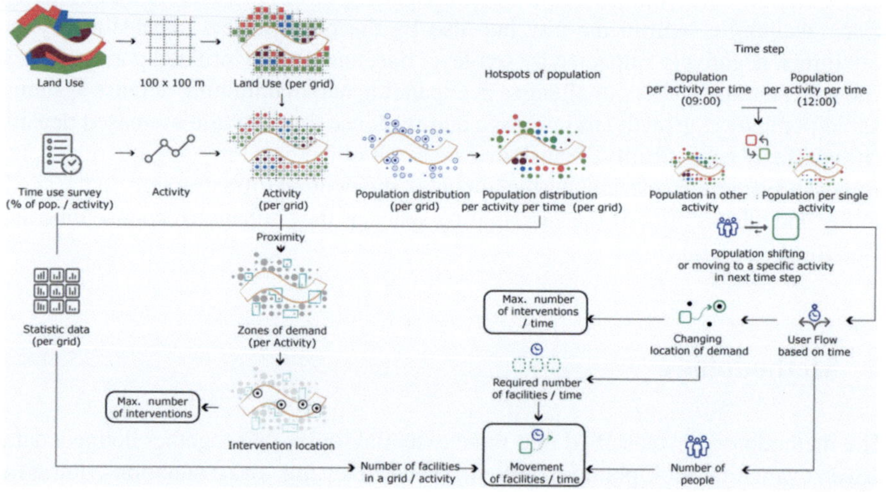

Fig. 2 An overview of the planning system based on frequency of use (translation of user demands to object movement)

2.1 Planning System: Translation of User Demands to Object Movement

Defining User Location Based on Land-Use. The planning system is established through several critical steps: defining user location based on land use, determining user position for object distribution, and scheduling user flow. The initial step involves the spatial harmonization of all data sets to a 100 m × 100 m grid to ensure data interoperability. Each grid cell collects land use information correlated with activities defined in Time Use Surveys. Lastly, activities are clustered into categories such as personal care corresponding to residential areas and paid work to industrial areas and offices, creating a detailed activity map.

In parallel, population density data is also converted into 100 m grids to identify activity based hot spots throughout a typical working day. The study visualizes user distribution and activity patterns by plotting the percentage of the population engaged in different activities at ten-minute intervals over 24 h. This analysis uses a typical working day (Tuesday) to show how population density shifts across the urban landscape.

National statistical data is overlaid on the standardized grid to evaluate proximity to specific amenities e.g. household necessities, with a 300 m threshold considered based on national standards. A circular packing system arranges circles within a given space to optimize the use of a particular functional space, minimizing gaps and maximizing efficiency. This system determines intervention locations on land and water, satisfying these services' 300 m proximity requirement. For research purposes, only water-based locations are analyzed. This approach identifies zones where new infrastructure can meet demand, such as satisfying 20,200 people across 32 locations for household activities on a typical day.

Defining User Location to Distribution. The next step involves verifying user positions at various time steps to predict user flow between locations for specific activities. The study identifies shifting activity patterns by analyzing the proportion of the population engaged in different activities over time. For example, every three-hour interval is examined to determine population movements, helping to pinpoint where interventions are necessary at different times of the day. This process involves calculating the proportion of people moving from one activity to another, using this data to define the distribution of intervention locations based on time steps. By overlaying this information with the demand zones identified in the previous step, the planning system determines where and when floating infrastructures should be deployed to meet changing user demands.

Defining User Schedule to Flow. The final step involves defining a schedule of demand locations and the number of people shifting to these activities. Using data on population movements, the study calculates the number of people near these demand locations at specific times. This information is combined with previous results to determine the placement and timing of movable floating infrastructures, ensuring that demand is met efficiently throughout the day. For instance, at different times of the day, the number of demand locations and people involved can vary significantly. One scenario may show a single-demand location with 825 people, while another might involve 15 locations with 4474 people. This variability demonstrates the necessity for flexible, mobile infrastructures that can adapt to changing urban dynamics.

2.2 Testing for Different Scenarios

To evaluate the performance of the planning system, an agent-based model is used to understand the behavioral patterns of floating infrastructures under different scenarios. This model employs Reinforcement Learning to train agents (i.e. the floating infrastructures) to respond to varying demand locations over time. Reinforcement Learning is a type of machine learning where an agent learns to make decisions by performing actions in an environment to maximize cumulative rewards through trial and error. The agents' actions are guided by three primary parameters: context (current status), actions (agent behavior), and new states (outcomes based on reward functions).

The context is defined by the current state of demand and supply, using a matrix of travel times between locations and a demand table for each location over time. Agents then determine their following location based on higher demand and minimal travel time, balancing the need to satisfy demand with efficient movement. Reward functions categorize demand satisfaction and distance traveled, ensuring agents prioritize high-demand locations while minimizing travel distances.

Weekdays Scenario. The initial iteration begins with six platforms (mean value of the identified spots) acting as agents to understand their movement patterns over time. The agents start at the highest recurring demand locations and move according to the calculated time steps. Evaluating the capacity of these floating infrastructures against the number of people in each location reveals the dynamics of demand satisfaction. In the first iteration, agents can meet up to 33.43% of demand, with extra space availability reaching 83.11% during specific intervals (see Fig. 3). Further iterations with increased agents show improved performance, with twelve agents satisfying 56.24% of demand and maintaining 72.78% extra space availability. These results highlight the effectiveness of deploying multiple agents to meet urban demands dynamically.

Seasonal Hinderance Scenario. The tool is also tested under seasonal flooding scenarios to evaluate its robustness. Flood zones are converted into 100 m grids to identify new demand locations. This data reveals how seasonal changes impact infrastructure needs within existing demand zones. In this scenario, agents' movement patterns shift significantly, with a higher demand satisfaction rate of 93.6% and 26.23% extra space availability during peak flooding periods. These results underscore the tool's adaptability to extreme conditions, demonstrating its potential to manage urban infrastructure efficiently under various scenarios.

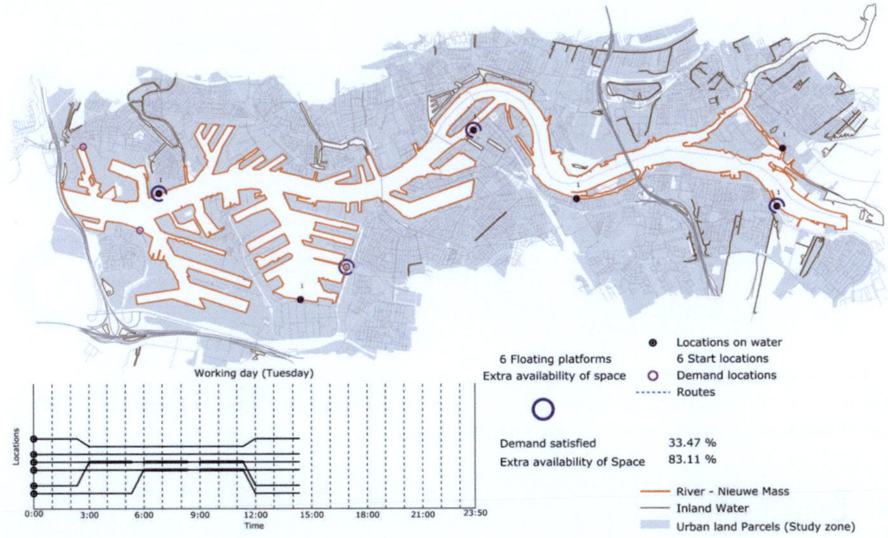

Fig. 3 Agents (floating infrastructure) behavior at a time interval on a weekday [Iteration—1]

3 Evaluation Criteria and Impact Assessment

The evaluation process focuses on the capacity of the system to reduce infrastructure needs by dynamically responding to population flows based on the concepts of de-densification and proximity. Performance is measured by demand satisfaction and extra space availability, providing insights into the tool's effectiveness.

Initial calculations show that the dynamic response to demand variations over time can significantly reduce the required infrastructure. The tool validates the need for mobile infrastructures that adapt to changing urban conditions by ensuring efficient use of functional spaces throughout the day. The planning tool's performance is evaluated by comparing demand satisfaction and extra space availability across different scenarios. Iterations with varying numbers of agents reveal an optimal balance between infrastructure deployment and demand fulfillment. For example, the second iteration with twelve agents shows a marked performance improvement, with higher demand satisfaction and efficient use of space. Testing under seasonal hindrance conditions further validates the tool's adaptability, with agents successfully meeting increased demand despite environmental challenges (see Fig. 4).

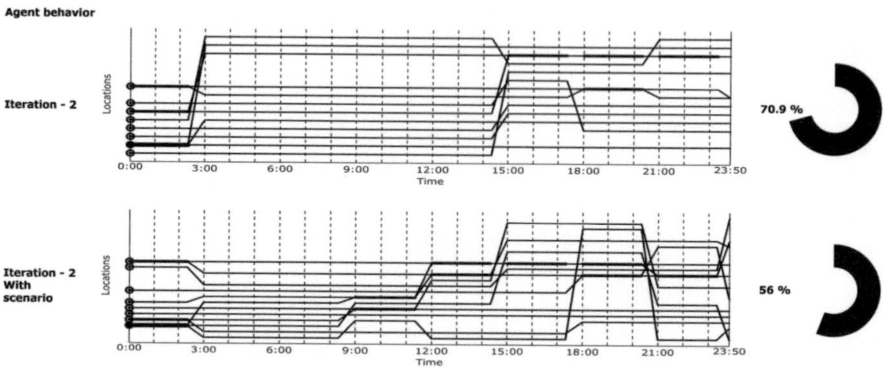

Fig. 4 Agents' movement patterns and performance comparisons between iteration-2 with 12 agents and testing under scenario—seasonal hindrance

4 Conclusion

4.1 Transitional Phases

The defined planning process and performance evaluation identify four transition phases for developing urban areas into the water (see Fig. 5). Initially, floating infrastructures are deployed to dynamically respond to city demands based on flow patterns and needs. In the second phase, the increased spatial capacity produced by the dynamic system allows refurbishment for immediate concerns on land, such as housing in Rotterdam. The third phase involves developing floating infrastructures to support additional activities, increasing urban capacity, and freeing land space. In the final phase, a comprehensive toolbox of functional spaces on modular floating platforms integrates with urban areas, reducing land dependency and accommodating sea-level rise, eventually allowing certain land regions to accommodate sea-level rise and rejuvenate natural resources. These phases enable urban planning systems to shift from land to water, preparing for uncertain future conditions.

4.2 Discussions and Future Steps

This research defines delta urbanism resilience through three core attributes: reflective, learning from experiences to avoid fighting against water; robust, managing systems through behavioral/activity changes; and flexible, adapting to disruptions and evolving circumstances via scenario-based planning.

The study validates the receptivity of floating infrastructures in urban areas to address rising sea levels. The ability to dynamically respond to uncertain situations

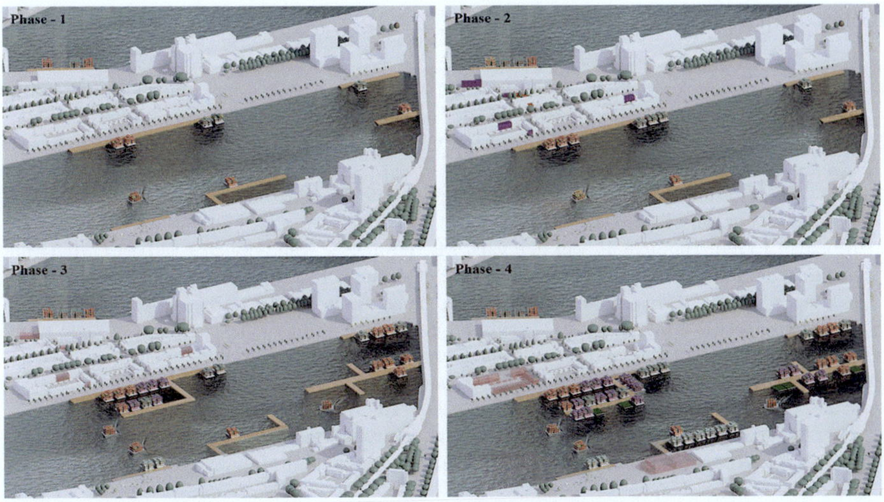

Fig. 5 Four transitional phases for developing urban functions from land to water

demonstrates flexibility in urban planning by incorporating mobile floating infrastructures. The performance system enables decision-makers to strategically plan the transition of infrastructures from land to water, simulating various scenarios. This methodology establishes a foundation for a planning system with a coefficient of floating infrastructure.

The primary goal was to develop a planning tool based on user activity frequency, integrating floating solutions to adapt and grow amidst sea-level rise uncertainties. Currently, the method relies on the Time Use Survey to understand human flows. Expanding this to include GPS data and other data collection methods could refine the data used in the research. The urban infrastructure systems are not integrated into this study, by incorporating other city systems, the transitional phases could be enriched. Each section of the methodology can be further studied. For instance, parameters such as the life expectancy of systems, economic factors, and life cycle are yet to be tackled. Also, future analysis could compare the proposed approach with existing defense systems, as well as research into legality, stakeholder involvement, and public receptivity could enhance its urban strategies. Finally, the introduced methodology highlights the potential of floating solutions, providing a leading case study for the ongoing research on urban development on water as an adaptive response to sea-level rise.

Acknowledgements This paper is a result of **Frequencity**—a data-driven tool for resilience of delta urbanism, a project of IAAC, Institute for Advanced Architecture of Catalonia, developed at Master in City & Technology in **2021/22** by student: **Sridhar Subramani** and faculty: **Mathilde Marengo, Gonzalo Delacámara & Iacopo Neri**.

References

1. Global Center on Adaptation, World Resources Institute (2019) Adapt now: a global call for leadership on climate resilience. Rotterdam, The Netherlands, Washington, DC, USA
2. IPCC (2021) Summary for policymakers. In: Climate change 2021: the physical science basis. Cambridge, UK
3. IPCC (2022) Summary for policymakers. In: Climate change 2022: impacts, adaptation, and vulnerability. Cambridge, UK
4. Bloemen PJTM, Hammer F, van der Vlist MJ, Grinwis P, van Alphen J (2019) DMDU into practice: adaptive delta management in The Netherlands. In: Marchau V, Walker W, Bloemen P, Popper S (eds) Decision making under deep uncertainty. Springer, Cham, pp 321–351
5. Deltacommissaris (2021) The 2022 delta programme, every new development climate proof. Ministry of Infrastructure and Water Management, The Hague, The Netherlands
6. Deltacommissaris (2010) The 2011 delta programme, working on the delta. Ministry of Transport, Public Works and Water Management, The Hague, The Netherlands
7. Rotterdam Climate Programme (2009) Adaptation programme
8. City of Rotterdam (2013) Rotterdam climate change adaptation strategy

Wave Tank Tests

Investigation on Wave Attenuation Performance of a Flexible Porous Floating Breakwater

Zitan Zhang, Kwok Wing Chow, and Ye Li

Abstract With the rapid increase of human exploration and utilization of ocean space, awareness of threats posed by extreme wave events elevates significantly. To protect man-made structures, breakwaters are widely used and they can be classified according to bottom fixed and floating ones. Compared to bottom fixed breakwaters, floating breakwaters can have the advantages such as low cost, easy installation, and environmental friendliness. The relevant technology has not been thoroughly studied yet. Recently, we performed a sequence of preliminary experimental and numerical investigations. The experimental tests were conducted in a wave flume at the Shanghai Jiao Tong University with image processing techniques. The numerical simulations were implemented with a Reynolds averaged Navier–Stokes equation code. We investigated the performance of porous floating breakwaters with different input wave states. In addition, we examined and discussed the wave attenuation performance of floating breakwaters with different moment of inertia and length scale. The results will enhance the scientific understanding and guide the feasibility studies of using floating breakwaters to protect the coast of Asian cities.

Keywords Floating breakwater · Coast protection · Transmission coefficient · Laboratory experiment · RANS equation

---

Z. Zhang · Y. Li (✉)
Multifunction Towing Tank Laboratory, Shanghai Jiao Tong University, Shanghai, China
e-mail: liye@sustech.edu.cn

K. W. Chow
Department of Mechanical Engineering, University of Hong Kong, Hong Kong, China

Y. Li
Department of Ocean Science and Engineering, Southern University of Science and Technology, Shenzhen, China

School of Engineering, The University of Edinburgh, The King's Buildings, Edinburgh, UK

Department of Civil and Mechanical Engineering, Technical University of Denmark, Nils Koppels Allé, Kongens Lyngby, Denmark

Department of Wind and Energy Systems, Technical University of Denmark, Nils Koppels Allé, Kongens Lyngby, Denmark

X. L. Zhao et al. (eds.), *Proceedings of the Fourth World Conference on Floating Solutions: WCFS 2024; 2–4 December; Hong Kong*, Lecture Notes in Civil Engineering 597, https://doi.org/10.1007/978-981-96-4569-5_99

1 Introduction

Climate change and global warming lead to sea level rise and an increased likelihood of 'super' typhoons [1]. At present, large and unexpected waves are the main threat to marine vessels and offshore structures in ocean and coastal engineering [2]. These extreme wave events have negative impact on the ocean tidal energy [3], wave energy [4], offshore wind [5], ocean thermal energy conversion [6], and coastal protection [7]. As innovative coastal structures, breakwaters can reduce the effects of waves in offshore waters and the erosion of coastal areas by waves. However, the wave attenuation performance of floating breakwaters may be affected by various factors such as (translational and rotational) motion, hull shape, width-to-wavelength ratio, and intrinsic period of the structure [8].

Laboratory experiments and numerical simulations play a crucial role in revealing the wave interactions, hydrodynamic properties, and structural response of floating breakwaters [9]. Advanced experimental techniques are widely used to quantify wave heights and investigate the hydrodynamic properties of different breakwater configurations [10, 11]. Numerical simulations are widely used to study complex fluid–structure interactions [12, 13]. The existing research has extensively studied the factors that affect the wave attenuation performance of various types of floating breakwaters [14]. These factors include anchor chain stiffness, breakwater shape, size, mass (moment of inertia), and even hydrodynamic interference between floating bodies. However, there is a lack of research efforts on the wave attenuation performance factors of flexible porous floating breakwaters. The influence of wave attenuation performance by various factors is not clear since the shape of such floating breakwaters is complex. Consequently, it is necessary to study the wave attenuation performance of flexible porous floating breakwaters.

In this paper, we investigate the hydrodynamic properties of a flexible porous floating breakwater experimentally and numerically, with the aim of providing a deeper understanding of the wave attenuation performance of flexible porous floating breakwater.

2 Methods

2.1 Laboratory Experiment

The laboratory experiments were performed in the wind tunnel-circulating water flume at the State Key Laboratory of Ocean Engineering in Shanghai Jiao Tong University [15]. The flume is 27.0 m in length, 7.5 m in width, and 9.0 m in depth, as shown in Fig. 1, and consists of four essential systems: the wave maker, the flow generator, the wave absorbers, and the movable truss trailer. The working test section measures 8 m in length, 3 m in width, and has a depth of 1.95 m. The water depth can be adjusted within the range of 1.6–1.63 m through the installation of artificial structures. There is a single paddle wave at the inlet of the test section within the circulating water flume to generate the various waves, such as regular linear waves and diverse spectral types of random waves. At the opposite end of the test section a wave absorber is located to reduce the wave reflections.

In this paper, a model with a scale factor of 1:20 was selected based on the dimensions of the experimental facility and the wave conditions, as shown in Fig. 2. In the experiment, two capacitive wave probes were used to measure the surface elevation of incident and transmitted waves, with a sampling frequency of 50 Hz. In the calibration experiment, the measuring instruments were calibrated to ensure their accuracy. The absolute error of the wave probe is 1 mm. In the seakeeping model tests, it is necessary to ensure that the Froude number F_r and the Strouhal number S_t are similar. Hence, the relationship between laboratory waves and the actual sea conditions is given by:

$$H_{\mathrm{m}} = H_{\mathrm{p}}/\lambda \tag{1}$$

$$T_{\mathrm{m}} = T_{\mathrm{p}}/\sqrt{\lambda} \tag{2}$$

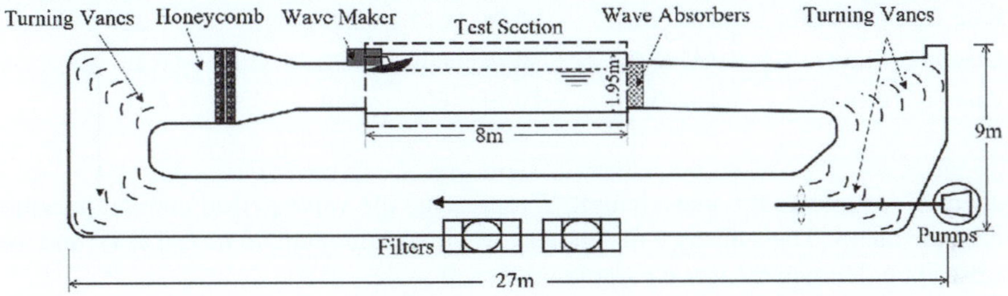

Fig. 1 Diagram of the wind tunnel-circulating water flume at the State Key Laboratory of Ocean Engineering in Shanghai Jiao Tong University

Fig. 2 Schematic diagram of the experiment setup

Table 1 The parameter matrix of the experiment

Case	Wave height (m)	Wave period (t)	Wave length (m)	Wave steepness	Relative width
Case 1	0.02 (0.4)	0.40 (1.79)	0.25 (5.0)	0.08	0.60
Case 2	0.02 (0.4)	0.49 (2.19)	0.37 (7.4)	0.05	0.40
Case 3	0.02 (0.4)	0.56 (2.50)	0.49 (9.8)	0.04	0.31
Case 4	0.02 (0.4)	0.60 (2.68)	0.56 (11.2)	0.04	0.27
Case 5	0.02 (0.4)	0.67 (3.00)	0.70 (14.0)	0.03	0.21
Case 6	0.03 (0.6)	0.49 (2.19)	0.37 (7.4)	0.08	0.40
Case 7	0.03 (0.6)	0.56 (2.50)	0.49 (9.8)	0.06	0.31
Case 8	0.03 (0.6)	0.60 (2.68)	0.56 (11.2)	0.05	0.27
Case 9	0.03 (0.6)	0.67 (3.00)	0.70 (14.0)	0.04	0.21
Case 10	0.04 (0.8)	0.49 (2.19)	0.37 (7.4)	0.11	0.40
Case 11	0.04 (0.8)	0.56 (2.50)	0.49 (9.8)	0.08	0.31
Case 12	0.04 (0.8)	0.60 (2.68)	0.56 (11.2)	0.07	0.27
Case 13	0.04 (0.8)	0.67 (3.00)	0.70 (14.0)	0.06	0.21
Case 14	0.05 (1.0)	0.49 (2.19)	0.37 (7.4)	0.13	0.40
Case 15	0.05 (1.0)	0.56 (2.50)	0.49 (9.8)	0.10	0.31
Case 16	0.05 (1.0)	0.60 (2.68)	0.56 (11.2)	0.09	0.27
Case 17	0.05 (1.0)	0.67 (3.00)	0.70 (14.0)	0.07	0.21
Case 18	0.10 (2.0)	0.67 (3.00)	0.70 (14.0)	0.14	0.21

where H represents the wave height, T represents the wave period and λ represents the scale factor. The subscript m denotes the wave parameters in model scale and the subscript p denotes the wave parameters in full scale.

We consider a series of experimental cases with various wave states to investigate the hydrodynamic properties of the flexible porous floating breakwater. The wave parameter matrix in the experiments is listed in Table 1. We show the input parameters of the experiment and (in brackets) the wave parameters in full scale. In this experiment, we focus on the regular waves and the working water depth is 1.6 m. We perform a wave calibration to ensure that the waves within the test section meet the design requirements.

2.2 Numerical Simulation

In this paper, we use the Reynolds averaged Navier Stokes (RANS) equation to solve the flow field. The numerical simulations were conducted to calculate the wave attenuation capacity of the floating breakwater. The model is 0.14 m in breadth, 0.09 m in height and the draught is 0.02 m. We compare the experimental measurements and

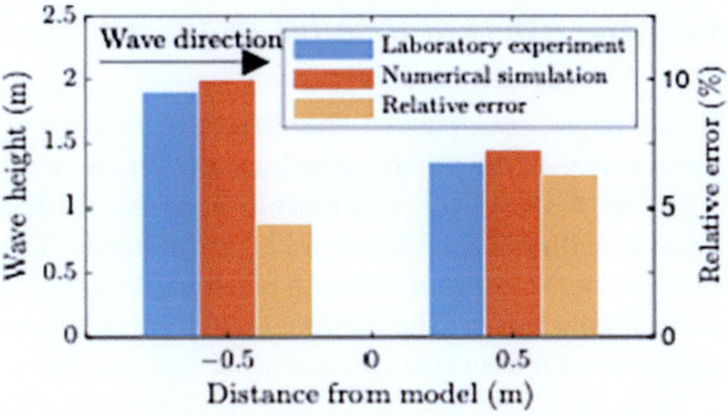

Fig. 3 Comparison between the results of laboratory experiment and numerical simulation

the numerical results to evaluate the validity and accuracy of the numerical model. Figure 3 highlights the results of the laboratory experiment and numerical simulation, and we also show the relative error as a reference. The relative error is approximately 5%. The numerical model developed in this paper has respectable performance in the wave attenuation simulation of floating breakwaters.

3 Results

3.1 Basic Parameters

There are two dimensionless parameters, relative width B/λ and wave steepness ε, used to evaluate the performance of the floating breakwater. The fundamental nonlinearity of a sea state, its steepness ε, for random deep-water surface gravity waves is defined as H/λ, where H the wave height and λ the incident wave length. The relative width is defined as B/λ, where B the width of the breakwater.

In this paper, we use the transmission coefficient K_t to evaluate the performance of the flexible porous floating breakwater. The wave transmission coefficient K_t represents the ratio of the transmitted wave height to the incident wave height, which is given by:

$$K_t = H_t/H_i \tag{3}$$

where H_t is the transmitted wave height and H_i is the incident wave height. The transmission coefficient K_t assesses the ability of a wave to pass through a boundary without being reflected or absorbed.

3.2 The Effect of Relative Width B/λ

In this paper, we consider the effects of relative width *B/λ* on wave attenuation performance of the floating breakwater with the regular waves.

Figure 4 shows the relationship between the transmission coefficient K_t with relative width *B/λ* at different wave height inputs. The results showed that the transmission coefficient K_t of floating breakwater significantly increases with the relative width *B/λ* increases. The floating breakwater has better performance on short-wave attenuation for the same incident wave height input.

We compare the porous floating breakwater investigated in the present paper with different types of breakwaters analyzed in previous work [8, 12, 14, 16], as shown in Fig. 5. Compared to other breakwaters, the transmission coefficient of the porous floating breakwater in this study is larger, and the wave attenuation performance is relatively weak. The porous floating breakwater in this paper has advantages in installation, movement, and maintenance. This type of breakwater can thus be a good option when the relative width *B/λ* is large.

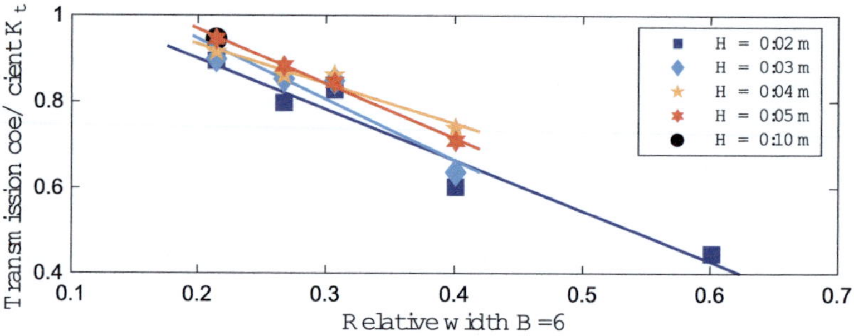

Fig. 4 The transmission coefficient K_t with different relative width *B/λ*

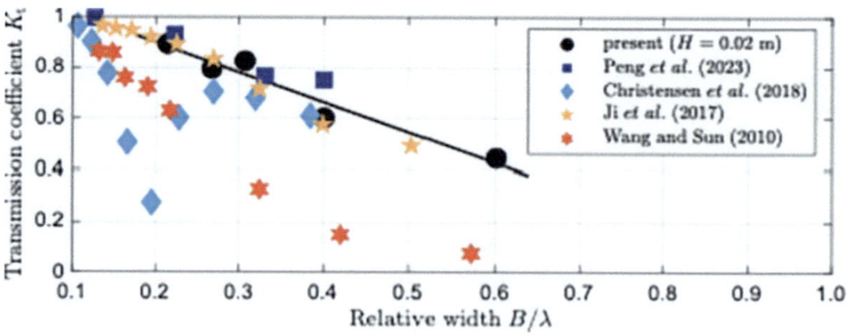

Fig. 5 Comparison of wave transmission coefficient K_t with different types of floating breakwaters

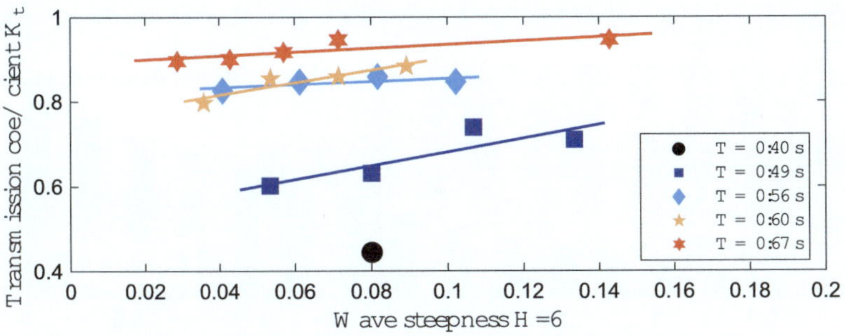

Fig. 6 The transmission coefficient K_t with different wave steepness ε

3.3 The Effect of Wave Steepness E

Figure 6 shows the relationship between the transmission coefficient K_t with wave steepness ε at different wave period inputs. The transmission coefficient K_t slightly increases with the wave steepness ε increases, indicating that the wave attenuation performance is enhanced for waves with lower steepness since the height of the breakwater decreases. However, the wave dissipation capacity of the breakwater decreases when the wave input is steeper and larger.

3.4 The Effect of Dimensionless Moment of Inertia

The rotational inertia of floating breakwaters has a significant impact on their wave attenuation performance since the floating breakwaters move in the waves. We calculated the wave attenuation performance of breakwaters with different moments of inertia.

Figure 7 shows the comparison of transmission coefficient K_t for floating breakwaters with different moments of inertia. The red line represents the transmission coefficient K_t of a fixed breakwater, and the dashed line represents the transmission coefficient K_t of the floating breakwaters with different moment of inertia. The results suggest that the wave attenuation performance of the floating breakwaters is better than that of the fixed breakwater. For the floating breakwater, the smaller the moment of inertia of the floating breakwater, the better will its wave attenuation performance be.

3.5 The Effect of Floating Breakwater Scale

We considered the effect of breakwater scale on the wave attenuation performance of the floating breakwater. We calculated the wave transmission coefficients K_t of three

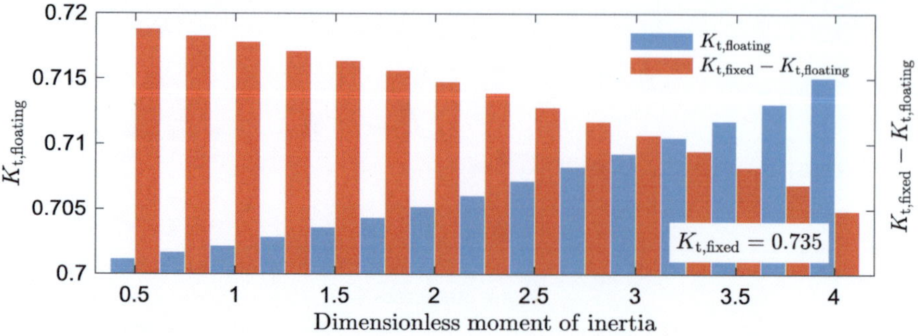

Fig. 7 The transmission coefficient K_t as a function of dimensionless moment of inertia. The dimensionless moment of inertia of fixed breakwater is $+\infty$, and the transmission coefficient of fixed breakwater $K_{t,fixed}$ is 0.735

different scales (small scale, baseline scale, and large scale) of floating breakwaters, with the baseline box having the same scale as the floating breakwater model used in the experiment. Table 2 presents the dimensions of three breakwater models, as well as the non-dimensional incident wave height and wavelength.

Figure 8 shows the transmission coefficient K_t with different scales of floating breakwater. The results suggest that the larger the scale of a floating breakwater, the better will its wave attenuation performance be. The movement of larger breakwaters results in higher wave energy dissipation, which leads to significant reduction in wave height. The increase in the scale of the breakwater has a weak impact on its wave attenuation performance, but it will significantly increase the anchoring forcing.

Table 2 Calculation conditions of floating breakwater with different scale

Case	Scale (Length × Breadth × Height) [m]	Incident wave height (m)	Incident wave length (m)
Small case (−10%)	0.45 × 0.13 × 0.81	0.155	3.81
Baseline case	0.50 × 0.14 × 0.09	0.141	3.46
Large case (+10%)	0.55 × 0.16 × 0.10	0.127	3.11

Fig. 8 The transmission coefficient K_t with different scale of floating breakwater

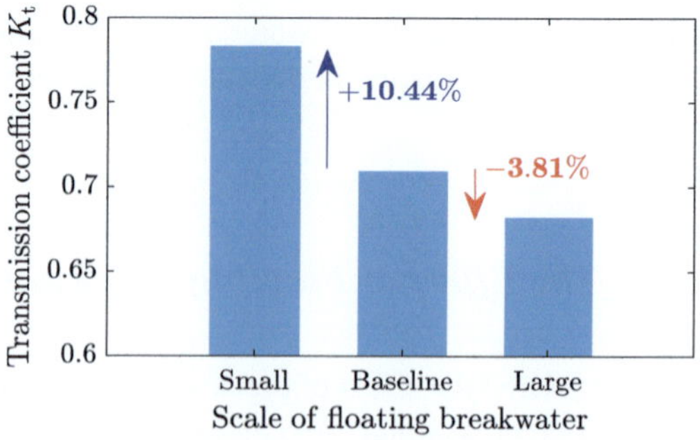

4 Conclusions

In this project we performed a series of laboratory experiments in a wave tank at the Shanghai Jiao Tong University to evaluate the wave attenuation performance of a flexible porous floating breakwater with different wave states. In addition, we investigate the effects of scale and inertia moment on the wave attenuation performance of floating breakwaters using the Reynolds-averaged Navier–Stokes equations. The results suggest that the flexible porous floating breakwater has superior performance in short wave attenuation compared to the cases with long waves (incident wave). In addition, the wave attenuation performance of floating breakwaters with smaller moments of inertia is significant. The increase in the scale of the breakwater has a weak impact on its wave attenuation performance.

The application of floating breakwaters in very large floating structures (VLFS) and floating islands may have special significance. Breakwaters can effectively mitigate the impact of natural disasters on these floating structures, such as storms, tsunamis, and extreme waves. The floating breakwaters not only protect the coastlines but can also be combined with wave energy converters to achieve efficient utilization of resources. The floating breakwaters can integrate renewable energy converters such as wave and tidal energy, providing clean energy for floating islands. The traditional breakwaters reduce the wave impact on the coast and port by reflecting and dissipating wave energy. However, these sources wave energy are not fully utilized. Integrating wave energy converter into breakwater structures can effectively utilize these wave energies and convert them into electrical energy. Compared to the traditional wave energy converter, this floating breakwater integrated with wave energy converters does not require additional occupation of ocean space.

Acknowledgements This work is partially supported by the Drainage Services Department of the Hong Kong Special Administrative Region Government.

References

1. Mayo TL, Lin N (2022) Climate change impacts to the coastal flood hazard in the northeastern United States. Weather Clim Extremes 36:100453
2. Zhang Z, Tang T, Zheng X, Xu W, Zhang L, Lee J, Adcock T, Monty J, Slunyaev A, van den Bremer T, Li Y (2024) Laboratory study of wind impact on steep unidirectional waves in a long tank. Phys Rev Fluids 9:104801
3. Li Y, Yi JH, Song HM, Wang Q, Yang ZQ, Kelley ND, Lee KS (2014) On the natural frequency of tidal current power systems—a discussion of sea testing. Appl Phys Lett 105(2):023902
4. Xu QL, Li Y, Yu YH, Ding BY, Jiang ZY, Lin ZL, Cazzolato B (2019) Experimental and numerical investigations of a two-body floating-point absorber wave energy converter in regular waves. J Fluids Struct 91:102613
5. Gao ZT, Feng XY, Zhang ZT, Liu ZL, Gao XX, Zhang LJ, Li S, Li Y (2022) A brief discussion on offshore wind turbine hydrodynamics problem. J Hydrodyn 34:15–30
6. Zhang W, Li Y, Wu XN, Guo SH (2018) Review of the applied mechanical problems in ocean thermal energy conversion. Renew Sustain Energy Rev 93:231–244

7. Li Y, Calisal SM, Landsburgh AC (2017) An integrated approach to increase marine transportation safety in harbor areas. Mar Technol Soc J 51(14):72–85
8. Wang HY, Sun ZC (2010) Experimental study of a porous floating breakwater. Ocean Eng 37(5):520–527
9. Dai J, Wang CM, Utsunomiya T, Duan WH (2018) Review of recent research and developments on floating breakwaters. Ocean Eng 158:132–151
10. Loukogeorgaki E, Lentsiou EN, Aksel M, Yagci O (2017) Experimental investigation of the hydro-elastic and the structural response of a moored pontoon-type modular floating breakwater with flexible connectors. Coast Eng 121:240–254
11. Liang JM, Liu Y, Chen YL, Li AJ (2022) Experimental study on hydrodynamic characteristics of the box-type floating breakwater with different mooring configurations. Ocean Eng 254:111296
12. Christensen ED, Bingham HB, Friis APS, Larsen AK, Jensen KL (2018) An experimental and numerical study of floating breakwaters. Coast Eng 137:43–58
13. Cebada-Relea AJ, López M, Aenlle M (2022) Time-domain numerical modelling of the connector forces in a modular pontoon floating breakwater under regular and irregular oblique waves. Ocean Eng 243:110263
14. Peng NN, Lau WK, Wai OWH, Chow KW (2023) Computational and experimental studies of wave-structure interaction: Wave attenuation by a floating breakwater. Phys Fluids 35(4):045112
15. Li S, Wei FY, Xu H, Li Y, Zhang LJ (2024) Experiment study on wave attenuation performance of a new type of porous floating breakwater. Ocean Eng 309:118334
16. Ji CY, Cheng Y, Yang K, Oleg G (2017) Numerical and experimental investigation of hydrodynamic performance of a cylindrical dual pontoon-net floating breakwater. Coast Eng 129:1–16

Experimental Study on Wave Attenuation Performance of Pontoon Type Floating Breakwater and Structure Optimization

Yusheng Shen, Junning Pan, and Yiren Zhou

Abstract Breakwaters are an important structure of port and coastal engineering. Compared to traditional rubble mound breakwaters or caisson breakwaters, floating breakwaters can adapt to situations with deeper water, soft foundations, large tidal ranges, and the needs of water exchanges. Floating breakwaters have broad application prospects in port and coastal engineering, offshore engineering, aquaculture, and many other fields. This paper starts with the structure of a traditional single pontoon floating breakwater anchored by chains. By 2D physical model experiments conducted in a wave flume, the wave-damping performance of structures of single pontoon, free twin pontoons, rigidly connected twin pontoons, and twin pontoons combined with plates was compared and analyzed with the test results of wave transmission coefficients. The research results indicate that, the wave transmission coefficients of the rigidly connected twin pontoons are less than those of the single pontoon and free twin pontoons, and the floating breakwater structure combining twin pontoons with vertical plates can further enhance the wave-damping performance.

Keywords Pontoon · Floating breakwater · Wave transmission coefficient · Different structures · Physical model test

1 Introduction

The floating breakwater can adapt to conditions with deep water, poor seabed foundation, macro-tidal range and requirement of water exchange and, etc., and have the advantages of cheap cost and rapid construction, having board application prospects in many fields [1]. However, there are a lot of factors limiting the application of the floating breakwater, such as less shielding effectiveness than bottom type breakwater, low wave attenuation effectiveness for relatively long period waves, and difficult to ensure the structure safety.

Y. Shen (✉) · J. Pan · Y. Zhou
Nanjing Hydraulic Research Institute, Nanjing, China
e-mail: ysshen@nhri.cn

X. L. Zhao et al. (eds.), *Proceedings of the Fourth World Conference on Floating Solutions: WCFS 2024; 2–4 December; Hong Kong*, Lecture Notes in Civil Engineering 597, https://doi.org/10.1007/978-981-96-4569-5_100

For the floating breakwaters, many efforts are made to study the pontoon type floating breakwaters worldwide. To improve the wave attenuation performance of the single pontoon type floating breakwater, the floating breakwater composed of twin pontoons, multi pontoons, and plates are proposed and studied. Williams et al. [2] investigated theoretically the hydrodynamic properties of a pair of long floating pontoon breakwaters of rectangular section by the boundary integral equation method using an appropriate Green's function, and it is found that the wave reflection properties of the structures depend strongly on their width, draft and spacing and the mooring line stiffness, while their excess buoyancy is of lesser importance. Syed et al. [3] analyzed a floating system consisting of three pontoons equispaced, rigidly interconnected and taut moored to act as a single breakwater, and the results show that the spaces between the pontoons have a definite effect on transmission and reflection characteristics of floating breakwater system. Ruol et al. [4] defined a simple and useful formula to predict wave transmission for π-type of floating breakwater (FB). Alizadeh et al. [5] studied the performance of different floating breakwater shapes using the physical modeling method, with the result that π-type floating breakwater is considerably more efficient than T-type floating breakwater. Chen et al. [6] studied theoretically the hydrodynamic behaviors of a floating breakwater consisting of a rectangular pontoon and horizontal plates, and the results indicate that the transmission coefficient decreases obviously as the relative width of the pontoon increases and the horizontal plates help to reduce the wave transmission over the floating breakwater. Ji et al. [7] proposed a new type of floating breakwater consisting of two parts: a main body of rigid cylinders, and a flexible mesh cage containing a number of suspending balls, and results showed that wave transmission was significantly reduced by the mesh cage with the balls. Ji et al. [8] studied a dual pontoon floating breakwater when attached to one or more rows of plane net by using physical and numerical models, and it is found that the new floating breakwater exhibits a better performance with the optimal design parameters. It can be concluded that the combination of pontoons and wave attenuation elements is a good way to improve the wave attenuation performance of the floating breakwater.

In this paper, the wave attenuation performance of pontoon type floating breakwater is studied based on the wave transmission coefficient results derived from the 2D physical model tests, including the structures of single pontoon, free twin pontoons, rigidly connected twin pontoons, twin pontoons combined with horizontal plates and twin pontoons combined with vertical plates.

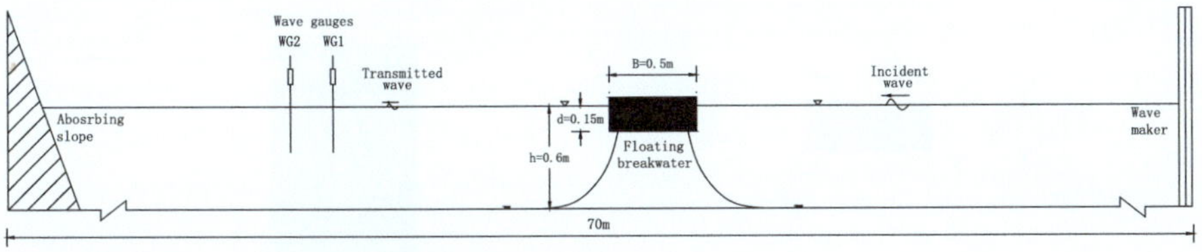

Fig. 1 Model layout in the wave flume

2 Physical Mode Tests

2.1 Testing Facilities and Model Set up

The 2D physical model tests for the floating breakwater were conducted in a wave flume of Nanjing Hydraulic Research Institute, and the wave flume is 70 m long, 0.8 m wide and 1.8 m deep. The wave flume is equipped with a paddle-type wave maker with the wave generation and the active reflection compensation system.

The model layout in the wave flume is shown in Fig. 1. The floating breakwater model was arranged in the middle of the wave flume, and the incident waves were calibrated in the wave flume before the installation of the floating model. The pontoons of the floating breakwater were modelled by rigid plastics and were loaded with stones to the target draft. Waves behind the floating breakwater were measured by wave gauges.

2.2 Testing Conditions

For the pontoon type breakwater, the structures of single pontoon (Fig. 2), free twin pontoons (Fig. 3), rigidly connected twin pontoons (Fig. 4), twin pontoons combined with horizontal plates (Fig. 5) and twin pontoons combined with vertical plates (Fig. 6) were tested. For the floating breakwater of twin pontoons combined with horizontal plates, the horizontal plates were set at the draft of 0.15 m with the porosity of 50% between the twin pontoons. For the floating breakwater of twin pontoons combined with vertical plates, two layers of vertical plates with the porosity of 11% were set between the twin pontoons. For all types of structures, the pontoon draft is all 0.15 m and the floating breakwater model was all anchored with chains.

For the 2D physical model tests for each type structure, the water depth was 0.60 m, and the incident wave heights (significant wave height *Hs*) were 0.075 and 0.125 m (two incident wave heights), and the incident wave periods (mean wave period *Tm*) were in the range of 0.89–2.01 s with the incident wave length *L* in the range of 1.23–4.39 m (six incident wave periods). The irregular waves with the Jonswap spectrum were applied for the tests.

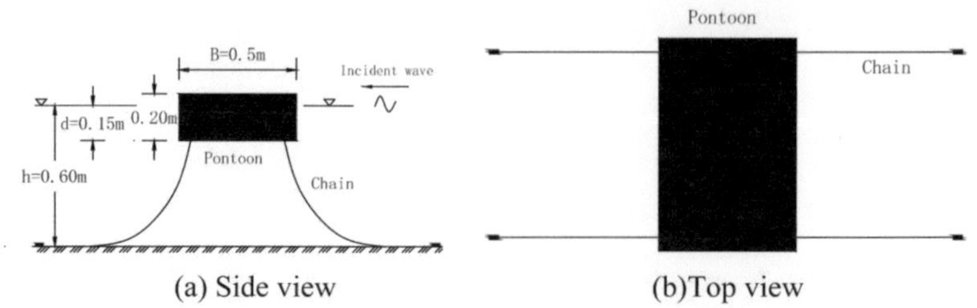

Fig. 2 Floating breakwater of single pontoon

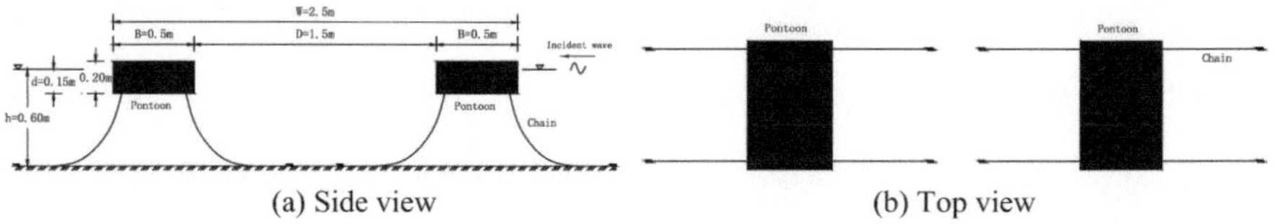

Fig. 3 Floating breakwater of free twin pontoons

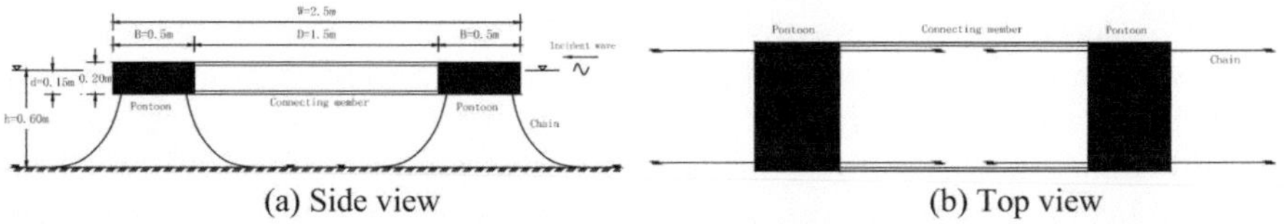

Fig. 4 Rigidly connected twin pontoons

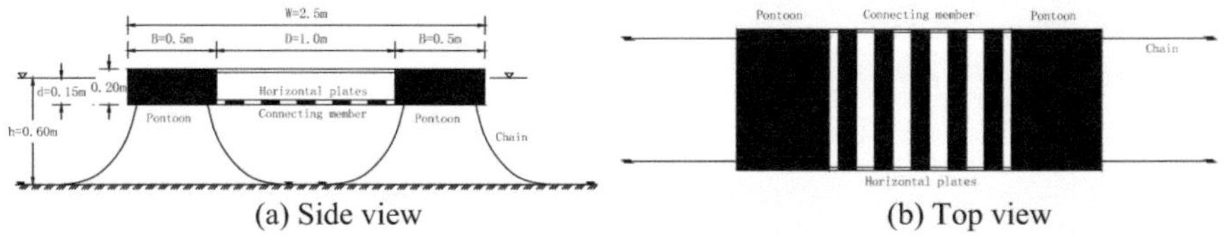

Fig. 5 Floating breakwater of twin pontoons combined with horizontal plates

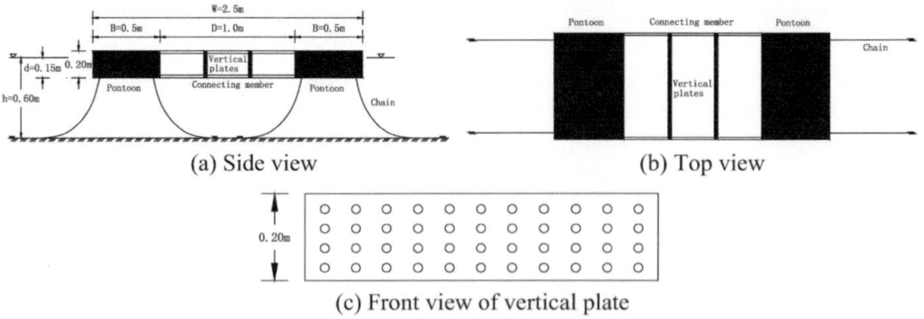

Fig. 6 Floating breakwater of twin pontoons combined with vertical plates

The wave transmission coefficient C_t is the ratio of the measured significant wave height behind the floating breakwater to the incident significant wave height.

3 Results and Analysis

3.1 Single Pontoon

The tested single pontoon (Fig. 2) is 0.5 m wide ($B = 0.5$ m) with the relative width B/L in the range of 0.11–0.41.

Wave transmission coefficient result of the floating breakwater of single pontoon is shown in Fig. 7. It can be seen from the results that, the wave transmission coefficients of the floating breakwater of single pontoon increase significantly with the decrease of the relative width B/L of the pontoon, just with the increase of the incident wave period and wave length. The different between wave transmission coefficients under two incident wave heights is not large generally, and the wave transmission coefficients are slightly smaller for larger incident wave heights or larger wave steepness Hs/L when the relative width B/L is small. In the range of relative width B/L of 0.11–0.41, the wave transmission coefficients of the floating breakwater of single pontoon are in the range of 0.42–0.97.

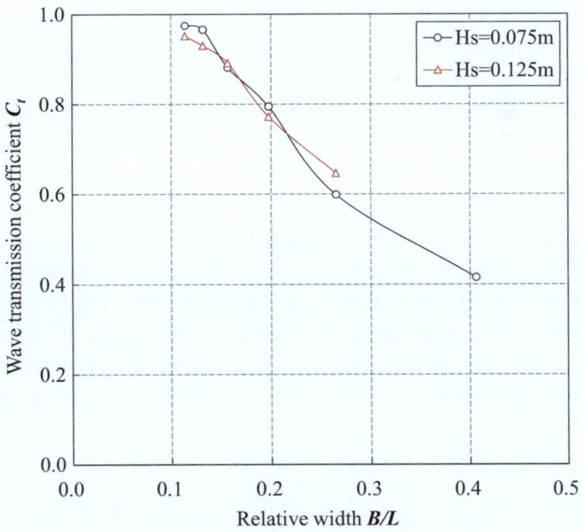

Fig. 7 Wave transmission coefficient result of the floating breakwater of single pontoon

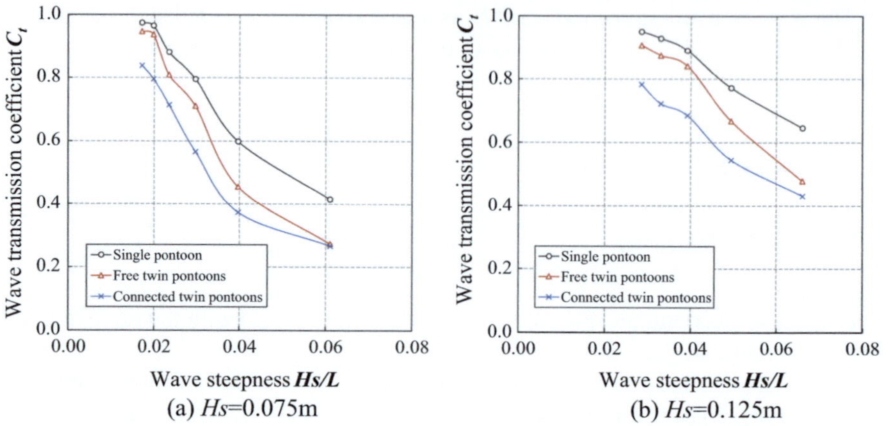

Fig. 8 Comparison results of wave transmission coefficients of twin pontoons and single pontoon

3.2 *Twin Pontoons*

The tested twin pontoons included the free twin pontoons (Fig. 3) and rigidly connected twin pontoons (Fig. 4) with the distance between the twin pontoons of 1.5 m ($D/2B = 1.5$), and the wave transmission coefficient results of the floating breakwater of free twin pontoons and rigidly connected twin pontoons are compared with those of the single pontoon.

Comparison results of wave transmission coefficients of the floating breakwaters of twin pontoons and single pontoon are shown in Fig. 8. It can be seen from the results that, under the same incident wave conditions, the wave attenuation performance of the twin pontoons is better than those of the single pontoon with smaller wave transmission coefficients of the twin pontoons for both free twin pontoons and rigidly connected twin pontoons, and wave transmission coefficients of the rigidly connected twin pontoons are smaller than those of the free twin pontoons especially for relatively small incident wave steepness, due to the observed less motions with waves and more interactions between waves for the rigidly connected twin pontoons during the tests. Compared to the floating breakwater of single pontoon, wave transmission coefficients of the floating breakwaters of free twin pontoons and rigidly connected twin pontoons are reduced by about 9% and 25% respectively.

3.3 *Twin Pontoons Combined with Plates*

Different types of plates were combined with twin pontoons, including the perforated horizontal plates (Fig. 5) and porous vertical plates (Fig. 6) between the twin pontoons, with the distance between the twin pontoons of 1.0 m ($D/2B = 1.0$) and the twin pontoons and plates rigidly connected, to investigate the effect of different

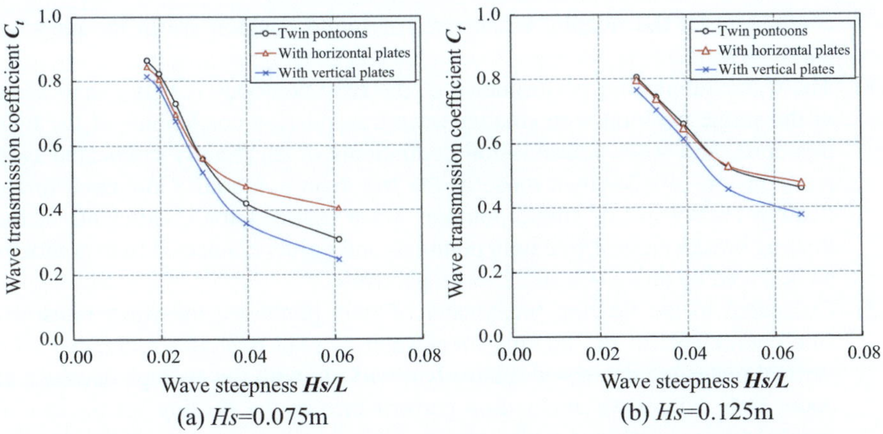

Fig. 9 Comparison results of wave transmission coefficients of twin pontoons combined with plates and twin pontoons

types of plates on the wave attenuation performance of the floating breakwater of twin pontoons.

Comparison results of wave transmission coefficients of the floating breakwaters of twin pontoons combined with plates and twin pontoons are shown in Fig. 9. It can be seen from the results that, compared to the floating breakwater of twin pontoons, the wave transmission coefficients of the floating breakwater of twin pontoons combined with horizontal plates are only decreased slightly when the incident wave steepness is relatively small with relatively long incident wave periods and wave lengths, and under the conditions of relatively large incident wave steepness with relatively short incident wave periods and wave lengths, the wave transmission coefficients are even increased for the horizontal plates. Compared to the floating breakwater of twin pontoons, the wave transmission coefficients of the floating breakwater of twin pontoons combined with vertical plates are decreased relatively obviously with the average decrease of about 10%, as the fact that more water turbulence excited between the twin pontoons and more wave energy dissipated between the twin pontoons thanks to the porous vertical plates between the twin pontoons were observed during the tests. The wave attenuation performance of the floating breakwater of twin pontoons combined with vertical plates is better than that of the floating breakwaters of twin pontoons and twin pontoons combined with horizontal plates.

4 Conclusions

(1) For the floating breakwater of single pontoon, the wave transmission coefficients increase significantly with the decrease of the relative width *B/L* of the pontoon. In the range of relative width *B/L* of 0.11–0.41, the wave transmission

coefficients of the floating breakwater of single pontoon are in the range of 0.42–0.97.

(2) The wave attenuation performance of the twin pontoons is better than those of the single pontoon with smaller wave transmission coefficients of the twin pontoons, and wave transmission coefficients of the rigidly connected twin pontoons are smaller than those of the free twin pontoons. Compared to the floating breakwater of single pontoon, wave transmission coefficients of the floating breakwaters of free twin pontoons and rigidly connected twin pontoons are reduced by about 9% and 25% respectively.

(3) Compared to the floating breakwater of twin pontoons, the wave transmission coefficients of the floating breakwater of twin pontoons combined with vertical plates are decreased relatively obviously with the average decrease of about 10%. The wave attenuation performance of the floating breakwater of twin pontoons combined with vertical plates is better than that of the floating breakwaters of twin pontoons and twin pontoons combined with horizontal plates.

References

1. McCartney BL (1985) Floating breakwater design. J Waterw Port Coast Ocean Eng 111(2):304–318
2. Williams AN, Lee HS, Huang Z (2000) Floating pontoon breakwaters. Ocean Eng 27(3):221–240
3. Syed SA, Mani JS (2006) Performance of multiple pontoons floating breakwater-a numerical approach. In: The International conference on civil engineering in the oceans VI, pp 342–353
4. Ruol P, Martinelli L, Pezzutto P (2013) Formula to predict transmission for π-type floating breakwaters. J Waterw Port Coast Ocean Eng 139(1):1–8
5. Alizadeh MJ, Kolahdoozan M, Tahershamsi A, Abdolali A (2014) Experimental study of the performance of floating breakwaters with heave motion. Civ Eng Infrastruct J 47(1):59–70
6. Chen ZJ, Wang YX, Dong HY, Zheng BX (2012) Time-domain hydrodynamic analysis of pontoon-plate floating breakwater. Water Sci Eng 5(3):291–303
7. Ji CY, Chen X, Cui J, Yuan MZ (2015) Experimental study of a new type of floating breakwater. Ocean Eng 105(1):295–303
8. Ji CY, Cheng Y, Yang K et al (2017) Numerical and experimental investigation of hydrodynamic performance of a cylindrical dual pontoon-net floating breakwater. Coast Eng 129(11):1–16

Wave Tank Model Testing Analysis for Lindeman Island Floating Jetty

H. L. Han

Abstract This study presents comprehensive wave tank model testing and analysis for the proposed Lindeman Island Floating Jetty in Queensland, Australia. Using a 1:25 scale model, we investigated wave-structure interactions under extreme scenarios combining 500-year and 50-year Average Recurrence Interval (ARI) wind and tide events. Our methodology integrated global computer modelling, local Computational Fluid Dynamics (CFD) simulations, and physical wave tank tests, incorporating high-resolution bathymetric data. Results indicate that the 50-year ARI wind combined with 500-year ARI water level produces the worst-case wave loads, with maximum residual wave heights reaching 3.06 m in full-scale equivalent. The study revealed significant load variability, with peak horizontal forces (Fx) of 1288.59 kN and vertical moments (Mz) up to 15,070.73 kN·m in the worst-case scenario. The unique Z-shaped bathymetry was found to play a crucial role in wave transformation, leading to a 39.2% increase in horizontal force under submerged conditions. This research provides critical insights into load variability, submerged state effects, and the impact of tidal variations on wave dynamics, offering valuable design inputs for the floating jetty. The hybrid approach employed demonstrates the effectiveness of combining numerical modelling with physical testing for complex marine environments, contributing significantly to the field of coastal engineering and floating structure design.

Keywords Floating jetty · Wave tank testing · Computational fluid dynamics · Bathymetric effects · Coastal infrastructure · Extreme wave loads

H. L. Han (✉)
Paya Lebar Link, Singapore, Singapore
e-mail: henrylhan@hann-ocean.com

X. L. Zhao et al. (eds.), *Proceedings of the Fourth World Conference on Floating Solutions: WCFS 2024; 2–4 December; Hong Kong*, Lecture Notes in Civil Engineering 597, https://doi.org/10.1007/978-981-96-4569-5_101

1 Introduction

The Lindeman Island Floating Jetty project is a crucial component of Well Smart Group's redevelopment initiative for Lindeman Island in the Whitsunday region of Queensland, Australia. Commissioned in December 2023, this project aims to enhance the island's tourism infrastructure and accessibility by designing a floating jetty capable of withstanding challenging marine conditions while providing safe and reliable access across wave and tide conditions.

1.1 Project Background and Objectives

The proposed floating jetty is designed to serve as the primary point of arrival and departure for visitors and supplies. Its location on the island's eastern side exposes it to a range of wave conditions influenced by both local wind patterns and long-range swell.

The primary objectives of this study were to:

- Perform small-scale wave flume tests to validate global climate computational simulation results provided by the client.
- Investigate wave transformation in different water depths based on site bathymetry.
- Conduct physical scaled-down wave tank model tests to measure dynamic forces acting on the jetty's piles.

1.2 Significance of the Study

This study is significant because it combines physical model testing with advanced numerical simulations to provide a comprehensive understanding of wave dynamics at the Lindeman Island site. The global model simulation considers two extreme design scenarios:

- 500-year ARI wind combined with 50-year ARI tide
- 50-year ARI wind combined with 500-year ARI tide.

These scenarios are analyzed for 2024 and 2074, including a projected 0.6 m sea level rise. By integrating multiple approaches, this study aims to overcome the limitations of each method and provide a more accurate assessment of the wave loads and dynamics that the Lindeman Island Floating Jetty must withstand.

2 Methodology

2.1 Wave Load Study Approach

To derive the dynamic loads on the floating jetty platform and its securing piles, Hann-Ocean Technology recommended a hybrid approach consisting of the following three steps, shown in Fig. 1.

Step 1: Global computer modelling for wind, wind-induced waves, tide, and current using 2D spectrum simulations (provided by Well Smart's consultant).
Step 2: Local computer modelling for wave and tide using 3D time-domain CFD simulation.
Step 3: Physical Model Testing in a wave tank, combined with 3D time-domain Computational Fluid Dynamics (CFD) simulation for wave and tide effects.

This approach was chosen as a more economical and efficient alternative to a full 3D time-domain CFD simulation with fluid–structure interaction, which would be highly resource-intensive and time-consuming.

2.2 Physical Model Testing Setup

The wave tank tests were conducted using a 1:25-scale physical model of the floating jetty and its surrounding bathymetry. Key features of the setup included:

- 120 kW wave maker with 4 wave paddles, 5 m long × 0.75 m wide each.
- 90 m long U-shape wave flume with a width varying from 3 to 5.5 m and a depth varying from 5.5 to 3.5 m.
- 9.09 m long and 3 m wide seabed platform accommodating up to a net 0.4 m water depth above the model seabed.
- Z-shaped seabed profile based on sonar data of the Lindeman Island bathymetry.
- 1:25-scale jetty model made of 20 mm-thick wood panels.

Two primary water depths were tested: 341 mm and 356 mm, corresponding to the real sea depths of 8.525 m (AHD 3.7 m) and 8.9 m (AHD 4.08 m) respectively.

Figure 2 features a 1:25 scale wave tank setup that includes a Z-shaped seabed profile. This profile is based on high-resolution sonar data, effectively illustrating the unique bathymetric conditions at the Lindeman Island site.

Figure 3 illustrates the scaled wooden jetty model to represent the proposed floating jetty design. Key components include:

- One Platform: A 1:25 scale of the jetty platform was created, with provisions for floating and submerged conditions.
- Two Piles: The model included representations of the two piles that secure the floating platform.

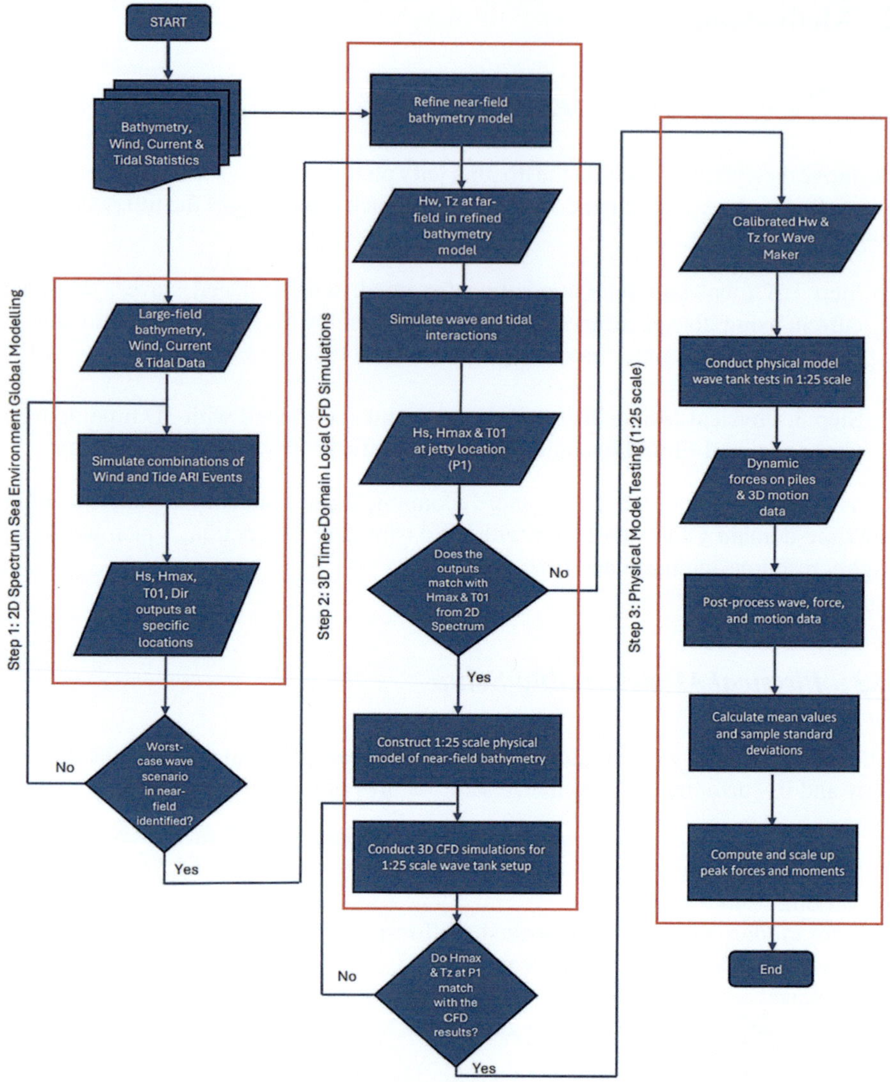

Fig. 1 Flowchart of three-step hybrid approach

- Eight Force Sensors: 8 force sensors were installed to measure loads in different directions.

2.3 *Measurement Equipment and Procedures*

The test setup included various measurement devices:

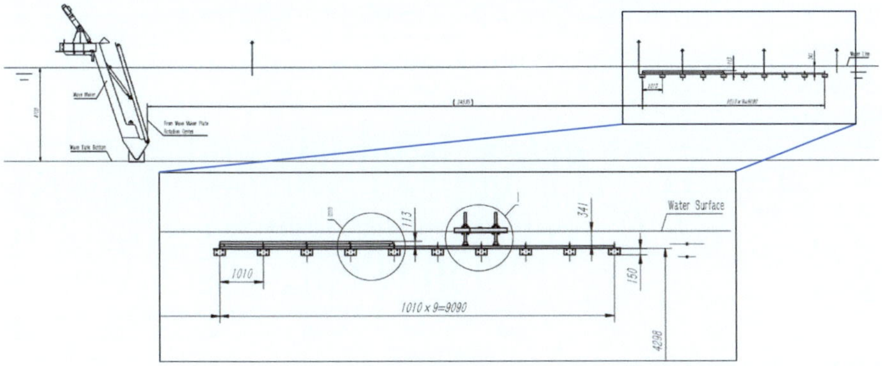

Fig. 2 Cross-section of the wave tank with the Z-shaped seabed and jetty model

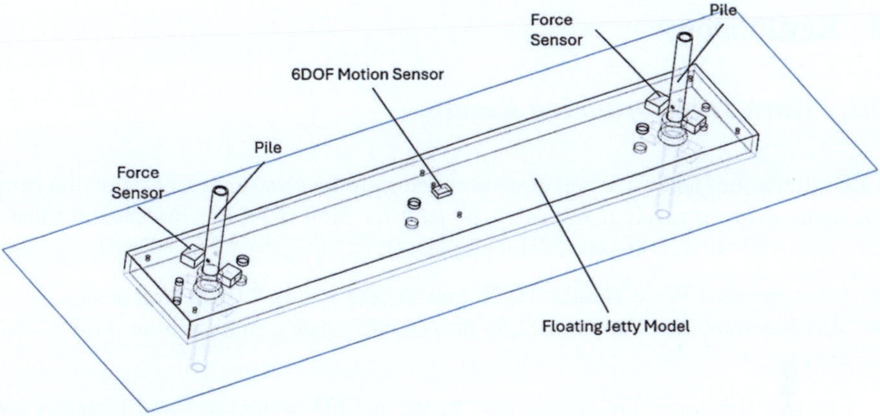

Fig. 3 Floating jetty model configuration

- Wave Gauges (Model No. MTS-1200MM-S/A-FS): Placed at strategic locations to measure wave heights and periods.
- Force Transducers (Model No. ZNLBS-VII): Secured to the platform and in contact with the piles to measure dynamic forces in different directions.
- Data Acquisition System (Model No. DATAQ 710 & 808): To record and process the measurements from wave gauges and force transducers.
- 3D Motion Sensor (Model No. HWT901B-TTL/232): To record the floating jetty's 6 degrees of freedom motion.

Multiple 60 s duration tests were conducted for each scenario to ensure statistical reliability. Tests were conducted with both the original approximated flat seabed and the corrected Z-shaped seabed to assess the impact of bathymetry on wave transformation.

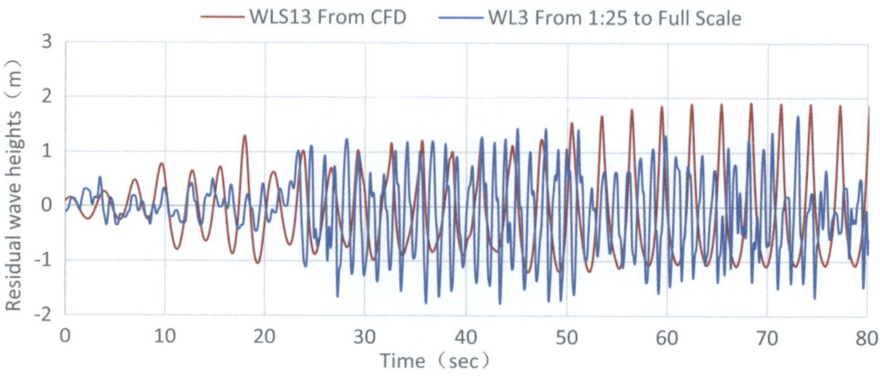

Fig. 4 CFD predictions versus physical test results for residual wave heights

3 Key Findings

3.1 *Wave Transformation Results*

CFD simulations and wave tank tests revealed significant wave height reduction from incident waves to residual waves at the jetty location (P1). For the critical Case J (50-year ARI wind, 500-year ARI water level):

- Average Mean Wave Height: 80.28 mm (model scale), 2.01 m (full scale).
- Max Residual Wave at P1: 122.53 mm (model scale), equivalent to 3.06 m (full scale).

Figure 4 illustrates the comparison between CFD predictions and physical test results for residual wave heights at the jetty location (P1), demonstrating the complementary nature of numerical and experimental approaches.

3.2 *Tidal Level and Submerging Effects*

Higher tidal levels (deeper water) allowed slightly larger waves to reach the jetty. The submerged state experienced higher overall loads, particularly in horizontal force and vertical moment:

- Floating Mode: Mean horizontal force (Fx) at P1 = 31.86 N.
- Submerged Mode: Mean horizontal force (Fx) at P1 = 44.35 N.

Figure 5 provides a comparison of the forces and moments exerted on the piles in both floating and fully submerged modes at a critical water depth of 356 mm, highlighting the differences in load distribution between these conditions.

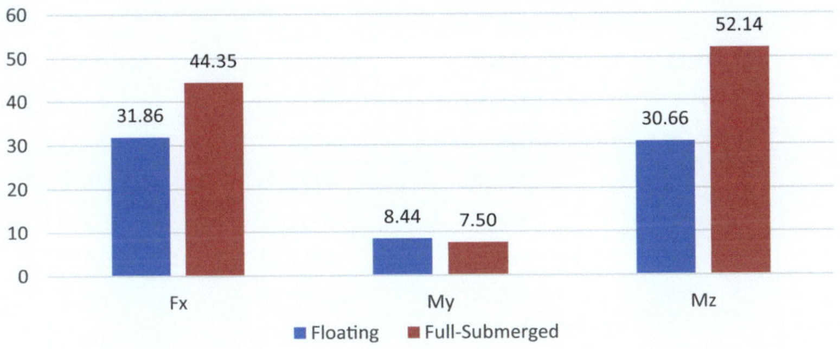

Fig. 5 Forces and moments on piles in floating and fully submerged modes at 356 mm water depth

3.3 Bathymetry Effects on Wave Heights

The corrected Z-shaped bathymetry profile led to significant wave height reduction as waves approached the jetty location. This resulted in lower wave heights at the jetty than might be expected from deep-water conditions. The unique bathymetry, particularly the 2.5 m step identified in the sonar data, played a crucial role in wave transformation and energy dissipation.

The fully submerged condition resulted in a 39.2% increment of horizontal force (Fx), an 11.1% decrement of the horizontal moment (My) and a 70.1% increment of Vertical moment (Mz).

4 Dynamic Load Measurement and Analysis

Full-scale design loads were derived, with high variability observed in all load components. A Peak Factor of 3.5 was used for design load calculations, capturing approximately 99.97% of potential load events. Table 1 presents a summary of the full-scale design loads on piles.

These loads represent the critical design scenarios for the jetty structure and its piles, accounting for the most severe conditions observed during testing.

Table 1 Summary of full-scale design loads on piles

Dynamic loads on piles	Mean	Max peak	Max trough
Horizontal force (Fx), kN	639.52	1288.59	– 608.50
Moment about Y-axis (My), kN-m	3801.98	6122.30	– 6385.81
Moment about Z-axis (Mz), kN-m	4444.61	15,070.73	– 34,511.77

Table 2 Full-scale wave-induced design loads on piles

Case No	Peak	Fx (kN)	My (kN-m)	Mz (kN-m)
1	+Fx	**1288.59**	3631.99	– 6434.53
2	−Fx	*– 608.50*	– 4214.53	– 5365.61
3	+My	597.59	**6122.30**	3630.23
4	−My	– 398.72	*– 6385.81*	– 21,249.62
5	+Mz	714.18	4821.61	**15,070.73**
6	−Mz	400.20	– 2742.19	*– 34,511.77*

5 Discussion

5.1 Implications for Jetty Design

The study provides crucial inputs for the jetty design:

- Load Resistance: Design piles to withstand six typical phases in the worst-case scenario, where a dominant peak load and two associated loads are applied simultaneously at the jetty-pile interfaces. See Table 2.
- Peak Factor: Implement a peak factor of 3.5 for extreme structural design conditions, considering the high load variability observed.
- Fatigue Analysis: Conduct comprehensive fatigue analysis using the full range of observed loads to ensure long-term structural integrity.

5.2 Comparison of CFD and Tank Test Results

While both CFD and tank tests demonstrated significant wave height reduction, the physical model showed more conservative results. This underscores the importance of physical testing in capturing real-world wave behaviours not fully represented in numerical simulations. The combination of CFD simulations and physical model tests provides a robust basis for understanding these effects and informing the jetty design.

5.3 Model Testing Limitations

Key limitations of the study include:

- Scaling Effects: The 1:25 scale model may not perfectly represent all force interactions, particularly for complex phenomena like wave breaking.

- Simplifications: The model simplifies the jetty and piles interfaces of the full-scale jetty, which should be considered when interpreting results.
- Environmental Factors: Factors like wind and currents, which weren't replicated in the wave tank, may influence loads in real-world conditions.

6 Conclusion

The wave tank model testing provides crucial data for the design of the Lindeman Island Floating Jetty, highlighting the complex interaction between the proposed structure and site-specific wave conditions. The study underscores the importance of considering multiple modelling approaches and the value of physical testing in capturing real-world wave behavior that may not be fully represented in numerical simulations.

Key findings include the identification of the worst-case scenario (50-year ARI wind combined with 500-year ARI water level), resulting in maximum residual wave heights of 3.06 m in full-scale equivalent. The study also revealed significant load variability and the critical impact of the unique Z-shaped bathymetry on wave transformation.

These results have important implications not only for the Lindeman Island project but also for similar coastal infrastructure developments worldwide. The hybrid approach employed in this study, combining global and local computer modelling with physical testing, offers a robust methodology for addressing the complex challenges posed by coastal environments.

References

1. Hann-Ocean Technology Pte Ltd (2024) Wave tank model testing analysis for Lindeman island floating jetty. Full Technical Report
2. Port & Coastal Solutions (2023) Global climate computational simulation for Lindeman Island. Technical Report
3. Hughes SA (1993) Physical models and laboratory techniques in coastal engineering. World Scientific
4. Chakrabarti SK (1987) Hydrodynamics of offshore structures. WIT Press
5. PIANC (1994) Guidelines for the design of floating breakwaters. Report No. 13
6. Faltinsen OM (1990) Sea loads on ships and offshore structures. Cambridge University Press
7. Holthuijsen LH (2007) Waves in oceanic and coastal waters. Cambridge University Press

Wave Response Experiment on a Floating Model with a Hanging Weight Aimed at Reducing the Response

Yoshitsugu Kawakami, Takumi Ishihara, and Ryuji Endo

Abstract This paper is the result of an experiment on a model aimed at reducing the response of a floating body to waves. A feature of the floating model is that a weight is hung from the bottom with a wire. The response reduction system of this floating model was inspired by oyster farming rafts. The experiments included fluid-coupled free vibration experiments, wave response experiments using regular waves, and irregular wave response experiments caused by wind. The floating model varied the type, shape, and hanging position of the weights. The frequency response function is obtained from the experimental results, and the reduction of the response is considered.

Keywords Floating architecture · Water tank experiment · Hanging weight model

1 Introduction

Floating structures are subject to complex responses due to the external environmental forces acting on them that are different from those on land. For this reason, accurate understanding of responses to currents and waves is extremely important in design. As an element that may contribute to reducing the response of floating structures, we propose a weight suspended from the bottom of the floating structure by a wire or the like (hereafter referred to as a suspended weight). It is believed that appropriately placed hanging weights can reduce the tilting and swaying of a structure. In this experiment, a response experiment was conducted when a hanging weight was attached to a floating model placed in rough waves.

Y. Kawakami (✉) · T. Ishihara
Department of Architectural Engineering, Hiroshima Institute of Technology, Hiroshima, Japan
e-mail: y.kawakami.4i@cc.it-hiroshima.ac.jp

R. Endo
Polytechnic University, Tokyo, Japan

X. L. Zhao et al. (eds.), *Proceedings of the Fourth World Conference on Floating Solutions: WCFS 2024; 2–4 December; Hong Kong*, Lecture Notes in Civil Engineering 597, https://doi.org/10.1007/978-981-96-4569-5_102

2 Overview of Floating Model and Free Vibration Experiment

There were two types of floating models: A was 450 mm long and B was 600 mm long. Both floats A and B are rectangular prisms with a thickness of 40 mm and a width of 100 mm. And are made of Styrofoam. The draft was set to 20 mm. To prevent bending deformation, aluminum plates were attached to the long side and bottom surfaces, and rigid body motion was assumed. Table 1 shows an overview of the floating model. Figure 1 shows the measurement method used in the experimental tank. The tank is 13.0 m long and 0.3 m wide.

Free vibration experiments were conducted to determine the natural frequency of the floating model. Acceleration sensors were attached to the front and back of the floating model, and the floating model was placed in still water and flicked by hand from above to cause free vibration, and the response acceleration was measured. We then performed an FFT analysis to calculate the fluid-coupled natural frequency. The results for models A and B were almost the same (in Table 1). Figure 2 shows the response acceleration of Type B with a hanging weight. From this measurement result, we can see that a different low-frequency response occurs while the acceleration is attenuating. This result was not obtained with the model without a hanging weight. The results of the FFT analysis are shown in Fig. 3. Two peaks are observed

Table 1 Overview of the floating models

		Length (mm)	Width (mm)	Draft (mm)	Fluid-coupled natural frequency (Hz)
Model A	Weight-mounted	450	100	20	2.20
	Weight-hanging	450	100	20	2.17
Model B	Weight-mounted	600	100	20	2.21
	Weight-hanging	600	100	20	2.17

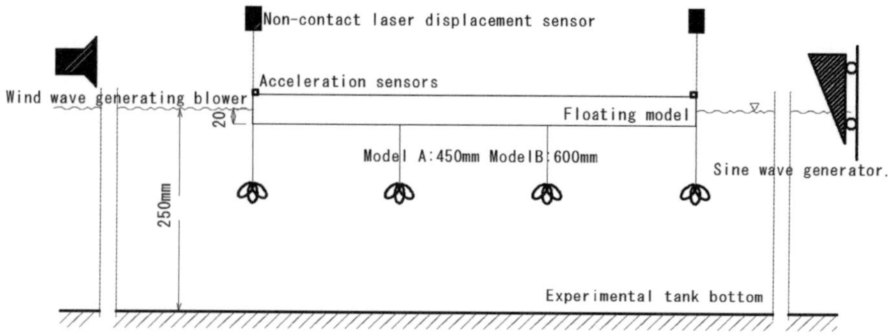

Fig. 1 A figure is the measurement method used in the experimental tank

Picture 1 A picture is the wave response experiment model B (Weight-hanging model)

Fig. 2 A figure is the response acceleration of Type B with a hanging weight

(0.72 Hz, 0.94 Hz) at frequencies lower than the fluid-coupled natural frequency of the floating body model. Picture 1 shows the experiment in the tank.

3 Wave Response Experiment

First, a response experiment was conducted in which a sine wave was applied to grasp the response characteristics of the floating model. The waves generated by a sine wave generator were 0.5 to 2.0 Hz, and because the wave height and wave frequency decay as the waves travel, the wave height and wave frequency were measured near the floating model, as shown in Fig. 1. In addition, a non-contact laser

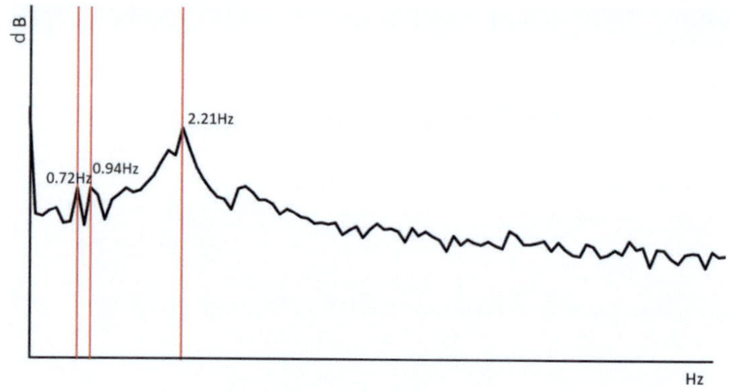

Fig. 3 A figure is the results of the FFT analysis of Type B with a hanging weight

displacement meter and an accelerometer were installed on the upper and lower sides of the waves, respectively, to measure the response.

Next, we generated irregular waves using a fan to generate wind waves, and we measured the wave response. In this experiment, we also measured the wave height and wave frequency near the floating model. Figure 4 shows the results of an experiment on the irregular wave response for the weight-hanging model, and Fig. 5 shows the results of an experiment on the weight-mounted model under the same conditions.

It can be seen that the response of the floating model is large on the upper side of the waves and small on the lower side of the waves. Also, no significant difference

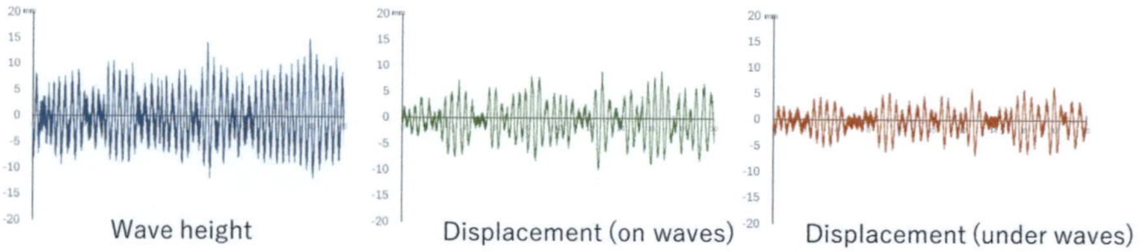

Fig. 4 A figure is the results of an experiment on the irregular wave response for the weight-hanging model

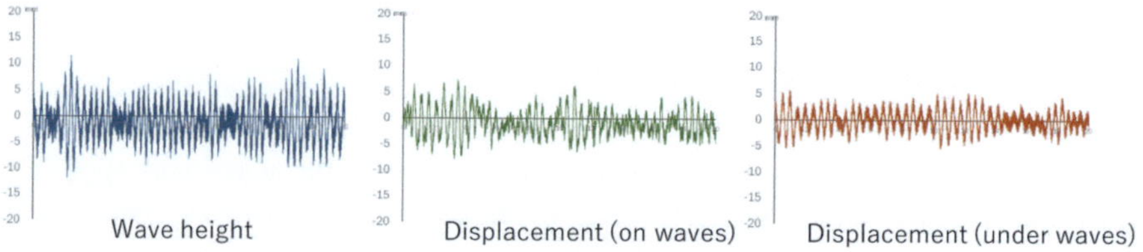

Fig. 5 A figure is the results of an experiment on the weight-mounted model

was observed in the waveforms between the hanging weight model and the weight-mounted model. Each experiment was performed three times, with measurements taken for 30 s after the waves had fully developed. The wave height and response displacement obtained were treated as significant waves.

4 Results and Discussion

The obtained floating body response acceleration and floating body response displacement were normalized by wave height. In the figure, the horizontal axis indicates the vibration of the incident wave, and the vertical axis indicates the response displacement divided by the wave height of the incident wave. Figure 6 shows the experimental results for the upper wave side of model A, and Fig. 7 shows the results for the lower wave side. Similarly, Fig. 8 shows the upper wave side of model B, and Fig. 9 shows the results for the lower wave side.

The wave response magnification on the up-wave side of both models is large, and when the natural frequency of 2.20 Hz is exceeded, it becomes smaller. This is the same for both the hanging weight model and the weight-mounted model. Analyzing the four graphs, we can see that the response reduction effect of the drooping model was not observed. The reason for this is thought to be that the floating model is assumed to behave as a rigid body, and it is a model in which elastic deformation does not occur. Although the characteristics of the hanging model were observed in the free vibration experiment as shown in the Fig. 2, this behavior was not observed in the wave response experiment.

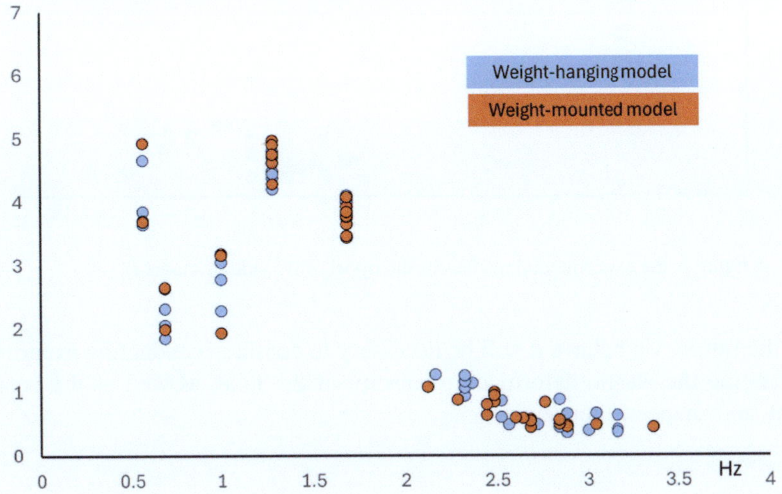

Fig. 6 A figure is the experimental results for the upper wave side of model A

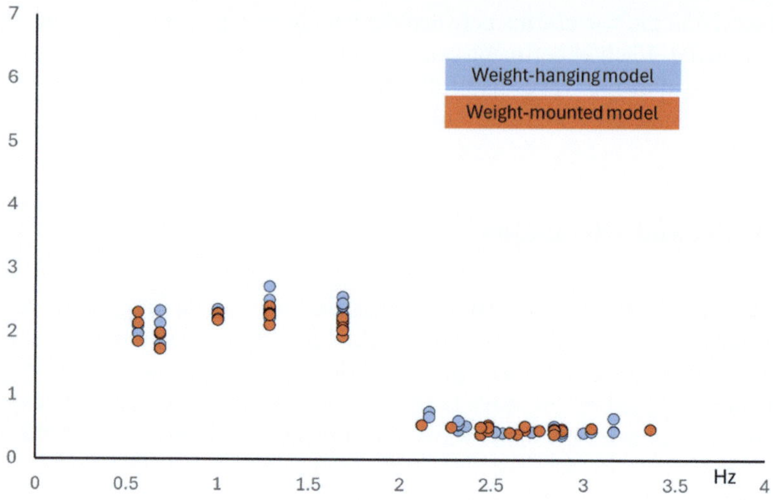

Fig. 7 A figure is the experimental results for the lower wave side of model A

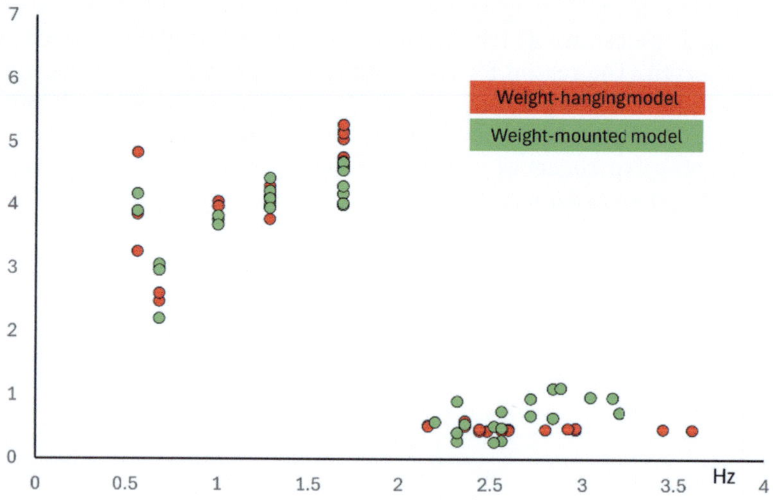

Fig. 8 A figure is the experimental results for the upper wave side of model B

In the future, we believe it will be necessary to continue conducting experiments by changing the elastic deformation capacity of the float, as well as the position, weight, and shape of the weight.

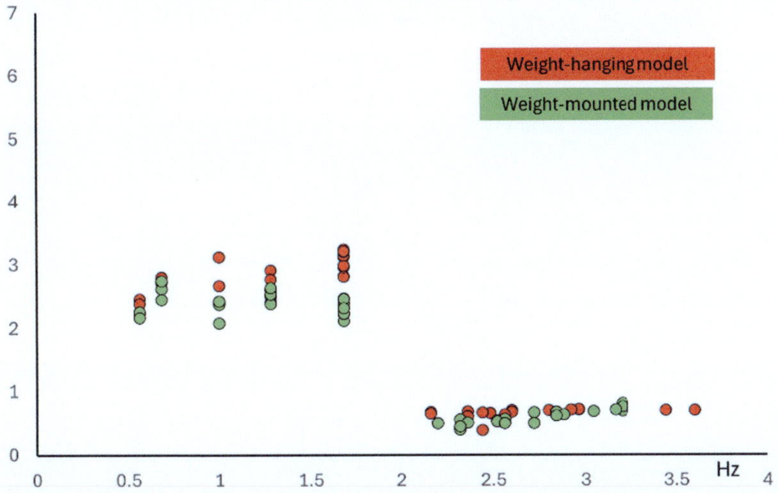

Fig. 9 A figure is the experimental results for the lower wave side of model B

5 Conclusion

In this study, a floating body with a hanging weight was used to conduct experiments aimed at reducing wave responses. Results obtained through comparison with a model with a weight of the same weight mounted on top did not show any effect in reducing the response. However, the response waveforms of the free vibration experiment confirmed behaviors not seen in the model with a weight mounted on top. In the future, we plan to continue conducting experiments by changing the elastic deformation capacity of the float, as well as the position, weight, and shape of the weight.

Acknowledgements We gratefully acknowledge the work of present members of our laboratory Mr. Hirai, Mio, Bisaga, Tomokawa, Ono, Hasebe, Ino, Fujimoto, Fukumoto, Koike, Miyazato and Imamura.

References

1. Endo R, Kawakami Y (2013) Water tank experiments on harmonic wave response focusing on dynamic pressure distributions of elastic floating structures. J Struct Constr Eng 78(684):395–404
2. Endo R, Kawakami Y, Imai T, Tosaka N (1999) Identification analysis of structural damage on unit-linked offshore floating models. Int J Offshore Polar Eng 9(3)

Wave Absorbance System for Towing Tanks

Ofer Berman, Ezri Tarazi, Thijs Roumen, and Gil Wang

Abstract Much of the research on floating solutions relies on physical experiments conducted in towing tanks, flumes, and wave basin facilities, in which specimens are exposed to wave loads, simulating oceanic conditions at scale. 98.5% of towing tank setups are closed on both ends (aka dead-ended systems). Consequently, the conducted experiments are significantly impacted by wave reflections from the end of the tank, there situated a wave absorbance structure commonly referred to as the "beach". These reflections require post-processing on the recorded signals to filter out the noise, potentially reducing the accuracy of the analysis. To minimize this problem, it is common to install a beach at the end of the tank for wave-breaking purposes—essentially a slope (linear or parabolic) designed to break the incident wave energy. Currently, conventional beach designs are made from commonly available materials and structures and therefore limited in its wave absorption capacity. This is especially present in short towing facilities or flumes where there is not enough length to efficiently absorb waves using longer beaches, making these facilities vulnerable to reflective waves. This study aims to improve beach design and reduce the influence of reflecting waves in testing facilities through a biomimetic noise reduction principle driven by computational design and digital fabrication techniques. The proposed design integrates vertical floating structures (resembling kelp blades) and ceramic porous structures (resembling corals) to achieve better wave absorbance compared to the existing solutions. This study presents the design methodology and preliminary results of the proposed solution.

Keywords Biomimicry design · Physical testing · Wave absorbance

O. Berman (✉) · T. Roumen
Matter of Tech Lab, Cornell Tech, New York, NY, USA
e-mail: ofer.berman@technion.ac.il

O. Berman · E. Tarazi
Design-Tech Lab, Industrial Design Department at the Faculty of Architecture and Town Planning, Israel Institute of Technology, Haifa, Israel

G. Wang
CAMERI—Coastal and Marine Engineering Research Institute, Haifa, Israel

X. L. Zhao et al. (eds.), *Proceedings of the Fourth World Conference on Floating Solutions: WCFS 2024; 2–4 December; Hong Kong*, Lecture Notes in Civil Engineering 597, https://doi.org/10.1007/978-981-96-4569-5_103

1 Introduction

Towing tank facilities are the cornerstones of naval architecture and physical studies of marine engineering to simulate oceanic conditions [1]. These versatile facilities are used for different types of research studies, both fundamental and applied [2, 3]. They are composed of three fundamental components: (1) Wavemaker which generates monochromatic and irregular waves. (2) Towing car, which is responsible for dragging or positioning an object along the tank and measure its drag in that time, for example, ship hulls at different speeds and in various conditions, floating, or completely submerged [4]. (3) The beach which is situated at the end of the tank (opposite the wavemaker), to passively break the wave train and reduce the impact of the reflective waves [5]. The effectiveness of the beach in damping waves and shortening cooling times of the tank after runs is proportional to its length together with its physical characteristics of it [6, 7]. Elongating towing tank beaches increases costs and reduces the space in the tank for experiments.

1.1 Surveying Existing Towing Tank Facilities

Almost 4 decades has pass from the last survey of towing tanks facilities, focusing especially on their beach design properties [7]. To get a current overview of the topic we surveyed the International Towing Tank Convention (ITTC) (https://ittc.info/) which has more than 100 members around the world, with a wide variety of beach parameters. Surveying the convention members can outline the general setups of existing towing tank facilities globally. At the ITTC members documentation we found that 98.5% of the ITTC towing tank setups are dead-ended systems, ending with a front beach structure, rather than a 'U' shaped wave absorber. Most ITTC towing tanks have a minimum length of approximately 50 m, while many of them easily cross the 150 m length benchmark as shown in Fig. 1 Up. The length of the tank enables an extensive and gradient-slope beach, which helps dissipate the propagating (incident) wave energy and reduces the impact of reflecting waves toward the measuring devices. While the length of the beach plays a crucial role as longer beaches provide more space for the dissipation of wave energy, reducing noise in experiments, and shortening cooling time between runs [8]. Moreover, out of the facilities with registered beach parameters, the length of most beaches is approximately 2–6% of the overall tank length as given in Fig. 1 Down. Even though there no much data quantity, its fits with the findings of Ouellet & Datta [7], who surveyed towing tanks facilities back in 1986. Furthermore, Ouellet and Datta [7], surveyed the type of wave absorbers existed in 76 water flumes and wave basins around the world. They have found that most of the beaches are using materials such as wire screens, transversal bars, horsehair, sand, gravel and stones, wooden laths, and perforated plywood. Since then, the practice has not changed dramatically, as current facilities still uses similar materials to build their newly towing tank beaches [9–11].

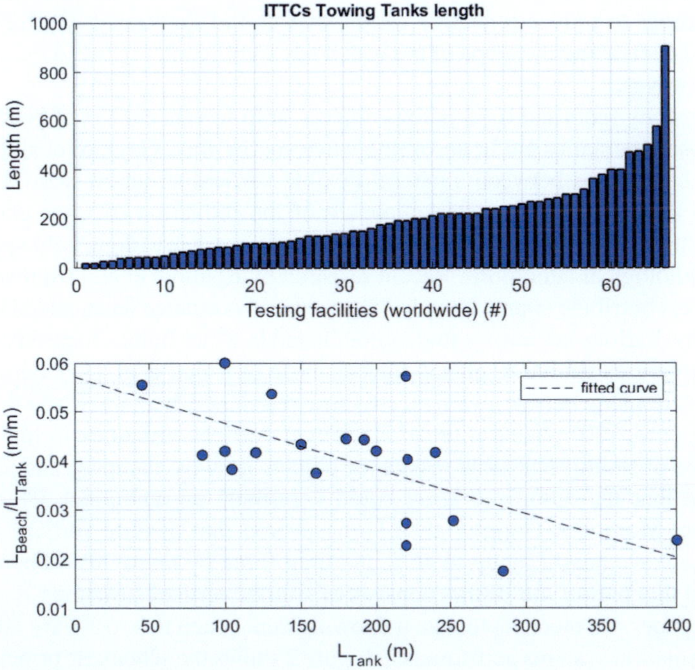

Fig. 1 **Up**, List of facilities, members of the ITTC, with their equivalent length of the towing tank. **Down**, tank length (x-axis), and beach length to total length ratio (y-axis)

1.2 The Absorbance of Waves Over Natural Systems

To better understand wave absorbance, we overlook mechanisms occurring in natural shallow marine ecosystems, such as coral reefs and kelp forests. These ecosystems exhibit intriguing interactions with wave formation and energy dissipation. Coral reefs have been studied for their role in shoreline protection against wave flooding and erosion [12, 13]. Several studies highlight the ability of the reef geometry to reduce the energy of waves by up to 99.9% during low tides [14], with the reef crest alone contributing to the energy reduction by 84% [13]. The rugosity of stony coral reefs introduces friction with the wave mass, effectively dissipating its movement. In some laboratory-based studies, researchers try to absorb waves using artificial objects, focusing on the field of porous materials as guidance. Chwang and Chan [15] define porous material as rigid body containing random pores distributed throughout the body, thus forming a homogenous matrix allowing fluid to pass through. Cho and Kim [10] study the interaction of monochromatic incident waves with a series of inclined punched steel plates, and found that a 0.1 porosity was the optimal in their setup. Later, George et al. [11] found that a wave absorber made of series of horizontal plates could be further optimized using artificial neural network model. Moreover, Sulisz [16] used plastic doormat to build and measure the efficiency of wave absorber to reduce reflective waves, they found that increasing the absorber length and decreasing the slope of the beach has significantly reduced the amount of reflective waves.

Unlike stony reefs and rigid porous objects, kelp forests consist of flexible structures whose kinematics aid in absorbing wave energy flux. Gaylord et al. [17] identified the correlation between a natural flexible vertical structure such as kelp and the depth of the water, with the efficiency of the reduction of wave energy. The study highlighted differences in wave absorbing between various kelp species and their morphological alterations. Recent research by Elsmore et al. [18] revealed that kelp forests contribute approximately 7% to wave absorbance when added to benthic topography. Various laboratory studies conducted in water flumes focused on vertical flexible structures similar to plants, such as those of Luhar et al. [19], demonstrated a direct relationship between vertical structure resistance, buoyancy, and flux. Nove-Josserand et al. [20] placed a set of simplified vertical structures to highlight the importance of multiple flexible resistance objects in wave damping and absorption and found that the optimized array could contribute an additional 5% to energy dissipation. Moreover, Gong et al. [21] discovered that flexible vertical structures can reduce wave height by up to 46% and drag force by up to 90%. Therefore, we concluded that part of our design consists of similar vertical structures.

In this paper, we reconceptualize the towing tank beach to boost wave absorbance following natural systems performance. Figure 2 shows the schematic principle of the beach design, containing a combination of porous rigid structures and vertical floating structures. We hypothesize that replicating the absorbance mechanisms found in coral reefs and kelp forests will produce superior results to the current state-of-the-art. Today, there are different studies which separately test wave absorbance in flexible or rigid structures. We believe that combining flexible and rigid absorbance mechanisms and utilizing fabrication processes to closely resemble nature might bring improved results than current state-of-the-art wave absorbance-based studies. Accordingly as a cut out of flexible PVC sheet designed to mimic vegetation [21] or as parabolic beach structure partially paneled with slabs [9]. Our main contribution sits in the design-space of structured materials and shapes using digital fabrication techniques to enable high wave absorbing capabilities.

2 Methodology

This paper methodology is directed toward bio-mimetic beach design consisting of rigid porous tiles and vertical floating structures. The goal is to reproduce the natural mechanisms existing in coral reefs and kelp forests which are relevant to wave absorbance. The major morphological element in corals responsible for wave absorbing besides overall topographic condition, is their permeability, allowing water to dissipate into multiple streams between branches and canopies. The major morphological element responsible for wave absorbance in kelps is their kinematic ability to force movement in the water column. Therefore, as described in the literature review,

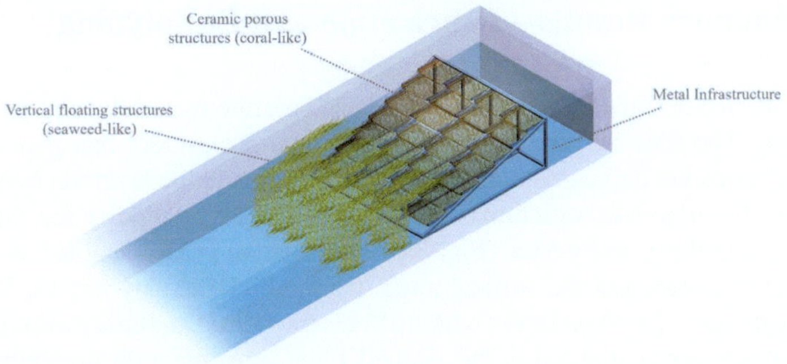

Fig. 2 Scheme of the suggested layout of the novel hybrid biomimetic noise reduction beach, the seaweed-like structures are positioned before the beach, the horizontal ceramics porous panels are distributed in stapes, and in the rear end of the slope there is quieting pool

we look for two separate sets of properties in each system, one rigid and permeable and the other vertical and flexible with a decent amount of surface area facing the water.

The rigid beach wave interface is made of ceramic, divided into 60 × 45 cm tiles that fit nicely inside the ceramic kiln. Designing permeable ceramic structures can benefit from fabrication process that uses continuous deposition path [22]. This way, we can better control the generated pores and strength of the structure. Based on an earlier experience, that was a development of a free-formed lattice structure inspired by the complexity of stony corals [23], we choose one of the readily available materials and techniques—paste-based extrusion of clay. This technique involves deposition of a soft clay mixture which after hydration fires to become ceramic. Generally, ceramic materials can achieve good structural ability and are suitable for underwater environments [24].

In natural ecosystems seaweed and kelps have evolved to cope with pressure and friction drag loads [25]. Here, we seek to generate drag using vertical floating structure, although the motivations contrast, the design can benefit from several formerly investigated fluid interaction qualities [25]. Such as the combination of porous and elastic features (poroelasticity), buoyancy, and skin friction in the material domain; and flutter, torsion, chirality, and vortices creation in the structural domain. Achieving these characteristics can assist in attenuating the water energy by its movement requiring a flexible structure and an interface that can capture the water movement. To realize how to fabricate these complex structures artificially, we used common materials like plywood, PLA, Mylar and fabric & metal rod, and widely used techniques such as laser-cutting, 3D printing, and embroidery & wire bending.

3 Preliminary Results—Fabrication and Prototyping

In this study, we designed a porous tile made of ceramic material using paste-based 3D printing. The tiles are made through an original design tool that generates clay loops. The loops are deposited over the building plate and the layered clay and form a complex of bridge-like structures. The system enables to modify the intensity of the loops along the x and y axes (Fig. 3a). This system is also enabled in choosing the number of layers and the vertical angle from which the clay moves. It can also change the distance between layers which affects the falling and chiral arrangement of the clay, consequently disrupting the aligned water movement through the medium. The final tiles can differ by their volumetric compactness, porosity, and surface complexity, as shown in Fig. 3b-b1. The distribution of the clay pasty coil over the build plate forming the thickness of the structure, generated collection of gravitated bridges arranged in the tile boundary (Fig. 3c). Finally, these tiles can be assembled on a metal infrastructure transforming into a fully functional beach, situated in a towing tank termination (Fig. 3d).

To facilitate mimicking the kelp blades, we focused on the discovery of the kelp genus *Alaria* which holds about 17 species [26]. This genus species could usually be found in mid-to-low rocky shores on the northeast Pacific coast coping with a highly energetic environment. As present in Fig. 4a, this genus blade is composed of two major elements—the midrib which gives its structural ability and the blade membrane which responsible mainly for wave absorbance (among other important things) [18]. To mimic these two structural elements in our blade design, we generated four archetypes that give strength and provide the blade surface area, each in general size of 120 mm (L) and 50 mm (W). We prototyped the blades using common

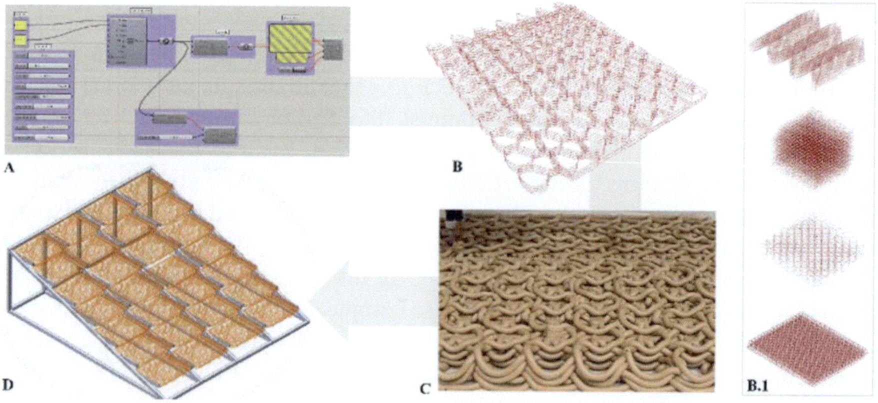

Fig. 3 **a** Parametric design tool that controls the movements of the 3D printer and is possible to generate multiple outcomes. **b** Dedicated tool path designed to dissipate wave energy. **b1** Some of the possible tool-path variations generated for making tiles. **c** The printed path was deposited out of soft clay, exhibiting various passages for the water. After the printing process, the clay is hydrated and fired in a kiln. **d** Scheme of the ceramic beach assembly containing 28 tiles

Fig. 4 **a** *Alaria Marginata* image retrieved from the Beaty Museum algae collection (https://explore.beatymuseum.ubc.ca/algae/). The morphology of this genus can typically be described as a long constructive midrib laminated with a thin blade membrane. **b** Presentation of the four-blade types connected to aluminum bar. **c** Initial reaction test of the blades manually pulled inside a small water tank

materials and techniques as presented in Fig. 4b. That includes (from left to right) Blade membrane made of manually cut fabric sewed using digital embroidering to 1 mm wire-bended metal rod. Laser-cutting 2.5 mm plywood. 3D printed PLA (thermoplasty polymer) blade integrating membrane and structural midrib. Laser-cutting PET (mylar) sheet in thickness of 0.254 mm, after the cut the blade was manually bent along the mid-perforation.

To get a kinematic behavioral sense of the different materials and techniques used to fabricate the blades we came up with a simple water pull-reaction (Fig. 4c). Out of this test we have found that the mylar blade deformed the most, then the fabric & metal rod blade and the PLA blade and plywood blade were too stiff for any noticeable change.

4 Discussion and Conclusions

Although wave absorbance effect of natural ecosystems such as coral reefs and kelp forests has been studied broadly, little attention has been given to these mechanisms in towing tank beach design. Our study aims to establish a holistic set of criteria for the design of towing tank beaches, by replicating natural ecosystems and developing fabrication workflows. Our preliminary results suggest a feasibility to scale up this technology into a full beach assembly. In wave absorbance structures ceramic is not a common material, however, this material and the associated techniques offer the creation of complex yet stable medium with the benefit of multi-scale permeable pores. Entering the next phase of physical wave absorbance tests in towing tank requires fabricating a series of tiles with gradient porosity to evaluate the respective absorbance. As for the kelp blade designs, we demonstrated it is possible to make variations using different materials and techniques. Future studies should include physical performance evaluations. The four blade designs currently vary by buoyancy and stiffness, future research should include the exploration of other important characteristics, and the assembly into full functional structures.

Acknowledgements We would like to thank the support given by AOL, Inc. Fund of the Cornell University Foundation.

References

1. Gao L (2022) Towing tank test. In: Cui W, Fu S, Hu Z (eds.) Encyclopedia of ocean engineering. Springer Nature Singapore, Singapore, pp 1978–1986. https://doi.org/10.1007/978-981-10-6946-8_264
2. Jiao J, Ren H, Guedes Soares C (2021) A review of large-scale model at-sea measurements for ship hydrodynamics and structural loads. Ocean Eng 227:108863
3. Wang G, Bar D, Schreier S (2024) The potential of end-of-life ships as a floating seawall and the methodical use of gap resonance for wave attenuation. Ocean Eng 298:117246
4. Olivieri A, Pistani F, Penna R (2001) Towing tank experiments of boundary layer, wake and free surface flow around a naval combatant. Presented at the SNAME 26th American towing tank conference, July 23
5. Urrego-Pabón J, Mercado J, Obando-Vega F, Rubio-Clemente A, Chica E (2024) Numerical and experimental study of several passive wave absorber behavior in a wave channel. Results Eng 23:102455
6. ITTC (2017) Q.S.G. of the 28th: recommended procedures and guidelines; Guideline on best practices for the applications of PIV/SPIV in towing tanks and cavitation tunnels. https://ittc.info/downloads/archive-of-recommended-procedures/
7. Ouellet Y, Datta I (1986) A survey of wave absorbers. J Hydraul Res 24:265–280
8. Klamo JT, Jones LM, Kwon YW, Didoszak JM (2017) Upgrades to and current research efforts involving the tow tank facility at the naval postgraduate school. Presented at the SNAME American towing tank conference
9. Malas B, Creasey L, Buckland D, Turnock SR (2023) Design, development and commissioning of the Boldrewood towing tank—a decade of endeavour. Int. J. Marit. Eng. 165:255–272
10. Cho IH, Kim MH (2008) Wave absorbing system using inclined perforated plates. J Fluid Mech 608:1–20. https://doi.org/10.1017/S0022112008001845
11. George A, Poguluri SK, Kim J, Cho IH (2022) Design optimization of a multi-layer porous wave absorber using an artificial neural network model. Ocean Eng 265:112666
12. Elliff CI, Silva IR (2017) Coral reefs as the first line of defense: shoreline protection in face of climate change. Mar Environ Res 127:148–154. https://doi.org/10.1016/j.marenvres.2017.03.007

13. Ferrario F, Beck MW, Storlazzi CD, Micheli F, Shepard CC, Airoldi L (2014) The effectiveness of coral reefs for coastal hazard risk reduction and adaptation. Nat Commun 5
14. Costa MBSF, Araújo M, Araújo TCM, Siegle E (2016) Influence of reef geometry on wave attenuation on a Brazilian coral reef. Geomorphology 253:318–327
15. Chwang AT, Chan AT (1998) Interaction between porous media and wave motion. Annu Rev Fluid Mech 30:53–84. https://doi.org/10.1146/annurev.fluid.30.1.53
16. Numerical modeling of wave absorbers for physical wave tanks. https://ascelibrary.org/doi/epdf/10.1061/(ASCE)0733-950X(2003)129:1(5). Last accessed 12 Jul 2024
17. Gaylord B, Denny MW (1997) Flow and flexibility I. Effects of size, shape and stiffness in determining wave forces on the stipitate kelps Eisenia arborea and Pterygophora californica. J Exp Biol 200:3141–3164
18. Elsmore K, Nickols KJ, Miller LP, Ford T, Denny MW, Gaylord B (2023) Wave damping by giant kelp, Macrocystis pyrifera. Ann Bot. https://doi.org/10.1093/aob/mcad094
19. Luhar M, Nepf HM (2011) Flow-induced reconfiguration of buoyant and flexible aquatic vegetation. Limnol Oceanogr 56:2003–2017. https://doi.org/10.4319/lo.2011.56.6.2003
20. Nove-Josserand C, Godoy-Diana R, Thiria B (2018) Array optimisation for wave damping and wave absorption. In: 4th Asian wave and tidal energy conference (AWTEC 2018)
21. Gong S, Xu S, Yin K, Chen Y, Chen J, Lei J (2024) Experimental study on wave attenuation and cross-shore profiles by submerged flexible vegetation. Ocean Eng 307:118218
22. Hergel J, Hinz K, Lefebvre S, Thomaszewski B (2019) Extrusion-based ceramics printing with strictly-continuous deposition. ACM Trans Graph 38:194:1–194:11. https://doi.org/10.1145/3355089.3356509
23. Berman O, Weizman M, Oren A, Neri R, Parnas H, Shashar N, Tarazi E (2023) Design and application of a novel 3D printing method for bio-inspired artificial reefs. Ecol Eng 188:106892
24. Hu Y, Zhao Y, Zhao M, Feng M (2022) Three phases fluid-structure interactive simulations of the deep sea ceramic sphere's failure and underwater implosion. Ocean Eng 245:110494
25. Gosselin FP (2019) Mechanics of a plant in fluid flow. J Exp Bot 70:3533–3548
26. Lane CE, Lindstrom SC, Saunders GW (2007) A molecular assessment of northeast Pacific *Alaria* species (Laminariales, Phaeophyceae) with reference to the utility of DNA barcoding. Mol Phylogenet Evol 44:634–648

Experimental and Numerical Studies on the Hydrodynamics of a Novel Floating Solar-Wave Converter Hub

Hongbin Hao, Jinghua Wang, Huan-Feng Duan, and Senqin Zhang

Abstract The sustainable offshore energy resources (OER) (e.g., solar, wind, wave and tide) are considered an alternative and supplement to meet the massive energy demands as well as to promote the global carbon mitigation. Due to the intermittent and low energy generation capacity of stand-alone OER systems, the concept of hybrid energy technology combing multiple OERs has become a promising and hopeful alternative towards sustainable energy exploitation. This paper introduces the design of a new Floating Solar Converter Hub (FloSCH) which is supported by four floaters and proposes the integration of a OWC-WEC concept to create a novel hybrid Floating Solar-Wave Converter Hub (FloSWACH). Then, the heave and pitch responses of FloSCH with three types of floaters due to regular waves are analyzed through a combination of model tests and numerical simulations. A noticeable agreement between the numerical and experimental results is observed. Compared to the FloSCH, the motion response of FloSWACH due to the wave around Hong Kong shows a significant improvement in predictability, potentially enhancing the conversion efficiency of solar energy.

Keywords Floating solar-wave converter hub · Hydrodynamics · Model test · Numerical simulation

H. Hao · J. Wang · H.-F. Duan (✉) · S. Zhang
Department of Civil and Environmental Engineering, The Hong Kong Polytechnic University, Hong Kong, China
e-mail: hf.duan@polyu.edu.hk

J. Wang · H.-F. Duan
Research Institute of Sustainable Urban Development, The Hong Kong Polytechnic University, Hong Kong, China

X. L. Zhao et al. (eds.), *Proceedings of the Fourth World Conference on Floating Solutions: WCFS 2024; 2–4 December; Hong Kong*, Lecture Notes in Civil Engineering 597, https://doi.org/10.1007/978-981-96-4569-5_104

1 Introduction

Greenhouse gas (GHG) emissions from fossil fuels and land use have continuously grown in the past decade. The present primary source of global energy consumption largely depends on fossil fuels (i.e., coal, petroleum, and natural gas), which has contributed nearly 70% of the greenhouse gas (GHG) emissions [1]. Such critical situation of global energy demands and GHG emissions will be continuously intensified due to industrialization and climate change, especially in the populated tropical and mid-latitude areas (e.g., the US, Europe and China). In this context, many governments and authorities have placed their targets on clean and renewable energy. Compared to the onshore, the offshore energy resources (OERs) (e.g., wave, wind, tide and solar) are more abundant and save land, with less impact on human life.

In recent years, offshore solar energy has been rapidly developing, with the frequent adoption of Floating Photovoltaic (FPV) concepts. The offshore environment offers benefits such as lower temperatures and high wind speed that assist in cooling the Photovoltaic (PV) panels, thus enhancing efficiency. Liu et al. conducted a comparative study of traditional terrestrial PV systems and an FPV system, revealing that the FPV systems exhibit a higher electricity generation capacity. Simulation tests indicate that the relative annual average energy output is approximately 12.96% higher at sea (reaching 18% in specific months) compared to land-based systems [2]. This improvement is attributed to the lower sea temperatures, which facilitate natural cooling processes [3].

According to DNV GL, the most common types of Floating Photovoltaic (FPV) arrays are categorized structurally into three main groups: Pure floats (or floats with PV modules), Modular rafts and Membranes, as depicted in Fig. 1 [4]. In Pure floats, the PV modules are directly mounted on floats. In Modular rafts, the structural framework is supported by floats, while Membranes utilize a reinforced membrane supported by additional structures.

Pure floats are commonly utilized in inland lakes. The offshore application originated with Zon-op-Zee [5], which marks the commencement of offshore Floating Photovoltaic (FPV) projects, operational since 2017 [6]. This concept was also adopted in China's first FPV project located in Zhoushan [6]. A notable example of a membrane system is the Ocean Sun, implemented in the Shandong Peninsula, representing the first deep-sea "wind + solar" project [6]. Featuring modified solar PV modules on a thin, flexible membrane, it is adaptable to various sea conditions, from

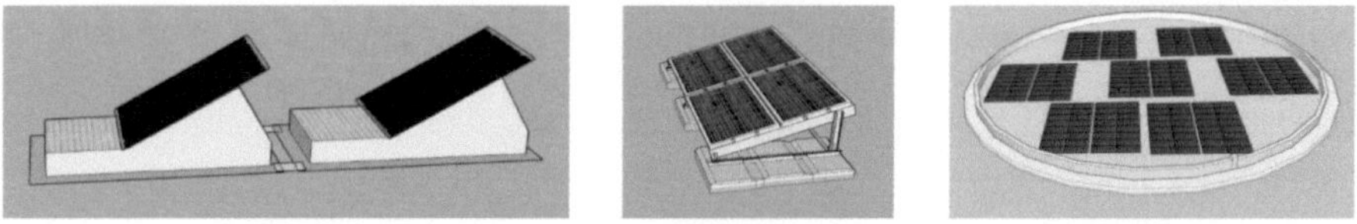

(a) Pure floats (b) Modular rafts (c) Membranes

Fig. 1 The schematic of three categories of FPV [4]

offshore locations to reservoirs and lakes [7]. Modular rafts have seen widespread use in offshore projects, incorporating raised PV modules to mitigate wave impact. In 2022, SolarDuck introduced a triangular FPV module design [8], positioning the PV modules 3 m above the sea surface. Similarly, CIMC Solar launched China's first semi-submersible marine FPV venture [9], comprising four FPV modules capable of withstanding open seas with wave heights up to 6.5 m, tides reaching 4.6 m, and winds up to 34 m/s [6].

Due to the highly variable and unpredictable nature of stand-alone OER systems (e.g., stability and survivability), which can result in intermittent and low energy generation capacity, a novel concept of hybrid energy technology combining multiple OERs has emerged as a promising and sustainable alternative for energy exploitation. This paper introduces the design of a new Floating Solar Converter Hub (FloSCH) and proposes the integration of a wave energy converter device to create a novel hybrid Floating Solar-Wave Converter Hub (FloSWACH). Additionally, the hydrodynamic performance of FloSWACH is examined through experimental and numerical studies.

2 Conceptual Design of FloSWACH

In this design, the Floating Solar Converter Hub (FloSCH) is proposed to function in a modular manner, with the expected power rating of a single FloSCH around 108 kW. As illustrated in Fig. 2, the PV panels are mounted to a square steel sub-truss measuring 1.0 m in height and 30 m in side length, with a total mass of approximately 50 tons for the SCH. In future practical applications, modifications to the structure and materials of the sub-truss can be made to accommodate marine environmental loads. To prevent wave impact, the PV panels are supported above the sea surface using four floaters. The floaters have a height and draught of 9.0 m and 4.0 m, respectively, positioning the PV panels approximately 5.0 m above the sea surface.

The floater consists of two cylinders. The larger cylinder, measuring 2.0 m in height and 4.0 m in diameter, is positioned at the bottom to provide necessary buoyancy and increase motion response damping. At the top, the smaller cylinder, with dimensions of 7.0 m in height and 1.4 m in diameter, connects and supports the sub-truss. The waterline is situated on the smaller cylinder to reduce wave loads due to its

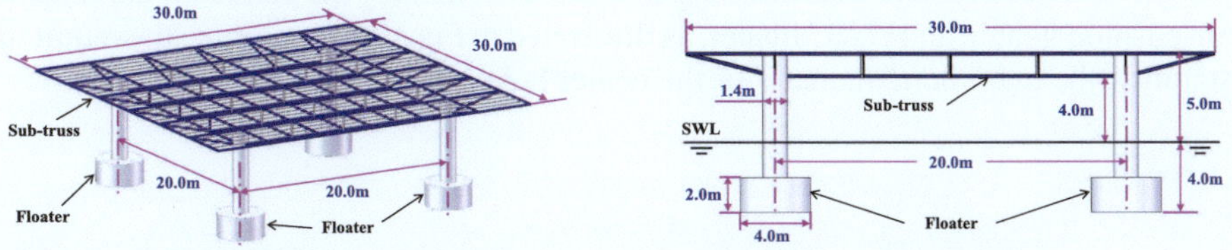

(a) 3-D view (b) Front view

Fig. 2 The structure diagram of Floating Solar Converter Hub

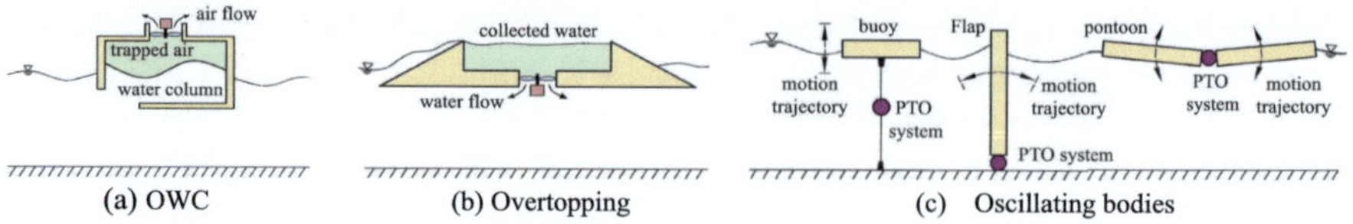

Fig. 3 Wave energy devices classified by working principle [11]

smaller waterplane area. The simple design of the floater allows for construction with concrete or FRP, reducing construction costs and the impact of seawater corrosion.

In this study, the wave energy device is introduced to operate in conjunction with FloSCH. According to the working principle, Wave Energy Converters (WECs) mainly fall into three categories: Oscillating Water Column (OWC), overtopping, and oscillating body (OB) [10, 11]. The OWC-WEC system (Fig. 3a) utilizes a semi-submerged chamber to capture wave energy, converting it into the kinetic energy of the water or trapped air column. In contrast, the overtopping WEC (Fig. 3b) employs a reservoir-like structure to absorb wave energy, transforming it into the potential energy of the collected water above the outside water surface. The OB-WEC system (Fig. 3c) utilizes a floater to capture wave energy, converting it into the kinetic energy between oscillating bodies or between an oscillating body and a fixed body (e.g., a seabed foundation). Due to its simplified and reliable nature with a single major moving component, the OWC-WEC device is considered straightforward. Consequently, the integration of the OWC concept is proposed to collaborate with FloSCH.

In real sea conditions, pitch motion increases reflection losses and decreases the effective area on PV panels, thereby reducing the efficiency of the FloSCH. The greater the amplitude of pitch motion, the more power decreases. To enhance the seakeeping ability of FloSCH and optimize the absorption and conversion of wave energy, a round hole is strategically positioned at the center of the floater. This hole, similar to the concept of OWC, allows for the absorption of wave energy. The harvested wave energy can be converted into electricity using turbines or utilized to cool the PV panels by transporting cold seawater close to SCH. This innovative design is referred to as the Floating Solar-Wave Converter Hub (FloSWACH), and its operational state lies between that of an Air Cushion Barge Platform [12] and a opening OWC system. Consequently, this study delves into the hydrodynamic performance of FloSCH with three types of floaters, namely the conventional floater, air-cushion floater, and OWC floater, as illustrated in Fig. 4. To guarantee the required stability, the maximum diameter of the center hole is 1.2 m.

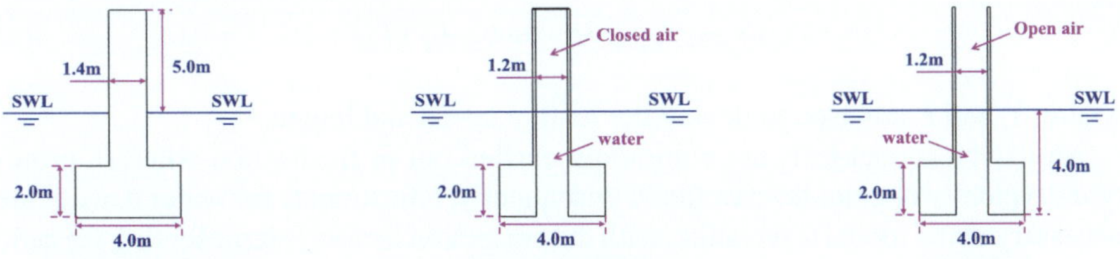

Fig. 4 The structure diagram of different floaters for FloSCH

3 Study of Hydrodynamic Performance

The hydrodynamic performance is analyzed through a combination of model tests and numerical simulations. According to the symmetry, an approximate 2D model is employed for the study. The model test is carried out in the wave tank at the hydraulics laboratory of PolyU, while the numerical simulations are performed using the commercial software Star-CCM+ [16]. The solar converter hub model is supported by three types of floaters respectively.

3.1 Model Test

Generally, the geometric similarity is a crucial aspect in model testing, requiring a consistent ratio between linear scales of both the prototype and the scale model. If L, B and d denote length, width and draft respectively, then

$$L_F/L_M = B_F/B_M = d_F/d_M = \lambda \quad (1)$$

In which, λ is the scale factor. The subscripts F and M denote the full scale and model scale, respectively.

When conducting model tests in a wave tank, it is essential for the water depth h_M, wave height H_M and wavelength λ_M to satisfy the geometric similarity requirements with the actual sea conditions, that is

$$h_F/h_M = H_F/H_M = \lambda_F/\lambda_M = \lambda \quad (2)$$

For traditional floating structures like surface ships and offshore platforms, wave force represents the primary external load, with gravity and inertia forces being the key influencing factors. Hence, Froude similarity is commonly applied in hydrodynamic model test of conventional floating structures at sea [13–15]. To guarantee the accurate similarity of gravity and inertial forces between the scaled model system and the prototype system, the Froude numbers (Fr) of the two systems must remain equal, i.e.

$$Fr = V_F/\sqrt{gL_F} = V_M/\sqrt{gL_M} \tag{3}$$

where, V and L are used to denote the feature speed and length.

Moreover, model tests are commonly carried out in freshwater, whereas prototypes typically operate in seawater. Consequently, adjustments for water density are necessary in the model test results, with the correction factor (γ) representing the ratio of seawater density to freshwater density, conventionally set at $\gamma = 1.025$. The conversion relationship between the physical parameters of the scaled model system and the prototype system under geometric and Froude similarities is detailed in Table 1.

The model test is carried out in the wave tank of the hydraulics laboratory at PolyU, with dimensions of 27 m × 1.5 m × 1.5 m, as illustrated in Fig. 5. Following to the principal dimension of FloSWACH, a scale factor of 1:20 is applied, positioning the experimental model 15 m from the wave maker.

Constrained by the wave tank dimensions, the model test is conducted as an approximate two-dimensional experiment, with the mooring system not taken into account. Consequently, to confine the motion responses of FloSWACH to heave and pitch only, a heave and pitch limiting device is utilized. Referring to the setup depicted in Fig. 5, the model is tested in regular waves detailed in Table 2. The heave and pitch motions are determined from the coordinates of marked points captured using a 3-dimensional motion capture system.

Table 1 Conversion relationship of the quantity between model scale and full scale

Items	Relationship	Items	Relationship
Length and angle	$L_M = L_F/\lambda,\ \phi_M = \phi_F$	Linear and angular velocity	$V_M = V_F/\lambda^{1/2},\ \varphi_M = \varphi_F \cdot \lambda^{1/2}$
Area and volume	$A_M = A_F/\lambda^2,\ \nabla_M = \nabla_F/\lambda^3$	Mass and inertia moment	$\Delta_M = \Delta_F/\gamma\lambda^3,\ I_M = I_F/\gamma\lambda^5$
Period and frequency	$T_M = T_F/\lambda^{1/2},\ f_M = f_F \cdot \lambda^{1/2}$	Force and torque	$F_M = F_F/\gamma\lambda^3,\ M_M = M_F/\gamma\lambda^4$

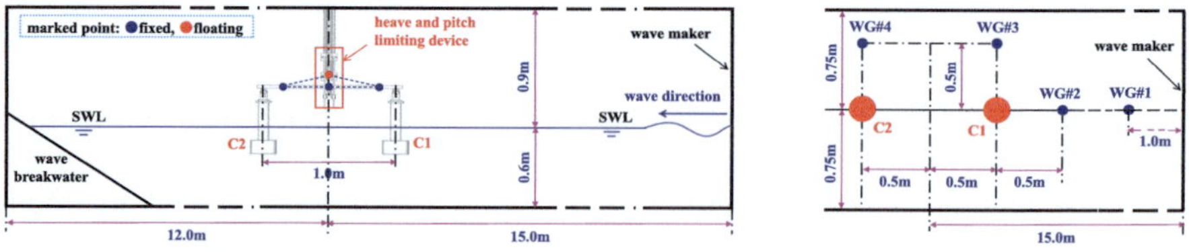

(a) Arrangement of model

(b) arrangement of wave gauge

Fig. 5 Setup of model test

Table 2 Regular waves of model test

No	Period (s)	Height (mm)	No	Period (s)	Height (mm)	No	Period (s)	Height (mm)	No	Period (s)	Height (mm)
1	0.50	20	5	0.90	30	9	1.30	50	13	1.80	60
2	0.60	30	6	1.00	40	10	1.40	50	14	2.00	60
3	0.70	30	7	1.10	40	11	1.50	60	15	2.20	60
4	0.80	30	8	1.20	40	12	1.60	60	16	2.40	60

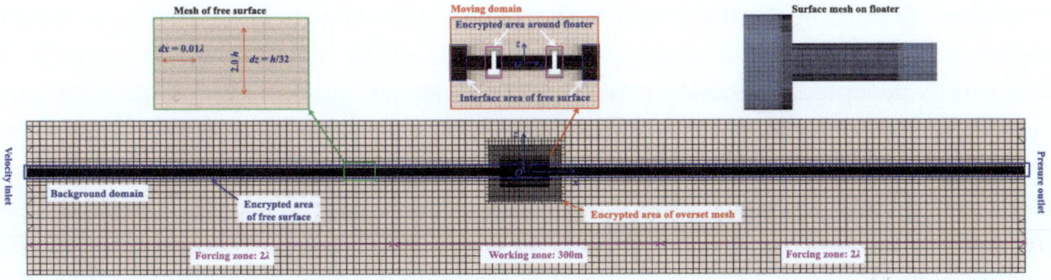

Fig. 6 Setup and mesh of numerical simulation

3.2 *Numerical Simulations*

Based on the model test, the hydrodynamic of full-scale FloSWACH model is numerically simulated in the commercial software Star-CCM+. The Volume of Fluid (VOF) method and overset mesh technique are used to calculate the heave and pitch due to regular waves of the FloSCH. The computational domain comprises a background domain and a moving domain, as illustrated in Fig. 6.

The background domain is primarily utilized for numerical wave generation, divided into two forcing zones on both sides and a central working zone. The working zone in the middle is approximately 300m long, while the forcing zones located after the velocity inlet and before the pressure outlet span about 2λ, where λ represents the wave length. An encrypted mesh region of $2h$ in height surrounds the free surface, with a mesh size (dx) in the wave length direction of 0.01λ and a mesh size (dz) in the wave height direction of $h/32$, where h represents the wave height. To enhance accuracy in data transfer between the two domains, an encrypted area is established within the background domain. This overset mesh area features a similar mesh size and complete coverage of the moving domain.

The moving domain primarily serves to simulate the heave and pitch motion responses, utilizing the motion specifications of DFBI rotation and translation. It is essential to encrypt the mesh near the free surface in the moving domain to ensure full coverage of the background domain. To enhance calculation accuracy and model resolution, the maximum mesh size on the floater's surface is constrained to less than 0.2m. Moreover, the mesh surrounding the floater is similarly encrypted with the same size as the surface mesh. Similar to the model test, mooring is also not considered, and the motion responses are confined to heave and pitch.

4 Results and Discussions

The motion response of free floating device in regular waves can be described by the Response Amplitude Operator (RAO), which represents the dimensionless motion response. According to the linear wave theory, the amplitude of floater's motion response is directly proportional to the amplitude of the incident regular wave [17]. It is widely accepted that ocean wave height follows a Gaussian random process with zero mean and ergodic states. Additionally, it is believed that the floater's response to each wave component is independent of the response to other wave components [18]. Specifically, for translation motion (e.g., heave, surge), the RAO is defined as the ratio of motion response amplitude to wave amplitude. On the other hand, for rotational motion (e.g., pitch, roll), the RAO is defined as the ratio of motion response amplitude to the maximum slope of the wave θ_m [19], which can be approximately obtained by

$$\theta_m = ar\tan\left[\left(\frac{\partial\eta}{\partial x}\right)_m\right] = ar\tan(2a\pi/\lambda_w) \tag{4}$$

where, η is the equation of wave elevation, a and λ_w are the amplitude and length of wave respectively.

According to the experimental and numerical time-history curves of heave, pitch and wave, the heave and pitch RAOs of different models are depicted in Fig. 7. In which, the experimental RAOs have been translated to prototype. The value of dimensionless RAOs remain constant when compared to the real scale. However, the wave periods in the experiments have been scaled to the real scale using Froude similarity, as outlined in Table 1. A noticeable agreement between the numerical and experimental results is observed in the overlapping duration (7.0–10.7s), indicating the accuracy and reliability of the numerical simulations. As the wave periods increase, the heave and pitch RAOs progressively peak during their respective natural periods and subsequently stabilize. The air-cushion FloSCH model demonstrates motion responses similar to the conventional FloSCH; however, the OWC FloSCH exhibits increased natural periods and peak values of heave and pitch RAOs due to its smaller waterline area. Consequently, the heave and pitch RAOs of the OWC FloSCH are significantly lower than those of the conventional FloSCH when wave periods are in the ranges of 8-14s and 9-15s respectively. As mentioned above, the operational state of FloSWACH will lie between the air-cushion FloSCH and OWC FloSCH, it is predictable that the pitch RAO of FloSWACH will be significantly smaller than that of conventional FloSCH when the wave period is in 9–12 s, potentially enhancing the efficiency of FloSCH systems in the seas around Hong Kong. Additionally, it is also helpful to reduce the fatigue of FloSCH.

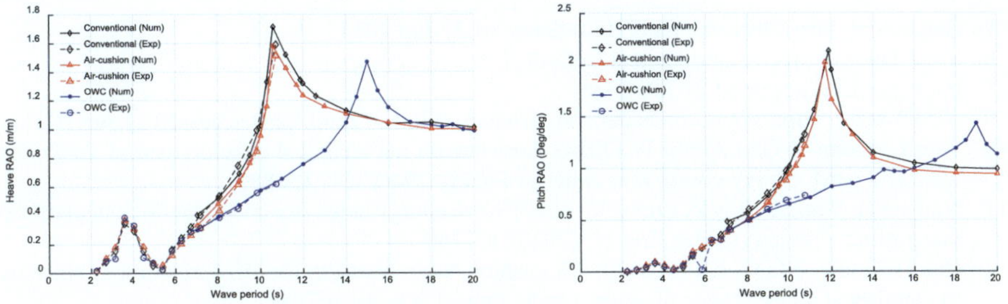

Fig. 7 RAOs of FloSCH with different types of floaters

5 Conclusions

To enhance the efficiency of floating offshore solar energy, a novel concept of hybrid Floating Solar-Wave Converter Hub (FloSWACH) is proposed, and its motion response due to waves is evaluated through a combination of model tests and numerical simulations.

Initially, a new FloSCH design, supported by four floaters to elevate the PV panels above the sea surface, is conceptually formulated. Subsequently, the integration of the OWC-WEC concept into the FloSCH involves a modification by adding a central hole on one of the floaters, thus giving rise to the innovative FloSWACH concept.

Following this design alteration, the hydrodynamic performance of the FloSWACH system is assessed. Model tests in a wave tank and numerical simulations using Star-CCM+ are conducted to analyze the heave and pitch responses of the different types of floaters (conventional, air-cushion, and OWC) under regular wave conditions. A noticeable agreement between the numerical and experimental results is observed. Specifically, it is predictable that the pitch RAO of FloSWACH is comparable to, yet significantly smaller than, that of the conventional FloSCH when subjected to wave periods less than 9 s and between 9 and 12 s.

References

1. Zheng X, Dalia S, Tomas B, Abbas M, Fausto C, Liao H (2019) A review of greenhouse gas emission profiles, dynamics, and climate change mitigation efforts across the key climate change players. J Clean Prod 234(2019):1113–1133
2. Liu L, Wang Q, Lin H, Li H, Sun Q, Ronald W (2017) Power generation efficiency and prospects of floating photovoltaic systems. Energy Procedia 105:1136–1142
3. Zahra Golroodbari S, van Sark W (2020) Simulation of performance differences between offshore and land-based photovoltaic systems. Prog Photovolt Res Appl 28:873–886
4. DNV GL (2021) Design, development and operation of floating solar photovoltaic systems. DNVGL-RP-0584
5. Ocean of Energy. https://oceansofenergy.blue/projects/. Last accessed 21 Jun 2024
6. Shi W, Yan C, Ren Z, Yuan Z, Liu Y, Zheng S, Li X, Han X (2023) Review on the development of marine floating photovoltaic systems. Ocean Eng 286:115560

7. Ocean Sun. https://oceansun.no/. Last accessed 21 Jun 2024
8. Solar Duck. https://solarduck.tech/solarduck-unveils-design-of-its-offshore-floating-solar-technology/. Last accessed 21 Jun 2024
9. CIMC Solar. http://www.cimcsolar.com/product_details/4.html. Last accessed 21 Jun 2024
10. Wan L, Moan T, Gao Z, Shi W (2024) A review on the technical development of combined wind and wave energy conversion systems. Energy 294(2024):130885
11. Nguyen H, Wang C, Tay Z, Luong V (2020) Wave energy converter and large floating platform integration: a review. Ocean Eng 213(2020):107768
12. Hao H, Guo Z, Ma Q, Xu G (2020) Air cushion barge platform for offshore wind turbine and its stability at a large range of angle. Ocean Eng 217(2020):107886
13. Chakrabarti S (1998) Physical model testing of floating offshore structures. In: Dynamic positioning conference, pp 1–33
14. Faltinsen O (1993) Sea loads on ships and offshore structures. Cambridge University Press
15. Sarpkaya T, Isaacson M, Wehausen JV (1982) Mechanics of wave forces on offshore structures. J Appl Mech 49:466–467
16. Siemens (2020) User manual: Star-CCM+ V15.02.007-R8
17. Yu K, Dai Y, Meng F, Mao D, Liu Y, Liang J (2021) Numerical simulation research on hydrodynamic characteristics of new-style buoys. Port Eng Technol 58(5):1–4
18. Shi H, Liu D, Liu Z (2011) Calculation of the motion response of a novel wave energy converter. Periodic Ocean Univ China 41(10):111–116
19. Ni B, Ye Y, Hu W, Xue Y (2020) Study on motion response of ice floe under Stokes wave. J Huazhong Univ Sci Technol (Nat Sci Ed) 48(10):75–81

MIX
Papier aus verantwortungsvollen Quellen
Paper from responsible sources
FSC® C105338

If you have any concerns about our products,
you can contact us on
ProductSafety@springernature.com

In case Publisher is established outside the EU,
the EU authorized representative is:
Springer Nature Customer Service Center GmbH
Europaplatz 3, 69115 Heidelberg, Germany

Printed by Libri Plureos GmbH
in Hamburg, Germany